Order Number: T-POWER-0101



# AIRCRAFT POWERPLANT MAINTENANCE

*Publisher* Karl D. Stoltzfus, Jr.
*Series Editor* Dale Hurst

**Production Staff**

*Lead Illustrator/Designer* Mary Evans
*Illustrator/Designer* Holly Curry



International Standard Book Number 0-9708109-3-8

*For Sale by:* Avotek Information Resources
A Select Aerospace Industries company

*Mail to:*
P.O. Box 219
Weyers Cave, Virginia 24486

*Ship to:*
200 Packaging Drive
Weyers Cave, Virginia 24486

*Toll Free:* 1-800-828-6835
*Telephone:* 1-540-234-9090
*Fax:* 1-540-234-9399

www.avotekbooks.com

Printed in the USA

# *Preface*

This textbook, the fourth in a series of four, was written for the Aviation Maintenance Technician student of today. It is based on the real-world requirements of today's aviation industry. At the same time, it does not eliminate the traditional subject areas taught since the first A&E schools were certificated.

This series of textbooks has evolved through careful study and gathering of information offered by the Federal Aviation Administration, the Blue Ribbon Panel, the Joint Task Analysis report, industry involvement and AMT schools nationwide.

The series is designed to fulfill both current and future requirements for a course of study in Aviation Maintenance Technology.

Textbooks, by their very nature, must be general in their overall coverage of a subject area. As always, the aircraft manufacturer is the sole source of operation, maintenance, repair and overhaul information. Their manuals are approved by the FAA and must always be followed. You may not use any material presented in this or any other textbook as a manual for actual operation, maintenance or repairs.

The writers, individuals and companies which have contributed to the production of this textbook have done so in the spirit of cooperation for the good of the industry. To the best of their abilities, they have tried to provide accuracy, honesty and pertinence in the presentation of the material. However, as with all human endeavors, errors and omissions can show up in the most unexpected places. If any exist, they are unintentional. Please bring them to our attention. ➔

**Email us at corrections@avotek.com for comments or suggestions.**

***Avotek® Aircraft Maintenance Series:***

*Introduction to Aircraft Maintenance*
*Aircraft Structural Maintenance*
*Aircraft System Maintenance*
*Aircraft Powerplant Maintenance*

***Other Books by Avotek®:***

*Aircraft Structural Technician*
*Avionics: Systems and Troubleshooting*

# *Acknowledgements*


Dave Stanley and Mike Leasure — *Purdue University*

*Academy of Infrared Training, Inc.*

*Air Methods (Rocky Mountain Helicopters)*

Al Dibble — *Snap-On Tools*

Alan Bandes — *UE Systems, Inc.*

Andy Wilson — *B/E Aerospace*

Bob Blouin — *National Business Aircraft Association*

*Boeing Commercial Airplane Co.*

Brian Stoltzfus — *Priority Air Charter*

Cal Crowder

*Champlin Fighter Aircraft Museum*

Charlie Witman — *Avotek*

Chris McGee — *USAF Museum*

David Jones — *AIM*

David Posavec — *Barry Controls*

Debbie Jones — *MD Helicopters, Inc.*

*De-Ice Systems International*

*Dynamic Solution Systems, Inc.*

*Eaton Aerospace Fluid Systems (Aeroquip)*

Fred Workley — *Workley Aviation Mx*

Greg Campbell, Dennis Burnett, Sherman Showalter, Stacey Smith — *Shenandoah Valley Regional Airport*

Harry Moyer, Virgil Gottfried — *Samaritan's Purse*

Jack Knox — *Evergreen Air Center*

Jean Watson — *FAA*

Jim Akovenko — *JAARS*

Karl Stoltzfus, Sr., Michael Stoltzfus, Aaron Lorson & staff — *Dynamic Aviation Group, Inc.*

Ken Hyde, Weldon Britton — *The Wright Experience*

Ken Stoltzfus, Jr. — *Preferred Airparts*

Lee Helm — *RAM Aircraft Corporation*

*Lilbern Design*

Lori Johnson, Larry Bartlett — *Duncan Aviation*

Mary Ellen Gubanic — *Alcoa*

*Mckee Foods Flight Department*

Michelle Moyer

*Micro-Mesh*

Pat Colgan — *Colgan Air*

Paul Geist, J.R. Dodson — *Dodson International Parts*

*Phoenix Composites*

*Pilgrim's Pride Aviation Department*

*Precision Airmotive Corp.*

*Precision Instruments*

Raymond Goldsby — *AIA*

Rich Welsch — *Dynamic Solutions Systems, Inc.*

Richard Kiser, Steve Bradley — *Classic Aviation Services*

Richard Milburn — *Northrop Grumman Corporation*

Robert Kafales — *Frontier Airlines*

*Scott Aviation Enterprises*

*Select Aerospace Industries, Inc*

*Select Airparts*

*Stern Technologies*

Steve Hanson, Jeff Ellis, Tim Travis — *Raytheon Aircraft*

Susan Timmons — *JRA Executive Air*

Tom Brotz — *U.S. Industrial Tools*

Vern Raburn, Andrew Broom — *Eclipse Aviation*

*Virginia State Police Med Flight III*

*Vought Aircraft Industries, Inc.*

# Contents

Continental
Continental
Continental

Chapter 1

# RECIPROCATING *engines*

Section 1

## Reciprocating Engine Theory

**Learning Objectives:**

- *Theory*
- *Design*
- *Construction*
- *Maintenance and and Operation*
- *Instrument Systems*
- *Lubrication Systems*
- *Ignition and Starting Systems*
- *Electronics Systems*
- *Fuel and Fuel Metering Systems*
- *Induction and Airflow Systems*
- *Cooling Systems*
- *Exhaust Systems*

For an aircraft to remain in level, un-accelerated flight, a thrust must be provided that is equal to the aircraft drag but in the opposite direction. A suitable type of heat engine provides this propulsive force, or thrust.

All heat engines have in common the ability to convert heat energy into mechanical energy, by the flow of some fluid mass through the engine. In all cases, the heat energy is released at a point in the cycle where the pressure is high, relative to the atmospheric pressure.

These engines are customarily divided into groups or types depending upon:

- The working fluid used in the engine cycle
- The means by which the mechanical energy is transmitted into a propulsive force
- The method of compressing the engine working fluid

The propulsive force is obtained by the displacement of a working fluid (not necessarily the same fluid used within the engine) in a direction opposite to that in which the airplane is propelled. This is an application of Newton's third law of motion. Air is the principal fluid used for propulsion in every type of powerplant except the rocket, in which only the by-products of combustion are accelerated and displaced.

The types of engines are illustrated in Table 1-1-1 (top of next page).

**_Left._ Modern reciprocating engines come in many sizes and horsepower ratings to meet a variety of applications.**

| ENGINE TYPES | MAJOR MEANS OF COMPRESSION | ENGINE WORKING FLUID | PROPULSIVE WORKING FLUID |
|---|---|---|---|
| Reciprocating | Reciprocating action of pistons | Fuel/air mixture | Ambient air |
| Turboprop | Turbine-driven compressor | Fuel/air mixture | Ambient air |
| Turbofan | Turbine-driven compressor | Fuel/air mixture | Ambient air through fan/engine working fluid |
| Turbojet | Turbine-driven compressor | Fuel/air mixture | Same as the engine working fluid |
| Ramjet | Ram compression due to high flight speeds | Fuel/air mixture | Same as the engine working fluid |
| Pulse-jet | Compression due to combustion | Fuel/air mixture | Same as the engine working fluid |
| Rocket | Compression due to combustion | Oxidizer/fuel mixture | Same as the engine working fluid |

**Table 1-1-1. Types of engines**

**Figure 1-1-1. Most large, commercial airplanes are powered by turbine engines**

**Figure 1-1-2. A reciprocating engine in a Beechcraft A36 Bonanza airplane**

Propellers of aircraft powered by reciprocating or turboprop engines accelerate a large mass of air through a small velocity change. The fluid (air) used for the propulsive force is a different quantity than that used within the engine to produce the mechanical energy. Turbojets, ramjets and pulsejets accelerate a smaller quantity of air through a large velocity change. They use the same working fluid for propulsive force that is used within the engine. A rocket carries its own oxidizer rather than using ambient air for combustion. It discharges the gaseous byproducts of combustion through the exhaust nozzle at an extremely high velocity.

Engines are further characterized by the means of compressing the working fluid before the addition of heat. The basic methods of compression are:

- The turbine-driven compressor (turbine engine), as shown in Figure 1-1-1
- The positive displacement, piston-type compressor (reciprocating engine), as shown in Figure 1-1-2
- Ram compression due to forward flight speed (ramjet), as shown in Figure 1-1-3
- Pressure rise due to combustion (pulse-jet and rocket), as shown in Figure 1-1-4

A more specific description of the major engine types used in commercial aviation is given later in this chapter.

## Comparison of Powerplants

In addition to the differences in the methods employed by the various types of powerplants for producing thrust, there are differences in their suitability for different types of aircraft. The following discussion points out some of the important characteristics, which determine their suitability.

### *General Requirements*

All engines must meet certain general requirements of efficiency, economy and reliability. Besides being economical in fuel consumption, an aircraft engine must be economical (the cost of original procurement and the cost of maintenance) and it must meet exacting requirements of efficiency and have a low weight per horsepower ratio. It must be capable of sustained high-power output with no sacrifice in reliability. It must also have the durability to operate for long periods of time between overhauls. It needs to be as compact as possible, yet have easy accessibility for maintenance. It is required to be as vibration-free as possible and be able to cover a wide range of power output at various speeds and altitudes.

These requirements dictate the use of ignition systems that will deliver the firing impulse to the spark plugs at the proper time in all kinds of weather and under adverse conditions. Fuel-metering devices are needed that will deliver fuel in the correct proportion to the air ingested by the engine, regardless of the altitude or type of weather in which the engine is operated. The engine needs an oil system that delivers oil under the proper pressure when it is running. Also, it must have a system of damping units to damp out the vibrations of the engine.

### *Power and Weight*

The useful output of all aircraft powerplants is *thrust*: the force that propels the aircraft. Since the reciprocating engine is rated in b.hp. (brake horsepower) and the gas turbine engine is rated in pounds of thrust; no direct comparison can be made. However, since the reciprocating engine/propeller combination receives its thrust from the propeller, a comparison can be made by converting the horsepower developed by the reciprocating engine to thrust.

If desired, the thrust of a gas turbine engine can be converted into t.hp. (thrust horsepower), but it is necessary to consider the speed of the aircraft. This conversion can be accomplished by using the formula:

$$\text{t.hp.} = \frac{\text{thrust x aircraft speed (mph)}}{\text{375 mile-pounds per hour}}$$

Figure 1-1-3. An early example of a ramjet engine

Photo courtesy of NASA

The value 375 mile-lbs. per hour is derived from the basic horsepower formula, as follows:

1 hp. = 33,000 ft-lb per minute

33,000 x 60 = 1,980,000 ft-lbs per hour

$$\frac{1{,}980{,}000}{5{,}280} = \text{375 mile-lbs. per hour}$$

One horsepower equals 33,000 ft.-lb. per minute or 375 mile-lbs. per hour. Under static

Figure 1-1-4. A rocket booster at takeoff illustrates the rise in pressure and thrust resulting from compression

Photo courtesy of NASA

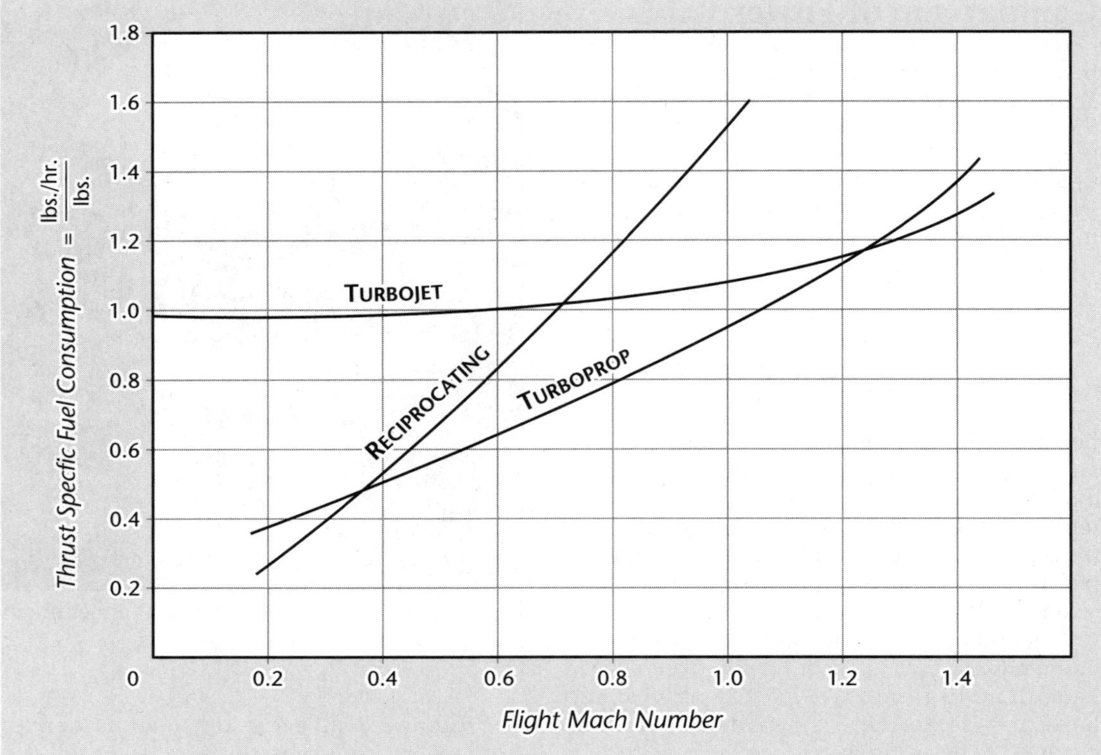


**Figure 1-1-5. Comparison of fuel consumption for three types of engines at rated power at sea level**

conditions, thrust is figured as equivalent to approximately 2.6 lbs. per hour.

If a gas turbine is producing 4,000 lbs. of thrust and the aircraft in which the engine is installed is traveling at 500 m.p.h., the t.hp. will be:

$$\frac{4{,}000 \times 500}{375} = 5{,}333.33 \text{ t.hp.}$$

It is necessary to calculate the horsepower for each speed of an aircraft, since the horsepower varies with speed. Therefore, it is not practical to try to rate or compare the output of a turbine engine on a horsepower basis.

The aircraft engine operates at a relatively high percentage of its maximum power output throughout its service life. The aircraft engine is at full power output whenever a takeoff is made. It may hold this power for a period of time up to the limits set by the manufacturer. The engine is seldom held at maximum power for more than 2 minutes, and usually not that long. Within a few seconds after lift-off, the power is reduced to a power that is used for climbing and that can be maintained for longer periods of time. After the aircraft has climbed to cruising altitude, the power of the engine is further reduced to a cruise power that can be maintained for the duration of the flight.

If the weight of an engine per b.hp. (called the specific weight of the engine) is decreased, the useful load that an aircraft can carry and the performance of the aircraft obviously is increased. Every excess pound of weight carried by an aircraft engine reduces its performance. Tremendous gains in reducing the weight of the aircraft engine through improvement in design and metallurgy have resulted in reciprocating engines now producing approximately 1 hp. for each pound of weight.

## *Fuel Economy*

The basic parameter for describing the fuel economy of aircraft engines is specific fuel consumption. Specific fuel consumption for turbojets and ramjets is the fuel flow (lbs./hr.) divided by thrust (lbs.), and for reciprocating engines the fuel flow (lbs./hr.) divided by brake horsepower. These are called thrust specific fuel consumption and brake specific fuel consumption, respectively. Equivalent specific fuel consumption is used for the turboprop engine and is the fuel flow in pounds per hour divided by a turboprop's equivalent shaft horsepower. Comparisons can be made between the various engines on a specific fuel consumption basis.

At low speed, the reciprocating and turbopropeller engines have better economy than the turbojet engines. However, at high speed, because of losses in propeller efficiency, the reciprocating or turbo-propeller engine's efficiency becomes less than that of the turbojet. Figure 1-1-5 shows a comparison of average thrust specific fuel consumption of three types of engines at rated power at sea level.

### *Durability and Reliability*

Durability and reliability are usually considered identical factors since it is difficult to mention one without including the other. An aircraft engine is reliable when it can perform at the specified ratings in widely varying flight attitudes and in extreme weather conditions. The FAA, the engine manufacturer and the airframe manufacturer agree upon standards of powerplant reliability. The engine manufacturer ensures the *reliability* of the product by design, research and testing. Close control of manufacturing and assembly procedures is maintained to ensure that all engines produced conform to their appropriate Type Certificates. From these collaborations, the standards and instructions for maintaining airworthiness are developed.

*Durability* is the amount of engine life obtained while maintaining the desired reliability. The fact that an engine has successfully completed its type or proof test indicates that it can be operated in a normal manner over a long period before requiring overhaul. However, no definite time interval between overhauls is specified or implied in the engine rating. The *TBO* (time between overhauls) varies with the operating conditions such as engine temperatures, amount of time the engine is operated at high-power settings and the maintenance received. In essence, for a private reciprocating-engined airplane, it is the choice of the operator when to overhaul the engine. However, should an operator operate an engine significantly past TBO and have an accident, there are many insurance companies that will not cover the accident because of the excess time on the engine.

Airplanes used in commercial service are another matter. They must follow the manufacturer's recommended TBO times and mandatory replacement schedules. These schedules and time limits are part of the operator's operations manual. The regulations concerning their TBO times are covered in FAR part 91.403 and FAR part 125.247(d)(1).

Reliability and durability are thus built into the engine by the manufacturer, but the maintenance, overhaul and operating personnel determine the continued reliability of the engine. Careful maintenance and overhaul methods, thorough periodical and preflight inspections and strict observance of the operating limits established by the engine manufacturer will make engine failure a rare occurrence.

### *Operating Flexibility*

Operating flexibility is the ability of an engine to run smoothly and give desired performance at all speeds, from idling to full-power output. The aircraft engine must also function efficiently through all the variations in atmospheric conditions encountered in widespread operations.

### *Compactness*

The shape and size of the engine must be as compact as possible to allow proper streamlining and balancing of an aircraft. In single-engine aircraft, the shape and size of the engine also affect the view of the pilot. This makes a smaller engine better from this standpoint, in addition to reducing the drag created by a large frontal area.

Weight limitations, naturally, are closely related to the compactness requirement. The more elongated and spread out an engine is, the more difficult it becomes to keep the specific weight within the allowable limits.

## Powerplant Selection

Engine weight and specific fuel consumption were discussed in the previous paragraphs. For certain design requirements, powerplant selection may be based on factors other than those which can be discussed from an analytical point of view. For that reason, a general discussion of powerplant selection is included here.

For aircraft whose cruising speeds will not exceed 250 m.p.h., the reciprocating engine is the usual choice (Figure 1-1-6). When economy is required in the low-speed range, the conventional reciprocating engine is chosen because of its excellent efficiency. When high-altitude performance is required, the turbo-supercharged reciprocating engine may be chosen because it

**Figure 1-1-6. A reciprocating engine is the powerplant of choice for airplanes whose cruising speeds generally fall below 250 m.p.h.**

**Figure 1-1-7. This turboprop plane is a typical example of engines found in aircraft that require more power than is provided by a reciprocating engine.**

**Figure 1-1-8. Turbofans are quieter and more efficient than turbojets, and have become the standard for commercial and business airlines.**

**Figure 1-1-9. Turbojet engines can power an airplane as fast as Mach 2 at high altitudes.**

is capable of maintaining rated power to a high altitude (above 30,000 ft.).

Reciprocating engines can be used in many different types of installations. For instance, helicopter installations usually have the engine mounted vertically, with the drive end up. This produces a cooling problem that is solved by attaching a multi-bladed fan to the crankshaft. With modified baffles, the lack of forward speed is not a problem.

In the range of cruising speeds from 180-350 m.p.h., the turbo-propeller engine (Figure 1-1-7) performs better than other types of engines. It develops more power per pound of weight than does the reciprocating engine. This allows a greater fuel load or payload for engines of a given power. The maximum overall efficiency of a turboprop powerplant is less than that of a reciprocating engine at low speed. Turboprop engines operate most economically at high altitudes, but may have a slightly lower service ceiling than some turbo-supercharged reciprocating engines. Economy of operation of turboprop engines, in terms of cargo per ton per miles-per-pound of fuel, will usually be poorer than that of reciprocating engines. This is because cargo-type aircraft are usually designed for low-speed operation. However, because it burns cheaper fuel, the cost of operating a turboprop engine may be similiar to that of a reciprocating engine.

Turbofan engines (Figure 1-1-8) have emerged as the compromise between the efficiency of the turboprop and the high-speed ability of the turbojet. They also offer much quieter operation than turbojets. At the beginning of the 21st century, the turbofan engine has emerged as the clear choice for commercial air carriers and business jets.

Aircraft intended to cruise from high subsonic speeds up to Mach 2.0 are powered by turbojet engines (Figure 1-1-9). Like the turboprop and turbofan, the turbojet operates most efficiently at high altitudes. High-speed, turbojet-propelled aircraft fuel economy, in terms of miles per pound of fuel, is poorer than that attained at low speeds with reciprocating engines.

However, reciprocating engines are more complex in operation than other engines. Correct operation of reciprocating engines requires about twice the instrumentation required by turbojets or turboprops, and it requires several more controls. A change in power setting on some reciprocating engine installations may require the adjustment of five controls, but a change in power on a turbojet requires only a change in throttle setting. Furthermore, there is a greater number of critical temperatures and pressures to be watched on reciprocating engine installations than on turbojet or turboprop installations.

## Maintenance Publications

Many different sources of information are used in maintaining a reciprocating engine. A partial list is:

- **Type Certificate Data Sheets.** These give all information necessary for a conformity check. Data Sheets contain the information used to certify the engine.

- **Engine Manufacturer's Service Manuals and Service Bulletins.** While Service Manuals contain mandatory procedures and information, Service Bulletins do not. They are not mandatory, but it is prudent to follow them.
- **Engine Manufacturer's FAA-approved Overhaul Manual.**
- **Engine Manufacturer's Parts Manuals.**
- **Component Manufacturer's Operations and Overhaul Manuals.**
- **FAA Advisory Circulars (FAA AC 43.13-1B).** FAA AC 43.13-1B is approved information for airplanes that do not have manuals, some repairs and some inspection procedures that have no reference in other places.
- **FAA Airworthiness Directives.** Study the procedures for Airworthiness Directives. They are mandatory. Compliance must be listed in the log book in specified manners. AD listings and compliance methods are required to be kept with the log books.
- **Service Manuals for expendable aftermarket parts (spark plugs, etc.).**
- **Operational Instructions for various pieces of shop equipment.** In some cases, an engine manufacturer will specify a specific shop tool or procedure. In most of those cases, that is exactly what you must use.
- **Instructions provided by a manufacturer holding Parts Manufacturing Authority.**
- **FAA Regulations concerning certification and operation.** These FARs are important places to obtain the scope and detail of required inspection items.
- **FAA Regulation part 43.** FAR part 43 is a section of the regulations that contains procedural instructions, as well as some performance standards. One of these standards has to do with how to correctly sign a log book: Title 14 Codified Federal Regulations — Chapter I — Part 43, Paragraph § 43.9 Content, form and disposition of maintenance, preventive maintenance, rebuilding and alteration records.

The FAA will test on the contents of this list. Learn it verbatim.

## Principles of Energy Transformation

A study of this section will help in understanding the basic operating principles of reciprocating engines. The principles that govern the

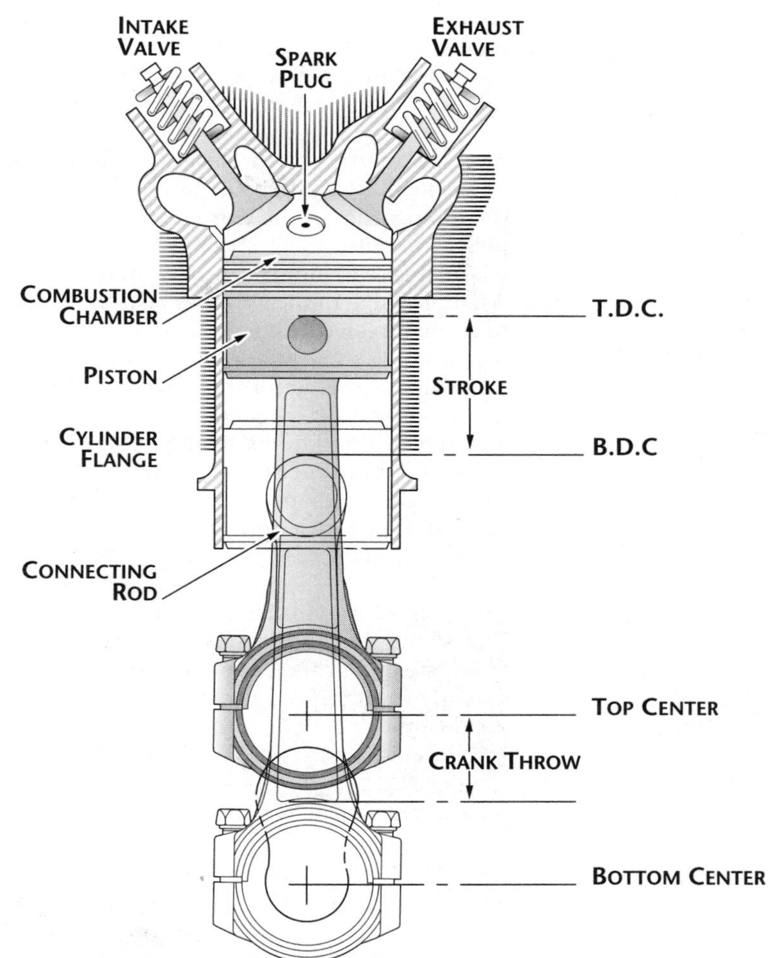


Figure 1-1-10. Components and terminology of engine operation

relationship between the pressure, volume and temperature of gases are the basic principles of engine operation.

An internal-combustion engine is a device for converting heat energy into mechanical energy. Gasoline is vaporized and mixed with air, forced or drawn into a cylinder, compressed by a piston, and then ignited by an electric spark. The conversion of the resultant heat energy into mechanical energy and then into work is accomplished in the cylinder. Figure 1-1-10 illustrates the various engine components necessary to accomplish this conversion and also presents the principal terms used to indicate engine operation.

The operating cycle of an internal combustion reciprocating engine includes the series of events required to induct, compress, ignite, burn and expand the fuel/air charge in the cylinder and to scavenge or exhaust the byproducts of the combustion process.

When the compressed mixture is ignited, the resultant gases of combustion expand very rapidly and force the piston to move away from the cylinder head. This downward motion of the piston, acting on the crankshaft through the connecting rod, is converted to a circular or rotary motion by the crankshaft.

A valve in the top or head of the cylinder opens to allow the burned gases to escape, and the momentum of the crankshaft and the propeller forces the piston back up in the cylinder, where it is ready for the next event in the cycle. Another valve in the cylinder head then opens to let in a fresh charge of the fuel/air mixture.

The valve allowing for the escape of the burning exhaust gases is called the exhaust valve, and the valve that lets in the fresh charge of the fuel/air mixture is called the intake valve. These valves are opened and closed mechanically at the proper times by the valve-operating mechanism.

The bore of a cylinder is its inside diameter. The stroke is the distance the piston moves from one end of the cylinder to the other, specifically from top dead center (TDC) to bottom dead center (BDC) or vice versa (see Figure 1-1-10, previous page).

## Operating Cycles

There are two operating cycles in use:

- The two-stroke cycle
- The four-stroke cycle

As the name implies, two-stroke-cycle engines require only one upstroke and one downstroke of the piston to complete the required series of events in the cylinder. Thus, the engine completes the operating cycle in one revolution of the crankshaft.

Most aircraft reciprocating engines operate on the four-stroke cycle, sometimes called the *Otto cycle*, after its originator, German physicist Nikolaus August Otto (1832-1891). The four-stroke-cycle engine has many advantages for use in aircraft. One advantage is that it lends itself readily to high performance through supercharging.

In this type of engine, four strokes are required to complete the required series of events or operating cycle of each cylinder. Two complete revolutions of the crankshaft (720°)

Figure 1-1-11. One of the most popular two-cycle engines used in experimental aircraft is the Rotax.The two-stroke cycle engine is actually simpler than the four-stroke cycle engine, but it is less fuel efficient.

are required for the four strokes; thus, each cylinder in an engine of this type fires once in every two revolutions of the crankshaft.

## *Two-stroke Cycle*

Four-stroke cycle engines power most of today's aircraft, but two-stroke cycle engines are used on a number of home-built aircraft and to operate auxiliary equipment (see Figure 1-1-11).

The two-stroke cycle engine (Figure 1-1-12) contains most of the same basic components of the four-stroke cycle engine, such as the crankcase, crankshaft, camshaft, pistons and valves. The main difference is in the valve arrangement and the fuel intake system.

As the piston moves up in the cylinder, a low pressure is created in the crankcase. This low pressure draws the fuel air mixture into the crankcase, from the carburetor, through a check valve. When the piston reaches TDC, the check valve is closed and the crankcase is filled with fuel-air mixture. As the piston starts downward in the cylinder, the fuel-air mixture in the crankcase is compressed. Then, when the piston reaches BDC, the intake port is open and allows the compressed fuel-air mixture to enter the cylinder head. This is the *intake event.*

As the piston moves back up the cylinder, the intake port is closed and the fuel-air mixture is compressed even more. During this event, more fuel-air is being drawn into the crankcase at the same time. This is called the *compression event.*

Just prior to the piston reaching TDC, the spark plug fires and drives the piston downward. This is the *power event.*

As the piston reaches the lower section of the cylinder, the exhaust port is opened to allow the exhaust gases to escape. The exhaust port is opened immediately prior to the intake port being opened. When the intake port is opened, the fresh air coming in the cylinder helps to scavenge the exhaust gases out.

In the two-stroke cycle engine, it takes one complete revolution of the crankshaft to complete the cycle of operation.

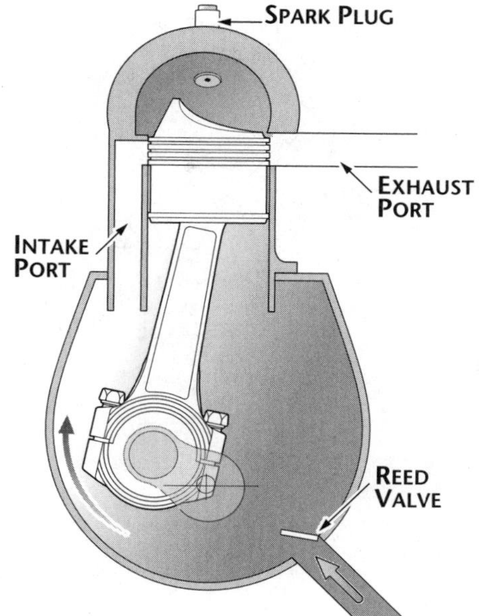


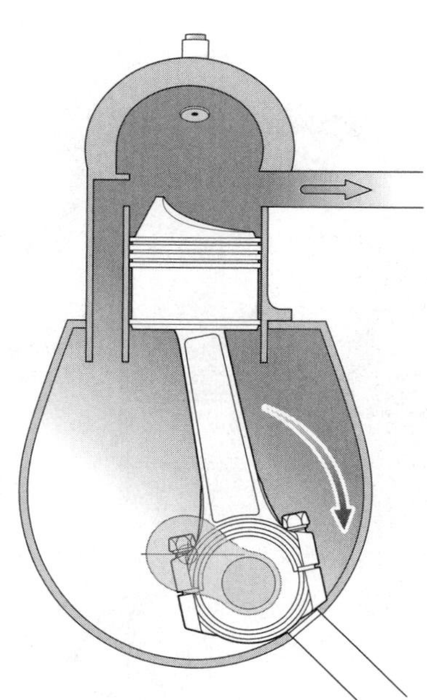


**Figure 1-1-12. Two-stroke cycle**

## *Four-stroke Cycle*

In the four-stroke-cycle engine operation, the timing of the ignition and the valve events will vary considerably in different engine designs. Many factors influence the timing of a specific engine, and it is most important that the engine manufacturer's recommendations be followed in maintenance and overhaul. The timing of the valve and ignition events is always specified in degrees of crankshaft travel. The four-stroke cycle is illustrated in Figure 1-1-13 (next page).

In the following paragraphs, the timing of each event is specified in terms of degrees of crankshaft travel on the stroke during which the event occurs. A certain amount of crankshaft

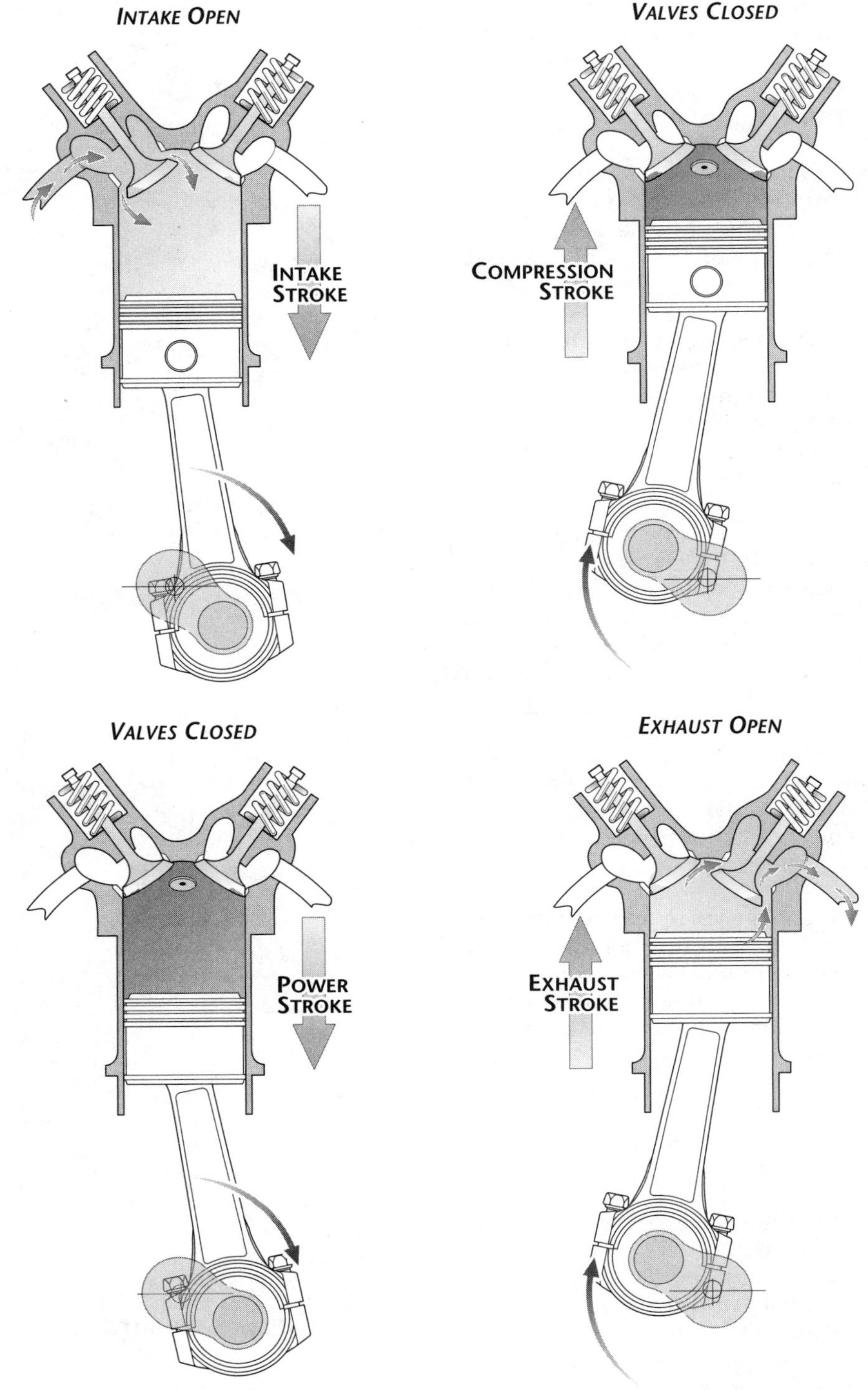


**Figure 1-1-13. Four-stroke cycle**

travel is required to open a valve fully; therefore, the specified timing represents the start of opening rather than the full-open position of the valve.

**Intake stroke.** During the intake stroke, the piston is pulled downward in the cylinder by the rotation of the crankshaft. This reduces the pressure in the cylinder and causes air under atmospheric pressure to flow through the carburetor, which meters the correct amount of fuel. The fuel/air mixture passes through the intake pipes and intake valves into the cylinders. The quantity or weight of the fuel/air charge depends upon the degree of throttle opening.

The intake valve is opened considerably before the piston reaches TDC on the exhaust stroke, in order to induce a greater quantity of the fuel/air charge into the cylinder and thus increase the horsepower. The distance the valve may be opened before TDC, however, is limited by several factors, such as the possibility that hot gases remaining in the cylinder from the previous cycle may flash back into the intake pipe and the induction system.

In all high-power aircraft engines, both the intake and the exhaust valves are off the valve seats at TDC at the start of the intake stroke. As mentioned above, the intake valve opens before TDC on the exhaust stroke (valve lead), and the closing of the exhaust valve is delayed considerably after the piston has passed TDC and has started the intake stroke (valve lag). This timing is called *valve overlap* and is designed to aid in cooling the cylinder internally by circulating the cool incoming fuel/air mixture, to increase the amount of the fuel/air mixture induced into the cylinder and to aid in scavenging the by products of combustion.

The intake valve is timed to close about 50°-75° past BDC on the compression stroke, depending upon the specific engine, to allow the momentum of the incoming gases to charge the cylinder more completely. Because of the comparatively large volume of the cylinder above the piston when the piston is near BDC, the slight upward travel of the piston during this time does not have a great effect on the incoming flow of gases. This late timing can be carried too far, because the gases may be forced back through the intake valve and defeat the purpose of the late closing.

**Compression stroke.** After the intake valve is closed, the continued upward travel of the piston compresses the fuel/air mixture to obtain the desired burning and expansion characteristics.

The charge is fired by means of an electric spark as the piston approaches TDC. The time of ignition will vary from 20° to 35° before TDC, depending upon the requirements of the specific engine, to ensure complete combustion of the charge by the time the piston is slightly past the TDC position.

Many factors affect ignition timing, and the engine manufacturer has expended considerable time in research and testing to determine the best setting. All engines incorporate devices for adjusting the ignition timing, and it is most important that the ignition system be timed according to the engine manufacturer's recommendations.

**Power stroke.** As the piston moves through the TDC position at the end of the compression stroke and starts down on the power stroke, it is pushed downward by the rapid expansion of the burning gases within the cylinder head with a force that can be greater than 15 tons (30,000 p.s.i.) at maximum power output of the engine. The temperature of these burning gases may be 3,000°-4,000°F.

As the piston is forced downward during the power stroke by the pressure of the burning gases exerted upon it, the downward movement of the connecting rod is changed to rotary movement by the crankshaft. Then the rotary movement is transmitted to the propeller shaft to drive the propeller. As the burning gases are expanded, the temperature drops to within safe limits before the exhaust gases flow out through the exhaust port.

The timing of the exhaust valve opening is determined by, among other considerations, the desirability of using as much of the expansive force as possible and of scavenging the cylinder as completely and rapidly as possible. The valve is opened considerably before BDC on the power stroke (on some engines, at 50° and 75° before BDC) while there is still some pressure in the cylinder. This timing is used so that the pressure can force the gases out of the exhaust port as soon as possible. This process frees the cylinder of waste heat after the desired expansion has been obtained and avoids overheating the cylinder and the piston. Thorough scavenging is very important. because any exhaust products remaining in the cylinder will dilute the incoming fuel/air charge at the start of the next cycle.

**Exhaust stroke.** As the piston travels through BDC at the completion of the power stroke and starts upward on the exhaust stroke, it will begin to push the burned exhaust gases out the exhaust port. The speed of the exhaust gases leaving the cylinder creates a low pressure in the cylinder. This low or reduced pressure speeds the flow of the fresh fuel/air charge into the cylinder as the intake valve is beginning to open. The intake valve opening is timed to occur at 8° to 55° before TDC on the exhaust stroke on various engines.

## Diesel Power for Aircraft

While not new, diesel-powered aircraft engines are being given serious consideration at the opening of this century. The first flight of a diesel-powered aircraft was made on Sept. 19, 1928, in a Stinson Detroiter, which was powered by a radial diesel engine built by Packard Motor Corporation. It was reported to have performed admirably at lower altitudes, but did not do so well up high. The project was discontinued, and diesel power for aircraft was not given much further consideration for a very long time.

**Figure 1-1-14. A modern diesel aircraft engine, the Snecma Renault 230 hp. four-cylinder engine has been type-certified in the U.S.**

Prompted by the promise that leaded aviation fuels are rapidly moving toward extinction, several companies are again taking a look at diesel power for aircraft engines. Using essentially the same fuel as turbine engines gives the industry a one-fuel possibility (see Figure 1-1-14).

Diesel engines may be built to operate with either a two-stoke cycle or a four-stroke cycle. Besides the obvious fuel differences, diesel engines are characterized by higher compression ratios and lower operating temperatures. While gasoline-powered aircraft engines employ compression ratios of 7.5:1 to 9:1, diesels may be as high as 20:1

Fuel is injected directly into the cylinder of a diesel engine, and the heat of compression provides the ignition. The high compression ratios produce more complete combustion and better thermal efficiency than is possible from gas engines. This reduces environmental pollution, especially carbon dioxide.

## Reciprocating Engine Work and Power

All aircraft engines are rated according to their ability to do work and produce power. This section presents an explanation of work and power and how they are calculated. Also discussed are the various efficiencies that govern the power output of a reciprocating engine.

**Work.** In the study of physics, work is defined as the amount of force applied to an object multiplied by the distance the object moves.

Work (W) = Force (F) x Distance (D)

The actual work achieved is measured by several standards. One of the most common is the foot-pound. That is, when a 1-lb. mass is moved 1 ft., 1 ft.-lb. of work has been performed. The greater the weight of the mass and the greater the distance it is moved, the greater the work accomplished.

**Power.** Power is defined as the time-rate of doing work. In the English system, power is expressed in foot-pounds per second. In the metric system, it is expressed in meter-kilograms per second. The formula to find power is:

$$\text{Power (P)} = \frac{\text{Work (w)}}{\text{Time (T)}}$$

For many applications in our industry, horsepower is used as measurement of power.

**Horsepower.** The common unit of mechanical power is the *horsepower* (hp.). Late in the 18th century, Scottish engineer James Watt (1736-1819), the inventor of the steam engine, found that an English workhorse could labor at the rate of 550 ft.-lbs. per second, or 33,000 ft.-lbs. per minute, for a reasonable length of time. From his observations came the term horsepower, which is the standard unit of power in the English system of measurement. To calculate the horsepower rating of an engine, divide the power developed in foot-pounds per minute by 33,000, or the power in foot-pounds per second by 550.

$$\text{hp.} = \frac{\text{feet-pounds per minute}}{33{,}000}$$

or

$$\text{hp.} = \frac{\text{feet-pounds per second}}{550}$$

Work is the product of force and distance, and power is work per unit of time.

**EXAMPLE:** If a 33,000-lb. weight is lifted through a vertical distance of 1 ft. in 1 min., the power expended is 33,000 ft.-lbs. per min., or exactly 1 hp.

Work is performed not only when a force is applied for lifting; force may be applied in any direction. If a 100-lb. weight is dragged along the ground, a force is still being applied to perform work, although the direction of the resulting motion is approximately horizontal. The amount of this force would depend upon the roughness of the ground.

**EXAMPLE:** If the weight were attached to a spring scale graduated in pounds, then dragged by pulling on the scale handle, the amount of force required could be measured. Assume that the force required is 90 lbs., and the 100-lb. weight is dragged 660 ft. in 2 min. The amount of work performed in that 2 min. will be 59,400 ft.-lbs., or 29,700 ft.-lbs. per min. Since 1 hp. is 33,000 ft.-lbs. per min., the hp. expended in this case will be 29,700 divided by 33,000, or 0.9 hp.

## *Piston Displacement*

When other factors remain equal, the greater the piston displacement, the greater the maximum horsepower an engine will be capable of developing. When a piston moves from BDC to TDC, it displaces a specific volume. The volume displaced by the piston is known as piston displacement, which is expressed in cubic inches for most American-made engines and cubic centimeters for foreign-made engines.

The piston displacement of one cylinder may be obtained by multiplying the area of the cross section of the cylinder by the total distance the piston moves in the cylinder in one stroke. For multi-cylinder engines, this product is multiplied by the number of cylinders to get the total piston displacement of the engine.

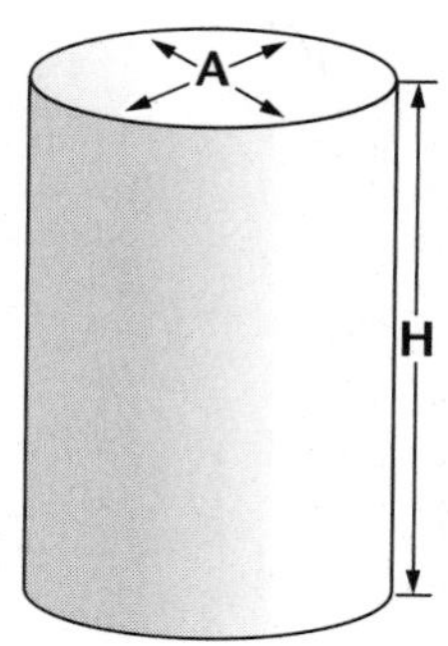

**Figure 1-1-15. A cylinder**

Since the volume (V) of a geometric cylinder equals the area (A) of the base multiplied by the altitude (H), it is expressed mathematically as:

$$V = A \times H$$

For our purposes, the area of the base is the area of the circular cross section of the cylinder or of the piston top (Figure 1-1-15).

**Area of a circle.** To find the area of a circle, it is necessary to use a number called Pi ($\pi$). This number represents the ratio of the circumference to the diameter of any circle. Pi cannot be found with exact accuracy, because it is a never-ending decimal, but expressed to four decimal places it is 3.1416, which is accurate enough for most computations.

The area of a circle, as in a rectangle or triangle, must be expressed in square units. The distance that is one-half the diameter of a circle is known as the radius. The area of any circle is found by squaring the radius and multiplying by pi. The formula is expressed thus:

$$A = \pi R^2$$

Where A is the area of a circle, pi is the given constant and r is the radius, which is equal to half the diameter, or D divided by two.

**EXAMPLE:** Compute the piston displacement of a four-cylinder aircraft engine that has a bore of 3.780 inches and a stroke of 4.000 inches. Start by finding the volume of a single cylinder.

Substitute values into these formulas and complete the calculation.

$$V = \pi R^2 H$$

$$V = 3.1416 \times (1.89)^2 \times 4$$

$$V = 44.880 \text{ cubic inches per cylinder}$$

$$44.880 \times 4 \text{ pistons} = 179.520 \text{ cubic inches}$$

Rounded off to the next whole number, total piston displacement equals 180 cubic inches.

## *Compression Ratio*

All internal-combustion engines must compress the fuel/air mixture to receive a reasonable amount of work from each power stroke. The fuel/air charge in the cylinder can be compared to a coil spring: The more it is compressed, the more work it is potentially capable of doing.

The *compression ratio* of an engine (Figure 1-1-16) is a comparison of the volume of space in a cylinder when the piston is at the bottom of the stroke to the volume of space when the piston is at the top of the stroke. This comparison is expressed as a ratio, hence the term compression ratio. Compression ratio is a controlling factor in the maximum horsepower developed by an engine, but it is limited by present-day fuel grades and the high engine speeds and manifold pressures required for takeoff.

**EXAMPLE:** If there is 140 cu. in. of space in the cylinder when the piston is at the bottom and there are 20 cu. in. of space when the piston is at the top of the stroke, the compression ratio would be 140 to 20. If this ratio were expressed in fraction form, it would be 140/20, or 7 to 1, usually represented as 7:1.

To grasp more thoroughly the limitation placed on compression ratios, manifold pressure and its effect on compression pressures should be understood. Manifold pressure is the average absolute pressure of the air or fuel/air charge in the intake manifold and is measured in units of inches of mercury (Hg). Manifold pressure is dependent on engine speed (throttle setting) and supercharging. The engine-driven internal supercharger (blower) and the external exhaust-driven supercharger (turbo) are actually centrifugal-type air compressors. The operation of these superchargers increases the weight of the charge entering the cylinder. When either one or both are used with the aircraft engine, the manifold pressure may be considerably higher than the pressure of the outside atmosphere. The advantage of this condition is that a greater amount of charge is forced into a given cylinder volume, and a greater power results.

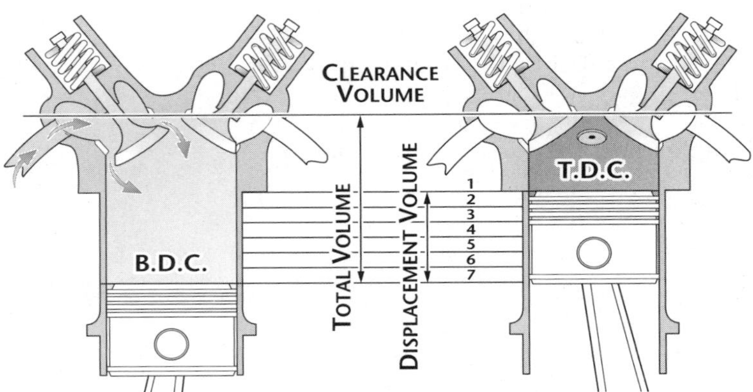


Figure 1-1-16. Compression ratio

Compression ratio and manifold pressure determine the pressure in the cylinder in that portion of the operating cycle when both valves are closed. The pressure of the charge before compression is determined by manifold pressure, while the pressure at the height of compression (just prior to ignition) is determined by manifold pressure times the compression ratio.

**EXAMPLE:** If an engine were operating at a manifold pressure of 30 inches Hg with a compression ratio of 7:1, the pressure at the instant before ignition would be approximately 210 inches Hg. At a manifold pressure of 60 inches Hg, the pressure would be 420 inches Hg.

Without going into great detail, it has been shown that the compression event magnifies the effect of varying the manifold pressure, and the magnitude of both affects the pressure of the fuel charge just before the instant of ignition. If the pressure at this time becomes too high, premature ignition or knock will occur and produce overheating.

One of the reasons for using engines with high compression ratios is to obtain long-range fuel economy, that is, to convert more heat energy into useful work than is done in engines of low compression ratio. Since more heat of the charge is converted into useful work, less heat is absorbed by the cylinder wall. This factor promotes cooler engine operation, which in turn increases the thermal efficiency.

Here, again, a compromise is needed between the demand for fuel economy and the demand for maximum horsepower without knocking. Some manufacturers of high-compression engines suppress knock at high manifold pressures by injecting an antiknock fluid into the fuel/air mixture. The fluid acts primarily as a coolant so that the engine can deliver more power for short periods, such as at takeoff and during emergencies, when power is critical. This high power should be used for short periods only.

## *Indicated Horsepower*

The indicated horsepower produced by an engine is the horsepower calculated from the indicated *mean effective pressure* and the other factors that affect the power output of

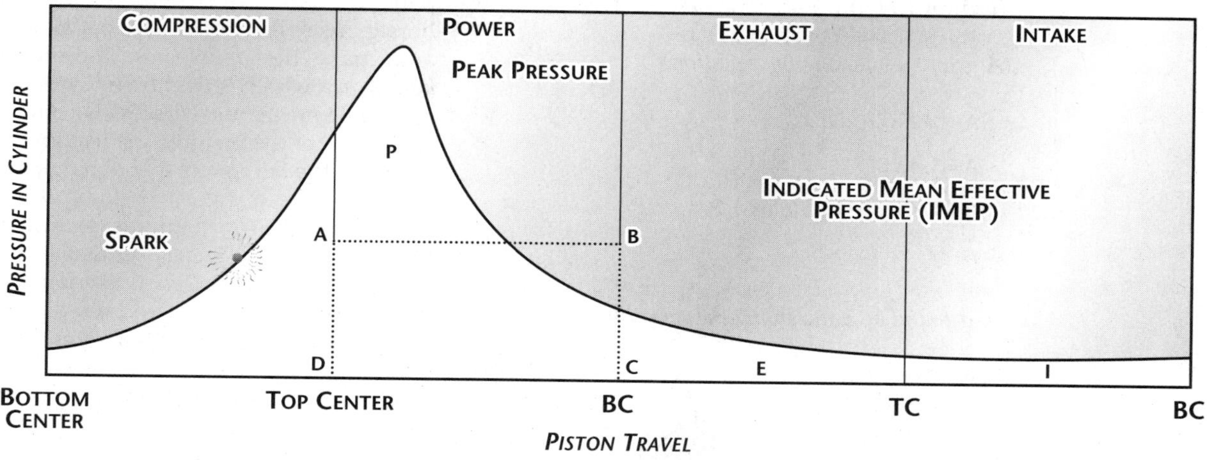


Figure 1-1-17. Cylinder pressure during power cycle

an engine. Indicated horsepower is the power developed in the combustion chambers without reference to friction losses within the engine.

This horsepower is calculated as a function of the actual cylinder pressure recorded during engine operation. To facilitate the indicated horsepower calculations, a mechanical indicating device, attached to the engine cylinder, scribes the actual pressure existing in the cylinder during the complete operating cycle. This pressure variation can be represented by the kind of graph shown in Figure 1-1-17. Notice that the cylinder pressure rises on the compression stroke, reaches a peak after TDC then decreases as the piston moves down on the power stroke.

Since the cylinder pressure varies during the operating cycle, an average pressure, line AB, is computed. This average pressure, if applied steadily during the time of the power stroke, would do the same amount of work as the varying pressure during the same period. This average pressure is known as indicated mean effective pressure and is included in the indicated horsepower calculation with other engine specifications. If the characteristics and the indicated mean effective pressure of an engine are known, it is possible to calculate the indicated horsepower rating.

The indicated horsepower for a four-stroke-cycle engine can be calculated from the following formula, in which the letter symbols in the numerator are arranged to spell the word PLANK to assist in memorizing the following formula:

P = indicated mean effective Pressure, in p.s.i.

L = Length of the stroke in feet or in fractions of a foot

A = Area of the piston head or cross-sectional area of the cylinder, in square inches

N = Number of power strokes per minute, (r.p.m. divided by 2)

K = number of Cylinders

$$\text{indicated hp.} = \frac{PLANK}{33{,}000}$$

In the formula above, the area of the piston times the indicated mean effective pressure gives the force acting on the piston in pounds. This force, multiplied by the length of the stroke in feet, gives the work performed in one power stroke, which, multiplied by the number of power strokes per minute, gives the number of foot-pounds per minute of work produced by one cylinder. Multiplying this result by the number of cylinders in the engine gives the amount of work performed, in foot-pounds, by the engine. Since horsepower is defined as work done at the rate of 33,000 ft/lb. per minute, the total number of foot-pounds of work performed by the engine is divided by 33,000 to find the indicated horsepower.

Given:

indicated mean effective pressure (P) = 165 lbs./sq. in.

stroke (L) = 6 in. or 0.5 ft.

bore = 5.5 in.

r.p.m. = 3,000

no. of cylinders (K) = 12

$$\text{indicated hp.} = \frac{PLANK}{33{,}000 \text{ feet-pounds per minute}}$$

Find indicated hp.:

*(A is found by using the equation):*

$A = 1/2\ \pi\ (1/2D)^2$

$A = 1/4\pi\ D^2$

$A = 1/4 \times 3.1416 \times 5.5 \times 5.5$

A = 23.76 sq inches

*(N is found by multiplying the r.p.m. by 1/2):*

$N = 1/2 \times 3{,}000$

N = 1500 r.p.m.

Now, substituting in the formula:

$$\text{indicated hp.} = \frac{165 \times .5 \times 23.76 \times 1{,}500 \times 12}{33{,}000}$$

indicated hp. = 1069.20

## *Brake Horsepower*

The indicated horsepower calculation discussed in the preceding paragraph is the theoretical power of a frictionless engine. The total horsepower lost in overcoming friction must be subtracted from the indicated horsepower to arrive at the actual horsepower delivered to the propeller. The power delivered to the propeller for useful work is known as b.hp. (brake horsepower). The difference between indicated and brake horsepower is known as *friction horsepower,* which is the horsepower required to overcome mechanical losses such as the pumping action of the pistons, the friction of the pistons and the friction of all moving parts.

In practice, the measurement of an engine's b.hp. involves the measurement of a quantity known as torque, or twisting moment. Torque is the product of a force and the distance of the force from the axis about which it acts, or:

torque = force x distance (at right angles to the force)

Torque is a measure of load and is properly expressed in pound-inches (lb.-in.) or pound-feet (lb.-ft.) and should not be confused with work, which is expressed in inch-pounds (in.-lbs.) or foot-pounds (ft.-lbs.).

There are a number of devices for measuring torque, of which the Prony brake, dynamometer and torquemeter are examples. Typical of these devices is the Prony brake (see Figure 1-1-18), which measures the usable power output of an engine on a test stand. It consists essentially of a hinged collar, or brake, which can be clamped to a drum, splined to the propeller shaft. The collar and drum form a friction brake, which can be adjusted by a wheel. An arm of a known

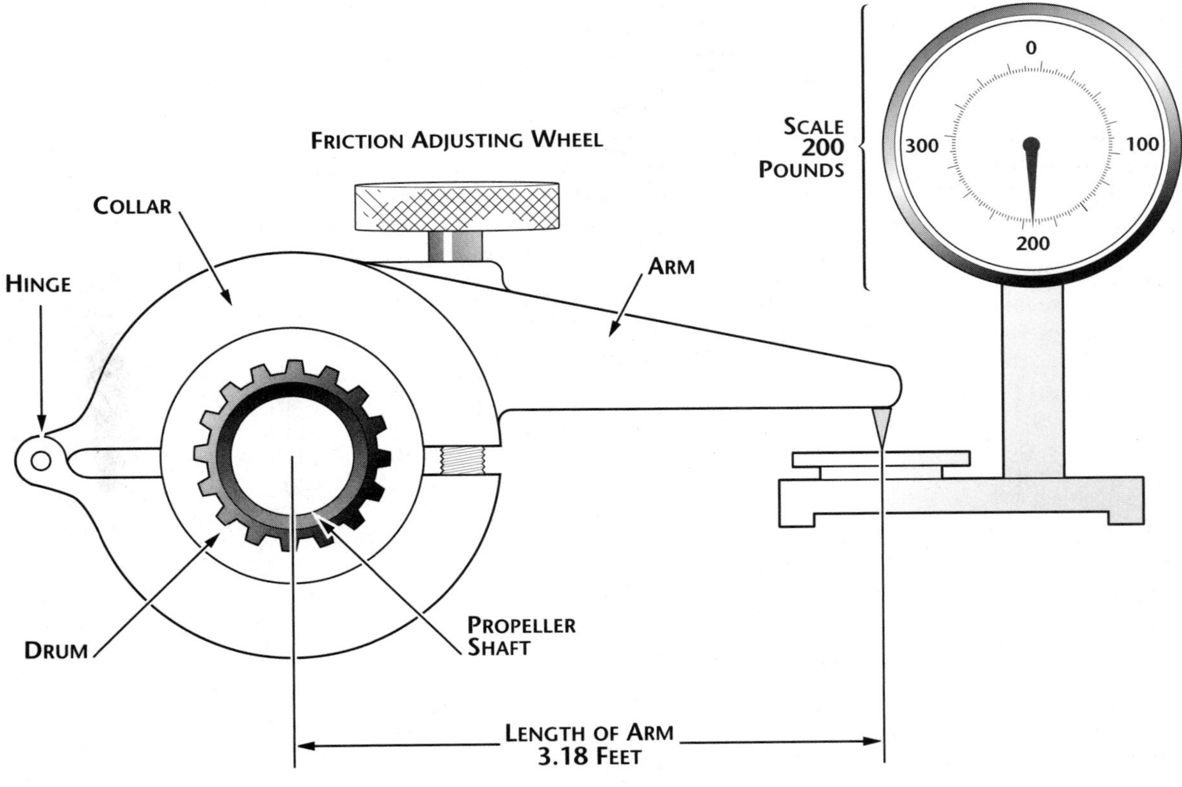


Figure 1-1-18. Typical Prony brake

length is rigidly attached to or is a part of the hinged collar and terminates at a point that bears on a set of scales.

As the propeller shaft rotates, it tries to carry the hinged collar of the brake with it and is prevented from doing so only by the arm that bears on the scale. The scale reads the force necessary to arrest the motion of the arm. If the resulting force registered on the scale is multiplied by the length of the arm, the resulting product is the torque exerted by the rotating shaft.

**EXAMPLE:** If the scale registers 200 lbs. and the length of the arm is 3.18 ft., the torque exerted by the shaft is:

200 lb. x 3.18 ft. = 636 lb.-ft.

Once the torque is known, the work done per revolution of the propeller shaft can be computed without difficulty by the equation:

work per revolution = $2\pi$ x torque

If work per revolution is multiplied by the r.p.m., the result is work per minute, or power. If the work is expressed in ft.-lbs. per min., this quantity is divided by 33,000; the result is the brake horsepower of the shaft. In other words:

power = work per revolution x r.p.m.

and

$$\text{b.hp.} = \frac{\text{work per revolution x r.p.m.}}{33{,}000}$$

or

$$\text{b.hp.} = \frac{2\pi \text{ x force on scales x length of arm x r.p.m.}}{33{,}000}$$

Given:

force on scales = 200 lbs.

length of arm = 3.18 ft.

r.p.m. = 3,000

$\pi$ = 3.1416

Find b.hp.:

$$\text{b.hp.} = \frac{6.2832 \text{ x } 200 \text{ x } 3.18 \text{ x } 3{,}000}{33{,}000}$$

$$= 363 \text{ b.hp.}$$

As long as the friction between the brake collar and propeller shaft drum is great enough to impose an appreciable load on the engine, but is not great enough to stop the engine, it is not necessary to know the amount of friction between the collar and drum to compute the b.hp. If there were no load imposed, there would be no torque to measure, and the engine would run away. If the imposed load is so great that the engine stalls, there may be considerable torque to measure, but there will be no r.p.m. In either case, it is impossible to measure the b.hp. of the engine. However, if a reasonable amount of friction exists between the brake drum and the collar and the load is then increased, the tendency of the propeller shaft to carry the collar and arm about with it becomes greater, thus imposing a greater force upon the scales.

As long as the torque increase is proportional to the r.p.m. decrease, the horsepower delivered at the shaft remains unchanged. This can be seen from the equation in which $2\pi$ and 33,000 are constants and torque and r.p.m. are variables. If the change in r.p.m. is inversely proportional to the change in torque, their product will remain unchanged. Therefore, b.hp. remains unchanged. This is important, as it shows that horsepower is the function of both torque and r.p.m. and can be changed by changing either torque, r.p.m. or both.

## *Friction Horsepower*

As explained in the previous paragraphs, friction horsepower is the indicated horsepower minus brake horsepower. It is the horsepower used by an engine in overcoming the friction of moving parts, drawing in fuel, expelling exhaust, driving oil and fuel pumps and the like. On modern aircraft engines, this power loss through friction may be as high as 10-15 percent of the indicated horsepower.

## *Friction and Brake Mean Effective Pressures*

The *IMEP* (*indicated mean effective pressure*) discussed previously is the average pressure produced in the combustion chamber during the operating cycle and is an expression of the theoretical, frictionless power known as *indicated horsepower.* In addition to completely disregarding power lost to friction, indicated horsepower gives no indication as to how much actual power is delivered to the propeller shaft for doing useful work. However, it is related to actual pressures that occur in the cylinder and can be used as a measure of these pressures.

To compute the friction loss and net power output, the indicated horsepower of a cylinder may be thought of as two separate powers, each producing a different effect. The first power overcomes internal friction, and the horsepower thus consumed is known as friction horsepower. The second power, known as brake horsepower, produces useful work at

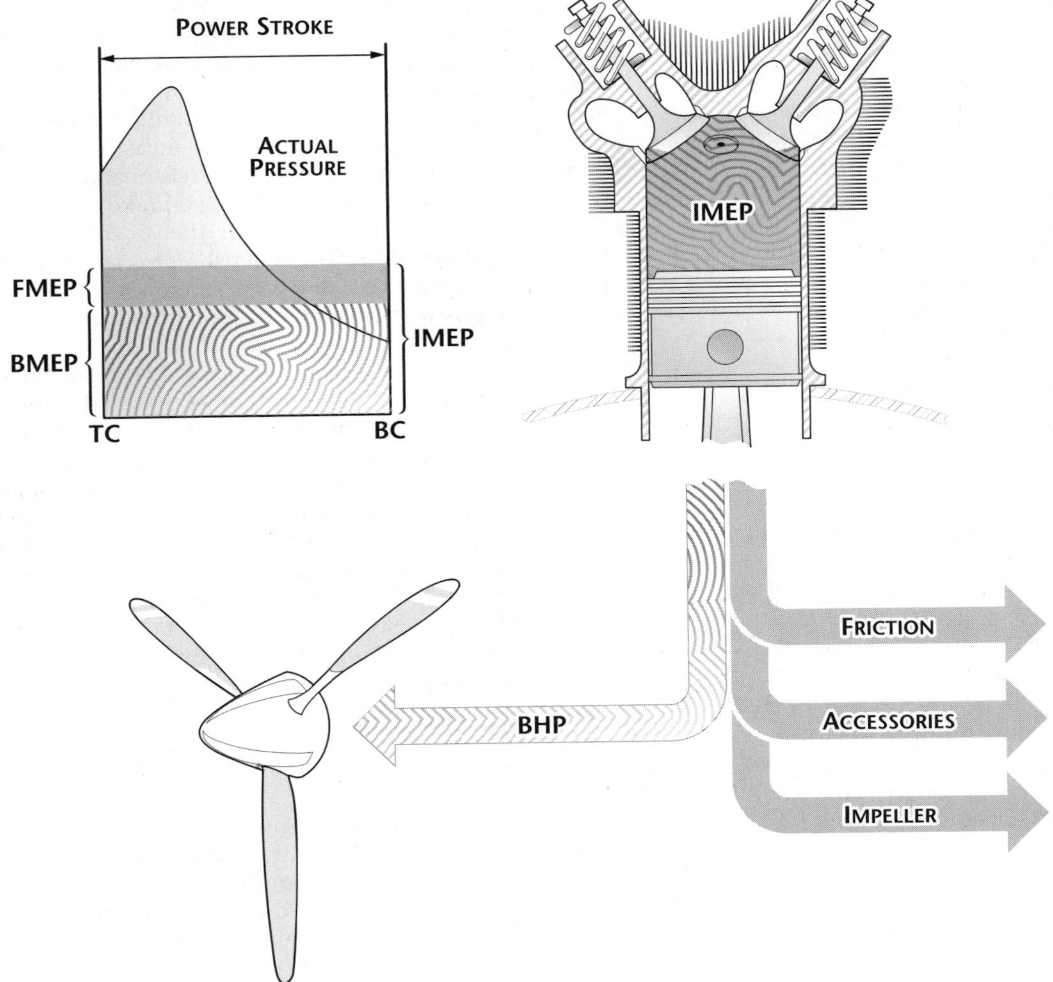


Figure 1-1-19. Power and pressures

the propeller. Logically, therefore, that portion of IMEP that produces brake horsepower is called *BMEP* (*brake mean effective pressure*). The remaining pressure used to overcome internal friction is called *FMEP* (*friction mean effective pressure*). This is illustrated in Figure 1-1-19. IMEP is a useful expression of total cylinder power output, but is not a real physical quantity; likewise, FMEP and BMEP are theoretical but useful expressions of friction losses and net power output.

Although BMEP and FMEP have no real existence in the cylinder, they provide a convenient means of representing pressure limits or rating engine performance throughout its entire operating range. This is true since there is a relationship between IMEP, BMEP and FMEP.

One of the basic limitations placed on engine operation is the pressure developed in the cylinder during combustion. In the discussion of compression ratios and indicated mean effective pressure, it was found that, within limits, the increased pressure resulted in increased power. It was also noted that if the cylinder pressure was not controlled within close limits, it would impose dangerous internal loads that might result in engine failure. It is therefore important to have a means of determining these cylinder pressures as a protective measure and for efficient application of power.

If the b.hp. is known, the BMEP can be computed by means of the following equation:

$$\text{BMEP} = \frac{\text{b.hp.} \times 33{,}000}{\text{PLANK}}$$

## Thrust Horsepower

Thrust horsepower (t.hp.) can be considered as the result of the engine and the propeller working together. If a propeller could be designed to be 100 percent efficient, the thrust- and the brake horsepower would be the same. However, the efficiency of the propeller varies with the engine speed, attitude, altitude, temperature and airspeed, thus the ratio of the thrust horsepower and the brake horsepower delivered to the propeller shaft will never be equal.

**EXAMPLE:** If an engine develops 1,000 b.hp., and it is used with a propeller having 85 percent efficiency, the thrust horsepower of that engine-propeller combination is 85 percent of 1,000, or 850 thrust hp. Of the four types of horsepower discussed, it is the thrust horsepower that determines the performance of the engine-propeller combination.

## Efficiencies

**Thermal efficiency.** Any study of engines and power involves consideration of heat as the source of power. The heat produced by the burning of gasoline in the cylinders causes a rapid expansion of the gases in the cylinder and this, in turn, moves the pistons and creates mechanical energy.

It has long been known that mechanical work can be converted into heat and that a given amount of heat contains the energy equivalent of a certain amount of mechanical work. Heat and work are theoretically interchangeable and bear a fixed relation to each other. Heat can therefore be measured in work units (for example, ft.-lbs.) as well as in heat units. The BTU (British thermal unit) of heat is the quantity of heat required to raise the temperature of 1 lb. of water 1°F. It is equivalent to 778 ft.-lbs. of mechanical work. A pound of petroleum fuel, when burned with enough air to consume it completely, gives up about 20,000 BTU, the equivalent of 15,560,000 ft.-lbs. of mechanical work. These quantities express the heat energy of the fuel in heat and work units, respectively.

The ratio of useful work done by an engine to the heat energy of the fuel it uses, expressed in work or heat units, is called the thermal efficiency of the engine. If two similar engines use equal amounts of fuel, the engine that converts into work the greater part of the energy in the fuel (higher thermal efficiency) will deliver the greater amount of power. Furthermore, the engine that has the higher thermal efficiency will have less waste heat to dispose of to the valves, cylinders, pistons and cooling system of the engine. A high thermal efficiency also means a low specific fuel consumption and, therefore, less fuel for a flight of a given distance at a given power. Thus, the practical importance of a high thermal efficiency is threefold, and it constitutes one of the most desirable features in the performance of an aircraft engine.

Of the total heat produced, 25-30 percent is utilized for power output; 15-20 percent is lost in cooling (heat radiated from cylinder head fins); 5-10 percent is lost in overcoming friction of moving parts; and 40-45 percent is lost through the exhaust. Anything that increases the heat content that goes into mechanical work on the piston, which reduces the friction and pumping losses or which reduces the quantity of unburned fuel or the heat lost to the engine parts, increases the thermal efficiency.

The portion of the total heat of combustion that is turned into mechanical work depends to a great extent upon the compression ratio. Compression ratio is the ratio of the piston displacement plus combustion chamber space to the combustion chamber space. Other things being equal, the higher the compression ratio, the larger the proportion of the heat energy of combustion turned into useful work at the crankshaft. On the other hand, increasing the compression ratio increases the cylinder head temperature. This is a limiting factor, for the extremely high temperature created by high compression ratios causes the material in the cylinder to deteriorate rapidly and the fuel to detonate.

The thermal efficiency of an engine may be based on either b.hp. or i.hp. and is represented by the formula:

$$\text{indicated thermal efficiency} = \frac{\text{i.hp.} \times 33{,}000}{\text{weight of fuel burned per minute} \times \text{heat value} \times 778}$$

The formula for brake thermal efficiency is the same as shown above except the value for b.hp. is inserted instead of the value for i.hp.

Reciprocating engines are only about 34 percent thermally efficient; that is, they transform only about 34 percent of the total heat produced by the burning fuel into mechanical energy. The remainder of the heat is lost through the exhaust gases, the cooling system and the friction within the engine.

**Mechanical efficiency.** Mechanical efficiency is the ratio that shows how much of the power developed by the expanding gases in the cylinder is actually delivered to the output shaft. It is a comparison between the b.hp. and the i.hp. It can be expressed by the formula:

$$\text{mechanical efficiency} = \frac{\text{b.hp.}}{\text{i.hp.}}$$

As discussed earlier, brake horsepower is the useful power delivered to the propeller shaft. Indicated horsepower is the total horsepower developed in the cylinders. The difference between the two is friction horsepower (f.hp.): the power lost in overcoming friction.

The factor that has the greatest effect on mechanical efficiency is the friction within the engine itself. The friction between moving parts in an engine remains practically constant

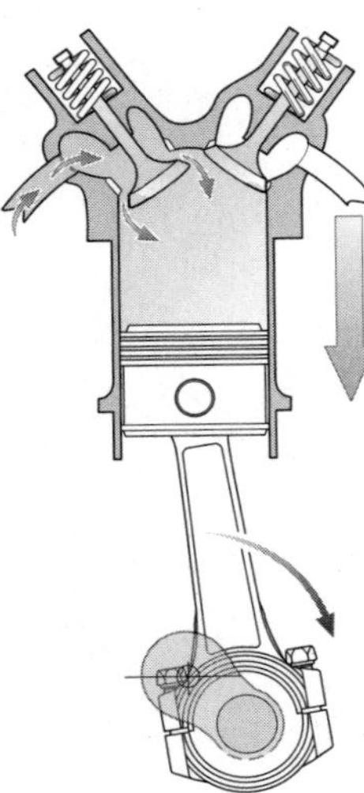


Figure 1-1-20. The intake stroke

throughout an engine's speed range. Therefore, the mechanical efficiency of an engine will be highest when the engine is running at the r.p.m. at which maximum b.hp. is developed. Mechanical efficiency of the average aircraft reciprocating engine approaches 90 percent.

**Volumetric efficiency.** Volumetric efficiency, another engine efficiency, is a ratio expressed in terms of percentages. It is a comparison of the volume of fuel/air charge (corrected for temperature and pressure) inducted into the cylinders to the total piston displacement of the engine. Various factors cause departure from a 100 percent volumetric efficiency.

The pistons of an un-supercharged engine displace the same volume each time they sweep the cylinders from top center to bottom center. The amount of charge that fills this volume on the intake stroke depends on the existing pressure and temperature of the surrounding atmosphere. Therefore, to find the volumetric efficiency of an engine, standards for atmospheric pressure and temperature had to be established. The *U.S. standard atmosphere* was established in 1958, providing the necessary pressure and temperature values to calculate volumetric efficiency.

The standard sea-level temperature is 59°F or 15°C. At this temperature, the pressure of one atmosphere is 14.69 lbs./sq. in., and this pressure will support a column of mercury 29.92 in. high. These standard sea-level conditions determine a standard density, and if the engine draws in a volume of charge of this density exactly equal to its piston displacement, it is said to be operating at 100 percent volumetric efficiency. An engine drawing in less volume than this has a volumetric efficiency lower than 100 percent. An engine equipped with a high-speed internal or external blower may have a volumetric efficiency greater than 100 percent. The equation for volumetric efficiency is:

$$\text{volumetric efficiency} = \frac{\text{volume of charge (corrected for temperature and pressure)}}{\text{piston displacement}}$$

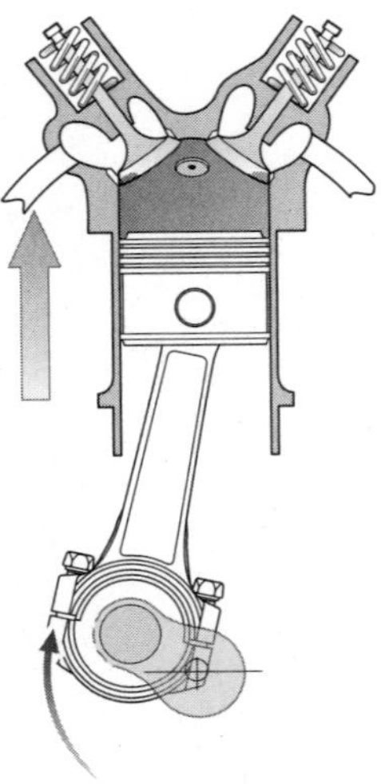


Figure 1-1-21. The compression stroke

Many factors decrease volumetric efficiency. Some of these are:

- Part-throttle operation
- Long intake pipes of small diameter
- Sharp bends in the induction system
- Carburetor air temperature too high
- Cylinder-head temperature too high
- Incomplete scavenging
- Improper valve timing

**Propulsive efficiency.** A propeller is used with an engine to provide thrust. The engine supplies b.hp. through a rotating shaft, and the propeller absorbs the b.hp. and converts it into thrust horsepower. In this conversion, some power is wasted. Since the efficiency of any machine is the ratio of useful power output to the power input, propulsive efficiency (in this case, propeller efficiency) is the ratio of thrust horsepower to b.hp.. On the average, t.hp. constitutes approximately 80 percent of the b.hp.. The other 20 percent is lost in friction and slippage. Controlling the blade angle of the propeller is the best method of obtaining maximum propulsive efficiency for all conditions encountered in flight.

During takeoff, when the aircraft is moving at low speeds and when maximum power and thrust are required, a low propeller blade angle will give maximum thrust. For high-speed flying or diving, the blade angle is increased to obtain maximum thrust and efficiency. The constant-speed propeller is used to give required thrust at maximum efficiency for all flight conditions.

## Basic Engine Operating Principles

A thorough understanding of the basic principles on which a reciprocating engine operates and the many factors that affect its operation is necessary to diagnose engine malfunctions. Some of these basic principles are reviewed not as a mere repetition of basic theory, but as a concrete, practical discussion of what makes for good or bad engine performance.

The conventional reciprocating aircraft engine operates on the four-stroke-cycle principle. Pressure from burning gases acts upon a piston, causing it to reciprocate back and forth in an enclosed cylinder. This reciprocating motion of the piston is changed into rotary motion by a crankshaft, to which the piston is coupled by means of a connecting rod. The crankshaft, in turn, is attached or geared to the aircraft propeller. Therefore, the rotary motion of the crankshaft causes the propeller to revolve. Thus, the motion of the propeller is a direct result of the forces acting upon the piston as it moves back and forth in the cylinder.

Four strokes of the piston, two up and two down, are required to provide one power impulse to the crankshaft. Each of these strokes is considered an event in the cycle of engine operation. Ignition of the gases (fuel/air mixture) at the end of the second, or compression, stroke makes a fifth event. Thus, the five events that make up a cycle of operation occur in four strokes of the piston.

As the piston moves downward on the first stroke (*intake stroke*), the intake valve is open and the exhaust valve is closed (Figure 1-1-20). As air is drawn through the carburetor gasoline is introduced into the stream of air, forming a combustible mixture.

On the second stroke, the intake closes and the combustible mixture is compressed as the piston moves upward. This is the *compression stroke* (Figure 1-1-21).

At the correct instant, an electric spark jumps across the terminals of the spark plug and ignites the fuel/air mixture. The ignition of the fuel/air mixture is timed to occur just slightly before the piston reaches TDC.

As the mixture burns, temperature and pressure rise rapidly. The pressure reaches maximum just after the piston has passed TDC. The expanding and burning gas forces the piston downward, transmitting energy to the crankshaft. This is the *power stroke* (Figure 1-1-22). Both intake and exhaust valves are closed at the start of the power stroke.

Near the end of the power stroke, the exhaust valve opens, and the burned gases start to escape through the exhaust port. On its return stroke, the piston forces out the remaining gases. This stroke, the *exhaust stroke* (Figure 1-1-23), ends the cycle. With the introduction of a new charge through the intake port, the action is repeated and the cycle of events occurs over and over again as long as the engine is in operation.

Ignition of the fuel charge must occur at a specific time in relation to crankshaft travel. The igniting device is timed to ignite the charge just before the piston reaches top center on the compression stroke. Igniting the charge at this point permits maximum pressure to build up at a point slightly after the piston passes over TDC. For ideal combustion, the point of ignition should vary with engine speed and with degree of compression, mixture strength and other factors governing the rate of burning. However, certain factors, such as the limited range of operating r.p.m. and the dangers of operating with incorrect spark settings, prohibit the use of variable spark control in most instances. Therefore, most aircraft ignition system units are timed to ignite the fuel/air charge at one fixed position (advanced).

On early models of the four-stroke-cycle engine, the intake valve opened at TDC (beginning of the intake stroke). It closed at BDC (end of intake stroke). The exhaust valve opened at BDC (end of power stroke) and closed at TDC (end of exhaust stroke). More efficient engine operation can be obtained by opening the intake valve several degrees before TDC and closing it several degrees after bottom center. Opening the exhaust valve before BDC and closing it after TDC, also improves engine performance. Since the intake valve opens before TDC exhaust stroke and the exhaust valve closes after TDC intake stroke, there is a period where both the intake and exhaust valves are open at the same time. This is known as *valve lap* or *valve overlap*. In valve timing, reference to piston or crankshaft position is always made in terms of before or after the TDC and BDC points, e.g.: ATC, BTC, ABC and BBC.

Opening the intake valve before the piston reaches top center starts the intake event while the piston is still moving up on the exhaust stroke. This aids in increasing the volume of charge admitted into the cylinder. The selected point at which the intake valve opens depends on the r.p.m. at which the engine normally operates. At low r.p.m. this early timing results in poor efficiency since the incoming charge is not drawn into and the exhaust gases are not expelled out of the cylinder with sufficient speed to develop the necessary momentum. Also, at low r.p.m. the cylinder is not well scavenged, and residual gases mix with the incoming fuel and are trapped during the compression stroke. Some of the incoming mixture is lost through the open exhaust port. However, the advantages obtained at normal operating r.p.m. will more than make up for the poor efficiency at low r.p.m. Another advantage of this valve timing is the increased fuel vaporization and beneficial cooling of the piston and cylinder.

Delaying the closing of the intake valve takes advantage of the inertia of the rapidly moving fuel/air mixture entering the cylinder. This ramming effect increases the charge over that which would be taken in if the intake valve closed at bottom center (end of intake stroke). The intake valve is actually open during the latter part of the exhaust stroke, all of the intake stroke and the first part of the compression stroke. Fuel/air mixture is taken in during all this time.

The *early opening and late closing* of the exhaust valve goes along with the intake valve timing to improve engine efficiency. The exhaust valve opens on the power stroke, several crankshaft degrees before the piston reaches BDC. This early opening aids in obtaining better scavenging of the burned gases. It also results in improved cooling of the cylinders, because of the early escape of the hot gases. Actually, on aircraft engines, the major portion of the exhaust gases and the unused heat escape before the piston reaches BDC. The burned gases continue to escape as the piston passes BDC, moves upward on the exhaust stroke and starts the next intake stroke. The late closing of the exhaust valve still further improves scavenging by taking advantage of the inertia of the rapidly moving outgoing gases. The exhaust

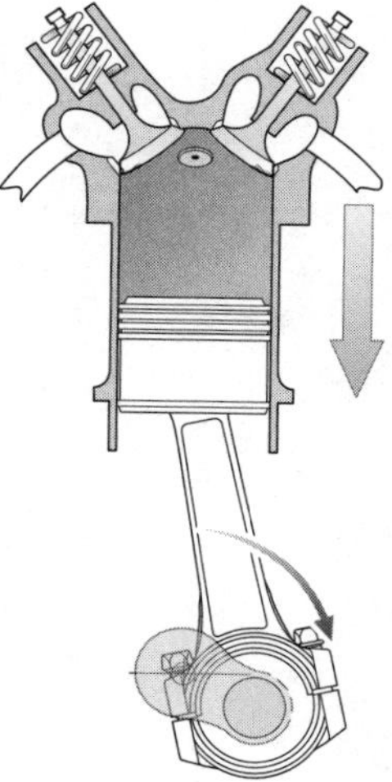


Figure 1-1-22. The power stroke

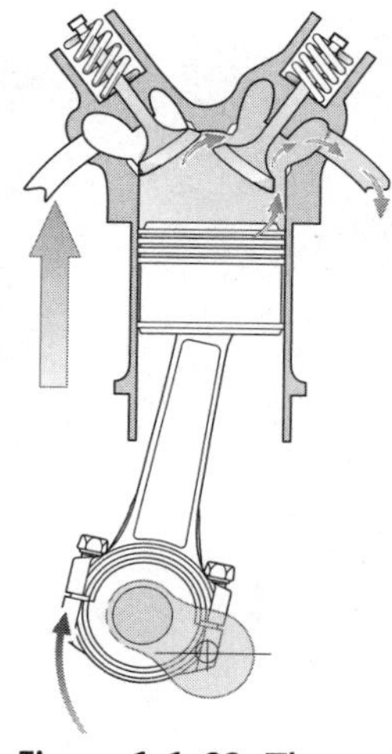


Figure 1-1-23. The exhaust stroke

valve is actually open during the latter part of the power stroke, all of the exhaust stroke and the first part of the intake stroke.

From this description of valve timing, it can be seen that the intake and exhaust valves are open at the same time on the latter part of the exhaust stroke and the first part of the intake stroke. During this valve overlap period, the last of the burned gases are escaping through the exhaust port while the fresh charge is entering through the intake port.

Supercharging an aircraft engine increases the pressure of the air or fuel/air mixture before it enters the cylinder. In other words, the air or fuel/air mixture is forced into the cylinder, rather than being drawn in. Supercharging increases engine efficiency and makes it possible for the engine to maintain its efficiency at high altitudes. This is true because the higher pressure packs more fuel/air charge into the cylinder during the intake event. This increase in weight of the charge results in a corresponding increase in power. In addition, the higher pressure of the incoming gases more forcibly ejects the burned gases out through the exhaust port. This results in better scavenging of the cylinder.

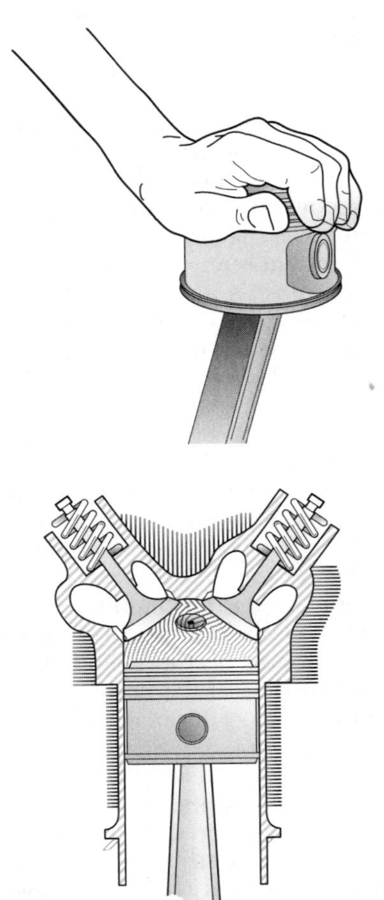

Figure 1-1-24. Normal combustion in a cylinder

## Combustion Process

Normal combustion occurs when the fuel/air mixture ignites in the cylinder and burns progressively at a relatively uniform rate across the combustion chamber. When ignition is properly timed, maximum pressure is built up just after the piston has passed TDC at the end of the compression stroke.

The flame fronts start at each spark plug and burn in more or less wavelike forms (Figure 1-1-24). The velocity of the flame travel is influenced by several factors:

- Type of fuel
- Ratio of the fuel/air mixture
- Pressure and temperature of the fuel mixture

**Normal combustion.** With normal combustion, the flame travel is about 100 ft./sec. The temperature and pressure within the cylinder rise at normal rates as the fuel/air mixture burns.

There is a limit, however, to the amount of compression and the degree of temperature rise that can be tolerated within an engine cylinder while still permitting normal combustion. All fuels have critical limits of temperature and compression. Beyond this limit, they will ignite spontaneously and burn with explosive violence. This instantaneous and explosive burning of the fuel/air mixture or, more accurately, of the latter portion of the charge, is called *detonation*.

As previously mentioned, during normal combustion, the flame fronts progress from the point of ignition across the cylinder. These flame fronts compress the gases ahead of them. At the same time, the upward movement of the piston is also compressing the gases. If the total compression on the remaining unburned gases exceeds the critical point, detonation occurs.

**Detonation.** Detonation (see Figure 1-1-25) then, is the spontaneous *combustion* of the unburned charge ahead of the flame fronts after ignition of the charge.

The explosive burning during detonation results in an extremely rapid pressure rise. This rapid pressure rise and the high instantaneous temperature combined with the high turbulence generated cause a scrubbing action on the cylinder and the piston. This can burn a hole completely through the piston.

The critical point of detonation varies with the ratio of fuel to air in the mixture. Therefore varying the fuel/air ratio can control the detonation characteristic of the mixture. At high power output combustion pressures and temperatures are higher than they are at low or medium power. Therefore, at high power the fuel/air ratio is made richer than is needed for good combustion at medium or low power output. This is done because, in general, a rich mixture will not detonate as readily as a lean mixture.

Unless detonation is heavy, there is no cockpit evidence of its presence. Light to medium detonation does not cause noticeable roughness, temperature increase or loss of power. As a result, it can be present during takeoff and high-power climb without being known to the crew.

In fact, the effects of detonation are often not discovered until after teardown of the engine. When the engine is overhauled, the presence of severe detonation during its operation is indicated by dished piston heads, collapsed valve heads, broken ring lands or eroded portions of valves, pistons or cylinder heads.

Basic protections from detonation are provided in the design of the engine carburetor setting, which automatically supplies the rich mixtures required for detonation suppression at high power; the rating limitations, which include the maximum operating temperatures; and selection of the correct grade of fuel. The design factors, cylinder cooling, magneto timing, mixture distribution, supercharging and carburetor setting are taken care of in the design and development of the engine and its method of installation in the aircraft.

The remaining responsibility for prevention of detonation, which often leads to pre-ignition, rests squarely in the hands of the ground and flight crews. They are responsible for observance of engine speed and manifold pressure limits. Proper use of supercharger and fuel mixture, and consistent maintenance of suitable cylinder head and carburetor-air temperatures are entirely within their control. Normal carbon deposits can also cause pre-ignition.

**Pre-ignition.** Pre-ignition, as the name implies, means that combustion takes place within the cylinder before the timed spark jumps across the spark plug terminals. Do not confuse pre-ignition, though, with the spark that occurs too early in the cycle. *Pre-ignition is caused by a hot spot in the combustion chamber, not by incorrect ignition timing.* This condition can often be traced to excessive carbon or other deposits which cause local hot spots. Detonation often leads to pre-ignition, which may also be caused by high-power operation on excessively lean mixtures. Abnormal carbon deposits can also cause pre-ignition, which are usually indicated in the cockpit by engine roughness, backfiring and by a sudden increase in cylinder head temperature.

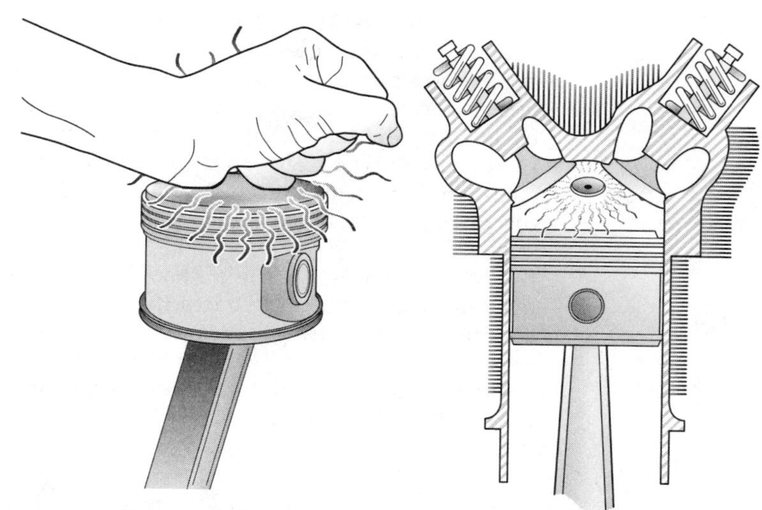

Figure 1-1-25. Detonation within a cylinder

Any area within the combustion chamber that becomes incandescent may serve as an igniter in advance of normal timed ignition and cause combustion earlier than desired. An area roughened and heated by detonation erosion may cause pre-ignition. A cracked valve or piston, or a broken spark plug insulator, may create a hot spot that functions as a glow plug. The hot spot can be caused by deposits on the chamber surfaces resulting from the use of leaded fuels.

The most obvious method of correcting pre-ignition is to reduce the cylinder temperature, which can be done immediately by retarding the throttle. This reduces the amount of fuel charge and the amount of heat generated. If the engine is at high power when pre-ignition occurs, retarding the throttle for a few seconds may provide enough cooling to chip off some of the lead, or other deposit, within the combustion chamber. These chipped-off particles pass out through the exhaust.

**Backfiring.** When a fuel/air mixture does not contain enough fuel to consume all the oxygen, it is called a *lean mixture.* Conversely, a charge that contains more fuel than required is called a *rich mixture.* An extremely lean mixture either will not burn at all or will burn so slowly that combustion is not complete at the end of the exhaust stroke. The flame lingers in the cylinder and then ignites the contents in the intake manifold or the induction system when the intake valve opens. This causes an explosion known as *backfiring,* which can damage the carburetor and other parts of the induction system.

A point worth stressing is that backfiring rarely involves the whole engine. Therefore, it is seldom the fault of the carburetor. In practically all cases, backfiring is limited to one or two cylinders. Usually, it is the result of faulty valve clearance setting, defective fuel injector nozzles or other conditions that cause these cylinders to operate leaner than the engine as a whole. There can be no permanent cure until these defects are discovered and corrected. Because these backfiring cylinders will fire intermittently and therefore run cool, they can be detected by the cold-cylinder check. The cold-cylinder check is discussed later in this chapter.

In some instances, an engine backfires in the idle range but operates satisfactorily at medium- and high-power settings. The most likely cause, in this case, is an excessively lean idle mixture. Proper adjustment of the idle fuel/air mixture usually corrects this difficulty.

Figure 1-1-26. Piston rings

Figure 1-1-27. Valve seat

**Afterfiring.** Afterfiring, sometimes called *afterburning*, often results when the fuel/air mixture is too rich. Overly rich mixtures are also slow burning. Because they contain more fuel than is required to burn the oxygen. Therefore, charges of unburned fuel are present in the exhausted gases. Air from outside the exhaust stacks mixes with this unburned fuel and it ignites, causing an explosion in the exhaust system. Afterfiring is perhaps more common where long exhaust ducting retains greater amounts of unburned charges. As in the case of backfiring, the correction for afterfiring is the proper adjustment of the fuel/air mixture.

Afterfiring can also be caused by cylinders not firing because of faulty spark plugs, defective fuel-injection nozzles or incorrect valve clearance. The unburned mixture from these dead cylinders passes into the exhaust system, where it ignites and burns. Unfortunately, the resultant torching or afterburning can easily be mistaken for evidence of a rich carburetor. Cylinders which are firing intermittently, can cause a similar effect. Again, only discovering the real cause and correcting the defect can remedy the malfunction. Either dead or intermittent cylinders can be located by the cold-cylinder check.

## Factors Affecting Engine Operation

**Compression.** To prevent loss of power, all openings to the cylinder must close and seal completely on the compression and power strokes. In this respect, there are three items in the proper operation of the cylinder that must be right for maximum efficiency. First, the piston rings (Figure 1-1-26) must be in good condition to provide maximum sealing during the stroke of the piston. There must be no leakage between the piston and the walls of the combustion chamber. Second, the intake and exhaust valves (Figure 1-1-27) must close tightly so that there will be no loss of compression at these points. Third, and very important, the timing of the valves must be such that highest efficiency is obtained when the engine is operating at its normal rated r.p.m. A failure at any of these points results in greatly reduced engine efficiency.

**Fuel metering.** The induction system is the distribution and fuel metering part of the engine. Obviously, any defect in the induction system seriously affects engine operation. For best operation, each cylinder of the engine must be provided with the proper fuel/air mixture, usually metered by the carburetor (Figure 1-1-28). On some fuel-injection engines, fuel is metered by the fuel injector flow divider and fuel injection nozzles.

Figure 1-1-28. Carburetor

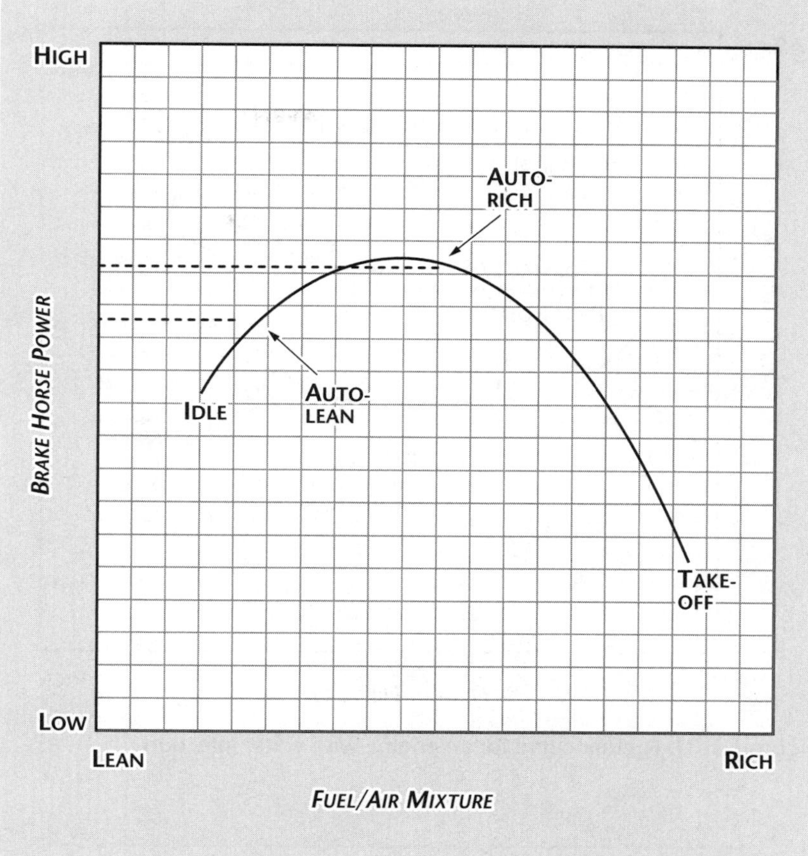


Figure 1-1-29. Power-versus-fuel/air mixture curve

The relation between fuel/air ratio and power is illustrated in Figure 1-1-29. Note that as the fuel mixture is varied from lean to rich, the power output of the engine increases until it reaches a maximum. Beyond this point, the power output falls off as the mixture is further enriched. This is because the fuel mixture is now too rich to provide perfect combustion. Note that maximum engine power can be obtained by setting the carburetor for one point on the curve.

In establishing the carburetor settings for an aircraft engine, the design engineers run a series of curves similar to the one shown. A curve is run for each of several engine speeds.

**EXAMPLE:** The idle speed is 600 r.p.m., and the first curve might be run at this speed. Another curve might be run at 700 r.p.m., another at 800 r.p.m. and so on, in 100-r.p.m. increments, up to takeoff r.p.m. The points of maximum power on the curves are then joined to obtain the best-power curve of the engine for all speeds. This best-power curve establishes the automatic rich setting of the carburetor.

In establishing the detailed engine requirements regarding carburetor setting, the fact that the cylinder head temperature varies with fuel/air ratio must be considered. This variation is illustrated in the curve shown in Figure 1-1-30. Note that the cylinder head temperature is lower with the auto-lean setting than it is with the auto-rich mixture. This is exactly opposite common belief, but it is true.

Knowledge of this fact can be used to advantage by flight crews. If, during cruise, it becomes difficult to keep the cylinder head temperature within limits, the fuel/air mixture may be leaned out to get cooler operation.

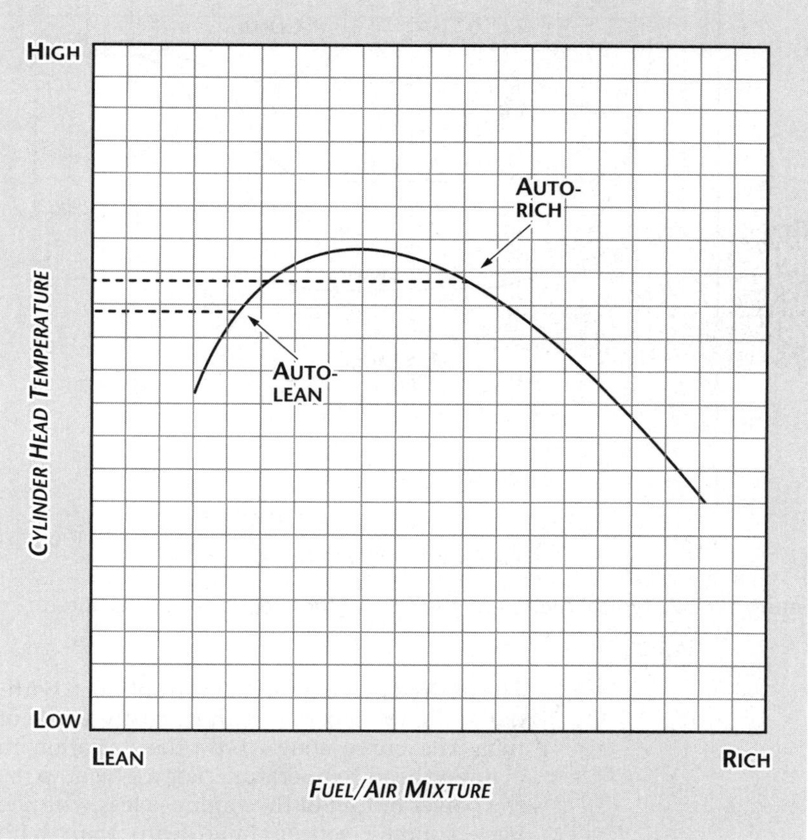


Figure 1-1-30. Variation in head temperature with fuel/air mixture

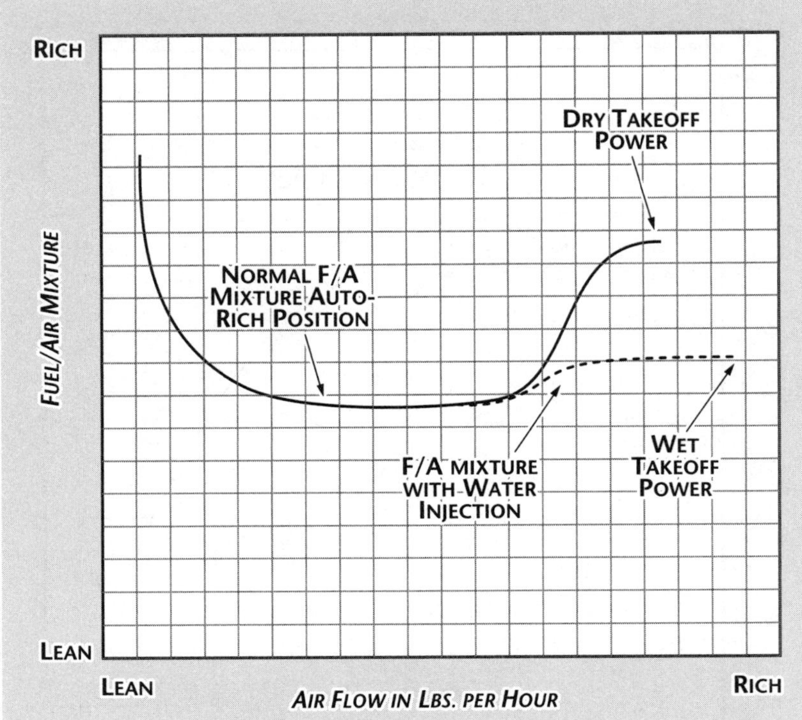


**Figure 1-1-31. Fuel/air curve for an engine with water injection**

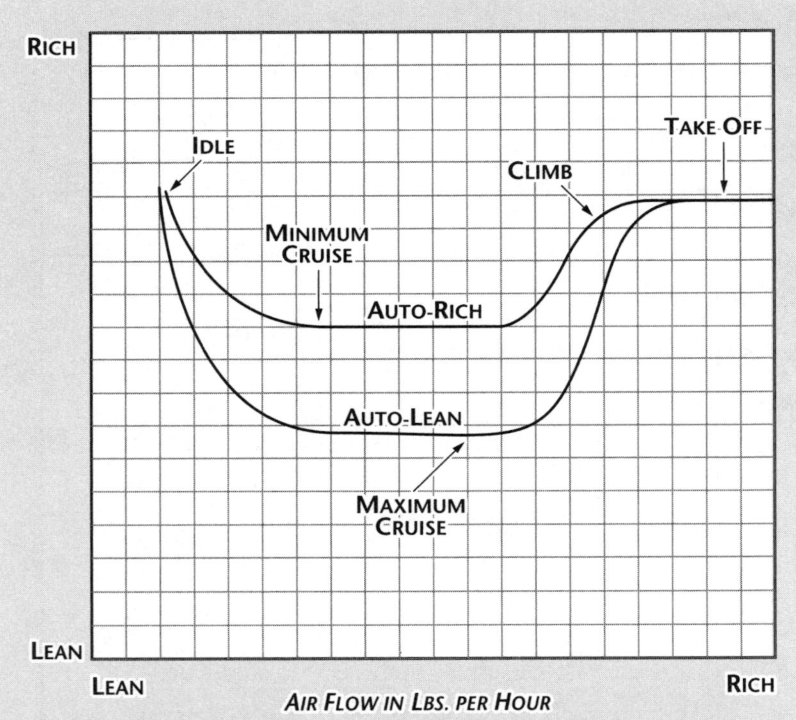


**Figure 1-1-32. Typical fuel/air mixture curve for injection-type carburetor**

The desired cooling can then be obtained without going to auto-rich with its costly waste of fuel. The curve shows only the variation in cylinder head temperature. For a given r.p.m., the power output of the engine is less with the best-economy setting (auto-lean) than with the best-power mixture.

The decrease in cylinder head temperature with a leaner mixture holds true only through the normal cruise range. At higher power settings, cylinder temperatures are higher with the leaner mixtures. The reason for this reversal hinges on the cooling ability of the engine. As higher powers are approached, a point is reached where the airflow around the cylinders will not provide sufficient cooling. At this point, a secondary cooling method must be used. This secondary cooling is done by enriching the fuel/air mixture beyond the best-power point. Although enriching the mixture to this extent results in a power loss, both power and economy must be sacrificed for engine cooling purposes.

To further investigate the influence of cooling requirements on fuel/air mixture, the effects of water injection must be examined. Figure 1-1-31 shows a fuel/air curve for a water-injection engine. The dotted portion of the curve shows how the fuel/ air mixture is leaned out during water injection. This leaning is possible because water, rather than extra fuel, is used as a cylinder coolant.

This permits leaning out to approximately best-power mixture without danger of overheating or detonation. This leaning out gives an increase in power. The water does not alter the combustion characteristics of the mixture. Fuel added to the auto-rich mixture in the power range during dry operation is solely for cooling. A leaner mixture would give more power.

Actually, water, or more accurately, the anti-detonant (water/alcohol) mixture is a better coolant than extra fuel. Therefore, water injection permits higher manifold pressures and a still further increase in power.

In establishing the final curve for engine operation, the engine's ability to cool itself at various power settings is, of course, taken into account. Sometimes the mixture must be altered for a given installation to compensate for the effect of cowl design, cooling airflow or other factors on engine cooling.

The final fuel/air mixture curves take into account economy, power, engine cooling, idling characteristics and all other factors that affect combustion.

The chart in Figure 1-1-32 shows a typical final curve for injection-type carburetors. Note that the fuel/air mixture at idle and at takeoff power is the same in auto-rich and auto-lean. Beyond idle, a gradual spread occurs as cruise power is approached. This spread is maximal in the cruise range. The spread decreases toward takeoff power. This spread between the two curves in the cruise range is the basis for the cruise metering check.

Figure 1-1-33 shows a typical final curve for a float-type carburetor. Note that the fuel/air mixture at idle is the same in rich and in manual lean.

The mixture remains the same until the low cruise range is reached. At this point, the curves separate and then remain parallel through the cruise and power ranges.

Note the spread between the rich and lean setting in the cruise range of both curves. Because of this spread, there will be a decrease in power when the mixture control is moved from auto-rich to auto-lean with the engine operating in the cruise range. This is true because the auto-rich setting in the cruise range is very near the best-power mixture ratio. Therefore, any leaning out will give a mixture that is leaner than best power.

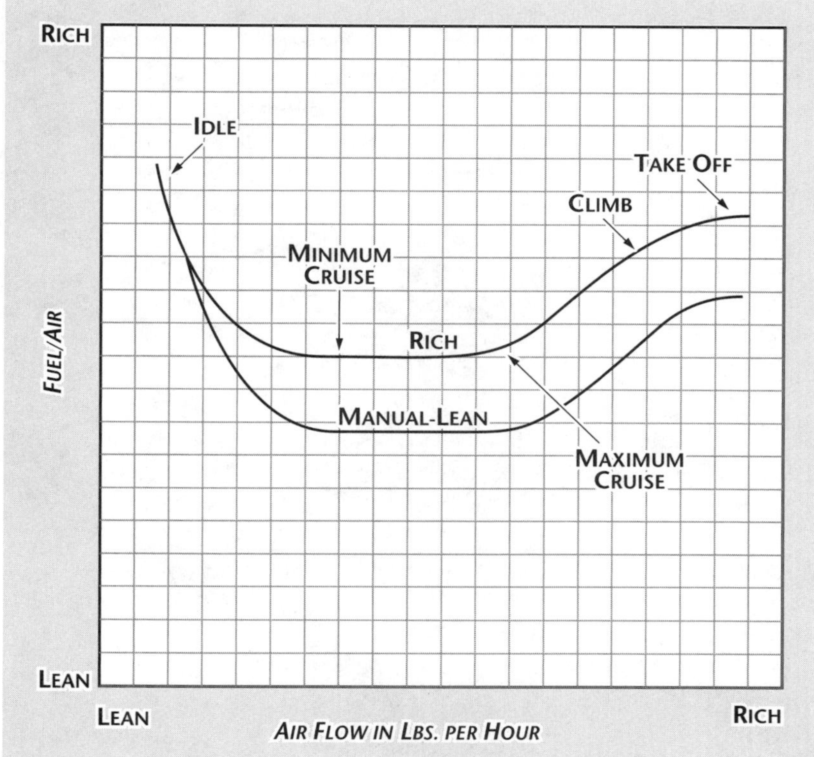


Figure 1-1-33. Typical fuel/air mixture curve for a float-type carburetor

## Idle Mixture

The *idle mixture curve* (Figure 1-1-34) shows how the mixture changes when the idle mixture adjustment is changed. Note that the greatest effect is at idling speeds. However, there is some effect on the mixture at airflows above idling. The airflow at which the idle adjustment effect cancels out varies from minimum cruise to maximum cruise. The exact point depends on the type of carburetor and the carburetor setting. In general, the idle adjustment affects the fuel/air mixture up to medium cruise on most engines having pressure-injection-type carburetors on up to low cruise on engines equipped with float-type carburetors. This means that incorrect idle mixture adjustments can easily give faulty cruise performance as well as poor idling.

There are variations in mixture requirements between one engine and another because of the fuel distribution within the engine and the ability of the engine to cool. Remember that a carburetor setting must be rich enough to supply a combustible mixture for the leanest cylinder. If fuel distribution is poor, the overall mixture must be richer than would be required for the same engine if distribution were good. The engine's ability to cool depends on such factors as cylinder design (including the design of the cooling fins), compression ratio, accessories on the front of the engine which cause individual cylinders to run hot, and the design of the baffling used to deflect airflow around the cylinder. At takeoff power, the mixture must be rich enough to supply sufficient fuel to keep the hottest cylinder cool.

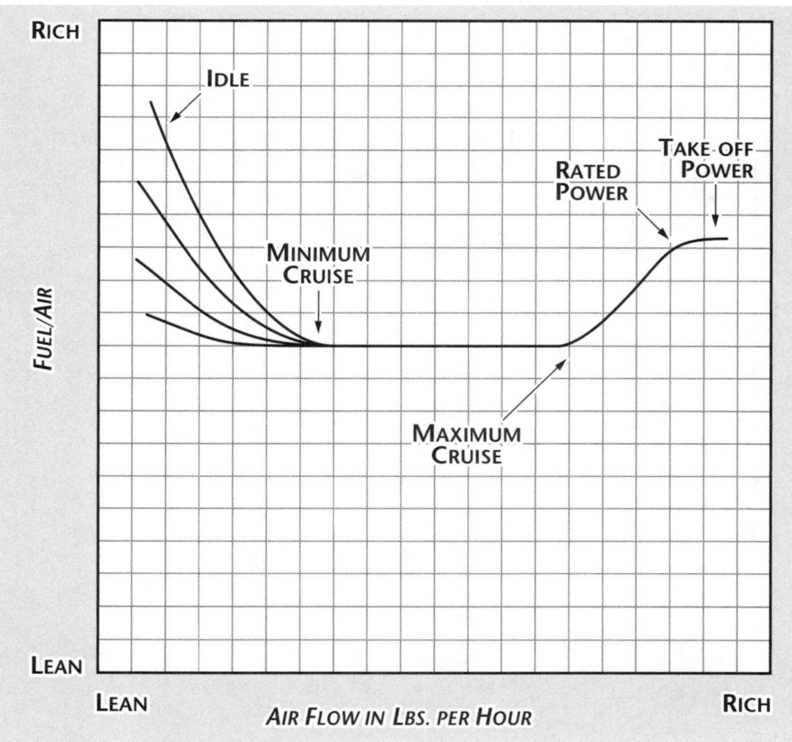


Figure 1-1-34. Idle mixture curve

## The Induction Manifold

The *induction manifold* (Figure 1-1-35, top of next page) provides the means of distributing air, or the fuel/air mixture, to the cylinders. Whether the manifold handles a fuel/air mix-

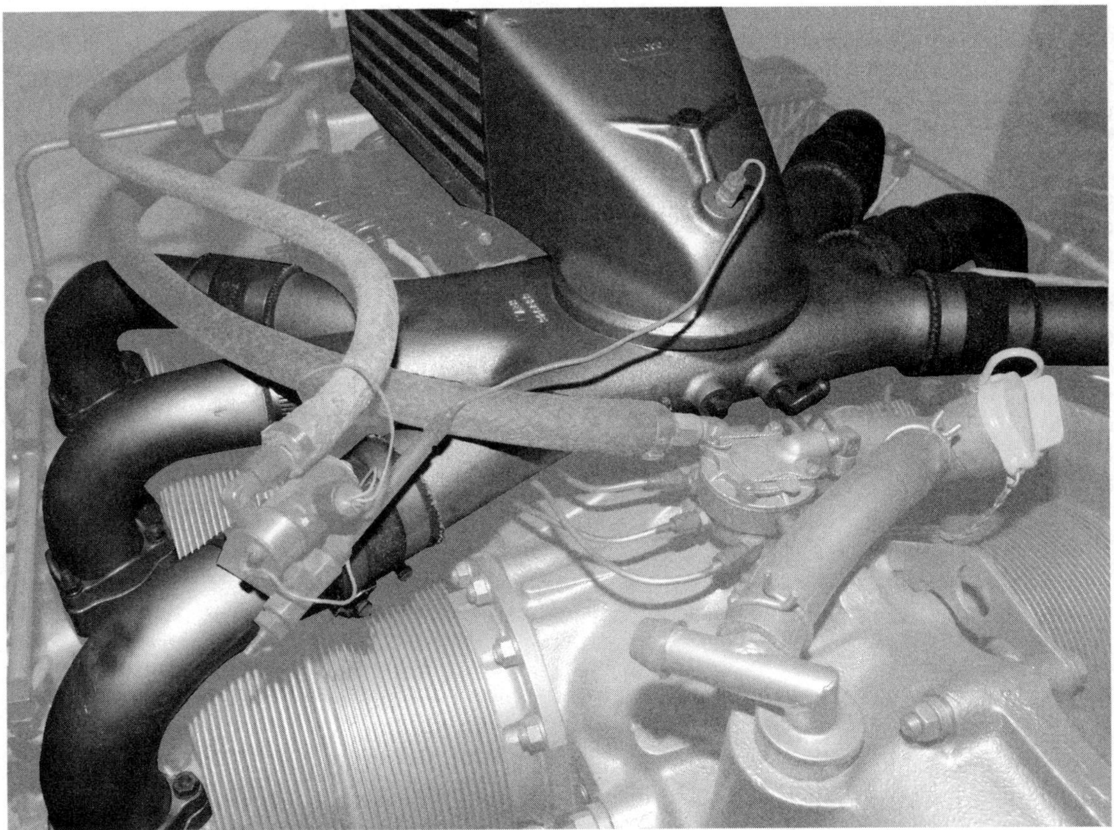
Figure 1-1-35. Induction manifold

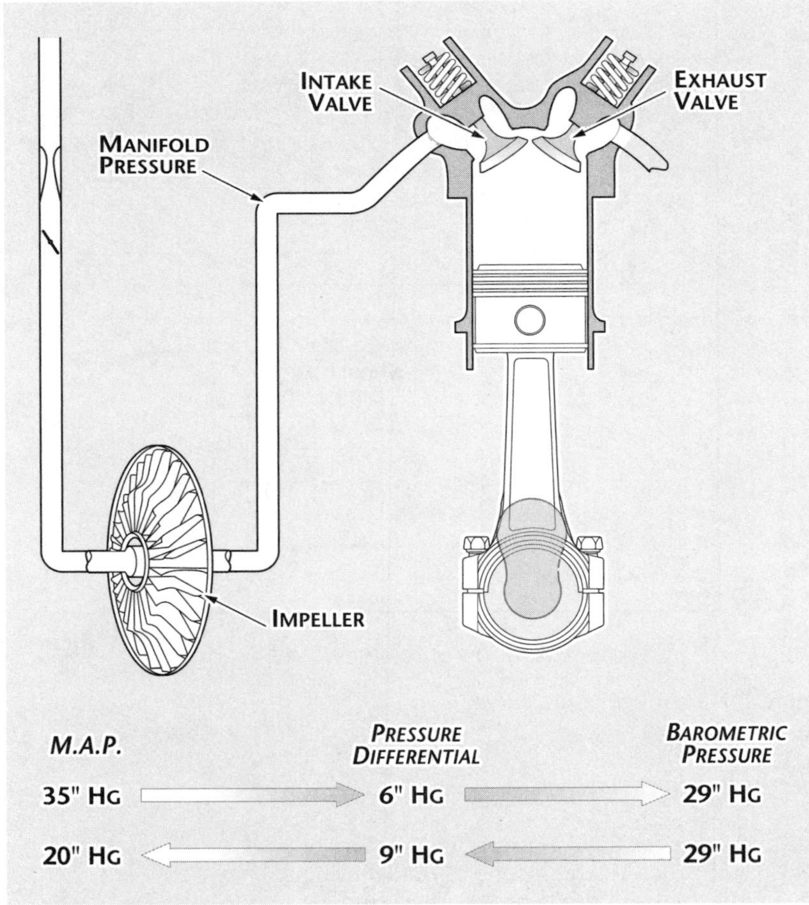

Figure 1-1-36. Effect of valve overlap

ture or air alone depends on the type of fuel metering system used. On an engine equipped with a carburetor, the induction manifold distributes a fuel/air mixture from the carburetor to the cylinders. On most fuel-injected engines, the fuel is delivered to injection nozzles, one per cylinder, which provides the proper spray pattern for efficient burning. Thus, the mixing of fuel and air takes place in the cylinders or at the inlet port to the cylinder. On these engines, the induction manifold handles only air.

The induction manifold is an important item because of the effect it can have on the fuel/air mixture that finally reaches the cylinder. Fuel is introduced into the air stream by the carburetor in a liquid form. To become combustible, the fuel must be vaporized in the air. This vaporization takes place in the induction manifold, which includes the internal supercharger if one is used. Any fuel that does not vaporize will cling to the walls of the intake pipes. Obviously, this affects the effective fuel/air ratio of the mixture, which finally reaches the cylinder in vapor form. This explains the reason for the apparently rich mixture required to start a cold engine. In a cold engine, some of the fuel in the air stream condenses out and clings to the walls of the manifold. This is in addition to that fuel that never vaporized in the first place. As the engine warms up, less fuel is required because less fuel is condensed out of the air stream and more of the fuel is vaporized, thus giving the cylinder the required fuel/air mixture for normal combustion.

Any leak in the induction system has an effect on the mixture reaching the cylinders. This is particularly true of a leak at the cylinder end of an intake pipe. At manifold pressures below atmospheric pressure, such a leak will lean out the mixture. This occurs because additional air is drawn in from the atmosphere at the leaky point. The affected cylinder may overheat, fire intermittently or even cut out altogether.

## Operational Effect of Valve Clearance

While considering the operational effect of *valve clearance*, keep in mind that all aircraft reciprocating engines of current design use *valve overlap*.

Figure 1-1-36 shows the pressures at the intake and exhaust ports under two different sets of operating conditions. In one case, the engine is operating at a manifold pressure of 35 inches Hg. Barometric pressure (exhaust back pressure) is 29 inches Hg. This gives a pressure differential of 6 inches Hg (3 p.s.i.) acting in the direction indicated by the arrow.

During the valve overlap period, this pressure differential forces the fuel/air mixture across the combustion chamber toward the open exhaust. This flow of fuel/air mixture forces ahead of it the exhaust gases remaining in the cylinder, resulting in complete scavenging of the combustion chamber. This, in turn, permits complete filling of the cylinder with a fresh charge on the following intake event. This is the situation in which valve overlap gives increased power.

In a situation where the manifold pressure is below atmospheric pressure, 20 inches Hg, for example, there is a pressure differential of 9 inches Hg (4.5 p.s.i.) in the opposite direction. This causes air or exhaust gas to be drawn into the cylinder through the exhaust port during valve overlap.

In engines with collector rings, this inflow through the exhaust port at low power settings consists of burned exhaust gases. These gases are pulled back into the cylinder and mix with the incoming fuel/air mixture. However, these exhaust gases are inert; they do not contain oxygen. Therefore, the fuel/air mixture ratio is not affected much. With open exhaust stacks, the situation is entirely different. Here, fresh air containing oxygen is pulled into the cylinders through the exhaust. This leans out the mixture. Therefore, the carburetor must be set to deliver an excessively rich idle mixture so

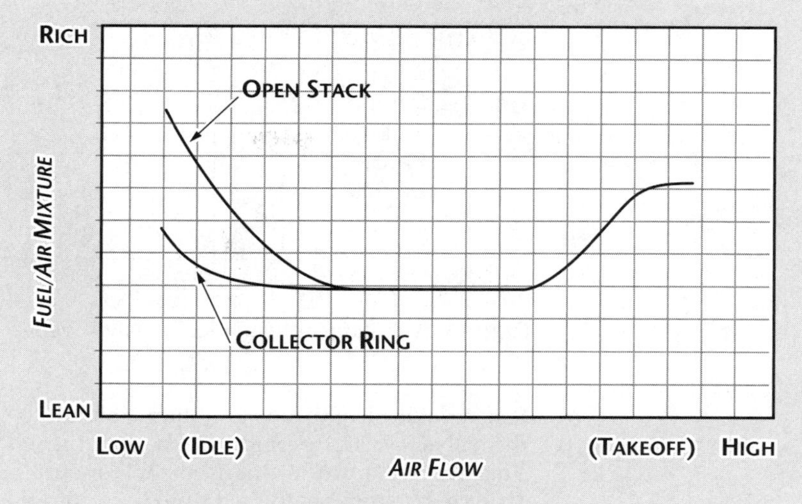


**Figure 1-1-37. Comparison of fuel/air mixture curves for open stack and collector ring installations**

that, when this mixture is combined with the fresh air drawn in through the exhaust port, the effective mixture in the cylinder will be at the desired ratio.

At first thought, it does not appear possible that the effect of valve overlap on fuel/air mixture is sufficient to cause concern. However, the effect of valve overlap becomes apparent when considering idle fuel/air mixtures. These mixtures must be enriched 20-30 percent when open stacks instead of collector rings are used on the same engine. This is shown graphically in Figure 1-1-37. Note the spread at idle between an open stack and an exhaust collector ring installation for engines that are otherwise identical. The mixture variation decreases as the engine speed or airflow is increased from idle into the cruise range.

Engine, airplane and equipment manufacturers provide a powerplant installation that will give satisfactory performance. Cams are designed to give best valve operation and correct overlap, but valve operation will be correct only if valve clearances are set and remain at the value recommended by the engine manufacturer. If valve clearances are set wrong, the valve overlap period will be longer or shorter than the manufacturer intended. The same is true if clearances get out of adjustment during operation.

Where there is too much valve clearance, the valves will not open as wide or remain open as long as they should. This reduces the overlap period. At idling speed, it will affect the fuel/air mixture, since a less-than-normal amount of air or exhaust gases will be drawn back into the cylinder during the shortened overlap period. As a result, the idle mixture will tend to be too rich.

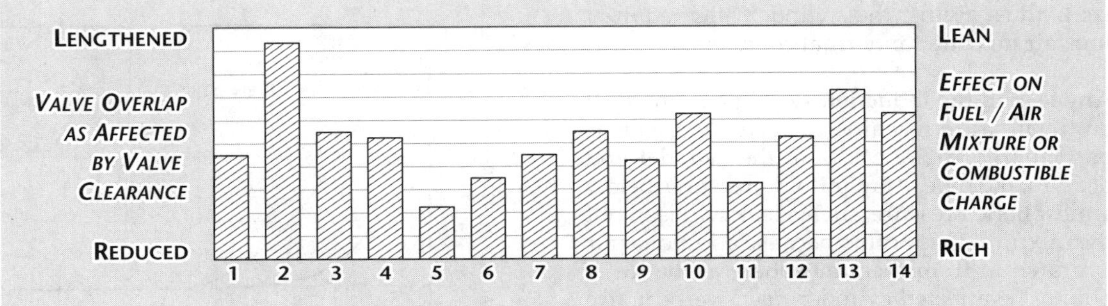


**Figure 1-1-38. Effect of variation in valve overlap on fuel/air mixture between cylinders**

When valve clearance is less than it should be, the valve overlap period will be lengthened. This permits a greater than-normal amount of air or exhaust gases to be drawn back into the cylinder at idling speeds. As a result, the idle mixture will be leaned out at the cylinder. The carburetor is adjusted with the expectation that a certain amount of air or exhaust gases will be drawn back into the cylinder at idling. If more or less air or exhaust gases are drawn into the cylinder during the valve-overlap period, the mixture will be too lean or too rich.

When valve clearances are wrong, it is unlikely that they will all be wrong in the same direction. Instead, there will be too much clearance on some cylinders and too little on others. Naturally, this gives a variation in valve overlap between cylinders. This, in turn, results in a variation in fuel/air ratio at idling and lower-power settings, since the carburetor delivers the same mixture to all cylinders. The carburetor cannot tailor the mixture to each cylinder to compensate for variation in valve overlap.

The effect of variation in valve clearance and valve overlap on the fuel/air mixture between cylinders is illustrated in Figure 1-1-38. Note how the cylinders with too little clearance run rich and those with too much clearance run lean. Note also the extreme mixture variation between cylinders. On such an engine, it would be impossible to set the idle adjustment to give correct mixtures on all cylinders, nor can all cylinders of such an engine be expected to produce the same power. Variations in valve clearance of as little as 0.005-inch have a definite effect on mixture distribution between cylinders.

Another aspect of valve clearance is its effect on volumetric efficiency. Considering the intake valve first, suppose valve clearance is greater than that specified. As the cam lobe starts to pass under the cam roller, the cam step or ramp takes up part of this clearance. However, it doesn't take up all the clearance, as it should. Therefore, the cam roller is well up on the lobe proper before the valve starts to open. As a result, the valve opens later than it should. In a similar way, the valve closes before the roller has passed from the main lobe to the ramp at its end. With excessive clearance, then, the intake valve opens late and closes early. This produces a throttling effect on the cylinder. The valve is not open long enough to admit a full charge of fuel and air. This will cut down the power output particularly at high-power settings.

Insufficient intake valve clearance has the opposite effect. The clearance is taken up and the valve starts to open while the cam roller is still on the cam step. The valve doesn't close until the riser at the end of the lobe has almost completely passed under the roller. Therefore, the intake valve opens early, closes late and stays open longer than it should. At low power, early opening of the intake valve can cause backfiring because of the hot exhaust gases backing out into the intake manifold and igniting the mixture there.

Excessive exhaust valve clearance causes the exhaust valve to open late and close early. This shortens the exhaust event and causes poor scavenging. The late opening may also lead to cylinder overheating. The hot exhaust gases are held in the cylinder beyond the time specified for their release.

When exhaust valve clearance is insufficient, the valve opens early and closes late. It remains open longer than it should. The early opening causes a power loss by shortening the power event. The pressure in the cylinder is released before all the useful expansion has worked on the piston. The late closing causes the exhaust valve to remain open during a larger portion of the intake stroke than it should. This may result in good mixture being lost through the exhaust port.

As mentioned before, there will probably be too little clearance on some cylinders and too much on others whenever valve clearances are incorrect. This means that the effect of incorrect clearances on volumetric efficiency will usually vary from cylinder to cylinder. One cylinder will take in a full charge while another receives only a partial charge. As a result, cylinders will not deliver equal power.

One cylinder will backfire or run hot while another performs satisfactorily.

On some direct fuel-injection engines, variations in valve clearance will affect only the amount of air taken into the cylinders. This is true when the induction manifold handles only air. In this case there will be no appreciable effect on the distribution of fuel to the individual cylinders. This means that, when clearances vary between cylinders, air charges will also vary but fuel distribution will be uniform. This faulty air distribution, coupled with proper fuel distribution, will cause variations in mixture ratio.

In all cases, variations in valve clearance from the value specified have the effect of changing the valve timing from that obtained with correct clearance. This is certain to give something less than perfect performance.

## Ignition System

The next item to be considered regarding engine operation is the ignition system. Although simple at its essence, it is sometimes not understood clearly.

An ignition system consists of four main parts:

1. Magneto
2. Distributor (often built into the magneto)
3. Ignition harness
4. Spark plug

The *magneto* (Figure 1-1-39) is a high-voltage generating device. It must be adjusted to give maximum voltage at the time the points open and ignition occurs. It must also be synchronized accurately to the firing position of the engine. The magneto generates a series of peak voltages that are released by the opening of the breaker points. A distributor is necessary to distribute these peak voltages from the magneto to the cylinders in the proper order. The ignition harness constitutes the insulated and shielded high-tension lines that carry the high voltages from the distributor to the spark plugs.

The magnetos used on aircraft engines are capable of developing voltages as high as 15,000 volts. The voltage required to jump the specified gap in a spark plug will usually be about 4,000-5,000 volts maximum. The spark plugs serve as safety valves to limit the maximum voltage in the entire ignition system. As spark plug gaps open up as a result of erosion, the voltage at the plug terminals increases. A higher voltage is required to jump the larger gap. This higher voltage is transmitted through the secondary circuit. The increased voltage in the circuit becomes a hazard. It is a possible source of breakdown in the ignition harness and can cause flashover in the distributor.

Figure 1-1-39. Magneto

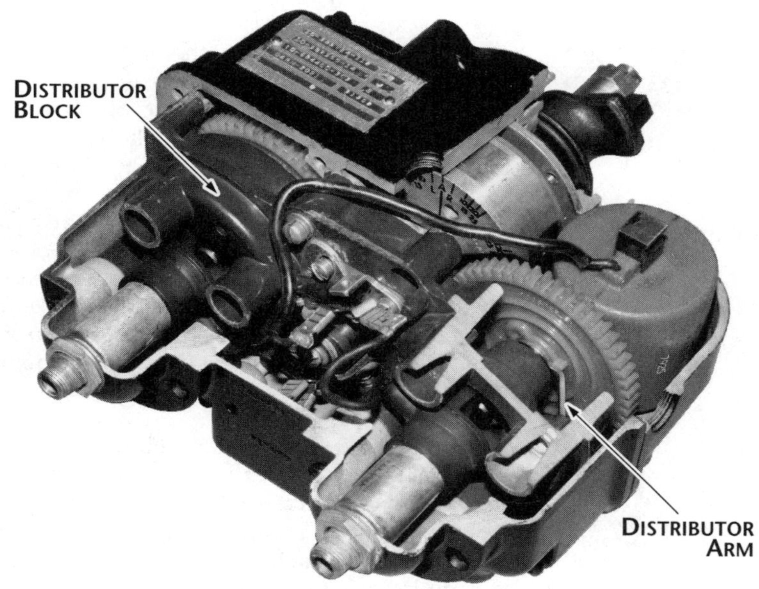


Figure 1-1-40. A cutaway magneto with exposed distributor

The *distributor* (Figure 1-1-40) directs the firing impulses to the various cylinders. It must be timed properly to both the engine and the magneto. The distributor finger must align with the correct electrode on the distributor block at the time the magneto points break. Any misalignment may cause the high voltage to jump to a cylinder other than the one intended. This will cause severe backfiring and general malfunctioning of the engine.

The manufacturer has selected the best compromise and specified an alignment with the No. 1 electrode for timing. However, even with

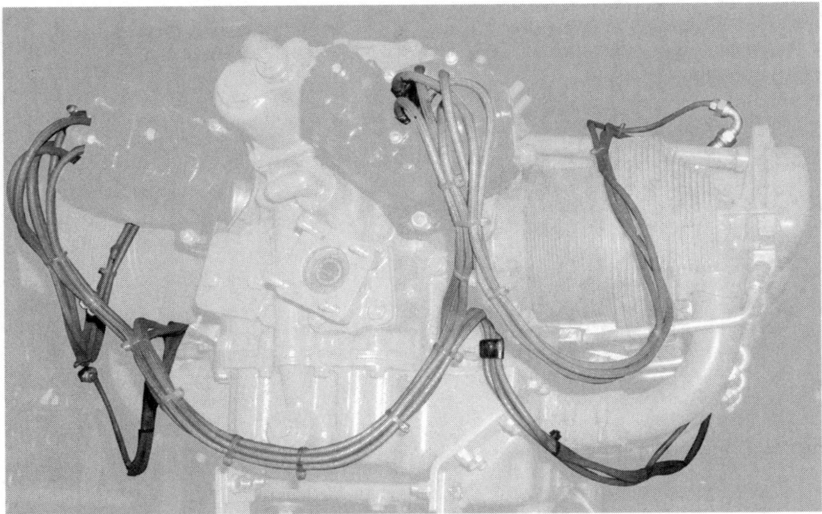

Figure 1-1-41. An ignition harness

perfect distributor timing, the finger is behind on some electrodes and ahead on others. For a few electrodes (cylinders) the alignment is as far from perfect as it can safely be. A slight error in timing, added to this already imperfect alignment, may put the finger so far from the electrode that the high voltage will not jump from finger to electrode, or the high voltage may be routed to the wrong cylinder. Therefore, the distributor must be timed perfectly. The finger must be aligned with the No. 1 electrode exactly as prescribed in the maintenance manual for the particular engine and airplane.

Although the *ignition harness* (Figure 1-1-41) is simple, it is a critical part of the ignition system. A number of things can cause failure in the ignition harness. Insulation may break down on a wire inside the harness and allow the high voltage to leak through to the shielding (and to ground), instead of going to the spark plug. Open circuits may result from broken wires or poor connections. A bare wire may be in contact with the shielding, or two wires may be shorted together.

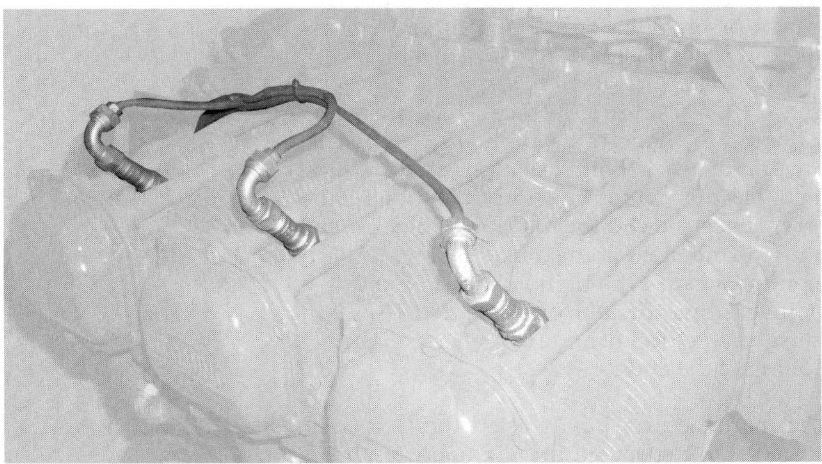

Figure 1-1-42. Spark plugs

Any serious defect in an individual lead prevents the high-voltage impulse from reaching the spark plug to which the lead is connected. As a result, only one spark plug in the cylinder will fire. This causes the cylinder to operate on single ignition. This is certain to result in detonation, since dual ignition is required to prevent detonation at takeoff and during other high-power operation. Two bad leads to the same cylinder will cause the cylinder to go completely dead. On engines with separate distributors, a faulty magneto-to-distributor lead can cut out half the ignition system.

Among the most common ignition harness defects, and the most difficult to detect, are high-voltage leaks. However, a complete harness check will reveal these and other defects.

Although the *spark plug* (Figure 1-1-42 is simple both in construction and in operation, it is nevertheless the direct or indirect cause of a great many malfunctions encountered in aircraft engines. Proper precaution begins with plug selection. Be sure to select and install the plug specified for the particular engine. One of the reasons a particular plug is specified is its heat range. The heat range of the spark plug determines the temperature at which the nose end of the plug operates. It also affects the ability of the spark plug to ignite mixtures that are borderline from the standpoint of high oil content or excessive richness or leanness.

A great many troubles attributed to spark plugs are the direct result of malfunctions somewhere else in the engine. Some of these are excessively rich idle mixtures, improperly adjusted valves and impeller oil seal leaks.

## Propeller Governor

The final item to be considered regarding engine operation is the effect of the *propeller governor* on engine operation. In the curve shown in Figure 1-1-43, note that the manifold pressure change with r.p.m. is gradual until the propeller governor cut-in speed is reached. Beyond this point, the manifold pressure increases, but no change occurs in the engine r.p.m. as the carburetor throttle is opened wider.

A true picture of the power output of the engine can be determined only at speeds below the propeller governor cut-in speed. The propeller governor (Figure 1-1-44) is set to maintain a given engine r.p.m. Therefore, the relationship between engine speed and manifold pressure as an indication of power output is lost, unless it is known that all cylinders of the engine are functioning properly.

In fact, on a multi-engine aircraft, an engine can fail and still produce every indication that it is developing power. The propeller governor will flatten out the propeller blade angle and windmill the propeller to maintain the same engine r.p.m. Heat of compression within the cylinder will prevent the cylinder head temperature from falling rapidly. The fuel pressure will remain constant and the fuel flow will not change unless the manifold pressure is changed. On an engine not equipped with a turbocharger, the manifold pressure will remain where it was. On a turbocharged engine, the manifold pressure will not drop below the value that the mechanical supercharger can maintain. This may be well above atmospheric pressure, depending upon the blower ratio of the engine and the specific conditions existing. Thus, it is difficult to recognize a sudden failure unless the engines are equipped with torque meters, or the pilot notices the fluctuation in r.p.m. at the time the engine cuts out.

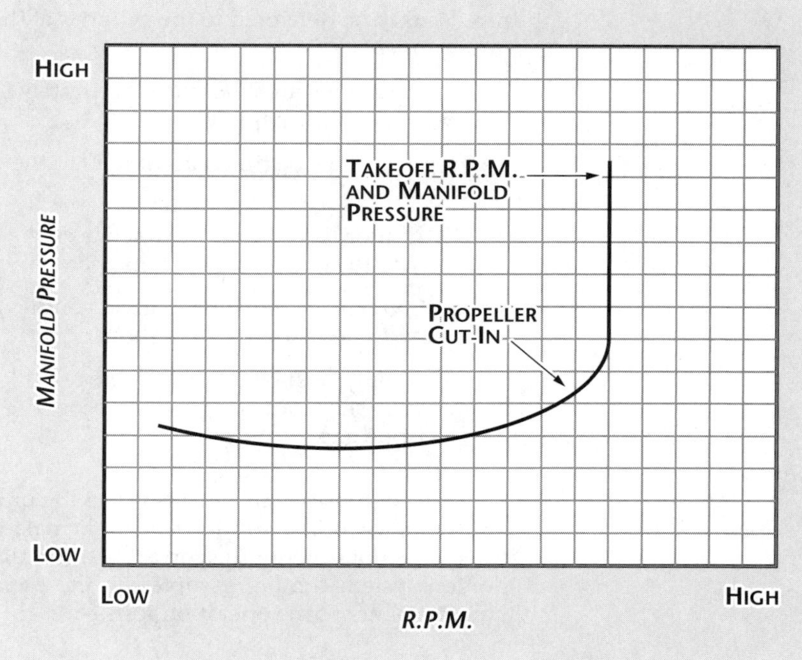


Figure 1-1-43. Effect of propeller governor on manifold pressure

## Overlapping Phases of Engine Operation

Up to this point, the individual phases of engine operation have been discussed. The relationship of the phases and their combined effect on engine operation will now be considered. Combustion within the cylinder is the result of fuel metering, compression and ignition. Since valve overlap affects fuel metering, proper combustion in all the cylinders involves correct valve adjustment in addition to the other phases. When all conditions are correct, there is a burnable mixture. When ignited, this mixture will give power impulses of the same intensity from all cylinders.

The system that ignites the combustible mixture requires that the following five conditions occur simultaneously if the necessary spark

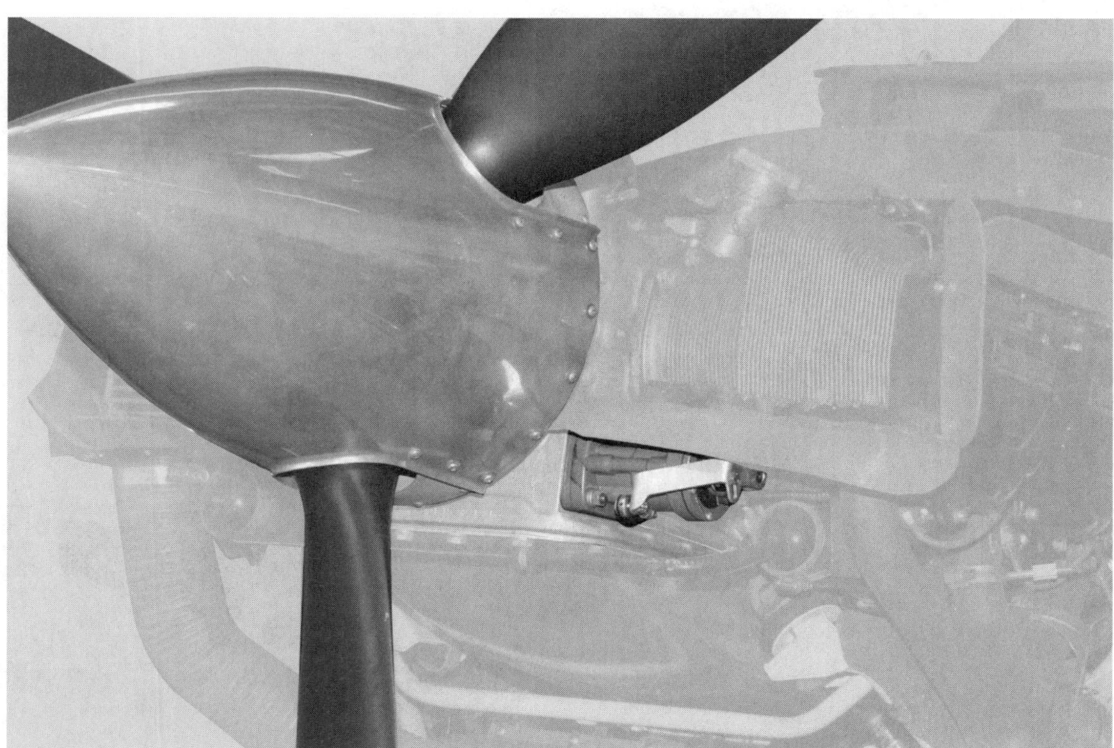

Figure 1-1-44. Propeller governor

impulse is to be delivered to the cylinder at the proper time:

- Breaker points must be timed accurately to the magneto (E-gap)
- Magneto must be timed accurately to the engine
- Distributor finger must be timed accurately to the engine and the magneto
- Ignition harness must be in good condition with no tendency to flashover
- Spark plug must be clean, have no tendency to short out and have the proper electrode gap

If any one of these requirements is lacking or if any one phase of the ignition system is maladjusted or is not functioning correctly, the entire ignition system can be disrupted to the point that improper engine operation results.

As an example of how one phase of engine operation can be affected by other phases, consider spark plug fouling. Spark plug fouling causes malfunctioning of the ignition system, but the fouling seldom results from a fault in the plug itself. Usually some other phase of operation is not functioning correctly, causing the plug to foul out. If excessively rich fuel/air mixtures are being burned because of either basically rich carburetion or improperly adjusted idle mixture, spark plug fouling will be inevitable. Generally, these causes will result in fouled spark plugs appearing over the entire engine, not necessarily confined to one or a few cylinders.

If the fuel/air mixture is too lean or too rich on any one cylinder because of a loose intake pipe or improperly adjusted valves, improper operation of that cylinder will result. The cylinder will probably backfire. Spark plug fouling will occur continually on that cylinder until the defect is remedied.

Impeller oil seal leaks, which can be detected only by removal of intake pipes, will cause spark plug fouling. Here, the fouling is caused by excess oil being delivered to one or more cylinders. Stuck or broken rings will cause oil pumping in the affected cylinders with consequent plug fouling and high oil consumption. Improperly adjusted cylinder valves cause spark plug fouling, hard starting and general engine malfunctioning. They may also cause valve failure as a result of high-seating velocities or of the valve holding open, with subsequent valve burning.

Whenever the true cause of engine malfunctioning is not determined and whenever the real disorder is not corrected, the corrective

**Figure 1-2-1. Opposed engine**

Figure 1-2-2. V-type engine

measure taken will provide only temporary relief. For example, the standard fix for engine backfiring is to change the carburetor. However, as a result of many tests, it is now known that the usual cause of engine backfiring is an improperly adjusted or defective ignition system or improperly adjusted engine valves.

Backfiring is usually caused by one cylinder, not all the cylinders. To remedy backfiring, first locate which cylinder is causing it, and then find out why that cylinder is backfiring.

## Section 2

# Reciprocating Engine Design and Construction

## Types of Reciprocating Engines

Many types of reciprocating engines have been designed. However, manufacturers have developed some designs that are used more commonly than others and are therefore recognized as conventional. Reciprocating engines may be classified according to cylinder arrangement with respect to the crankshaft (in-line, V-type, radial and opposed) or according to the method of cooling (liquid-cooled or air-cooled). Actually, all engines are cooled by transferring excess heat to the surrounding air. In air-cooled engines, this heat transfer is direct from the cylinders to the air. In liquid-cooled engines, the heat is transferred from the cylinders to the coolant, which is then sent through tubing and cooled within a radiator placed in the airstream. The radiator must be large enough to cool the liquid efficiently. Heat is transferred to air more slowly than it is to a liquid. Therefore, it is necessary to provide thin metal fins on the cylinders of an air-cooled engine in order to have increased surface for sufficient heat transfer. Most aircraft engines are air-cooled.

**In-line engines.** In-line engines normally have an even number of cylinders arranged in one row or bank. The pistons of these cylinders are connected to one common crankshaft. The in line engine may be either liquid or air-cooled. Because the in-line engine has a very high power-to-weight ratio and its shape is very streamlined, it has been used in many high-horsepower reciprocating engine fighter-type aircraft.

The primary draw back of the in-line engine is the ability to cool the rear cylinders. Because of this cooling problem, it is normally not used in modern aircraft.

**O-type engines.** The *opposed-type engine,* also called an O-type engine (shown in Figure 1-2-1) has two banks of cylinders directly opposite each other with a crankshaft in the center. The pistons of both cylinder banks are connected to the single crankshaft. Although the engine can be either liquid-cooled or air-cooled, the air-cooled version is used predominantly in aviation. It can be mounted with the cylinders in either a vertical or horizontal position.

The opposed-type engine has a low weight-to-horsepower ratio, and its narrow silhouette makes it ideal for horizontal installation on the aircraft wings. Another advantage is its comparative freedom from vibration.

**V-type engines.** In the V-type engines (Figure 1-2-2), the cylinders are arranged in two in-

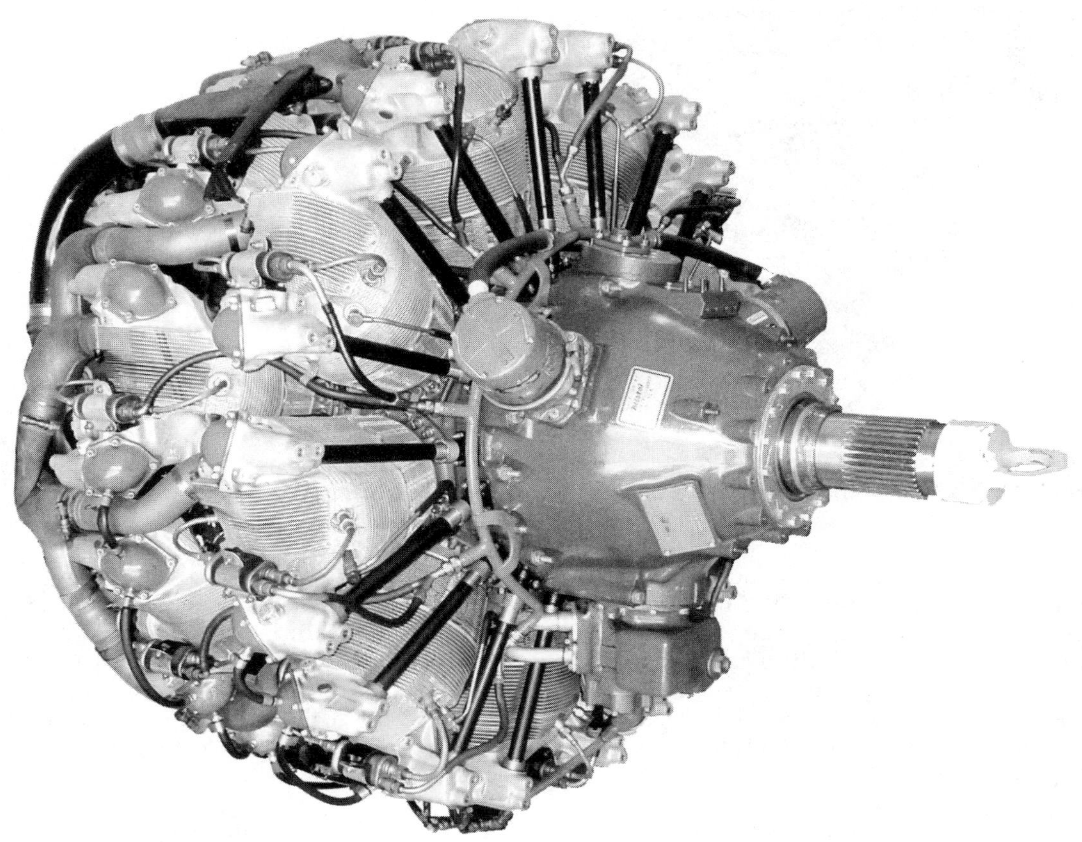

Figure 1-2-3. Radial engine

line banks generally set 60° apart. Most of these engines have 8-12 cylinders and may be either liquid-cooled or air-cooled. The engines are designated by a V, followed by a dash and the piston displacement in cubic inches, for example, V-1710.

**Radial engines.** The radial engine consists of a row or rows of cylinders arranged radially about a central crankcase (see Figure 1-2-3). This type of engine has proven to be very rugged and dependable. The number of cylinders composing a row may be either three, five, seven or nine. Some radial engines have two rows of seven or nine cylinders arranged radially about the crankcase. One type has four rows of cylinders with seven cylinders in each row.

The power output from the different sizes of radial engines varies from 100 hp. to 3,800 hp.

## Reciprocating Engine Construction

The basic parts of a reciprocating engine are the crankcase, cylinders, pistons, connecting rods, valves, valve-operating mechanism and crankshaft. In the head of each cylinder are the valves and spark plugs. One of the valves is in a passage leading from the induction system; the other is in a passage leading to the exhaust system. Inside each cylinder is a movable piston connected to a crankshaft by a connecting rod. Figure 1-2-4 (next page) illustrates the basic parts of a reciprocating engine.

## Crankcase Sections

The foundation of an engine is the *crankcase*. It contains the bearings in which the crankshaft revolves. Besides supporting itself, the crankcase must provide a tight enclosure for the lubricating oil and must support various external and internal mechanisms of the engine. It also provides support for attachment of the cylinder assemblies and the powerplant to the aircraft. It must be sufficiently rigid and strong to prevent misalignment of the crankshaft and its bearings. Cast or forged aluminum alloy is generally used for crankcase construction because it is light and strong (Figure 1-2-5).

The crankcase is subjected to many variations of vibrational and other forces. Since the cylinders are fastened to the crankcase, the tremendous expansion forces tend to pull the

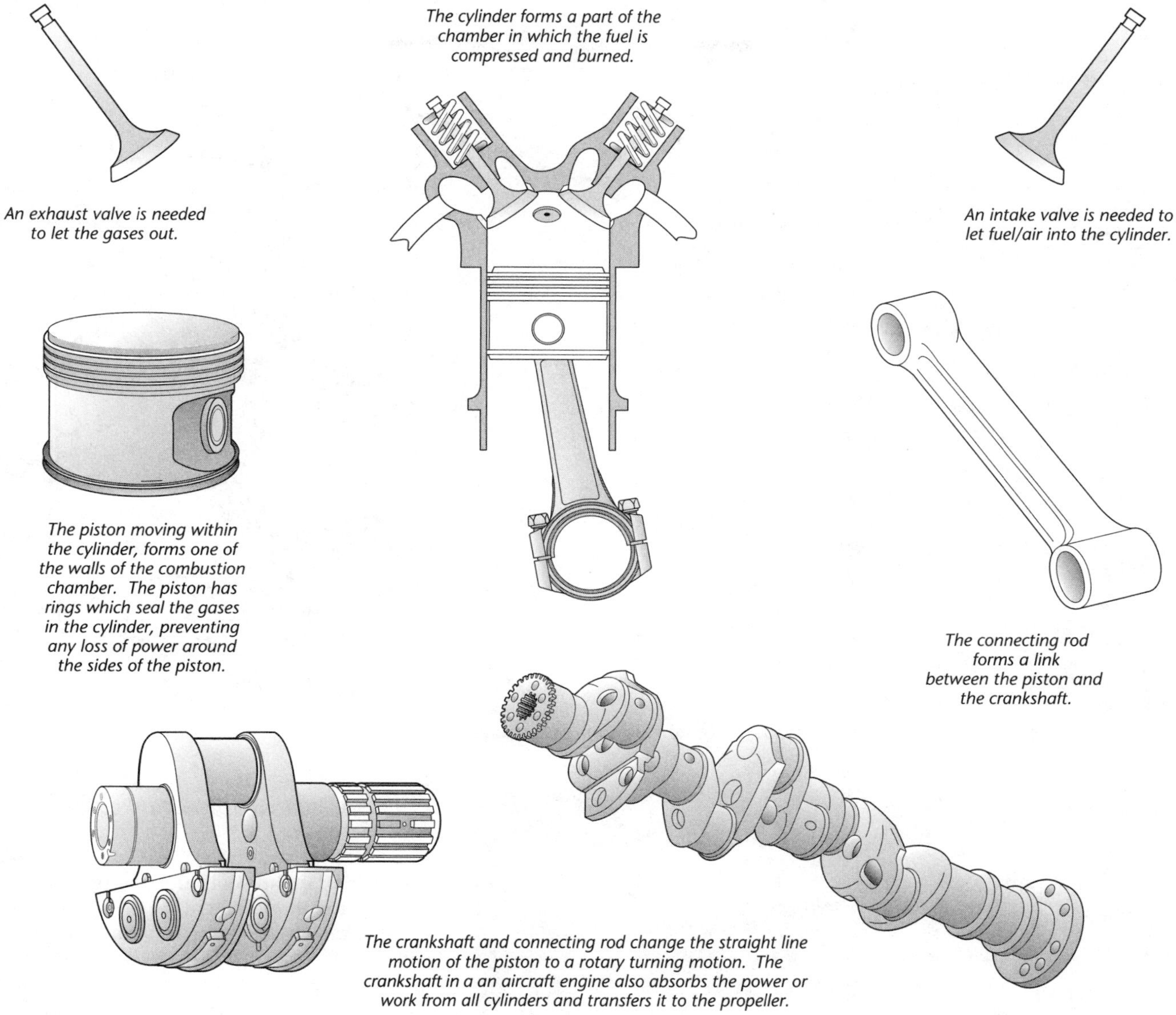

*The cylinder forms a part of the chamber in which the fuel is compressed and burned.*

*An exhaust valve is needed to let the gases out.*

*An intake valve is needed to let fuel/air into the cylinder.*

*The piston moving within the cylinder, forms one of the walls of the combustion chamber. The piston has rings which seal the gases in the cylinder, preventing any loss of power around the sides of the piston.*

*The connecting rod forms a link between the piston and the crankshaft.*

*The crankshaft and connecting rod change the straight line motion of the piston to a rotary turning motion. The crankshaft in a an aircraft engine also absorbs the power or work from all cylinders and transfers it to the propeller.*

**Figure 1-2-4. Basic parts of a reciprocating engine**

cylinder off the crankcase. The unbalanced centrifugal and inertia forces of the crankshaft acting through the main bearing subject the crankcase to bending moments which change continuously in direction and magnitude. The crankcase must have sufficient stiffness to withstand these bending moments. If the engine is equipped with a propeller reduction gear, the front, or drive, end will be subjected to additional forces.

In addition to the thrust forces developed by the propeller under high-power output, there are severe centrifugal and gyroscopic forces applied to the crankcase due to sudden changes in the direction of flight, such as those occurring during maneuvers of the airplane. Gyroscopic forces are, of course, particularly severe when a heavy propeller is installed.

**Figure 1-2-5. A cast aluminum crankcase for a four-cylinder engine**

*Photo courtesy of Engine Components, Inc.*

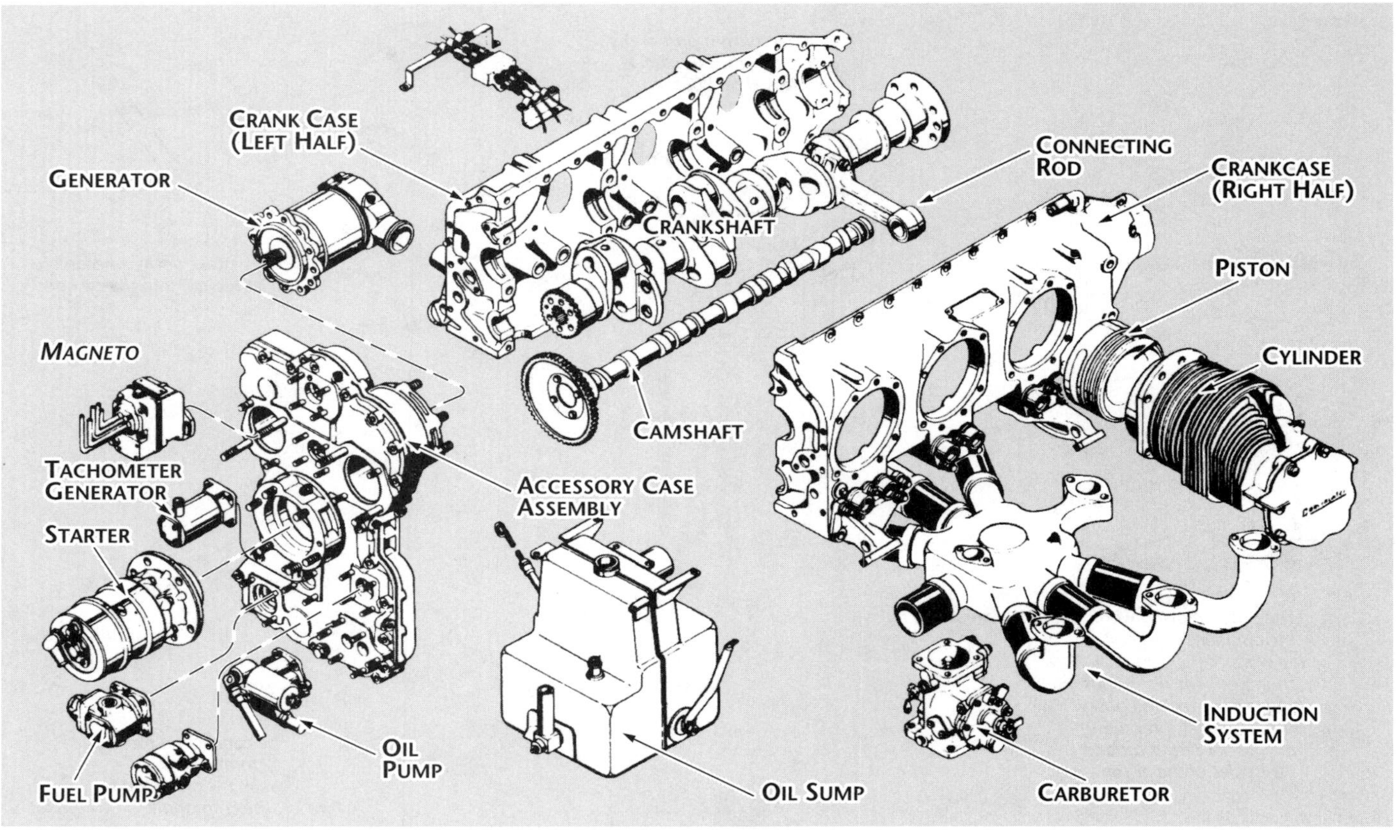

**Figure 1-2-6. Typical opposed engine, exploded into component assemblies**

**Figure 1-2-7. Crankshaft forgings awaiting the crankshaft grinder**

*photo courtesy of Engine Components, Inc*

## *Opposed and In-line Crankcases*

The crankcases used on engines having opposed or in-line cylinder arrangements vary in form for the different types of engines, but in general they are approximately cylindrical. One or more sides are surfaced to serve as a base to which the cylinders are attached by means of cap screws, bolts or studs. These accurately machined surfaces are frequently referred to as cylinder pads.

The crankshaft is carried in a position parallel to the longitudinal axis of the crankcase and is generally supported by a main bearing between each throw. The crankshaft main bearings must be supported rigidly in the crankcase. This usually is accomplished by means of transverse webs in the crankcase, one for each main bearing. The webs form an integral part of the structure and, in addition to supporting the main bearings, add to the strength of the entire case.

The crankcase is divided into two sections in a longitudinal plane. This division may be in the plane of the crankshaft so that one-half of the main bearing (and sometimes camshaft bearings) is carried in one section of the case and the other half in the opposite section (see Figure 1-2-6). Another method is to divide the case in such a manner that the main bearings are secured to only one section of the case

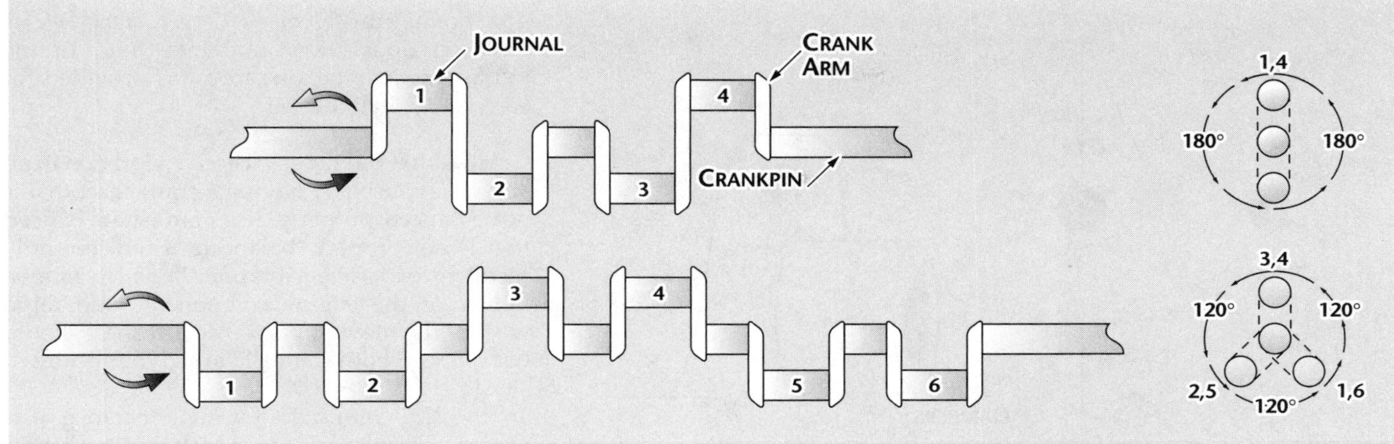

Figure 1-2-8. Solid types of crankshafts

on which the cylinders are attached, thereby providing means of removing a section of the crankcase for inspection without disturbing the bearing adjustment.

# Engine Components

**Crankshafts.** The crankshaft is the backbone of the reciprocating engine. It is subjected to most of the forces developed by the engine. Its main purpose is to transform the reciprocating motion of the piston and connecting rod into rotary motion for rotation of the propeller. The crankshaft, as the name implies, is a shaft composed of one or more cranks located at specified points along its length. The *cranks,* or throws, are formed by forging offsets into a shaft before it is machined. Since crankshafts must be very strong, they generally are forged from a very strong alloy, such as chromium-nickel-molybdenum steel (see Figure 1-2-7).

Figure 1-2-8 shows two representative types of solid crankshafts used in aircraft engines. The *four-throw crankshaft* construction may be used either on four-cylinder horizontal opposed or four-cylinder in-line engines. The throws are 180° apart.

The *six-throw crankshaft shift* is used on six-cylinder in-line engines, 12-cylinder V-type engines and six-cylinder opposed engines. The throws are 120° apart.

Crankshafts of radial engines may be the single-throw, two-throw or four-throw type, depending on whether the engine is the single-row, twin-row or four-row type. A two-throw radial engine crankshaft is shown in Figure 1-2-9. A crankshaft may be of single-piece or multi-piece construction.

No matter how many throws it may have, each crankshaft has three main parts:

- Journals
- Crankpins
- Crank cheeks

*Counterweights* and *dampers,* although not true parts of a crankshaft, are usually attached to it to reduce engine vibration.

The *journal* is supported by, and rotates in, a main bearing. It serves as the center of rotation of the crankshaft. The journal is surface-hardened to reduce wear.

The *crankpin* is the section to which the connecting rod is attached. It is off-center from the main journals and is often called the throw. Two crank cheeks and a crankpin make a *throw.* When a force is applied to the crankpin in any direction other than parallel or perpendicular to and through the center line of the crankshaft, it will cause the crankshaft to rotate.

The outer surface is hardened by *nitriding* to increase its resistance to wear and to provide the required bearing surface. The crankpin is

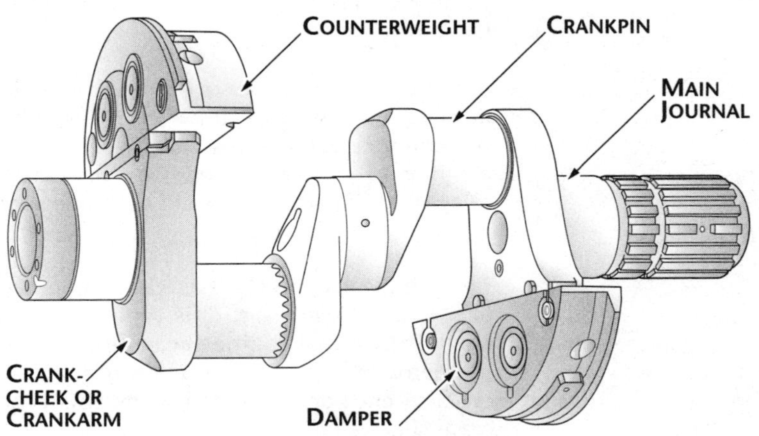

Figure 1-2-9. A two-throw crankshaft for a radial engine

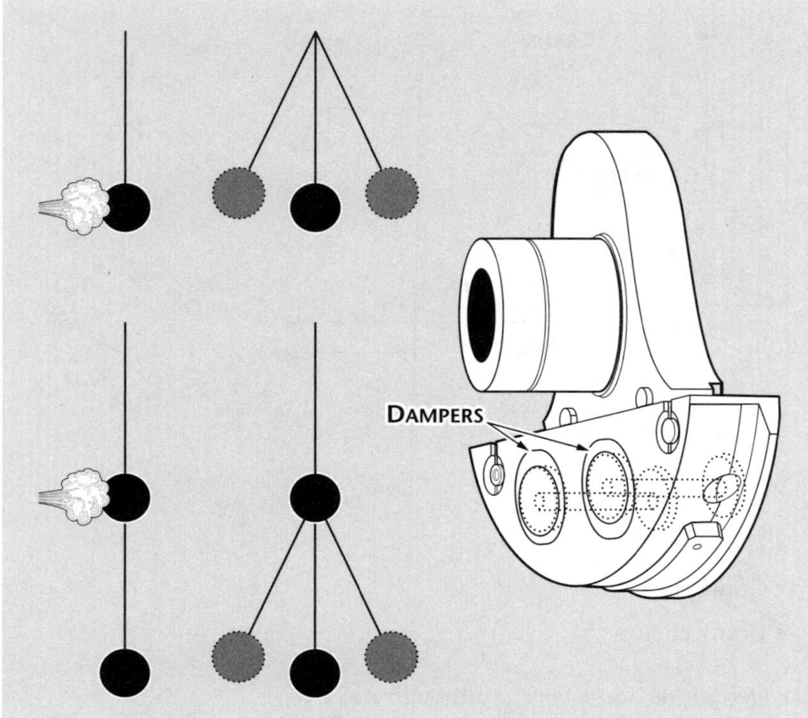


Figure 1-2-10. Principles of a dynamic damper

usually hollow. This reduces the total weight of the crankshaft and provides a passage for the transfer of lubricating oil. The hollow crankpin also serves as a chamber for collecting sludge, carbon deposits and other foreign material. Centrifugal force throws these substances to the outside of the chamber and thus keeps them from reaching the connecting-rod bearing surface. On some engines, a passage is drilled in the crank cheek to allow oil from the hollow crankshaft to be sprayed on the cylinder walls.

The *crank cheek* connects the crankpin to the main journal. In some designs, the cheek extends beyond the journal and carries a counterweight to balance the crankshaft. The crank cheek must be of sturdy construction to obtain the required rigidity between the crankpin and the journal.

In all cases, the type of crankshaft and the number of crankpins must correspond with the cylinder arrangement of the engine. The position of the cranks on the crankshaft in relation to the other cranks of the same shaft is expressed in degrees.

The simplest crankshaft is the single-throw. or *360°-type crankshaft*. This type is used in a single-row radial engine. It can be constructed in one or two pieces. Two main bearings (one on each end) are provided when this type of crankshaft is used.

The double-throw or *180°-type crankshaft* is used on double-row radial engines. In the radial-type engine, one throw is provided for each row of cylinders.

**Crankshaft balance.** Excessive vibration in an engine is caused by having a crankshaft that is not balanced properly. If a crankshaft is used that is not properly balanced, it will not only result in excessive vibration, but also fatigue failure of the engine components and rapid wear of the moving parts. A crankshaft should be balanced both statically and dynamically.

To check a crankshaft for static balance, it is placed on two knife-edges with the crank pins, counterweights and crank cheeks installed. If the crankshaft has a tendency to rotate to any one position, it is not statically balanced. A statically balanced crankshaft is balanced around its plane of rotation.

A crankshaft is dynamically balanced when all the forces created by crankshaft rotation and power impulses are balanced within themselves so that little or no vibration is produced when the engine is operating. To reduce vibration to a minimum during engine operation, *dynamic dampers* are incorporated on the crankshaft. A dynamic damper is merely a pendulum that is fastened to the crankshaft so that it is free to move in a small arc. It is incorporated in the counterweight assembly. Some crankshafts incorporate two or more of these assemblies, each being attached to a different crank cheek. The distance the pendulum moves and its vibrating frequency correspond to the frequency of the power impulses of the engine. When the vibration frequency of the crankshaft occurs, the pendulum oscillates out of time with the crankshaft vibration, thus reducing vibration to a minimum.

**Dynamic dampers.** The construction of the dynamic damper used in one engine consists of a movable slotted-steel counterweight attached to the crank cheek. Two spool-shaped steel pins extend into the slot and pass through oversized holes in the counterweight and crank cheek. The difference in the diameter between the pins and the holes provides a pendulum effect. An analogy of the functioning of a dynamic damper is shown in Figure 1-2-10.

## Connecting Rods

The *connecting rod* is the link that transmits forces between the piston and the crankshaft. Connecting rods must be strong enough to remain rigid under load and yet be light enough to reduce the inertia forces which are produced when the rod and piston stop, change direction and start again at the end of each stroke.

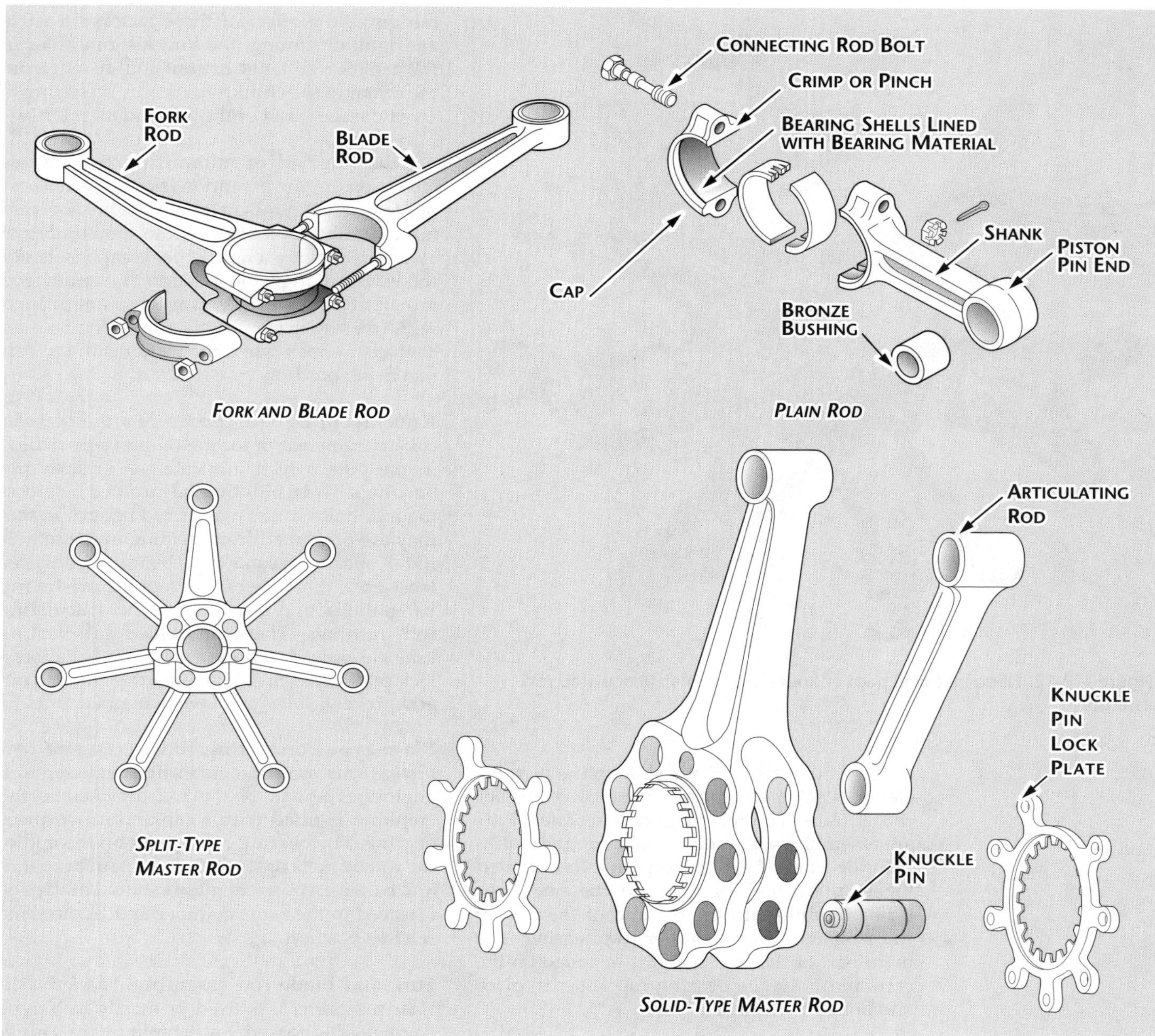


Figure 1-2-11. Connecting-rod assemblies

There are three types of connecting-rod assemblies:

- Plain-type connecting rod
- Fork-and-blade connecting rod
- Master-and-articulated rod assembly (see Figure 1-2-11)

**Master-and-articulated rod assembly.** The *master-and-articulated rod assembly* is commonly used in radial engines. In a radial engine the piston in one cylinder in each row is connected to the crankshaft by a master rod. All other pistons in the row are connected to the master rod by an articulated rod. An 18-cylinder engine has two rows of cylinders and uses two master rods and 16 articulated rods. The articulated rods are constructed of forged steel alloy in either the I- or H-shape, denoting the cross-sectional shape. Bronze bushings are pressed into the bores in each end of the articulated rod to provide knuckle-pin and piston-pin bearings.

The *master rod* serves as the connecting link between the piston pin and the crankpin. The crankpin end, or the big end, contains the crankpin or master rod bearing. Flanges around the big end provide for the attachment of the articulated rods. The *articulated rods* are attached to the master rod by knuckle pins, which are pressed into holes in the master rod flanges during assembly. A plain bearing, usually called a piston-pin bushing, is installed in the piston end of the master rod to receive the piston pin.

Figure 1-2-12. Elliptical travel path of knuckle pins in an articulated rod assembly

When a crankshaft of the split-spline or split-clamp type is employed, a one-piece master rod is used. The master and articulated rods are assembled and then installed on the crankpin; the crankshaft sections are then joined together. In engines that use the one-piece type of crankshaft, the big end of the master rod is split, as is the master rod bearing. The main part of the master rod is installed on the crankpin; then the bearing cap is set in place and bolted to the master rod.

The centers of the knuckle pins do not coincide with the center of the crankpin. Thus, while the crankpin center describes a true circle for each revolution of the crankshaft, the centers of the knuckle pins describe an elliptical path (see Figure 1-2-12).

The elliptical paths are symmetrical about a center line through the master rod cylinder. It can be seen that the major diameters of the ellipses are not the same. Thus, the link rods will have varying degrees of angularity relative to the center of the crank throw.

Because of the varying angularity of the link rods and the elliptical motion of the knuckle pins, all pistons do not move an equal amount in each cylinder for a given number of degrees of crank throw movement. This variation in piston position between cylinders can have considerable effect on engine operation. To minimize the effect of these factors on valve and ignition timing, the knuckle-pin holes in the master rod flange are not equidistant from the center of the crankpin, thereby offsetting to an extent the effect of the link rod angularity.

Another method of minimizing the adverse effects on engine operation is to use a *compensated magneto.* In this magneto the breaker cam has a number of lobes equal to the number of cylinders on the engine. To compensate for the variation in piston position due to link rod angularity, the breaker cam lobes are ground with uneven spacing. This allows the breaker contacts to open when the piston is in the correct firing position.

**Knuckle pins.** The *knuckle pins* are of solid construction except for the oil passages drilled in the pins, which lubricate the knuckle-pin bushings. These pins may be installed by pressing into holes in the master rod flanges so that they are prevented from turning in the master rod. Knuckle pins may also be installed with a loose fit so that they can turn in the master rod flange holes and also turn in the articulating rod bushings. These are called full-floating knuckle pins. In either type of installation a lock plate on each side retains the knuckle pin and prevents a lateral movement of it.

**Plain-type connecting rods.** *Plain-type connecting rods* are used in in-line and opposed engines. The end of the rod attached to the crankpin is fitted with a cap and a two-piece bearing. The bearing cap is held on the end of the rod by bolts or studs. To maintain proper fit and balance, connecting rods should always be replaced in the same cylinder and in the same relative position.

**Fork and blade rod assembly.** The *fork-and-blade rod assembly* is used primarily in V-type engines. The forked rod is split at the crankpin end to allow space for the blade rod to fit between the prongs. A single two-piece bearing is used on the crankshaft end of the rod.

## Pistons

The *piston* of a reciprocating engine is a cylindrical member that moves back and forth within a steel cylinder. The piston acts as a moving wall within the combustion chamber. As the piston moves down in the cylinder, it draws in the fuel/air mixture. As it moves upward, it compresses the charge, ignition occurs, and the expanding gases force the piston downward. This force is transmitted to the crankshaft through the connecting rod. On the return upward stroke, the piston forces the exhaust gases from the cylinder.

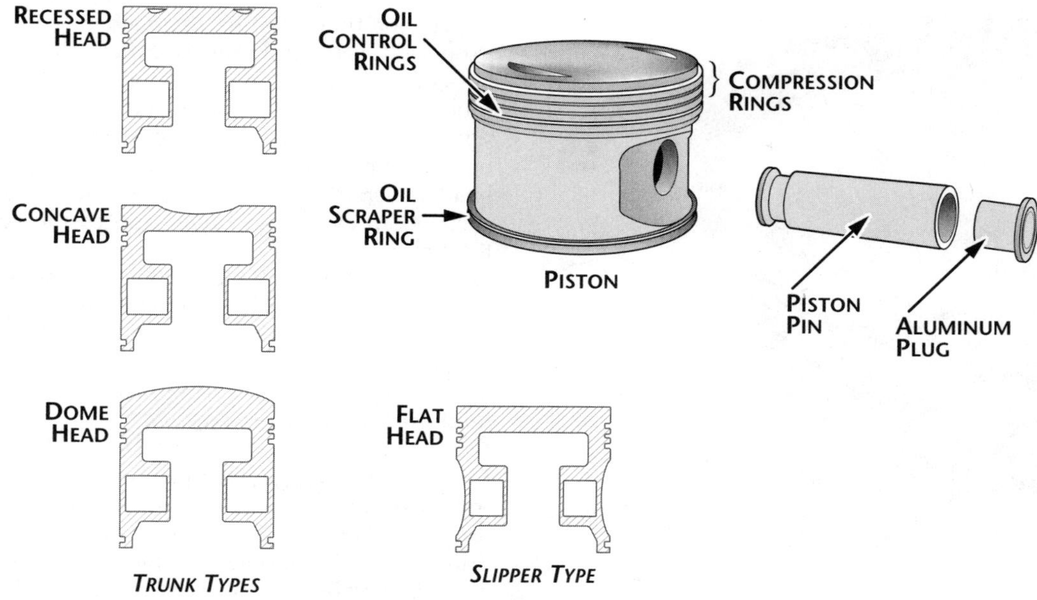


**Figure 1-2-13. Piston assembly and types of pistons**

## *Piston construction*

The majority of aircraft engine pistons are machined from aluminum alloy forgings. Grooves are machined in the outside surface of the piston to receive the piston rings, and cooling fins are provided on the inside of the piston for greater heat transfer to the engine oil.

Pistons may be either the *trunk-type* or the *slipper-type;* both are shown in Figure 1-2-13. Slipper-type pistons are not used in modern, high-powered engines because they do not provide adequate strength or wear resistance. The top face of the piston, or head, may either be flat, convex or concave. Recesses may be machined in the piston head to prevent interference with the valves.

As many as six grooves may be machined around the piston to accommodate the compression rings and oil rings (see Figure 1-2-13). The compression rings are installed in the three uppermost grooves; the oil control rings are installed immediately above the piston pin. The piston is usually drilled at the oil control ring grooves to allow surplus oil scraped from the cylinder walls by the oil control rings to pass back into the crankcase. An oil scraper ring is installed at the base of the piston wall or skirt to prevent excessive oil consumption. The portions of the piston walls that lie between each pair of ring grooves are called the ring lands.

In addition to acting as a guide for the piston head, the piston skirt incorporates the piston-pin bosses. The piston-pin bosses are of heavy construction to enable the heavy load on the piston head to be transferred to the piston pin.

**Piston pin.** The piston pin joins the piston to the connecting rod. It is machined in the form of a tube from a nickel steel alloy forging, casehardened and ground. The piston pin is sometimes called a wristpin because of the similarity between the relative motions of the piston and the articulated rod and that of the human arm.

The piston pin used in modern aircraft engines is the full-floating type, so called because the pin is free to rotate in both the piston and in the connecting rod piston-pin bearing.

The piston pin must be held in place to prevent the pin ends from scoring the cylinder walls. In earlier engines, spring coils were installed in grooves in the piston-pin bores at either end of the pin. The current practice is to install a plug of relatively soft aluminum in the pin ends to provide a good bearing surface against the cylinder wall.

**Piston rings.** The piston rings prevent leakage of gas pressure from the combustion chamber and reduce to a minimum the seepage of oil into the combustion chamber. The rings fit into the piston grooves but spring out to press against the cylinder walls; when properly lubricated, the rings form an effective gas seal.

Most piston rings are made of high-grade cast iron. After the rings are made, they are ground to the cross section desired. They are then split so that they can be slipped over the outside of the piston and into the ring grooves that are

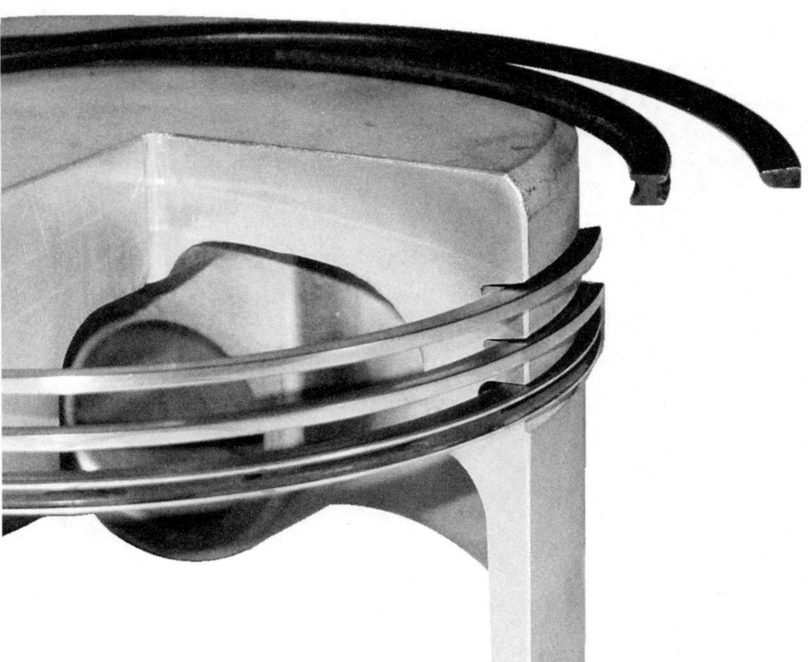

Figure 1-2-14. This cutaway of a three-ring piston shows two wedge-shaped compression rings and an oil control ring. The groove that allows the oil to return to the engine interior is also visible.

machined in the piston wall. Since their purpose is to seal the clearance between the piston and the cylinder wall, they must fit the cylinder wall snugly enough to provide a gastight fit; they must exert equal pressure at all points on the cylinder wall; and they must make a gastight fit against the sides of the ring grooves.

Gray cast iron is most often used in making piston rings. However, many other materials have been tried. In some engines, chrome-plated mild steel piston rings are used in the top compression ring groove because these rings can better withstand the high temperatures present at this point (see Figure 1-2-14).

**Compression ring.** The purpose of the compression rings is to prevent the escape of gas past the piston during engine operation. They are placed in the ring grooves immediately below the piston head. The number of compression rings used on each piston is determined by the type of engine and its design, although most aircraft engines use two compression rings plus one or more oil control rings.

The cross section of the ring is either rectangular or wedge shaped with a tapered face. The tapered face presents a narrow bearing edge to the cylinder wall, which helps to reduce friction and provide better sealing.

**Oil control rings.** Oil control rings are placed in the grooves immediately below the compression rings and above the piston pin bores. There may be one or more oil control rings per piston; two rings may be installed in the same groove, or they may be installed in separate grooves. Oil control rings regulate the thickness of the oil film on the cylinder wall. If too much oil enters the combustion chamber, it will burn and leave a thick coating of carbon on the combustion chamber walls, the piston head, the spark plugs and the valve heads. This carbon can cause the valves and piston rings to stick if it enters the ring grooves or valve guides. In addition, the carbon can cause spark plug misfiring as well as detonation, pre-ignition or excessive oil consumption. To allow the surplus oil to return to the crankcase, holes are drilled in the piston ring grooves or in the lands next to these grooves.

**Oil scraper ring.** The oil scraper ring is installed in the bottom ring groove on the piston skirt. The scraper ring will normally contain a beveled face. The direction in which this bevel is installed is very important. The direction of the bevel will determine whether the oil is held up onto the cylinder walls or scraped off into the crankcase. The required direction of this bevel installation is determined by several factors such as the type of material used in the cylinder wall and the location of the cylinder on the crankcase. The manufacturers instructions should always be consulted and followed when installing the oil scraper rings.

## Cylinders

The portion of the engine in which the power is developed is called the cylinder. The cylinder provides a combustion chamber where the burning and expansion of gases take place, and it houses the piston and the connecting rod.

There are four major factors that need to be considered in the design and construction of the cylinder assembly:

1. It must be strong enough to withstand the internal pressures developed during engine operation.
2. It must be constructed of a lightweight metal to keep down engine weight.
3. It must have good heat-conducting properties for efficient cooling.
4. It must be comparatively easy and inexpensive to manufacture, inspect and maintain.

The cylinder used in the air-cooled engine is the overhead valve-type shown in Figure 1-2-15. Each cylinder is an assembly of two major parts:

- Cylinder head
- Cylinder barrel

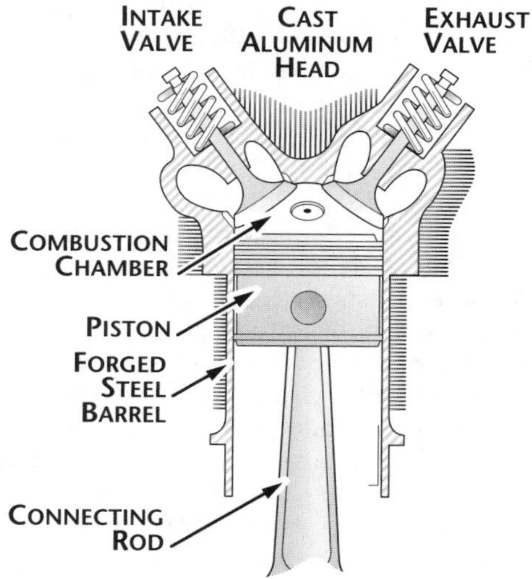


**Figure 1-2-15. Cutaway view of the cylinder assembly**

**Figure 1-2-16. A new cylinder head in its as-cast condition**

*Photo courtesy of Engine Components, Inc.*

**Cylinder heads.** The head is either produced singly for each cylinder in air-cooled engines or is cast in-block (all cylinder heads in one block) for most liquid-cooled engines. The *cylinder head* of an air-cooled engine is generally made of aluminum alloy, because aluminum alloy is a good conductor of heat and is light, reducing the overall engine weight. Cylinder heads are forged or die-cast for greater strength (see Figure 1-2-16).

At assembly, the cylinder head is expanded by heating and then screwed down on the cylinder barrel, which has been chilled; thus, when the head cools and contracts and the barrel warms up and expands, a gas-tight joint results (see Figure 1-2-17).

The inner shape of a cylinder head may be flat, semispherical or peaked, in the form of a house roof. The semispherical type has proved most satisfactory because it is stronger and aids in a more rapid and thorough scavenging of the exhaust gases. The purpose of the cylinder head is to provide a place for combustion of the fuel/air mixture and to give the cylinder more heat conductivity for adequate cooling. The fuel/air mixture is ignited by the spark in the combustion chamber and commences burning as the piston travels toward TDC on the compression stroke. The ignited charge is rapidly expanding at this time, and pressure is increasing so that as the piston travels through the TDC position, it is driven downward on the power stroke. The intake and exhaust valve ports are located in the cylinder head along with the spark plugs and the intake and exhaust valve actuating mechanisms.

After casting, the spark plug bushings, valve guides, rocker arm bushings and valve seats are installed in the cylinder head. Spark plug openings may be fitted with bronze or steel bushings that are shrunk and screwed into the openings. Stainless steel Heli-Coil spark plug inserts are used in many engines currently manufactured.

Bronze or steel valve guides are usually shrunk or screwed into drilled openings in the cylinder head to provide guides for the valve stems (see Figure 1-2-18). The valve seats are circular rings of hardened metal that protect the relatively soft metal of the cylinder head from the hammering action of the valves and from the exhaust gases.

**Figure 1-2-17. After chilling the barrel and heating the cylinder head, the head is screwed onto the barrel.**

*Photo courtesy of Engine Components, Inc.*

**Figure 1-2-18. New cylinder head being reamed for installation of valve guides**

*Photo courtesy of Engine Components, Inc.*

**Figure 1-2-19. Cylinder forgings before being machined**

*Photo courtesy of Engine Components, Inc.*

The cylinder heads of air-cooled engines are subjected to extreme temperatures; it is therefore necessary to provide adequate fin area and to use metals that conduct heat rapidly. Cylinder heads of air-cooled engines are usually cast or forged singly. Aluminum alloy is used in the construction for a number of reasons. It is well adapted for casting or for the machining of deep, closely spaced fins, and it is more resistant than most metals to the corrosive attack of tetraethyl lead in gasoline. The greatest improvement in air-cooling has resulted from reducing the thickness of the fins and increasing their depth. In this way the fin area has been increased from approximately 1,200 sq. inches to more than 7,500 sq. inches per cylinder in modern engines. Cooling fins taper from 0.090 inches at the base to 0.060 inches at the tip end. Because of the difference in temperature in the various sections of the cylinder head, it is necessary to provide more cooling-fin area on some sections than on others. The exhaust valve region is the hottest part of the internal surface; therefore, more fin area is provided around the outside of the cylinder in this section.

**Cylinder barrels.** In general, the cylinder barrel in which the piston operates must be made of a high-strength steel. It must be as light as possible, yet have the proper characteristics for operating under high temperatures. It must be made of a good bearing material and have high tensile strength.

The cylinder barrel is made from a steel alloy forging (see Figure 1-2-19).

The inner surface hardened to resist wear of the piston and the piston rings which bear against it. This hardening is usually done by exposing the steel to ammonia or cyanide gas while the steel is very hot. The steel soaks up nitrogen from the gas and forms iron nitrides on the exposed surface. As a result of this process, the metal is said to be nitrided.

The barrel will have threads on the outside surface at one end so that it can be screwed into the cylinder head (see Figure 1-2-20).

Some air-cooled cylinder barrels have replaceable aluminum cooling fins attached to them, while others have the cooling fins machined as an integral part of the barrel.

**Cylinder numbering.** Occasionally it is necessary to refer to the left or right side of the engine or to a particular cylinder. Therefore, it is necessary to know the engine directions and how cylinders of an engine are numbered.

The propeller shaft end of the engine is always the front end, and the accessory end is the rear end, regardless of how the engine is mounted in an aircraft. When referring to the right side or left side of an engine, always assume you are viewing it from the rear or accessory end. As seen from this position, crankshaft rotation is referred to as either clockwise or counterclockwise.

Radial engine cylinders are numbered clockwise as viewed from the accessory end. In-line and V-type engine cylinders are usually numbered from the rear. In V-engines, the cylinder banks are known as the right bank and the left bank, as viewed from the accessory end.

The numbering of engine cylinders is shown in Figure 1-2-21. Note that the cylinder numbering of the opposed engine shown begins with the right rear as No. 1, and the left rear as No. 2. The one forward of No. 1 is No. 3; the one forward of No. 2 is No. 4; and so on. The numbering of opposed engine cylinders is by no means standard. Some manufacturers number their cylinders from the rear and others from the front of the engine. Always refer to the appropriate engine manual to determine the correct numbering system used by the manufacturer.

**Figure 1-2-20. This cylinder is finish-machined, threaded and ready for assembly of the cylinder head.**

*Photo courtesy of Engine Components, Inc.*

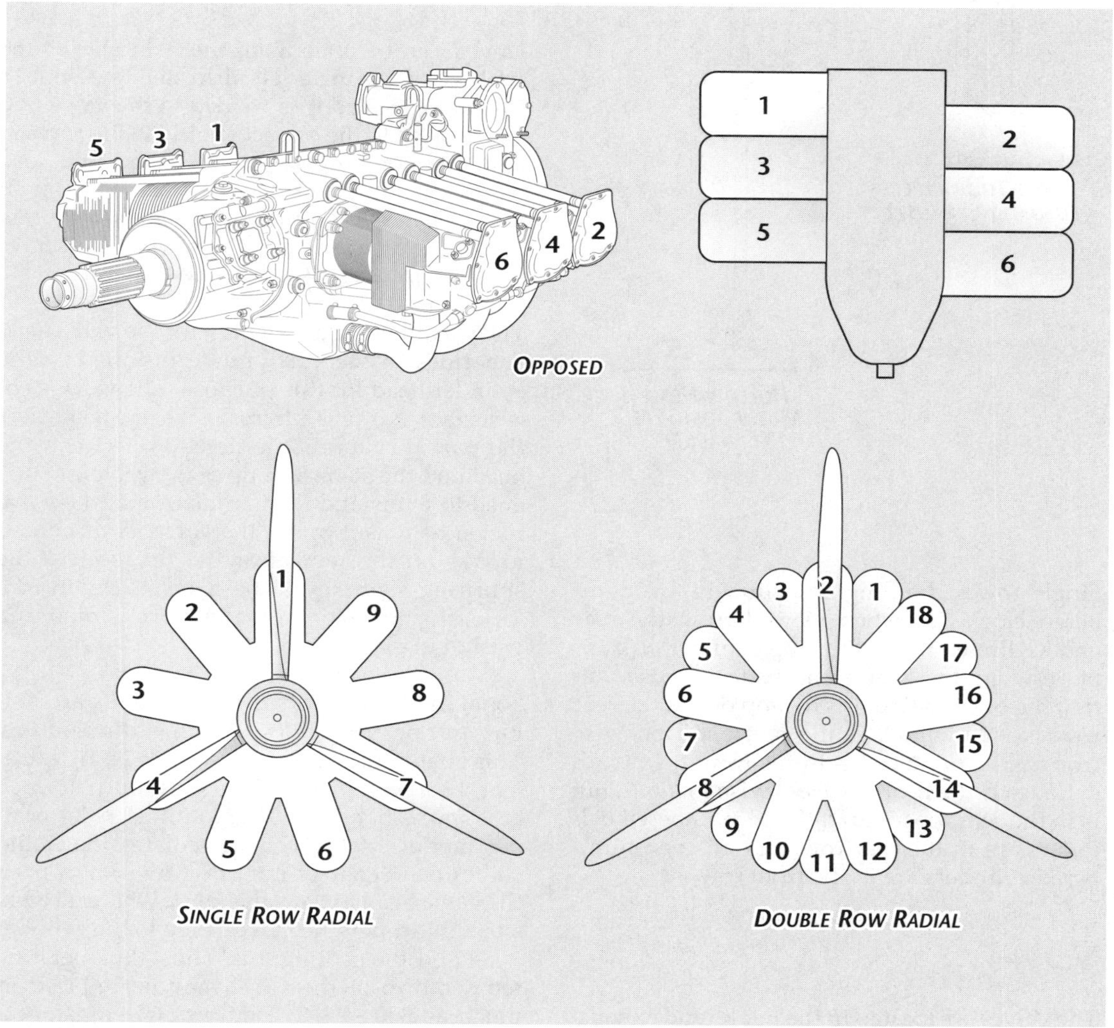

*Opposed*

*Single Row Radial*

*Double Row Radial*

Figure 1-2-21. Numbering of engine cylinders

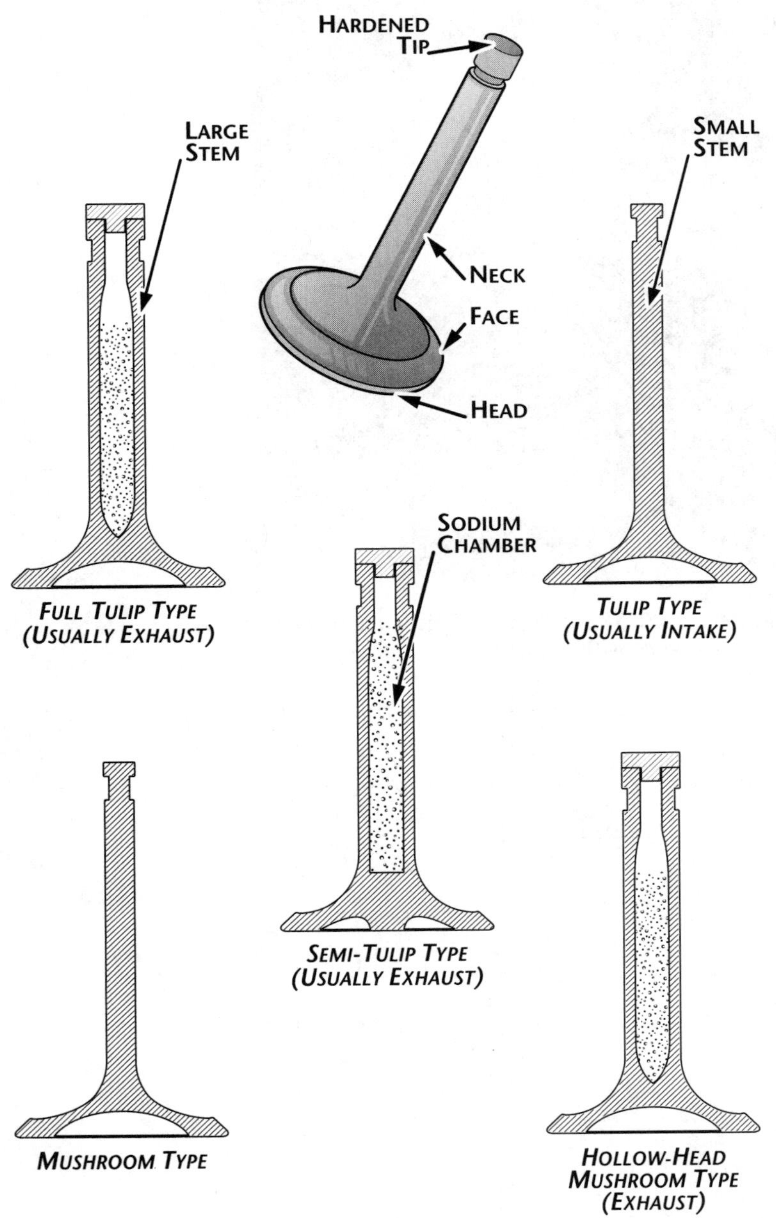


**Figure 1-2-22. Valve types**

Single-row radial engine cylinders are numbered clockwise when viewed from the rear end. Cylinder No. 1 is the top cylinder. In double-row engines, the same system is used, in that the No. 1 cylinder is the top one in the rear row. No. 2 cylinder is the first one clockwise from No. 1, but No. 2 is in the front row. No. 3 cylinder is the next one clockwise to No. 2, but is in the rear row. Thus, all odd-numbered cylinders are in the rear row, and all even-numbered cylinders are in the front row.

## Valves

The valves are located in the intake and exhaust ports of the cylinder head. The fuel/air mixture enters the cylinder head through the intake valve ports and the burned exhaust gases are exhausted through exhaust valve ports. The conventional poppet type valves are normally used in aircraft engines. Valves are not only typed by their intended use but also their size and shape. Figure 1-2-22 shows various shapes and types of these valves.

**Valve construction.** The valves in the cylinders of an aircraft engine are subjected to high temperatures, corrosion and operating stresses. To provide reliable service the metal alloy in the valves must be able to resist all these factors.

Because intake valves operate at lower temperatures than exhaust valves, they can be made of chrome-nickel steel. Exhaust valves are usually made of nichrome, silchrome or cobalt-chromium steel.

The valve head has a ground face that forms a seal against the ground valve seat in the cylinder head when the valve is closed. The face of the valve is usually ground to an angle of either 30° or 45°. In some engines, the intake-valve face is ground to an angle of 30° and the exhaust valve face is ground to a 45° angle.

Valve faces are often made more durable by the application of a material called stellite. About 1/16-inch of this alloy is welded to the valve face and ground to the correct angle. Stellite is resistant to high-temperature corrosion and also withstands the shock and wear associated with valve operation. Some engine manufacturers use a nichrome facing on the valves. This serves the same purpose as the stellite material.

The valve stem acts as a pilot for the valve head and rides in the valve guide installed in the cylinder head for this purpose. The valve stem is surface-hardened to resist wear. The neck is the part that forms the junction between the head and the stem. The tip of the valve is hardened to withstand the hammering of the valve rocker arm as it opens the valve. A machined groove on the stem near the tip receives the split-ring stem keys. These stem keys form a lock ring to hold the valve spring retaining washer in place.

Some intake and exhaust valve stems are hollow and partially filled with metallic sodium. This material is used because it is an excellent heat conductor. The sodium will melt at approximately 208°F, and the reciprocating motion of the valve circulates the liquid sodium and enables it to carry away heat from the valve head to the valve stem, where it is dissipated through the valve guide to the cylinder head and the cooling fins. Thus, the operating temperature of the valve may be reduced as much as 300°-400°F. Under no circumstances should a sodium-filled valve be cut open or subjected to treatment that may cause it to rupture. Exposure of the sodium in these valves to the outside air will result in fire or explosion with possible personal injury.

The most commonly used intake valves have solid stems, and the head is either flat or tulip shaped. Intake valves for low-power engines are usually flat headed.

In some engines, the intake valve may be the tulip type and have a smaller stem than the exhaust valve, or it may be similar to the exhaust valve but have a solid stem and head. Although these valves are similar, they are not interchangeable since the faces of the valves are constructed of different material. The intake valve will usually have a flat milled on the tip to identify it.

**Valve seats.** The aluminum alloy that the cylinder head is made of is not hard enough to withstand the forces of the valve being pulled closed by the valve spring. To provide for an adequate long lasting seal around the valve, a valve seat of either a steel or bronze alloy is used. The valve seat is either screwed or shrunk (tight fit) in the intake or exhaust port of the cylinder head. If the seat is installed using the shrink (tight fit) method, the cylinder must be heated to between 500°-600°F. The seat is then chilled in dry ice and pressed into place.

The valve and the seat are normally ground to a slightly different angle to produce an interference fit. The valve face is normally ground to an angle of 1° less then the angle of the seat. This interference fit provides for positive seating of the valve on the seat.

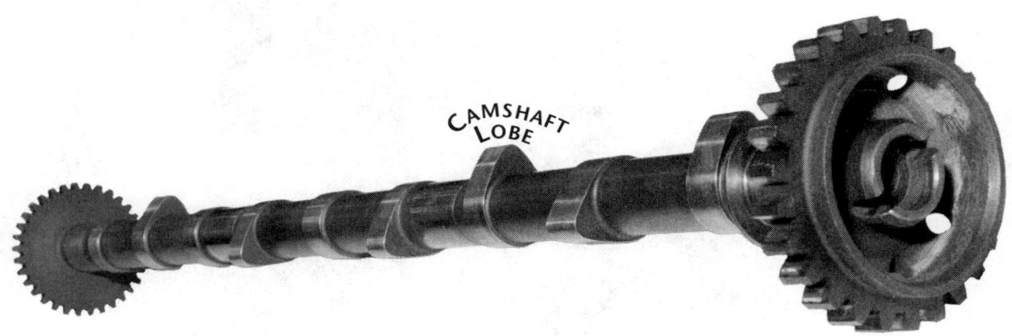

Figure 1-2-23. Typical cam lobes

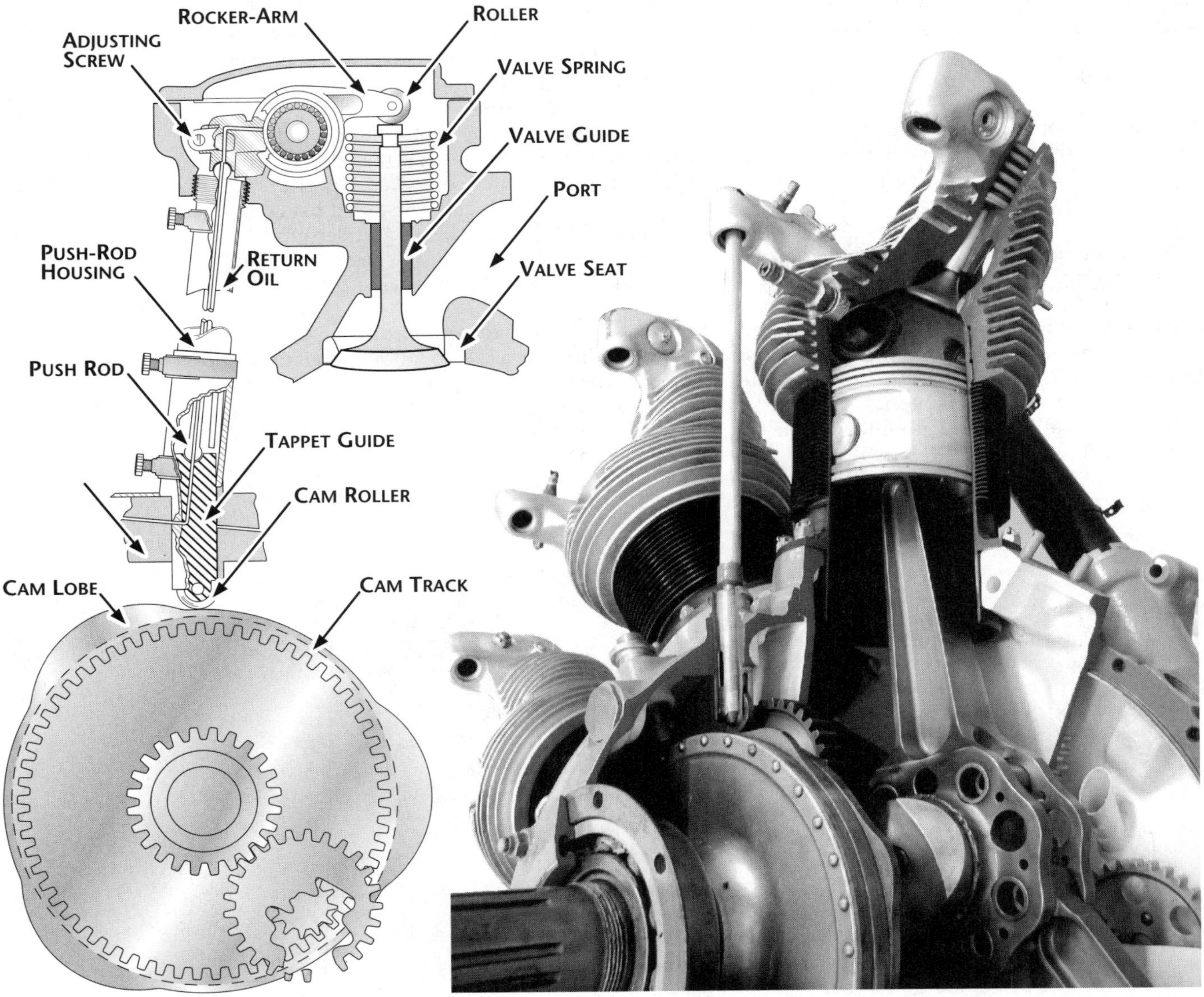

**Figure 1-2-24. Valve-operating mechanism for a radial engine**

**Figure 1-2-25. Valve-operating mechanism for an opposed engine**

## Valve-operating Mechanism

For operation of a reciprocating engine, each valve must open at the proper time, stay open for the required time and close at the proper time. Intake valves are opened just before the piston reaches TDC, and exhaust valves remain open after TDC. At a particular instant, both valves are open (end of the exhaust stroke, beginning of the intake stroke). This overlap permits better volumetric efficiency and lowers the cylinder operating temperature and is controlled by the valve-operating mechanism.

The *valve lift* (distance that the valve is lifted off its seat) and the *valve duration* (length of time the valve is held open) are both determined by the shape of the cam lobes.

Typical cam lobes are illustrated in Figure 1-2-23 (see page 1-49). The portion of the lobe that

| CAM RING TABLE | | | | | | |
|---|---|---|---|---|---|---|
| 5 Cylinders | | 7 Cylinders | | 9 Cylinders | | Direction of Rotation |
| Number of Lobes | Speed | Number of Lobes | Speed | Number of Lobes | Speed | |
| 3 | 1/6 | 4 | 1/8 | 5 | 1/10 | With crankshaft |
| 2 | 1/4 | 3 | 1/6 | 4 | 1/8 | Opposite crankshaft |

**Table 1-2-1. Radial engine cam ring table**

gently starts the valve-operating mechanism moving is called a *ramp,* or *step.* The ramp is machined on each side of the cam lobe to ease the rocker arm into contact with the valve tip, thus reducing the shock load.

The valve-operating mechanism consists of a cam ring or camshaft equipped with lobes, which work against a *cam roller,* or *cam follower* (see Figures 1-2-24 and 1-2-25). The cam follower then pushes a push rod and ball socket which, in turn, actuates a rocker arm that opens the valve. Springs, which slip over the stem of the valves and are held in place by the valve-spring retaining washer and stem key, close each valve and push the valve mechanism in the opposite direction when the cam roller or follower rolls along a low section of the cam ring.

**Cam ring.** The valve mechanism of a radial engine is operated by one or two cam rings, depending upon the number of rows of cylinders. In a single-row radial engine one ring with a double cam track is used. One track operates the intake valves; the other, the exhaust valves. The cam ring is usually located between the propeller reduction gearing and the front end of the power section.

The *cam ring* is a circular piece of steel with a series of cams or lobes on the outer surface. The surface of these lobes and the space between them (on which the cam rollers ride) is known as the *cam track.* As the cam ring revolves, the lobes cause the cam roller to raise the tappet in its guide, transmitting the force through the push rod and rocker arm to open the valve.

The cam ring is mounted concentrically with the crankshaft and is driven by the crankshaft at a reduced rate of speed through the cam intermediate drive gear assembly. The cam ring has two parallel sets of lobes spaced around the outer periphery, one set (cam track) for the intake valves and the other for the exhaust valves. The cam rings used may have four or five lobes on both the intake and the exhaust tracks. The timing of the valve events is determined by the spacing of these lobes and the speed and direction at which the cam rings are driven in relation to the speed and direction of the crankshaft.

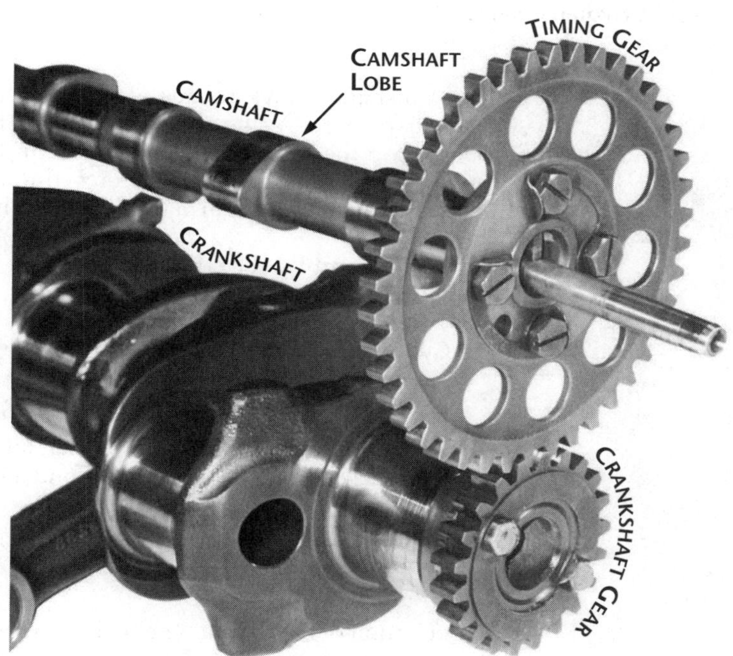


**Figure 1-2-26. Cam drive mechanism used on opposed-type aircraft engines**

The method of driving the cam varies on different engine makes. The cam ring can be designed with teeth on either the inside or outside periphery. If the reduction gear meshes with the teeth on the outside of the ring, the cam will turn in the direction of rotation of the crankshaft. If the ring is driven from the inside, the cam will turn in the opposite direction from the crankshaft. This method is illustrated in Table 1-2-1.

**Camshaft.** A camshaft is used to operate the valve mechanism of an opposed engine. It is driven by a gear that mates with another gear attached to the crankshaft (see Figure 1-2-26).

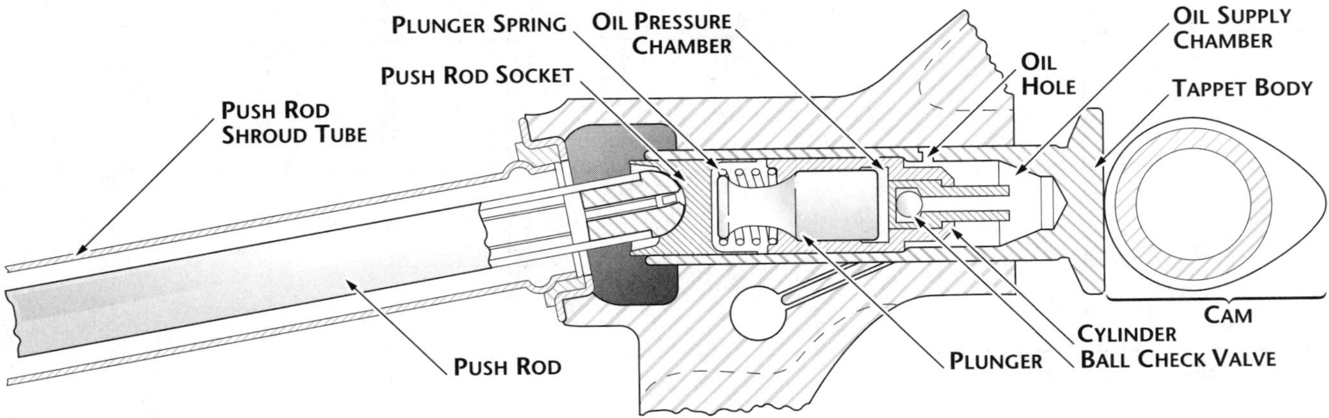


**Figure 1-2-27. Hydraulic valve lifters**

The camshaft always rotates at one-half the crankshaft speed. As the camshaft revolves, the lobes cause the tappet assembly to rise in the tappet guide, transmitting the force through the push rod and rocker arm to open the valve.

**Tappet assembly.** The tappet assembly, or *lifter assembly,* consists of:

- Cylindrical tappet, which slides in and out in a tappet guide installed in one of the crankcase sections around the cam ring
- Cam follower or tappet roller, which follows the contour of the cam ring and lobes
- Tappet ball socket or push rod socket
- Tappet spring

The function of the lifter assembly is to convert the rotational movement of the cam lobe into reciprocating motion, transmitting this motion to the push rod, rocker arm and then to the valve tip, opening the valve at the proper time.

**Solid lifters.** The solid lifter uses the tappet spring to take up the clearance between the rocker arm and the valve tip. This reduces the shock load when the valve is opened. An oil passage drilled through the tappet allows engine oil to flow through the hollow push rods to lubricate the rocker assemblies.

When an engine is equipped with solid lifters the valve clearance must be set at the correct operating temperature. This results in the engine running rough until the operating temperature and the clearance are correct.

**Hydraulic valve lifters.** Most modern aircraft engines incorporate hydraulic lifters that automatically keep the valve clearance at zero, eliminating the necessity for any valve clearance adjustment mechanism. A typical hydraulic tappet (zero-lash valve lifter) is shown in Figure 1-2-27.

When the engine valve is closed, the face of the tappet body (cam follower) is on the base circle or back of the cam, as shown in Figure 1-2-26. The light plunger spring lifts the hydraulic plunger so that its outer end contacts the push rod socket, exerting a light pressure against it, thus eliminating any clearance in the valve linkage. As the plunger moves outward, the ball check valve moves off its seat. Oil from the supply chamber, which is directly connected with the engine lubrication system, flows in and fills the pressure chamber. As the camshaft rotates, the cam pushes the tappet body and the hydraulic lifter cylinder outward. This action forces the ball check valve onto its seat; thus, the body of oil trapped in the pressure chamber acts as a cushion. During the interval when the engine valve is off its seat, a predetermined leakage occurs between plunger and cylinder bore which compensates for any expansion or contraction in the valve train. Immediately after the engine valve closes, the amount of oil required to fill the pressure chamber flows in from the supply chamber, preparing for another cycle of operation.

Hydraulic valve lifters are normally adjusted at the time of overhaul. They are assembled dry (no lubrication), clearances are checked and adjustments are usually made by use of pushrods having different lengths. Minimum and maximum valve clearances are established. Any measurement between these extremes is acceptable, but approximately half way between the extremes is desired. Hydraulic valve lifters require less maintenance, are better lubricated and operate more quietly than the screw adjustment type.

**Push rod.** The push rod is a hollow tubular-shaped component that transmits the force of the valve tappet to the rocker arm. Each end of the tube is fitted with a hardened steel ball that is allowed to rest in a socket in the tappet and the rocker arm. The hollow tube and the oil passage in the hardened steel balls allows for lubrication of the rocker arm, the rocker arm bearing and in some cases the valve stem and guide. For most modern reciprocating engines, the pushrods are available in different lengths. The length of the pushrod will determine the rocker arm/valve clearance. The pushrod is enclosed in a tubular housing that extends from the crankcase to the cylinder head.

**Rocker arms.** The rocker arms transmit the lifting force from the cams to the valves. Rocker arm assemblies are supported by a plain, roller or ball bearing, or a combination of these, which serves as a pivot. Generally one end of the arm bears against the push rod and the other bears on the valve stem. One end of the rocker arm is sometimes slotted to accommodate a steel roller. The opposite end is constructed with either a threaded split clamp and locking bolt or a tapped socket.

Some rocker arms have an adjusting screw for adjusting the clearance between the rocker arm and the valve stem.

**Valve springs.** Each valve is closed by two or three helical-coiled springs. If a single spring were used, it would vibrate or surge at certain speeds. To eliminate this difficulty, two or more springs (one inside the other) are installed on each valve. Each spring will therefore vibrate at a different engine speed, and rapid damping out of all spring-surge vibrations during engine operation will result. Two or more springs also reduce danger of weakness and possible failure by breakage due to heat and metal fatigue.

The springs are held in place by split locks installed in the recess of the valve spring upper retainer or washer and engage a groove machined into the valve stem (see Figure 1-2-28). The functions of the valve springs are to close the valve and to hold the valve securely on the valve seat.

Figure 1-2-28. Installing split locks on a valve — while one person compresses the valve springs, another installs the keepers.

A *bearing* is any surface that supports, or is supported by, another surface. A good bearing must be composed of material that is strong enough to withstand the pressure imposed on it and should permit the other surface to move with a minimum of friction and wear. The parts must be held in position within very close tolerances to provide efficient and quiet operation, yet allow freedom of motion. To accomplish this, and at the same time reduce friction of moving parts so that power loss is not excessive, lubricated bearings of many types are used. Bearings are required to take radial loads, thrust loads or a combination of the two.

There are two ways in which bearing surfaces move in relation to each other. One is by the sliding movement of one metal against the other, and the second is for one surface to roll over the other. The three different types of bearings in general use are plain, roller and ball (see Figure 1-2-29).

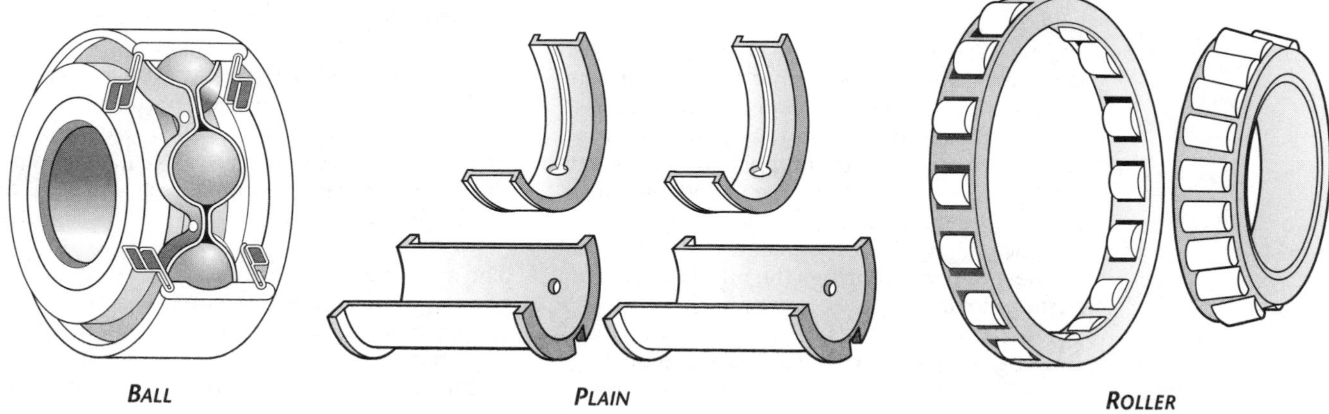


Figure 1-2-29. Bearings

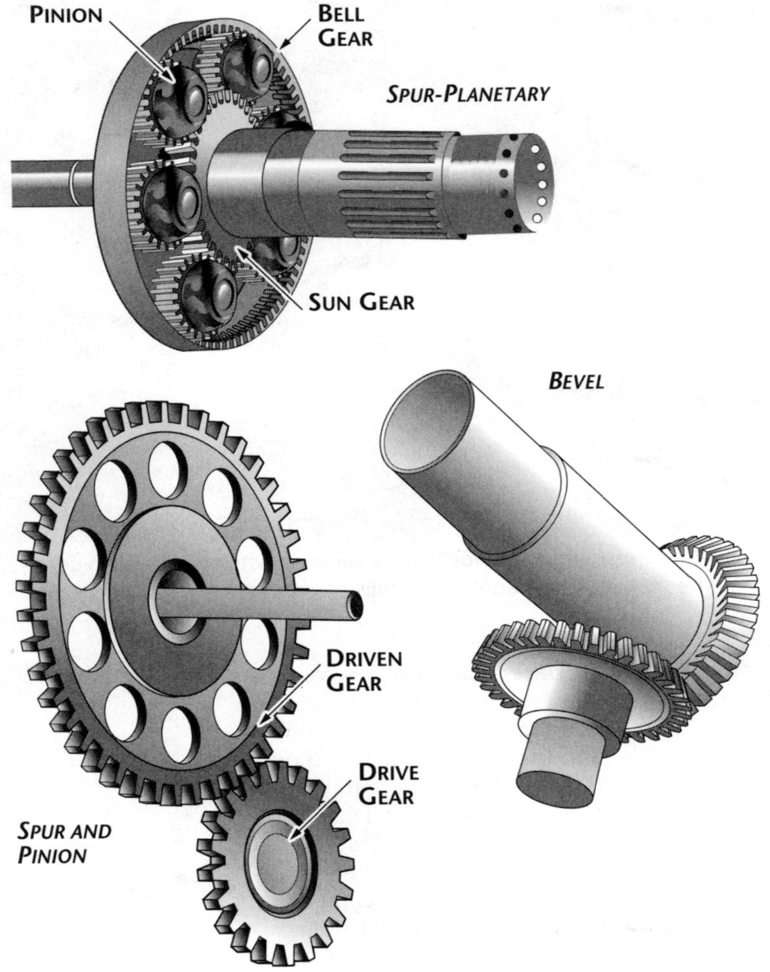


Figure 1-2-30. The three most commonly used types of reduction gears

**Plain bearings.** Plain bearings are generally used for the crankshaft, cam ring, camshaft, connecting rods and the accessory drive shaft bearings. Such bearings are usually subjected to radial loads only, although some have been designed to take thrust loads.

Plain bearings are usually made of non-ferrous (having no iron) metals, such as silver, bronze, aluminum and various alloys of copper, tin or lead. Master rod or crankpin bearings in some engines are thin shells of steel, plated with silver on both the inside and the outside surfaces and with lead-tin plated over the silver on the inside surface only. Smaller bearings, such as those used to support various shafts in the accessory section, are called bushings. Porous Oilite bushings are widely used in this instance. They are impregnated with oil so that the heat of friction brings the oil to the bearing surface during engine operation.

**Ball bearings.** A ball bearing assembly consists of grooved inner and outer races, one or more sets of balls and, in bearings designed for disassembly, a bearing retainer. They are used for supercharger impeller shaft bearings and rocker arm bearings in some engines. Special deep-groove ball bearings are used in aircraft engines to transmit propeller thrust to the engine nose section.

**Roller bearings.** Roller bearings are made in many types and shapes, but the two types generally used in the aircraft engine are the straight roller and the tapered roller bearings. Straight roller bearings are used where the bearing is subjected to radial loads only. In tapered roller bearings, the inner- and outer-race bearing surfaces are cone shaped. Such bearings will withstand both radial and thrust loads. Straight roller bearings are used in high-power aircraft engines for the crankshaft main bearings. They are also used in other applications where radial loads are high.

## Propeller Reduction Gearing

The increased brake horsepower delivered by a high-horsepower engine results partly from increased crankshaft r.p.m. It is therefore necessary to provide reduction gears to limit the propeller rotation speed to a value at which efficient operation is obtained. Whenever the speed of the blade tips approaches the speed of sound, the efficiency of the propeller decreases rapidly. The general practice has been to provide reduction gearing for propellers of engines whose speeds are above 2,000 r.p.m., because propeller efficiency decreases rapidly above this speed.

Since reduction gearing must withstand extremely high stresses, the gears are machined from steel forgings. Many types of reduction gearing systems are in use. The three types (Figure 1-2-30) most commonly used are:

- Spur planetary
- Bevel planetary
- Spur and pinion

Figure 1-2-31 is a cutaway component, revealing a spur-type planetary reduction gear.

The planetary reduction gear systems are used with radial and opposed engines, and the spur-and-pinion system is used with opposed, in-line and V-type engines. Two of these types, the spur planetary and the bevel planetary, are discussed here.

The spur planetary reduction gearing consists of a large driving gear or sun gear splined (and sometimes shrunk) to the crankshaft, a large stationary gear, called a bell gear, and a set of small spur planetary pinion gears mounted on a carrier ring. The ring is fastened to the propeller shaft, and the planetary gears mesh

**Figure 1-2-31. Spur planetary reduction gears**

with both the sun gear and the stationary bell or ring gear. The stationary gear is bolted or splined to the front-section housing. When the engine is operating, the sun gear rotates. Because the planetary gears are meshed with this ring, they also must rotate. Since they also mesh with the stationary gear, they will walk or roll around it as they rotate, and the ring in which they are mounted will rotate the propeller shaft in the same direction as the crankshaft but at a reduced speed.

In some engines, the bell gear is mounted on the propeller shaft, and the planetary pinion gear cage is held stationary. The sun gear is splined to the crankshaft and thus acts as a driving gear. In such an arrangement, the propeller travels at a reduced speed, but in opposite direction to the crankshaft.

In the bevel planetary reduction gearing system, the driving gear is machined with beveled external teeth and is attached to the crankshaft. A set of mating bevel pinion gears is mounted in a cage attached to the end of the propeller shaft. The pinion gears are driven by the drive gear and walk around the stationary gear, which is bolted or splined to the front-section housing. The thrust of the bevel pinion gears is absorbed by a thrust ball bearing of special design. The drive and the fixed gears are generally supported by heavy-duty ball bearings. This type of planetary reduction assembly is more compact than the other one described and can therefore be used where a smaller propeller gear step-down is desired.

## Propeller Shafts

Propeller shafts are one of three major types:

- Tapered
- Splined
- Flanged

Figure 1-2-32. A tapered propeller crankshaft

Figure 1-2-33. A splined propeller crankshaft

Figure 1-2-34. A flange-type propeller shaft

Tapered shafts are identified by taper numbers, splined and flanged shafts by SAE numbers.

**Tapered.** The propeller shaft of many low-power output engines is forged as part of the crankshaft. It is tapered and a milled slot is provided so that the propeller hub can be keyed to the shaft. The keyway and key index of the propeller are in relation to the No. 1 cylinder TDC. The end of the shaft is threaded to receive the propeller-retaining nut. Tapered propeller shafts (Figure 1-2-32) are common on older opposed and in-line engines.

**Splined.** The propeller shaft of a high-output engine generally is splined (Figure 1-2-33). It is threaded on one end for a propeller hub nut. The thrust bearing, which absorbs propeller thrust, is located around the shaft and transmits the thrust to the nose-section housing. The shaft is threaded for attaching the thrust-bearing retaining nut. On the portion protruding from the housing (between the two sets of threads), splines are located to receive the splined propeller hub. The shaft is generally machined from a steel-alloy forging throughout its length. The propeller shaft may be connected by reduction gearing to the engine crankshaft, but in smaller engines the propeller shaft is simply an extension of the engine crankshaft. To turn the propeller shaft, the engine crankshaft must revolve.

**Flanged.** Flanged propeller shafts (Figure 1-2-34) are used on low- to medium-horsepower reciprocating and small turboprop engines. One end of the shaft is flanged with drilled holes to accept the propeller mounting bolts. The installation may be a short shaft with internal threading to accept the distributor valve to be used with a controllable propeller. The flanged propeller shaft is a normal installation on most modern reciprocating engines.

## Section 3

# Engine Maintenance and Operation

## Reciprocating Engine Operation

The operation of the powerplant is controlled from the cockpit. All installations have numerous control handles and levers connected to the engine by rods, cables, bellcranks, pulleys, etc. The control handles, in most cases, are mounted on quadrants in the cockpit. In order to help prevent operational errors, the location, placards, colors and shapes on the quadrant indicate the functions and positions of the levers. The shape of each handle is regulated by its usage. Both lever and push-pull controls have specific shapes for each different function. FAR part 23.781 shows the shapes for all cockpit controls (Figure 1-3-1). The other FAA regulations that cover cockpit control shapes and operations are FAR parts 23.777 and 23.779. The same sections apply for turbine aircraft, with the prefix of part 25 instead of part 23.

Figure 1-3-2 is a picture of the throttle quadrant of a Beechcraft Baron. It clearly shows the differing handle shapes, making confusion over what control performs what action practically a non-issue. Memorize the shapes of at least the basic controls before doing a run-up. When a flight instructor checks you out in starting and run-up procedures, he or she will ask for a demonstration of your understanding of the control shapes and operation.

Manifold pressure, r.p.m., engine temperature, oil temperature, carburetor air temperature and the fuel/air ratio can be controlled by manipulating the cockpit controls. Coordinating the movement of the controls with the instrument readings protects against exceeding operating limits.

Engine operation is usually limited by specified operating ranges of the following:

- Crankshaft speed (r.p.m.).
- Manifold pressure
- Cylinder head temperature
- Carburetor air temperature
- Oil temperature
- Oil pressure
- Fuel pressure
- Fuel/air mixture setting

The procedures, pressures, temperatures and engine speeds used throughout this section are solely for the purpose of illustration and do not have general application. The operating procedures and limits used on individual makes and models of aircraft engines vary considerably from the values shown here. For exact information regarding a specific engine model, consult the engine manufacturer's instructions.

**Engine instruments.** The term *engine instruments* includes all those instruments required to measure and indicate the functioning of the powerplant. The engine instruments are generally installed on the instrument panel so that all of them can easily be observed at one time.

Some of the simpler, light aircraft may be equipped only with a tachometer, oil pressure

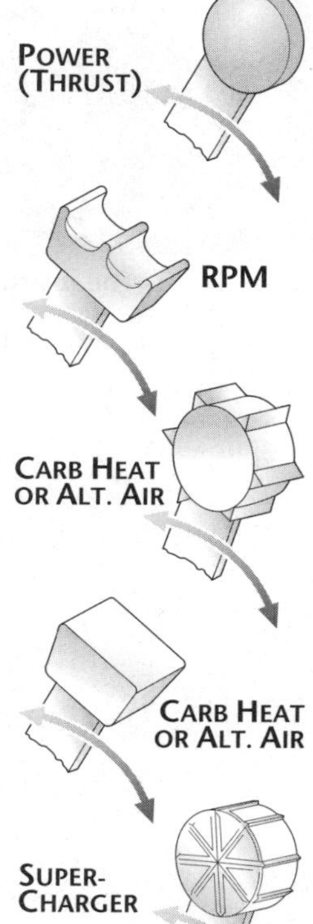


PANEL MOUNTED CONTROL KNOBS

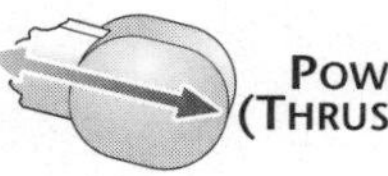


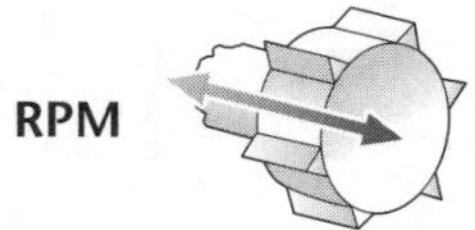

Figure 1-3-1. Handle shapes for cockpit controls are uniquely designed.

Figure 1-3-2. This photo shows the throttle quadrant of a Beechcraft Baron. Following the regulations, propeller controls are on the left, throttles are in the center and mixture controls are on the right. Notice that each lever has its specified control knob shape. In addition, they are clearly placarded.

**Figure 1-3-3. The instrument panel of a small aircraft**

**Figure 1-3-4. This instrument panel, from a Beechcraft Baron, shows some of the many indicator gauges necessary for each engine.**

gauge and oil temperature gauges (Figure 1-3-3). The heavier, more complex aircraft instrument panels (Figure 1-3-4) will have all or part of the following engine instruments:

- Oil pressure indicator and warning system
- Oil temperature indicator
- Fuel pressure indicator and warning system
- Carburetor air temperature indicator
- Cylinder head temperature indicator for air-cooled engines
- Manifold pressure indicator
- Tachometer
- Fuel quantity indicator
- Fuel flowmeter or fuel mixture indicator
- Oil quantity indicator
- Augmentation liquid quantity indicator
- Fire-warning indicators
- A means to indicate when the propeller is in reverse pitch
- BMEP (brake mean effective pressure) indicator

**Engine starting.** Correct starting technique is an important part of engine operation. Improper procedures often are used because some of the basic principles involved in engine operation are misunderstood. It is important for the technician to understand that starting procedures will be dictated by the type of fuel metering system used on the engine, as well as by ambient conditions. The specific manufacturer's procedures for a particular engine and aircraft combination should always be followed. Always use the proper checklist for the aircraft.

Many airports require engine run-ups to be performed in a specific part of the airport. Generally, this requirement has to do with noise pollution standards and must be followed closely. If using the run-up apron requires the airplane to be taxied, a technician should be checked out by a Certified Flight Instructor first. Never try to taxi an airplane for which you have not been checked out and/or approved to taxi. Additionally, you may be required to contact ground control by radio before moving the airplane (see Figure 1-3-5).

**Engine warm-up.** Proper engine warm-up is important, particularly when the condition of the engine is unknown. Improperly adjusted idle mixture, intermittently firing spark plugs and improperly adjusted engine valves all have an overlapping effect on engine stability. Therefore, the warm-up should be made at the engine speed where maximum engine stability is obtained. Experience has shown that the optimum warm-up speed is from 1,000-1,600 r.p.m. The actual speed selected should be the speed at which engine operation is the smoothest, since the smoothest operation is an indication that all phases of engine operation are the most stable.

During warm-up, watch the instruments associated with engine operation. This will aid in making sure that all phases of engine operation are normal. For example, engine oil pressure should be indicated within 30 sec. after the start. Furthermore, if the oil pressure is not up to or above normal within 1 min. after the engine starts, the engine should be shut down. Cylinder head or coolant temperatures should be observed continually to see that they do not exceed the maximum allowable limit.

**Figure 1-3-5. Many airports have very specific rules about performing run-ups in the airport.**

An overly lean mixture should not be used to hasten the warm-up. Actually, at the warm-up r.p.m. there is very little difference in the mixture supplied to the engine, whether the mixture is in a rich or lean position, since metering in this power range is governed by throttle position.

Carburetor heat can be used as required under conditions leading to ice formation. For engines equipped with a float-type carburetor, it may be desirable to raise the carburetor air temperature during warm-up to prevent ice formation and to ensure smooth operation.

> **NOTE:** *Caution should be exercised when using carburetor heat during ground operations, as normally the heated air does not pass through the air filter.*

The magneto safety check can be performed during warm-up. Its purpose is to ensure that all ignition connections are secure and that the ignition system will permit operation at the higher power settings used during later phases of the ground check. The time required for proper warm-up gives ample opportunity to perform this simple check, which may disclose a condition that would make it inadvisable to continue operation until after corrections have been made.

The magneto safety check is conducted with the propeller in the high r.p.m. (low-pitch) position, at approximately 1,000 r.p.m. Move the ignition switch from BOTH to RIGHT, then return to BOTH; from BOTH to LEFT and return to BOTH; from BOTH to OFF momentarily, then return to BOTH.

While switching from BOTH to a single magneto position, a slight but noticeable drop in r.p.m. should occur. This indicates that the opposite magneto has been properly grounded out. Complete cutting out of the engine when switching from BOTH to OFF indicates that both magnetos are grounded properly. Failure to obtain any drop while in the single magneto position, or failure of the engine to cut out while switching to OFF, indicates that one or both ground connections are not secured.

**Ground check.** A ground check is performed to evaluate the function of the engine by comparing power input, measured by manifold pressure, with power output, measured by r.p.m.

The engine may be capable of producing a prescribed power, even rated takeoff, and not be functioning properly. Only by comparing the manifold pressure required during the check against a known standard will an unsuitable condition be disclosed. The magneto check can also fail to show up shortcomings, since the allowable r.p.m. drop-off is only a measure of an improperly functioning ignition system and is not necessarily affected by other factors. Conversely, it is possible for the magneto check to prove satisfactory with an unsatisfactory condition present elsewhere in the engine.

The ground check should be made after the engine is thoroughly warm. It consists of checking the operation of the powerplant and accessory equipment by ear, by visual inspection and by proper interpretation of instrument readings, control movements and switch reactions.

During the ground check, the aircraft should be headed into the wind, if possible, to take advantage of the cooling airflow. A ground check may be performed as follows:

Control ............. Position Check

Cowl flaps .......... Open

Mixture ............ Rich

Figure 1-3-6. Moving the propeller governor control lever from one position to another allows a technician to check operation of the propeller governor.

Propeller ........... High r.p.m.

Carburetor heat ..... Cold

Carburetor air filter .. As required

Procedure:

- Check the propeller, according to propeller manufacturer's instruction
- Open the throttle to manifold pressure equal to field barometric pressure
- Move mag switch from both to right and return to both. Switch from both to left and return to both. Observe the r.p.m. drop while operating on the right and left positions. The maximum drop should not exceed that specified by the engine manufacturer.

Figure 1-3-7. Opening and closing the throttle during a power check causes shifts in the manifold pressure gauge readings, allowing the technician to detect cylinder and timing malfunctions.

- Check the fuel pressure and oil pressure. They must be within the established tolerance for the subject engine.
- Take note of the r.p.m.
- Retard the throttle

In addition to the operations outlined above, check the functioning of various items of aircraft equipment, such as generator systems, hydraulic systems, etc.

**Propeller pitch check.** The propeller is checked to ensure proper operation of the pitch control and the pitch-change mechanism. The operation of a controllable-pitch propeller is checked by the indications of the tachometer and manifold pressure gauge when the propeller governor control is moved from one position to another (Figure 1-3-6). Because each type of propeller requires a different procedure, the applicable manufacturer's instructions should be followed.

**Power check.** Specific r.p.m. and manifold pressure relationship should be checked during each ground check. This can be done at the time the engine is run-up to make the magneto check. The basic idea of this check is to measure the performance of the engine against an established standard. Calibration tests have determined that the engine is capable of delivering a given power at a given r.p.m. and manifold pressure. The original calibration, or measurement of power, is made by means of a dynamometer. During the ground check, power is measured with the propeller. With constant conditions of air density, the propeller, at any fixed-pitch position, will always require the same r.p.m. to absorb the same horsepower from the engine. This characteristic is used in determining the condition of the engine.

With the governor control set for full low pitch, the propeller operates as a fixed-pitch propeller. Under these conditions, the manifold pressure for any specific engine, with the mixture control in auto-rich, indicates whether all the cylinders are operating properly. With one or more dead or intermittently firing cylinders, the operating cylinders must provide more power for a given r.p.m. Consequently, the carburetor throttle must be opened further (Figure 1-3-7), resulting in higher manifold pressure. Different engines of the same model using the same propeller installation and in the same geographical location should require the same manifold pressure, within 1 inch Hg, to obtain a certain r.p.m. when the barometer and temperature are at the same readings. A higher-than-normal manifold pressure usually indicates a dead cylinder or late ignition timing. An excessively low manifold pressure for a particular r.p.m. usually indicates that the ignition timing is early. Early ignition can cause detonation and loss of power at takeoff power settings.

Before starting the engine, observe the manifold pressure gauge, which should read approximately atmospheric (barometric) pressure when the engine is not running. At sea level this is approximately 30 inches Hg, and at fields above sea level, the atmospheric pressure will be less, depending on the height above sea level.

When the engine is started and then accelerated, the manifold pressure will decrease until about 1,600-1,700 r.p.m. is reached, and then it will begin to rise. At approximately 2,000 r.p.m., with the propeller in low-pitch position, the manifold pressure should be the same as the field barometric pressure. If the manifold pressure-gauge reading (field barometric pressure) was 30 inch Hg before starting the engine, the pressure reading should return to 30 inch Hg at approximately 2,000 r.p.m. If the manifold pressure gauge reads 26 inch Hg before starting, it should read 26 inch Hg again at approximately 2,000 r.p.m. The exact r.p.m. will vary with various models of engines or because of varying propeller characteristics. In certain installations, the r.p.m. needed to secure field barometric pressure may be as high as 2,200 r.p.m. However, once the required r.p.m. has been established for an installation, any appreciable variation indicates some malfunctioning. This variation may occur because the low-pitch stop of the propeller has not been properly set or because the carburetor or ignition system is not functioning properly.

The accuracy of the power check may be affected by the following variables:

- *Wind.* Any appreciable air movement (5 m.p.h. or more) will change the air load on the propeller blade when it is in the fixed-pitch position. A headwind will increase the r.p.m. obtainable with a given manifold pressure. A tailwind will decrease the r.p.m.
- *Atmospheric temperatures.* The effects of variations in atmospheric temperature tend to cancel each other. Higher carburetor intake and cylinder temperatures tend to lower the r.p.m., but the propeller load is lightened because of the decreased density of the air.
- *Engine and induction system temperature.* If the cylinder and carburetor temperatures are high because of factors other than atmospheric temperature, a low r.p.m. will result, since the power will be lowered without a compensating lowering of the propeller load.
- *Oil temperature.* Cold oil will tend to hold down the r.p.m., since the higher viscosity results in increased friction horsepower losses.

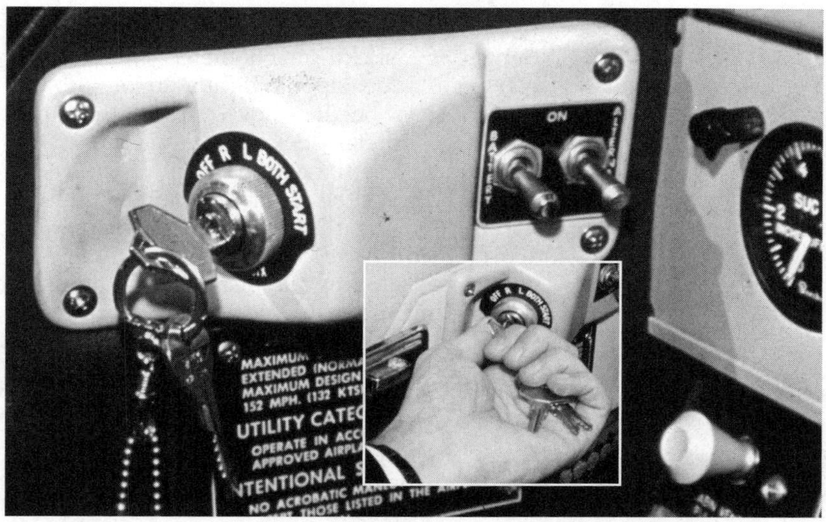


**Figure 1-3-8. A switch can be flipped to cut individual magnetos, making detection of faulty ignition easier.**

**Ignition system operational check.** In performing the ignition system operational check (magneto check), the power-absorbing characteristics of the propeller in the low fixed-pitch position are utilized. In switching to individual magnetos (Figure 1-3-8), cutting out the opposite plugs results in a slower rate of combustion, which gives the same effect as retarding the spark advance. The drop in engine speed is a measure of the power loss at this slower combustion rate.

By comparing the r.p.m. drop with a known standard. The following may be determined:

- Proper timing of each magneto
- General engine performance as evidenced by smooth operation
- Additional check of the proper connection of the ignition leads

Any unusual roughness on either magneto is an indication of faulty ignition caused by plug fouling or by malfunctioning of the ignition system. The operator should be very sensitive to engine roughness during this check. Lack of drop-off in r.p.m. may be an indication of faulty grounding of one side of the ignition system. Complete cutting out when switching to one magneto is definite evidence that its side of the ignition system is not functioning. Excessive difference in r.p.m. drop-off between the left and right switch positions can indicate a difference in timing between the left and right magnetos.

Sufficient time should be given the check on each single switch position to permit complete stabilization of engine speed and manifold pressure. There is a tendency to perform this check too rapidly with resultant wrong indications. Single ignition operation for as long as 1 minute is not excessive.

Another point that must be emphasized is the danger of a sticking tachometer. The tachometer should be tapped lightly to make sure the indicate needle moves freely. In some cases, tachometer sticking has caused errors in indication to the extent of 100 r.p.m. Under such conditions the ignition system could have had as much as a 200-r.p.m. drop with only a 100-r.p.m. drop indicated on the instrument. In most cases, tapping the instrument eliminates the sticking and results in accurate readings.

In recording the results of the ignition system check, record the amount of the total r.p.m. drop which occurs rapidly and the amount that occurs slowly. This breakdown in r.p.m. drop provides a means of pinpointing certain troubles in the ignition system. It can save a lot of time and unnecessary work by confining maintenance to the specific part of the ignition system that is responsible for the trouble.

Fast r.p.m. drop is usually the result of either faulty spark plugs or faulty ignition harness. This is true because faulty plugs or leads take effect at once. The cylinder goes dead or starts firing intermittently the instant the switch is moved from both to the right or left position.

Slow r.p.m. drop usually is caused by incorrect ignition timing or faulty valve adjustment. With late ignition timing, the charge is fired too late (with relation to piston travel) for the combustion pressures to build up to the maximum at the proper time. The result is a power loss greater than normal for single ignition because of the lower peak pressures obtained in the cylinder. However, this power loss does not occur as rapidly as that which accompanies a dead spark plug. This explains the slow r.p.m. drop as compared to the instantaneous drop with a dead plug or defective lead. Incorrect valve clearances, through their effect on valve overlap, can cause the mixture to be too rich or too lean. The too-rich or too-lean mixture may affect one plug more than another, because of the plug location, and show up as a slow r.p.m. drop on the ignition check.

**Cruise mixture check.** The cruise mixture check is a check of carburetor metering. Checking the carburetor metering characteristics at 200- to 300-r.p.m. intervals, from 800 r.p.m. to the ignition system check speed, gives a complete pattern for the basic carburetor performance.

To perform this test, set up a specified engine speed with the propeller in full low pitch. The first check is made at 800 r.p.m. With the carburetor mixture control in the automatic-rich position, read the manifold pressure. With the throttle remaining in the same position, move the mixture control to the automatic lean position. Read and record the engine speed and manifold pressure readings. Repeat this check at 1,000, 1,200, 1,500, 1,700 and 2,000 r.p.m. or at the engine speed specified by the manufacturer. Guard against a sticking instrument by tapping the tachometer.

Moving the mixture control from the auto-rich position to the auto-lean position checks the cruise mixture. In general, the speed should not increase more than 25 r.p.m. or decrease more than 75 r.p.m. during the change from auto-rich to auto-lean.

For example, suppose that the r.p.m. change is above 100 for the 800-1,500 r.p.m. checks; it is obvious that the probable cause is an incorrect idle mixture. When the idle is adjusted properly, carburetion will be correct throughout the range.

**Idle speed and idle mixture checks.** Plug fouling difficulty is the inevitable result of failure to provide a proper idle mixture setting. The tendency seems to be to adjust the idle mixture on the extremely rich side and to compensate for this by adjusting the throttle stop to a relatively high r.p.m. for minimum idling. With a properly adjusted idle mixture setting, it is possible to run the engine at idle r.p.m. for long periods. Such a setting will result in a minimum of plug fouling and exhaust smoking and it will pay dividends from the savings on the aircraft brakes after landing and while taxiing.

If the wind is not too strong, the idle mixture setting can be checked easily during the ground check as follows:

- Close the throttle.
- Move the mixture control to the idle cutoff position and observe the change in r.p.m. Return the mixture control back to the rich position before engine cutoff.

As the mixture control lever is moved into idle cutoff, and before normal drop-off, one of two things may occur momentarily:

- The engine speed may increase. An increase in r.p.m. but less than that recommended by the manufacturer (usually 20 r.p.m.), indicates proper mixture strength. A greater increase indicates that the mixture is too rich.
- The engine speed may not increase or may drop immediately. This indicates that the idle mixture is too lean.

The idle mixture should be set to give a mixture slightly richer than best power, resulting in a 10- to 20-r.p.m. rise after idle cutoff.

The idle mixture of engines equipped with electric primers can be checked by flicking the primer switch momentarily and noting any change in manifold pressure and r.p.m. A decrease in r.p.m. and an increase in manifold pressure will occur when the primer is energized if the idle mixture is too rich. If the idle mixture is adjusted too lean, the r.p.m. will increase and manifold pressure will decrease.

**Acceleration and deceleration checks.** The acceleration check is made with the mixture control in both auto-rich and auto-lean. Move the throttle from idle to takeoff smoothly and rapidly. The engine speed should increase without hesitation and with no evidence of engine backfire.

This check will, in many cases, show up borderline conditions that will not be revealed by any of the other checks. This is true because the high cylinder pressures developed during this check put added strain on both the ignition system and the fuel metering system. This added strain is sufficient to point out certain defects that otherwise go unnoticed. Engines must be capable of rapid acceleration (since in an emergency), such as a go-around during landing, the ability of an engine to accelerate rapidly is sometimes the difference between a successful go-around and a crash landing.

The deceleration check is made while retarding the throttle from the acceleration check. Note the engine behavior. The speed should decrease smoothly and evenly. There should be little or no tendency for the engine to *afterfire*.

**Engine stopping.** With each type of carburetor installation, specific procedures are used in stopping the engine. The general procedure outlined in the following paragraphs reduces the time required for stopping, minimizes backfiring tendencies and, most importantly, prevent overheating of tightly baffled air-cooled engines during operation on the ground.

In stopping any aircraft engine, the controls are set as follows, irrespective of carburetor type or fuel system installation:

- Cowl flaps are always placed in the full open position, to avoid overheating the engine, and are left in that position after the engine is stopped, to prevent engine residual heat from deteriorating the ignition system
- Carburetor air-heater control is left in the cold position to prevent damage that may occur from backfire
- Turbocharger waste gates are set in the full open position

No mention is made of the throttle, mixture control, fuel selector valve and ignition switches in the preceding set of directions because the operation of these controls varies with the type of fuel metering system used with the engine.

Generally, an engine equipped with an idle cutoff is stopped as follows:

- Idle the engine by setting the throttle for 800-1,000 r.p.m.
- Move the mixture control to the idle cutoff position.
- After the propeller has stopped rotating, place the ignition switch in the off position.

## Engine Troubleshooting

The need for troubleshooting (Figure 1-3-9) normally is dictated by poor operation of the complete powerplant. Power settings for the type of operation at which any difficulty is encountered in many cases will indicate which part of the powerplant is the basic cause of difficulty.

The cylinders of an engine, along with the supercharger (when installed), form an air pump. Furthermore, the power developed in the cylinders varies directly with the rate of air consumption. Therefore, a measure of air consumption or airflow into the engine is a measure of power input. Ignoring for the moment such factors as humidity and exhaust back pressure, the manifold pressure gauge and the engine tachometer provide a measure of engine air consumption. Thus, for a given r.p.m. any change in power input will be reflected by a corresponding change in manifold pressure.

**Figure 1-3-9. A technician making a carburetor adjustment during engine troubleshooting**

The power output of an engine is the power absorbed by the propeller. Therefore, propeller load is a measure of power output. Propeller load, in turn, depends on the propeller r.p.m., blade angle and air density. For a given angle and air density, propeller load (power output) is directly proportional to engine speed.

The basic power of an engine is related to manifold pressure, fuel flow and r.p.m. Because the r.p.m. of the engine and the throttle opening directly control manifold pressure, the primary engine power controls are the throttle and the r.p.m. control. An engine equipped with a fixed-pitch propeller has only a throttle control. In this case, the throttle setting controls both manifold pressure and engine r.p.m.

With proper precautions, manifold pressure can be taken as a measure of power input, and r.p.m. can be taken as a measure of power output. However, the following factors must be considered:

- Atmospheric pressure and air temperature must be considered, since they affect air density.
- These measures of power input and power output should be used only for comparing the performance of an engine with its previous performance or for comparing identical powerplants.
- With a controllable propeller, the blades must be against their low-pitch stops, since this is the only blade position in which the blade angle is known and does not vary. Once the blades are off their low-pitch stops, the propeller governor takes over and maintains a constant r.p.m., regardless of power input or engine condition. This precaution means that the propeller control must be set to maximum or takeoff r.p.m. and the checks made at engine speeds below this setting.

Having relative measures of power input and power output, the condition of an engine can be determined by comparing input and output. This is done by comparing the manifold pressure required to produce a given r.p.m. with the manifold pressure required to produce the same r.p.m. at a time when the engine (or an identical powerplant) was known to be in top operating condition.

An example will best show the practical application of this method of determining engine condition. With the propeller control set for takeoff r.p.m. ( full low blade angle), an engine may require 32 inches of manifold pressure to turn 2,200 r.p.m. for the ignition check. On previous checks, this engine required only 30 inches of manifold pressure to turn 2,200 r.p.m. at the same altitude and under similar atmospheric conditions. Obviously, something is wrong; a higher power input (manifold pressure) is now required for the same power output (r.p.m.). There is a good chance that one cylinder has cut out.

There are several standards against which engine performance can be compared. The performance of a particular engine can be compared with its past performance, provided adequate records are kept, and with that of other engines on the same aircraft or aircraft having identical installations.

If a fault does exist, it may be assumed that the trouble lies in one of the following systems:

- Ignition system
- Fuel metering system
- Induction system
- Power section (valves, cylinders, etc.)
- Instrumentation

If a logical approach to the problem is taken and the instrument readings properly utilized, the malfunctioning system can be pinpointed and the specific problem in the defective system can be singled out.

The more information available about any particular problem, the better will be the opportunity for a rapid repair. Information that is of value in locating a malfunction includes:

- Was any roughness noted? Under what conditions of operation?
- What is the time on the engine and spark plugs? How long since last inspection?
- Were the ignition system operational check and power check normal?
- When did the trouble first appear?
- Was backfiring or afterfiring present?
- Was the full throttle performance normal?

From a different point of view, the powerplant is, in reality, a number of small engines turning a common crankshaft and being operated by two common phases:

1. Fuel metering
2. Ignition

When backfiring, low power output or other powerplant difficulty is encountered, first find out which system (fuel metering or ignition) is involved and then determine whether the entire engine or only one cylinder is at fault.

For example, backfiring normally will be caused by:

- Valves holding open or sticking open in one or more of the cylinders
- Lean mixture
- Intake pipe leakage
- An error in valve adjustment which causes individual cylinders to receive too small a charge or one too large, even though the mixture to the cylinders has the same fuel/air ratio

Ignition system reasons for backfiring might be a cracked distributor block or a high-tension leak between two ignition leads. Either of these conditions could cause the charge in the cylinder to be ignited during the intake stroke. Ignition system troubles involving backfiring normally will not be centered in the basic magneto since a failure of the basic magneto would result in the engine not running, or it would run well at low speeds but cut out at high speeds. On the other hand, replacement of the magneto would correct a difficulty caused by a cracked distributor where the distributor is a part of the magneto.

## Trouble — Cause — Remedy

Troubleshooting is a systematic analysis of the symptoms which indicate engine malfunction. since it would be impractical to list all the malfunctions that could occur in a reciprocating engine, only the most common malfunctions are discussed. A thorough knowledge of the engine systems, applied with logical reasoning, will solve any problems which may occur.

Table 1-3-1 lists general conditions or troubles which may be encountered on reciprocating engines, such as engine fails to start. They are further divided into the probable causes contributing to such conditions. Corrective actions are indicated in the remedy column. The items are presented with consideration given to frequency of occurrence, ease of accessibility and complexity of the corrective action indicated.

## Cylinder Maintenance

Each cylinder of the engine is, in reality, an engine in itself. In most cases the cylinder receives its fuel and air from a common source such as the carburetor. Every phase of cylinder operation, such as compression, fuel mixture and ignition must function properly, since even one type of malfunctioning will cause engine difficulty. Engine backfiring, for example, may be caused by a lean fuel/air mixture in one of the cylinders. The lean mixture may be caused by such difficulties as an improper valve adjustment, a sticking intake or exhaust valve or a leaking intake pipe. Most engine difficulties can be traced to one cylinder or a small number of cylinders. Therefore, engine difficulty can be corrected only after malfunctioning cylinders have been located and defective phases of cylinder operation brought up to normal.

### *Hydraulic Lock*

Whenever a radial engine remains shut down for any length of time beyond a few minutes, oil or fuel may drain into the combustion chambers of the lower cylinders or accumulate in the lower intake pipes ready to be drawn into the cylinders when the engine starts (Figure 1-3-10). As the piston approaches top center of the compression stroke (both valves closed), this liquid, being incompressible, stops piston movement. If the crankshaft continues to rotate, something must give. Therefore, starting or attempting to start an engine with a hydraulic lock of this nature may cause the affected cylinder to blow out or, more likely, may result in a bent or broken connecting rod.

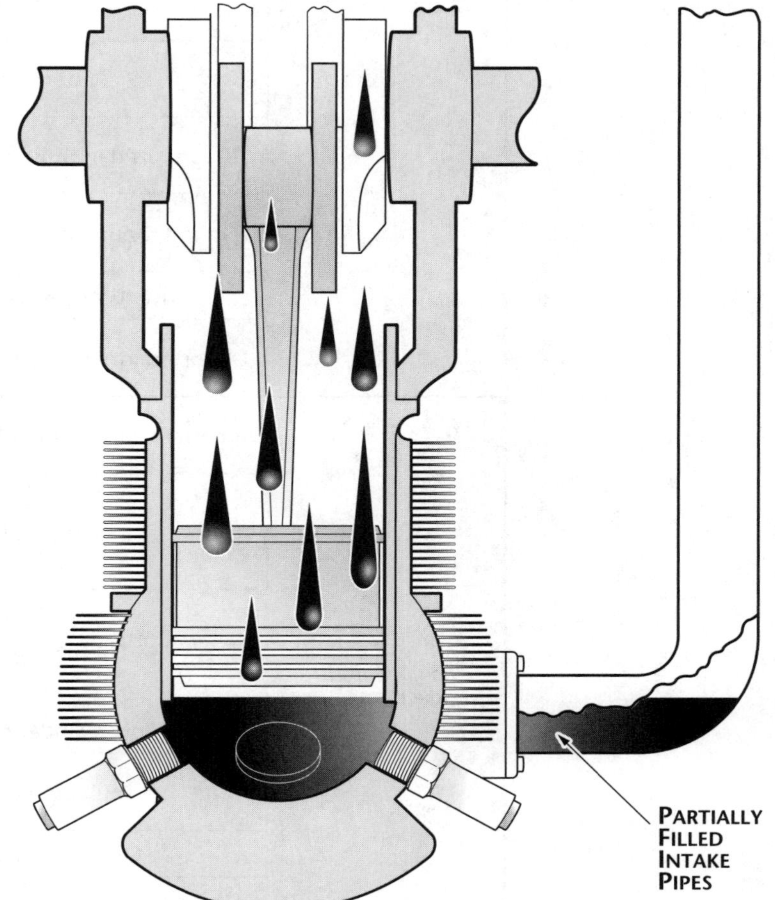


Figure 1-3-10. Initial step in developing a hydraulic lock

**Table 1-3-1. Troubleshooting opposed engines**

| TROUBLESHOOTING OPPOSED ENGINES | | |
|---|---|---|
| SYMPTOM/TROUBLE | PROBABLE CAUSE | REMEDY |
| *Engine fails to start* | Lack of fuel | Check fuel system for leaks. Fill fuel tank. Clean dirty lines, strainers or fuel valves. |
| | Underpriming | Use correct priming procedure. |
| | Overpriming | Open throttle and "unload" engine by rotating propeller. |
| | Incorrect throttle setting | Open throttle to 1/10 of its range. |
| | Defective spark plugs | Clean and re-gap or replace spark plugs. |
| | Defective ignition wire | Test and replace any defective wires. |
| | Defective or weak battery | Replace with charged battery. |
| | Improper operation of magneto or breaker points | Check internal timing of magnetos. |
| | Water in carburetor | Drain carburetor and fuel lines. |
| | Internal failure | Check oil sump strainer for metal particles. |
| | Magnetized impulse coupling, if installed | Demagnetized impulse coupling. |
| | Frozen spark plug electrodes | Replace or dry out spark plugs. |
| | Mixture contol in idle cutoff | Open mixture control. |
| | Shorted ignition switch or loose ground | Check and replace or repair. |
| *Engine fails to idle properly* | Incorrect carburetor idle speed adjustment | Adjust throttle stop to obtain correct idle. |
| | Incorrect idle mixture | Adjust mixture (refer to engine manufacturer's handbook for proper procedure). |
| | Leak in the induction system | Tighten all connections in the induction system. Replace any defective parts. |
| | Low cylinder compression | Check cylinder compression. |
| | Faulty ignition system | Check entire ignition system. |
| | Open or leaking primer | Lock or repair primer. |
| | Improper spark plug setting for altitude | Check spark plug gap. |
| | Dirty air filter | Clean or replace. |

(Table 1-3-1 cont'd on next two pages)

Table 1-3-1. Troubleshooting opposed engines (cont'd from previous page)

| TROUBLESHOOTING OPPOSED ENGINES | | |
|---|---|---|
| SYMPTOM/TROUBLE | PROBABLE CAUSE | REMEDY |
| *Low power and engine running unevenly* | Mixture too rich; indicated by sluggish engine operation, red exhaust flame and black smoke | Check primer. Re-adjust carburetor mixture. |
| | Mixture too lean; indicated by overheating or backfiring | Check fuel lines for dirt or other restrictions. Check fuel supply. |
| | Leaks in induction system | Tighten all connections. Replace defective parts. |
| | Defective spark plugs | Clean or replace spark plugs. |
| | Improper grade of fuel | Fill tank with recommended grade. |
| | Magneto breaker points not working properly | Clean points. Check internal timing of magneto. |
| | Defective ignition wire | Test and replace any defective wires. |
| | Defective spark plug terminal connectors | Replace connectors on spark plug wire. |
| | Incorrect valve clearance | Adjust valve clearance. |
| | Restriction in exhaust system | Remove restriction. |
| | Improper ignition timing | Check magnetos for timing and synchronization. |
| *Engine fails to develop full power* | Throttle lever out of adjustment | Adjust throttle lever. |
| | Leak in induction system | Tighten all connections. Replace defective parts. |
| | Restriction in carburetor airscoop | Examine airscoop and remove restriction. |
| | Improper fuel | Fill tank with recommended fuel. |
| | Propeller governor out of adjustment | Adjust governor. |
| | Faulty ignition | Tighten all connections. Check system. Check ignition timing. |
| *Rough-running engine* | Cracked engine mount(s) | Repair or replace engine mount(s) |
| | Unbalanced propeller | Remove propeller. Have it checked for balance. |
| | Defective mounting bushings | Install new mounting bushings. |
| | Lead deposit on spark plugs | Clean or replace spark plugs. |
| | Primer unlocked | Lock primer. |

(Table 1-3-1 cont'd on next page)

**Table 1-3-1. Troubleshooting opposed engines (cont'd from previous pages)**

| TROUBLESHOOTING OPPOSED ENGINES | | |
|---|---|---|
| SYMPTOM/TROUBLE | PROBABLE CAUSE | REMEDY |
| *Low oil pressure* | Insufficient oil | Check oil supply. |
| | Dirty oil strainers | Remove and clean oil strainers. |
| | Defective pressure gauge | Replace gauge. |
| | Air lock or dirt in relief valve | Remove and clean oil pressure relief valve. |
| | Leak in suction line or pressure line | Check gasket between accessory housing crankcase. |
| | High oil temperature | See "*High oil temperature*" in SYMPTOM/TROUBLE column. |
| | Stoppage in oil pump intake passage | Check line for obstruction. Clean suction strainer. |
| | Worn or scored bearings | Overhaul engine. |
| *High oil temperature* | Insufficient air cooling | Check air inlet/outlet for deformation or obstruction. |
| | Insufficient oil supply | Fill oil tank to proper level. |
| | Clogged oil lines or strainers | Remove and clean oil lines or strainers. |
| | Failing or failed bearings | Examine sump for metal particles and, if found, overhaul engine. |
| | Defective thermostats | Replace thermostats. |
| | Defective temperature gauge | Replace gauge. |
| | Excessive blow-by | Usually caused by weak or stuck rings. Overhaul engine. |
| *Excessive oil consumption* | Failing or failed bearing | Check sump for metal particles and, if found, an overhaul of engine is indicated. |
| | Worn or broken piston rings | Install new rings. |
| | Incorrect installation of piston rings | Install new rings. |
| | External oil leakage | Check engine carefully for leaking gaskets or O-rings. |
| | Leakage through engine fuel pump vent | Replace fuel pump seal. |
| | Engine breather or vacuum, pump breather | Check engine and overhaul or replace vacuum pump. |

A complete hydraulic lock, one that stops crankshaft rotation, can result in serious damage to the engine. Still more serious, however, is the slight damage resulting from a partial hydraulic lock that goes undetected at the time it occurs. The piston meets extremely high resistance but is not completely stopped. The engine falters but starts and continues to run as the other cylinders fire. The slightly bent connecting rod resulting from the partial lock also goes unnoticed at the time it is damaged but is sure to fail later. The eventual failure is almost certain to occur at a time when it can be least tolerated, since it is during such critical operations as takeoff and go-around that maximum power is demanded of the engine and maximum stresses are imposed on its parts. A hydraulic lock and some possible results are shown in Figure 1-3-11.

Before starting any radial engine that has been shut down for more than 30 minute, check the ignition switches for OFF and then pull the propeller through in the direction of rotation a minimum of two complete turns to make sure that there is no hydraulic lock or to detect the hydraulic lock if one is present. Any liquid present in a cylinder will be indicated by the abnormally great effort required to rotate the propeller. However, never use force when a hydraulic lock is detected.

When engines that employ direct-drive or combination inertia and direct-drive starters are being started and an external power source is being used, a check for hydraulic lock may be made by intermittently energizing the starter and watching for a tendency of the engine to stall. Use of the starter in this way will not exert sufficient force on the crankshaft to bend or break a connecting rod if a lock is present.

To eliminate a lock, remove either the front or rear spark plug of the lower cylinders and pull the propeller through in the direction of rotation. The piston will expel any liquid that may be present.

If the hydraulic lock occurs as a result of over-priming prior to initial engine start, eliminate the lock in the same manner, i.e., remove one of the spark plugs from the cylinder and rotate the crankshaft through two turns.

Never attempt to clear the hydraulic lock by pulling the propeller through in the direction opposite to normal rotation, since this tends to inject the liquid from the cylinder into the intake pipe with the possibility of a complete or partial lock occurring on the subsequent start.

## *Valve Blow-by*

Valve blow-by is indicated by a hissing or whistle when pulling the propeller through prior to starting the engine, when turning the engine with the starter or when running the engine at slow speeds. It is caused by a valve sticking open or being warped to the extent that compression is not built up in the cylinder as the piston moves toward TDC on the compression stroke. Blow-by past the exhaust valve can be heard at the exhaust stack, and blow-by past the intake valve is audible through the carburetor.

Correct valve blow-by immediately to prevent valve failure and possible engine failure by taking the following steps:

- Perform a cylinder compression test to locate the faulty cylinder.
- Check the valve clearance on the affected cylinder. If the valve clearance is incorrect, the valve may be sticking in the valve guide. To release the sticking valve, place a fiber drift on the rocker arm immediately over the valve stem and strike the drift several times with a mallet. Sufficient hand pressure should be exerted on the fiber drift to remove any space between the rocker arm and the valve stem prior to hitting the drift.
- If the valve is not sticking and the valve clearance is incorrect, adjust it as necessary.

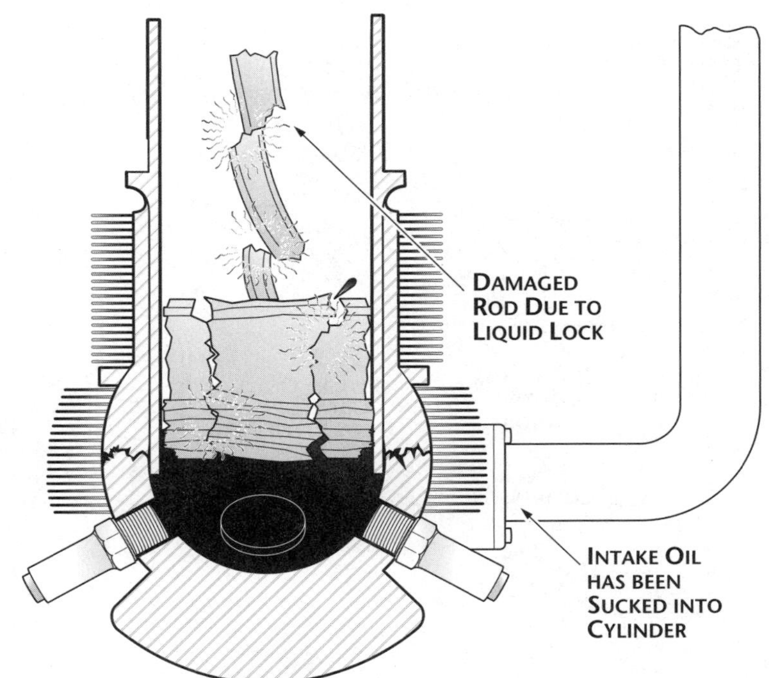


Figure 1-3-11. Results of a hydraulic lock

- Determine whether blow-by has been eliminated by again pulling the engine through by hand or turning it with the starter. If blow-by is still present, it may be necessary to replace the cylinder.

## Cylinder Compression Tests

The cylinder compression test determines if the valves, piston rings and pistons are adequately sealing the combustion chamber. If pressure leakage is excessive, the cylinder cannot develop its full power. The purpose of testing cylinder compression is to determine whether cylinder replacement is necessary. The detection and replacement of defective cylinders will prevent a complete engine change because of cylinder failure. It is essential that cylinder compression tests be made periodically.

Although it is possible for the engine to lose compression for other reasons, low compression for the most part can be traced to leaky valves. Conditions which affect engine compression are:

- Incorrect valve clearances
- Worn, scuffed or damaged piston
- Excessive wear of piston rings and cylinder walls
- Burned or warped valves
- Carbon particles between the face and the seat of the valve or valves
- Early or late valve timing

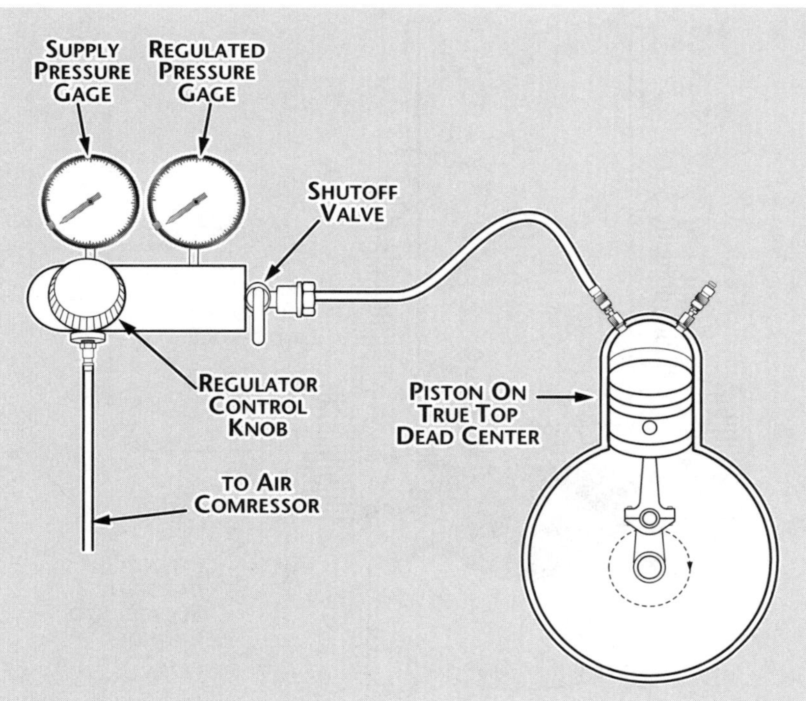


Figure 1-3-12. Differential compression tester

Perform a compression test as soon as possible after the engine is shut down so that piston rings, cylinder walls and other parts are still freshly lubricated. However, it is not necessary to operate the engine prior to accomplishing compression checks during engine buildup or on individually replaced cylinders. In such cases, before making the test, spray a small quantity of lubricating oil into the cylinder or cylinders and turn the engine over several times to seal the piston and rings in the cylinder barrel.

Be sure that the ignition switch is in the off position so that there will be no accidental firing of the engine. Remove necessary cowling and the most accessible spark plug from each cylinder. When removing the spark plugs, identify them to coincide with the cylinder. Close examination of the plugs will aid in diagnosing problems within the cylinder. Review the maintenance records of the engine being tested. Records of previous compression checks help in determining progressive wear conditions and in establishing the necessary maintenance actions.

**Differential compression tester.** The differential pressure tester is designed to check the compression of aircraft engines by measuring the leakage through the cylinders caused by worn or damaged components. Operation of the compression tester is based on the principle that, for any given airflow through a fixed orifice, a constant pressure drop across that orifice will result. The restrictor orifice dimensions in the differential pressure tester should be sized for the particular engine as follows:

- Engines up to 1,000 cubic inch displacement: 0.040-inch orifice diameter, 0.250 inch long, 60-degree approach angle.
- Engines in excess of 1,000 cubic inch displacement: 0.060-inch orifice diameter, 0.250 inch long, 60° approach angle.

The differential pressure tester (Figure 1-3-12) requires the application of air pressure to the cylinder being tested with the piston at top-center compression stroke.

To ensure accuracy, the differential compression tester must be tested using a *master orifice*. The master orifice is a calibration standard that establishes the acceptable cylinder pressure leakage limit for the test equipment being used and the atmospheric conditions at the time of the test. Some newer model compression testers have a master orifice built in.

As with other maintenance operations, the recommendations of the engine manufacturer must be followed closely. Specific instructions regarding compression testing and evaluating the test results may be found in Service Bulletins or other publications issued by the manufacturer.

General guidelines for performing a differential compression test are:

1. Perform the compression test as soon as possible after engine shutdown to provide uniform lubrication of cylinder walls and rings.
2. Remove the most accessible spark plug from the cylinder or cylinders and install a spark plug adapter in the spark plug insert.
3. Connect the compression tester assembly to a 100- to 150-p.s.i. compressed air supply. With the shutoff valve on the compression tester closed, adjust the regulator of the compression tester to obtain 80 p.s.i. on the regulated pressure gauge.
4. Open the shutoff valve and attach the air hose quick-connect fitting to the spark plug adapter. The shutoff valve, when open, will automatically maintain a pressure of 15 to 20 p.s.i. in the cylinder when both the intake and exhaust valves are closed.
5. By hand, turn the engine over in the direction of rotation until the piston in the cylinder being tested comes up on the compression stroke against the 15 p.s.i. Continue turning the propeller slowly in the direction of rotation until the piston reaches TDC. TDC can be detected by a decrease in force required to move the propeller. If the engine is rotated past TDC, the 15 to 20 p.s.i. will tend to move the propeller in the direction of rotation. If this occurs, back the propeller up at least one blade prior to turning the propeller again in the direction of rotation. This backing up is necessary to eliminate the effect of backlash in the valve-operating mechanism and to keep the piston rings seated on the lower ring lands.
6. Close the shutoff valve in the compression tester and re-check the regulated pressure to see that it is 80 p.s.i. with air flowing into the cylinder. If the regulated pressure is more or less than 80 p.s.i., readjust the regulator in the test unit to obtain 80 p.s.i. When closing the shutoff valve, make sure that the propeller path is clear of all objects. There will be sufficient air pressure in the combustion chamber to rotate the propeller if the piston is not on TDC.
7. With regulated pressure adjusted to 80 p.s.i., if the cylinder pressure reading indicated on the cylinder pressure gauge is below the minimum specified for the engine being tested, move the propeller in the direction of rotation to seat the piston rings in the grooves. Check all the cylinders and record the readings.

If low compression is obtained on any cylinder, turn the engine through with the starter or restart and run the engine to takeoff power and re-check the cylinder or cylinders having low compression. If the low compression is not corrected, remove the rocker-box cover and check the valve clearance to determine if the difficulty is caused by inadequate valve clearance. If the low compression is not caused by inadequate valve clearance, place a fiber drift on the rocker arm immediately over the valve stem and tap the drift several times with a 1- to 2-pound hammer to dislodge any foreign material that may be lodged between the valve and valve seat. After staking the valve in this manner, rotate the engine with the starter and re-check the compression. Do not make a compression check after staking a valve until the crankshaft has been rotated either with the starter or by hand to re-seat the valve in the normal manner. The higher seating velocity obtained when staking the valve will indicate valve seating even though valve seats are slightly egged or eccentric.

Cylinders having compression below the minimum specified after staking should be further checked to determine whether leakage is past the exhaust valve, intake valve or piston. Excessive leakage can be detected:

- At the exhaust valve by listening for air leakage at the exhaust outlet
- At the intake valve by escaping air at the air intake
- Past the piston rings by escaping air at the engine breather outlets

The wheeze test is another method of detecting leaking intake and exhaust valves. In this test, as the piston is moved to TDC on the compression stroke, the faulty valve may be detected by listening for a wheezing sound in the exhaust outlet or intake duct.

Another method is to admit compressed air into the cylinder through the spark plug hole. The piston should be restrained at TDC of the compression stroke during this operation. A leaking valve or piston rings can be detected by listening at the exhaust outlet, intake duct, or engine breather outlets.

Next to valve blow-by, the most frequent cause of compression leakage is excessive leakage past the piston. This leakage may occur because of lack of oil. To check this possibility, squirt engine oil into the cylinder and around the piston. Then re-check the compression. If this procedure raises compression to or above the minimum required, continue the cylinder in service. If the cylinder pressure readings still do not meet the minimum requirement, replace the cylinder. When it is necessary to replace a cylinder as a result of low compression, record the cylinder number and the compression value of the newly installed cylinder on the compression check sheet.

## Cylinder Replacement

Reciprocating engine cylinders are designed to operate a specified time before normal wear will require their overhaul. If the engine is operated as recommended and proficient maintenance is performed, the cylinders normally will last until the engine is removed for high-time reasons. It is known from experience that materials fail and engines are abused through incorrect operation; this has a serious effect on cylinder life. Another reason for premature cylinder change is poor maintenance.

Figure 1-3-13. Chrome cylinder with orange markings ( shown in white on cooling fins)

Therefore, exert special care to ensure that all the correct maintenance procedures are adhered to when working on the engine.

Some reasons for cylinder replacement are:

- Low compression
- High oil consumption in one or more cylinders
- Excessive valve guide clearance
- Loose intake pipe flanges
- Loose or defective spark plug inserts
- External damage, such as cracks

When conditions like these are limited to one or a few cylinders, replacing the defective cylinders should return the engine to a serviceable condition.

When spare serviceable cylinders are available, replace cylinders when the man-hour requirement for changing them does not exceed the time required to make a complete engine change.

Except under certain conditions, do not attempt to replace individual parts, such as pistons, rings or valves. This precaution guarantees that clearance and tolerances are correct. Other parts, such as valve springs, rocker arms and rocker box covers, may be replaced individually.

Normally, all the cylinders in an engine are similar; that is, all are standard size or all a certain oversize, and all are steel bore or all are chrome-plated. In some instances, because of shortages at the time of overhaul, it may be necessary that engines have two different sizes of cylinder assemblies.

Replace a cylinder with an identical one, if possible. If an identical cylinder is not available. it is permissible to install either a standard or oversize cylinder and piston assembly, since this will not adversely affect engine operation.

In some instances, air-cooled engines will be equipped with chrome-plated cylinders. Chrome-plated cylinders are usually identified by a paint band around the barrel, between the attaching flange and the lower barrel cooling fin. This color band is usually international orange (Figure 1-3-13). When installing a chrome-plated cylinder, do not use chrome-plated piston rings. The matched assembly will, of course, include the correct piston rings. However, if a piston ring is broken during cylinder installation, check the cylinder marking to determine what ring, chrome plated or otherwise, is correct for replacement. Similar precautions must be taken to be sure that the correct size rings are installed.

Correct procedures and care are important when replacing cylinders. Careless work or the use of incorrect tools can damage the replacement cylinder or its parts. Incorrect procedures in installing rocker-box covers may result in troublesome oil leaks. Improper torque on cylinder hold down nuts or cap screws can easily result in a cylinder malfunction and subsequent engine failure.

The discussion of cylinder replacement in this book is limited to the removal and installation of air-cooled engine cylinders. The discussion is centered on radial and opposed engines, since these are the aircraft engines on which cylinder replacements are most often performed.

Since these instructions are meant to cover all air-cooled engines, they are necessarily of a general nature. The applicable manufacturer's maintenance manual should be consulted for torque values and special precautions applying to a particular aircraft and engine. However, always practice neatness and cleanliness and always protect openings so that nuts, washers, tools and miscellaneous items do not enter the engine's internal sections.

## *Cylinder Removal*

After all the cowling, brackets and baffles have been removed, the removal of the cylinder can continue. Start by first removing the ignition leads, then the intake and exhaust pipes (Figure 1-3-14). Remove the rocker box covers(Figure 1-3-15). Remove the rocker arms and the push rods. The rocker arm can be removed easily if the crankshaft is rotated so the respective piston in on TDC, this will remove the pressure on both the intake and exhaust rocker arms. After the push rods have been removed the push rod tubes may then be removed.

It is a good practice to mark all parts as they are removed to assure that they are reinstalled in the same place. If the engine is fuel injected, the fuel injection line should be removed and carefully repositioned out of the way.

Use a cylinder base wrench to loosen all of the cylinder holddown nuts (Figure 1-3-16). Remove all of the nuts except for two located on opposite sides of the cylinder.

When all of the nuts have been removed, including the last two, gently pull the cylinder away from the crankcase while holding it up. It is a good practice to have two people working together when removing the cylinder. One person should hold up on the rod, as soon as it is accessible, while the second person pulls the cylinder off the piston.

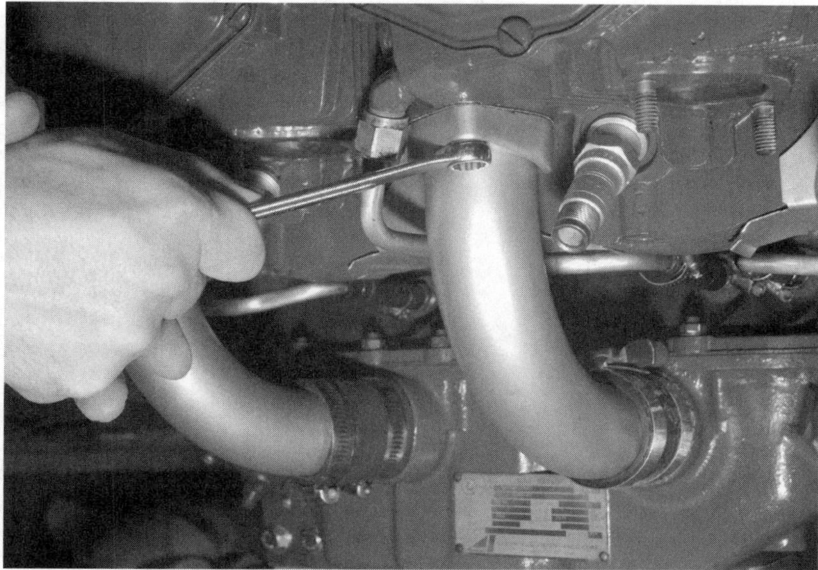

Figure 1-3-14. Removal of the intake pipes

Figure 1-3-15. Removal of the rocker box covers

Figure 1-3-16. Removing the the holddown nuts with a cylinder base wrench

Figure 1-3-17. A soft mallet may be used to drive out the pin when removing the piston from the rod. The rod is supported by a board to prevent bending.

Figure 1-3-18. Tying the connecting rod with a rubber band prevents any damage after the piston and cylinder have been removed.

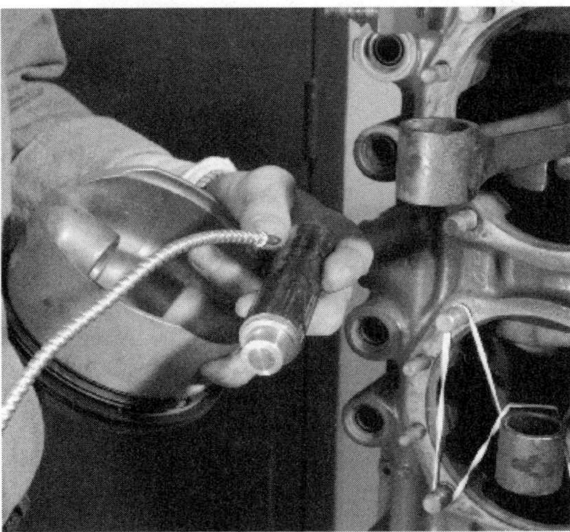

Figure 1-3-19. Thorough cleaning and lubrication of the piston pin makes reinstallation easier.

Care should be taken to prevent any foreign objects, such as broken piston rings, from falling into the engine as the cylinder is removed.

To remove the piston from the rod (Figure 1-3-17), a brass drift and soft mallet may be used to drive out the pin. Special care must be used when doing this to prevent bending of the rod.

After the cylinder and the piston have been removed, the connecting rod must be supported to prevent damage to the rod and the case. This can be accomplished by tying the rod up with a rubber band (Figure 1-3-18) or the cylinder base seal from the cylinder just removed.

When the cylinder and piston have been removed and the rod secured, the mounting flange and cap screws on the crankcase should be cleaned and inspected. Carefully examine the flange for cracks or defects. Examine the cap screws and studs for security and damaged threads. If a cap screw or stud is found to be loose, not only should it be changed but all fasteners for that cylinder should be replaced. The excessive load applied to them may be caused by the loose stud and may cause them to fail. The engine manufacturer will determine the number of studs that are required to be changed after a stud failure.

When it becomes necessary to remove a broken stud, take care to prevent metal shavings or pieces of the stud from entering the engine. Before reinstalling the cylinder, all washers and nuts should be cleaned and any roughness or burrs removed.

## *Cylinder Installation*

Before installing the cylinder, make certain that all parts are clean and dry. Install the piston and rings on the rod, and be certain to lubricate the piston pin before it is installed (Figure 1-3-19).

Next, apply oil to the piston skirt and ring lands (Figure 1-3-20). Rotate the rings so that the ring gaps are staggered.

If it is necessary to replace the rings on one or more of the pistons, check the side clearance against the manufacturer's specification using a thickness gauge. The method for checking side and end clearances is shown in Figure 1-3-21 (top of next page). A piston (without rings) may be inserted in the cylinder and the ring inserted in the cylinder bore. Insert the ring in the cylinder skirt below the mounting flange, since this is usually the smallest bore diameter. Pull the piston against the ring to align it properly in the bore.

If it is necessary to remove material to obtain the correct side clearance, it can be done either by turning the piston grooves a slight amount on each side or by lapping the ring on a surface plate.

If the end gap is too close, the excess metal can be removed by clamping a mill file in a vise, holding the ring in proper alignment and dressing off the ends. In all cases, the engine manufacturer's procedures must be followed.

Before installing the cylinder, check the flange to see that the mating surface is smooth and clean. Coat the inside of the cylinder barrel generously with oil. Be sure that the cylinder oil-seal ring is in place and that only one seal ring is used.

To start the cylinder onto the head of the piston, the rings must be compressed with a ring compressor, as shown in Figure 1-3-22. Be careful not to rotate the cylinder after it is started over the piston. Continue to carefully slide the cylinder over the piston until all of the rings are covered by the cylinder walls. Remove the ring compressor and place it out of the way. Remove the rod support at this time and continue to slide the cylinder onto the mounting studs. Install the cylinder holddown plates if used and two nuts, making them finger tight.

If the cylinder is secured to the crankcase by conical washers and nuts or cap screws, position the cylinder on the crankcase section with two special locating nuts or cap screws. These locating nuts or cap screws do not remain on the engine, but are removed and replaced with regular nuts or cap screws and conical washers after they have served their purpose and the other nuts or cap screws have been installed and tightened to the prescribed torque.

Install the remaining nuts or cap screws with their conical washers, and tighten the nuts or cap screws until they are snug. Make sure that the conical side of each washer is toward the

**Figure 1-3-20.** When installing a cylinder, oil should be applied to the piston skirt and ring lands.

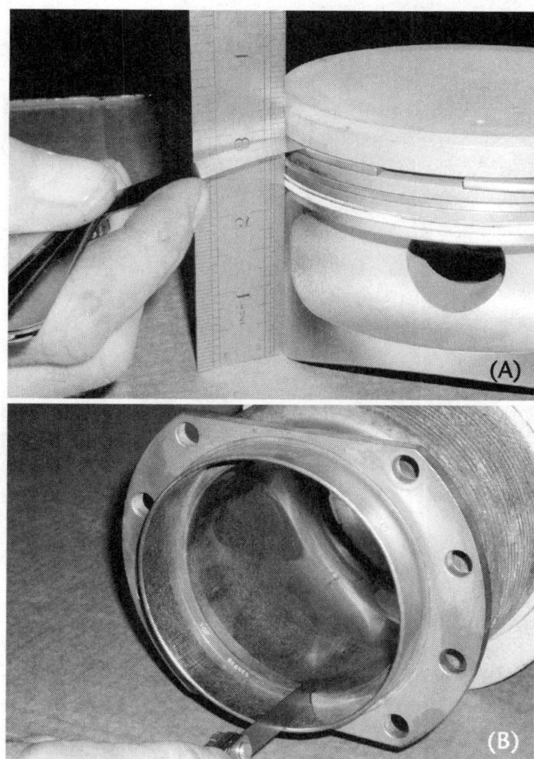

**Figure 1-3-21.** (A) Measuring piston ring side clearance and (B) end gap

**Figure 1-3-22.** (A) A ring compressor is used (B) to start the cylinder onto the head of the piston.

Figure 1-3-23. Torquing cylinder holddown nuts

Figure 1-3-24. Re-installing the push rods

cylinder mounting flange. Before inserting cap screws, coat them with a good sealer to prevent oil leakage. Generally, studs fit into holes and the fit is tight enough to prevent leakage.

The holddown nuts should now be torqued to the specified value. When torquing the nuts (Figure 1-3-23), the proper sequence for tightening is found in the manufacturer's service or overhaul manual. This sequence must be followed.

If locating nuts or cap screws are being used, they should be torqued first. The tightening of the remaining screws or nuts should be alternated as the torquing continues around the cylinder. Apply the torque with a slow, steady motion until the prescribed value is reached. Hold the tension on the wrench for a sufficient length of time to ensure that the nut or cap screw will tighten no more at the prescribed torque value. In many cases, additional turning of the cap screw or nut as much as one-quarter turn can be done by maintaining the prescribed torque on the nut for a short period of time. After tightening the regular nuts or cap screws, remove the two locating nuts or cap screws, install regular nuts or cap screws and tighten them to the prescribed torque.

After the stud nuts or cap screws have been torqued to the prescribed value, safety them in the manner recommended in the engine manufacturer's service manual.

Re-install the push rods (Figure 1-3-24), push rod housings, rocker arms, barrel deflectors, intake pipes, ignition harness lead clamps and brackets, fuel injection line clamps and fuel injection nozzles, exhaust stack, cylinder head deflectors and spark plugs. Remember that the push rods must be installed in their original locations and must not be turned end to end.

Make sure, too, that the push rod ball end seats properly in the tappet. If it rests on the edge or shoulder of the tappet during valve clearance adjustment and later drops into place, valve clearance will be off. Furthermore, rotating the crankshaft with the push rod resting on the edge of the tappet may bend the push rod.

After installing the push rods and rocker arms, set the valve clearance.

Before installing the rocker box covers, lubricate the rocker arm bearings and valve stems (Figure 1-3-25). Check the rocker box covers for flatness, and re-surface them if necessary. After installing the gaskets and covers, tighten the rocker box cover nuts to the specified torque.

Safety those nuts, screws and other fasteners which require safetying. Follow the recommended safetying procedures.

## Valves and Valve Mechanisms

Valves open and close the ports in the cylinder head to control the entrance of the combustible mixture and the exit of the exhaust gases. It is important that they open and close properly and seal tight against the port seats to secure maximum power from the burning fuel/air mixture for the crankshaft and to prevent valve burning and warping. The motion of the valves is controlled by the valve-operating mechanism.

The valve mechanism (Figure 1-3-26) includes cam plates or shafts, cam followers, pushrods, rocker arms, valve springs and retainers. All parts of a valve mechanism must be in good condition and valve clearances must be correct if the valves are to operate properly.

Checking and adjusting the valve clearance is perhaps the most important part of valve inspection, and certainly it is the most difficult part. However, the visual inspection should not be slighted. It should include a check for the following major items:

- Metal particles in the rocker box are indications of excessive wear or partial failure of the valve mechanism. Locate and replace the defective parts.
- Excessive side clearance or galling of the rocker arm side. Replace defective rocker arms. Add shims when permitted, to correct excessive side clearance.
- Insufficient clearance between the rocker arm and the valve spring retainer. Follow the procedure outlined in the engine service manual for checking this clearance, and increase it to the minimum specified.
- Replace any damaged parts, such as cracked, broken or chipped rocker arms, valve springs or spring retainers. If the damaged part is one which cannot be replaced in the field, replace the cylinder.
- Excessive valve stem clearance. A certain amount of valve stem wobble in the valve guide is normal. Replace the cylinder only in severe cases.
- Evidence of incorrect lubrication. Excessive dryness indicates insufficient lubrication. However, the lubrication varies between engines and between cylinders in the same model engine. For example, the upper boxes of radial engines will normally run drier than the lower rocker boxes. These factors must be taken into account in determining whether or not ample lubrication is being obtained. Wherever improper lubrication is indicated, determine the cause and correct it. For example, a dry rocker may be caused by a plugged oil passage in the pushrod. Excessive oil in the rocker box may be caused by plugged drains between the rocker box and the crankcase. If the push rod drains become clogged the oil forced to the rocker arm and other parts of the valve mechanism cannot drain back to the crankcase. This may result in oil leakage at the rocker box cover or in oil seepage along valve stems into the cylinder or exhaust system, causing excessive oil consumption on the affected cylinder and smoking in the exhaust.
- Excessive sludge in the rocker box. This indicates an excessive rocker box temperature, which, in turn, may be caused by improper positioning of cowling or exhaust heat shields or baffles. After correcting the cause of the difficulty, spray the interior of the rocker box with dry cleaning solvent, blow it dry with compressed air, then coat the entire valve mechanism and interior of the rocker box with clean engine oil.
- Variation in valve clearance not explained by normal wear. If there is excessive valve clearance, check for bent push rods. Replace any that are defective. Check also for valve sticking. If the push rod is straight and the valve opens and closes when the propeller is pulled through by hand, check the tightness of the adjusting screw to determine whether the clearance was set incorrectly or the adjusting screw has loosened.

Warped rocker box covers are a common cause of oil leakage. Therefore, the box covers should be checked for flatness at each valve inspection. Re-surface any warped covers by lapping them on emery cloth laid on a surface plate.

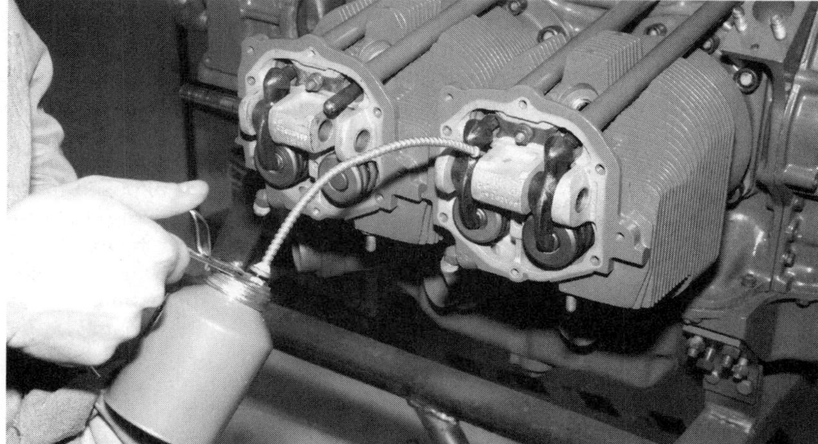

Figure 1-3-25. Lubricate the rocker arm bearings and valve stems before installing the rocker box covers

Figure 1-3-26. Valve mechanism

Rocker box cover warpage is often caused by improper tightening of the rocker box cover nuts. Eliminate further warpage by torquing the nuts to the values specified in the manufacturer's service manual.

## *Valve Clearance*

The amount of power that can be produced by a cylinder depends primarily on the amount of heat produced in that cylinder without destructive effects on the components. Any limits to the amount of heat in the cylinder also limits the power that cylinder can produce.

The manufacturer, in determining valve timing and establishing the maximum power setting at which the engine will be permitted to run, considers the amount of heat at which cylinder components such as spark plugs and valves can operate efficiently. The heat level of the exhaust valve must be below that at which pitting and warping of the valve occurs.

The head of the exhaust valve is exposed to the heat of combustion at all times during the combustion period. In addition, the head of this valve and a portion of the stem are exposed to hot exhaust gases during the exhaust event. Under normal operation, the exhaust valve remains below the critical heat level because of its contact with the valve seat when closed and because of the heat dissipated through the stem. Any condition that prevents the valve from seating properly for the required proportion of time will cause the valve to exceed the critical heat limits during periods of high power output. In cases of extremely poor valve seat contact, the exhaust valve can warp during periods of low power output.

Figure 1-3-27. If the exhaust valve is correctly adjusted and seated, most of the heat is transferred into the cylinder head.

Normally, the exhaust valve is closed (Figure 1-3-27) and in contact with its seat about 65 percent of the time during the four-stroke cycle. If the valve adjustment is correct, and if the valve seats firmly when closed, much of the heat is transferred from the valve, through the seat, into the cylinder head.

In order for a valve to seat, the valve must be in good condition, with no significant pressure being exerted against the end of the valve by the rocker arm. If the expansion of all parts of the engine including the valve train were the same, the problem of ensuring valve seating would be very easy to solve. Practically no free space would be necessary in the valve system. However, since there is a great difference in the amount of expansion of various parts of the engine, there is no way of providing a constant operating clearance in the valve train. The clearance in the valve-actuating system is very small when the engine is cold but is much greater when the engine is operating at normal temperature. The difference is caused by differences in the expansion characteristics of the various metals and by the differences in temperature of various engine parts.

There are many reasons why proper valve clearances are of vital importance to satisfactory and stable engine operation. For example, when the engine is operating, valve clearances establish valve timing. Since all cylinders receive their fuel/air mixture (or air) from a common supply, valve clearance affects both the amount and the richness or leanness of the fuel/air mixture. Therefore, it is essential that valve clearances be correct and uniform between each cylinder.

**Valve adjustment on opposed engines.** The valve adjustments on opposed engines are normally made by replacing the push rod with one of a different length. In other words, if the valve clearance is too large a push rod of greater length should be installed and visa-versa.

**Valve spring replacement.** A single broken valve spring seldom affects engine operation — in part, because multiple springs are used — and can, therefore, be detected only during careful inspection. But when a broken valve spring is discovered, it can be replaced without removing the cylinder. During valve spring replacement, the important precaution to remember is not to damage the spark plug hole threads. The complete procedure for valve spring replacement is as follows:

- Remove one spark plug from the cylinder.
- Turn the propeller in the direction of rotation until the piston is at the top of the compression stroke (Figure 1-3-28).
- Remove the rocker arm.

- Using a valve spring compressor, compress the spring and remove the valve keepers (Figure 1-3-29A). During this operation, it may be necessary to insert a piece of brass rod through the spark plug hole to decrease the space between the valve and the top of the piston head to break the spring-retaining washer loose from the keepers. The piston, being at the top position on the compression stroke, prevents the valve from dropping down into the cylinder once the spring retaining washers are broken loose from the keepers on the stem.
- Remove the defective spring and any broken pieces from the rocker box (Figure 1-3-29B).
- Install a new spring and correct washers. Then, using the valve spring compressor, compress the spring and, if necessary, move the valve up from the piston by means of a brass rod inserted through the spark plug hole.
- Re-install the keepers and rocker arms. Then check and adjust the valve clearance.
- Re-install the rocker box cover and the spark plug.

**Figure 1-3-28. Turn the propeller until the piston is at the top dead center of the compression stroke, which looks like this internally.**

## Cold-cylinder Check

The cold-cylinder check determines the operating characteristics of each cylinder of an air-cooled engine. The tendency for any cylinder or cylinders to be cold or to be only slightly warm indicates lack of combustion or incomplete combustion within the cylinder. This must be corrected if best operation and power conditions are to be obtained. The cold-cylinder check is made with a cold-cylinder indicator (*Magic Wand*), an *infrared laser thermometer* or a temperature probe on a *multimeter.* Engine

**Figure 1-3-29. After removing the rocker arm (A), a valve spring compressor helps to remove the keepers (B).**

difficulties that can be analyzed by use of the cold-cylinder indicator (Figure 1-3-30) are:

- Rough engine operation
- Excessive r.p.m. drop during the ignition system check
- High manifold pressure for a given engine r.p.m. during the ground check when the propeller is in the full low-pitch position
- Faulty mixture ratios caused by improper valve clearance

In preparation for the cold-cylinder check, head the aircraft into the wind to minimize irregular cooling of the individual cylinders and to ensure even propeller loading during engine operation.

Open the cowl flaps. Do not close the cowl flaps under any circumstances, as the resulting excessive heat radiation will affect the readings obtained and can damage the ignition leads.

Start the engine with the ignition switch in the BOTH position. After the engine is running, place the ignition switch in the position in which any excessive r.p.m. drop is obtained. When excessive r.p.m. drop is encountered on both right and left switch positions, or when excessive manifold pressure is obtained at a given engine r.p.m., perform the check twice, once on the left and once on the right switch position.

Operate the engine at its roughest speed, between 1,200-1,600 r.p.m., until a cylinder head temperature reading of 150°-170°C (302°-338°F) is obtained, or until temperatures stabilize at a lower reading. If engine roughness is encountered at more than one speed, or if there is an indication that a cylinder ceases operating at idle or higher speeds, run the engine at each of these speeds and perform a cold cylinder check to pick out all the dead or intermittently operating cylinders. When low-power output or engine vibration is encountered at speeds above 1,600 r.p.m. with the ignition switch on BOTH, run the engine at the speed where the difficulty is encountered until the cylinder head temperatures are up to 150°-170°C or until the temperatures have stabilized at a lower value.

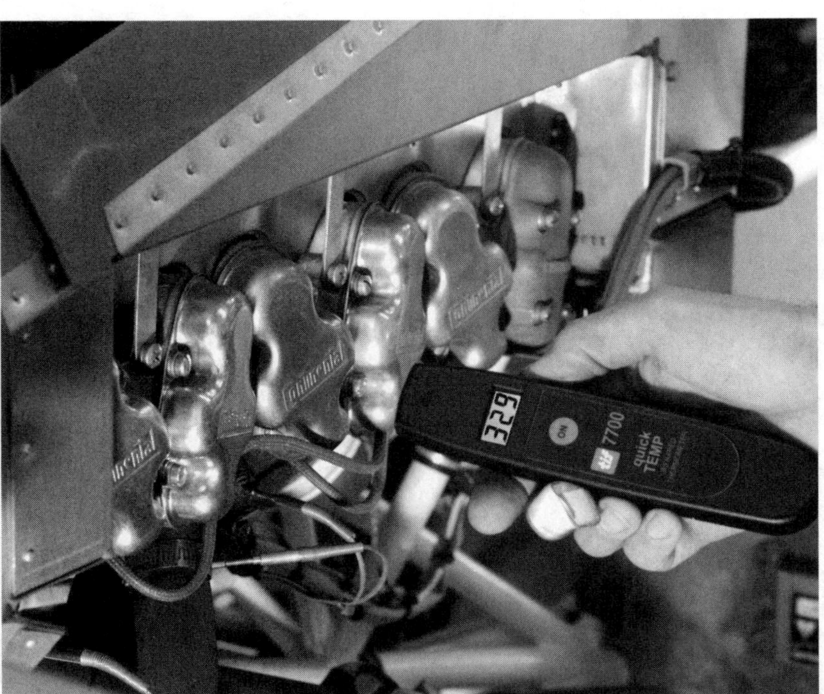

Figure 1-3-30. An infrared laser thermometer used to find cold cylinders

Infrared laser thermometers have become the tool of choice for the AMT because they do not require physical contact with the engine. They can even be used while the engine is operating.

Difficulties that may cause a cylinder to be inoperative (dead) on both right and left magneto positions are:

- Defective spark plugs
- Incorrect valve clearances
- Leaking impeller oil seal
- Leaking intake pipes
- Lack of compression
- Plugged push rod housing drains
- Faulty operation of the fuel-injection nozzle (on fuel-injection engines)

Before changing spark plugs or making an ignition harness test on cylinders that are not operating or are operating intermittently, check the magneto ground leads to determine that the wiring is connected correctly.

Repeat the cold cylinder test for the other magneto positions on the ignition switch, if necessary. Cooling the engine between tests is unnecessary. The airflow created by the propeller and the cooling effect of the incoming fuel/air mixture will be sufficient to cool any cylinders that are functioning on one test and not functioning on the next.

In interpreting the results of a cold cylinder check, remember that the temperatures are relative. A cylinder temperature taken alone means little, but when compared with the temperatures of other cylinders on the same engine, it provides valuable diagnostic information.

**For example:** A review of the temperature readings shows that cylinder No. 3 runs cold. This indicates that cylinder six is firing intermittently and cylinders two and three are dead during engine operation on the right magneto.

Cylinders three and five are dead during operation on the plugs fired by the left magneto. Cylinder three is completely dead.

An ignition system operational check would not disclose this dead cylinder since the cylinder is inoperative on both right and left switch positions.

A dead cylinder can be detected during runup, since an engine with a dead cylinder will require a higher-than-normal manifold pressure to produce any given r.p.m. below the cut-in speed of the propeller governor.

Defects within the ignition system that can cause a cylinder to go completely dead are:

- Both spark plugs inoperative
- Both ignition leads grounded, leaking or open
- A combination of inoperative spark plugs and defective ignition leads

Faulty fuel-injection nozzles, incorrect valve clearances and other defects outside the ignition system can also cause dead cylinders.

In interpreting the readings obtained on a cold-cylinder check, the amount the engine cools during the check must be considered. To determine the extent to which this factor should be considered in evaluating the readings, re-check some of the first cylinders tested and compare the final readings with those made at the start of the check. Another factor to be considered is the normal variation in temperature between cylinders and between rows. This variation results from those design features that affect the airflow past the cylinders.

## Reciprocating Engine Overhaul

The intervals in which maintenance and overhaul are performed on a reciprocating engine are normally by the number of hours that the engine has been in operation. This time span is called Time Between Overhaul (TBO). These intervals are determined by the manufacturer of the engine and approved by the FAA. While these TBO figures are not mandatory for privately operated general aviation airplanes, they are mandatory for commercial operators. Tests by manufactures have shown that the operation of an engine beyond these limits will cause the wear rate to increase, thus creating further wear at a faster rate. In short, it makes economic sense to not go past TBO.

During overhaul, all parts must be carefully inspected to make certain that they meet the manufacturer's service limits. The primary purpose of overhauling an engine is to inspect the engine parts. It is also important to remember that the inspection process is the most important phase of the engine overhaul. The inspection of the engine parts should always be accomplished with utmost care and attention to detail.

Each engine manufacturer provides very specific tolerances to which the engine parts must conform and provides general instructions to aid in determining the airworthiness of the part. However, in many cases, the final decision is left up to the mechanic. The AMT must determine if the part is serviceable, repairable or should be rejected. A knowledge of the operating principles, strength and stresses applied to a part is essential in making this decision. When the powerplant mechanic signs for the overhaul of an engine, it certifies that the work has been performed using methods, techniques and practices acceptable to the FAA Administrator.

It is normal practice to overhaul all accessories at the same time that the engine is overhauled.

**Top overhaul.** The term top overhaul refers to those parts ont the crankcase that can be repaired or replaced with out the dismantling of the crankcase (Figure 1-3-31).

**Figure 1-3-31. An engine is dissassembled to this point for a top overhaul. It is often done without removing the engine from the aircraft.**

A typical top overhaul would include the removal, inspection, repair and replacement of such items as:

- The engine cylinder(s)
- The pistons and rings
- Valves, valve guides, springs and spring locks
- Magnetos, ignition harness and spark plugs
- Intake and exhaust pipes
- Fuel injection system, including discharge nozzles or the carburetor

Top overhaul is not recommended by all aircraft engine manufacturers. Many stress that if an engine requires this much dismantling it should be completely disassembled and receive a major overhaul.

**Major overhaul.** Major overhaul consists of the complete reconditioning of the powerplant. The actual overhaul period for a specific engine will generally be determined by the manufacturer's recommendations or by the maximum hours of operation between overhauls, as approved by the FAA.

At regular intervals, an engine should be completely dismantled, thoroughly cleaned and inspected. Each part should be overhauled in accordance with the manufacturer's instructions and tolerances for the engine involved. At this time all accessories are removed, overhauled and tested. Here again, instructions of the manufacturer of the accessory concerned should be followed.

There are many terms that are associated with the overhaul of aircraft engines. At the least they tend to cause confusion. It is very important for the technician to understand the sometime subtle differences that have a large impact on the work that must be done and the person or agency that is permitted to return the engine to service.

**Overhauled engine**. An overhauled engine is one that has been disassembled, cleaned, inspected, repaired as necessary and tested using FAA approved procedures. The engine may be overhauled to New Limits or Service Limits and still be considered an FAA Approved overhaul. The engine's previous operating history is maintained, and it is returned to service with zero time since major overhaul and a total time since new that includes all previous operating hours. Both times must be maintained in the engine log book.

**Rebuilt engine**. A rebuilt engine is one that has been overhauled using new and used parts to New Limits by the manufacturer or a facility approved by the manufacturer. The engine's previous operating history is expunged and it is returned to service with zero hours total time in service.

**New limits**. These are the FAA approved fits and tolerances to which the engine was manufactured to. This may be accomplished using standard or approved undersized and oversized tolerances.

**Service limits**. These are the FAA approved tolerances that a part may wear to and still be a useable component. This may also be accomplished using standard and approved undersized and oversized tolerances.

## *General Overhaul Procedures*

The manufacturer's overhaul instruction should always be followed when conducting a major overhaul of an engine. Make certain that you are using the latest service information available. This will always include the list of parts that must be replaced during the overhaul.

During the overhaul of the engine a few general procedures should be followed:

1. Disassemble the engine
2. Clean all parts
3. Inspect and measure all parts
4. Replace all parts that do not meet manufacturer's specification
5. Test the engine for proper operation

**Disassembly.** Inasmuch as visual inspection immediately follows disassembly, all individual parts should be laid out in an orderly manner on a workbench as they are removed. To guard against damage and to prevent loss, suitable containers should be available in which to place small parts, nuts, bolts, etc., during the disassembly operation.

Other practices during disassembly include:

- Dispose of all safety devices as they are removed. Never use safety wire, cotter pins, etc., a second time. Always replace them with new safety devices.
- All loose studs and loose or damaged fittings should be carefully tagged to prevent their being overlooked during inspection.
- Always use the proper tool for the job and the one that fits. Use sockets and box end

wrenches wherever possible. If special tools are required, use them rather than improvising.

- Drain the engine oil pumps and remove the oil filter. Drain into a suitable container, strain it through a clean cloth, and check both for metal particles.
- Before disassembly, wash the exterior of the engine thoroughly.

Cleaning. Procedures and materials used to clean engine components must follow the recommendations of the manufacturer. Failure to follow the established procedure could result in damage to expensive engine components.

Some non-recommended cleaning compounds may contain a caustic compound that will corrode aluminum and magnesium. Also, compounds from soaps and household cleaners may contaminate the pores in aluminum cases and cause oil foaming.

**Inspection.** The inspection of engine parts during overhaul is divided into three categories:

- Visual
- Non-destructive
- Dimensional

The first two methods are aimed at determining structural failures in the parts, while the last method deals with the size and shape of each part.

Visual inspection should precede all other inspection procedures. Parts should not be cleaned before a preliminary visual inspection, since indications of a failure often may be detected from the residual deposits of metallic particles in some recesses in the engine.

Structural failures can be determined using several different non-destructive inspection methods. Non-austenitic steel parts can readily be examined by the magnetic particle method. Aluminum and other parts that cannot be magnetized may be inspected using the dye penetrant method. Other non-destructive methods such as ultra-sonic, X-ray or etching can also be used if recommended by the engine manufacturer.

Several terms are used to describe defects detected in engine parts during inspection. Some of the more common terms and definitions are:

- **Abrasion.** An area of roughened scratches or marks usually caused by foreign matter between moving parts or surfaces.
- **Brinelling.** One or more indentations on bearing races usually caused by high static loads or application of force during installation or removal. Indentations are rounded or spherical due to the impression left by the contacting balls or rollers of the bearing.
- **Burning.** Surface damage due to excessive heat. It is usually caused by improper fit, defective lubrication or over-temperature operation.
- **Burnishing.** Polishing of one surface by sliding contact with a smooth, harder surface (Figure 1-3-32). Usually no displacement nor removal of metal.
- **Burr.** A sharp or roughened projection of metal usually resulting from machine processing.
- **Chafing.** Describes a condition caused by a rubbing action between two parts under light pressure which results in wear.
- **Chipping.** The breaking away of pieces of material usually caused by excessive stress concentration or careless handling.
- **Corrosion.** Loss of metal by a chemical or electrochemical action. The corrosion products generally are easily removed by mechanical means. Iron rust is an example of corrosion.
- **Crack.** A partial separation of material usually caused by vibration, overloading, internal stresses, defective assembly or fatigue. Depth may be a few thousandths to the full thickness of the piece.
- **Cut loss of metal.** Usually to an appreciable depth over a relatively long, narrow area, mechanically, as would occur with

Figure 1-3-32. A crankcase bearing showing burnishing marks at one end.

Figure 1-3-33. Heavy scoring on a piston sidewall

Figure 1-3-34. Staining

the use of a saw blade, chisel or sharp-edged stone striking a glancing blow.

- **Dent.** A small rounded depression in a surface usually caused by the part being struck with a rounded object.
- **Erosion.** Loss of metal from the surface by mechanical action of foreign objects, such as grit or fine sand. The eroded area will be rough and may be lined in the direction in which the foreign material moved relative to the surface.
- **Flaking.** The breaking loose of small pieces of metal or coated surfaces, which is usually caused by defective plating or excessive loading.
- **Fretting.** A condition of surface erosion caused by minute movement between two parts usually clamped together with considerable unit pressure.
- **Galling.** A severe condition of chafing or fretting in which a transfer of metal from one part to another occurs. It is usually caused by a slight movement of mated parts having limited relative motion and under high loads.
- **Gouging.** A furrowing condition in which a displacement of metal has occurred (a torn effect). It is usually caused by a piece of metal or foreign material between close moving parts.
- **Grooving.** A recess or channel with rounded and smooth edges usually caused by faulty alignment of parts.
- **Inclusion.** Presence of foreign or extraneous material wholly within a portion of metal. Such material is introduced during the manufacture of rod, bar or tubing by rolling or forging.
- **Nick.** A sharp-sided gouge or depression with a V-shaped bottom that is generally the result of careless handling of parts.
- **Peening.** A series of blunt depressions in a surface.
- **Pick-up** or **scuffing.** A buildup or rolling of metal from one area to another, which is usually caused by insufficient lubrication, clearances or foreign matter.
- **Pitting.** Small hollows of irregular shape in the surface, usually caused by corrosion or minute mechanical chipping of surfaces.
- **Scoring.** A series of deep scratches caused by foreign particles between moving parts or careless assembly or disassembly techniques (Figure 1-3-33).
- **Scratches.** Shallow, thin lines or marks, varying in degree of depth and width caused by presence of fine foreign particles during operation or contact with other parts during handling.
- **Staining.** A change in color, locally, causing a noticeably different appearance from the surrounding area (Figure 1-3-34).
- **Upsetting.** A displacement of material beyond the normal contour or surface (a local bulge or bump). Usually indicates no metal loss.

Examine all gears for evidence of pitting or excessive wear. These conditions are of particular importance when they occur on the teeth; deep pit marks in this area are sufficient cause to reject the gear. Bearing surfaces of all gears should be free from deep scratches. However, minor abrasions usually can be dressed out with a fine abrasive cloth.

All bearing surfaces should be examined for scores, galling and wear. Considerable scratching and light scoring of aluminum bearing surfaces in the engine will do no harm and should not be considered a reason for rejecting the part, provided it falls within the clearances set forth in the Table of Limits in the engine manufacturer's overhaul manual. Even though the part comes within the specific clearance limits, it will not be satisfactory for re-assembly in the engine unless inspection shows the part to be free from other serious defects.

Ball bearings should be inspected visually and by feel for roughness, flat spots on balls, flaking or pitting of races or scoring on the outside of races. All journals should be checked for galling, scores, misalignment or out-of-round condition. Shafts, pins, etc., should be checked for straightness. This may be done in most cases by using V-blocks and a dial indicator.

Pitted surfaces in highly stressed areas resulting from corrosion can cause ultimate failure of the part. The following areas should be examined carefully for evidence of such corrosion:

- Interior surfaces of piston pins
- The fillets at the edges of crankshaft main and crankpin journal surfaces (Figure 1-3-35)
- Thrust-bearing races

If pitting exists on any of the surfaces mentioned to the extent that it cannot be removed by polishing with crocus cloth or other mild abrasive, the part usually must be rejected.

Parts, such as threaded fasteners or plugs, should be inspected to determine the condition of the threads. Badly worn or mutilated threads cannot be tolerated; the parts should be rejected. However, small defects such as slight nicks or burrs may be dressed out with a small file, fine abrasive cloth or stone. If the part appears to be distorted, badly galled or mutilated by over-tightening, or from the use of improper tools, replace it with a new one.

Figure 1-3-35. Crankshaft main journals require special attention during inspections

Figure 1-3-36. Degreasing with solvents

**Cleaning.** After visually inspecting engine recesses for deposits of metal particles, it is important to clean all engine parts thoroughly to facilitate inspection. Two processes for cleaning engine parts are:

- Degreasing to remove dirt and sludge (soft carbon)(Figure 1-3-36)
- The removal of hard carbon deposits by de-carbonizing, brushing or scraping and grit-blasting

Degreasing can be done by immersing or spraying the part in a suitable commercial solvent. Extreme care must be used if any water-mixed degreasing solutions containing caustic compounds or soap are used. Such compounds, in addition to being potentially corrosive to aluminum and magnesium, may become impregnated in the pores of the metal and cause oil foaming when the engine is returned to service. When using water-mixed solutions, therefore, it is imperative that the parts be rinsed thoroughly and completely in clear boiling water after degreasing. Regardless of the method and type of solution used, coat or spray all parts with lubricating oil immediately after cleaning to prevent corrosion.

While the degreasing solution will remove dirt, grease and soft carbon, deposits of hard carbon will almost invariably remain on many interior surfaces. To remove these deposits, they must be loosened first by immersion in a tank containing a de-carbonizing solution (usually heated).

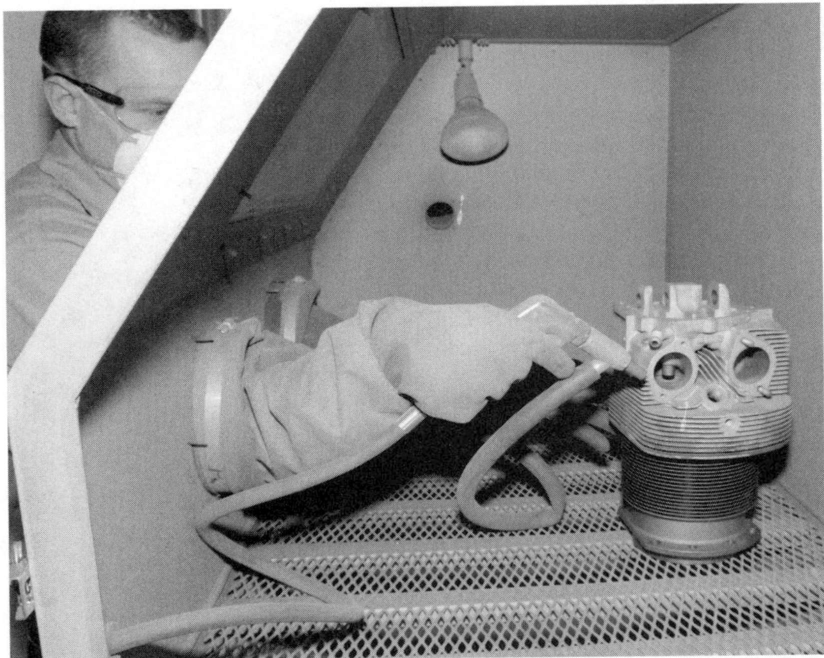
Figure 1-3-37. Grit-blasting

A variety of commercial de-carbonizing agents are available. De-carbonizers, like the degreasing solutions previously mentioned, fall generally into two categories, water-soluble and hydrocarbons. The same caution concerning the use of water-soluble degreasers is applicable to water-soluble de-carbonizers.

Extreme caution should be followed when using a de-carbonizing solution on magnesium castings. Avoid immersing steel and magnesium parts in the same de-carbonizing tank, because this practice often results in damage to the magnesium parts from corrosion.

De-carbonizing usually will loosen most of the hard carbon deposits remaining after degreasing; the complete removal of all hard carbon, however, generally requires brushing, scraping or grit-blasting. In all of these operations, be careful to avoid damaging the machined surfaces. In particular, wire brushes and metal scrapers must never be used on any bearing or contact surface.

When grit-blasting parts (Figure 1-3-37), follow the manufacturer's recommendations for the type of abrasive material to use. Sand, rice, baked wheat, plastic pellets, glass beads or crushed walnut shells are examples of abrasive substances that are used for grit-blasting parts.

All machined surfaces must be masked properly and adequately and all openings tightly plugged before blasting. The one exception to this is the valve seats, which may be left unprotected when blasting the cylinder head combustion chamber. It is often advantageous to grit-blast the seats, since this will cut the glaze that tends to form (particularly on the exhaust valve seat), thus facilitating subsequent valve seat reconditioning. Piston ring grooves may

Figure 1-3-38. Dimensional inspection

be grit-blasted if necessary; extreme caution must be used, however, to avoid the removal of metal from the bottom and sides of the grooves. When grit-blasting housings, plug all drilled oil passages with rubber plugs or other suitable material to prevent the entrance of foreign matter.

The de-carbonizing solution generally will remove most of the enamel on exterior surfaces. All remaining enamel should be removed by grit-blasting, particularly in the crevices between cylinder cooling fins.

At the conclusion of cleaning operations, rinse the part in petroleum solvent, dry, remove any loose particles of carbon or other foreign matter by air-blasting, and apply a liberal coating of preservative oil to all surfaces.

**Visual inspection.** Visual inspection of all engine parts should be accomplished with at least a 10-power magnifying glass.

**Fluorescent penetrant inspection.** All parts that cannot be tested using the magnetic particle inspection method should be inspected with fluorescent penetrant. This inspection must be accomplished in accordance with the procedures established for the type of penetrant used.

**Magnetic particle inspection.** All steel engine parts should be inspected using the magnetic particle inspection method. Special care should be taken to clean and demagnetize these parts after the inspection is completed. The type of magnetization (circular or longitudinal) and the magnetizing current are usually specified in the engine overhaul manual.

**Dimensional inspection.** A complete dimensional inspection (Figure 1-3-38) of all the engine parts should be completed at the time of overhaul. Careful consideration should be given to the calibration of the measuring equipment that you are using. The manufacturer's overhaul instruction will include the dimensional tolerances of the engine parts. These instructions should always be used to determine the acceptability or serviceability of all engine parts. Be certain that you are using the proper set of measurements (New Limits or Service Limits).

## *Repair*

FAA Part 43 contains explanations concerning major versus minor engine repairs. It is a good place to look for additional guidanceif the manufacturer's manual do not provide sufficient information.

**Crankcase.** A dimensional inspection should be completed to determine if the crankcase is still in limits. Then a dye penetrant inspection should be performed. After careful inspection, all defects should be repaired. Small gouges or scratches may be repaired by polishing. If any cracks are found, the crankcase should be sent to a FAA certified repair station to be welded.

**Camshaft.** The camshaft (Figure 1-3-39) is often replaced at overhaul, but if it's necessary to re-use the old camshaft, the same basic inspection and repair procedures should be used as recommended with the crankshaft. Always check the camshaft lobes for flat spots before re-installation. If found to have flat spots on the lobes, the camshaft should be replaced.

**Piston ring.** Piston rings should normally be replaced at overhaul. If it is necessary to re-use an old piston ring, it should always be lapped to fit the cylinder in which it will be placed.

## *Reassembly*

Reassembly of the engine should not be started until the cleaning, inspection and repair of all the engine parts have been completed.

The engine manufacturer's overhaul procedures should always be followed when reassembling the engine. The purpose of this section is only to provide basic reassembly information.

**Cylinder.** The cylinder reassembly consists mainly of installing the valve springs and valve spring locks after the valves and seats have been reconditioned (Figure 1-3-40, next page). Lubricate the valve stems before installation. It is a good practice to leak-test the valves after the valve springs have been installed. Wipe the cylinder dry and apply the recommended lubricant to the cylinder walls.

Figure 1-3-39. Camshaft

Figure 1-3-40. Reinstalling the valve springs and locks after reconditioning

Figure 1-3-41. The crankshaft is often reassembled upright on a stand

**Piston and piston ring.** The piston rings should be placed into the proper ring land. Care should be taken that the rings are installed with all bevels facing the proper direction. After the rings are installed, it is a good idea to re-measure the ring gap clearance and the ringside clearance to make certain the fit is correct. The rings and the ring lands should be well lubricated with the proper lubricant.

**Crankshaft.** The crankshaft is normally held upright on a stand for reassembly (Figure 1-3-41). The crankshaft journals should be cleaned and lubricated with the proper lubricant. The connecting rods and rod bearing should then be installed. The rod bolts should be torqued following the manufacturer's procedures.

**Crankcase.** The crankcase is laid down horizontally for reassembly. All seals and bearings should be installed at this time. The bearings should be lubricated with the recommended lubricant. The camshaft and the crankshaft, with the connecting rods installed, are then positioned in the bearings (Figure 1-3-42). The other half of the crankcase is then installed and torqued following the manufacturer's procedures.

### *Final Assembly*

The final assembly includes the installation of the cylinders on the crankcase. The installation of the rocker box covers, intake and exhaust systems and the engine accessories.

The final engine assembly is very important and the proper instructions should always be followed.

## Repair and Replacement

Damage such as burrs, nicks, scratches, scoring or galling should be removed with a fine oil stone, crocus cloth or any similar abrasive substance. Following any repairs of this type, the part should be cleaned carefully to be certain that all abrasive has been removed, and then checked with its mating part to assure that the clearances are not excessive. Flanged surfaces that are bent, warped or nicked can be repaired by lapping to a true surface on a surface plate. Again, the part should be cleaned to be certain that all abrasive has been removed. Defective threads can sometimes be repaired with a suitable die or tap. Small nicks can be removed satisfactorily with Swiss pattern files or small, edged stones. Pipe threads should not be tapped deeper to clean them, because this practice will result in an oversized tapped hole. If galling or scratches are removed from a bearing surface of a journal, it should be buffed to a high finish.

**Welding**. In general, welding of highly-stressed engine parts is not recommended for un-welded parts. However, welding may be accomplished if it can be reasonably expected that the welded repair will not adversely affect the airworthiness of the engine. A part may be welded when:

- The weld is located externally and can be inspected easily.
- The part has been cracked or broken as the result of unusual loads not encountered in normal operation.

- A new replacement part for an obsolete type of engine is unavailable.
- The welder's experience and the equipment used will ensure a first-quality weld and the restoration of the original heat treatment in heat-treated parts.

Many minor parts not subjected to high stresses may be safely repaired by welding. Mounting lugs, cowl lugs, cylinder fins, rocker box covers and many parts originally fabricated by welding are in this category.

Some certified repair stations have been approved for welded repairs to crankcases and other highly stressed parts. To obtain FAA approval of their process, the CRS must establish not only that they can do the repair, but that the repair will be serviceable (see Figure 1-3-43).

**Re-painting**. Parts requiring use of paint for protection or appearance should be re-painted according to the engine manufacturer's recommendations. One procedure is outlined in the following paragraphs.

Aluminum alloy parts should have original exterior painted surfaces sanded smooth to provide a proper paint base. See that surfaces to be painted are thoroughly cleaned. Care must be taken to avoid painting mating surfaces. Exterior aluminum parts should be primed first with a thin coat of zinc chromate primer. Each coat should be either air dried for 2 hours or baked at 177°C (350°F) for one-half hour. After the primer is dry, parts should be painted with engine enamel, which should be air-dried until hard, or baked for one-half hour at 82°C (180°F). Aluminum parts from which the paint has not been removed may be repainted without the use of a priming coat, provided no bare aluminum is exposed.

Parts requiring a black gloss finish should be primed first with zinc chromate primer and then painted with glossy black cylinder enamel. Each coat should be baked for 1-1/2 hours at 177°C (350°F). If baking facilities are not available, cylinder enamel may be air dried; however, an inferior finish will result. All paint preferably should be sprayed; however, if it is necessary to use a brush, use care to avoid an accumulation of paint pockets.

Magnesium parts should be cleaned thoroughly with a dichromate treatment prior to painting. This treatment consists of cleaning all traces of grease and oil from the part by using a neutral, non-corrosive degreasing medium followed by a rinse, after which the part is immersed for at least 45 minutes in a hot dichromate solution (three-fourths of a pound of sodium dichromate to 1 gallon of water at 180-200°F). Then the part should be washed thoroughly in cold running water, dipped in hot water and dried in an air blast. Immediately thereafter, the part should be painted with a primer coat and engine enamel in the same manner as that suggested for aluminum parts.

Any studs which are bent, broken, damaged or loose must be replaced. After a stud has been removed, the tapped stud hole should be examined for size and condition of threads. If it is necessary to re-tap the stud hole, it also will be necessary to use a suitable oversize stud. Studs that have been broken off flush with the case must be drilled and removed with suitable stud remover. Be careful not to damage any threads. When replacing studs, coat the coarse threads of the stud with anti-seize compound.

## Cylinder Assembly Reconditioning

Cylinder and piston assemblies are inspected according to the procedures contained in the engine manufacturer's manuals, charts and

Figure 1-3-42. Positioning the crankshaft into one half of the crankcase

Figure 1-3-43. This illustration shows a welded repair to a bearing flange on an engine crankcase. This type of repair can only be performed by a CRS which has the approval.

*Photo courtesy of Engine Components, Inc.*

**Figure 1-3-44. This crack, extending from the exhaust valve seat to the spark plug boss, is an example of something you can expect to find on high-time engines.**

service bulletins. A general procedure for inspecting and reconditioning cylinders will be discussed in the following section to provide an understanding of the operations involved.

**Cylinder head.** Inspect the cylinder head for internal and external cracks. Carbon deposits must be cleaned from the inside of the head, and paint must be removed from the outside for this inspection.

Exterior cracks will show up on the head fins where they have been damaged by tools or contact with other parts because of careless handling. Cracks near the edge of the fins are not dangerous if the portion of the fin is removed and contoured properly. Cracks at the base of the fin are a reason for rejecting the cylinder. Cracks may also occur on the rocker box or in the rocker bosses.

Interior cracks almost always will radiate from the valve seat bosses or the spark plug bushing boss (Figure 1-3-44). They may extend completely from one boss to the other. These cracks are usually caused by improper installation of the seats or bushings.

Use a bright light to inspect for cracks, and investigate any suspicious areas with a magnifying glass or microscope. Cracks in aluminum alloy cylinder heads generally will be jagged because of the granular nature of the metal. Do not mistake casting marks or laps for cracks. One of the best methods to double-check your findings is to inspect by means of the dye penetrant process. Any crack in the cylinder head, except those on the fins which can be worked out, is reason for rejecting the cylinder.

Inspect the head fins for other damage besides cracks. Dents or bends in the fins should be left alone unless there is danger of cracking. Where pieces of fin are missing, the sharp edges should be filed to a smooth contour. Fin breakage in a concentrated area will cause dangerous local hot spots. Fin breakage near the

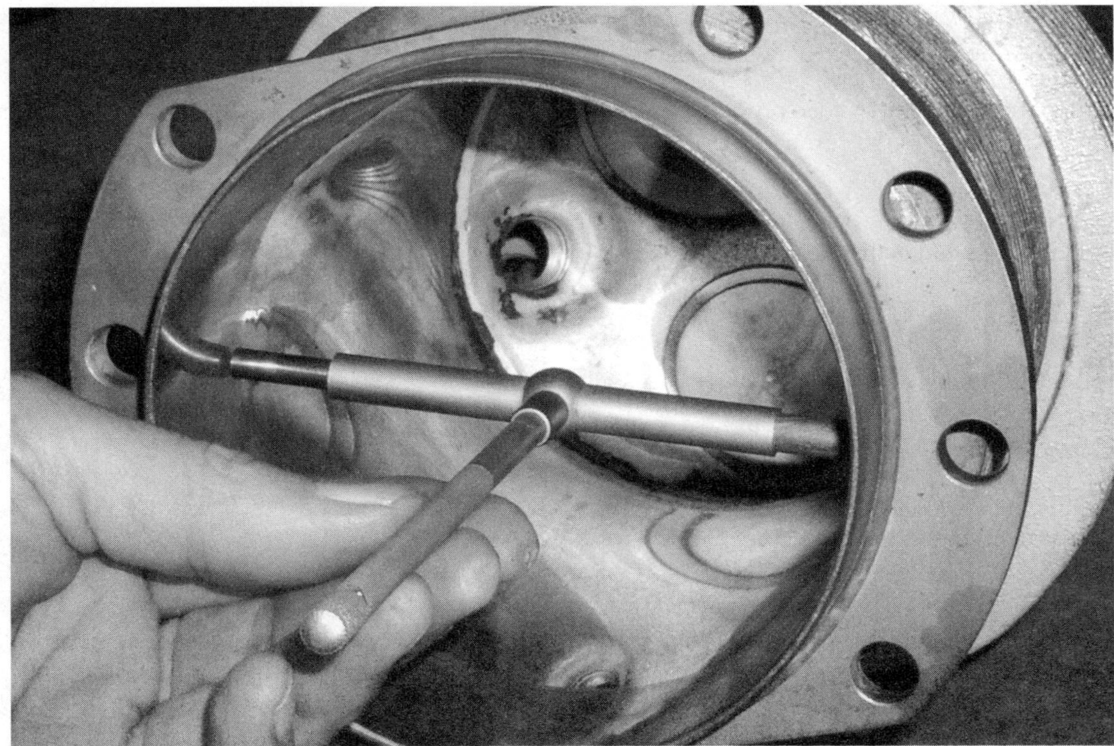

**Figure 1-3-45. Checking cylinder bore with a telescoping gauge**

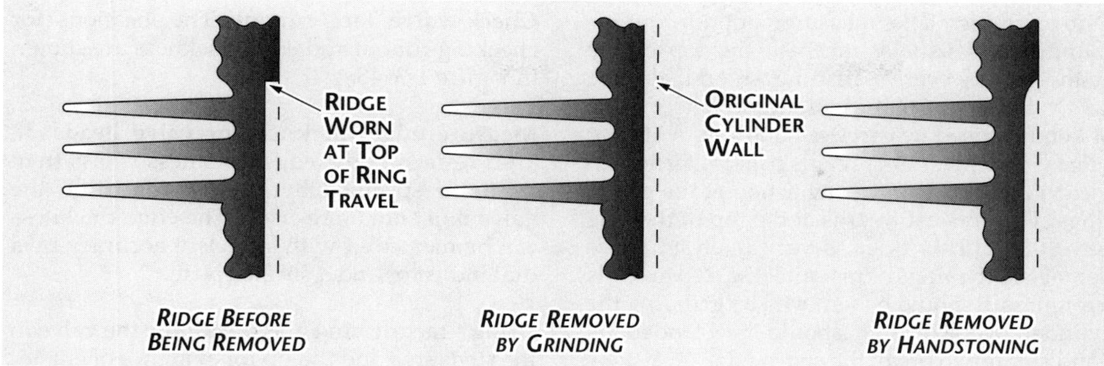


**Figure 1-3-46. Ridge or step formed in an engine cylinder**

spark plug bushings or on the exhaust side of the cylinder is obviously more dangerous than in other areas. When removing or re-profiling a cylinder fin, follow the instructions and the limits in the manufacturer's manual.

Inspect all the studs on the cylinder head for looseness, straightness, damaged threads and proper length. Slightly damaged threads may be chased with the proper die. The length of the stud should be correct within 1/32 inch (0.03125) to allow for proper installation of pal nuts or other safety devices.

Be sure the valve guides are clean before inspection. Very often, carbon will cover pits inside the guide. If a guide in this condition is put back in service, carbon will again collect in the pits, and valve sticking will result. Besides pits, scores and burned areas inside the valve guide, inspect them for wear or looseness. Most manufacturers provide a maximum wear gauge to check the dimension of the guide. This gauge should not enter the guide at all at either end. Do not confuse this gauge with the go and no-go gauge used to check new valve guides after reaming.

Inspection of valve seat inserts before they are re-faced is mostly a matter of determining if there is enough of the seat left to correct any pitting, burning, scoring or out-of-trueness.

Inspect spark plug inserts for the condition of the threads and for looseness. Run a tap of the proper size through the bushing. Very often, the inside threads of the bushing will be burned. If more than one thread is missing, the bushing should be rejected. Tighten a plug in the bushing to check for looseness.

Inspect the rocker shaft bosses for scoring, cracks, oversize or out-of-roundness. Scoring is generally caused by the rocker shaft turning in the bosses, which means either the shaft was too loose in the bosses or a rocker arm was too tight on the shaft. Out-of-roundness is usually caused by a stuck valve. If a valve sticks, the rocker shaft tends to work up and down when the valve offers excessive resistance to opening. Inspect for out-of-roundness and oversize using a telescopic gauge and a micrometer.

**Cylinder barrel.** Inspect the cylinder barrel for wear, using a dial indicator, a telescopic gauge and micrometer or an inside micrometer. Dimensional inspection of the barrel consists of the following measurements:

- Maximum taper of cylinder walls
- Maximum out-of-roundness
- Bore diameter
- Step
- Fit between piston skirt and cylinder

All measurements involving cylinder barrel diameters must be taken at a minimum of two positions 90° apart in the particular plane being measured. It may be necessary to take more than two measurements to determine the maximum wear. The use of a telescoping gauge to check a cylinder bore is shown in Figure 1-3-45.

**Taper**. Taper of the cylinder walls is the difference between the diameter of the cylinder barrel at the bottom and the diameter at the top. The cylinder is usually worn larger at the top than at the bottom. This taper is caused by the natural wear pattern. At the top of the stroke, the piston is subjected to greater heat and pressure and more erosive environment than at the bottom of the stroke. Also, there is greater freedom of movement at the top of the stroke. Under these conditions, the piston will wear the cylinder wall. In most cases, the taper will end with a ridge (see Figure 1-3-46) that must be removed during overhaul. Where cylinders are built with an intentional choke, measurement of taper becomes more complicated. It is necessary to know exactly how the size indicates wear or taper. Taper can be measured in any cylinder by a cylinder dial gauge as long as there is not a sharp step. The dial gauge tends to ride up on the step and causes inaccurate readings at the top of the cylinder.

**Out-of-round**. The measurement for out-of-roundness is usually taken at the top of the cylinder. However, a reading should also be taken at the skirt of the cylinder to detect dents or bends caused by careless handling. A step or ridge (Figure 1-3-46 previous page) is formed in the cylinder by the wearing action of the piston rings. The greatest wear is at the top of the ring travel limit. This ridge is very likely to cause damage to the rings or piston. If the step exceeds tolerances, it should be removed by grinding the cylinder oversize or it should be blended by hand stoning to break the sharp edge.

A step also may be found where the bottom ring reaches its lowest travel. This step is very rarely found to be excessive, but it should he checked.

Inspect the cylinder walls for rust, pitting or scores. Mild damage of this sort can be removed when the rings are lapped. With more extensive damage, the cylinder will have to be reground or honed. If the damage is too deep to be removed by either of these methods, the cylinder usually will have to be rejected. Most engine manufacturers have an exchange service on cylinders with damaged barrels.

**Warpage**. Check the cylinder flange for warpage by placing the cylinder on a suitable jig. Check to see that the flange contacts the jig all the way around. The amount of warp can be checked by using a thickness gauge (Figure 1-3-47). A cylinder whose flange is warped beyond the allowable limits should be rejected.

**Valves and valve springs.** Remove the valves from the cylinder head and clean them to remove soft carbon. Examine the valve visually for physical damage and damage from burning or corrosion. Do not re-use valves that indicate damage of this nature.

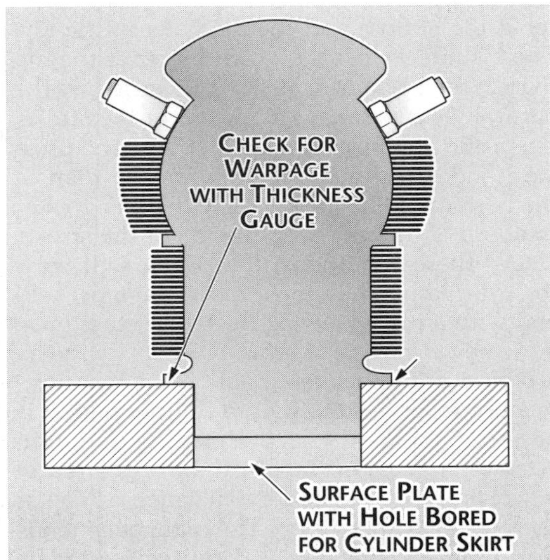


Figure 1-3-47. A method for checking cylinder flange warpage

**Check valve face runout**. The locations for checking runout and edge thickness are shown in Figure 1-3-48.

**Measure edge thickness of valve heads**. If, after re-facing, the edge thickness is less than the limit specified by the manufacturer, the valve must not be re-used. The edge thickness can be measured with sufficient accuracy by a dial indicator and a surface plate.

Using a magnifying glass, examine the valve in the stem area and the tip for evidence of cracks, nicks or other indications of damage. This type of damage seriously weakens the valve, making it susceptible to failure. If superficial nicks and scratches on the valve indicate that it might be cracked, inspect it using the magnetic particle or dye penetrant method.

Critical areas of the valve include the face and tip, both of which should be examined for pitting and excessive wear. Minor pitting on valve faces can sometimes be removed by grinding.

Inspect the valve for stretch and wear, using a micrometer or a valve radius gauge. If a micrometer is used, stretch will be found as a smaller diameter of the valve stem near the neck of the valve. Measure the diameter of the valve stem and check the fit of the valve in its guide (Figure 1-3-49).

Examine the valve springs for cracks, rust, broken ends and compression. Cracks can be located by visual inspection or the magnetic particle method. Compression is tested with a valve spring tester. The spring is compressed until its total height is that specified by the manufacturer. The dial on the tester should indicate the pressure (in pounds) required to compress the spring to the specified height. This must be within the pressure limits established by the manufacturer.

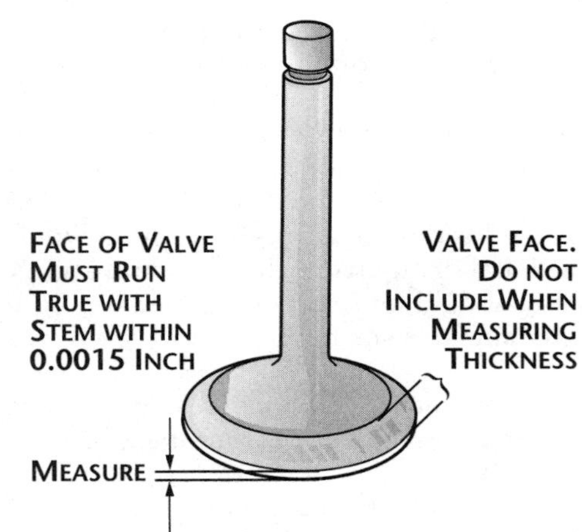


Figure 1-3-48. Valve, showing locations for checking run-out and section for measuring edge thickness

## Rocker Arms and Shafts

Inspect the valve rockers for cracks and worn, pitted or scored tips. See that all oil passages are free from obstructions.

Inspect the shafts for correct size with a micrometer. Rocker shafts very often are found to be scored and burned because of excessive turning in the cylinder head. Also, there may be some pickup on the shaft (bronze from the rocker bushing transferred to the steel shaft). Generally this is caused by overheating and too little clearance between shaft and bushing.

Inspect the rocker arm bushing for correct size. Check for proper clearance between the shaft and the bushing. Very often the bushings are scored because of mishandling during disassembly. Check to see that the oil holes line up. At least 50 percent of the hole in the bushing should align with the hole in the rocker arm.

On engines that use a bearing rather than a bushing, inspect the bearing to make certain it has not been turning in the rocker arm boss, as well as to determine its serviceability.

## Piston and Piston Pin

Inspect the piston for cracks. As an aid to this, heat the piston carefully with a blow torch. If there is a crack, the heat will expand it and will force out residual oil that remains in the crack no matter how well the piston has been cleaned. Cracks are more likely to be formed at the highly stressed points; therefore, inspect carefully at the base of the pin bosses, inside the piston at the junction of the walls and the head and at the base of the ring lands, especially the top and bottom lands.

When applicable, check for flatness of the piston head using a straightedge and thickness gauge as shown in Figure 1-3-50. If a depression is found, double check for cracks on the inside of the piston. A depression in the top of the piston usually means that detonation has occurred within the cylinder.

Inspect the exterior of the piston for scores and scratches. Scores on the top ring land are not cause for rejection unless they are excessively deep. Deep scores on the side of the piston are usually a reason for rejection. Examine the piston for cracked skirts, broken ring lands and scored piston pin holes. As a rule, faulty pistons are replaced, rather than repaired.

Measure the outside of the piston with a micrometer. Measurements must be taken in several directions and on the skirt, as well as on the lands. Check these sizes against the cylinder size. Several engines now use cam-ground pistons to compensate for the greater expansion parallel to the pin during engine operation. The diameter of these pistons measures several thousandths of an inch larger at an angle to the pin hole than parallel to the pin hole.

Inspect the ring grooves for evidence of a step. If a step is present, the groove will have to be machined to an oversize width. Use a standard piston ring and check side clearance with a

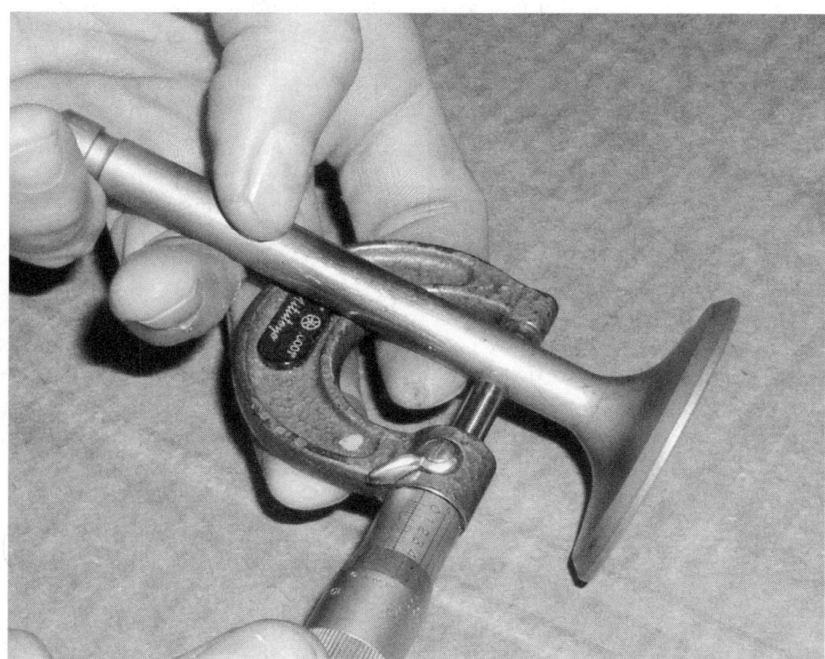

Figure 1-3-49. Checking valve stretch with a manufacturer's gauge

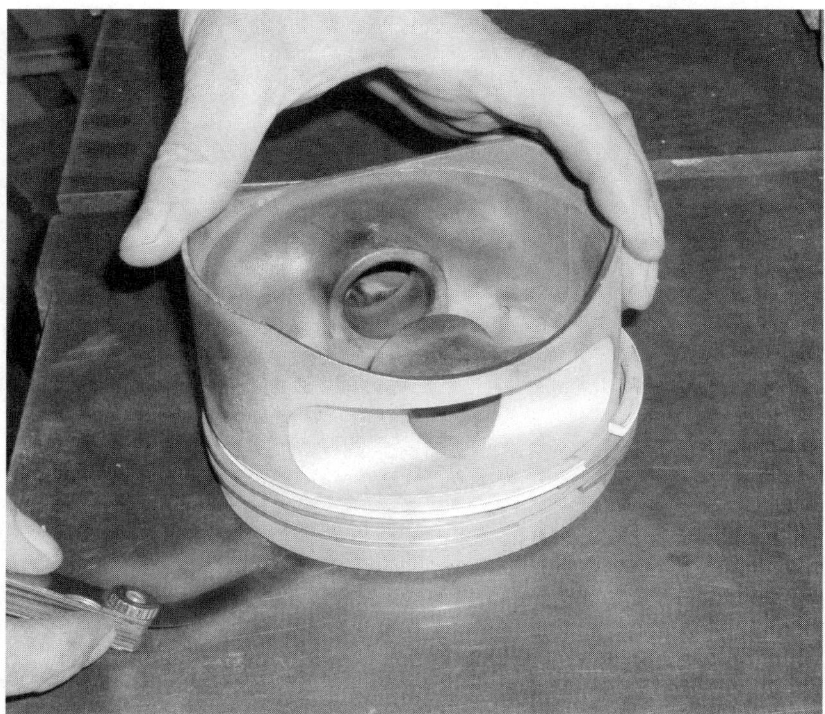

Figure 1-3-50. Checking a piston head for flatness

feeler gauge to locate wear in the grooves or to determine if the grooves have already been machined oversize. The largest allowable width is usually 0.020-inch oversize — any further machining weakens the lands excessively.

Examine the piston pin for scoring, cracks, excessive wear and pitting. Check the clearance between the piston pin and the bore of the piston pin bosses using a telescopic gauge and a micrometer. Use the magnetic particle method to inspect the pin for cracks. Since the pins are often case hardened, cracks will show up inside the pin more often than they will on the outside.

Check the pin for bends (Figure 1-3-51) using V-blocks and a dial indicator on a surface plate. Measure the fit of the plugs in the pin.

## Re-facing Valve Seats

The valve seat inserts of aircraft engine cylinders usually are in need of re-facing at every overhaul. They are re-faced to provide a true, clean and correct size seat for the valve. When valve guides or valve seats are replaced in a cylinder, the seats must be trued-up to the guide.

Modern engines use either bronze or steel seats. Steel seats are commonly used as exhaust seats and are made of a hard, heat-resistant and often austenitic steel alloy. Bronze seats are used for intake or for both seats; they are made of aluminum bronze or phosphor bronze alloys.

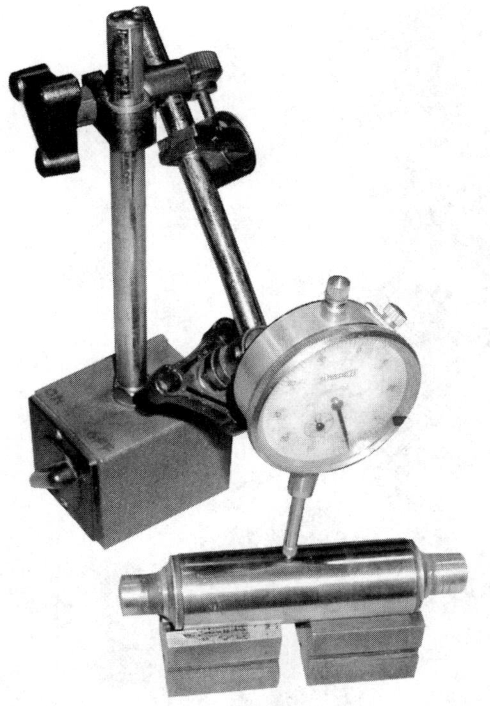

Figure 1-3-51. Checking a piston pin for bends

Steel valve seats are re-faced by grinding equipment. Bronze seats are re-faced preferably by the use of cutters or reamers, but they may be ground when this equipment is not available. The only disadvantage of using a stone on bronze is that the soft metal loads the stone to such an extent that much time is consumed in re-dressing the stone to keep it clean.

The equipment used on steel seats can be either wet or dry valve seat grinding equipment. Figure 1-3-52 illustrates a wet seat grinding table. The wet grinder uses a mixture of soluble oil and water to wash away the chips and to keep the stone and seat cool; this produces a smoother, more accurate job than the dry grinder. The stones may be either silicon carbide or aluminum oxide.

Before re-facing the seat, make sure that the valve guide is in good condition, is clean and will not have to be replaced.

Mount the cylinder firmly in the hold-down fixture. An expanding pilot is inserted in the valve guide from the inside of the cylinder, and an expander screw is inserted in the pilot from the top of the guide as shown in Figure 1-3-53. The pilot must be tight in the guide because any movement can cause a poor grind. The fluid hose is inserted through one of the spark plug inserts.

The three grades of stones available for use are classified as rough, finishing and polishing stones. The rough stone is designed to true and clean the seat. The finishing stone must follow the rough to remove grinding marks and produce a smooth finish. The polishing stone does just as the name implies and is used only where a highly polished seat is desired.

The stones are installed on special stone holders. The face of the stone is trued by a diamond dresser. The stone should be re-faced whenever it is grooved or loaded and when the stone is first installed on the stone holder. The diamond dresser also may be used to cut down the diameter of the stone. Dressing of the stone should be kept to a minimum as a matter of conservation; therefore, it is desirable to have sufficient stone holders for all the stones to be used on the job.

In the actual grinding job, considerable skill is required in handing the grinding gun. The gun must be centered accurately on the stone holder. If the gun is tilted off-center, chattering of the stone will result and a rough grind will be produced. It is very important that the stone be rotated at a speed that will permit grinding instead of rubbing. This speed is approximately 8,000-10,000 r.p.m. Excessive pressure on the stone can slow it down. It is not a good technique to let the stone grind at slow speed by

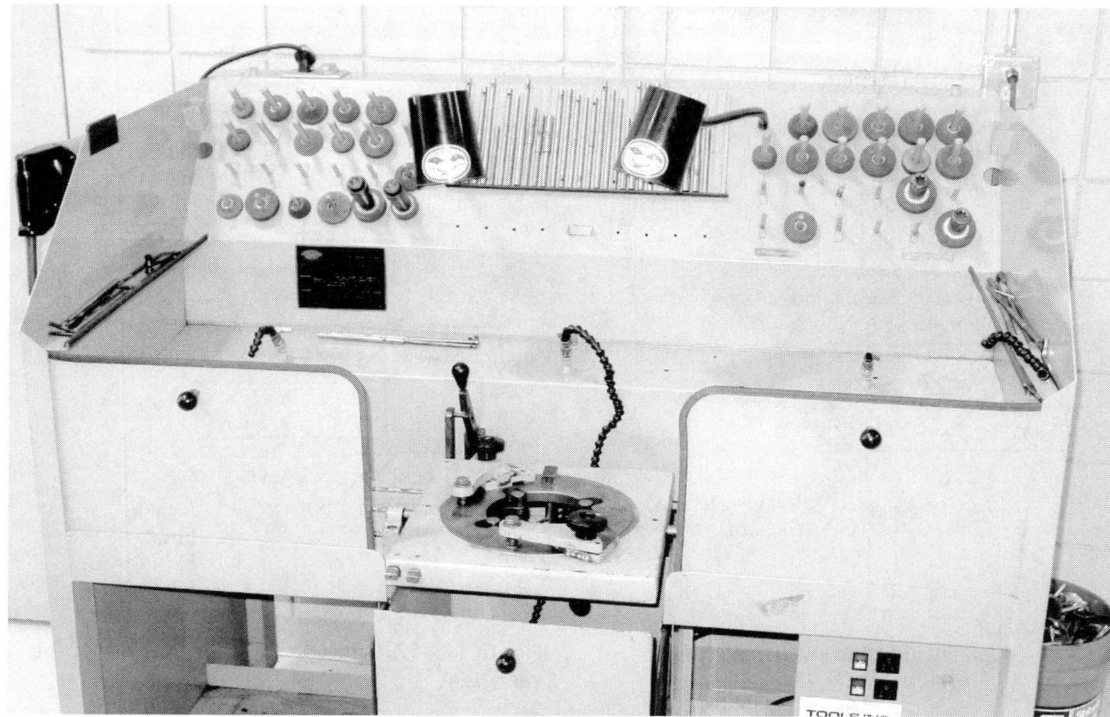

Figure 1-3-52. A wet seat grinding table, with cylinder hold-down clamps and recirculating cooling fluid, can help produce good, smooth valve seats.

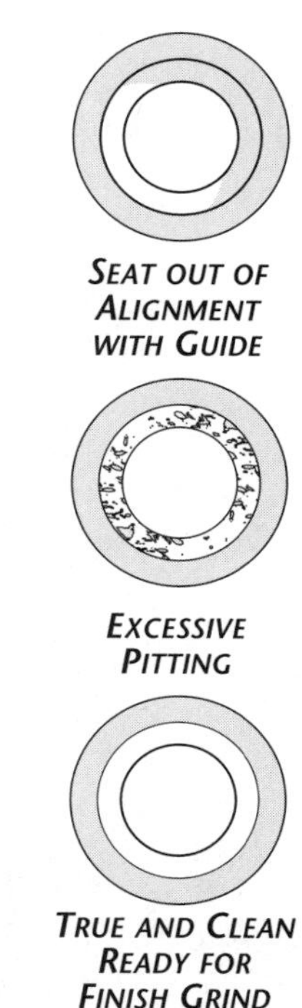


Figure 1-3-54. Valve seat grinding

putting pressure on the stone when starting or stopping the gun. The maximum pressure used on the stone at any time should be no more than that exerted by the weight of the gun.

Another practice which is conducive to good grinding is to ease off on the stone every second or so to let the coolant wash away the chips on the seat; this rhythmic grinding action also helps keep the stone up to its correct speed. Since it is quite a job to replace a seat, remove as little material as possible during the grinding. Inspect the job frequently to prevent unnecessary grinding.

The rough stone is used until the seat is true to the valve guide and until all pits, scores or burned areas (Figure 1-3-54) are removed. After re-facing, the seat should be smooth and true.

The finishing stone is used only until the seat has a smooth, polished appearance. Extreme caution should be used when grinding with the finishing stone to prevent chattering.

The size and trueness of the seat can be checked by several methods. Runout of the seat is checked with a special dial indicator and should not exceed 0.002 inches. The size of the seat may be determined by using Prussian blue. To check the fit of the seat, spread a thin coat of Prussian blue evenly on the seat. Press the valve onto the seat. The blue transferred to the valve will indicate the contact surface. The contact surface should be one-third to two-thirds the width of the valve face and in the middle of

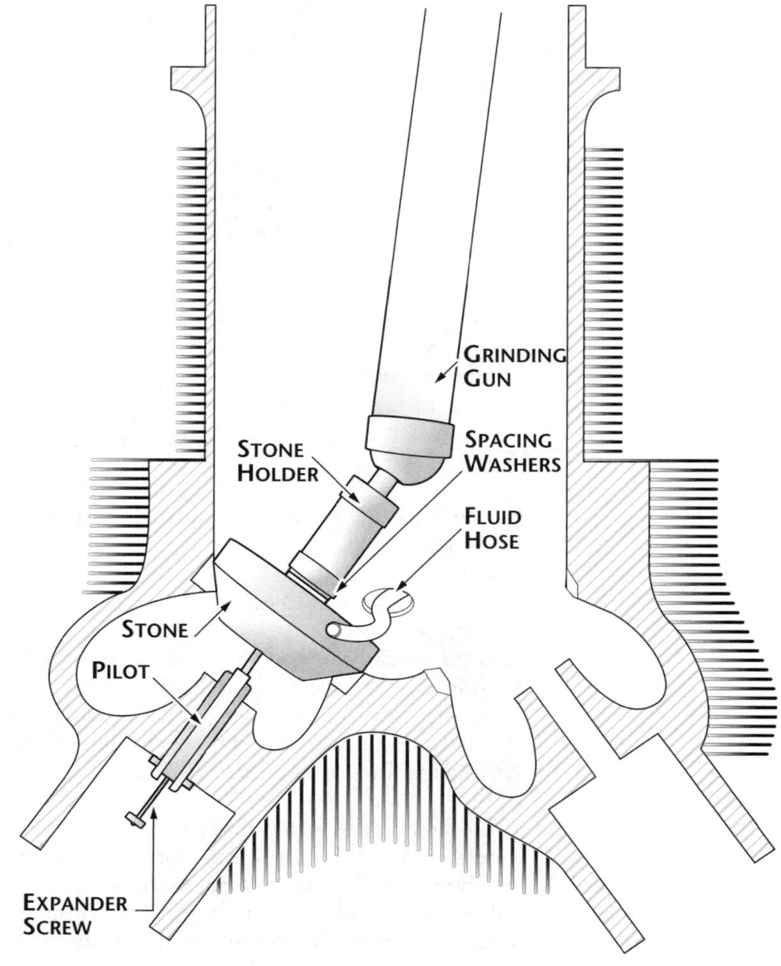


Figure 1-3-53. Valve seat grinding equipment

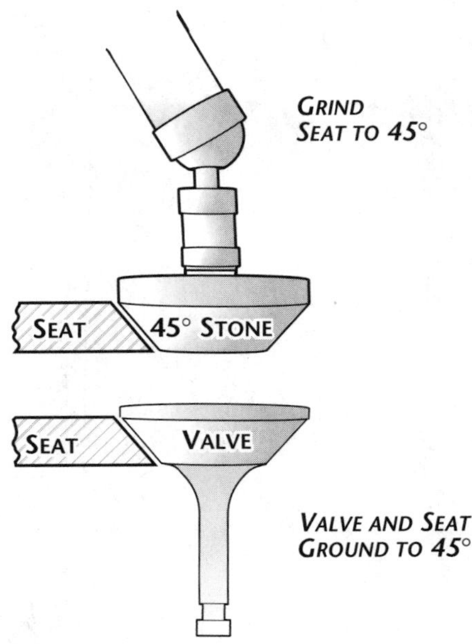


Figure 1-3-55. Fitting of the valve and seat

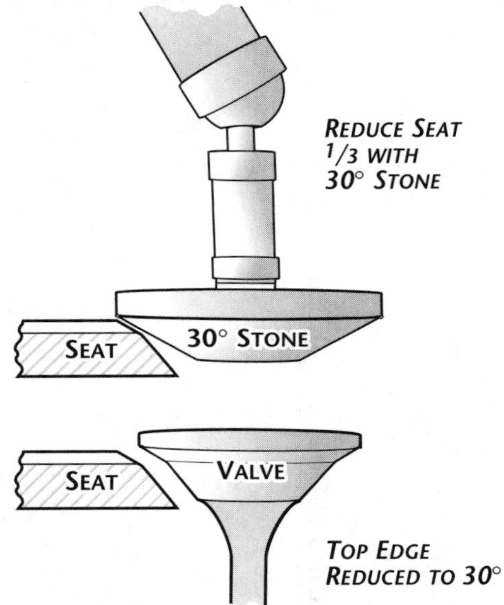


Figure 1-3-56. Grinding top surface of the valve seat

the face. In some cases, a go and no-go gauge is used in place of the valve when making the Prussian blue check. If Prussian blue is not used, the same check may be made by lapping the valve lightly to the seat. Examples of test results are shown in Figure 1-3-55.

If the seat contacts the upper third of the valve face, grind off the top corner of the valve seat as shown in Figure 1-3-56. Such grinding is called narrowing grinding. This permits the seat to contact the center third of the valve face without touching the upper portion of the valve face.

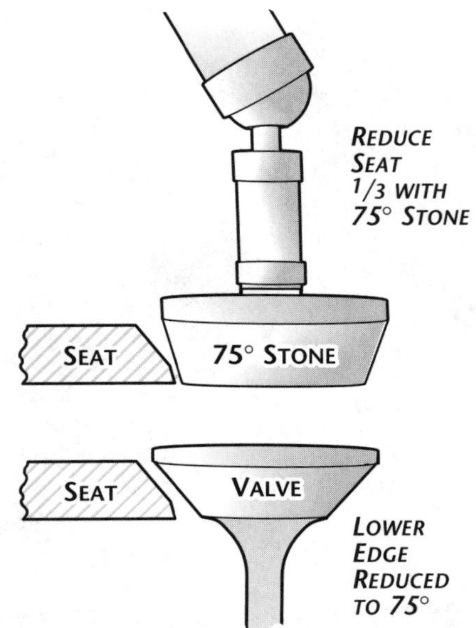


Figure 1-3-57. Grinding the inside corner of the valve seat

If the seat contacts the bottom third of the valve face, grind off the inner corner of the valve seat as shown in Figure 1-3-57.

The seat is narrowed by a stone other than the standard angle. It is common practice to use a 15° angle and 45° angle cutting stone on a 30° angle valve seat, and a 30° angle and 75° angle stone on a 45° angle valve seat (see Figure 1-3-58).

If the valve seat has been cut or ground too much, the valve will contact the seat too far up into the cylinder head, and the valve clearance, spring tension and the fit of the valve to the seat will be affected. To check the height of a valve, insert the valve into the guide and hold it against the seat. Check the height of the valve stem above the rocker box or some other fixed position.

Before re-facing a valve seat, consult the overhaul manual for the particular model engine. Each manufacturer specifies the desired angle for grinding and narrowing the valve seat.

**Valve reconditioning.** One of the most common jobs during engine overhaul is grinding the valves. The equipment used should preferably be a wet valve grinder. An example is shown in Figure1-3-59.

With this type of machine, a mixture of soluble oil and water is used to keep the valve cool and carry away the grinding chips.

Like many machine jobs, valve grinding is mostly a matter of setting up the machine. The following points should be checked or accomplished before starting a grind.

True the stone by means of a diamond nib. The machine is turned on, and the diamond is drawn across the stone, cutting just deep enough to true and clean the stone.

Determine the face angle of the valve being ground and set the movable head of the machine to correspond to this valve angle. Usually, valves are ground to the standard angles of 30° or 45°. However, in some instances, an interference fit of 0.5 or 1.5 less than the standard angle may be ground on the valve face.

The interference fit is used to obtain a more positive seal by means of a narrow contact surface. An example is shown in Figure 1-3-60. Theoretically, there is a line contact between the valve and seat. With this line contact, the entire load that the valve exerts against the seat is concentrated in a very small area, thereby increasing the unit load at any one spot. The interference fit is especially beneficial during the first few hours of operation after an overhaul. The positive seal reduces the possibility of a burned valve or seat that a leaking valve might produce. After the first few hours of running, these angles tend to pound down and become identical.

Notice that the interference angle is ground into the valve, not the seat. It is easier to change the angle of the valve grinder work head than to change the angle of a valve seat grinder stone. Do not use an interference fit unless the manufacturer approves it.

Install the valve into the chuck (Figure 1-3-61) and adjust the chuck so that the valve face is approximately 2 inches from the chuck. If the valve is chucked any further out, there is danger of excessive wobble and a possibility of grinding into the stem.

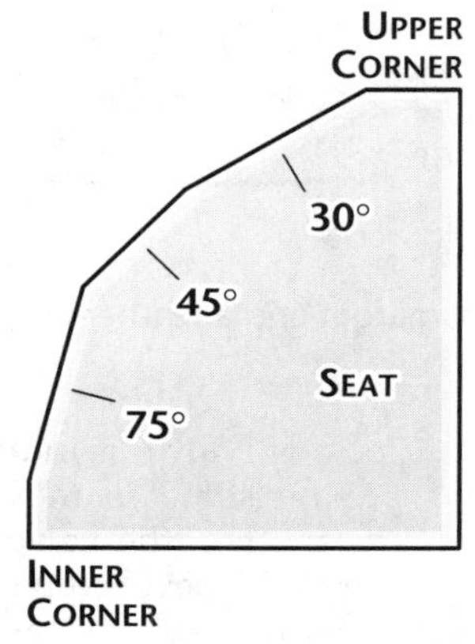


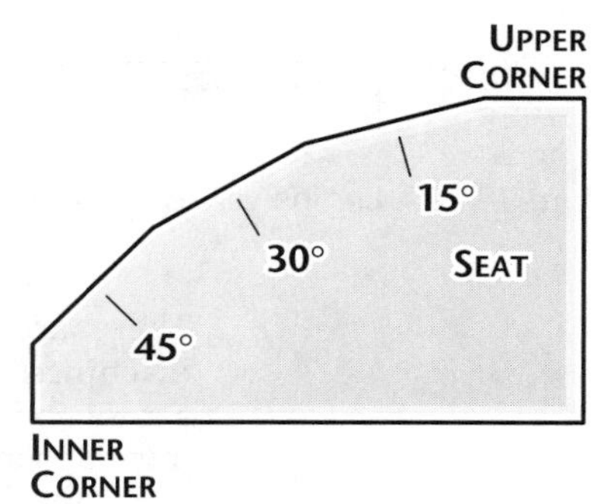


Figure 1-3-58. Valve seat angles

Figure 1-3-59. A wet valve-grinding machine for grinding valve faces

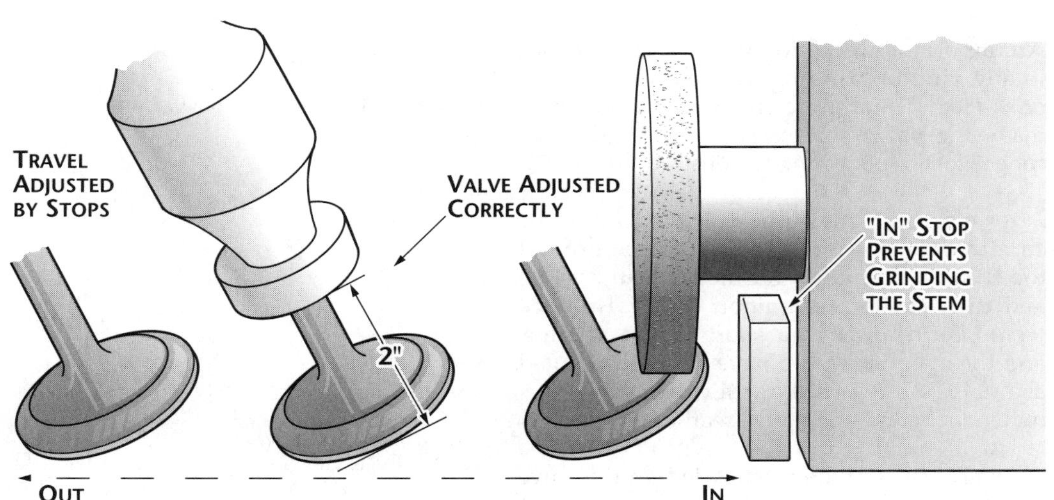


Figure 1-3-61. Valve installed in grinding machine

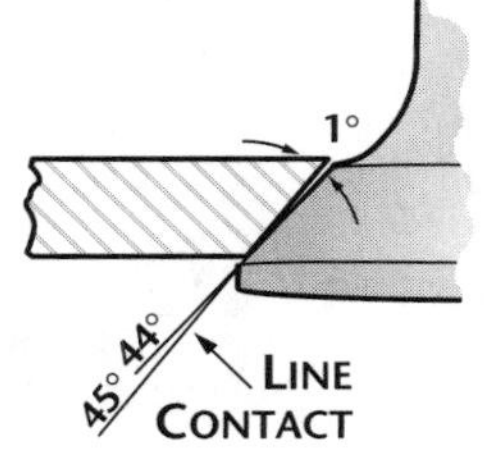


Figure 1-3-60. Interference fit of valve and valve seat

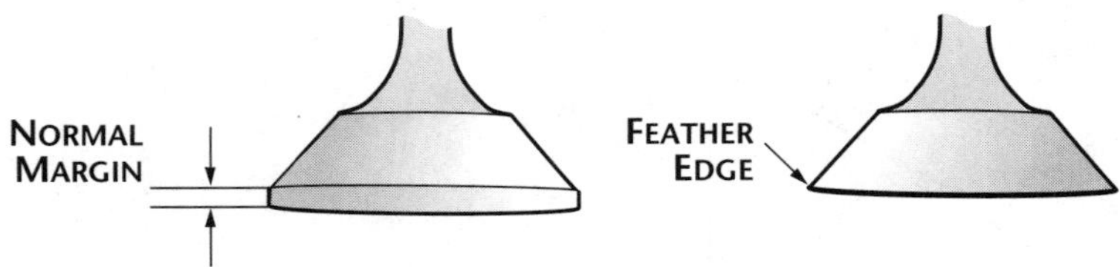


Figure 1-3-62. Engine valves showing a normal margin and a feather edge

There are various types of valve grinding machines. In one type the stone is moved across the valve face; in another, the valve is moved across the stone. Whichever type is used, the following procedures are typical of those performed when re-facing a valve.

Check the travel of the valve face across the stone. The valve should completely pass the stone on both sides and yet not travel far enough to grind the stem. There are stops on the machine that can be set to control this travel.

With the valve set correctly in place, turn on the machine and turn on the grinding fluid so that it splashes on the valve face. Back the grinding wheel off all the way. Place the valve directly in front of the stone. Slowly bring the wheel forward until a light cut is made on the valve. The intensity of the grind is measured by sound more so than any other way. Slowly draw the valve back and forth across the stone without increasing the cut. Move the work-head table back and forth using the full face of the stone but always keep the valve face on the stone. When the sound of the grind diminishes, indicating that some valve material has been removed, move the workhead table to the extreme left to stop rotation of the valve. Inspect the valve to determine if further grinding is necessary. If another cut must be made, bring the valve in front of the stone, then advance the stone out to the valve. Do not increase the cut without having the valve directly in front of the stone.

An important precaution in valve grinding, as in any kind of grinding, is to make light cuts only. Heavy cuts cause chattering, which may make the valve surface so rough that much time is lost in obtaining the desired finish.

After grinding, check the valve margin to be sure that the valve edge has not been ground too thin. A thin edge is called a feather edge and can lead to pre-ignition. The valve edge would burn away in a short period of time, and the cylinder would have to be overhauled again. Figure 1-3-62 shows a valve with a normal margin and one with a feather edge.

The valve tip may be re-surfaced on the valve grinder. The tip is ground to remove cupping or wear and also to adjust valve clearances on some engines.

The valve is held by a clamp (Figure 1-3-63) on the side of the stone. With the machine and grinding fluid turned on, the valve is pushed lightly against the stone and swung back and forth. Do not swing the valve stem off either edge of the stone. Because of the tendency for the valve to overheat during this grinding, be sure plenty of grinding fluid covers the tip.

Grinding of the valve tip may remove or partially remove the bevel on the edge of the valve. To restore this bevel, mount a vee-way approximately 45° to the grinding stone. Hold the valve onto the vee-way with one hand, then twist the valve tip onto the stone and, with a light touch, grind all the way around the tip. This bevel prevents scratching the valve guide when the valve is installed.

**Valve lapping and leak testing.** After the grinding procedure is finished, it is sometimes necessary that the valve be lapped to the seat. This is done by applying a small amount of lapping compound to the valve face, inserting the valve into the guide and rotating the valve with a lapping tool until a smooth, gray finish appears at the contact area. A correctly lapped valve is shown in Figure 1-3-64.

After the lapping process is finished, be sure that all lapping compound is removed from the valve face, seat and adjacent areas.

The final step is to check the valves seal by conducting a leak test. This is accomplished by

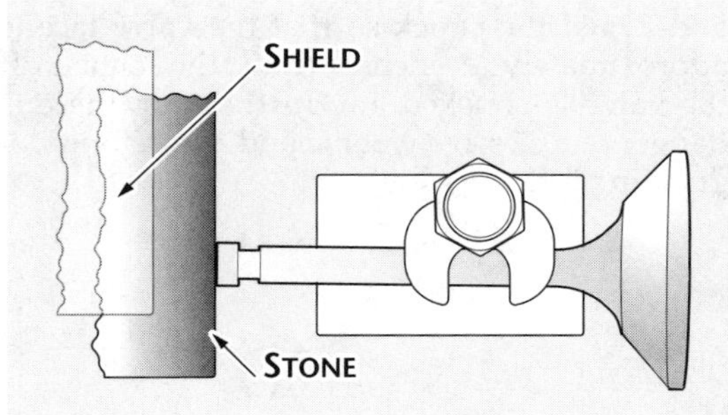


Figure 1-3-63. Grinding a valve tip

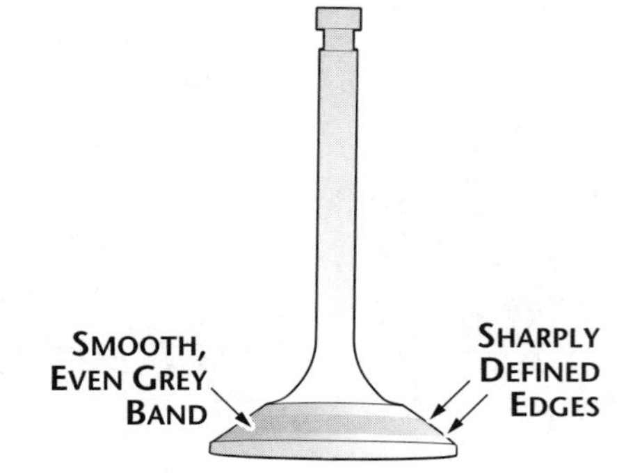


Figure 1-3-64. A correctly lapped valve

wiping the mating surfaces of both the valves and the seats with a dry cloth. Insert both the intake and exhaust valves in their respective guides and let them fall gently on their seats. Then pour enough Stoddard solvent into the cylinder to completely cover the valves. Watch the valve port for any signs of leakage. If any signs are detected, the valve should be re-lapped and the leak testing process repeated.

Incorrect examples of valve face surface appearance are of value in diagnosing improper valve and valve seat grinding. Incorrect indications, their cause and remedy, are shown in Figure 1-3-65.

## Piston Repairs

Piston repairs are not required as often as cylinder repairs since most of the wear is between the piston ring and cylinder wall, valve stem and guide and valve face and seat. A lesser amount of wear is encountered between the piston skirt and cylinder, ring and ring groove or piston pin and bosses.

The most common repair will be the removal of scores. Usually these may be removed only on the piston skirt if they are very light. Scores above the top ring groove may be machined or sanded out, as long as the diameter of the piston is not reduced below the specified minimum. To remove these scores, set the piston on a lathe. With the piston revolving at a slow speed, smooth out the scores with number 320 wet or dry sandpaper. Never use anything rougher than crocus cloth on the piston skirt.

On engines where the entire rotating and reciprocating assembly is balanced, the pistons must weigh within one-fourth ounce of each other. When a new piston is installed, it must be within the same weight tolerance as the one removed. It is not enough to have the pistons matched alone; they must be matched to the crankshaft, connecting rods, piston pins, etc. To make weight adjustments on new pistons, the manufacturer provides a heavy section at the base of the skirt. To decrease weight, file metal evenly off the inside of this heavy section. The piston weight can be decreased easily, but welding, metalizing or plating cannot be done to increase the piston weight.

If ring grooves are worn or stepped, they will have to be machined oversize so that they can accommodate an oversize width ring with the proper clearance. After machining, check to be sure that the small radius is maintained at the back of each ring groove. If it is removed, cracks may occur due to localization of stress. Ring groove oversizes are usually 0.005 inch, 0.010 inch or 0.020 inch. More than that would weaken the ring lands.

A few manufacturers sell 0.005-inch oversize piston pins. When these are available, it is permissible to bore or ream the piston-pin bosses to 0.005 inch oversize. However, these bosses must be in perfect alignment.

Small nicks on the edge of the piston-pin boss may be sanded down. Deep scores inside the boss or anywhere around the boss are definite reasons for rejection.

## Cylinder Grinding and Honing

If a cylinder has excessive taper, out-of-roundness, step or its maximum size is beyond limits, it can be re-ground to the next allowable oversize. If the cylinder walls are lightly rusted, scored or pitted, the damage may be removed by honing or lapping.

Regrinding a cylinder is a specialized job that the powerplant technician usually is not

| | INCORRECTLY LAPPED VALVES | | |
|---|---|---|---|
| INDICATION | 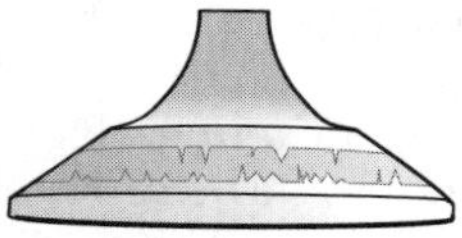 Fuzzy edge | 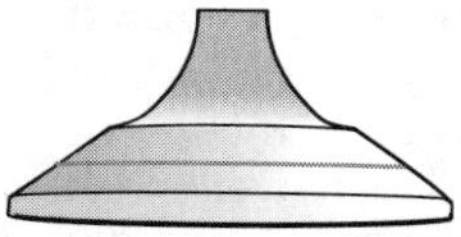 Too narrow contact | 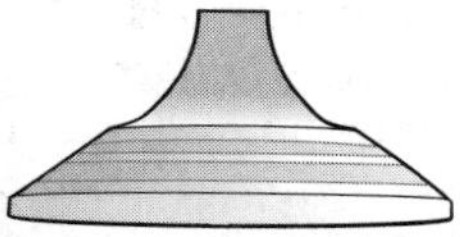 Two lap bands |
| CAUSE | Rough grind | Unintentional interference fit | Improper narrowing of seat |
| REMEDY | Regrind valve or continue lapping | Grind both valve and seat | Renarrow seat |

**Figure 1-3-65. Incorrectly lapped valves**

**Figure 1-3-66. A large cylinder bore grinding machine begins grinding on a new cylinder. Except for mounting, used cylinders are bored in the same manner.**

*Photo courtesy of Engine Components, Inc.*

expected to be able to do. However, the technician must be able to recognize when a cylinder needs re-grinding, and must know what constitutes a good or bad job.

Generally, standard aircraft cylinder oversizes are 0.010-inch, 0.015-inch, 0.020-inch or 0.030-inch. Unlike automobile engines that may be re-bored to oversizes of 0.075-inch to 0.100-inch, aircraft cylinders have relatively thin walls and may have a nitrided surface, which must not be ground away. Any one manufacturer usually does not allow all of the above oversizes. Some manufacturers do not allow re-grinding to an oversize at all. The manufacturer's overhaul manual or parts catalog usually lists the oversizes allowed for a particular make and model engine.

To determine the re-grind size, the standard bore size must be known. This usually can be determined from the manufacturer's specifications or manuals. The re-grind size is figured from the standard bore. For example, a certain cylinder has a standard bore of 3.875 inches. To have a cylinder ground to 0.015 inch oversize, it is necessary to grind to a bore diameter of 3.890 inches (3.875 + 0.015). A tolerance of ±0.0005 inch is usually accepted for cylinder grinding.

Another factor to consider when determining the size to which a cylinder must be re-ground is the maximum wear that has occurred. If there are spots in the cylinder wall that are worn larger than the first oversize, then obviously it will be necessary to grind to the next oversize to clean up the entire cylinder.

An important consideration when ordering a re-grind is the type of finish desired in the cylinder. Some engine manufacturers specify a fairly rough finish on the cylinder walls, which will allow the rings to seat even if they are not lapped to the cylinder. Other manufacturers desire a smooth finish to which a lapped ring will seat without much change in ring or cylinder dimensions. The latter type of finish is more expensive to produce.

The standard used when measuring the finish of a cylinder wall is known as microinch root-mean-square, or microinch r.m.s. In a finish where the depth of the grinding scratches are one-millionth (0.000001) of an inch deep, it is specified as 1 microinch r.m.s. Most aircraft cylinders are ground to a finish of 15 -20 microinch r.m.s. Several low-powered engines have cylinders that are ground to a relatively rough 20- to 30-microinch r.m.s. finish. On the other end of the scale, some manufacturers require a superfinish of approximately 4- to 6-microinch r.m.s.

Cylinder grinding (Figure 1-3-66) is accomplished by a firmly mounted stone that revolves around the cylinder bore, as well as up and down the length of the cylinder barrel. Either the cylinder, the stone or both may move to get this relative movement. The size of the grind is determined by the distance the stone is set away from the center line of the cylinder. Some cylinder bore grinding machines will produce a perfectly straight bore, while others are designed to grind a *choked bore*. A choked bore grind refers to the manufacturing process in which the cylinder walls are ground to produce a smaller internal diameter at the top than at the bottom. The purpose of this type grind or taper is to maintain a straight cylinder wall during operation. As a cylinder heats up during operation, the head and top of the cylinder are subjected to more heat than the bottom. This causes greater expansion at the top than at the bottom, thereby maintaining the desired straight wall.

After grinding a cylinder, it may be necessary to hone the cylinder bore to produce the desired finish. If this is the case, specify the cylinder re-grind size to allow for some metal removal during honing. The usual allowance for honing is 0.001 inch. If a final cylinder bore size of 3.890 inches is desired, specify the re-grind size of 3.889 inches, and then hone to 3.890 inches.

There are several different makes and models of cylinder hones. The burnishing hone is used only to produce the desired finish on the cylinder wall. The more elaborate micromatic hone can also be used to straighten out the cylinder wall. A burnishing hone should not be used in an attempt to straighten cylinder walls. Since the stones are only spring loaded, they will follow the contour of the cylinder wall and may aggravate a tapered condition. Figure 1-3-67 shows the desired honed pattern on a nickel process cylinder.

Another type of hone is the *brush hone,* illustrated in Figure 1-3-68. A brush hone looks like a giant bottle brush with diamond-impregnated bristles. They are designed to produce a crosshatched finish only — they will not resize a cylinder evenly and should not be used for that purpose.

**Figure 1-3-67. A cylinder honed with the desired crosshatch pattern**

*Photo courtesy of Engine Components, Inc.*

After the cylinders have been re-ground, check the size and wall finish and check for evidence of overheating or grinding cracks before installing on an engine.

## Crankshaft Inspection

The crankshaft should always be carefully inspected at overhaul. The crankshaft should be checked for cracks with either magnetic partial inspection or radiography. If the crankshaft is equipped with oil transfer tubes, they should be checked for tightness and security.

**Figure 1-3-68. Brush honing**

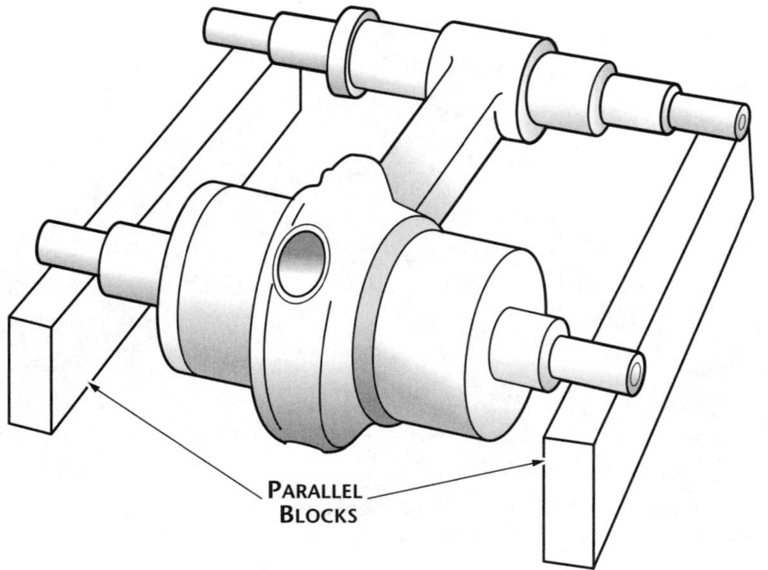


**Figure 1-3-69. Checking connecting rod squareness**

The crankshaft should be checked for straightness by the use of V-blocks and a dial indicator. If the crankshaft is found to be bent, it should be replaced.

Small scratches and scores may be removed from the crankshaft by polishing. If more extensive damage is found on the crankshaft, it should be sent to a certified repair station.

**Sludge chambers.** Some crankshafts are manufactured with hollow crankpins that serve as sludge removers. The sludge chambers may be formed by means of spool-shaped tubes pressed into the hollow crankpins or by plugs pressed into each end of the crankpin.

The sludge chamber or tubes must be removed for cleaning at overhaul. If these are not removed, accumulated sludge loosened during cleaning may clog the crankshaft oil passages and cause subsequent bearing failures. If the sludge chambers are formed by means of tubes pressed into the hollow crankpins, make certain they are re-installed correctly to avoid covering the ends of the oil passages.

After the sludge chambers have been cleaned and reinstalled, they should be checked for obstruction by blowing Stoddard solvent through the chambers with a small amount of compressed air.

## Connecting Rods

The inspection and repair of connecting rods include:

- Visual inspection
- Checking of alignment
- Re-bushing
- Replacement of bearings

Some manufacturers also specify a magnetic particle inspection of connecting rods.

**Visual inspection.** Visual inspection should be done with the aid of a magnifying glass or bench microscope. A rod which is obviously bent or twisted should be rejected without further inspection.

Inspect all surfaces of the connecting rods for cracks, corrosion, pitting, galling or other damage. Galling is caused by a slight amount of movement between the surfaces of the bearing insert and the connecting rod during periods of high loading, such as that produced during overspeed or excessive manifold pressure operation. The visual evidence produced appears as if particles from one contacting surface welded to the other. Evidence of any galling is sufficient reason for rejecting the complete rod assembly. Galling is a distortion in the metal and is comparable to corrosion in the way it weakens the metallic structure of the connecting rod.

To aid in inspection of the connecting rods for cracks magnetic partial inspection may also be used on some rods.

**Checking alignment.** Check bushings that have been replaced to determine if the bushing and rod bores are square and parallel to each other. The alignment of a connecting rod can be checked several ways. One method requires a push fit arbor for each end of the connecting rod, a surface plate and two parallel blocks of equal height.

To measure rod squareness (Figure 1-3-69) or twist, insert the arbors into the rod bores. Place the parallel blocks on a surface plate. Place the ends of the arbors on the parallel blocks. Check the clearance at the points where the arbors rest on the blocks, using a thickness gauge. This clearance, divided by the separation of the blocks in inches, will give the twist per inch of length.

To determine bushing or bearing parallelism (convergence), insert the arbors in the rod bores. Measure the distance between the arbors on each side of the connecting rod at points that are equidistant from the rod centerline. For exact parallelism, the distances checked on both sides should be the same. Consult the manufacturer's table of limits for the amount of misalignment permitted.

The preceding operations are typical of those used for most reciprocating engines and are included to introduce some of the operations involved in engine overhaul. It would be impractical to list all the steps involved in the overhaul of an engine. It should be understood that there are other operations and inspections that must be performed. For exact information regarding a specific engine model, consult the manufacturer's overhaul manual.

## Block Testing of Reciprocating Engines

According to FAR Part 33 Appendix A, the engine manufacturer must provide block testing instruction in the engine overhaul manual. Those are the instructions you must follow. The information presented here on block testing of engines is intended to familiarize you with the procedures and equipment. It cannot be construed to be the manufacturers directions.

Any new or recently overhauled aircraft engine must be in top mechanical condition. This condition must be determined after the engine has been newly assembled or completely overhauled. The method used is the *block test*, or *run-in*, which takes place at overhaul prior to delivery of the engine. Engine run-in is as vital as any other phase of engine overhaul, for it is the means by which the quality of a new or newly overhauled engine is checked, and it is the final step in the preparation of an engine for service.

Many times, an engine appears to be in perfect mechanical condition before the run-in tests, but tests show that it was actually in poor and unreliable mechanical condition. Thus, the reliability and potential service life of an engine is in question until it has satisfactorily passed the block test.

**Block test purpose.** The block test serves a dual purpose: first, it accomplishes piston ring run-in and bearing burnishing; and second, it provides valuable information that is used to evaluate engine performance and determine engine condition. To provide proper oil flow to the upper portion of the cylinder barrel walls with a minimum loss of oil, it is important that piston rings be properly seated in the cylinder in which they are installed. The process is called piston ring run-in and is accomplished chiefly by controlled operation of the engine in the high-speed range. Improper piston ring conditioning or run-in may result in unsatisfactory engine operation. A process called bearing burnishing creates a highly polished surface on new bearings and bushings installed during overhaul. The burnishing is usually accomplished during the first periods of the engine run-in at comparatively slow engine speeds.

**Testing of reciprocating engines.** Final testing of a newly overhauled reciprocating engine may be accomplished in a fixed test cell (block testing) or using a mobile test stand (Figure 1-3-70). They both have the same purpose: i.e., to ensure that the engine is fit to be installed on

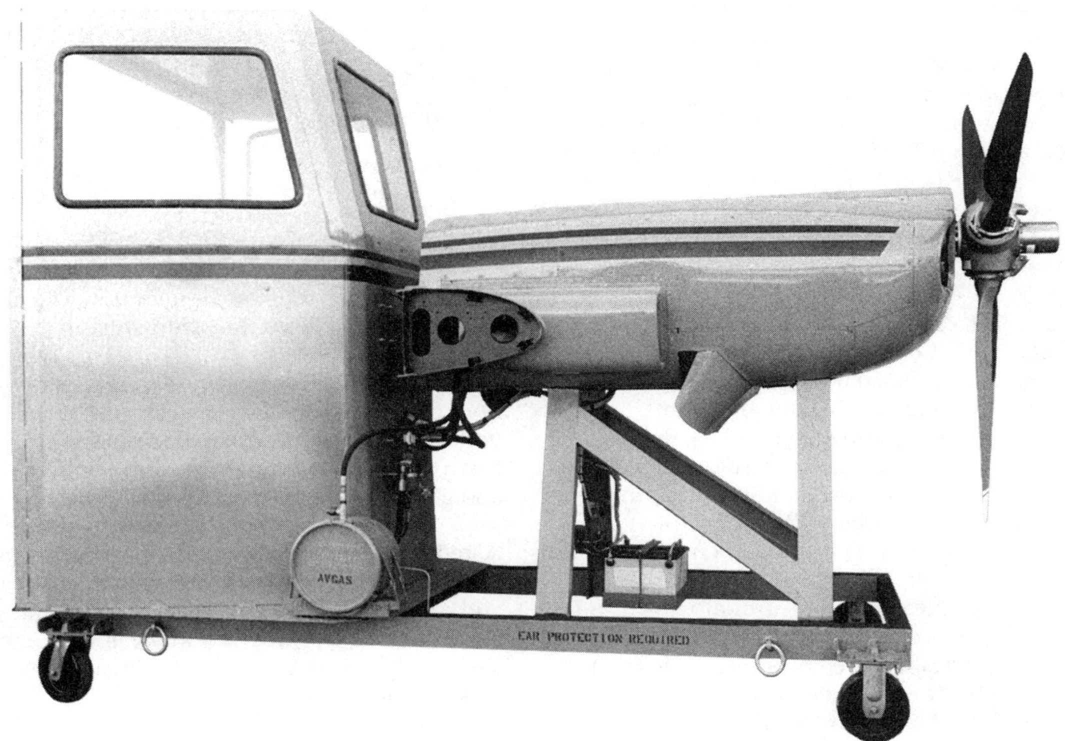


Figure 1-3-70. A GTSIO-520 engine in a mobile run-up stand

an aircraft. Once the engine has been operated on the test stand and any faults or troubles corrected, it is presumed that the engine will operate correctly on the aircraft.

The most important thing about positioning a mobile test stand is to face the propeller directly into the wind. If this is not done, engine testing will not be accurate.

**Block test instruments.** The block-test operator's control room houses the controls used to operate the engine and the instruments used to measure various temperatures and pressures, fuel flow and other factors. These devices are necessary in providing an accurate check and an evaluation of the operating engine. The control room is separate from, but adjacent to, the space (test cell) that houses the engine being tested.

The safe, economical and reliable testing of modern aircraft engines depends largely upon the use of instruments. In engine run-in procedures, the same basic engine instruments are used as when the engine is installed in the aircraft, plus some additional connections to these instruments and some indicating and measuring devices that cannot be practically installed in the aircraft. Instruments used in the testing procedures are inspected and calibrated periodically, as are instruments installed in the aircraft; thus, accurate information concerning engine operation is ensured.

**Test requirements.** The operational tests and test procedures vary with individual engines. The required tests and procedures will be detailed in the engine manufacturer's overhaul manual. The failure of any internal part during engine run-in requires that the engine be returned for replacement of the necessary units and then be completely re-tested. If any component part of the basic engine should fail, a new unit is installed; a minimum operating time is used to check the engine with the new unit installed.

After an engine has successfully completed block-test requirements, it may then be specially treated to prevent corrosion. This is normally accomplished if the engine is not going to be immediately installed on an aircraft. During the final run-in period at block test, the engines are operated on the proper grade of fuel prescribed for the particular kind of engine. The oil system is serviced with a mixture of corrosion-preventive compound and engine oil. The temperature of this mixture is maintained at 105°-121°C. Near the end of final run-in, *CPM (corrosion-preventive mixture)* is used as the engine lubricant; the engine induction passages and combustion chambers are also treated with CPM by an aspiration method (CPM is drawn or breathed into the engine).

## Section 4

# Engine Instrument Systems

The safe, economical and reliable operation of modern aircraft engines depends largely upon the use of instruments. In engine run-in procedures, the same basic engine instruments are used as when the engine is installed in the aircraft, plus some additional connections to these instruments and some indicating and measuring devices that cannot be practically installed in the aircraft. Instruments used in the testing procedures are inspected and calibrated periodically, as are instruments installed in the aircraft; thus, accurate information concerning engine operation is ensured.

Engine instruments are operated in several different fashions: some mechanically, some electrically and some by the pressure of a liquid. This chapter will not discuss how they operate, but rather the information they give, their common names and the markings on them. Newer style electronic instrumentation and displays will be covered in another chapter. The instruments to be covered here are:

- Carburetor air temperature gauge
- Fuel pressure gauge
- Fuel flowmeter
- Manifold pressure gauge
- Oil temperature gauge
- Oil pressure gauge
- Tachometer
- Cylinder head temperature gauge
- Torquemeter
- Suction gauge

Instrument markings and the interpretation of these markings will be discussed before considering the individual instruments.

Instrument markings indicate ranges of operation or minimum and maximum limits, or both. Generally, the instrument marking system consists of four colors (red, yellow, blue and green) and intermediate blank spaces.

A red line or mark indicates a point beyond which a dangerous operating condition exists, and a red arc indicates a dangerous operating range. Of the two, the red mark is used more commonly and is located radially on the cover glass or dial face.

The yellow arc covers a given range of operation and is an indication of caution. Generally, the yellow arc is located on the outer circumference of the instrument cover glass or dial face.

The blue arc, like the yellow, indicates a range of operation. The blue arc might indicate, for example, the manifold pressure gauge range in which the engine can be operated with the carburetor control set at automatic lean. The blue arc is used only with certain engine instruments, such as the tachometer, manifold pressure, cylinder head temperature and torquemeter.

The green arc shows a normal range of operation. When used on certain engine instruments, however, it also means that the engine must be operated with an automatic rich carburetor setting when the pointer is in this range.

When the markings appear on the cover glass, a white line is used as an index mark, often called a slippage mark. The white radial mark indicates any movement between the cover glass and the case, a condition that would cause dislocation of the other range and limit markings.

The instruments illustrated in Figures 1-4-1 through 1-4-8 are range-marked.

Figure 1-4-1. Carburetor air temperature (CAT) gauge

## Carburetor Air Temperature Indicator

Measured at the carburetor entrance, carburetor air temperature (CAT) is regarded by many as an indication of induction system ice formation. Although it serves this purpose, it also provides many other important items of information.

The powerplant is a heat machine, and the temperature of its components or the fluids flowing through it affects the combustion process either directly or indirectly. The temperature level of the induction air affects not only the charge density but also the vaporization of the fuel.

In addition to the normal use of CAT, it will be found useful for checking induction system condition. Backfiring will be indicated as a momentary rise on the gauge, provided it is of sufficient severity for the heat to be sensed at the carburetor air-measuring point. A sustained induction system fire will show a continuous increase of carburetor air temperature.

The CAT should be noted before starting and just after shutdown. The temperature before starting is the best indication of the temperature of the fuel in the carburetor body and tells whether vaporization will be sufficient for the initial firing or whether the mixture must be augmented by priming. If an engine has been shut down for only a short time, the residual heat in the carburetor may make it possible to rely on the vaporizing heat in the fuel and powerplant, and priming would then be unnecessary. After shutdown, a high CAT is a warning that the fuel trapped in the carburetor will expand, producing high internal pressure. When a high temperature is present at this time, the fuel line and manifold valves should be open so that the pressure can be relieved by allowing fuel passage back to the tank.

The carburetor air temperature gauge indicates the temperature of the air before it enters the carburetor. The temperature reading is sensed by a bulb. In the test cell, the bulb is located in the air intake passage to the engine. In an aircraft, it is located in the ram-air intake duct. Figure 1-4-1 shows a typical carburetor air temperature gauge, or CAT. This gauge, like many other multiengine aircraft instruments, is a dual gauge; that is, two gauges, each with a separate pointer and scale, are used in the same case. Notice the range markings used. The yellow arc indicates a range wherein lies the danger of icing. The green range indicates the normal operating range. The red line indicates the maximum operating temperature of 40° C — any operation at a temperature over this value places the engine in danger of detonation.

## Fuel Pressure Indicator

The fuel pressure gauge (Figure 1-4-2) is calibrated in pounds per square inch of pressure. It is used during the block test run-in to measure engine fuel pressure at the carburetor inlet, the fuel feed valve discharge nozzle and the main fuel supply line. Fuel gauges are located in the operator's control room and are connected by flexible lines to the different points at which pressure readings are desired during the testing procedures.

Figure 1-4-2. This fuel pressure gauge is a twin-needle unit, with ranges for different altitudes.

In some aircraft installations, the fuel pressure is sensed at the carburetor inlet of each engine, and the pressure is indicated on individual or dual-needle gauges (Figure 1-4-2) on the instrument panel. The dial is calibrated in 1 p.s.i. graduations. The numbers range from 0-20. The green arc shows the desired range of operation, which is 18 p.s.i. The red line at the 19 p.s.i. graduation indicates the maximum allowable fuel pressure. Fuel pressures vary with the type of carburetor installation and the size of the engine. In most reciprocating engines that use pressure injection carburetion, the fuel pressure range is the same as illustrated in Figure 1-4-2.

When float-type carburetors or low-pressure carburetion systems are used, the fuel pressure range is of a much lower value; the minimum allowable pressure is 3 p.s.i., and the maximum is 5 p.s.i., with the desired range of operation between 3-5 p.s.i.

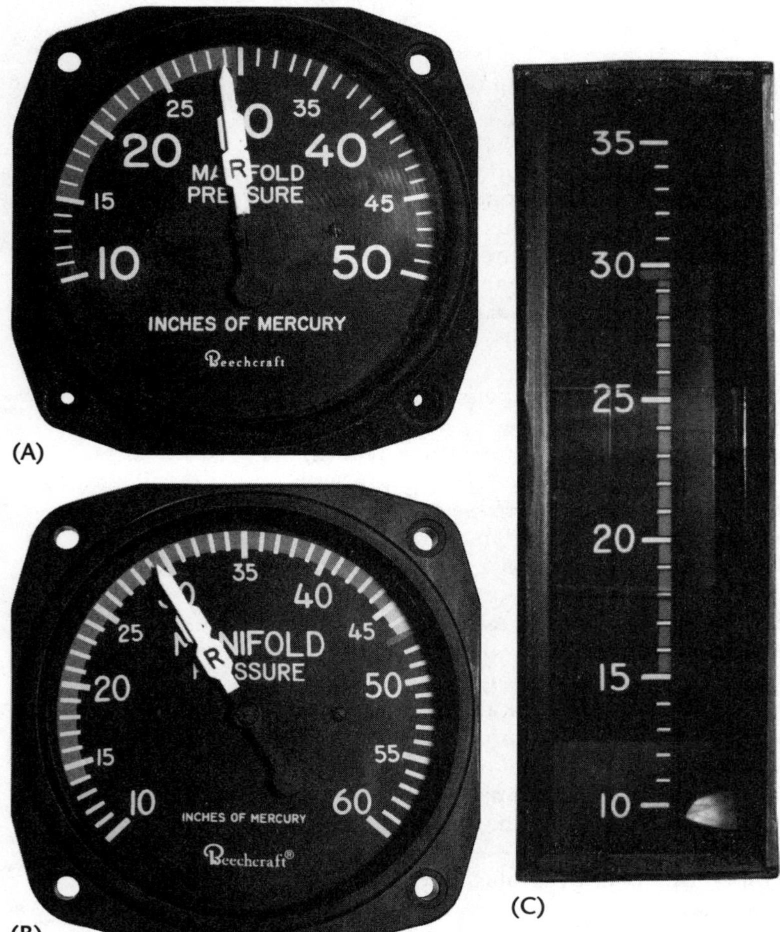


Figure 1-4-3. Manifold pressure gauges: (A) is normally used on a standard installation, while (B) would most likely be found on a supercharged engine. Gauge (C) is a vertical scale unit.

## Fuel Flowmeter

The fuel flowmeter measures the amount of fuel delivered to the carburetor. During engine block-test procedures, the fuel flow to the engine may be measured by a series of calibrated tubes located in the control room. The tubes are of various sizes to indicate different volumes of fuel flow. Each tube contains a float that can be seen by the operator, and as the fuel flow through the tube varies, the float is raised or lowered, indicating the amount of fuel flow. From these indications, the operator can determine whether an engine is operating at the correct fuel/air mixture for a given power setting.

In an aircraft installation, the fuel flow indicating system consists of a transmitter and an indicator for each engine. The fuel flow transmitter is conveniently mounted in the engine's accessory section, and measures the fuel flow between the engine-driven fuel pump and the carburetor. The transmitter is an electrical device that is connected electrically to the indicator located on the aircraft operator's panel. The reading on the indicator is calibrated to record the amount of fuel flow in pounds of fuel per hour. Some fuel flow systems are nothing more than a fuel pressure gauge calibrated in pounds per hour.

## Manifold Pressure Indicator

The preferred type of instrument for measuring the manifold pressure is a gauge that records the pressure as an absolute pressure reading (Figure 1-4-3). A mercury manometer, a tube calibrated in inches, is used during block-test procedures. It is partially filled with mercury and connected to the manifold pressure adapter located on the engine. Since it is impractical to install mercury manometers in an aircraft to record the manifold pressure of the engines, a specially designed manifold pressure gauge that indicates absolute manifold pressure in inches of mercury is used.

A blue arc represents the range within which operation with the mixture control in AUTOMATIC-LEAN (only on large radial engines) position is permissible. A green arc indicates the range within which the engine must be operated with the mixture control in the NORMAL or RICH position. The red arc indicates the maximum manifold pressure permissible during takeoff.

On installations where water injection is used, a second red line is located on the dial to indicate the maximum permissible manifold pressure for a wet takeoff.

Figure 1-4-4. Oil temperature gauge

Figure 1-4-5. Aircraft oil pressure gauges — (A) is described in the text.

## Oil Temperature Indicator

During engine run-in at block test, engine oil temperature readings are taken at the oil inlet and outlet. From these readings, it can be determined if the engine heat transferred to the oil is low, normal or excessive. This information is of extreme importance during the *breaking-in* process of large reciprocating engines. The oil temperature gauge line in the aircraft is connected at the oil inlet to the engine.

Three range markings are used on the oil temperature gauge. The yellow mark in Figure 1-4-4, at 40°C on the dial, shows the minimum oil temperature permissible for ground operational checks or during flight. The green mark between 40° and 110° shows the desired oil temperature for continuous engine operation. The red mark at 115°C indicates the maximum permissible oil temperature.

## Oil Pressure Indicator

The oil pressure on block-test engines is checked at various points. The main oil pressure reading is taken at the pressure side of the oil pump. Other pressure readings are taken from the nose section and blower section; and when internal supercharging is used, a reading is taken from the high- and low-blower clutch.

Generally, there is only one oil pressure gauge for each engine, and the connection is made at the pressure side (outlet) of the main oil pump.

The oil pressure gauge dial, marked as shown in Figure 1-4-5, does not show the pressure range or limits for all installations. The actual markings for specific aircraft may be found in the Aircraft Specifications or Type Certificate Data Sheets. The red line (at 40 p.s.i.) indicates the minimum oil pressure permissible in flight. The green arc between 65-85 p.s.i. shows the desired operating oil pressure range.

The oil pressure gauge indicates the pressure (in p.s.i.) that the oil of the lubricating system is being supplied to the moving parts of the engine. The engine should be shut down immediately if the gauge fails to register pressure when the engine is operating. Excessive oscillation of the gauge pointer indicates that there is air in the lines leading to the gauge or that some unit of the system is functioning improperly.

## Tachometer Indicator

The *tachometer* shows the engine crankshaft r.p.m. Figure 1-4-6, illustration A, shows a tachometer with range markings installed on the cover glass. The tachometer, often referred to as *tach,* is calibrated in hundreds with graduations at every 50-r.p.m. interval. The dial shown here starts at 0 and goes to 35 (3,500 r.p.m.).

The green arc indicates the operating r.p.m. range. The top of the green arc, 2,550 r.p.m., indicates maximum continuous power. All operation above this r.p.m. may be limited in time. The red line indicates the maximum r.p.m. permissible during takeoff; any r.p.m. beyond this value is an overspeed condition.

AC 43.13-1B contains standards for accuracy and inspection criteria for tachometer accuracy. On page 8-406, the FAA extablishes the need for tachometer accuracy and sets the error limit. Generally, the limit is ±2 percent. It the tachometer fails, the tach should be replaced.

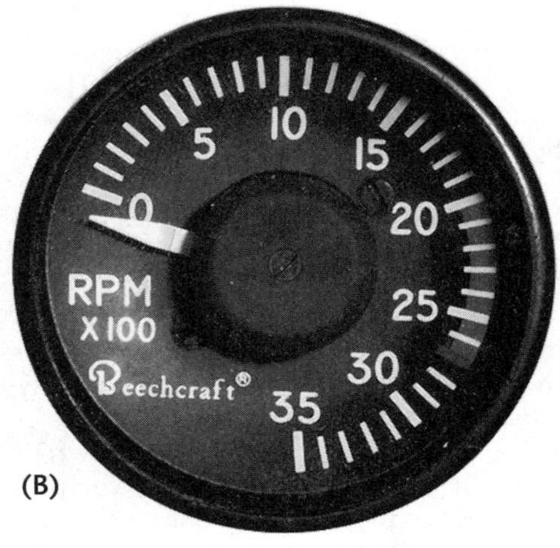


Figure 1-4-6. (A) is a mechanical tachometer with a built-in hour meter, while (B) is an electric unit that uses a tach generator.

## Cylinder Head Temperature Indicator

Cylinder head temperatures are indicated by a gauge connected to a thermocouple attached to the cylinder. Single-cylinder and multi-cylinder installations may be found. If a single thermocouple is used, it will be located on the cylinder determined by test to be the hottest. The thermocouple may be placed in a special gasket located under a spark plug or in a special well in the top or rear of the cylinder head.

The temperature recorded at either of these points is merely a reference or control temperature; as long as it is kept within the prescribed limits, the temperatures of the cylinder dome, exhaust valve and piston will be within a satisfactory range. When the thermocouple is attached to only one cylinder, it can do no more than give evidence of general engine temperature. While normally it can be assumed that the remaining cylinder temperatures will be lower, conditions such as detonation will not be indicated unless they occur in the cylinder that has the thermocouple attached.

The cylinder head temperature gauge range marking is similar to that of the manifold pressure and tachometer indicator. The cylinder head temperature gauge illustrated in Figure 1-4-7 is a dual gauge that incorporates two separate temperature scales. The scales are calibrated in increments of 20°F, with numerals at the 0°F, 200°F (2), 400°F (4) and 600°F (6) graduations.

The green arc describes the range within which operation is most desired. The top of this arc, 450°F, indicates maximum continuous power; all operation above this temperature may be limited in time. The red line indicates maximum permissible temperature, 550°F.

## Suction Gauge

The suction gauge is not classed as an engine instrument, since it does not indicate any information in determining efficient engine operation. The mechanic is concerned with it because in regards to adjusting the suction regulator and checking the suction gauge reading during the engine operational checks.

The suction gauge (Figure 1-4-8) is calibrated to indicate reduction of pressure below atmospheric pressure in inches of mercury; the space between the graduation lines represents 0.2 inches Hg. The range markings will be painted on when the gauge is installed. A red line at 3.75 inches Hg indicates minimum desirable suction. A green arc shows the desirable suction range, 3.75-4.25 inches Hg. A red line at 4.25 inches Hg indicates the maximum desirable suction.

## EGT Gauge

An *exhaust gas temperature gauge* (EGT) is principally an instrument designed for pilot use. Its purpose is to measure the temperature of the exhaust gas as it leaves the cylinder(s). From that reading, the fuel mixture can be leaned in-flight with no danger of overheating the engine. Some gauge systems use a multi-position switch with a single instrument. These are of the most value to a maintenance technician because they show the operating exhaust temperature of each cylinder in real time.

Figure 1-4-7. Cylinder head temperature gauge

*Section 5*

# *Lubrication Systems*

## Lubricating Oils

### *Oil Bases*

Vegetable oils. In early aviation, additives for mineral-based oils had not yet been invented. Many engines, particularly rotary radial engines, used a vegetable-based lubricant, mainly castor oil. While castor oil oxidized fairly readily and tended to gum up, it did not wash away with gasoline. Because rotary engines used a fuel-distribution system similar to a two-cycle (crankcase induction), castor oil worked for those engines. However, rapid improvement in mineral oil processing phased out its use.

**Mineral oils.** Mineral oils are conventional oils that have been used for decades, produced by crude oil distillation. Additives, like detergents, viscosity improvers, ash dispersants and antiwear agents are added to improve the usability.

**Synthetic oils.** Synthetic oils are completely man-made in laboratories by complex chemical processes, and they are more expensive. Synthetic oils have better thermal strength, meaning they are capable of maintaining their viscosity for longer periods of use and under much greater temperatures than conventional mineral oils.

## Principles of Engine Lubrication

The primary purpose of a lubricant is to reduce friction between moving parts. Because liquid lubricants (oils) can be circulated readily, they are used universally in aircraft engines.

In theory, fluid lubrication is based on the actual separation of the surfaces so that no metal-to-metal contact occurs. As long as the oil film remains unbroken, metallic friction is replaced by the internal fluid friction of the lubricant. Under ideal conditions, friction and wear are held to a minimum.

In addition to reducing friction, the oil film acts as a cushion between metal parts. This cushioning effect is particularly important for such parts as reciprocating engine crankshaft and connecting rods, which are subject to shock loading. As oil circulates through the engine, it absorbs heat from the parts. Pistons and cylinder walls in reciprocating engines are especially dependent on the oil for cooling. The oil also aids in forming a seal between the piston and the cylinder wall to prevent leakage of the gases from the combustion chamber. Oils also reduce abrasive wear by picking up foreign particles and carrying them to a filter, where they are removed.

Figure 1-4-8. Suction gauge

## *Requirements and Characteristics of Reciprocating Engine Lubricants*

While there are several important properties which a satisfactory reciprocating engine oil must possess, its viscosity is most important in engine operation. The resistance of an oil to flow is known as its viscosity. Oil which flows slowly is viscous or has a high viscosity. If it flows freely, it has a low viscosity.

Unfortunately, the viscosity of oil is affected by temperature. It is not uncommon for some grades of oil to become practically solid in cold weather. This increases drag and makes circulation almost impossible. Other oils may become so thin at high temperature that the oil film is broken, resulting in rapid wear of the moving parts. The viscosity index is the difference of the viscosity of the oil at two different temperatures. A low viscosity index would mean that a change in temperature would have a small effect on the pourability of the oil.

The oil selected for aircraft engine lubrication must be light enough to circulate freely, yet heavy enough to provide the proper oil film at engine operating temperatures. Since lubricants vary in properties and since no one oil is satisfactory for all engines and all operating conditions, it is extremely important that only the recommended grade be used.

Several factors must be considered in determining the proper grade of oil to use in a particular engine. The operating load, rotational speeds and operating temperatures are the most important. The operating conditions to be met in the various types of engines will determine the grade of the lubricating oil to be used.

The oil used in reciprocating engines has a relatively high viscosity because of:

- Large engine operating clearances due to the relatively large size of the moving parts, the different materials used and the different rates of expansion of the various materials
- High operating temperatures
- High bearing pressures

The following characteristics of lubricating oils measure their grade and suitability:

- *Flash point* and *fire point* are determined by laboratory tests that show the temperature at which a liquid will begin to give off ignitable vapors (flash) and the temperature at which there are sufficient vapors to support a flame (fire). These points are established for engine oils to determine that they can withstand the high temperatures encountered in an engine.
- *Cloud point* and *pour point* also help to indicate suitability. The cloud point of oil is the temperature at which its wax content, normally held in solution, begins to solidity and separate into tiny crystals, causing the oil to appear cloudy or hazy. The pour point of oil is the lowest temperature at which it will flow or can be poured.
- *Specific gravity* is a comparison of the weight of the substance to the weight of an equal volume of distilled water at a specified temperature. As an example, water weighs approximately 8 lbs. to the gallon; oil with a specific gravity of 0.9 would weigh 7.2 lbs. to the gallon.

Generally, commercial aviation oils are classified numerically, such as 80, 100, 140, etc., which are an approximation of their viscosity, as measured by a testing instrument called the *Saybolt Universal Viscosimeter.* In this instrument, a tube holds a specific quantity of the oil to be tested. The oil is brought to an exact temperature by a liquid bath surrounding the tube. The time in seconds required for exactly 60 cubic centimeters of oil to flow through an accurately calibrated orifice is recorded as a measure of the oil's viscosity.

If actual Saybolt values were used to designate the viscosity of oil, there probably would be several hundred grades of oil. To simplify the selection of oils, they often are classified under an SAE (Society of Automotive Engineers) system, which divides all oils into seven groups (SAE 10 to SAE 70) according to viscosity at either 130°F or 210°F.

SAE ratings are purely arbitrary and bear no direct relationship to the Saybolt or other ratings. The letter "W" occasionally is included in the SAE number giving a designation such as SAE 20W. This letter "W" indicates that the oil, in addition to meeting the viscosity requirements at the testing temperature specifications, is a satisfactory oil for winter use in cold climates. It does not stand for "weight," as in 40-weight oil.

Although the SAE scale has eliminated some confusion in the designation of lubricating oils, it must not be assumed that this specification covers all the important viscosity requirements. An SAE number indicates only the viscosity (grade) or relative viscosity; it does not indicate quality or other essential characteristics. It is well known that there are good oils and inferior oils that have the same viscosities at a given temperature and, therefore, are subject to classification in the same grade. The SAE letters on an oil container are not an endorsement or recommendation of the oil by the Society of Automotive Engineers.

Although each grade of oil is rated by an SAE number, depending on its specific use, it may be rated with a commercial aviation grade number or an Army and Navy specification number. The correlation between these grade-numbering systems is shown in Table 1-5-1.

| OIL GRADES BY NUMBER | | |
|---|---|---|
| Commercial Aviation Number | Commercial SAE Number | Army/Navy Specification Number |
| 65 | 30 | 1065 |
| 80 | 40 | 1080 |
| 100 | 50 | 1100 |
| 120 | 60 | 1120 |
| 140 | 70 | — |

**Table 1-5-1. Oil grades by SAE, commercial aviation and Army/Navy specification numbers**

## *Multi-viscosity Oil*

Airplane engines work in an environment that is more than a bit different than other engines. They are required to produce maximum power fairly soon after being started, hence the extra warm-up requirements. They tend to run consistently at about 65-percent power and above. Last, but not least, is their temperature operating environment. In the space of a few short minutes, airplanes can travel from desert heat to below-zero temperatures. All of these parameters add importance to the oil chosen for the lubricating system.

Viscosity is the most important property of an aviation oil. An oil with too low a viscosity can shear and lose film strength at high temperatures. Too high a viscosity, and the oil pump may not be able to push it through all the internal passages it needs to. Multi-grade oil is one answer to some of the problems presented by the changing environment of an operating aircraft.

Multi-viscosity oil works like this: Polymers are added to the base stock for the purpose of modifying the oil's viscosity by its temperature. At cold temperatures, the polymers are coiled up and allow the oil to flow as the low number on its identifier indicates. As the oil warms up, the polymers unwind into long chains that prevent the oil from thinning as much as it normally would. The result is that at 100°C, the oil has only thinned as much as the higher viscosity number indicates. Another way of looking at multi-viscosity oils is to think of a 20W50 as a 20W oil that will not thin more than a 50W oil would when hot.

Multi-grade oils are a major improvement in many operating environments.

## *Ash-dispersant Oil*

The lower the flash point of an oil, the greater its tendency to burn off on hot cylinder walls and pistons. Sulfated ash is the material left when oil burns internally. A high ash content will tend to form more sludge and deposits inside an engine. Low ash content tends to promote longer valve life.

The ash is removed from the inside of an engine by putting a dispersant additive in the oil. The dispersant allows the ash to stay in suspension and circulate with the oil through the system. Because of the relatively high rates of oil flow needed for air cooling, the oil is all circulated fairly rapidly. As it passes through the oil filter, the ash content is removed and thus is not allowed to settle in the engine crankcase or oil tank.

## *Synthetic Oils*

Synthetic oils offer the only truly significant differences in lubrication. This is due to their superior high-temperature oxidation resistance, high film strength, very low tendency to form deposits, stable viscosity base and low-temperature flow characteristics. Synthetics are generally superior lubricants compared to traditional mineral lubricants. Synthetics do not react to combustion and combustion by-products to the same extent that petroleum does.

These oils are not, however, without fault. Because they cannot suspend ash like a mineral oil, synthetics tend to build up more sludge. Rubber and plastic products may tend to soften, which can raise havoc with fuel systems. Some of them can even blister paint, if left exposed for any time.

Synthetic oils do not have an SAE rating. Viscosity is measured by a system called the *Kinematic Rating* in *centistokes* (cST). Regardless, a technician must not choose which oil to put in an engine but must use the oil specified, and it generally has either a MIL SPEC number, or a manufacturer's number to identify it.

> **CAUTION:** *Because of the vast differences in composition, mineral-based oil and synthetic oils may not be mixed.*

## *Dry-sump Systems*

Many reciprocating aircraft engines have pressure dry-sump lubrication systems. The oil supply in this type of system is carried in a tank. A pressure pump circulates the oil through the engine; scavenger pumps then return it to the

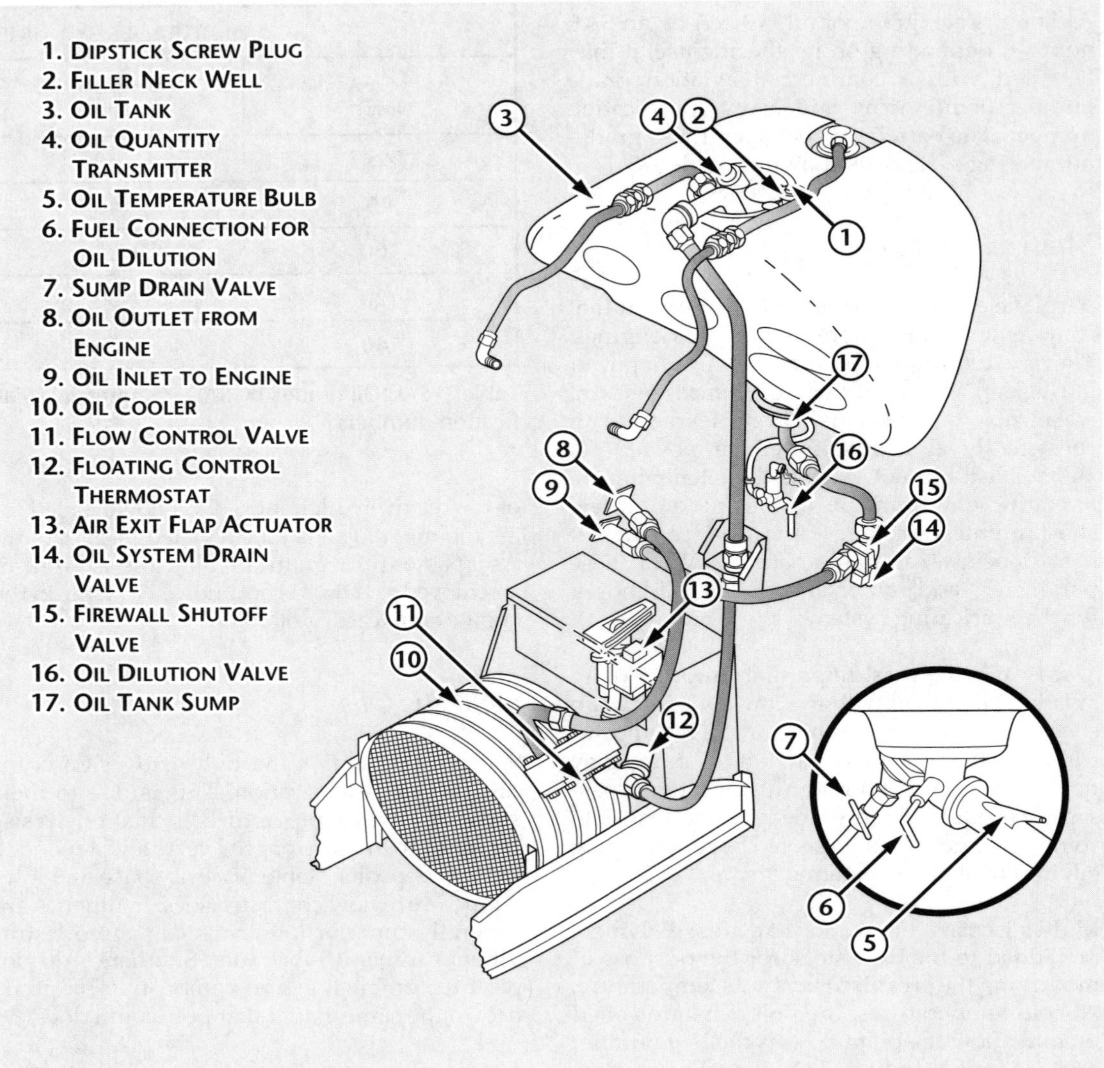


**Figure 1-5-1. Typical lubrication system**

tank as quickly as it accumulates in the engine sumps. The need for a separate supply tank is apparent when considering the complications that would result if large quantities of oil were carried in the engine crankcase. On multi-engine aircraft, each engine is supplied with oil from its own complete and independent system.

Although the arrangement of the oil systems in different aircraft varies widely and the units of which they are composed differ in construction details, the functions of all such systems are the same. A study of one system will clarify the general operation and maintenance requirements of other systems.

The principal units in a typical reciprocating engine dry-sump oil system include an oil supply tank, an engine oil pump, an oil cooler, an oil control valve, an actuator for an oil cooler air-exit control, a firewall shutoff valve, the necessary tubing and quantity, pressure and temperature indicators. Most of these units are shown in Figure 1-5-1.

## *Oil Tanks*

Oil tanks generally are constructed of aluminum alloy. The oil tank usually is placed close to the engine and high enough above the oil pump inlet to ensure gravity feed. Oil tank capacity varies with the different types of aircraft, but generally it is sufficient to ensure an adequate supply of oil for the total fuel supply. The tank filler neck is positioned to provide sufficient room for oil expansion and for foam to collect. The filler cap or cover is marked with the word OIL and the tank capacity. A drain in the filler cap well disposes of any overflow caused by the filling operation. Oil tank vent lines are provided to ensure proper tank ventilation in all attitudes of flight. These lines usually are connected to the engine crankcase to prevent the loss of oil through the vents. This indirectly vents the tanks to the atmosphere through the crankcase breather.

Some oil tanks have a built-in hopper (Figure 1-5-2), or temperature accelerating well, that

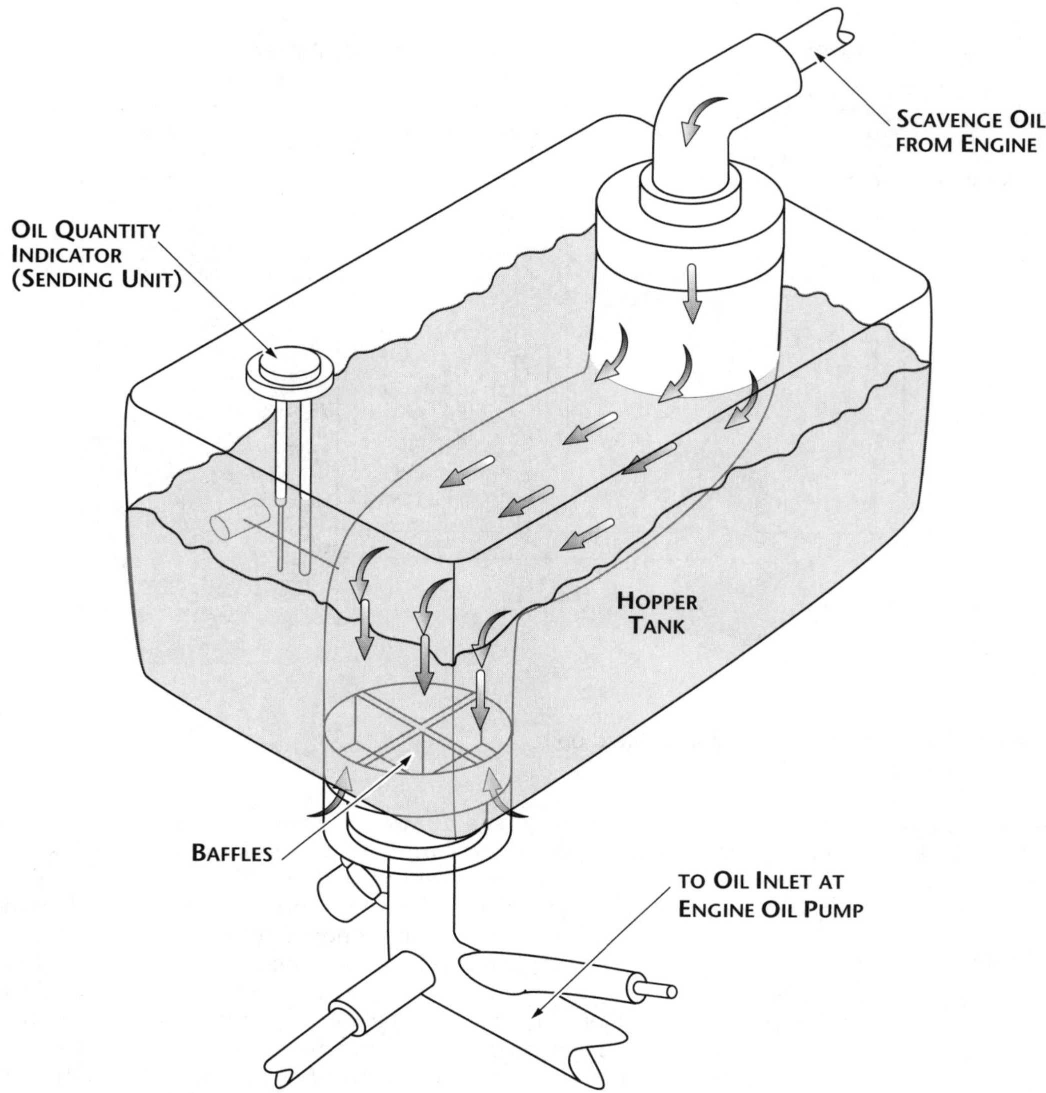


**Figure 1-5-2. Oil tank with hopper**

extends from the oil return fitting on top of the oil tank to the outlet fitting in the sump in the bottom of the tank. In some systems, the hopper tank is open to the main oil supply at the lower end; other systems have flapper-type valves that separate the main oil supply from the oil in the hopper.

The opening at the bottom of the hopper in one type and the flapper-valve-controlled openings in the other allow oil from the main tank to enter the hopper and replace the oil consumed by the engine. Whenever the hopper tank includes the flapper-valve-controlled openings, the valves are operated by differential oil pressure.

By separating the circulating oil from the surrounding oil in the tank, less oil is circulated. This hastens the warming of the oil when the engine is started. A hopper tank also makes oil dilution practical because only a relatively small volume of oil will have to be diluted. When it is necessary to dilute the oil, gasoline is added at some point in the inlet oil line to the engine, where it mixes with the circulating oil.

The return line in the top of the tank is positioned to discharge the returned oil against the wall of the hopper in a swirling motion. This method considerably reduces foaming. Baffles in the bottom of the hopper tank break up this swirling action to prevent air from being drawn into the line to the oil pressure pumps. In the case of oil-controlled propellers, the main outlet from the hopper tank may be in the form of a standpipe so that there will always be a reserve supply of oil for propeller feathering in case of engine failure. An oil tank sump, attached to the undersurface of the tank, acts as a trap for moisture and sediment (Figure 1-5-1). The water and sludge can be drained by manually opening the drain valve in the bottom of the sump.

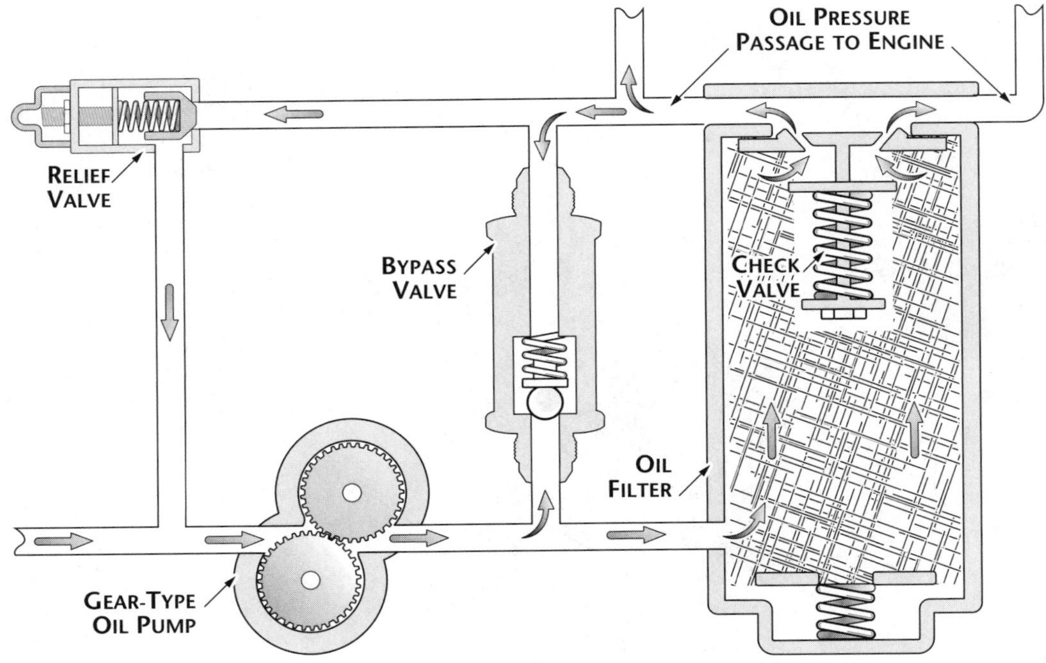


**Figure 1-5-3. Engine oil pump and associated units**

Most aircraft oil systems are equipped with the dipstick-type quantity gauge, often called a bayonet gauge.

## *Oil Pump*

Oil entering the engine is pressurized, filtered and regulated by units within the engine. They will be discussed, along with the external oil system, to provide a concept of the complete oil system.

As oil enters the engine (Figure 1-5-3), it is pressurized by a gear-type pump. This pump is a positive displacement pump that consists of two meshed gears that revolve inside a housing. The clearance between the teeth and housing is small. The pump inlet is located on the left, and the discharge port is connected to the engine's system pressure line. One gear is attached to a splined drive shaft that extends from the pump housing to an accessory drive shaft on the engine. Seals are used to prevent leakage around the drive shaft. As the lower gear is rotated counterclockwise, the driven (idler) gear turns clockwise.

As oil enters the gear chamber, it is picked up by the gear teeth, trapped between them and the sides of the gear chamber and is carried around the outside of the gears and discharged from the pressure port into the oil screen passage. The pressurized oil flows to the oil filter, where any solid particles suspended in the oil are separated from it, preventing possible damage to moving parts of the engine. Oil under pressure then opens the oil filter check valve mounted in the top of the filter. This valve is closed by a light spring loading of 1-3 lbs. when the engine is not operating to prevent gravity-fed oil from entering the engine and settling in the lower cylinders of radial engines. If oil were permitted to lie in the lower cylinders, it would gradually seep by the rings of the piston and fill the combustion chamber, contributing to a possible liquid lock.

The bypass valve, located between the pressure side of the oil pump and the oil filter, permits unfiltered oil to bypass the filter and enter the engine when the oil filter is clogged or during a cold engine start. The spring loading on the bypass valve allows the valve to open before the oil pressure collapses the filter or, in the case of cold, congealed oil, it provides a low-resistance path around the filter. It is felt that dirty oil in an engine is better than no lubrication at all.

## *Oil Filters*

The oil filters used on aircraft engines are usually one of four types:

1. Screen
2. Cuno
3. Air-Maze®
4. Automotive style filter cans

A *screen-type filter* (Figure 1-5-4) with its double-walled construction provides a large filtering area in a compact unit. As oil passes through the fine-mesh screen, dirt, sediment and other foreign matter are removed and settle to the bottom of the housing. At regular intervals, the cover is removed and the screen and housing cleaned with a solvent.

The *Cuno* or *edge filtration oil filter* has a cartridge made of disks and spacers. A cleaner blade fits between each pair of disks. When the cartridge is rotated, the cleaner blades comb the foreign matter from the disks. The cartridge of the manually operated Cuno filter is turned by an external handle. Cuno filters are old technology and only used on older airplanes that have large radial engines.

The *Air-Maze® filter* contains a series of round, fine-meshed screens mounted on a hollow shaft. The oil from the pump enters the well, surrounds the screens and then passes through them and the shaft before entering the engine. The carbon deposits that collect on the screens actually improve their filtering efficiency.

*Automotive-style filter cans* (Figure 1-5-5) have been the filter of choice for several years. They are manufactured specifically for aircraft engines and are not replaceable with any automotive equivalent. They are a full-filtration filter, normally with a built-in method for partial bypass should the element become clogged.

At each 100-hour inspection, the filters are to be replaced and cut open with a special cutter. The contents are then to be examined for any metal or foreign material. There can be quite a bit of foreign material in a filter, especially if the engine is new or freshly overhauled. Little pieces of gasket material, clumps of sealant and anything else small that made its way into the crankcase during the assembly process will show up in the filter.

Carbon chunks will be black and generally soft. They can generally be flattened between two fingernails.

## Oil Pressure-relief Valve

In all pressure oil lubrication systems, a pressure relief valve must be used to limit the oil pressure to a predetermined value. This setting will be predetermined by the engine manufacturer. The oil pressure must be regulated to allow for proper lubrication, but it must not be allowed to be too great because oil leakage or damage to the oil system may occur. Most aircraft reciprocating engines use a spring-loaded pressure relief valve. As the oil pressure reaches the predetermined setting the oil pressure pushes against the spring, opening the valve and relieving the

**Figure 1-5-4. A screen-type filter**

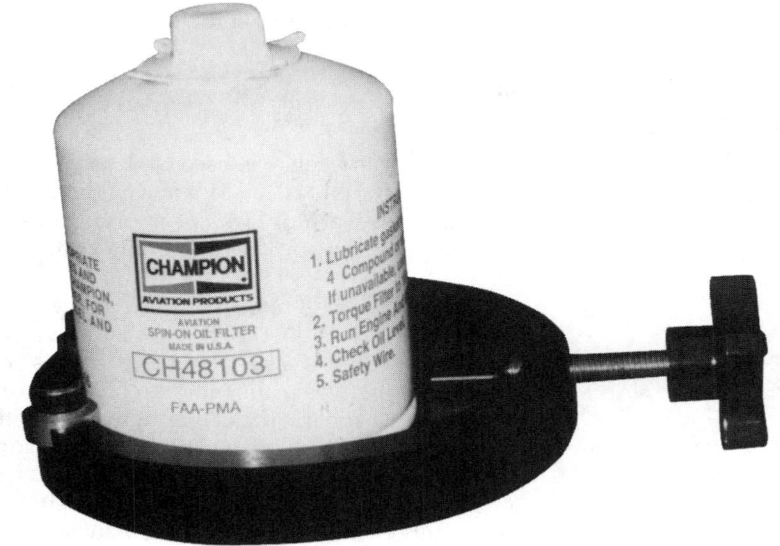


**Figure 1-5-5. An automotive-style filter and the special cutter necessary to cut it open**

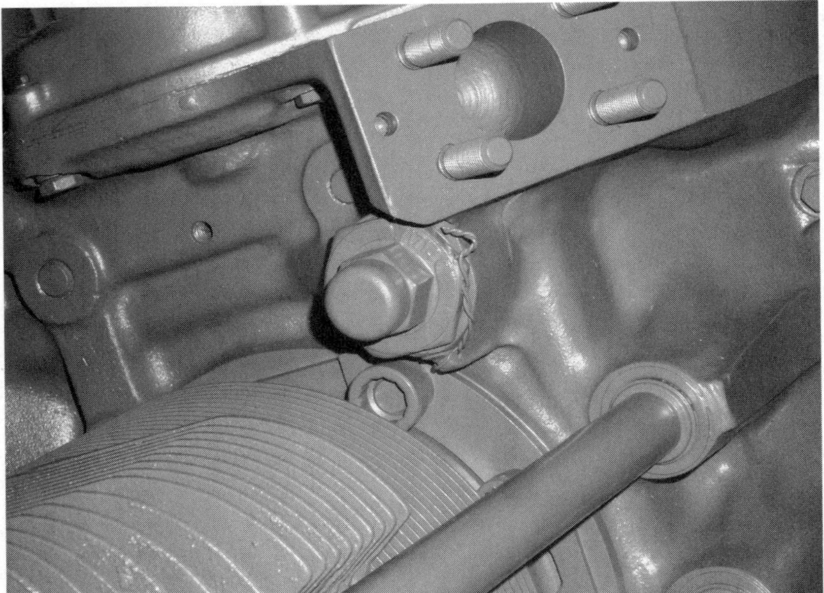

**Figure 1-5-6. Oil pressure relief-valve on the accessory section of a crankcase**

**Figure 1-5-7. Oil temperature regulator**

pressure (see Figure 1-5-6). The pressure relief valve is adjustable. This is done by either adding or removing washers under the spring or by the use of an adjusting screw. When adjusting engine oil pressure the technician should follow the manufacturer's instructions to determine the procedure and the exact setting.

## *Oil Pressure Gauge*

The oil pressure gauge is used to determine the pressure of the oil system. Monitoring the system pressure can warn of leaks or malfunctions in the oil system, or worn bearings in the engine.

One common type of oil pressure gauge used in small aircraft is the Bourdon-tube type. This type of oil pressure gauge measures the difference between the engine oil pressure and the atmospheric pressure in the cabin. The operation of the Bourdon-tube oil pressure gauge differs from other Bourdon-tube type of instruments only by the use of a restrictor in the tube. This restrictor is used to absorb the pulsations in pressure caused by the oil pump. Most oil pressure gauges have range markings between 0 p.s.i. and 200-300 p.s.i.

A dual-type oil pressure gauge is available for use on multi-engine aircraft. The dual indicator contains two Bourdon tubes, housed in a standard instrument case, with one tube being used for each engine. The connections extend from the back of the case to each engine. There is one common movement assembly, but the moving parts function independently. In some installations, the line leading from the engine to the pressure gauge is filled with light oil. Since the viscosity of this oil will not vary much with changes in temperature, the gauge will respond better to changes in oil pressure. In time, engine oil will mix with some of the light oil in the line to the transmitter; during cold weather, the thicker mixture will cause sluggish instrument readings. To correct this condition, the gauge line must be disconnected, drained and refilled with light oil.

Most new aircraft use an electrical oil pressure gauge. In this type of system, the oil pressure of the engine is measured by a transmitter at the engine. This transmitter will normally consist of a bellows that is used to vary a potentiometer. The potentiometer varies the amount of current to be sent through a wire to the indicator in the cockpit. Electrical oil pressure gauges are preferred over the Bourdon-tube type because they are lighter in weight.

## *Oil Temperature Regulator*

When the circulating oil has performed its function of lubricating and cooling the moving parts of the engine, it drains into the sumps in the lowest parts of the engine. Oil collected in these sumps is picked up by gear or *gerotor-type scavenger pumps* as quickly as it accumulates. These pumps have a greater capacity than the pressure pump. In dry-sump engines, this oil leaves the engine, passes through the oil temperature regulator and returns to the supply tank.

As discussed earlier, the viscosity of the oil varies with its temperature. Since the lubricating properties of the oil are affected by viscosity, the temperature of the oil must be regulated. Normally, the oil leaving the engine is hot and must be cooled before it reenters the engine. The oil temperature regulator (Figure 1-5-7) is used to regulate this temperature. If the oil is cool when it leaves the engine, as in conditions of operations before the engine has reached normal operating temperature, the regulator allows the oil to return directly to the oil sump. If the oil is hot, as in most normal operating conditions, the regulator directs the oil to flow through the oil cooler. As the oil flows through the oil cooler, it is cooled by air passing through the cooler fins. This action removes the heat from the oil before it returns to the oil sump (see Figure 1-5-1).

## Indicating Oil Temperature

In dry-sump lubricating systems, the oil temperature bulb (Figure 1-5-8) may be anywhere in the oil inlet line between the supply tank and the engine. Oil systems for wet-sump engines have the temperature bulb located where it senses oil temperature after the oil passes through the oil cooler. In either system, the bulb is located so that it measures the temperature of the oil before it enters the engine's hot sections.

In combination with the oil temperature regulating system, an oil temperature gauge is installed in the cockpit. This gauge is connected electrically to a temperature bulb. If the oil temperature regulating system malfunctions, it will be indicated on this gauge.

Oil flow in the system shown in Figure 1-5-1 can be traced from the oil outlet fitting of the tank. The next units through which oil must flow to reach the engine are the drain valve and the firewall shutoff valve. The drain valve in this installation is a manual, two-position valve. It is located in the lowest part of the oil inlet line to the engine to permit complete drainage of the tank and its inlet supply line. The firewall shutoff valve is an electric motor-driven gate valve installed in the inlet oil line at the firewall of the nacelle. This valve shuts off the oil supply when there is a fire, when there is a break in the oil supply line or when engine maintenance is performed that requires the oil to be shut off. In some aircraft, the firewall shutoff valves for the oil system, fuel system and hydraulic system are controlled by a single switch or mechanical linkage; other systems have separate control switches for each valve.

## Oil Cooler

Oil coolers on modern opposed engines are, almost without exception, flange mounted aluminum radiators. Most have built-in oil temperature control valves that double as clogged cooler bypass valves. They are normally considered as part of the engine and, if the engine is removed, they go with it.

The oil cooler system requires virtually no maintenance other than keeping its exterior clean and not allowing cowling or baffles to rub a hole in it. Figure 1-5-9 illustrates two types of aluminum flange mounted oil coolers.

The old-style oil coolers (Figure 1-5-10), either cylindrical or elliptical shaped, consist of a core enclosed in a double-walled shell. The core is built of copper or aluminum tubes with the tube ends formed to a hexagonal shape and joined together in the honeycomb effect shown in Figure 1-5-10. The ends of the tubes of the core are soldered or brazed. The tubes touch only at the ends so that a space exists between them along most of their lengths. This allows oil to flow through the spaces between the tubes while the cooling air passes through the tubes.

Figure 1-5-9 Flange-mounted aluminum oil coolers

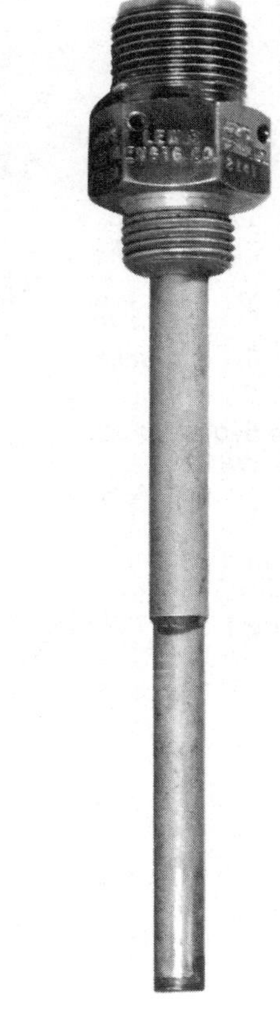

Figure 1-5-8. Oil temperature bulb

The space between the inner and outer shells is known as the annular or bypass jacket. Two paths are open to the flow of oil through a cooler. From the inlet it can flow halfway around the bypass jacket, enter the core from the bottom and then pass through the spaces between the tubes and out to the oil tank. This is the path of the oil when it is hot enough to require cooling. As the oil flows through the

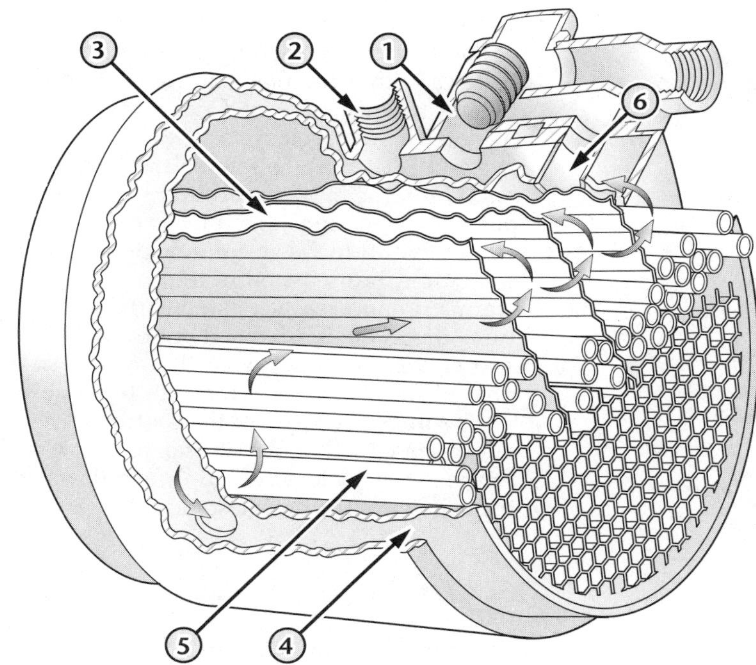


Figure 1-5-10. Oil cooler

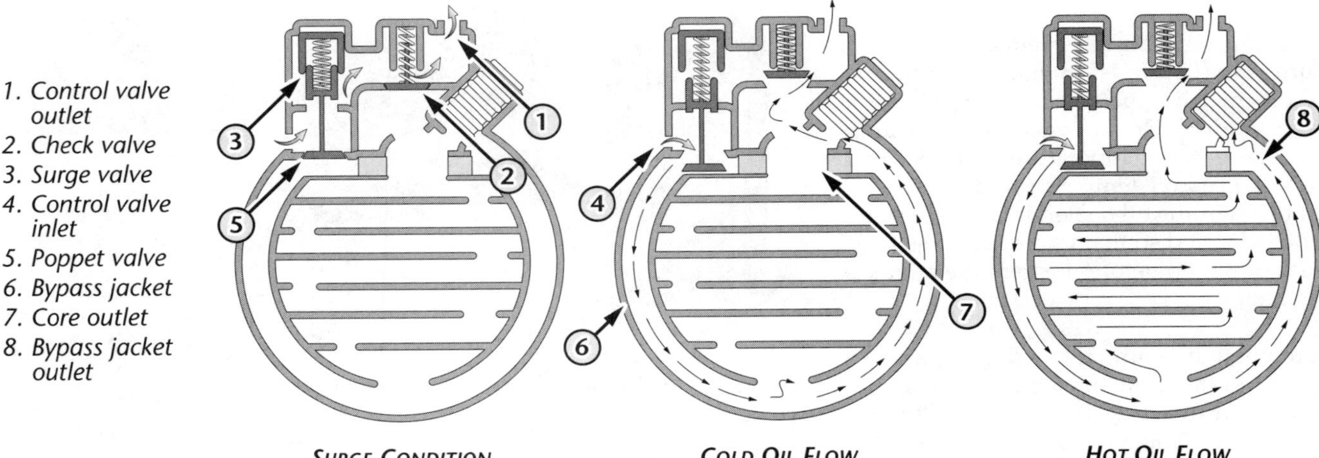


**Figure 1-5-11. Control valve with surge protection**

core, it is guided by baffles, which force the oil to travel back and forth several times before it reaches the core outlet. The oil can also pass from the inlet completely around the bypass jacket to the outlet without passing through the core. Oil follows this bypass route when the oil is cold or when the core is blocked with thick, congealed oil.

## *Flow Control Valve*

The flow control valve determines which of the two possible paths the oil will take through a cooler. There are two openings in a flow control valve which fit over the corresponding outlets at the top of the cooler. When the oil is cold, a bellows within the flow control contracts and lifts a valve from its seat. Under this condition, oil entering the cooler has a choice of two outlets and two paths. Following the path of least resistance, the oil flows around the jacket and out past the thermostatic valve to the tank. This allows the oil to warm up quickly and, at the same time, heats the oil in the core. As the oil warms up and reaches its operating temperature, the bellows of the thermostat expands and closes the outlet from the bypass jacket. The oil must now flow through the core into an opening in the base of the control valve and out to the tank. No matter which path it takes through the cooler, the oil always flows over the bellows of the thermostatic valve.

## *Surge-protection Valves*

When oil in the system is congealed, the scavenger pump may build up a very high pressure in the oil return line. To prevent this high pressure from bursting the oil cooler or blowing off the hose connections, some aircraft have surge-protection valves in the engine lubrication systems.

One type of surge valve (Figure 1-5-11) is incorporated in the oil cooler flow control valve; another type is a separate unit in the oil return line.

The surge-protection valve (Figure 1-5-11) incorporated in a flow control valve is the more common type. Although this flow control valve differs from the one just described, it is essentially the same except for the surge-protection feature. The high-pressure operation condition is shown in Figure 1-5-11, where the high oil pressure at the control valve inlet has forced the surge valve (3) upward. Note how this movement has opened the surge valve and, at the same time, seated the poppet valve (5). The closed poppet valve prevents oil from entering the cooler proper; therefore, the scavenge oil passes directly to the tank through outlet (1) without passing through either the cooler bypass jacket or the core. When the pressure drops to a safe value, the spring forces the surge and poppet valves downward, closing the surge valve (3) and opening the poppet valve (5). Oil then passes from the control valve inlet (4), through the open poppet valve and into the bypass jacket (5). The thermostatic valve, according to oil temperature, then determines oil flow either through the bypass jacket to port (8) or through the core to port (7). The check valve (2) opens to allow the oil to reach the tank return line.

## *Airflow Controls*

The temperature of the oil can be more accurately controlled by regulating the flow of air through the oil cooler. This can be controlled by two methods: with either a set of movable shutters behind the oil cooler or a flap installed in the air exit duct. In either case, the amount of air that flows through the oil cooler is controlled from inside the cockpit.

By regulating the airflow through the cooler, the temperature of the oil can be controlled to fit various operating conditions. For example, the oil will reach operating temperature more quickly if the airflow is cut off during engine warm-up. There are two methods in general use: one method employs shutters installed on the rear of the oil cooler, and the other uses a flap on the air-exit duct.

In some cases, the oil cooler air-exit flap is opened manually and closed by linkage attached to a cockpit lever. More often, the flap is opened and closed by an electric motor.

One of the most widely used automatic oil temperature control devices is the floating control thermostat that provides manual and automatic control of the oil inlet temperatures. With this type of control, the oil cooler air-exit door is opened and closed automatically by an actuator. Automatic operation of the actuator is determined by electrical impulses received from a controlling thermostat inserted in the oil pipe leading from the oil cooler to the supply tank. The actuator may be operated manually by an oil cooler air-exit door control switch. Placing this switch in the open or closed position produces a corresponding movement of the cooler door. Placing the switch in the auto position puts the actuator under the automatic control of the floating control thermostat (Figure 1-5-12). The thermostat is adjusted to maintain a normal oil temperature so that it will not vary more than 5°-8°C, depending on the installation.

During operation, the temperature of the engine oil flowing over the bimetal element (bottom illustration, Figure 1-5-12) causes it to wind or unwind slightly. This movement rotates the shaft (1) and the grounded center contact arm (3). As the grounded contact arm is rotated, it is moved toward either the open or closed floating contact arm (7). The two floating contact arms are oscillated by the cam (6), which is continuously rotated by an electric motor (4) through a gear train (5). When the grounded center contact arm is positioned by the bimetal element so that it will touch one of the floating contact arms, an electric circuit to the oil cooler exit-flap actuator motor is completed, causing the actuator to operate and position the oil cooler air-exit flap.

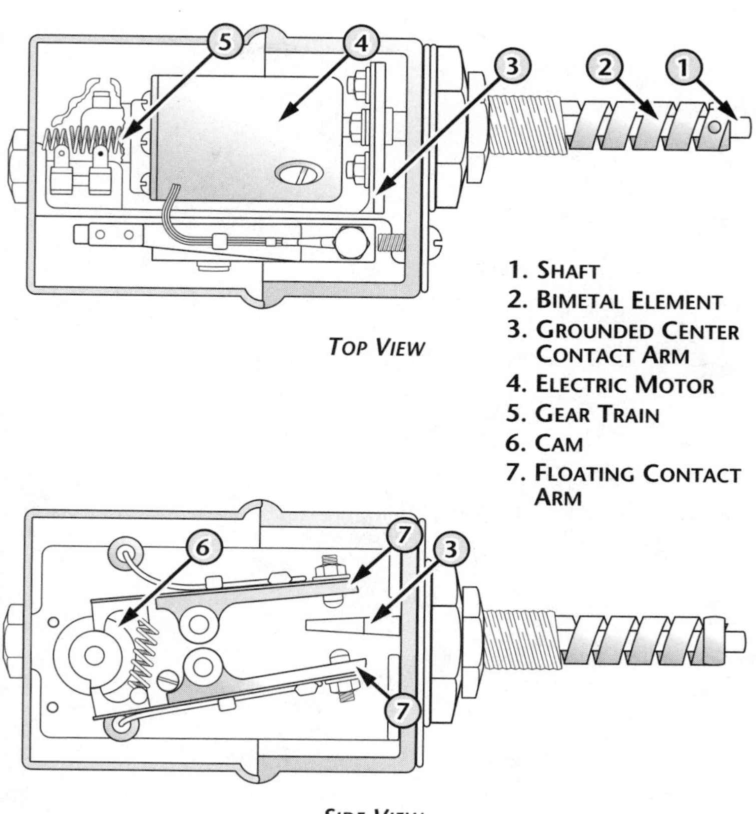


**Figure 1-5-12. Floating control thermostat**

## Internal Lubrication of Reciprocating Engines

The lubricating oil is distributed to the various moving parts of a typical internal-combustion engine by one of the three following methods:

1. Pressure
2. Splash
3. A combination of pressure and splash

### *Pressure Lubrication*

In a typical pressure-lubrication system (Figure 1-5-13, next page), a mechanical pump supplies oil under pressure to the bearings throughout the engine. The oil flows into the inlet or suction side of the oil pump through a line connected to the tank at a point higher than the bottom of the oil sump. This prevents sediment that falls into the sump from being drawn into the pump.

The pump forces the oil into a manifold that distributes the oil through drilled passages to the crankshaft bearings and other bearings throughout the engine.

Oil flows from the main bearings through holes drilled in the crankshaft to the lower connecting rod bearings. Each of these holes through which the oil is fed is located so that the bearing pressure at the point will be as low as possible.

Oil reaches the camshaft (in an in-line or opposed engine) or a cam plate or cam drum (in a radial engine) through a connection with

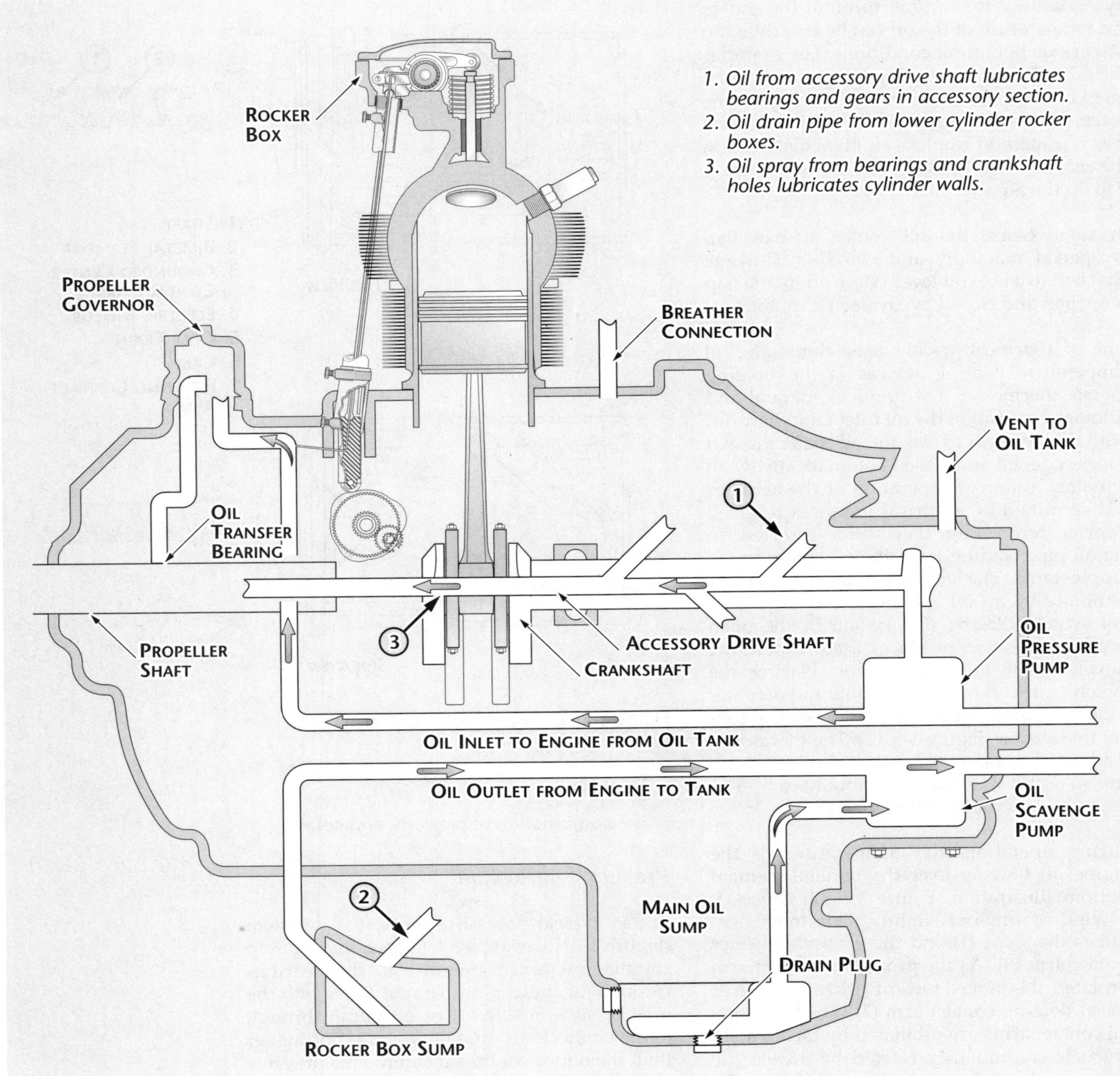


**Figure 1-5-13. Schematic showing pressure dry-sump lubrication system**

the end bearing or the main oil manifold; it then flows out to the various camshaft, cam drum or cam plate bearings and the cams.

The engine cylinder surfaces receive oil sprayed from the crankshaft and also from the crankpin bearings. Since oil seeps slowly through the small crankpin clearances before it is sprayed on the cylinder walls, considerable time is required for enough oil to reach the cylinder walls, especially on a cold day when the oil flow is more sluggish. This is one of the chief reasons for diluting the engine oil with gasoline for cold weather starting.

## *Splash-and-pressure Lubrication*

The pressure-lubrication system is the principal method of lubricating aircraft engines. Splash lubrication may be used in addition to pressure lubrication on aircraft engines, but it is never used by itself. Aircraft-engine lubrication systems are always either the pressure type or the combination pressure-and-splash type, usually the latter.

The advantages of pressure lubrication are:

- Positive introduction of oil to the bearings

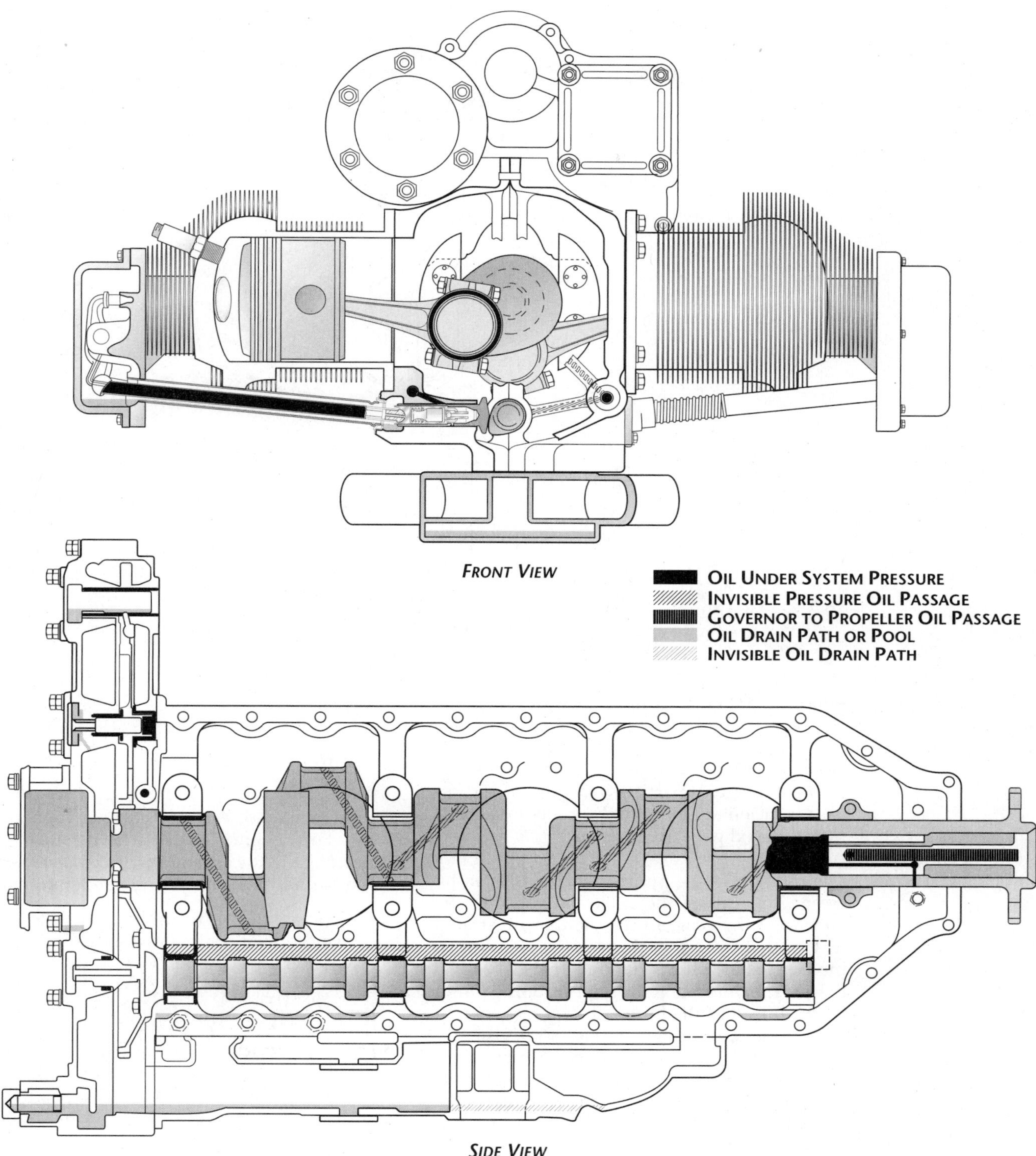


**Figure 1-5-14. Cross-sectional view of a typical wet-sump lubrication system**

- Cooling effect caused by the large quantities of oil which can be (pumped) circulated through a bearing
- Satisfactory lubrication in various attitudes of flight

## *Wet-sump Lubrication*

A simple form of a wet-sump system is shown in Figure 1-5-14. The system consists of a sump or pan in which the oil supply is contained. The level (quantity) of oil is indicated or measured by a vertical rod that protrudes into the oil from an elevated hole on top of the crankcase. In the bottom of the sump (oil pan) is a screen strainer

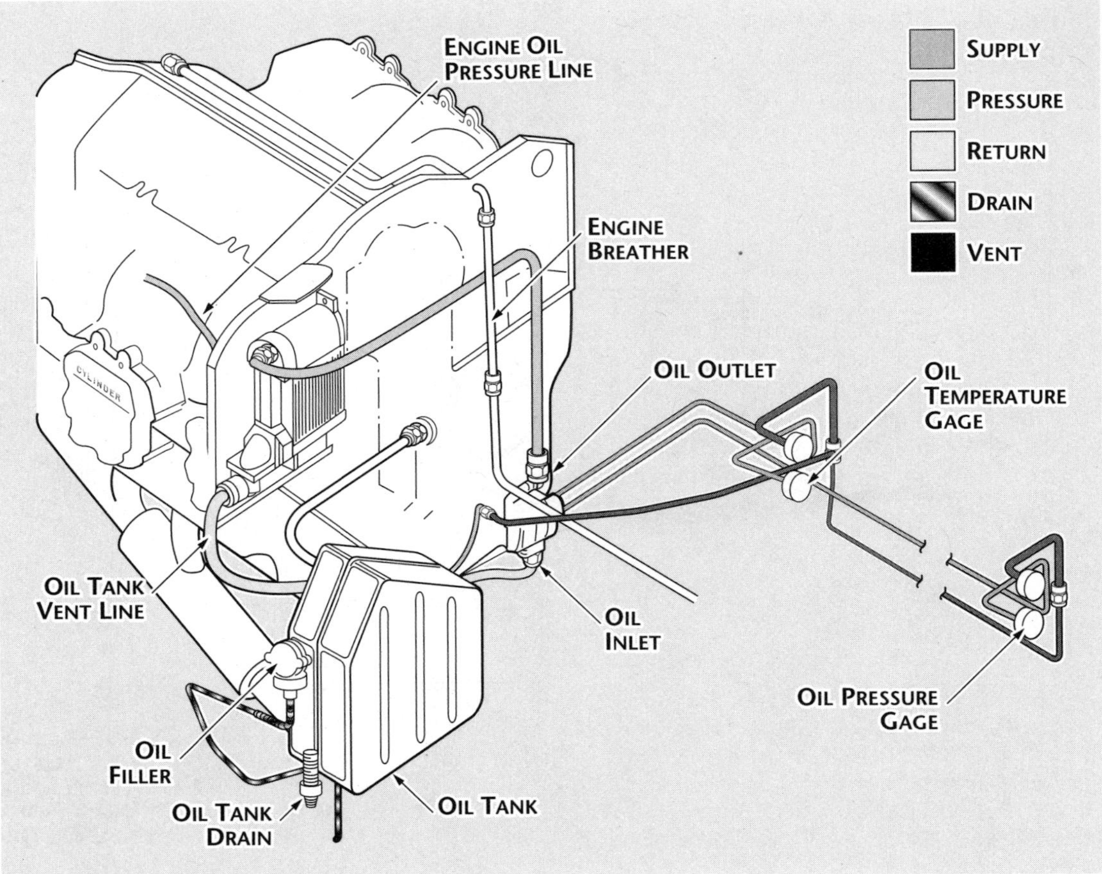


**Figure 1-5-15. Oil system scematic**

having a suitable mesh or series of openings to strain undesirable particles from the oil and yet pass sufficient quantity to the inlet or (suction) side of the oil pressure pump.

The rotation of the pump, which is driven by the engine, causes the oil to pass around the outside of the gears in the manner illustrated in Figure 1-5-3. This develops a pressure in the crankshaft oiling system (drilled passage holes). The variation in the speed of the pump from idling to full-throttle operating range of the engine and the fluctuation of oil viscosity because of temperature changes are compensated by the tension on the relief valve spring. The pump is designed to create a greater pressure than probably will ever be required to compensate for wear of the bearings or thinning out of oil. The parts oiled by pressure throw a lubricating spray into the cylinder and piston assemblies. After lubricating the various units on which it sprays, the oil drains back into the sump and the cycle is repeated.

The main disadvantages of the wet-sump system are:

- The oil supply is limited by the sump (oil pan) capacity.
- Provisions for cooling the oil are difficult to arrange because the system is a self-contained unit.
- Oil temperatures are likely to be higher on large engines, because the oil supply is so close to the engine and is continuously subjected to the operating temperatures.
- The system is not readily adaptable to inverted flying, since the entire oil supply will flood the engine.

## Lubrication System Maintenance Practices

The following lubrication system practices are typical of those performed on small, single-engine aircraft. The oil system and components are those used to lubricate a typical six-cylinder, horizontally opposed, air-cooled engine.

The oil system is the dry-sump type, using a pressure lubrication system sustained by engine-driven, positive-displacement, gear-type pumps. The system (Figure 1-5-15) consists of an oil cooler (radiator), a 3-gallon (U.S.) oil tank, oil pressure pump and scavenge pump, and the necessary interconnecting oil lines. Oil from the oil tank is pumped to the engine, where it circulates under pressure, then collects in the cooler and is returned to the oil tank. A thermostat in the cooler controls oil temperature by allowing part of the oil to flow through the cooler and part to flow directly into the oil supply tank. This arrangement allows hot engine oil, with a temperature still below 65°C. (150°F), to mix with the cold un-circulated oil in the tank. This raises the complete engine oil supply to operating temperature in a shorter period of time.

The oil tank, constructed of welded aluminum, is serviced through a filler neck located on the tank and equipped with a spring-loaded locking cap. Inside the tank a weighted, flexible rubber oil hose is mounted so that it is repositioned automatically to ensure oil pickup during inverted maneuvers. A dipstick guard is welded inside the tank for the protection of the flexible oil hose assembly. During normal flight, the oil tank is vented to the engine crankcase by a flexible line at the top of the tank. However, during inverted flight the normal vent is covered or submerged below the oil level within the tank. Therefore, a secondary vent and check-valve arrangement is incorporated in the tank for inverted operation. During an inversion, when air in the oil tank reaches a certain pressure, the check valve in the secondary vent line will unseat and allow air to escape from the tank. This assures an uninterrupted flow of oil to the engine.

The location of the oil system components in relation to each other and to the engine is shown in Figure 1-5-15.

## *Oil Tank*

Repair of an oil tank usually requires that the tank be removed. The removal and installation procedures normally remain the same regardless of whether the engine is removed or not.

First, the oil must be drained. Most light aircraft provide an oil drain similar to that shown in Figure 1-5-16. On some aircraft the normal ground attitude of the aircraft may prevent the oil tank from draining completely. If the amount of un-drained oil is excessive, the aft portion of the tank can be raised slightly after the tank straps have been loosened to complete the drainage.

After the oil tank has been drained, the cowl assembly is removed to provide access to the oil tank installation. After disconnecting the oil inlet and vent lines (Figure 1-5-17), the scupper drain hose and bonding wire can be removed.

The securing straps fitted around the tank can now be removed, as shown in Figure 1-5-18. Any safety wire securing the clamp must be removed before the clamp can be loosened and the strap disconnected.

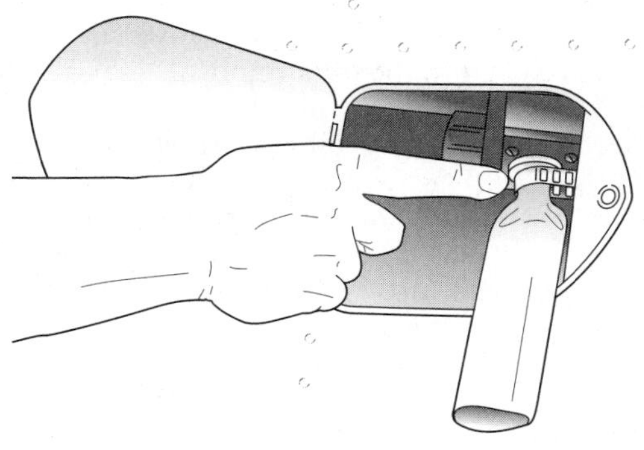

Figure 1-5-16. Oil tank drain

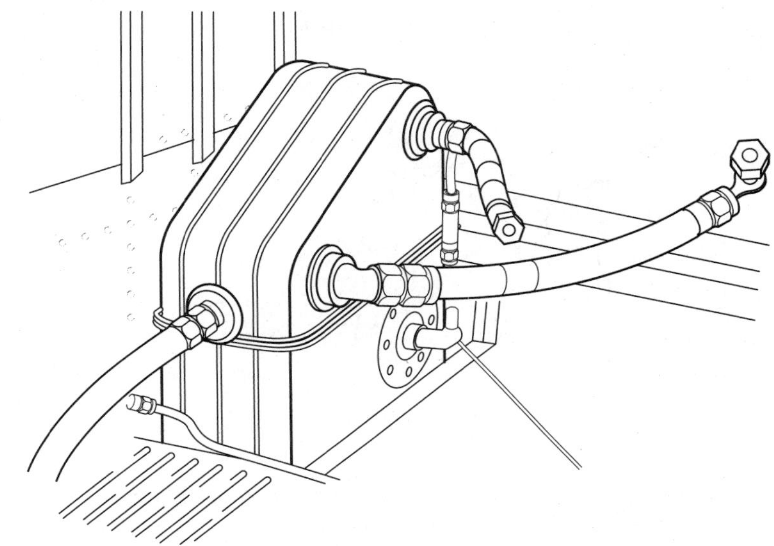

Figure 1-5-17. Disconnecting the oil lines

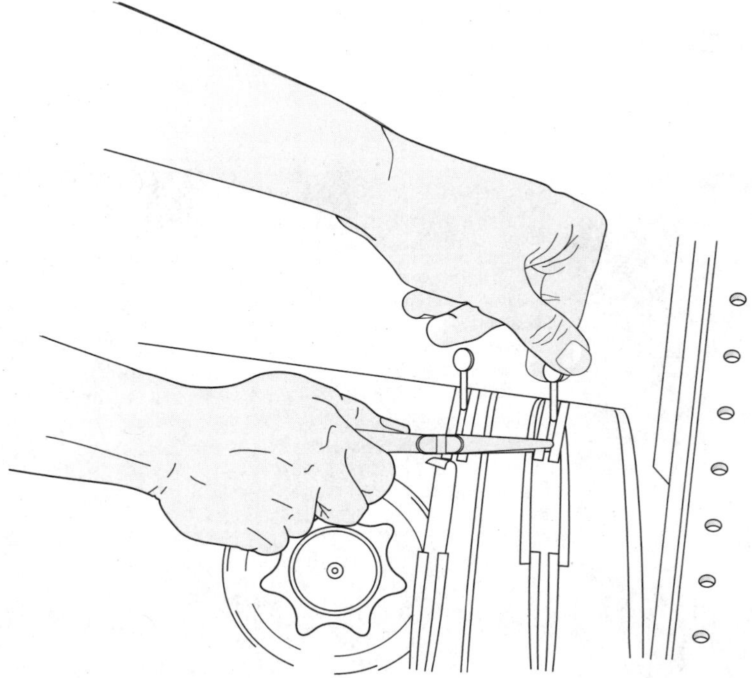

Figure 1-5-18. Removal of the securing straps

The tank can now be lifted out of the aircraft. The tank is re-installed by reversing the sequence used in the tank removal.

After installation, the oil tank should be filled to capacity (Figure 1-5-19), and the engine run for at least 2 minutes. Then the oil level should be checked and, if necessary, sufficient oil should be added to bring the oil up to the proper level on the dipstick.

## Oil Cooler

The oil cooler (Figure 1-5-20) used with this aircraft's opposed-type engine is the honeycomb type. With the engine operating and an oil temperature below 65°C (150°F), an oil cooler bypass valve opens, allowing oil to bypass the core. This valve begins to close when the oil temperature reaches approximately 65°C (150°F). When the oil temperature reaches 85°C (185°F), ±2°C, the valve is closed completely, diverting all oil flow through the cooler core.

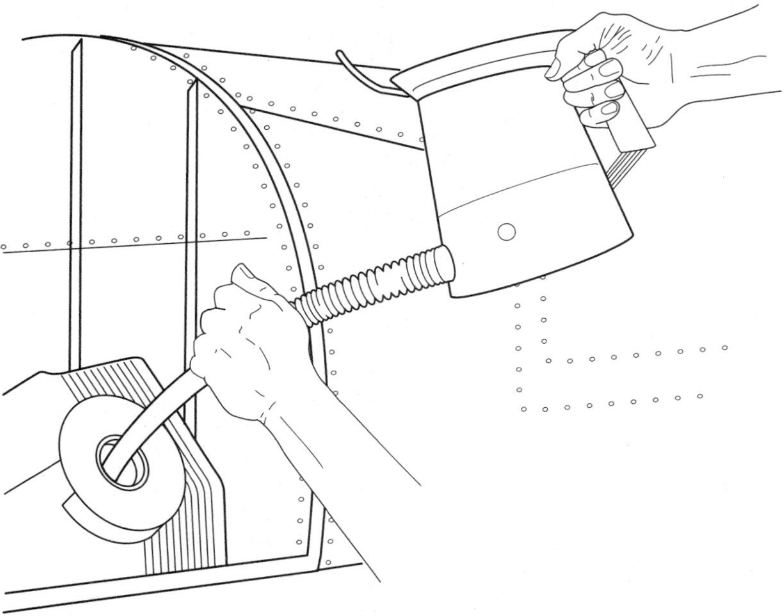

Figure 1-5-19. Filling the oil tank

Figure 1-5-20. Oil cooler

## Oil Temperature Bulbs

Most oil temperature bulbs are mounted in the pressure oil screen housing. They relay an indication of engine oil inlet temperature to the oil-temperature indicators mounted on the instrument panel. Temperature bulbs can be replaced by removing the safety wire and disconnecting the wire leads from the temperature bulbs; then remove the temperature bulbs, using the proper wrench, as shown in Figure 1-5-21.

## Pressure and Scavenge Oil Screens

During engine operation sludge will build up on the pressure and scavenge oil screen. (Figure 1-5-22) The screens must be removed, inspected and cleaned at intervals that are predetermined by the manufacturers.

Although the exact procedure for the removal of the screens will vary with each engine and installation, typical procedures will include the following:

- Remove the safety devices
- Loosen the screen housing or cover plate
- Use a clean container to catch the oil spilled as the screen is removed

After the screens have been removed they should be carefully inspected for metal particles that may indicate internal engine damage. After inspection the screens should be carefully cleaned and reinstalled.

When re-installing a filter, use new O-rings and gaskets, and tighten the filter housing or cover retaining nuts to the torque value specified in the applicable maintenance manual. Filters should be safetied as required.

## Oil Pressure-relief Valve

An oil pressure relief valve limits oil pressure to the value specified by the engine manufacturer. Oil pressure settings vary from 35 p.s.i. to 90 p.s.i., depending on the installation. The oil pressure must be high enough to ensure adequate lubrication of the engine and accessories at high speeds and powers. On the other hand, the pressure must not be too high, since leakage and damage to the system may result. The oil pressure is adjusted by removing a cover nut, loosening a locknut and turning the adjusting screw (see Figure 1-5-23). Turn the adjusting screw clockwise to increase the pressure or counterclockwise to decrease the pressure. Make the pressure adjustments while the engine is idling and tighten the adjustment screw locknut after each adjustment. Check the oil pressure reading while the engine is running at the r.p.m. specified in the maintenance manual (1,900-2,300 r.p.m.). The reading should be between the limits prescribed by the manufacturer.

### *Draining Oil*

In service, oil is constantly exposed to harmful substances that reduce its ability to protect moving parts. The main contaminants are:

- Gasoline
- Moisture
- Acids
- Dirt
- Carbon
- Metallic particles

Because of the accumulation of these harmful substances, common practice is to drain the entire lubrication system at regular intervals and refill with new oil. The time between oil changes varies with each make and model aircraft and engine combination.

### *Oil Filter Replacement*

If the engine uses a replaceable type oil filter, it should be changed at the same intervals that the oil is changed. The oil filter element should be carefully cut open and inspected for metal particles that might have come from the inside of the engine. If metal particles appear in the oil filter it might be an indication that internal engine damage is present and be cause for overhaul.

## Troubleshooting Oil Systems

The outline of malfunctions and their remedies listed in Table 1-5-2 (next page) can expedite troubleshooting of the lubrication system. The purpose of this section is to present typical troubles. It does not imply that any of the troubles are exactly as they may be in a particular airplane.

## FAA Requirements

FAR Part 23 and Part 25 require the oil system to meet the following standards:

Figure 1-5-21. Removing the oil temperature bulb

Figure 1-5-22. Removing scavenge screens

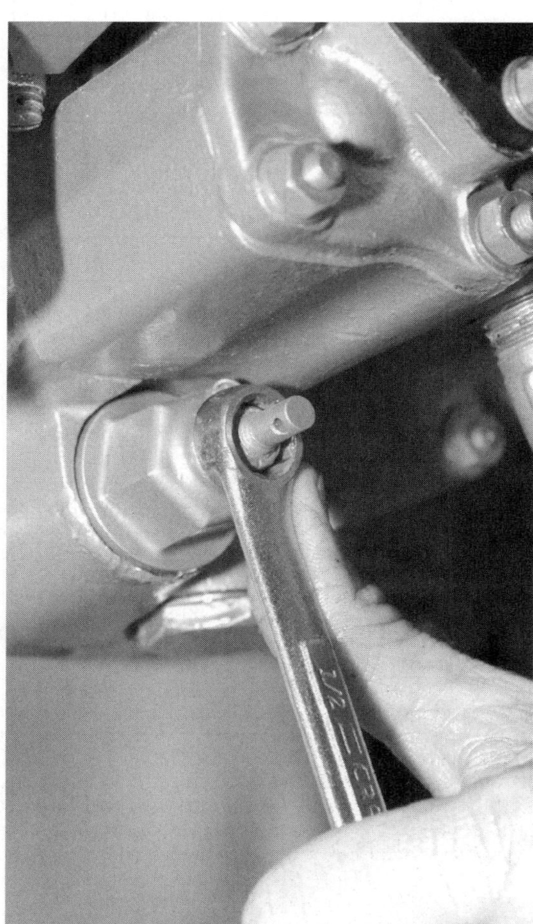

Figure 1-5-23. Oil pressure-relief valve adjustment

| TROUBLESHOOTING OIL SYSTEMS | | |
|---|---|---|
| Symptom/Trouble | Isolation Procedure | Remedy |
| *Excessve Oil Consumption* | | |
| Oil line leakage | Check external lines for evidence of oil leakage | Replace or repair defective lines. |
| Accessory seal leakage | Check for leak at accessories immediately after engine operation | Replace accessory and/or defective accessory oil seal. |
| Low grade of oil | N/A | Fill tank with proper grade of oil. |
| Failing or failed bearing | Check sump and oil pressure pump screen for metal particles | Replace engine, if metal particles are found. |
| *Indications of High or Low Oil Pressure* | | |
| Defective pressure gauge | Check indicator | Replace indicator, if found to be defective. |
| Improper operation of oil pressure-relief valve | Erratic pressure indications, either excessively high or low | Remove, clean and inspect relief valve. |
| Inadequate oil supply | Check oil quantity | Fill oil tank. |
| Diluted or contaminated oil | N/A | Drain engine and tank. Refill tank. |
| Clogged oil screen | N/A | Remove and clean oil screen. |
| Oil viscosity incorrect | Make sure correct oil is being used | Drain engine and tank. Refill tank. |
| Oil pump pressure-relief valve adjustment incorrect | Check pressure-relief valve adjustment | Make correct adjustment on oil pump pressure-relief valve. |
| *Indications of High or Low Oil Temperature* | | |
| Defective temperature gauge | Check indicator | Replace indicator, if found to be defective. |
| Inadequate oil supply | Check oil quality | Fill oil tank. |
| Diluted or contaminated oil | N/A | Drain engine and tank. Refill tank. |
| Obstruction in oil tank | Check tank | Drain oil and remove obstruction. |
| Clogged oil screen | N/A | Remove and clean oil screens. |
| Obstruction in oil cooler passages | Check cooler for blocked or deformed passages | Replace oil cooler, if found to be defective. |
| *Oil Foaming* | | |
| Diluted or contaminated oil | N/A | Drain engine and tank. Refill tank. |
| Oil level in tank too high | Check oil quantity | Drain excess oil from tank. |

**Table 1-5-2. Oil system troubleshooting procedures**

- The word "oil" and the system capacity must be located at the oil tank filler opening.
- A means of visually checking the oil quantity on the ground must be provided.
- An expansion space of 10 percent or 1/2 gallon, whichever is greater, must be built into the oil tank.
- The oil scavenge system must have at least twice the capacity of the pressure system.

## *Section 6*

# *Ignition and Starting Systems*

The basic purpose of an ignition system used on a reciprocating engine is to deliver a high-tension spark to each cylinder of the engine in the proper order at the proper time. The ignition event must occur in the cylinder head at the exact number of crankshaft degrees before TDC. The voltage output of the ignition system must be great enough to produce a spark across the gap of the spark plug under all operating conditions.

Ignition systems used on aircraft reciprocating engines are classified as dual-ignition systems. The dual ignition system contains two separate magnetos, two ignition harnesses and two spark plugs in each cylinder. Figure 1-6-1 shows two of the more common magnetos in service today.

## Battery Ignition System

Some older model aircraft do not use magnetos as a source of voltage for ignition. Instead, they use a battery or generator as the source. This battery ignition system is much like the ignition system of an automobile. This system uses a rotating cam, driven by the engine, to open and close a set of breaker points. This opening and closing of the breaker points interrupts the flow of electricity in a primary circuit, resulting in a continuously building and collapsing magnetic field, which in turn induces high voltage in the secondary ignition coil. This electricity is directed by a distributor to the proper cylinder (see Figure 1-6-2).

## Magneto Ignition System Operating Principles

The magneto, a special type of engine-driven AC generator, uses a permanent magnet as a source of energy. The magneto develops the

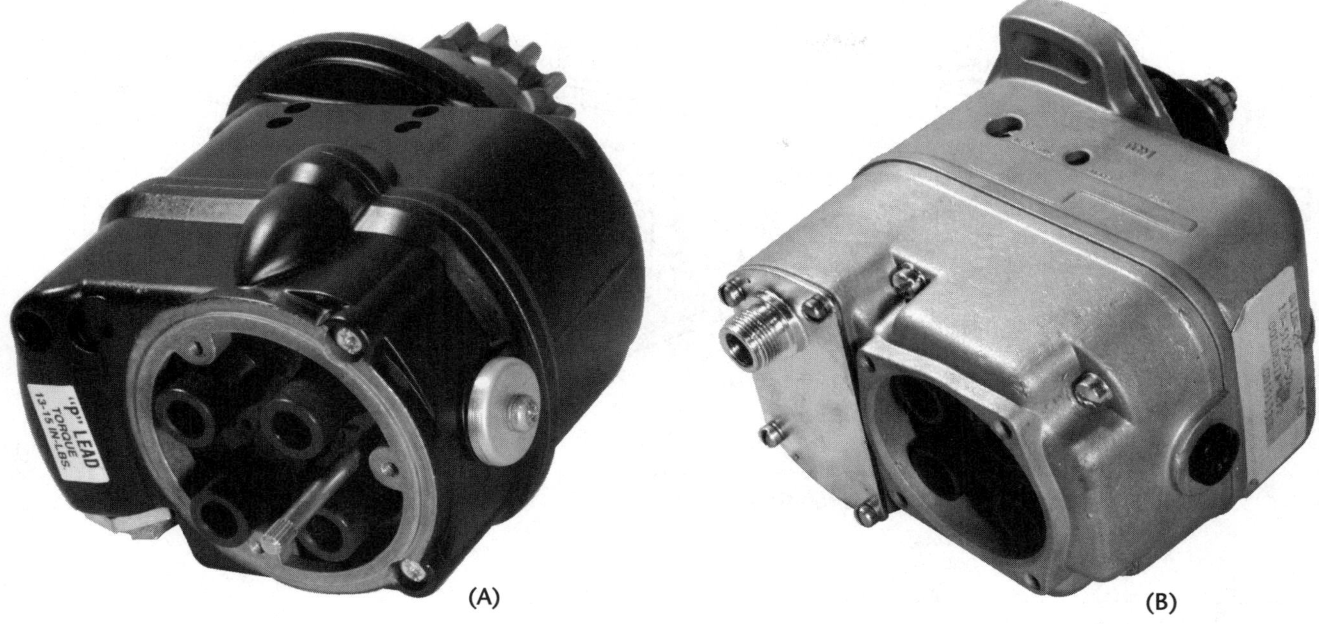


Figure 1-6-1. (A) A Slick magneto and (B) a Bendix magneto

high voltage that forces a spark to jump across the spark plug gap in each cylinder. Magneto operation is timed to the engine so that a spark occurs only when the piston is on the proper stroke at a specific position.

Aircraft magneto ignition systems can be classified as either high tension or low tension. The low-tension magneto system, covered in a later section of this chapter, generates a low voltage, which is distributed to a transformer coil near each spark plug. This system eliminates some problems inherent in the high-tension system.

The high-tension magneto system is the older of the two systems and, despite some disadvantages, is still the most widely used aircraft ignition system.

**High-tension magneto system.** For our purpose of discussion, the high-tension magneto is broken down into four different circuits:

1. Magnetic
2. Primary electrical
3. Secondary electrical
4. Mechanical

**The mechanical operating system.** The mechanical operating system is considered as the mechanical functions and components of the magneto that provide for proper magneto operation.

The magneto is engine-driven either by a splined or geared shaft. The shaft that contains the rotating magnet is supported on needle or ball bearings. The magneto is mounted to the engine either by a flange mount or a base mount on the magneto case.

**The magnetic circuit.** The magnetic circuit consists of a permanent multi-pole rotating magnet, a soft iron core and pole shoes. The magnet is geared to the aircraft engine and rotates in the gap between two pole shoes to furnish the magnetic lines of force (flux) necessary to produce an electrical voltage. The poles of the magnet are arranged in alternate polarity so that the flux can pass out of the north pole through the coil core and back to the south pole of the magnet. When the magnet is in the posi-

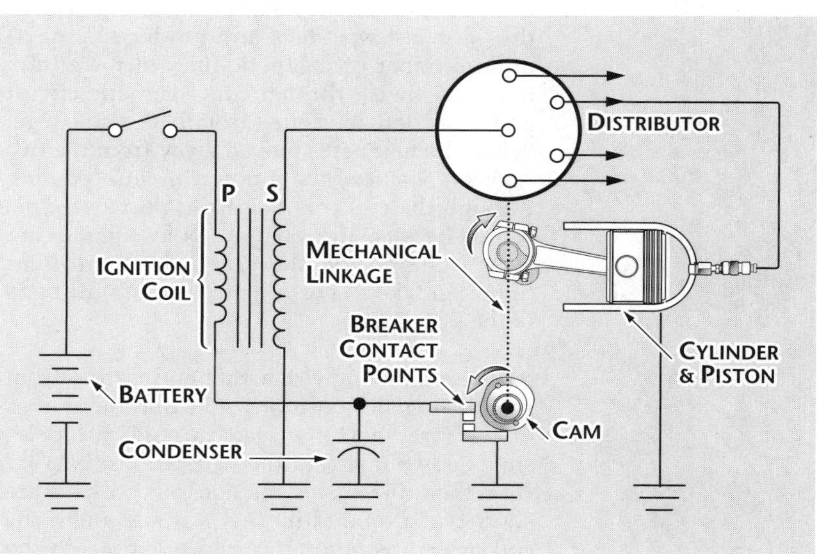


Figure 1-6-2. Battery ignition system

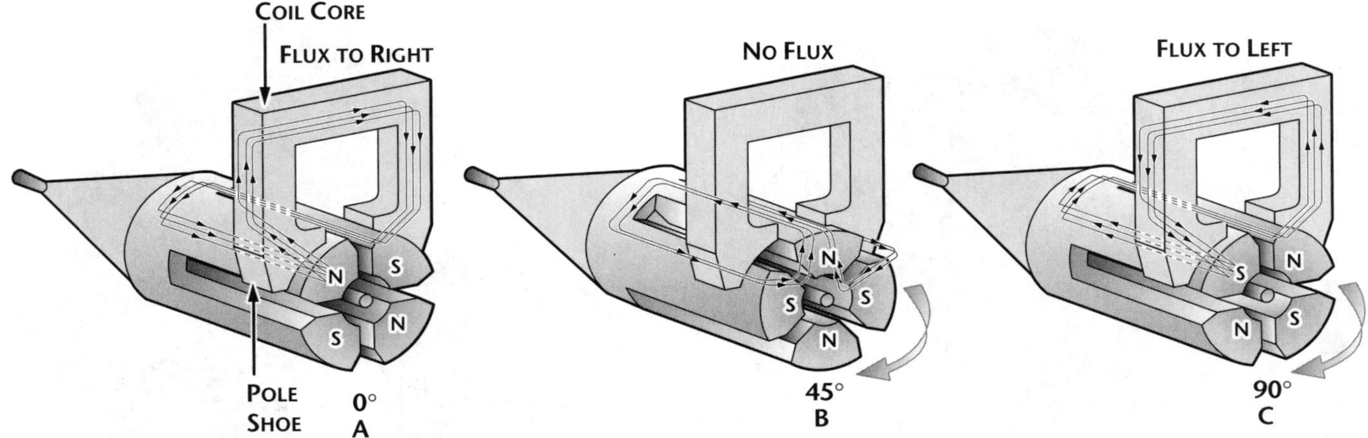


Figure 1-6-3. Magnetic flux at three points of the rotating magnet

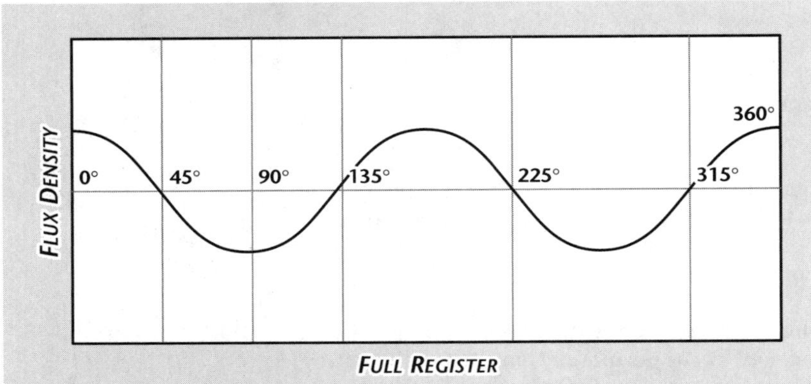


Figure 1-6-4. Change in flux density as magnet rotates

tion shown in illustration A of Figure 1-6-3, the number of magnetic lines of force through the coil core is maximum because two magnetically opposite poles are perfectly aligned with the pole shoes.

This position of the rotating magnet is called the *full-register position,* and produces a maximum number of magnetic lines of force (flux flow) clockwise through the magnetic circuit and from left to right through the coil core. When the magnet is moved away from the full register position, the amount of flux passing through the coil core begins to decrease. This results because the magnet's poles are moving away from the pole shoes, allowing some lines of flux to take a shorter path through the ends of the pole shoes.

As the magnet moves farther and farther from the full-register position, more and more lines of flux are short-circuited through the pole-shoe ends. Finally, at the neutral position (45° from the full-register position) all flux lines are short-circuited and no flux flows through the coil core (illustration B of Figure 1-6-3). As the magnet moves from full register to the neutral position, the number of flux lines through the coil core decreases in the same manner as the gradual collapse of flux in the magnetic field of an ordinary electromagnet.

The neutral position of the magnet is that position where one of the poles of the magnet is centered between the pole shoes of the magnetic circuit. As the magnet is moved clockwise from this position, the lines of flux that had been short-circuited through the pole-shoe ends begin to flow through the coil core again. But this time the flux lines flow through the coil core in the opposite direction, as shown in illustration C of Figure 1-6-3. The flux flow reverses as the magnet moves out of the neutral position, because the north pole of the rotating permanent magnet is opposite the right pole shoe instead of the left, as shown in illustration A of Figure 1-6-3.

When the magnet is again moved a total of 90°, another full-register position is reached, with a maximum flux flow in the opposite direction. The 90° of magnet travel is illustrated graphically in Figure 1-6-4, where a curve shows how the flux density in the coil core (without a primary coil around the core) changes as the magnet is rotated.

Figure 1-6-4 shows that as the magnet moves from the full-register position (0°), flux flow decreases and reaches a zero value as it moves into the neutral position (45°). While the magnet moves through the neutral position, flux flow reverses and begins to increase as indicated by the curve below the horizontal line. At the 90° position another position of maximum flux is reached. Thus, for one revolution (360°) of the four-pole magnet, there will be four positions of maximum flux, four positions of zero flux and four flux reversals.

This discussion of the magnetic circuit demonstrates how the coil core is affected by the rotating magnet. It is subjected to an increasing and decreasing magnetic field, and a change in polarity each 90° of magnet travel.

When a coil of wire as part of the magneto's primary electrical circuit is wound around the coil core, it is also affected by the varying magnetic field.

## *Primary and Secondary Circuits*

The primary electrical circuit (Figure 1-6-5) consists of a set of breaker contact points, a condenser and an insulated coil.

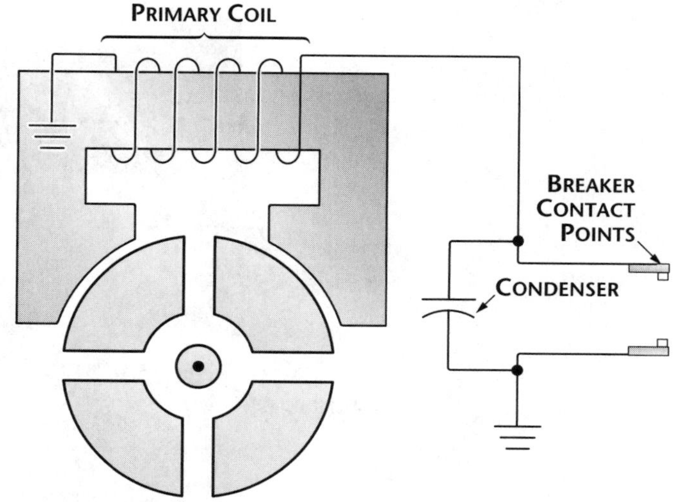


**Figure 1-6-5. Primary electrical circuit of a high-tension magneto**

The coil is made up of a few turns of heavy copper wire, one end of which is grounded to the coil core and the other end to the ungrounded side of the breaker points (see Figure 1-6-5). The primary circuit is complete only when the ungrounded breaker point contacts the grounded breaker point. The third unit in the circuit, the condenser (capacitor), is wired in parallel with the breaker points. The condenser prevents arcing at the points when the circuit is opened, hastening the collapse of the magnetic field about the primary coil.

The primary breaker closes at approximately full-register position. When the breaker points are closed, the primary electrical circuit is completed, and the rotating magnet will induce current flow in the primary circuit. This current flow generates its own magnetic field, which is in such a direction that it opposes any change in the magnetic flux of the permanent magnet's circuit.

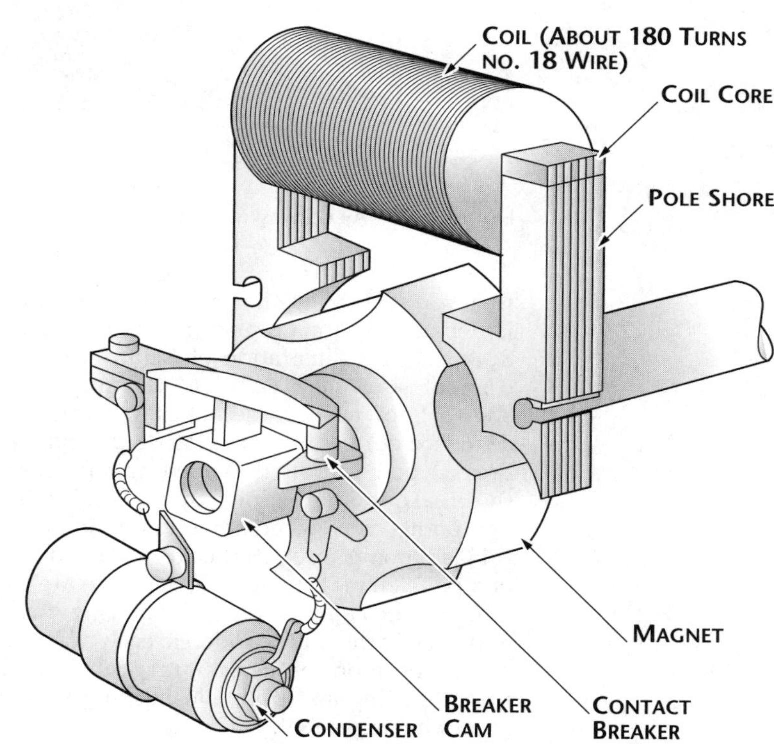


**Figure 1-6-6. Components of a high-tension magneto circuit**

While the induced current is flowing in the primary circuit, it will oppose any decrease in the magnetic flux in the core. This is in accordance with Lenz's law, stated as follows: *An inducted current always flows in such a direction that its magnetism opposes the motion or the change that induced it.* Thus, the current flowing in the primary circuit holds the flux in the core at a high value in one direction until the rotating magnet has time to rotate through the neutral position, to a point a few degrees beyond neutral. This position is called the *E-gap position* (E stands for *efficiency*).

With the magnetic rotor in E-gap position and the primary coil holding the magnetic field of the magnetic circuit in the opposite polarity, a very high rate of flux change can be obtained by opening the primary breaker points. Opening the breaker points stops the flow of current in the primary circuit and allows the magnetic rotor to quickly reverse the field through the coil core. This sudden flux reversal produces a high rate of flux change in the core, which cuts across the secondary coil of the magneto (wound over and insulated from the primary coil), inducing the pulse of high-voltage current in the secondary needed to fire a spark plug. As the rotor continues to rotate to approximately full-register position, the primary breaker points close again and the cycle is repeated to fire the next spark plug in firing order.

**Sequence of events**. With the breaker points, cam and condenser connected in the circuit as shown in Figures 1-6-6 and 1-6-7, the action

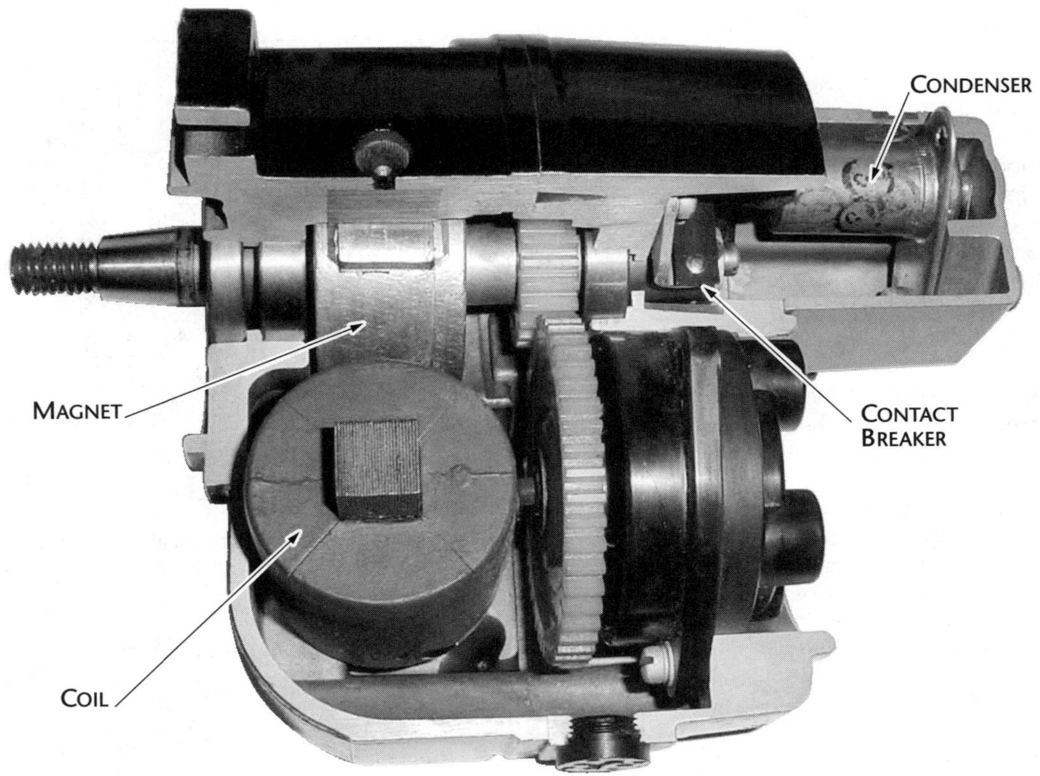


**Figure 1-6-7. A cutaway magneto, with contact breaker, coil, magnet and condenser highlighted**

that takes place as the magnetic rotor turns is depicted by the graph curve in Figure 1-6-8. At the top illustration of Figure 1-6-8 the original static flux curve of the magnets is shown. Shown below the static flux curve is the sequence of opening and closing the magneto breaker points. Note that opening and closing the breaker points is timed by the breaker cam. The points close when a maximum amount of flux is passing through the coil core and open at a position after neutral. Since there are four lobes on the cam, the breaker points will close and open in the same relation to each of the four neutral positions of the rotor magnet. Also, the point opening and point closing intervals are approximately equal.

Starting at the maximum flux position (0° at the top of Figure 1-6-8), the sequence of events in that occurs is described in the following paragraphs.

As the magnet rotor is turned toward the neutral position, the amount of flux through the core starts to decrease (illustration D of Figure 1-6-8). This change in flux linkages induces a current in the primary winding (illustration C of Figure 1-6-8). The induced current creates a magnetic field of its own. This magnetic field opposes the change of flux linkages inducing the current. Without current flowing in the primary coil, the flux in the coil core decreases to zero as the magnet rotor turns to neutral and starts to increase in the opposite direction (dotted static flux curve in illustration D of Figure 1-6-8). But the electromagnetic action of the primary current prevents the flux from changing and temporarily holds the field instead of allowing it to change (Resultant Flux line in illustration D of Figure 1-6-8).

As a result of the holding process, there is a very high stress in the magnetic circuit by the time the magnet rotor has reached the position where the breaker points are about to open.

The breaker points, when opened, function with the condenser to interrupt the flow of current in the primary coil, causing an extremely rapid change in flux linkages. The high voltage in the secondary winding discharges across the gap in the spark plug and ignites the fuel/air mixture in the engine cylinder. Each spark actually consists of one peak discharge, after which a series of small oscillations takes place. They continue to occur until the voltage becomes too low to maintain the discharge. Current flows in the secondary winding during the time that it takes for the spark to completely discharge. The energy or stress in the magnetic circuit is completely dissipated by the time the contacts close for the production of the next spark.

**Breaker assembly.** Breaker assemblies, used in high-tension magneto ignition systems, automatically open and close the primary circuit at

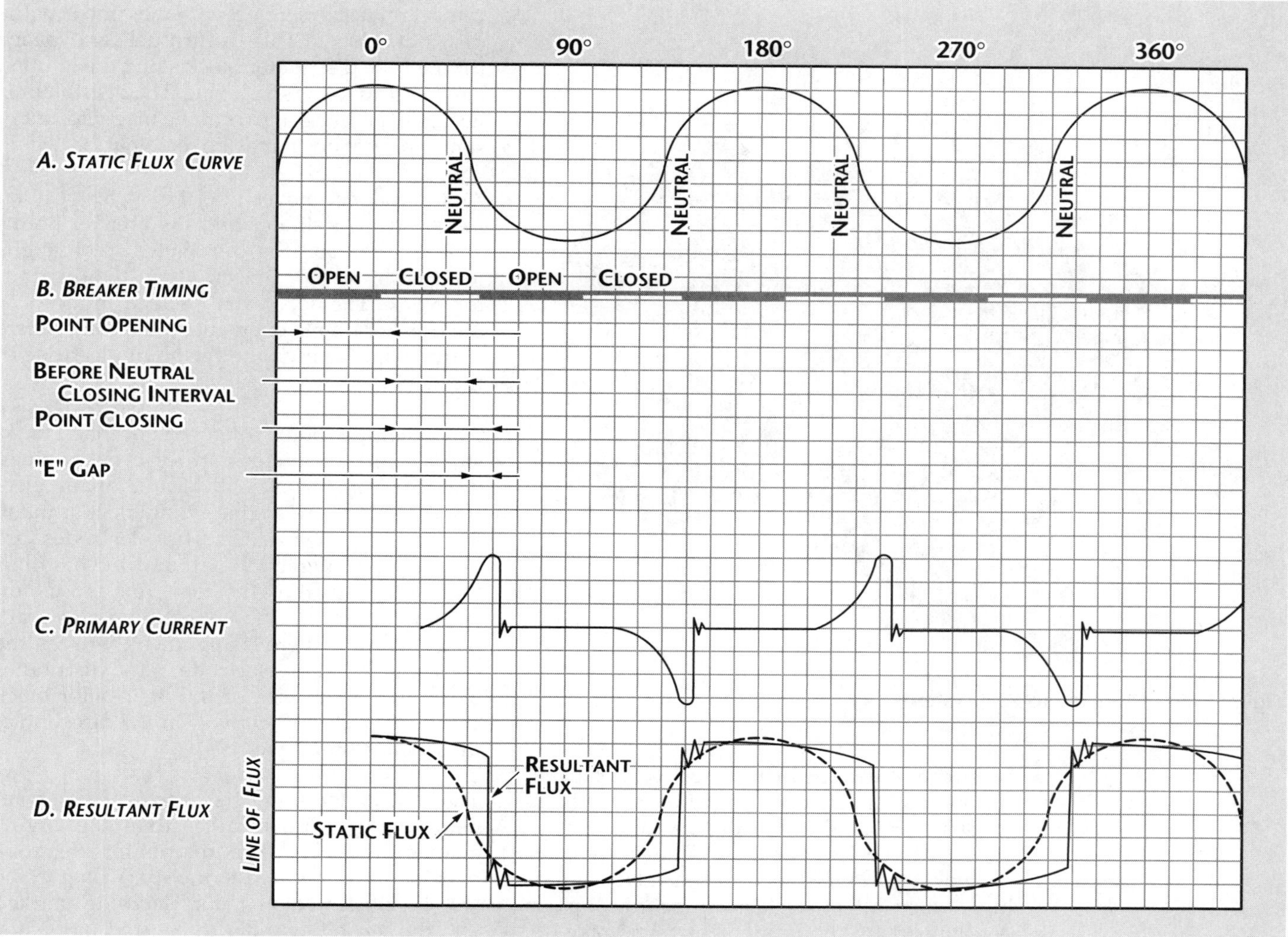


Figure 1-6-8. Magneto flux curves

the proper time in relation to piston position in the cylinder to which an ignition spark is being furnished. The interruption of the primary current flow is accomplished through a pair of breaker contact points, made of an alloy that resists pitting and burning.

A simple type of breaker assembly may be found on aircraft engines in the lower power range. This type, called the pivot type, has one hinged or pivoted arm with a contact point on the end opposite to the pivot or hinge point. The other contact point is secured to a stationary plate. A rubbing block, usually made of fibrous material, is located near the middle point of the movable breaker arm. When the engine rotates the cam, the lobes exert pressure against the rubbing block, causing the movable breaker arm to swing on its pivot point, opening the contact points.

The breaker-actuating cam may be directly driven by the magneto rotor shaft or through a gear train from the rotor shaft. Most large radial engines use a compensated cam, which is designed to operate with a specific engine and has one lobe for each cylinder to be fired by the magneto. The cam lobes are machine ground at unequal intervals to compensate for the top-dead-center variations of each position.

**Coil assembly.** The magneto coil assembly consists of the soft iron core around which is wound the primary coil and the secondary coil, with the secondary coil wound on top of the primary coil.

The secondary coil is made up of a winding containing approximately 13,000 turns of fine, insulated wire, one end of which is electrically grounded to the primary coil or to the coil core and the other end connected to the distributor rotor. The primary and secondary coils are encased in a non-conducting material of *bakelite, hard rubber* or *varnished cambric*. The whole assembly is then fastened to the pole shoes with screws and clamps.

When the primary circuit is closed, the current flow through the primary coil produces magnetic lines of force that cut across the secondary windings, inducing an electromo-

Figure 1-6-9. A magneto vent system

tive force. When the primary circuit is broken, the magnetic field about the primary windings collapses, causing the secondary windings to be cut by the lines of force. The strength of the voltage induced in the secondary windings, when all other factors are constant, is determined by the number of turns of wire. Since most high-tension magnetos have many thousands of turns of wire in the secondary, a very high voltage, often as high as 20,000 volts, is generated in the secondary circuit to jump the air gap of the spark plug in the cylinder.

**Safety gap.** Most magnetos have a safety gap to relieve the voltage in the secondary winding of the magneto, should the spark not be generated in the spark plug.

The actual mechanism to produce a safety gap is an electrode on the high-tension brush holder and a ground electrode. The gap between the two electrodes is greater than the gap between the distributor finger and the contacts in the cap. If a high voltage were left to dissipate by itself, it would damage coil insulation, causing distributor tracks.

**Distributor.** The voltage that is induced in the secondary coil is sent directly to the distributor. The distributor consists of two basic parts, the rotor and the distributor block. The rotor is a rotating conductor driven by the engine. The distributor block is a stationary component that consists of an electrical contact and wire terminals for each ignition lead. The voltage from the secondary winding is distributed as the rotor contact rotates to connect with the contact of the distributor block. This in turn delivers voltage through the distributor block along the ignition harness lead to the spark plug. The distributor is normally an integral part of the magneto, but in some applications it may be remotely located.

As the magnet moves into the E-gap position for the No. 1 cylinder and the breaker points start to separate, the distributor rotor aligns itself with the No. 1 electrode in the distributor block. The secondary voltage induced as the breaker points open enters the rotor where it arcs a small air gap to the No. 1 electrode in the block.

Since the distributor rotates at one-half crankshaft speed on all four-stroke-cycle engines, the distributor block will have as many electrodes as there are engine cylinders or as many electrodes as cylinders. The electrodes are located circumferentially around the distributor block so that, as the rotor turns, a circuit is completed to a different cylinder and spark plug each time there is alignment between the rotor finger and an electrode in the distributor block. The electrodes of the distributor block are numbered consecutively in the direction of distributor rotor travel.

The distributor numbers represent the magneto sparking order rather than the engine cylinder numbers. The distributor electrode marked 1 is connected to the spark plug in the No. 1 cylinder; distributor electrode marked 2 to the second cylinder to be fired (not No. 2 cylinder); distributor electrode marked 3 to the third cylinder to be fired (not No. 3 cylinder) and so forth.

## Magneto and Distributor Venting

Magneto and distributor assemblies are subjected to constant and sudden changes in temperature and altitude. These sudden changes cause condensation of moisture on the inside of these assemblies. This condensed moisture is an excellent conductor of electricity and can cause the high voltage current to be misdirected to the wrong spark plug. This condition is called flashover and results in cylinder misfiring. To prevent flashover, the internal components of the magneto such as the coils, distributor block condenser and rotor are protected from moisture contamination. Normally these parts are coated with a wax-like material which beads up the moisture preventing a solid conducting path.

Carbon tracking can be caused by flashover. It is indicated by fine, pencil-like lines that appear in the location in which the flashover has occurred, and is actually a trail of carbon

**Figure 1-6-10. An ignition harness installed on a Lycoming engine**

caused by the flashover burning dirt particles that contain hydrocarbon materials. The moisture is burned out of these hydrocarbon materials, leaving a conductive path of carbon. This path will conduct electricity and cause misdirection of the voltage even after the moisture is gone.

Magnetos cannot be sealed to prevent the contamination by moisture, because they are subjected to changes in temperature and pressure caused by varying altitude. For this reason, the magnetos must be adequately vented (Figure 1-6-9) and have sufficient moisture drains. Good circulation of the air through a magneto will reduce the possibility of moisture in the magneto. Circulation will also purge the magneto of corrosive gasses produced by normal arcing across the distributor block.

Magnetos that are used in high-altitude applications are pressurized to maintain a higher absolute pressure and eliminate flashover.

The vent system of the magneto must be kept free of obstruction. It is also important to make certain that the air circulating through the magneto is free of dirt and oil.

## Ignition Harness

The ignition harness consists of an insulated wire for each spark plug that the magneto fires. The insulation of the ignition harness serves a dual purpose. It supports the wire and protects it from the heat given off by the engine. It also provides a conductor for the magnetic fields that are produced by the high voltage electricity passing through the conductor. The insulation carries these magnetic fields directly to ground so magnetic interference to the aircraft radios is reduced. This type of magnetic field insulation is called shielding. Without harness shielding, radio use would be almost impossible.

On most installations the magnetos are mounted on the accessory section of the engine and individual wires are carefully routed from the magneto to the appropriate cylinder and installed with special clamps and stand-offs to prevent chafing and burning of the wires. In this type of ignition harness, the ignition wires are continuous from the distributor block to the spark plug. If trouble develops, the entire lead must be replaced.

**Construction.** Modern ignition harnesses are made of 5mm wire. The conductor of the wire is braided or a continuous spiral of copper or stainless steel wire. The insulation is rubber or silicone and is enclosed in a braided metal shield, which in turn is enclosed in a clear plastic outer cover. This plastic outer cover protects the wire from abrasion (Figure 1-6-10).

The spark plug end of the harness wire has a threaded fitting that screws onto the spark plug. The size of these fittings is either 5/8-24 or 3/4-20.

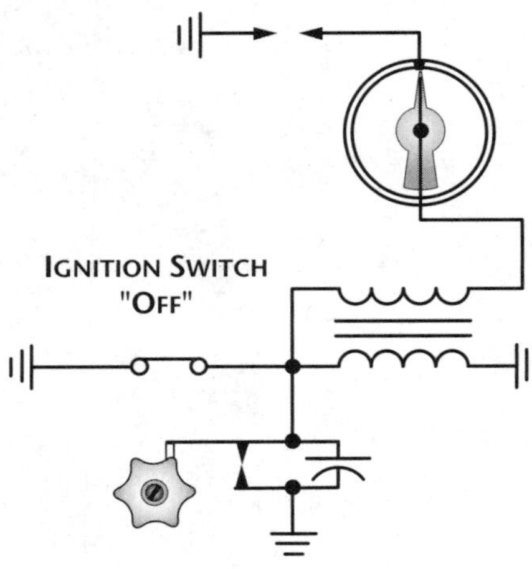


Figure 1-6-11. Typical ignition switch in the off position

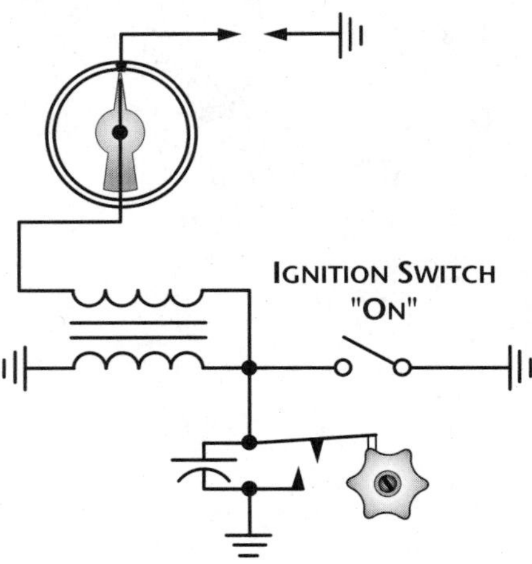


Figure 1-6-12. Typical ignition switch in the on position

The magneto end of this harness is not as standardized as the spark plug end. The magneto end is held to the distributor block by a variety of methods, some of the main ones are solder, threaded nuts, ferroles or crimped.

## Ignition Switches

All units in an aircraft ignition system are controlled by an ignition switch in the cockpit. The type of switch used varies with the number of engines on the aircraft and the type of magnetos used. All switches, however, turn the system off and on in much the same manner. The ignition switch is different in at least one respect from all other types of switches in that, when the ignition switch is in the OFF position, a circuit is completed through the switch to ground. In other electrical switches, the OFF position normally breaks or opens the circuit.

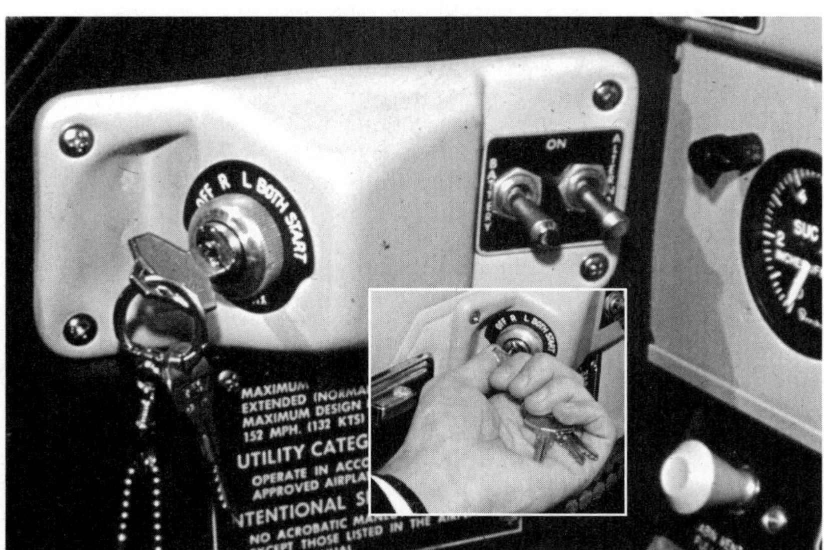


Figure 1-6-13. Switch positions for one ignition switch that controls two magnetos

The ignition switch has one terminal connected to the primary electrical circuit between the coil and the breaker contact points. The other terminal of the switch is connected to the aircraft ground (structure). As shown in Figure 1-6-11, two ways to complete the primary circuit are:

1. Through the closed breaker points to ground
2. Through the closed ignition switch to ground

In Figure 1-6-11, it can be seen that the primary current will not be interrupted when the breaker contacts open, since there is still a path to ground through the closed (OFF) ignition switch. Since primary current is not stopped when the contact points open (Figure 1-6-11), there can be no sudden collapse of the primary coil flux field and no high voltage induced in the secondary coil to fire the spark plug.

As the magnet rotates past the E-gap position, a gradual breakdown of the primary flux field occurs. But that breakdown occurs so slowly that the induced voltage is too low to fire the spark plug. Thus, when the ignition switch is in the OFF position (switch closed), the contact points are as completely short-circuited as if they were removed from the circuit, and the magneto is inoperative. When the ignition switch is placed in the ON position (switch open), as shown in Figure 1-6-12, the interruption of primary current and the rapid collapse of the primary coil flux field is once again controlled or triggered by the opening of the breaker contact points. When the ignition switch is in the ON position, the switch has absolutely no effect on the primary circuit.

Most single-engine aircraft ignition systems employ a dual-magneto system, in which the right magneto supplies the electric spark for the one plug in each cylinder, and the left magneto fires the other. One ignition switch is normally used to control both magnetos. An example of this type of switch is shown in Figure 1-6-13.

This switch has four positions: OFF, LEFT, RIGHT and BOTH. In the OFF position, both magnetos are grounded and thus are inoperative. When the switch is placed in the LEFT position, only the left magneto operates; in the RIGHT position, only the right magneto operates. In the BOTH position, both magnetos operate. The RIGHT and LEFT positions are used to check dual-ignition systems, allowing one system to be turned off at a time.

## Single and Dual High-tension System Magnetos

High-tension system magnetos are either single- or dual-type magnetos. The single-magneto design incorporates the distributor in the housing with the magneto breaker assembly, rotating magnet and coil. The dual magneto (Figure 1-6-14) incorporates two magnetos in one housing. One rotating magnet and a cam are common to two sets of breaker points and coils. Two separate distributor units are mounted on the engine apart from the magneto or within the magneto housing.

## Magneto Mounting Systems

Magnetos may be designed for either base mounting or flange mounting.

*Base-mounted magnetos* are secured to a mounting bracket on the engine. *Flange-mounted magnetos* are attached to the engine by a flange around the driven end of the rotating shaft of the magneto. Elongated slots in the mounting flange permit adjustment through a limited range to aid in timing the magneto to the engine.

## Low-tension Magneto System

High-tension ignition systems have been in use since the beginning of powered flight. Many refinements in the design have been made, but certain underlying problems have remained and others have intensified, such as:

- Increase in the number of cylinders per engine
- Requirement that all radio-equipped aircraft have their ignition wires enclosed in metal conduits
- Need for all-weather flying
- Operation at high altitudes

To overcome these problems, low-tension ignition systems were developed. Electronically, the low-tension system is different from the high-tension system. In the low-tension system, low-voltage is generated in the magneto and flows to the primary winding of a transformer coil located near the spark plug. There the voltage is increased to a high voltage by transformer action and conducted to the spark plug by very short high-tension leads. Figure 1-6-15 shows some of the primary components involved in a low tension ignition system. Figure 1-6-16 (next page) is a simplified schematic of a typical low-tension system.

Flashover is virtually eliminated in the low-tension system because the air gaps inside the distributor have been eliminated by the use of a brush type distributor. High voltage is only present in the short leads between the transformer and the spark plug.

Figure 1-6-14. Dual magnetos combine one magnet and cam with two sets of breaker points and coils.

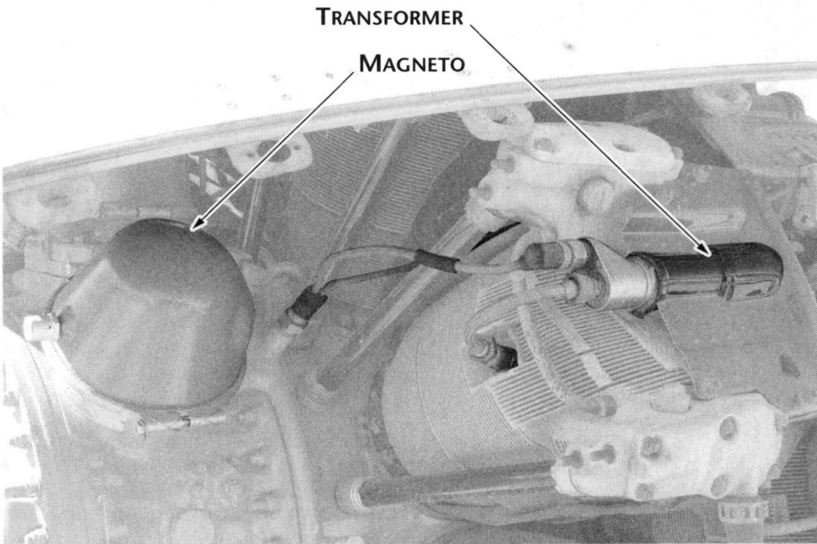


Figure 1-6-15. Low tension ignition system

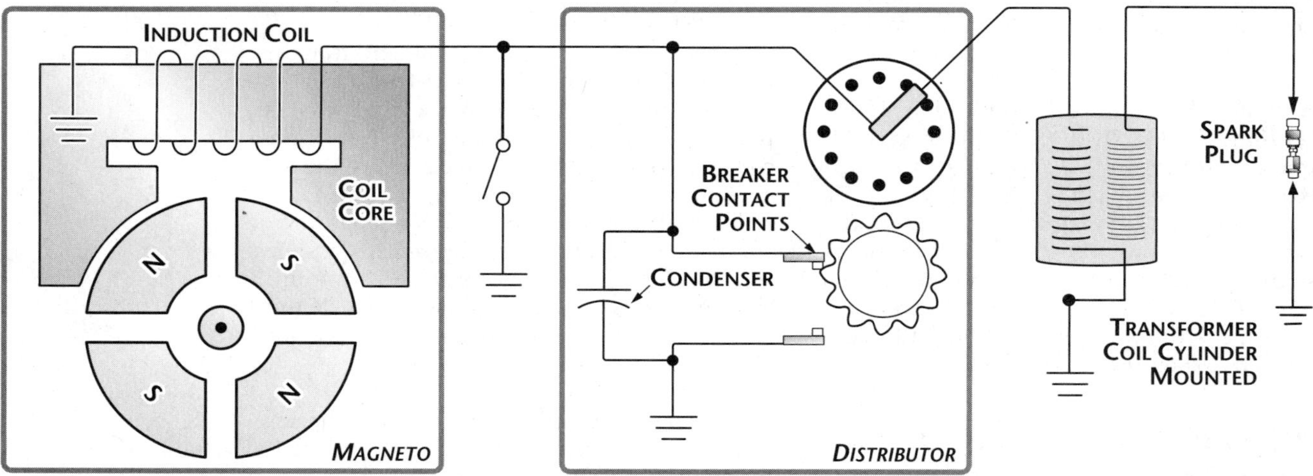


**Figure 1-6-16. Simplified low-tension ignition system schematic**

Electrical leakage occurs in all ignition systems. However, a greater amount of leakage occurs if the system uses a radio shielded ignition harness. Radio shielded harnesses increase the amount of leakage because the metal covering that leads to ground runs very close to the ignition wires for their entire length. This leakage can be reduced by the use of a low-tension system because the current flowing through the ignition wire is transmitted at a low voltage.

Despite the apparent advantages of low-tension systems, improvements in materials and construction have kept the high-tension magneto as the standard for reciprocating engines on aircraft. Recent developments and certification of electronic ignition units are setting the pace for the future of aircraft ignition systems. However, the low tension ignition system is not dead. Some current automobiles use low tension systems, primarily to meet extended warranty emission requirements.

## Auxiliary Ignition Units

During engine starting, the output of a magneto is low because the cranking speed of the engine is low. This is understandable when the factors that determine the amount of voltage induced in a circuit are considered.

To increase the voltage used for starting, one of three methods may be used:

1. Use a stronger magnet
2. Increase the number of turns in the coil
3. Increase the speed of relative motion between the magnet and the conductor

Since the strength of the rotating magnet and the number of turns in the coil are constant factors in both high- and low-tension magneto ignition systems, the voltage produced depends upon the speed at which the rotating magnet is turned. When the engine is being cranked for starting, the magnet is rotated at about 80 r.p.m. Since the value of the induced voltage is so low, a spark may not jump the spark plug gap. Thus, to facilitate engine starting, an auxiliary device is connected to the magneto to provide a high ignition voltage.

Ordinarily, such auxiliary ignition units are energized by the battery and connected to the right magneto or distributor. Reciprocating engine starting systems normally include one of the following types of auxiliary starting systems: booster coil, induction vibrator (sometimes called starting vibrator), impulse coupling or other specialized retard breaker and vibrator starting systems.

## Booster Coil

The booster coil assembly (Figure 1-6-17) consists of two coils wound on a soft iron core, a set of contact points and a condenser. The primary winding has one end grounded at the internal grounding strip and its other end connected to the moving contact point. The stationary contact is fitted with a terminal to which battery voltage is applied when the magneto switch is placed in the START position, or automatically applied when the starter is engaged. The secondary winding, which contains several times as many turns as the primary coil, has one end grounded at the internal grounding strip and the other terminated at a high-tension terminal. The high-tension terminal is connected to an electrode in the distributor by an ignition cable.

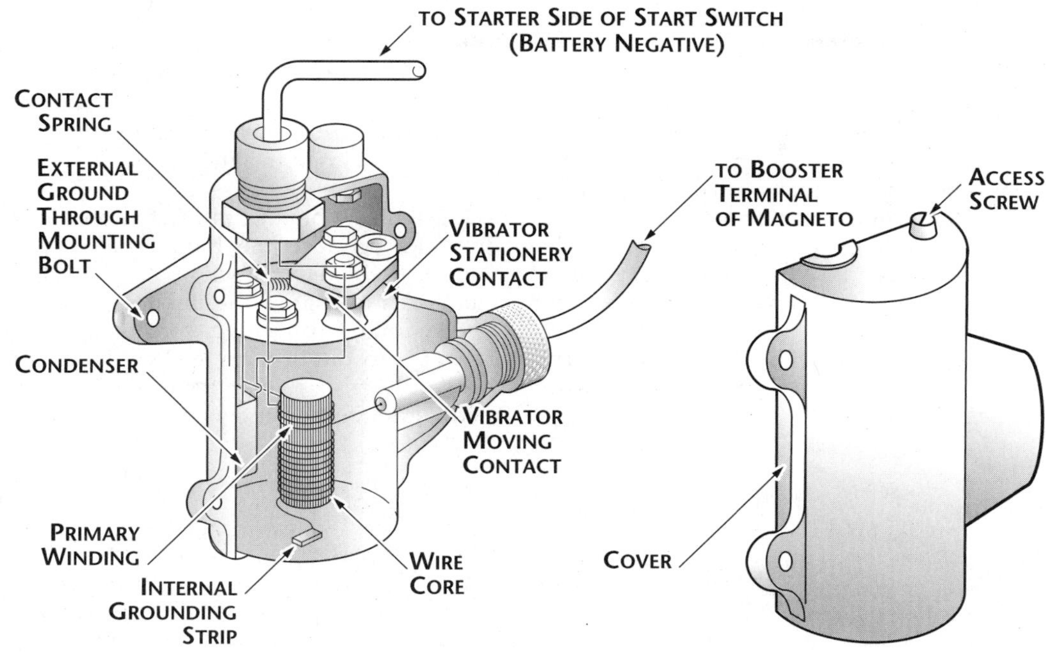


Figure 1-6-17. Booster coil

Since the regular distributor terminal is grounded through the primary or secondary coil of a high-tension magneto, the high voltage furnished by the booster coil must be distributed by a separate circuit in the distributor rotor. This is accomplished by using two electrodes in one distributor rotor. The main electrode, or *finger*, carries the magneto output voltage, and the auxiliary electrode distributes only the output of the booster coil. The auxiliary electrode is always located so that it trails the main electrode, thus retarding the spark during the starting period.

Figure 1-6-17 illustrates, in schematic form, the booster coil components shown in Figure 1-6-18. In operation, battery voltage is applied to the positive (+) terminal of the booster coil through the start switch. This causes current to flow through the closed contact points (Figure 1-6-18) to the primary coil and ground. Current flow through the primary coil sets up a magnetic field about the coil that magnetizes the coil core. As the core is magnetized, it attracts the movable contact point, which is normally held against the stationary contact point by a spring.

As the movable contact point is pulled toward the iron core, the primary circuit is broken, collapsing the magnetic field that extended about the coil core. Since the coil core acts as an electromagnet only when current flows in the primary coil, it loses its magnetism as soon as the primary coil circuit is broken. This permits the action of the spring to close the contact points and again complete the primary coil circuit. This, in turn, re-magnetizes the coil core, and again attracts the movable contact point, which again opens the primary coil circuit. This action causes the movable contact point to vibrate rapidly, as long as the start switch is held in the closed (on) position. The result of this action is a continuously expanding and collapsing magnetic field that links the secondary coil of the booster coil. With several times as many turns in the secondary as in the primary, the induced

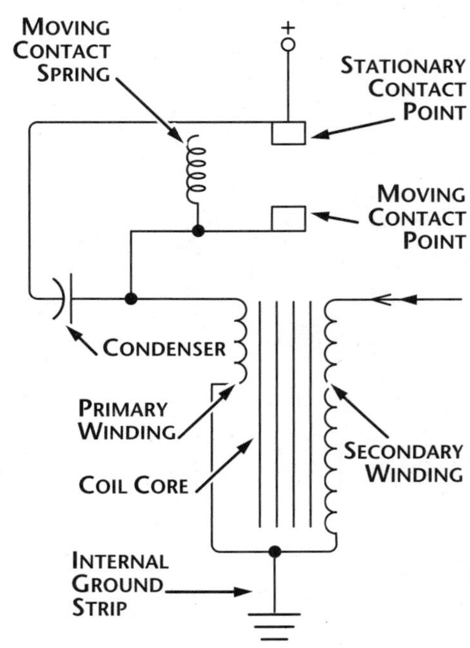


Figure 1-6-18. Booster coil schematic

voltage that results from lines of force linking the secondary is high enough to furnish ignition for the engine.

The condenser (Figure 1-6-18, previous page), which is connected across the contact points, has an important function in this circuit. As current flow in the primary coil is interrupted by the opening of the contact points, the high self-induced voltage that accompanies each collapse of the primary magnetic field surges into the condenser. Without a condenser, an arc would jump across the points with each collapse of the magnetic field. This would burn and pit the contact points and greatly reduce the voltage output of the booster coil.

## Induction Vibrator

The induction vibrator (or starting vibrator) shown in Figure 1-6-19 consists essentially of an electrically operated vibrator, a condenser, and a relay. These units are mounted on a base plate and enclosed in a metal case.

The starting vibrator, unlike the booster coil, does not produce the high ignition voltage within itself. The function of this starting vibrator is to change the direct current of the battery into a pulsating direct current and deliver it to the primary coil of the magneto. It also serves as a relay for disconnecting the auxiliary circuit when it is not in use.

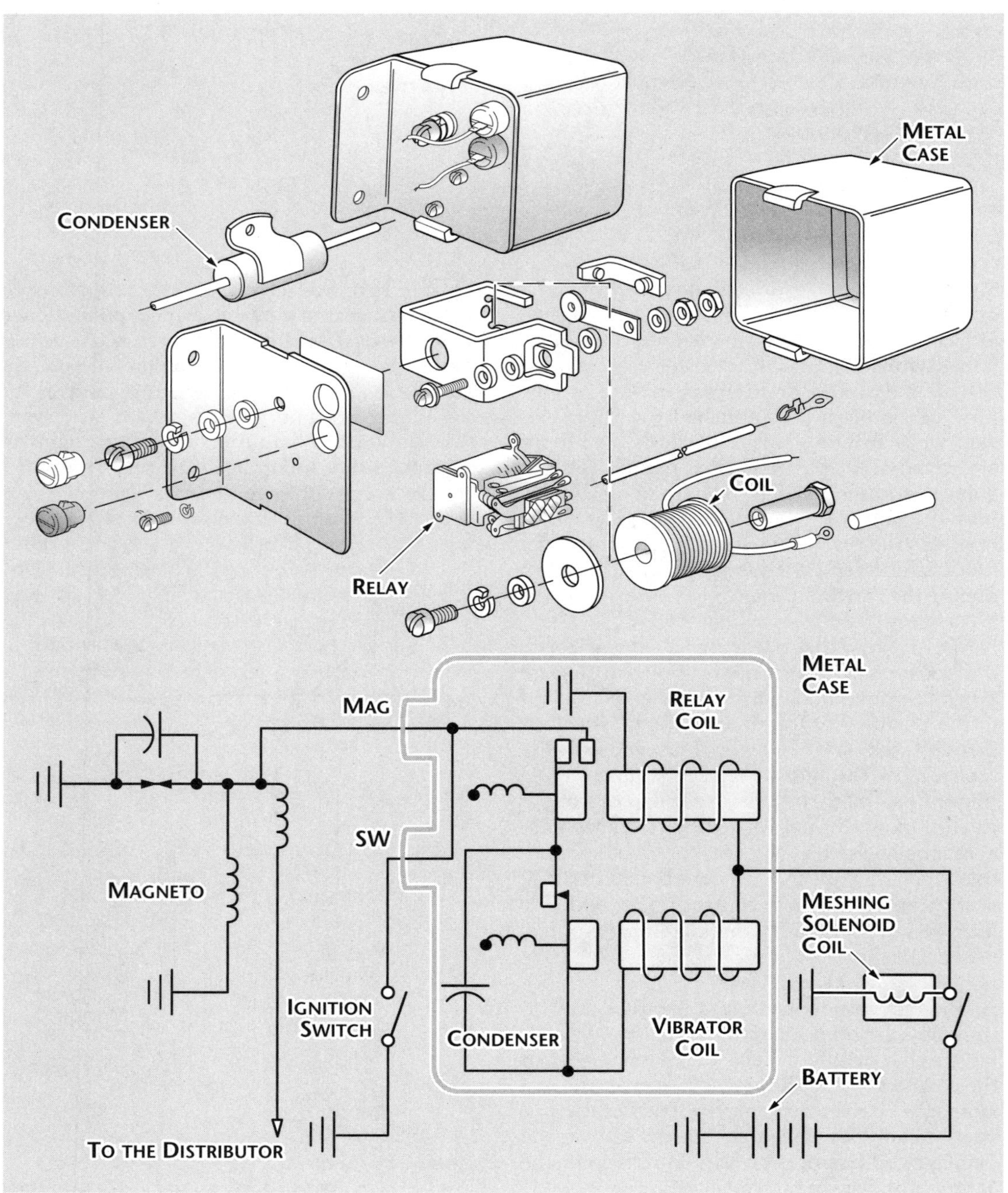


Figure 1-6-19. Induction vibrator

As shown in Figure 1-6-19, the positive terminal of the starting vibrator is connected into the starter meshing solenoid circuit. Closing this switch energizes the meshing solenoid and causes current to flow through the relay coil to ground. At the same time, current also flows through the vibrator coil and its contact points. Since current flow through the relay coil establishes a magnetic field that attracts and closes the relay points, the vibrator circuit is now complete to the magneto. The electrical path that battery current takes in the magneto is determined by the position of the primary breaker contact points; if the points are closed, current flows through them to ground; if the points are open, current flows through the primary coil to ground.

Current flow in the vibrator coil sets up a magnetic field that attracts and opens the vibrator points. When the vibrator points open, current flow in the coil stops, and the magnetic field that attracted the movable vibrator contact point disappears. This allows the vibrator points to close and again permits battery current to flow in the vibrator coil. This completes a cycle of operation. The cycle, however, occurs many times per second, so rapidly, in fact, that the vibrator points produce an audible buzz.

Each time the vibrator points close, current flows to the magneto. If the primary breaker contact points are closed, almost all the battery current passes to ground through them, and very little passes through the primary coil. Thus, no appreciable change in flux in the primary coil occurs. When the magneto breaker contact points open, current that had been flowing through the breaker points is now directed through the primary coil to ground. Since this current is being interrupted many times per second, the resulting magnetic field is building and collapsing across the primary and secondary coils of the magneto many times per second.

The rapid successions of separate voltages induced in the secondary coil will produce a *shower of sparks* across the selected spark plug air gap. The succession of separate voltages is distributed through the main distributor finger to the various spark plugs because the breaker contact points trigger the sparks just as they do when the magneto is generating its own voltage. Ignition systems that use an induction vibrator have no provision for retarding the spark; hence they will not have a trailing auxiliary distributor electrode.

When starting an engine equipped with an induction vibrator, the ignition switch must be kept in the OFF position until the starter has cranked the propeller through one revolution. Then, while the propeller is still turning, the ignition switch should be turned to the ON (or BOTH) position. If this precaution is not observed, engine kickback will probably result from ignition before top center and low cranking r.p.m. After the propeller has completed at least one revolution, it will have gained sufficient momentum to prevent kickback.

As soon as the engine begins firing and the starter switch is released, the electric circuit from the battery to the induction vibrator is opened. When battery current is cut off from the induction vibrator, the relay points open and break the connection between the induction vibrator and the magneto.

This connection must be broken to prevent the magneto from being grounded-out at the relay coil. If the relay points of the induction vibrator did not open when battery current was cut off, primary current in the magneto would not be interrupted when the breaker points open; instead, primary current would flow back through the relay and vibrator points of the induction vibrator and then to ground through the relay coil. In this event, the magneto would be just as inoperative as though the ignition switch were placed in the OFF position.

The latest development in this technology replaces the induction vibrator with solid-state circuitry that accomplishes the same thing, with no moving parts. These units are now certified to replace most induction vibrators.

## Impulse Coupling

Engines having a small number of cylinders are sometimes equipped with what is known as an impulse coupling (Figure 1-6-20). This is a unit which will, at the time of spark production, give one of the magnetos attached to the engine a brief acceleration and produce a hot spark for starting. This device consists of small flyweights and spring assemblies located within the housing that attaches the magneto to the accessory shaft.

**Figure 1-6-20. An impulse coupling on a magneto, showing the spring-loaded flyweights and stop pins**

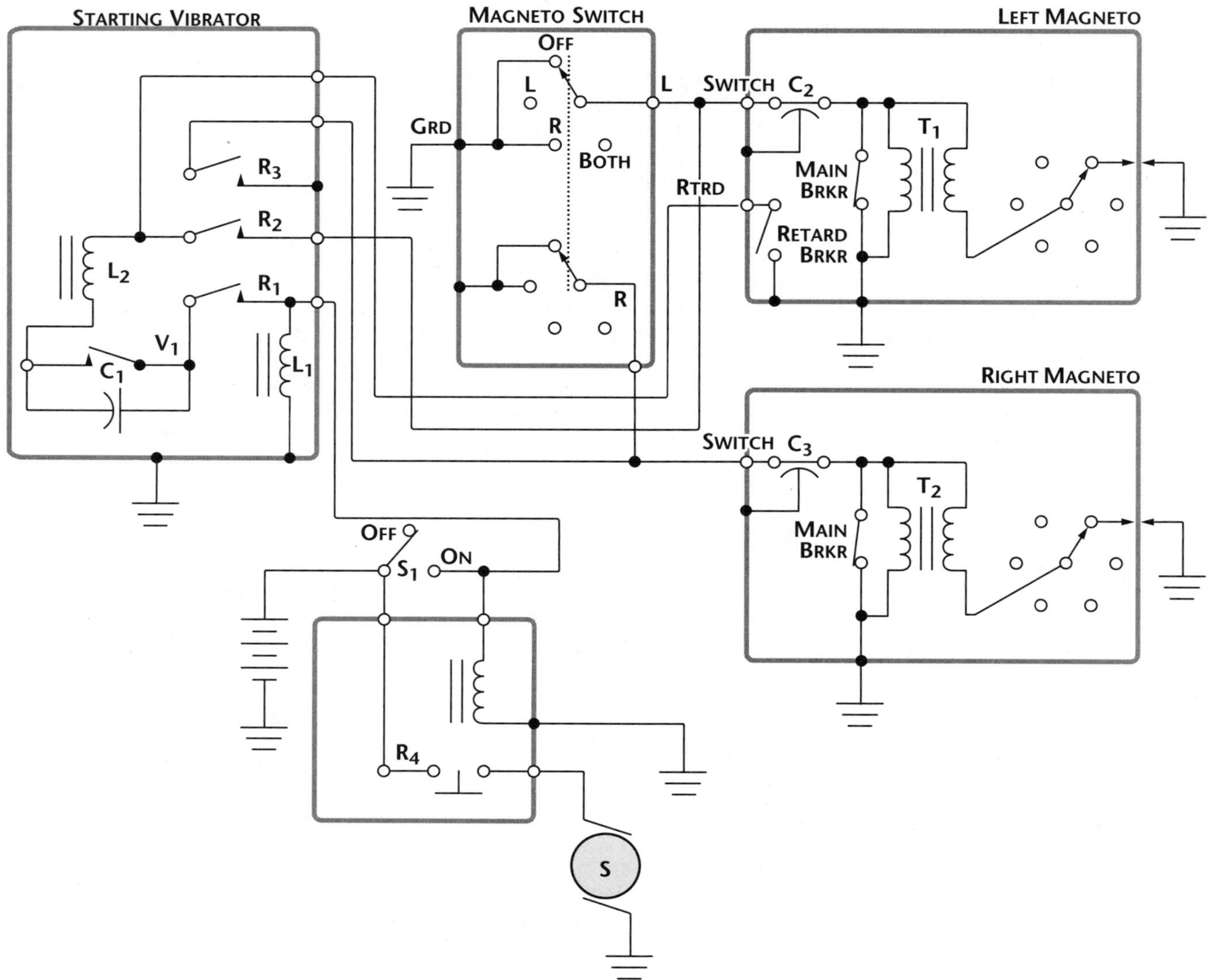


**Figure 1-6-21. High-tension, retard breaker magneto and starting vibrator circuit**

The magneto is flexibly connected through the impulse coupling by means of the spring so that at low speed the magneto is temporarily held while the accessory shaft is rotated until the pistons reach approximately a top center position. At this point the magneto is released and the spring kicks back to its original position, resulting in a quick twist of the rotating magnet. This, being equivalent to high-speed magneto rotation, produces a hot spark.

After the engine is started and the magneto reaches a speed at which it furnishes sufficient current, the flyweights in the coupling fly outward due to centrifugal force and lock the two coupling members together. That makes it a solid unit, returning the magneto to a normal timing position relative to the engine. The presence of an impulse coupling is identified by a sharp clicking noise as the crankshaft is turned at cranking speed past top center on each cylinder.

Use of the impulse coupling produces impact forces on the magneto, the engine drive parts, and various parts of the coupling unit. Often the flyweights become magnetized and do not engage the stop pins; congealed oil on the flyweights during cold weather may produce the same results. Another disadvantage of the impulse coupling is that it can produce only one spark for each firing cycle of the cylinder. This is a disadvantage especially during adverse starting conditions.

## High-tension Retard Breaker Vibrator

The retard breaker magneto and starting vibrator system is used as part of the high-tension system on some small aircraft. Designed for four- and six-cylinder ignition systems, the retard breaker magneto eliminates the need for the impulse coupling in light aircraft. This system uses an additional breaker to obtain retarded sparks for starting. The starting vibrator is also adaptable to many helicopter ignition systems. A schematic diagram of an ignition system using the retard breaker magneto and starting vibrator concept is shown in Figure 1-6-21.

With the magneto switch in the BOTH position (Figure 1-6-21) and the starter switch S1 in the ON position, starter solenoid L3 and coil L1 are energized, closing relay contacts R4, R1, R2 and R3. R3 connects the right magneto to ground, keeping it inoperative during starting operation. Electrical current flows from the battery through R1, vibrator points V1, coil L2, through both the retard breaker points, and through R2 and the main breaker points of the left magneto to ground.

The energized coil L2 opens vibrator points V1, interrupting the current flow through L2. The magnetic field about L2 collapses, and vibrator points V1 close again. Once more, current flows through L2, and again V1 vibrator points open. This process is repeated continuously, and the interrupted battery current flows to ground through the main and retard breaker points of the left magneto.

Since relay R4 is closed, the starter is energized and the engine crankshaft is rotated. When the engine reaches its normal advance firing position, the main breaker points of the left magneto open. The interrupted surges of current from the vibrator can still find a path to ground through the retard breaker points, which do not open until the retarded firing position of the engine is reached. At this point in crankshaft travel, the retard points open. Since the main breaker points are still open, the magneto primary coil is no longer shorted, and current produces a magnetic field around T1.

Each time the vibrator points V1 open, current flow through V1 is interrupted. The collapsing field about T1 cuts through the magneto coil secondary and induces a high-voltage surge of energy used to fire the spark plug. Since the V1 points are opening and closing rapidly and continuously, a shower of sparks is furnished to the cylinders when both the main and retard breaker points are open.

After the engine begins to accelerate, the manual starter switch is released, causing L1 and L3 to become de-energized. This action causes both the vibrator and retard breaker circuits to become inoperative. It also opens relay contact R3, which remove the ground from the right magneto. Both magnetos now fire at the advance (normal running) piston position.

# Spark Plugs

The function of the spark plug in an ignition system is to conduct a short impulse of high voltage current through the wall of the combustion chamber. Inside the combustion chamber it provides an air gap across which this impulse can produce an electric spark to ignite the fuel/air charge. While the aircraft spark plug is simple in construction and operation, it is nevertheless the direct or indirect cause of a great many malfunctions in aircraft engines. But spark plugs provide a great deal of trouble-free operation, considering the adverse conditions under which they operate.

In each cylinder of an engine operating at 2,100 r.p.m., approximately 17 separate and distinct high-voltage sparks bridge the air gap of a single spark plug each second. This appears to the naked eye as a continuous fire searing the spark plug electrodes at temperatures of over 3,000°F. At the same time the spark plug is subjected to gas pressures as high as 2,000 p.s.i. and electrical pressure as high as 15,000 volts.

The three main components of a spark plug (Figure 1-6-22) are the electrode, insulator and outer shell. The outer shell, threaded to fit into the cylinder, is usually made of finely machined steel and is often plated to prevent

Figure 1-6-22. A typical aircraft spark plug

corrosion from engine gases and possible thread seizure. Close-tolerance screw threads and a copper gasket prevent cylinder gas pressure from escaping around the plug. Pressure that might escape through the plug is retained by inner seals between the outer metal shell and the insulator, and between the insulator and the center electrode assembly.

The insulator provides a protective core around the electrode. In addition to affording electrical insulation, the ceramic insulator core also transfers heat from the ceramic tip, or nose, to the cylinder.

The types of spark plugs used in different engines vary in respect to heat range, reach, thread size or other characteristics of the installation requirements of different engines.

**Heat range.** The heat range of a spark plug is a measure of its ability to transfer heat to the cylinder head. The plug must operate hot enough to burn off deposits which can cause fouling, yet remain cool enough to prevent a pre-ignition condition. The length of the nose core is the principal factor in establishing the plug's heat range. *Hot plugs* have a long insulator nose that creates a long heat transfer path, whereas *cold plugs* have a relatively short insulator to provide a rapid transfer of heat to the cylinder head (Figure 1-6-23).

If an engine were operated at only one speed, spark plug design would be greatly simplified.

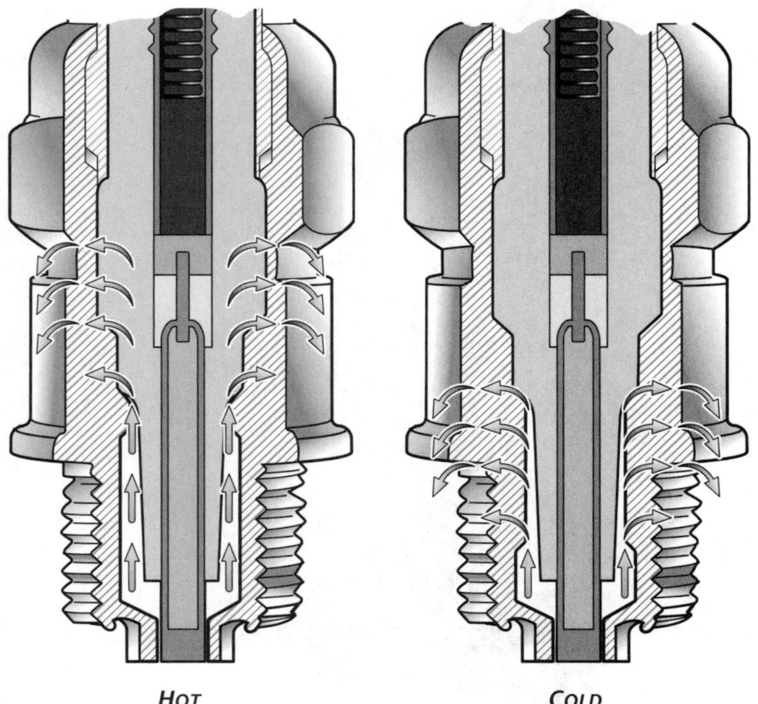


Figure 1-6-23. Hot and cold spark plugs

Because flight demands impose different loads on the engine, spark plugs must be designed to operate as hot as possible at slow speeds and light loads, and as cool as possible at cruise and takeoff power.

The choice of spark plugs to be used in a specific aircraft engine is determined by the engine manufacturer after extensive tests. When an engine is certified to use hot or cold spark plugs, the plug used is determined by how the engine is to be operated.

**Reach.** A spark plug with the proper reach (Figure 1-6-24) will ensure that the electrode end inside the cylinder is in the best position to achieve ignition. The *spark plug reach* is the threaded portion inserted in the spark plug bushing of the cylinder. Spark plug seizure and/or improper combustion within the cylinder will probably occur if a plug with the wrong reach is used.

Typically, the short plugs are installed in normally aspirated engines and the long plugs are installed in turbocharged engines. The extra length of the long reach plugs provides extra grip in the cylinder head to keep the plug from being blown out of the cylinder by the high pressures developed in supercharged engines.

Another feature of spark plugs that is determined by their application is the size of the barrel as defined by the outer diameter and pitch of the threads where the ignition lead attaches. The two sizes currently used are 5/8-24 and 3/4-20. The 5/8-24 thread is 5/8 inch in diameter with 24 threads per inch, and the 3/4-20 is 3/4 inch in diameter and has 20 threads per inch.

Generally, the 5/8-24 plugs are standard equipment on normally aspirated engines, and the 3/4-20 are used on turbocharged engines. The 3/4-20 plugs are also referred to as all-weather plugs, because features at the harness connection end provide for a better moisture seal than the small barreled plugs.

## Spark Plug Rotation

The policy of rotating spark plugs from top to bottom has been practiced by technicians and pilots for many years. It is common knowledge in the industry that the bottom plugs are always the dirty ones, and the top plugs are the clean ones. By periodically switching the plugs from top to bottom, you get a self-cleaning action from the engine, whereby the dirty plug placed in the top is cleaned, while the clean plug replaced in the bottom gradually becomes dirty. Based on this cleaning action, a rotational time period must be established.

Due to the ever-increasing cost of aircraft maintenance and a desire to get the maximum service life from your spark plugs, the following information is offered on the effects of constant polarity and how to rotate plugs to get maximum service life.

The polarity of an electrical spark — either positive or negative — and its effects on spark plug electrode erosion have long been known but have had little effect on spark plug life in the relatively low-performance engines of the past. In more recent, high-performance, normally aspirated and turbocharged engines, where cylinder temperature and pressure are much higher, the adverse effects of constant polarity are becoming more prevalent.

When a spark plug is installed in a cylinder that is fired negative and is allowed to remain there for a long period of time, more erosion occurs on the center electrode than on the ground electrode. When a spark plug is fired positive, more erosion occurs on the ground electrode than on the center electrode. From this, it follows that a periodic exchange of spark plugs fired positive with those fired negative will result in even wear and longer spark plug service life.

To get a polarity change, as well as switching the plugs from top to bottom, the following rotational sequence is suggested. First, when removing the spark plugs from the engine, keep them in magneto sets. After the plugs have been serviced and are ready to be reinstalled in the engine, make the following plug exchange.

- For six-cylinder engines, switch the plugs from the odd-numbered cylinders with the plugs from the even-numbered cylinders. For example, switch 1 with 6, 2 with 5 and 3 with 4.
- On four-cylinder engines, you must switch 1 with 4 and 2 with 3.

During the operating period following the rotation, each plug will be fired at reverse polarity to the former operating period. This will result in even spark plug wear and longer service life.

This rotational procedure works equally well on all four- and six-cylinder Lycoming engines, except four-cylinder engines equipped with the single-unit dual magneto. This is a constant-polarity magneto, and the only benefit to be gained by rotating the plugs is the reduction of lead deposit build-up on the spark plugs when a rotational time period is established and followed.

Another exception occurs on a few four-cylinder engines where one magneto will fire all of the top spark plugs and the other magneto will fire all of the bottom spark plugs. If the plugs are rotated as previously recommended, a polarity change will still result, but since the plugs do not get moved from top to bottom, no self-cleaning action by the engine will occur. This may result in the necessity to clean the dirtier bottom plugs at regular intervals.

Figure 1-6-24. spark plug reach

## Ignition System Maintenance and Inspection

An aircraft's ignition system is the result of careful design and thorough testing. In all probability, the ignition system will provide good, dependable service, provided it is maintained properly. However, difficulties can occur which will affect ignition system performance.

The most common of these maintenance difficulties, together with the most generally accepted methods of ignition inspection, are discussed in this section.

Breakdown of insulating materials, burning and pitting of breaker points and short circuits or broken electrical connections are not uncommon, and must be found and corrected.

Less common are the irregularities that involve human error. For example, ignition timing requires precise adjustment and painstaking care so that the following four conditions occur at the same instant:

1. The piston in the No. 1 cylinder is in a position a prescribed number of degrees before TDC on the compression stroke.
2. The rotating magnet of the magneto is in the E-gap position.
3. The breaker points are just opening on the No. 1 cam lobe.
4. The distributor finger is aligned with the electrode serving the No. 1 cylinder.

If one of these conditions is out of synchronization with any of the others, the ignition system is said to be out of time.

When ignition in the cylinder occurs before the optimum crankshaft position is reached, the timing is said to be *early*. If ignition occurs too early, the piston rising in the cylinder will be opposed by the full force of combustion. This condition results in a loss of engine power, overheating, and possible detonation and preignition. If ignition occurs at a time after the optimum crankshaft position is reached, the ignition timing is said to be *late*. If it occurs too late, not enough time will be allowed to consume the fuel/air charge, and combustion will be incomplete. As a result, the engine loses power, and a greater throttle opening will be required to carry a given propeller load.

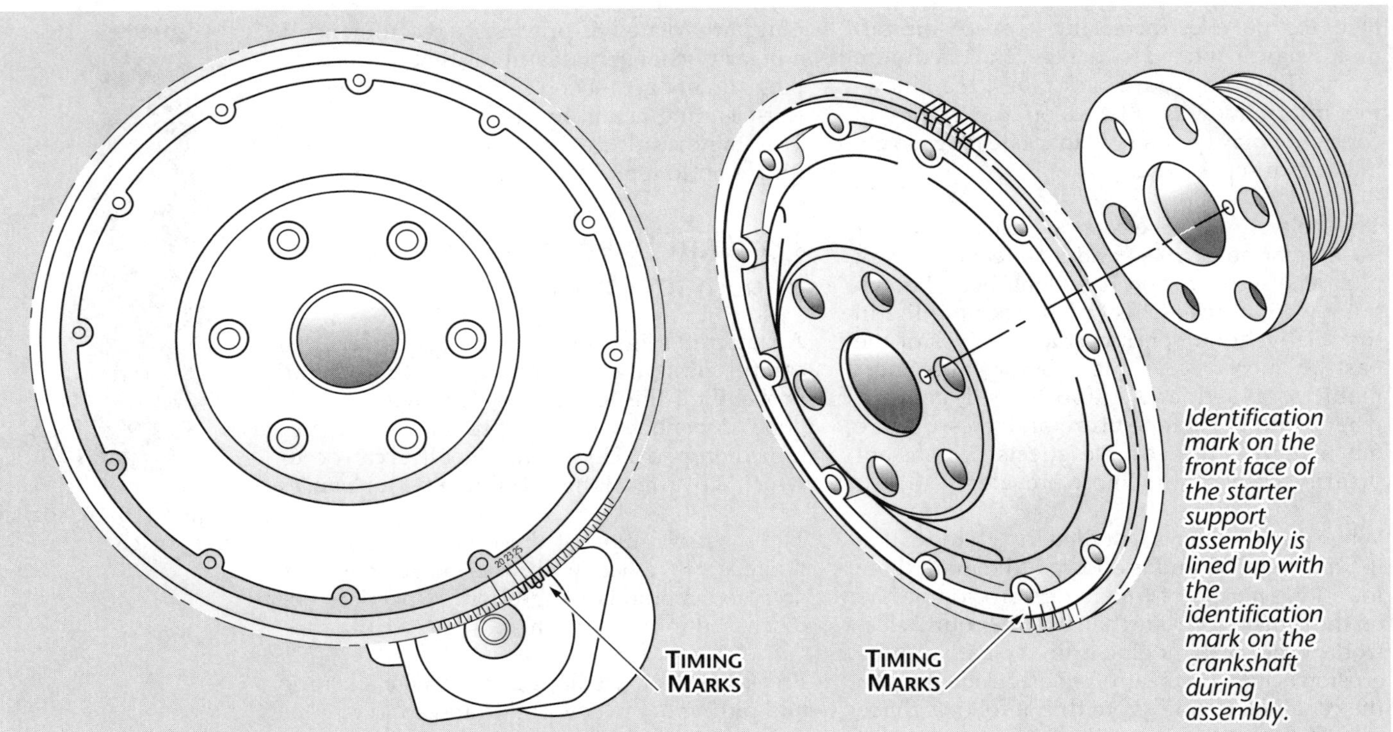


**Figure 1-6-25. Propeller flange timing marks**

More common irregularities are those caused by moisture forming on different parts of the ignition system. Moisture can enter ignition system units through cracks or loose covers, or it can result from condensation. *Breathing*, a situation that occurs during the readjustment of the system from low to high atmospheric pressure, can result in drawing in moisture-laden air.

Ordinarily, the heat of the engine is sufficient to evaporate this moisture, but occasionally the moist air condenses as the engine cools. The result is an appreciable moisture accumulation, which causes the insulation materials to lose their electrical resistance. A slight amount of moisture contamination may cause reduction in magneto output by short-circuiting to ground a part of the high-voltage current intended for the spark plug.

If the moisture accumulation is appreciable, the entire magneto output may be dissipated to ground by way of flashover and carbon tracking. Moisture accumulation during flight is extremely rare, because the high operating temperature of the system is effective in preventing condensation; hence difficulties from this cause will probably be more evident during ground operation.

Aircraft spark plugs take the blame unjustly for many ignition system malfunctions. Spark plugs are often diagnosed as being faulty when the real malfunction exists in some other system. Malfunctioning of the carburetor, poor fuel distribution, too much valve overlap, leaking primer system or poor idle speed and mixture settings will show symptoms that are the same as those for faulty ignition.

Unfortunately, many of these conditions can be temporarily improved by a spark plug change, but the trouble will recur in a short time because the real cause of the malfunction has not been eliminated. A thorough understanding of the various engine systems, along with meticulous inspection and good maintenance methods, can substantially reduce such errors.

## Magneto Ignition Timing

When the many opportunities for errors in timing the ignition system to the engine are considered, the emphasis placed on the correct use of timing devices that follows is easily justified. Errors can easily occur in positioning the piston in the timing cylinder. It can be placed at the wrong crankshaft degree or at the correct crankshaft degree but on the wrong stroke. When positioning the magneto's rotating magnet, an error can be made by not removing the backlash in the gear train. The breaker point assemblies may not be perfectly synchronized, or they may be synchronized but not opening at E-gap. Any other errors will alter the final spark timing. Because of the many chances for error, timing devices have been developed to ensure more consistent and accurate timing methods.

## Built-in Engine Timing Reference Marks

Most reciprocating engines have timing reference marks built in. On an engine that has no propeller reduction gear, the timing mark will normally be on the propeller flange edge (Figure 1-6-25). The *top center* (TC) mark stamped on the edge will align with the crankcase split line below the crankshaft when the No. 1 piston is at TDC. Other flange marks indicate degrees before TC. Some engines have degree markings on the propeller reduction drive gear. To time these engines, the plug provided on the exterior of the reduction gear housing must be removed to view the timing marks. On other engines, the timing marks are on a crankshaft flange and can be viewed by removing a plug from the crankcase. In every case, the engine manufacturer's instructions will give the location of built-in timing reference marks. Always check the manufacturer's maintenance manual for the location of timing marks.

When using built-in timing marks to position the crankshaft, be sure to sight straight across the stationary pointer or mark on the nose section, the propeller shaft, crankshaft flange or bell gear. Sighting at an angle will result in an error in positioning the crankshaft.

While many engines have timing reference marks, they leave something to be desired. The main drawback is the backlash factor. The amount of backlash in any system of gears will vary between installations and will even vary between two separate checks on the same piece of equipment. This results because there is no way of imposing a load on the gear train in a direction opposite the direction of crankshaft rotation. Another unfavorable aspect in the use of timing marks on the reduction gear is the small error that exists when sighting down the reference mark to the timing mark inside the housing on the reduction gear. Because there is depth between the two reference marks, each technician must be sighting from exactly the same plane. If not, each person will select a different crankshaft position for ignition timing.

## Timing Disks

The *timing disk* is a more accurate crankshaft positioning device than the timing reference marks. This device consists of a disk and a pointer mechanism mounted on an engine-driven accessory or its mounting pad. The pointer, which is indirectly connected to the accessory drive, indicates the number of degrees of crankshaft travel on the disk. The disk is marked off in degrees of crankshaft travel. By applying a slight torque to the accessory drive gear in a direction opposite that of the normal rotation, the *backlash* in the accessory gear train can be removed to the extent that a specific crankshaft position can be obtained with accuracy time after time.

Not all timing disks are marked off in the same number of degrees. For example, the disk designed for use on one type of engine is mounted on the fuel pump drive pad. Since the fuel pump is driven at the same speed as the crankshaft, the pointer will describe a complete circle when the crankshaft completes one revolution. Hence, the disk is laid out in one-degree increments throughout a total of 360°. However, the timing disk used on another engine is mounted on top of the magneto, which is driven at one-half crankshaft speed. With this arrangement, the crankshaft will move one degree while the pointer moves only one-half a degree. Therefore, the disk is marked off in 720 increments of 1/2°, each indicating one full degree of crankshaft travel.

## Piston Position Indicators

Any given piston position, whether it is to be used for ignition, valve or injection pump timing, is referenced to a piston position called *top dead center* (TDC). This piston position is not to be confused with a rather hazily defined piston position called top center (TC).

A piston in top center has little value from a timing standpoint, because the corresponding crankshaft position may vary from 1°-5° for this piston position. This is illustrated in Figure 1-6-26, which is exaggerated to emphasize the *no-travel zone* of the piston. Notice that the piston does not move while the crankshaft describes the small arc from position A to position B. This no-travel zone occurs between the time the crankshaft and connecting rod stop pushing the piston upward and continues until the crankshaft has swung the lower end of the connecting rod into a position where the crankshaft can start pulling the piston downward.

TDC is a piston and crankshaft position from which all other piston and crankshaft locations are referenced. When a piston is in the TDC position, it will be a maximum distance from the center of the crankshaft and in the center of the no travel zone. This places the piston in a position where a straight line can be drawn through the center of the crankshaft journal, the crankpin, and the piston pin, as shown in the lower diagram of Figure 1-6-26. With such an alignment, a force applied to the piston could not move the crankshaft.

Perhaps the earliest piston position indicator was a wooden rod or pencil. One end of this simple device was inserted at an angle through the spark plug hole of the timing cylinder until it came to rest on the top far edge of the piston,

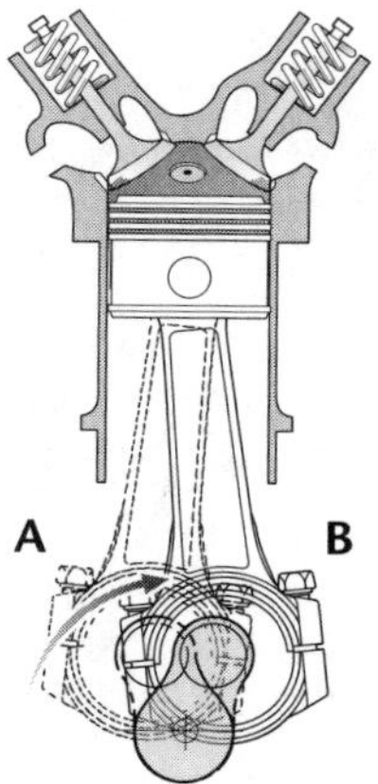


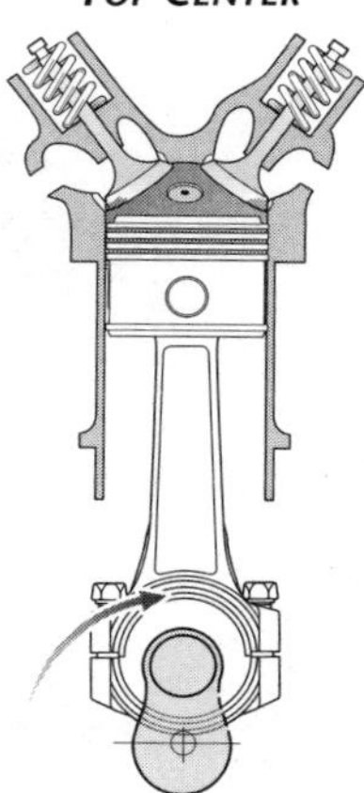


**Figure 1-6-26. Illustrating the difference between top center and top dead center (TDC)**

as shown in Figure 1-6-27. Then the crankshaft was rotated until the piston stopped moving the end of the rod outward. At this point, the mechanic would grasp the rod with a thumbnail resting at a point where the rod contacted the top edge of the spark plug hole. With the thumbnail still in this position, the rod would be extracted, and a notch cut about 1 inch above the thumbnail location. This notch provided the technician with an arbitrary reference point somewhere before TDC.

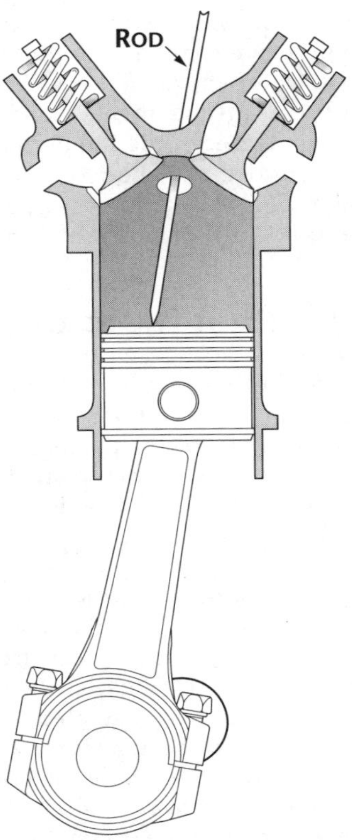


Figure 1-6-27. Simple piston position indicator

Figure 1-6-28. A Time-Rite® indicator

Such an inexact procedure cannot be relied on to find the same piston position each time. All piston position indicators in use today screw into the spark plug hole so that the indicator always enters the cylinder in exactly the same plane and its indicating rod always contacts the same part of the piston head.

One of the various piston position indicators in use today is a *Time-Rite® Indicator* (Figure 1-6-28). It serves the purpose of a piston position indicator and, to a limited degree, as a timing disk. The device consists of two parts, a body shell and a face. The shell is essentially an adapter which screws into the spark plug hole and supports the face. The face snaps into the adapter and contains a spring-loaded compensated indicator arm, a slide pointer, a removable scale calibrated in degrees, an indicator light, and a frame which extends behind the face to form a hinge point for the compensated indicator arm. One end of the compensated arm extends into the cylinder through the spark plug hole and is actuated by piston movement. The other end of the arm extends through a slot in the face and actuates the slide pointer over the scale.

A variety of arms and graduated scales is furnished with the Time-Rite®. Both arms and scales are compensated for the particular engine for which they are marked. Compensation is necessary because of variations in piston stroke and spark plug hole locations in different cylinders. The arms are compensated by varying their shapes and lengths, and the scales are compensated by the spacing of the degree markings. In this way, a particular arm and scale combination will indicate true piston position if used correctly.

To ensure even greater accuracy with the Time-Rite®, a small light, powered by a small dry cell battery, is mounted in its face. When the compensated arm contacts the slide pointer, an electrical circuit is completed and the light comes on. This light provides greater accuracy because the slide pointer can be positioned at the desired degree setting on the scale and the crankshaft slowly rotated by bumping the propeller shaft until the light just flashes on. The propeller shaft must be moved slowly and carefully so that the arm will not overshoot and move the slide pointer beyond the desired degree setting after the light flashes on.

There are two other common types of piston position indicators in use, both of which operate on the piston positioning principle. One features a scale of reference points. The other is simply a light that comes on when the piston touches the actuating arm and goes out when the piston moves below the reach of the arm.

## Timing Lights

Timing lights are used to help determine the exact instant that the magneto points open. There are two general types of timing lights in common use. Both have two lights and three external wire connections. Although both have internal circuits that are different, their function is much the same. One type of light and its internal circuit is shown in Figure 1-6-29.

Three wires plug into the top of the light box (top illustration of Figure 1-6-29). There are also two lights on the front face of the unit, and a switch to turn the unit on and off. The wiring diagram (bottom illustration of Figure 1-6-29) shows that the unit contains a battery, a vibrator coil and two transformers. To use the timing light, the center lead, marked *ground lead*, is connected to the case of the magneto being tested. The other leads are connected to the primary leads of the breaker point assembly of the magnetos being timed.

With the leads connected in this manner, it can be easily determined whether the points are open or closed by turning on the switch and observing the two lights. If the points are closed, most of the current will flow through the breaker points and not through the transformers, and the lights will not come on. If the points are open, the current will flow through the transformer and the lights will glow. Some models of timing lights operate in the reverse manner, i.e., the light goes out when the points open. Each of the two lights is operated separately by the set of breaker points to which it is connected. This makes it possible to observe the time, or point in reference to magneto rotor rotation, that each set of points opens.

Most timing lights use dry cell batteries that must be replaced after long use. Attempts to use a timing light with weak batteries may result in erroneous readings because of low current flow in the circuits.

## Checking the Internal Timing of a Magneto

When replacing a magneto or preparing a magneto for installation, the first concern is with the internal timing of the magneto. For each magneto model, the manufacturer determines how many degrees beyond the neutral position a pole of the rotor magnet should be to obtain the strongest spark at the instant of breaker point separation. This angular displacement from the neutral position, known as the *E-gap angle*, will vary with different magneto models.

On one model, a step is cut on the end of the breaker cam for checking internal timing of the magneto. When a straightedge is laid along this step and it coincides with the timing marks on the rim of the breaker housing, the magneto rotor is then in the E-gap position, and the breaker contact points should just begin to open.

Another method for checking E-gap is to align a timing mark with a pointed, chamfered tooth

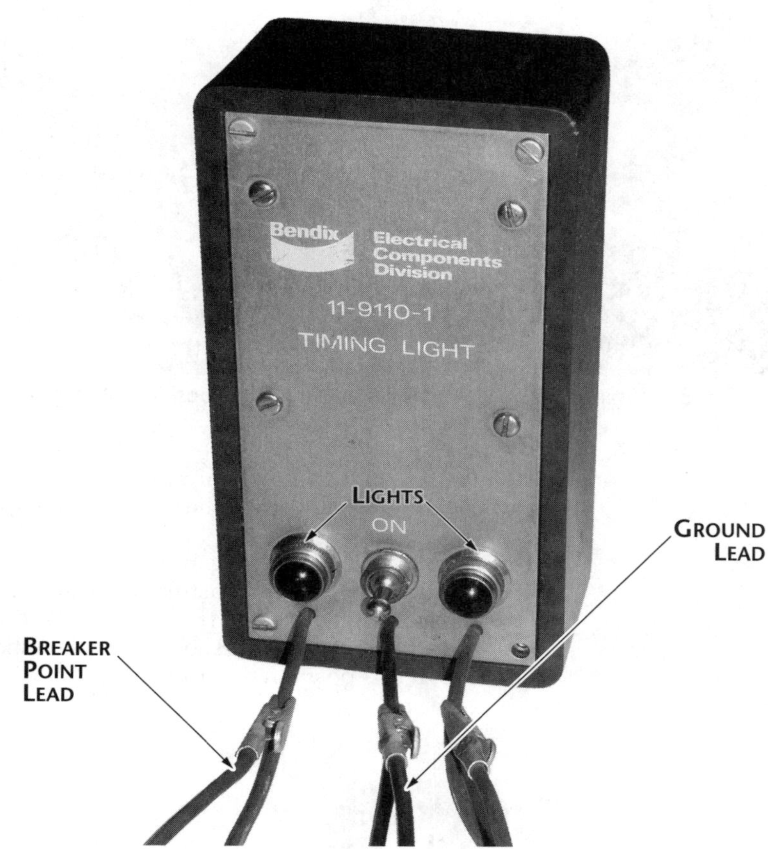


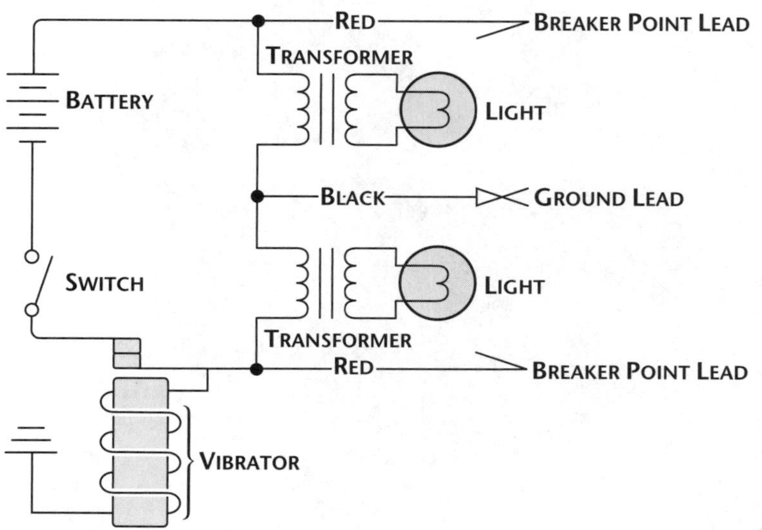


Figure 1-6-29. Timing light and wiring diagram

**Figure 1-6-30. Timing marks which indicate No. 1 firing position of magneto**

(Figure 1-6-30). The breaker points should be just starting to open when these marks line up.

In a third method, the E-gap is correct when the E gap gauge rests against the pole lamination in the magneto frame (Figure 1-6-31). The contact points should be just opening when the rotor is in this position described.

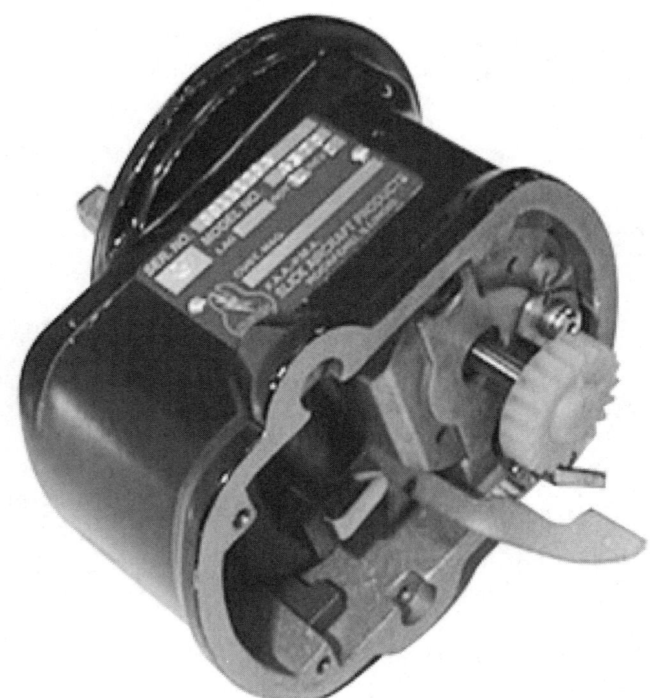

**Figure 1-6-31. Checking magneto E-gap**

*Bench timing* the magneto involves positioning the magneto rotor at the E-gap position and setting the breaker points to open when the timing lines or marks provided for that purpose are perfectly aligned.

## Timing the High-tension Magneto to the Engine

When replacing magnetos on aircraft engines, two factors must be considered:

1. The internal timing of the magneto, including breaker point adjustment, which must be correct to obtain maximum potential voltage from the magneto
2. The crankshaft position at which the spark occurs — some engines will use a staggered spark, and both magnetos will not fire at the same time

**NOTE:** *The purpose of staggering the spark is to fire the spark plug nearest the exhaust port first. This is becuase, in that area, the incoming mixture gets diluted with exhaust gases, and the flame front will travel more slowly.*

One breaker point gap should never be compared with another, since it would not be known if either set were opening at the proper number of degrees before TDC of the timing position of the engine. The magneto must be timed by first adjusting the internal timing of the magneto and then by checking and adjusting the ignition points to open at this position.

If the reference timing mark for the magneto lines up when the timing piston is at a prescribed number of degrees ahead of true TDC, both the right and left set of breaker points open at that instant and remain open for the prescribed number of degrees, the internal magneto timing is correct, proper magneto-to-engine timing exists, and all phases of magneto operation are synchronized. In no case should the breaker points be adjusted when the internal timing of the magneto, as designated by internal timing reference marks, is off in relation to the prescribed piston position.

For timing the magneto to the engine in the following example, a timing light is used. The timing light is designed in such a way that one of two lights will come on when the points open. The timing light incorporates two lights; hence, when connecting the timing light to the magneto, the leads should be connected so that the light on the right-hand side of the box represents the breaker points on the right side of the magneto, and the light on the left-hand side represents the left breaker points. The proper connection of wires can be established by turning the timing light on and then touching one of the red wires against the black wire.

If the right light goes out, the red lead used should be connected to the magneto housing or the engine to effect a ground. When using the timing light to check a magneto in a complete ignition system installed on the aircraft, the master ignition switch for the aircraft must be turned on and the ignition switch for the engine turned to both. Otherwise, the lights will not indicate breaker point opening. With the ignition switch on and the timing light connected, the magneto is rendered inoperative; hence, no firing impulses can occur when the propeller is turned.

After determining that the magneto internal timing is correct, turn the engine crankshaft until the piston of the No. 1 cylinder is in the firing position on the compression stroke. The firing position can be determined by referring to the engine manufacturer's service manual. Locate the firing position by using a piston position indicator.

While holding the magneto cam in the firing position for the No. 1 cylinder, as indicated by the alignment of the reference marks for the magneto, install the magneto drive into the engine drive.

Connect the timing light to the magneto and breaker points, and with the light and ignition switch turned on, rotate the magneto assembly, first in the direction of rotation and then in the opposite direction. Perform this procedure to determine that the timing light goes off and then on when the cam lifts the breaker points as the magneto is rotated on its mount.

Install the magneto attaching nuts on the studs and tighten slightly (Figure 1-6-32). The nuts must not be so tight as to prevent the movement of the magneto assembly when the magneto mounting flange is tapped with a mallet.

While holding the backlash out of the magneto gears and drive coupling, tap the magneto to advance or retard it until the timing marks align. This synchronizes the internal timing of the magneto to the prescribed number of degrees before TDC. After completing this adjustment, tighten the mounting nuts. Move the propeller opposite the direction of rotation one blade, and then pull it slowly in the direction of rotation until the crankshaft position is again at the prescribed number of degrees ahead of TDC. The purpose of this check is to eliminate the possibility of error because of backlash in the engine gear-train assembly and magneto gears. If the timing mark does not line up, loosen the mounting nuts and shift the magneto until the scale or straightedge will line up with the timing mark when the propeller is pulled through to the prescribed number of degrees.

**Figure 1-6-32. Installing a magneto's attaching nuts**

Reconnect the timing light. Move the propeller one blade opposite the direction of rotation and then, while observing the timing light, move the propeller in the direction of rotation until the prescribed number of degrees ahead of TDC is reached. Be sure that the lights for both sets of points come on within one-half-degree crankshaft movement of the prescribed crankshaft position.

After the points have been adjusted as necessary, recheck the point-adjusting lock screw for tightness. Always check the point opening after tightening the lock screw.

## Performing Ignition System Checks

There are normally three *ignition checks* performed on aircraft during engine run-up. The first is performed during engine warm-up. The second is performed at field barometric manifold pressure. The third is performed prior to engine shutdown.

The first ignition check is normally made during warm-up. Most manufacturers recommend this check during the warm-up period. Actually, this check is a combination of the ignition system check and ignition switch check and is used to check the ignition system for proper functioning before other checks are made. The second check is performed as the ignition system check and is used to check individual magnetos, harnesses and spark plugs. The third check is performed as the ignition switch check and checks the switch for proper grounding for ground safety purposes.

**Figure 1-6-33. Three types of spark plugs that indicate good operation. They may be cleaned, gapped, tested and reinstalled.**

The *ignition system check* is usually performed with the *power check*. The ignition check is sometimes referred to as the *field barometric check*, because on large engines it is performed with the engine operating at a manifold pressure equal to the field barometric pressure. The power check is also performed at the same manifold pressure setting. The ignition check should not be confused with the *full throttle check*. The exact r.p.m. and manifold pressure for making this check can be found in the applicable manufacturer's instructions.

The barometric pressure used as a reference will be the reading obtained from the manifold pressure gauge for the engine involved prior to starting the engine and after engine shutdown. After reaching the engine r.p.m. specified for the ignition system check, allow the r.p.m. to stabilize. Place the ignition switch in the right position and note the r.p.m. drop on the tachometer. Return the switch to the BOTH position. Allow the switch to remain in the BOTH position for a few seconds so that the r.p.m. will again stabilize. Place the ignition switch in the left position and again note the r.p.m. drop. Return the ignition switch to the BOTH position.

In performing this check, lightly tap the rim of the tachometer to ensure that the tach indicator pointer moves freely. A sticking tachometer pointer can conceal ignition malfunctions. There is also a tendency to perform this check too rapidly, which will result in wrong indications. Single ignition operation for as long as a minute is not considered excessive, but this time interval generally should not be exceeded.

Record the amount of the total r.p.m. drop which occurs immediately, and also the amount which occurs slowly for each switch position. This breakdown in r.p.m. drop provides useful information. This ignition system check is usually performed at the beginning of the engine run-up, because if the r.p.m. drops were not within the prescribed limits, it would affect all other later checks.

## Ignition Switch Check

The *ignition switch check* is usually made at 700 r.p.m. On those aircraft engine installations that will not idle at this low r.p.m., set the engine speed to the minimum possible to perform this check. When the speed to perform this check is obtained, momentarily turn the ignition switch to the OFF position. The engine should completely quit firing. After a drop of 200-300 r.p.m. is observed, return the switch to the BOTH position as rapidly as possible. Do this quickly to eliminate the possibility of afterfire and backfire when the ignition switch is returned to BOTH.

If the ignition switch is not returned quickly enough, the engine r.p.m. will drop off completely and the engine will stop. In this case, leave the ignition switch in the OFF position and place the mixture control in idle-cutoff position to avoid overloading the cylinders and exhaust system with raw fuel. When the engine has completely stopped, allow it to remain inoperative for a short time before restarting.

The ignition switch check is performed to see that all magneto ground leads are electrically grounded. If the engine does not cease firing in the OFF position, the magneto ground lead, more commonly referred to as the *P lead,* is open, and the trouble must be corrected.

### *Replacement of Ignition Leads*

When defective leads are revealed by an ignition harness test, continue the test to determine whether the leads or distributor block are defective. If the difficulty is in an individual ignition lead, determine whether the electrical leak is at the spark plug elbow or elsewhere. Remove the elbow, pull the ignition lead out of the manifold a slight amount and repeat the harness test on the defective lead. If this stops the leakage, cut away the defective portion of the lead, and re-install the elbow assembly, integral seal and terminal sleeve.

If the lead is too short to repair in the manner just described, or the electrical leak is inside the harness, replace the defective lead. If the harness is not the re-wirable type, the entire harness must be replaced.

## Spark Plug Inspection and Maintenance

Spark plug operation can often be a major source of engine malfunctions because of lead, graphite or carbon fouling and because of spark plug gap erosion. Most of these failures, which usually accompany normal spark plug operation, can be minimized by good operational and maintenance practices. Good spark plugs, as removed, are shown in Figure 1-6-33.

### *Carbon Fouling*

*Carbon fouling* (Figure 1-6-34) from fuel is associated with mixtures that are too rich to burn or mixtures that are so lean they cause intermittent firing. Each time a spark plug does not fire, raw fuel and oil collect on the non-firing electrodes and nose insulator. These difficulties are almost invariably associated with an improper idle mixture adjustment, a leaking primer or carburetor malfunction that causes too rich a mixture in the idle range. A rich fuel/air mixture is detected by soot or black smoke coming from the exhaust and by an increase in r.p.m. when the idling fuel/air mixture is leaned to *best power.* The soot that forms as a result of overly rich, idle fuel/air mixtures settles on the inside of the combustion chamber because the heat of the engine and the turbulence in the combustion chamber are slight. At higher engine speeds, the soot is swept out and does not condense out of the charge in the combustion chamber.

Even though the idling fuel/air mixture is correct, there is a tendency for oil to be drawn into the cylinder past the piston rings, valve guides and impeller shaft oil seal rings. At low engine speeds, the oil combines with the soot in the cylinder to form a solid that is capable of shorting-out the spark plug.

Spark plugs that are wet or covered with lubricating oil are usually grounded out during the engine start. These plugs may clear up and operate properly after a short period of engine operation.

Engine oil that has been in service for any length of time will hold in suspension minute carbon particles that are capable of conducting an electric current. Thus, a spark plug will not arc the gap between the electrodes when the plug is full of oil. Instead, the high-voltage impulse flows through the oil from one electrode to the other without a spark, just as surely as if a wire conductor were placed between the two electrodes.

Combustion in the affected cylinder does not occur until, at a higher r.p.m., increased airflow has carried away the excess oil. Then, when intermittent firing starts, combustion assists in emitting the remaining oil. In a few seconds, the engine is running clean with white fumes of evaporating and burning oil coming from the exhaust.

**Figure 1-6-34. Carbon-fouled spark plug**

## Lead Fouling

*Lead fouling* of aviation spark plugs is a condition likely to occur in any engine using "leaded fuels." Lead is added to aviation fuel to improve its anti-knock qualities. The lead, however, has the undesirable effect of forming lead oxide during combustion. This lead oxide forms as a solid with varying degrees of hardness and consistency. Lead deposits on combustion chamber surfaces are good electrical conductors at high temperatures and cause misfiring. At low temperatures, the same deposits may be good insulators. In either case, lead formations on aircraft spark plugs, such as those shown in Figure 1-6-35, prevent their normal operation. To minimize the formation of lead deposits, *ethylene dibromide* is added to the fuel as a scavenging agent that combines with the lead during combustion. Lead fouling is not as serious today as in the days before low-lead gasolines.

Lead fouling may occur at any power setting, but perhaps the power setting most conducive to lead fouling is cruising with lean mixtures. At this power, the cylinder head temperature is relatively low and there is an excess of oxygen over that needed to consume all the fuel in the fuel/air mixture. Oxygen, when hot, is very active and aggressive; when all the fuel has been consumed, some of the excess oxygen unites with some of the lead and some of the scavenger agent to form oxygen compounds of lead or bromine or both. Some of these undesirable lead compounds solidify and build up in layers as they contact the relatively cool cylinder walls and spark plugs.

To clean a lead- or mica-fouled spark plug can be a difficult job. The solidified material has to be dug out of the shell. If an entire set of plugs are lead-fouled, the expense will be considerable. A vibrating spark plug cleaner similar to that shown in Figure 1-6-36 uses replaceable chisel points and will clean a fouled spark plug in considerably less time than doing it by hand.

## Graphite Fouling

As a result of careless and excessive application of thread lubricant to the spark plug, the lubricant will flow over the electrodes and cause *shorting*. Shorting occurs because graphite is a good electrical conductor. The elimination of service difficulties caused by *graphite fouling* is up to the aviation technicians. Use care when applying the lubricant to make certain that smeared fingers, rags or brushes do not contact the electrodes or any part of the ignition system except the spark plug threads. Practically no success has been experienced in trying to burn off or dislodge the thread lubricant.

## Gap Erosion

Erosion of the electrodes takes place in all aircraft spark plugs as the spark jumps the *air gap* between the electrodes (see Figure 1-6-37).

The spark carries with it a portion of the electrode, part of which is deposited on the other electrode, and the remainder is blown off in the combustion chamber. As the air gap is enlarged by erosion, the resistance that the

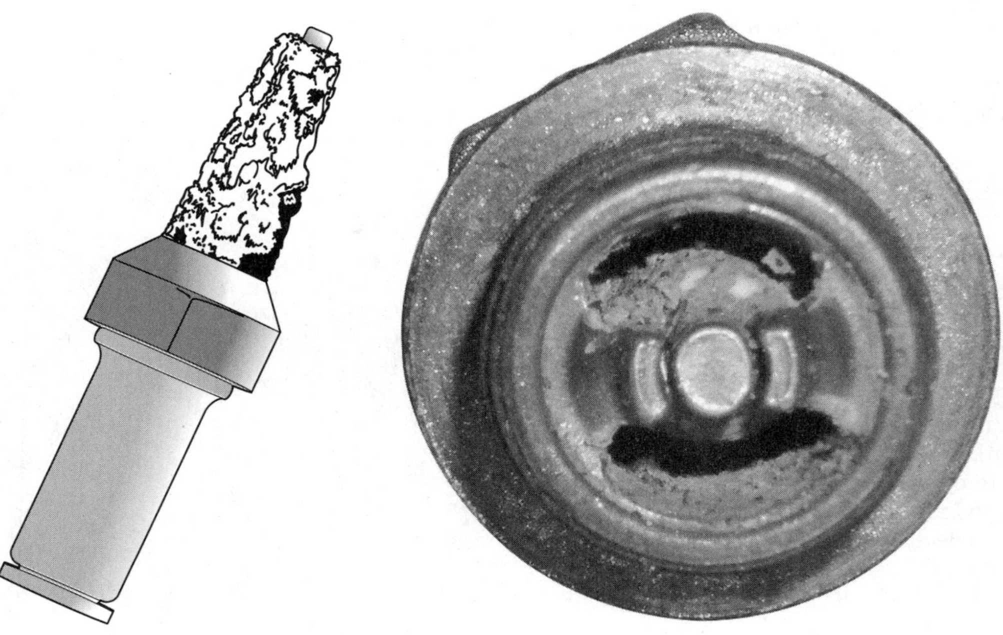

**Figure 1-6-35. Lead-fouled spark plugs**

spark must overcome in jumping the air gap also increases. This means that the magneto must produce a higher voltage to overcome the higher resistance. With higher voltages in the ignition system, a greater tendency exists for the spark to discharge at some weak insulation point in the ignition harness. Since the resistance of an air gap also increases as the pressure in the engine cylinder increases, a double danger exists at takeoff and during sudden acceleration with enlarged air gaps. *Insulation breakdown, premature flashover* and *carbon tracking* result in misfiring of the spark plug and go hand in hand with excessive spark-plug gap. Wide gap settings also raise the *coming in speed* of a magneto, and therefore cause hard starting.

Spark plug manufacturers have partially overcome the problem of gap erosion by using a hermetically sealed resistor in the center electrode of some spark plugs. This added resistance in the high-tension circuit reduces the peak current at the instant of firing. This reduced current flow aids in preventing metal disintegration in the electrodes. Also, due to the high erosion rate of steel or any of its known alloys, spark plug manufacturers are using tungsten or an alloy of nickel for their massive electrode plugs and platinum plating for their fine-wire electrode plugs.

## *Spark Plug Removal*

Spark plugs should be removed for inspection and servicing at the intervals recommended by the manufacturer. Since the rate of gap erosion varies with different operating conditions, engine models and type of spark plug, engine malfunction traceable to faulty spark plugs may occur before the regular servicing interval is reached. Normally, in such cases, only the faulty plugs are replaced.

**Figure 1-6-36. A vibrating cleaning tool for lead- and mica-fouled spark plugs**

**Figure 1-6-37. Spark plug gap erosion**

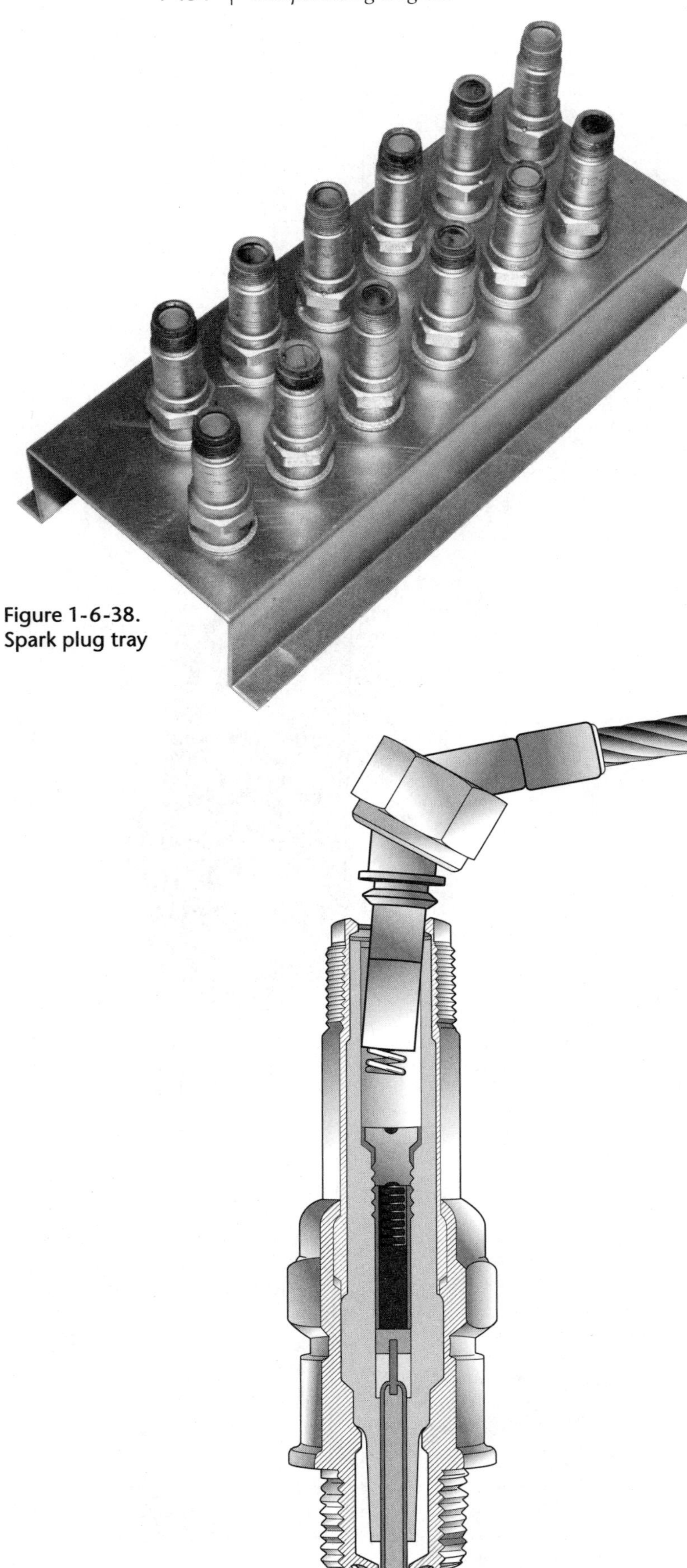

Figure 1-6-38. Spark plug tray

Figure 1-6-39. Improper lead-removal technique

Careful handling of the used and replacement plugs during installation and removal of spark plugs from an engine cannot be overemphasized, since spark plugs can be easily damaged. To prevent damage, spark plugs should always be handled individually, and new and reconditioned plugs should be stored in separate cartons. A common method of storage is illustrated in Figure 1-6-38. This is a *drilled tray*, which prevents the plugs from bumping against one another and damaging the fragile insulators and threads. If a plug is dropped on the floor or other hard surface, it should not be installed in an engine, since the shock of impact usually causes small, invisible cracks in the insulators. The plug should be tested under controlled pressure conditions before use.

Before a spark plug can be removed, the ignition harness lead must be disconnected. Using the special spark plug coupling elbow wrench, loosen and remove the spark plug-to-elbow coupling nut from the spark plug. Take care to pull the lead straight out and in line with the centerline of the plug barrel. If a side load is applied, as shown in Figure 1-6-39, damage to the barrel insulator and the ceramic lead terminal may result. If the lead cannot be removed easily in this manner, the neoprene collar may be stuck to the shielding barrel. Break loose the neoprene collar by twisting the collar as though it were a nut being unscrewed from a bolt.

After the lead has been disconnected, select the proper size deep socket for spark plug removal. A socket designed for the installation and removal of aircraft spark plugs is best. Standard deep sockets may allow enough sideways movement to damage the spark plug. Apply a steady pressure with one hand on the hinge handle, holding the socket in alignment with the other hand. Failure to hold the socket in correct alignment, as shown in Figure 1-6-40, will cause the socket to cock to one side and damage the spark plug.

In the course of engine operation, carbon and other products of combustion will be deposited across the spark plug and cylinder, and some carbon may even penetrate the lower threads of the shell. As a result, a high torque is generally required to break the spark plug loose. This factor imposes a *shearing load* on the shell section of the plug — if this load is great enough, the plug may break off, leaving the shell section in the cylinder spark plug hole.

## *Inspection and Maintenance Prior to Installation*

Before installing new or reconditioned spark plugs in the engine cylinders, clean the spark plug bushings or *Heli-Coil® inserts.*

**Figure 1-6-40. Proper spark plug removal technique**

Brass or stainless steel spark plug bushings are usually cleaned with a spark plug bushing *cleanout tap*. Before inserting the clean-out tap in the spark plug hole, fill the *flutes* of the tap (channels between threads) with clean grease to prevent hard carbon or other material removed by the tap from dropping into the inside of the cylinder. Align the tap with the bushing threads by sight where possible, and start the tap by hand until there is no possibility of the tap being cross-threaded in the bushing. To start the tap on some installations where the spark plug hole is located deeper than can be reached by a clenched hand, it may be necessary to use a short length of hose slipped over the square end of the tap to act as an extension. When screwing the tap into the bushing, be sure that the full cutting thread reaches the bottom thread of the bushing. This will remove carbon deposits from the bushing threads without removing bushing metal, unless the pitch diameter of the threads has contracted as the result of shrinkage or some other unusual condition.

If, during the thread-cleaning process, the bushing is found to be loose or is loosened in the cylinder or the threads are cross-threaded or otherwise seriously damaged, replace the cylinder. Spark plug Heli-Coil® inserts are cleaned with a round wire brush, preferably one having a diameter slightly larger than the diameter of the spark plug hole. A brush considerably larger than the hole may cause removal of material from the Heli-Coil® proper or from the cylinder head surrounding the insert. Also, the brush should not disintegrate with use, allowing wire bristles to fall into the cylinder. Clean the insert by carefully rotating the wire brush with a power tool.

When using the power brush, be careful that no material is removed from the spark plug gasket seating surface, since this may cause a change in the spark plug's heat range, combustion leakage and eventual cylinder damage. Never clean the Heli-Coil® inserts with a cleaning tap, since permanent damage to the insert will result. If a Heli-Coil® insert is damaged as a result of normal operation or while cleaning it, replace it according to the applicable manufacturer's instructions.

Using a lint-free rag and cleaning solvent, wipe the spark plug gasket seating surface of the cylinder to eliminate the possibility of dirt or grease being accidentally deposited on the spark plug electrodes at the time of installation.

Before the new or reconditioned plugs are installed, they must be inspected for each of the following conditions:

- Be sure the plug is of the approved type, as indicated by the applicable manufacturer's instructions.
- Check for evidence of rust-preventive compound on the spark plug exterior and core insulator and on the inside of the shielding barrel. Rust-preventive compound accumulations are removed by washing the plug with a brush and cleaning solvent. It must then be dried with a dry air blast.
- Check both ends of the plug for nicked or cracked threads and any indication of cracks in the nose insulator.
- Inspect the inside of the shielding barrel for cracks in the barrel insulator, and the center electrode contact for rust and foreign material that might cause poor electrical contact.
- Inspect the spark plug gasket. A gasket that has been excessively flattened, scarred, dented or distorted by previous use must not be used. Used, serviceable, copper spark plug gaskets should be annealed prior to reinstallation. When the thermocouple gasket is used, do not use an additional gasket.

The gap setting should be checked with a round wire thickness gauge, as shown in Figure 1-6-41. A flat-type gauge will give an incorrect clearance indication because the massive ground electrodes are contoured to the shape of the round center electrode. When using the wire thickness gauge, insert the gauge in each gap parallel to the centerline of the center electrode. If the gauge is tilted slightly, the indication will be incorrect. Do not install a plug that does not have an air gap within the specified clearance range.

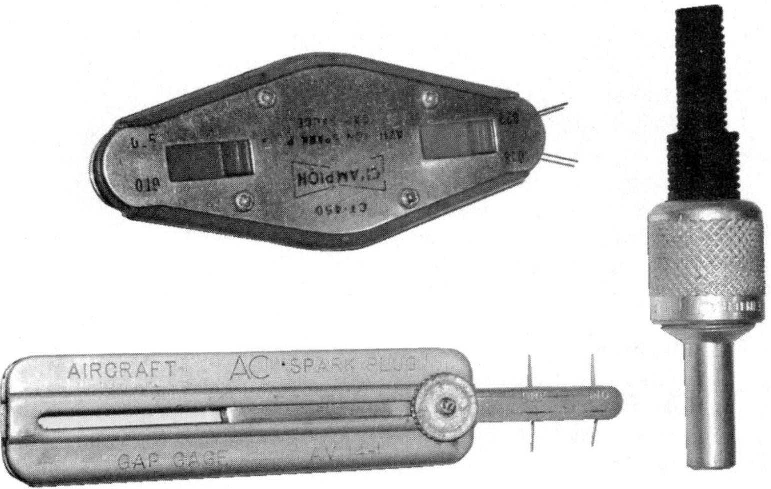


**Figure 1-6-41. Wire gap-measuring gauges, along with a stain-removal tool for spark plug barrels that uses an abrasive compound and a hand drill**

## *Spark Plug Installation*

Prior to spark plug installation, carefully coat the first two or three threads from the electrode end of the shell with a graphite-base *anti-seize compound*. Prior to application, stir the anti-seize compound to ensure thorough mixing. When applying the anti-seize compound to the threads, be extremely careful that none of the compound gets on the ground or center electrodes or on the nose of the plug, where it can spread to the ground or center electrode during installation. This precaution is mentioned because the graphite in the compound is an excellent electrical conductor and could cause permanent fouling.

To install a spark plug, start it into the cylinder without using a wrench of any kind, and turn it until the spark plug is seated on the gasket. If the plug can be screwed into the cylinder with comparative ease, using the fingers, this indicates good, clean threads. In this case, only a small additional tightening torque will be needed to compress the gasket to form a gas-tight seal.

If, on the other hand, a high torque is needed to install the plug, dirty or damaged threads on either the plug or plug bushing are indicated. The use of excessive torque might compress the gasket out of shape and distort the plug shell to a point where breakage would result during the next removal or installation. Shell stretching occurs as excessive torque continues to screw the lower end of the shell into the cylinder after the upper end has been stopped by the gasket shoulder. As the shell stretches (Figure 1-6-42), the seal between the shell and core insulator is opened, creating a loss of gas-tightness or damage to the core insulator. After a spark plug has been seated with the fingers, use a torque wrench and tighten to the specified torque.

## *Spark Plug Lead Installation*

Before installing the spark plug lead, carefully wipe the terminal sleeve (sometimes referred to as *cigarette*) and the integral seal with a cloth moistened with acetone, MEK or an approved solvent. After the plug lead is cleaned, inspect it for cracks and scratches. If the terminal sleeve is damaged or heavily stained, replace it.

After inspection of the spark plug lead, slip the lead into the shielding barrel of the plug with care. Then tighten the spark plug coupling elbow nut with the proper tool. Most manufacturer's instructions specify the use of a tool designed to help prevent an over-torque condition. After the coupling nut is tightened, avoid checking for tightness by twisting the body of the elbow.

After all plugs have been installed, torqued and the leads properly installed, start the engine and perform a complete ignition system operational check.

## Breaker Point Inspection

Inspection of the magneto consists essentially of a periodic breaker point and *dielectric inspection*. After the magneto has been inspected for security of mounting, remove the magneto cover or breaker cover, and check the cam for proper lubrication. Under normal conditions, there is usually ample oil in the *felt oiler pad* of the cam follower to keep the cam lubricated between overhaul periods. However, during the regular routine inspection, examine the felt pad on the cam follower to be sure it contains sufficient oil for cam lubrication. Make this check by pressing the thumbnail against the oiler pad. If oil appears on the thumbnail, the pad contains sufficient oil for cam lubrication. If there is no evidence of oil on the fingernail, apply one drop of a light aircraft engine oil to the bottom felt pad and one drop to the upper felt pad of the follower assembly, as shown in Figure 1-6-43.

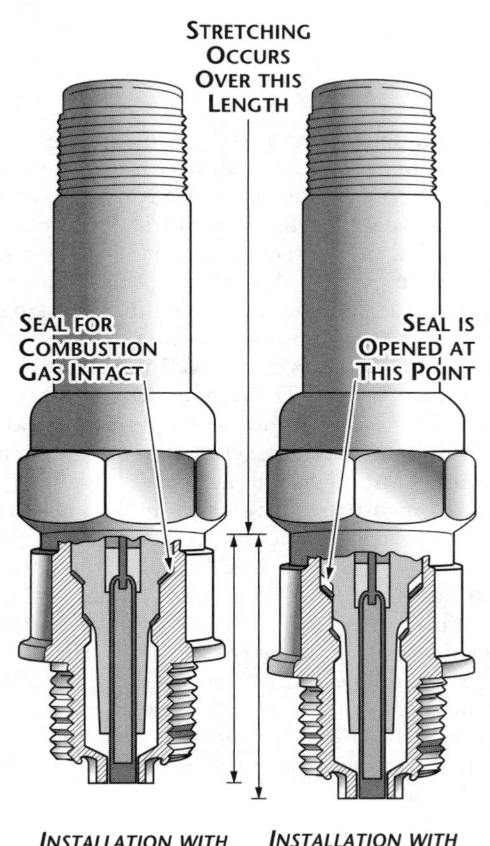


Figure 1-6-42. Effect of excessive torque when installing a spark plug

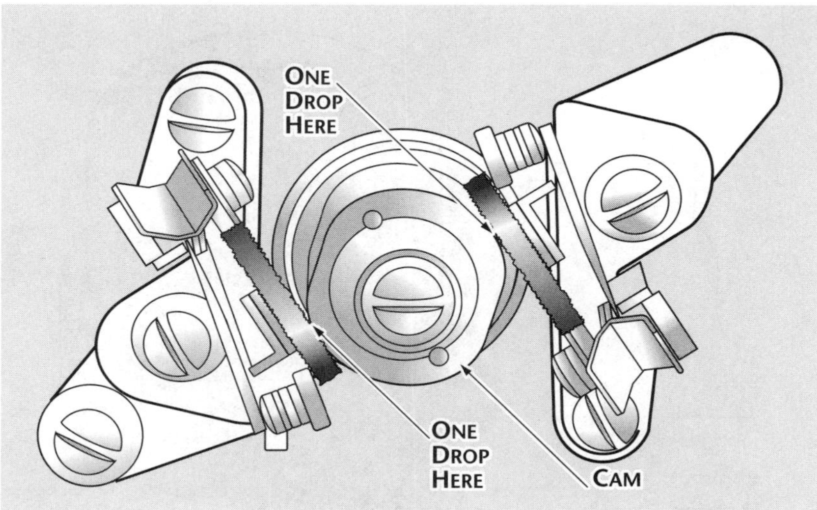


Figure 1-6-43. Applying engine oil to the felt oiler pad

After application, allow at least 15 minutes for the felt to absorb the oil. At the end of 15 minutes, blot off any excess oil with a clean, lint-free cloth. During this operation, or any time the magneto cover is off, use extreme care to keep the breaker compartment free of oil, grease or engine cleaning solvents, since each of these has an adhesiveness which collects dirt and grime that could foul an otherwise good set of breaker contact points.

After the felt oiler pad has been inspected, serviced and found to be satisfactory, visually inspect the breaker contacts for any condition that may interfere with proper operation of the magneto. If the inspection reveals an oily or gummy substance on the sides of the contacts, swab the contacts with a flexible wiper, such as a pipe cleaner dipped in acetone or some other approved solvent. By forming a hook on the end of the wiper, ready access can be gained to the back side of the contacts.

To clean the contact mating surfaces, force open the breaker points enough to admit a small swab. Whether spreading the points for purposes of cleaning or checking the surfaces for condition, always apply the opening force at the outer end of the mainspring and never spread the contacts more than 1/16 inch (0.0625 inch). If the contacts are spread wider than recommended, the *mainspring* (the spring carrying the movable contact point) is likely to take a permanent set. If the mainspring takes a permanent set, the movable contact point loses some of its closing tension, and the points will then either *bounce* or *float*, preventing the normal induction buildup of the magneto.

A swab can be made by wrapping a piece of linen tape or a small piece of lint-free cloth over one of the leaves of a clearance gauge and dipping the swab in an approved solvent. Then pass the swab between the carefully separated

Figure 1-6-44. Normal contact surface

Figure 1-6-45. Points with normal irregularities

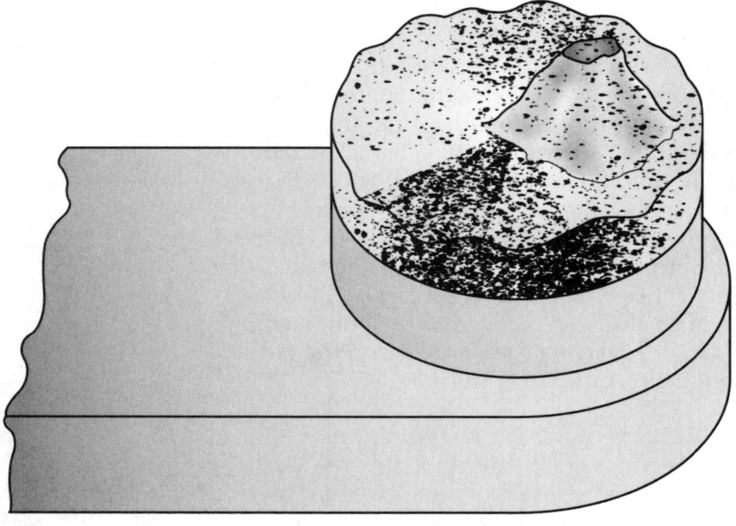

Figure 1-6-46. Points with well-defined peaks

contact surfaces until the surfaces are clean. During this entire operation, take care that drops of solvent do not fall on lubricated parts, such as the cam, follower block or felt oiler pad.

To inspect the breaker contact surfaces, it is necessary to know what a normal operating set of contacts looks like, what surface condition is considered as permissible wear and what surface condition is cause for dressing or replacement. The probable cause of an abnormal surface condition can also be determined from the contact appearance. The normal contact surface (Figure 1-6-44) has a dull gray, sandblasted (almost rough) appearance over the area where electrical contact is made. This gray, sandblasted appearance indicates that the points have worn in and have mated to each other and are providing the best possible electrical contact.

This is not meant to imply that this is the only acceptable contact surface condition. Slight, smooth-surfaced irregularities, without deep pits or high peaks, such as shown in Figure 1-6-45, are considered normal wear and are not cause for dressing or replacement.

However, when wear advances to a point where the slight, smooth irregularities develop into well-defined peaks extending noticeably above the surrounding surface, as illustrated in Figure 1-6-46, the breaker contacts must be dressed or replaced.

Unfortunately, when a peak forms on one contact, the mating contact will have a corresponding pit or hole. This pit is more troublesome than the peak because it penetrates the platinum pad of the contact surface. It is sometimes difficult to judge whether a contact surface is pitted deeply enough to require dressing, because in the final analysis this depends on how much of the original platinum is left on the contact surface. The danger arises from the possibility that the platinum pad may already be thin from a long service life and previous dressings. At overhaul facilities, a gauge is used to measure the remaining thickness of the pad, and no difficulty in determining the condition of the pad exists. At line maintenance activities, this gauge is generally unavailable. Therefore, if the peak is quite high or the pit quite deep, do not dress these contacts; instead, remove and replace them with a new or reconditioned assembly. A comparison between Figures 1-6-45 and 1-6-46 will help to draw the line between *minor irregularities* and *well-defined peaks*.

Some examples of possible breaker contact surface conditions are shown in Figure 1-6-47. Illustration A shows an example of erosion or wear called *frosting*. This condition results from an open-circuited condenser and is easily recognized by the coarse, crystalline surface and the black, sooty appearance of the sides of the points. The lack of effective condenser action results in an arc of intense heat being formed each time the points open. This, together with the oxygen in the air, rapidly oxidizes and erodes the platinum surface of the points, producing the coarse, crystalline, frosted appearance. Properly operating points have a fine-grained, frosted or silvery appearance and should not be confused with the coarse-grained and sooty point caused by faulty condenser action.

Illustrations B and C of Figure 1-6-47 illustrate badly pitted points. These points are identified by a fairly even contact edge (in the early stage) and minute pits or pocks in or near the center of the contact surface with a general overall smoky appearance. In more advanced stages, the pit may develop into a large, jagged crater, and eventually the entire contact surface will take on a burned, black, crumpled appearance. Pitted points, as a general rule, are caused by dirt and impurities on the contact surfaces. If points are excessively pitted, a new or reconditioned breaker assembly must be installed.

Illustration D of Figure 1-6-47 demonstrates a *crowned contact point* and can be readily identified by the concave center and a convex rim on the contact surface. This condition results from improper dressing, as may be the case when an attempt is made to dress points while they are still installed in the magneto. In addition to an uneven or un-square surface, the tiny particles of foreign material and metal that remain between the points after the dressing operation fuse and cause uneven burning of the inner contact surface. This burning differs from "frosting" in that a smaller arc produces less heat and less oxidation. In this instance the rate of burning is more gradual. Crowned points, if not too far gone, may be cleaned and returned to service. If excessive crowning has taken place, the points must be removed and replaced with a new or reconditioned set.

Illustration E of Figure 1-6-47 shows a built-up point that can be recognized by the mound of metal which has been transferred from one point to another. Buildup, like the other conditions mentioned, results primarily from the transfer of contact material by means of the arc as the points separate. But, unlike the others, there is no burning or oxidation in the process because of the closeness of the pit of one point and the buildup of the other. This condition may result from excessive breaker point spring tension, which retards the opening of the points or causes a slow, lazy break. It can also be caused by a poor primary condenser or a loose connection at the primary coil. If excessive buildup has occurred, a new or reconditioned breaker assembly must be installed.

Illustration F of Figure 1-6-47 shows *oily points,* which can be recognized by their smoked and smudged appearance and by the lack of any of the above-mentioned irregularities. This condition may be the result of excessive cam lubrication or of oil vapors, which may come from inside or outside of the magneto. A smoking or fuming engine, for example, could produce the oil vapors. These vapors then enter the magneto through the magneto ventilator and pass between and around the points. These conductive vapors produce arcing and burning on the contact surfaces. The vapors also adhere to the other surfaces of the breaker assembly and form the sooty deposit. Oily points can ordinarily be made serviceable by using a suitable cleaning procedure. However, the removal of smoke and smudge may reveal a need for dressing the points. If so, dress the points or install a new or reconditioned breaker assembly.

## *Dressing Breaker Points*

Generally speaking, disassembly and dressing of breaker points should not be a regular, routine step of magneto maintenance. By performing expensive and unnecessary maintenance on the point assemblies, many sets of points reach the scrap bin prematurely, with perhaps two-thirds to three-fourths of the platinum contact surface material filed away by repeated dressing operations. In a majority of the cases, breaker points will remain in satisfactory condition between overhaul periods with only routine inspection, cleaning and lubrication.

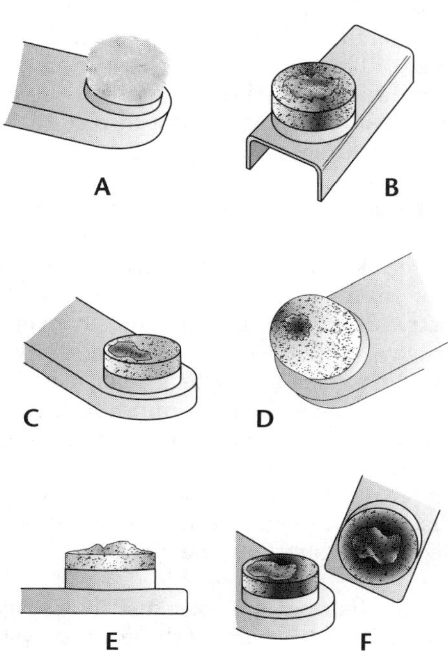


**Figure 1-6-47. Contact surface conditions**

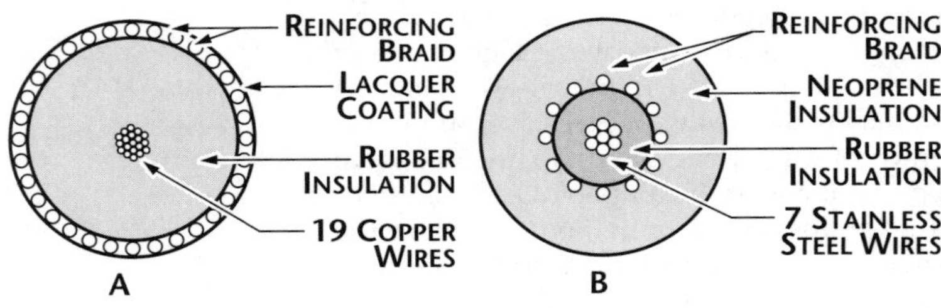


Figure 1-6-48. Cross-sectional view of typical high-tension ignition cable

If the breaker contacts have deep pits, mounds or burnt surfaces, they should be dressed or replaced according to the maintenance practice recommended by the manufacturer. If dressing of breaker contacts is approved, a special *contact point dressing kit* will normally be available. The kit includes a contact point dressing block and adapters to hold the contacts during the dressing operation, a special file to remove the peaks and mounds, and a very fine whetstone to be used in the final dressing operation to remove any ridges or burrs left by the file.

If the breaker contact points have been removed for any reason, the replacement or reconditioned points must be installed and precisely timed to open just as the rotating magnet of the magneto moves into the E-gap position for the No. 1 cylinder.

## Dielectric Inspection

Another phase of magneto inspection is the *dielectric inspection*. This inspection is a visual check for cleanliness and cracks. If inspection reveals that the coil cases, condensers, distributor rotor or blocks are oily or dirty or have any trace of carbon tracking, they will require cleaning and possibly waxing to restore their dielectric qualities.

Clean all accessible condensers and coil cases that contain condensers by wiping them with a lint-free cloth moistened with acetone. Many parts of this type have a protective coating. This protective coating is not affected by acetone, but it may be damaged by scraping or by the use of other cleaning fluids. Never use unapproved cleaning solvents or improper cleaning methods. Also, when cleaning condensers or parts which contain condensers, do not dip, submerge or saturate the parts in any solution because the solution used may seep inside the condenser and short out the plates.

Coil cases, distributor blocks, distributor rotors and other dielectric parts of the ignition system are treated with a wax coating when they are new, and again at overhaul. The waxing of dielectrics aids their resistance to moisture absorption, carbon tracking and acid deposits. When these parts become dirty or oily, some of the original protection is lost, and carbon tracking may result.

If any hairline carbon tracks or acid deposits are present on the surface of the dielectric, immerse the part in approved cleaning solvent and scrub it vigorously with a stiff-bristle brush. When the carbon track or acid deposits have been removed, wipe the part with a clean, dry cloth to remove all traces of the solvent used for cleaning. Then coat the part with a special ignition-treating wax. After wax treating the part, remove excess wax deposits and re-install the part in the magneto.

## Ignition Harness Maintenance

Although the ignition harness is simple, it is a vital link between the magneto and spark plug. Because the harness is mounted on the engine and exposed to the atmosphere, it is vulnerable to heat, moisture and the effects of changing altitude. These factors, plus aging insulation and normal gap erosion, work against efficient engine operation. The insulation may break down on a wire inside the harness and allow the high voltage to leak through the insulation to the harness shielding instead of going to the spark plug. Open circuits may result from broken wires or poor connections. A bare wire may be in physical contact with the shielding, or two wires may be shorted together.

Any serious defect in an individual lead prevents the high-tension impulse from reaching the spark plug to which the lead is connected. As a result, this plug will not fire. When only one spark plug is firing in a cylinder, the charge is not consumed as quickly as it would be if both plugs were firing. This factor causes the peak pressure of combustion to occur later on the power stroke. If the peak pressure occurs later than normal, a loss of power in that cylinder results. However, the power loss from a single cylinder becomes a minor factor when the effect of a longer burning time is considered. A longer burning time overheats the affected cylinder, causing detonation, possible pre-ignition and perhaps permanent damage to the cylinder.

The insulated wire that carries the electrical impulse is a special type of cable designed to prevent excessive losses of electrical energy. This wire is known as high-tension ignition cable and is manufactured in three diameters. The outside diameters of cables in current use are 5, 7 or 9 mm (millimeters). The reason for different cable diameters is that the amount and kind of insulation around the conducting core determines the amount of electrical loss during transmission of the high voltage. Since the conducting core carries only a weak current, the conductor is of a small diameter.

The 9-mm cable has only a limited application now because it is of early design and has a relatively thick layer of insulation. For the most part, present-day engines use the 7-mm size, but there are a few systems which are designed to use 5-mm cable. The increased use of the smaller size cable is largely due to improvements in the insulation material, which permits a thinner insulating sheath. Adapter sleeves have been designed for the ends of the smaller, improved cable so that it can be used in re-wirable harnesses where the distributor wells were originally designed for larger cable.

One type of cable construction uses a core consisting of 19 strands of fine copper wire covered with a rubber sheath. This, in turn, is covered by a reinforcing braid and an outside coat of lacquer (illustration A of Figure 1-6-48) A newer type of construction (illustration B, Figure 1-6-48) has a core of seven strands of stainless steel wire covered with a rubber sheath. Over this is woven a reinforcing braid, and a layer of neoprene is added to complete the assembly. This type of construction is superior to the older type, primarily because the neoprene has improved resistance to heat, oil and abrasion.

## *High-tension Ignition Harness Faults*

Perhaps the most common and most difficult high-tension ignition system faults to detect are high-voltage leaks. This is leakage from the core conductor through insulation to the ground of the shielded manifold. A small amount of leakage exists even in brand new ignition cable during normal operation. Various factors then combine to produce first a high rate of leakage and then complete breakdown. Of these factors, moisture in any form is probably the worst. Under high-voltage stress, an arc forms and burns a path across the insulator where the moisture exists. If there is gasoline, oil or grease present, it will break down and form carbon. The burned path is called a *carbon track*, since it is actually a path of carbon particles. With some types of insulation, it may be possible to remove the carbon track and restore the insulator to its former useful condition. This is generally true of porcelain, ceramics and some of the plastics, because these materials are not hydrocarbons, and any carbon track forming on them is the result of a dirt film that can be wiped away.

Differences in location and amount of leakage will produce different indications of malfunction during engine operation. Indications are generally misfiring or cross-firing. The indication may be intermittent, changing with manifold pressure or with climate conditions. An increase in manifold pressure increases the compression pressure and the resistance of the air across the air gap of the spark plugs.

An increase in the resistance at the air gap opposes the spark discharge and produces a tendency for the spark to discharge at some weak point in the insulation. A weak spot in the harness may be aggravated by moisture collecting in the harness manifold. With moisture present, continued engine operation will cause the intermittent faults to become permanent carbon tracks. Thus, the first indication of ignition harness un-serviceability may be engine misfiring or roughness caused by partial leakage of the ignition voltage.

Figure 1-6-49, showing a cross section of a harness, demonstrates four faults that may occur. Fault A shows a short from one cable conductor to another. This fault usually causes misfiring, since the spark is short-circuited to a plug in a cylinder where the cylinder pressure is low. Fault B illustrates a cable with a portion of its insulation scuffed away. Although the insulation is not completely broken down, more than normal leakage exists, and the spark plug to which this cable is connected may be lost during takeoff when the manifold pressure is quite high.

Fault C is the result of condensation collecting in the lowest portion of the ignition manifold. This condensation may completely evaporate during engine operation, but the carbon track that is formed by the initial flashover remains to allow continued flashover whenever high manifold pressure exists.

Fault D may be caused by a flaw in the insulation or the result of a weak spot in the insulation, aggravated by the presence of moisture.

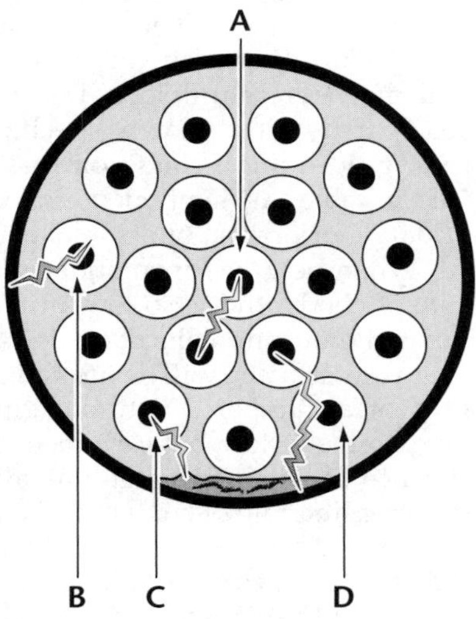


Figure 1-6-49. Cross-section of an ignition harness

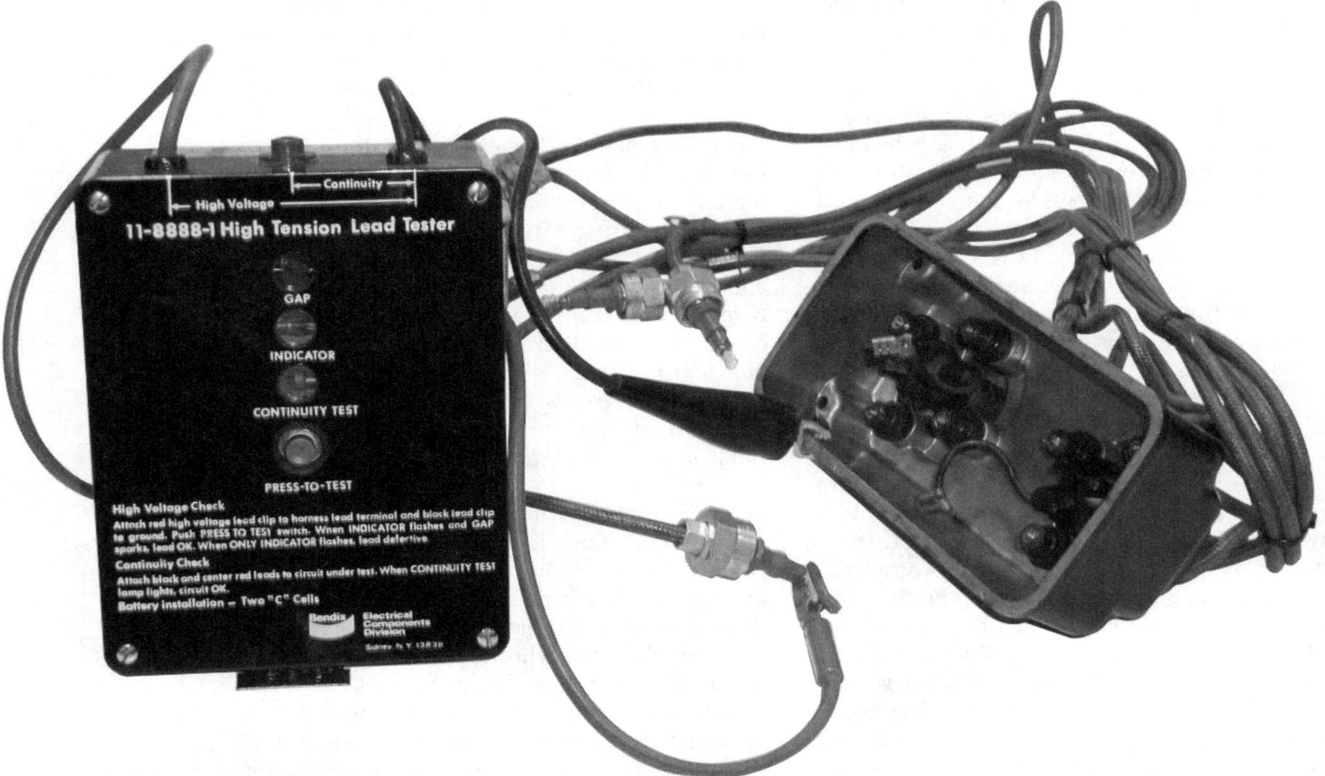


**Figure 1-6-50. A harness tester hooked up to a lead, which can actually be tested with the harness still on the magneto**

However, since the carbon track is in direct contact with the metal shielding, it will probably result in flashover under all operating conditions.

## Harness Testing

The electrical test of the ignition harness checks the condition or effectiveness of the insulation around each cable in the harness. The principle of this test involves application of a definite voltage to each lead and then measurement with a very sensitive meter of the amount of current leakage between the lead and the grounded harness manifold. This reading, when compared with known specifications, becomes a guide to the condition or serviceability of the cable. As mentioned earlier, there is a gradual deterioration of flexible insulating material. When new, the insulation will have a low rate of conductivity; so low, in fact, that under several thousand volts of electrical pressure the current leakage will be only a very few millionths of an ampere. Natural aging will cause an extremely slow, but certain, change in the resistance of insulating material, allowing an ever-increasing rate of current leakage.

Several different types of test devices are used for determining the serviceability of high-tension ignition harness. One is shown in Figure 1-6-50. Consult the applicable manufacturer's instructions before performing an ignition harness test.

## Engine Starting Systems

Most aircraft engines are started by a device called a *starter.* A starter is a mechanism capable of developing large amounts of mechanical energy that can be applied to an engine, causing it to rotate.

In the early stages of aircraft development, relatively low-powered engines were started by pulling the propeller through a part of a revolution by hand. Some difficulty was often experienced in cold-weather starting when lubricating oil temperatures were near the congealing point. In addition, the magneto systems delivered a weak starting spark at the very low cranking speeds. This was often compensated for by providing a hot spark, using such ignition system devices as the booster coil, induction vibrator or impulse coupling.

## Reciprocating Engine Starting Systems

Throughout the development of the aircraft reciprocating engine from the earliest use of starting systems to the present, a number of different starter systems have been used. These are:

- Cartridge starter (not in common use)
- Hand inertia starter (not in common use)
- Electric inertia starter (not in common use)
- Combination inertia starter (not in common use)
- Direct-cranking electric starter

Almost all reciprocating engine starters are of the direct-cranking electric type. A few older model aircraft may still be equipped with one of the types of *inertia starters (Figure 1-6-51)*.

Figure 1-6-51. Inertia starters can still be found on older aircraft

The most widely used starting system on all types of reciprocating engines utilizes the *direct-cranking electric starter* (Figure 1-6-52). This type of starter provides instant and continual cranking when energized. The direct-cranking electric starter consists of an electric motor, reduction gears and an automatic engaging and disengaging mechanism which is operated through an adjustable torque overload-release clutch. A typical circuit for a direct- cranking electric starter is shown in Figure 1-6-53.

The engine is cranked directly when the starter solenoid is closed. Since no flywheel is used in the direct-cranking electric starter, there is no preliminary storing of energy, as in the case of an inertia starter.

As shown in Figure 1-6-53, the main cables leading from the starter to the battery are heavy-duty to carry the high-current flow — maybe as high as 350 amperes, depending on the starting torque required. The use of solenoids and heavy wiring with a remote control switch reduces overall cable weight and total circuit voltage drop.

Figure 1-6-52. Most reciprocating engines use a direct-cranking starter

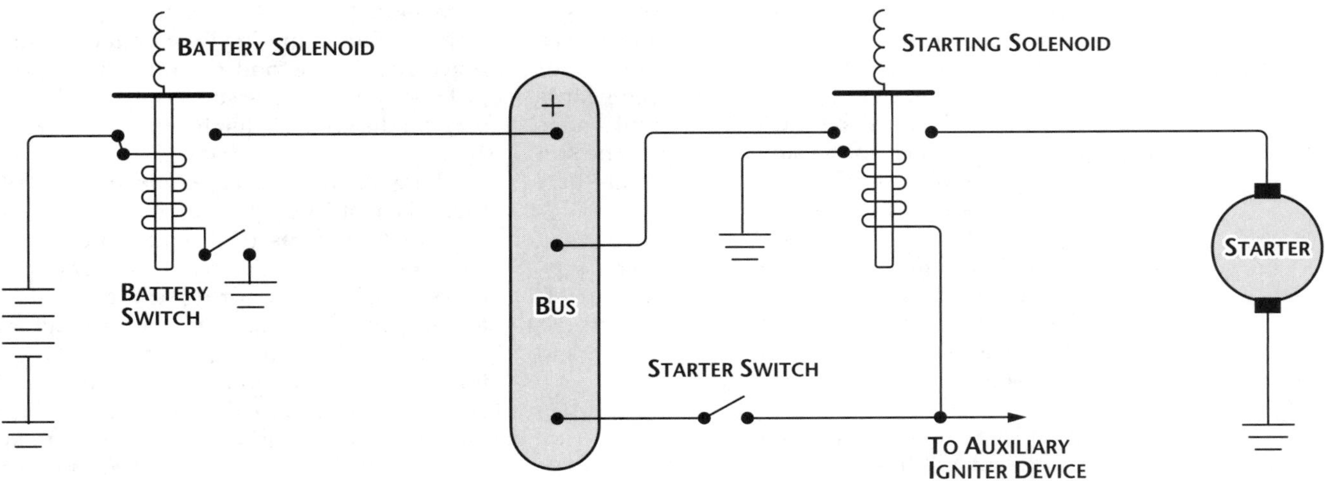


Figure 1-6-53. Typical starting circuit using a direct-cranking electric starter

The typical starter motor is a 12- or 24-volt series-wound motor, which develops high starting torque. The torque of the motor is transmitted through reduction gears to the overload-release clutch. Typically, this actuates a helically splined shaft, moving the starter jaw outward to engage the engine-cranking jaw before the starter jaw begins to rotate. After the engine reaches a predetermined speed, the starter automatically disengages.

## *Direct-cranking Electric Starting System for High-power Aircraft*

In a typical high-horsepower reciprocating engine starting system, the direct-cranking electric starter consists of two basic components: a *motor assembly* and a *gear section*. The gear section is bolted to the drive end of the motor to form a complete unit.

The motor assembly consists of the *armature and motor pinion assembly*, the *end bell assembly* and the *motor housing assembly*. The motor housing also acts as the magnetic yoke for the field structure.

The starter motor is a non-reversible, series interpole motor. Its speed varies directly with the applied voltage and inversely with the load.

The starter gear section, shown in Figure 1-6-54, consists of a housing with an integral mounting flange, planetary gear reduction, a sun and integral gear assembly, a torque-limiting clutch, and a jaw and cone assembly.

When the starter circuit is closed, the torque developed in the starter motor is transmitted to the starter jaw through the reduction gear train and clutch. The starter gear train converts the high-speed low-torque of the motor to the low-speed high-torque required to crank the engine.

In the gear section, the motor pinion engages the gear on the intermediate countershaft (see Figure 1-6-54). The pinion of the countershaft engages the internal gear. The internal gear is an integral part of the sun gear assembly and is rigidly attached to the sun gear shaft. The sun gear drives three planet gears, which are part of the planetary gear assembly.

The individual planet gear shafts are supported by the planetary carrying arm, a barrel-like part shown in Figure 1-6-54. The carrying arm transmits torque from the planet gears to the starter jaw as follows:

- The cylindrical portion of the carrying arm is splined longitudinally around the inner surface
- Mating splines are cut on the exterior surface of the cylindrical part of the starter jaw
- The jaw slides fore and aft inside the carrying arm to engage and disengage with the engine

The three planet gears also engage the surrounding internal teeth on the six steel clutch plates (Figure 1-6-54). These plates are interleaved with externally splined bronze clutch plates that engage the sides of the housing, preventing them from turning. The proper pressure is maintained upon the clutch pack by a *clutch spring retainer assembly*.

A cylindrical traveling nut inside the starter jaw extends and retracts the jaw. Spiral jaw-engaging splines around the inner wall of the nut mate with similar splines cut on an extension of the sun gear shaft (Figure 1-6-54). Being splined in this fashion, rotation of the shaft forces the nut out and the nut carries the jaw with it. A jaw spring around the traveling nut carries the jaw with the nut and tends to keep a conical clutch surface around the inner wall of the jaw head seated against a similar surface around the under side of the nut head. A return spring is installed on the sun gear shaft extension between a shoulder formed by the splines around the inner wall of the traveling nut and a jaw-stop retaining nut on the end of the shaft.

Because the conical clutch surfaces of the traveling nut and the starter jaw are engaged by jaw-spring pressure, the two parts tend to rotate at the same speed. However, the sun gear shaft extension turns six times faster than the jaw. The spiral splines on it are cut left-hand, and the sun gear shaft extension, turning to the right in relation to the jaw, forces the traveling nut and the jaw out from the starter its full travel (about 5/16 inch) in approximately half a rotation of the jaw. The jaw moves out until it is stopped either by engagement with the engine or by the jaw-stop retaining nut. The travel nut continues to move slightly beyond the limit of jaw travel, just enough to relieve some of the spring pressure on the conical clutch surfaces.

As long as the starter continues to rotate, there is just enough pressure on the conical clutch surfaces to provide torque on the spiral splines, which balances most of the pressure of the jaw spring. If the engine fails to start, the starter jaw will not retract, since the starter mechanism provides no retracting force. However, when the engine fires and the engine jaw overruns the starter jaw, the sloping ramps of the jaw teeth force the starter jaw into the starter against the jaw spring pressure. This disengages the conical clutch surfaces entirely, and the jaw-spring pressure

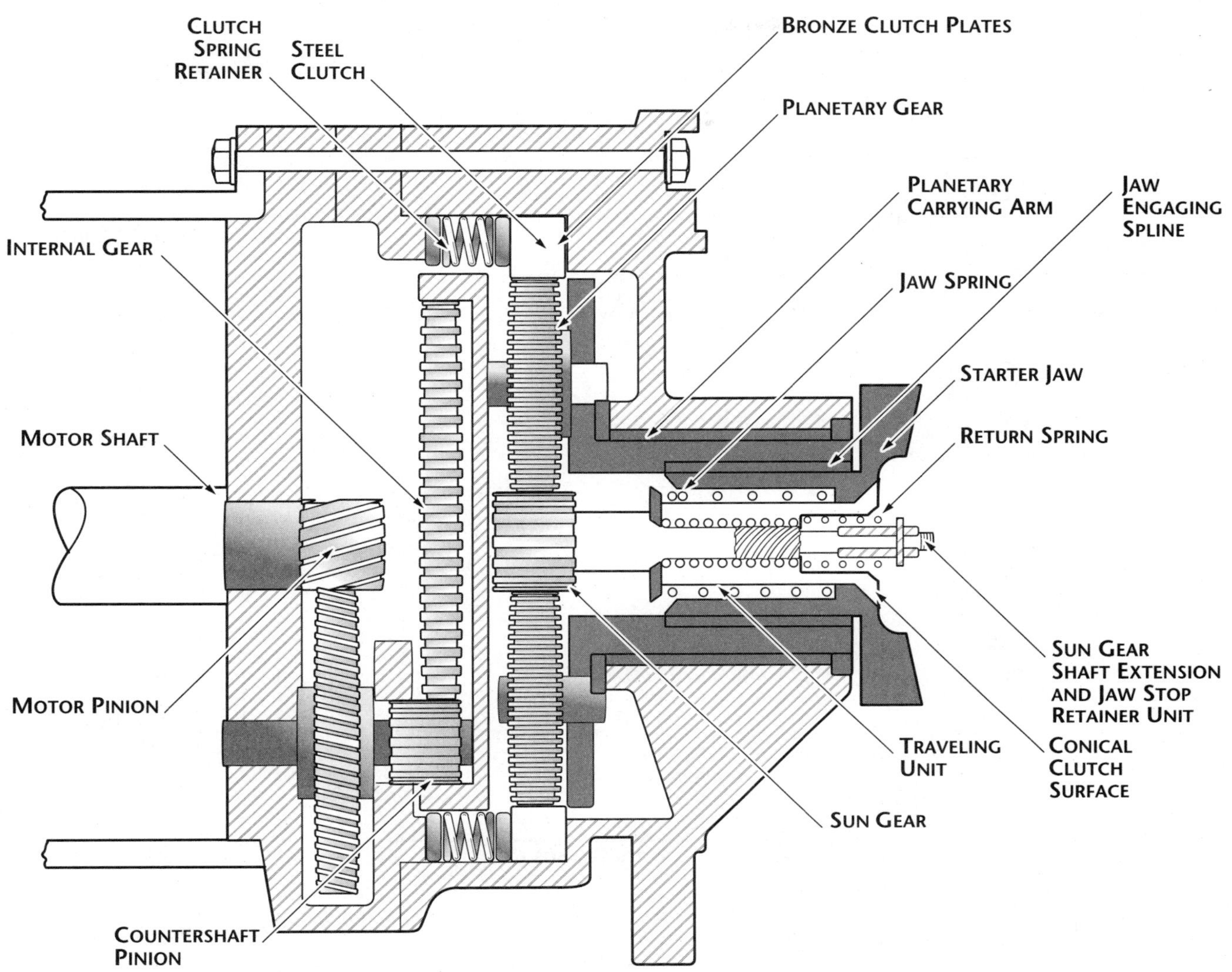


Figure 1-6-54. Starter gear section

forces the traveling nut to slide in along the spiral splines until the conical clutch surfaces are again in contact.

With the starter and engine both running, there will be an engaging force keeping the jaws in contact, which will continue until the starter is de-energized. However, the rapidly moving engine-jaw teeth, striking the slowly moving starter-jaw teeth, hold the starter jaw disengaged. As soon as the starter comes to rest, the engaging force is removed, and the small return spring will throw the starter jaw into its fully retracted position, where it will remain until the next start.

When the starter jaw first engages the engine jaw, the motor armature has had time to reach considerable speed because of its high starting torque. The sudden engagement of the moving starter jaw with the stationary engine jaw would develop forces high enough to severely damage the engine or the starter, were it not for plates in the clutch pack, which slip when engine torque exceeds clutch-slipping torque.

In normal direct-cranking action, the internal gear clutch plates (steel) are held stationary by the friction of the bronze plates, with which they are interleaved. When the torque imposed by the engine exceeds the clutch setting, however, the internal gear-clutch plates rotate against the clutch friction, allowing the planet gears to rotate while the planetary carrying arm and the jaw remain stationary. When the engine comes up to the speed at which the starter is trying to drive it, the torque drops off to a value less than the clutch setting, the internal gear clutch plates are again held stationary, and the jaw rotates at the speed at which the motor is attempting to drive it.

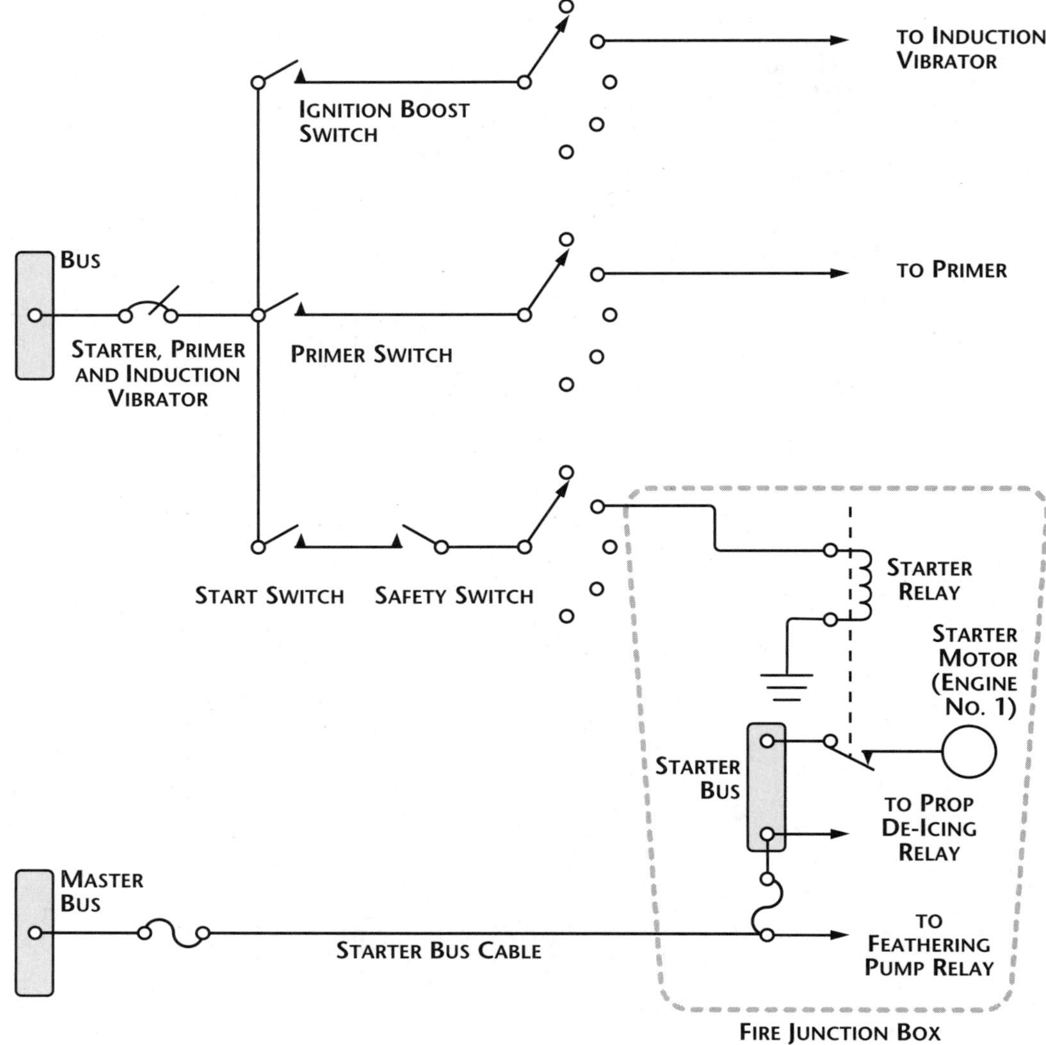


**Figure 1-6-55. A schematic of a starter control circuit**

The starter control switches are shown in Figure 1-6-55. The engine selector switch must be positioned, and both the starter switch and the safety switch (wired in series) must be closed before the starter can be energized.

Current is supplied to the starter-control circuit through a circuit breaker, labeled *Starter, Primer and Induction Vibrator* in Figure 1-6-55. When the engine selector switch is in position for the engine start, closing the starter and safety switches energizes the starter relay located in the firewall junction box.

Energizing the starter relay completes the power circuit to the starter motor. The current necessary for this heavy load is taken directly from the master bus through the starter bus cable. Starters tend to build internal heat rapidity. After one minute of operation it should be allowed to cool for an additional minute. If the starter is again operated for one minute, allow it to cool for five minutes.

## *Direct-cranking Electric Starting System for Low-power Aircraft*

Most small reciprocating-engine aircraft employ a direct-cranking electric starting system. The starters are very similar to an automotive starter with a splined Bendix drive. Most are mounted on a pad below the centerline of the engine and behind the propeller flange.

A starter solenoid is activated by either a push button or a key switch on the instrument panel. When the solenoid is activated, its contacts close and electrical energy energizes the starter motor, causing it to rotate. Initial rotation of the starter motor engages the starter pinion gear with a ring gear mounted on the crankshaft flange and the engine begins turning over.

As the engine fires the higher speed of the ring gear exceeds the speed of the pinion gear. Releasing the starter button or key switch causes the Bendix spring to withdraw the pinion gear and the starter stops turning, while the engine starts running. These types of units are common on Lycoming engines (Figure 1-6-56).

Other units operate similarly; only they mount on a gear driven starter adapter that attaches to the rear case or the engine. They also contain a similar Bendix that engages a starter jaw mounted in the accessory case on the rear of the engine. These types of starters are more common on Continental engines and are called angle drive starters.

## Starting System Maintenance Practices

Maintenance practices include replacing the starter brushes and brush springs, cleaning dirty commutators, and turning down burned or out-of-round starter commutators. As a matter of fact, condition of the brushes is a 100-hour inspection item.

As a rule, starter brushes should be replaced when worn down to approximately one-half their original length. Brush spring tension should be sufficient to give brushes a good firm contact with the commutator. Brush leads should be unbroken and lead terminal screws tight.

A glazed or dirty starter commutator can be cleaned by holding a strip of double-O sandpaper or a brush seating stone against the commutator as it is turned. The sandpaper or stone should be moved back and forth across the commutator to avoid wearing a groove. Emery paper or carborundum should never be used for this purpose because of their possible shorting action. Roughness, out-of-roundness, or high-mica conditions are reasons for turning down the commutator. In the case of a high-mica condition, the mica should be undercut after the turning operation is accomplished. All commutator segments and windings should be tested for continuity and resistance with a volt-ohm meter. All starter bearings should be replaced when the starter is disassembled for overhaul.

Most small airplane starters that fail prematurely do so because of neglect. As an example, over-lubrication causes dirt and debris to gum up the Bendix splines. This can cause the pinion not to disengage. The starter ring gear will then rapidly spin the starter once the engine starts. The starter will overheat rapidly and literally destroy itself. There will not be much left that is worth trying to overhaul.

Covering the starter with plastic bags or wraps when washing the engine is another item that is generally overlooked. This washes away the lubrication and seriously affects the service life of a starter.

A low battery can also bring on the premature failure of a starter. If system voltage is low and a cable connection or two is not tight and corrosion free, the starter can overheat while still not getting the engine started. Low voltage is also very common.

### *Inspection*

Routine inspections of the starter system should be conducted at the time periods recommended by the manufacturer. The starter system inspection should include a visual inspection of the brushes, commutator, bearings, electrical connections and mechanical linkage.

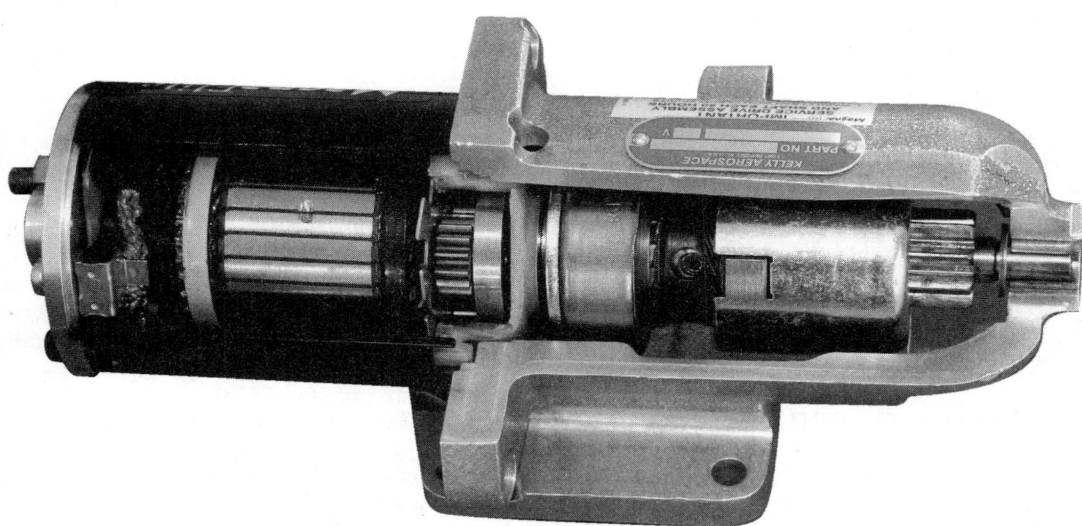


**Figure 1-6-56. Starter cutaway, showing motor and Bendix drive**

| SMALL AIRCRAFT TROUBLESHOOTING PROCEDURES | | |
|---|---|---|
| PROBABLE CAUSE | ISOLATION PROCEDURE | REMEDY |
| | *STARTER WILL NOT OPERATE* | |
| Defective master switch or circuit | Check master circuit | Repair circuit. |
| Defective starter switch or switch circuit | Check switch circuit continuity | Replace switch or wires. |
| Starter lever does not activate switch | Check starter lever adjustment | Adjust starter lever in accordance with manufacturer's instructions. |
| Defective starter | Check through items above. If another cause is not apparent, starter is defective. | Remove and repair/replace starter |
| | *STARTER MOTOR RUNS BUT DOES NOT TURN CRANKSHAFT* | |
| Starter lever adjusted to activate switch without engaging pinion with crankshaft gear | Check starter level adjustment | Adjust starter lever, in accordance with manufacturer's instructions. |
| Defective overrunning clutch or drive | Remove starter and check starter drive and overrunning clutch | Replace defective parts. |
| Damaged starter pinion gear or crankshaft gear | Remove and check pinion gear and crankshaft gear | Replace defective parts. |
| | *STARTER DRAGS* | |
| Low battery | Check battery | Charge/replace battery. |
| Starter switch or relay contacts burned or dirty | Check contacts | Replace with serviceable unit/ component. |
| Defective starter | Check starter brushes, brush spring tension for solder thrown on brush cover | Repair or replace starter. |
| Dirty, worn commutator | Clean and check visually | Turn down commutator. |
| | *STARTER EXCESSIVELY NOISY* | |
| Worn starter pinion | Remove and examine pinion | Replace starter drive. |
| Worn or broken teeth on crankshaft gears | Remove starter and turn over engine by hand to examine crankshaft gear | Replace crankshaft gear. |

**Table 1-6-1. Small-aircraft starter troubleshooting procedures**

### *Troubleshooting Small Aircraft Starting System*

The troubleshooting procedures listed in Table 1-6-1 are typical of those used to isolate malfunction in small aircraft starting systems.

# *Section 7*

# ***Engine Electronics Systems***

## Digital Instrumentation

There is a steady migration in the aviation industry from mechanical and analog electrical instrumentation to more sophisticated digital instrumentation. Mechanically operated instrumentation, often referred to as *steam gauges*, are being replaced by digital information displayed on various types of screens in front of the pilot. This type of display is called the *glass cockpit*.

Traditional instrumentation, such as the bourdon tube, requires a direct connection to the system being monitored. This means that the fuel pressure or oil pressure gauge must include a tube or hose running from the engine to the cockpit instrument. Using mechanical instrumentation places flammable liquid lines behind the aircraft's instrument panel, literally in the lap of the flight crew. Remote-reading analog instruments were developed to replace mechanical instrumentation, but each instrument represented a stand-alone system, with its own display, and was not capable of a high degree of integration.

Even with its limitations, electronic instrumentation began to replace the steam gauges. For the most part, certification and installation of instrumentation was done on a per-instrument basis. Early forms of integration allowed multiple temperature probes to connect to a single readout with a selector switch or automatic scanning circuitry. When aircraft manufacturers began installing electronic instrumentation on new aircraft, these systems started gaining acceptance (Figure 1-7-1).

While the idea of cockpit safety was appealing, there was a reluctance to accept the use of electronic instruments. This is mostly because the traditional mechanical instruments had performed so reliably. Pilots and technicians were somewhat skeptical of relying on instruments that required electrical power. The constant question was always, "What happens if the power fails?" It was a slow process for the industry to accept that the level of reliability of aircraft electrical systems and the simplicity of these instrument systems made them very trustworthy.

Fast-forward to the 1990s, and we can see another milestone in the development of engine electronic systems. Analog instrumentation began being replaced with new digital electronic systems.

Once again, this began with the introduction of replacement instruments. Individual units can be replaced with a new digital system that has provisions for interfacing with other electronic instrument and radio systems. Multiple functions can be monitored using a single flat-screen liquid crystal display (LCD), as shown in Figure 1-7-2. They are designed for readability and often show information in a format similiar to the older steam gauges.

Of course, large commercial aircraft and the business jets were the first to embrace the new technology, but smaller, less costly versions are being developed and marketed to the general aviation and reciprocating engine-powered fleet. Integration of digital instrument systems and engine controls occurred first with turbine engines. It was a number of years before the technology became available for reciprocating engine-powered aircraft.

Helping advance the industry toward digital technology were the efforts to introduce electronic ignition and electronic fuel injection to aircraft. These systems had already proven themselves in the automotive world, and manufacturers were eager to offer aircraft applications. FAA approvals and market acceptance came slowly. As engine and accessory manufacturers began to modify their products to accommodate new technologies, the framework was laid for further advancement.

Figure 1-7-1. Electronic display on an engine intsrument

# FADEC

The *FADEC system* (Full-authority Digital Engine Control) is where digital monitoring of critical engine parameters meets with engine controls. First developed for turbine-powered aircraft, this level of integration is now available for reciprocating engines. These systems provide computer-controlled electronic ignition and sequential direct port electronic fuel injection, and have the capability of interfacing with an electronic propeller control (Figure 1-7-3, next page). This places control of all engine functions with a single lever. Carburetor heat, mixture, primer and even the prop control can be eliminated.

PowerLink® FADEC, by Teledyne Continental, is a solid-state digital electronic system with only one moving part, the *pintel* that moves in the fuel injectors at each cylinder. This system

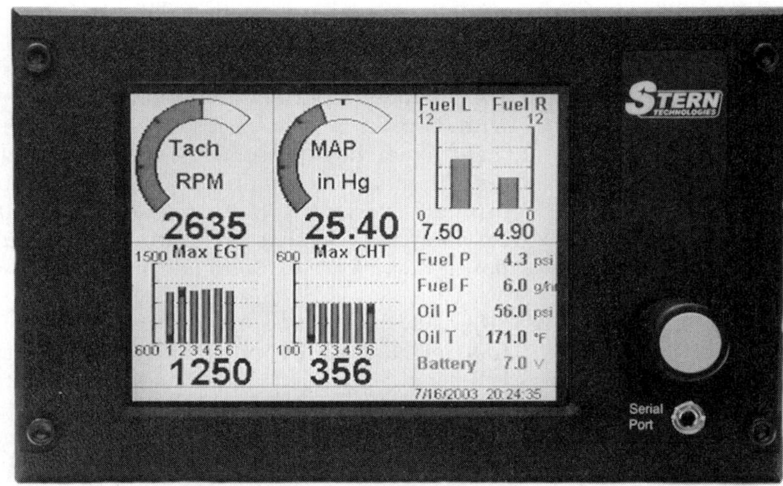


Figure 1-7-2. Flat-panel display of engine instruments

*Photo courtesy of Stern Technologies*

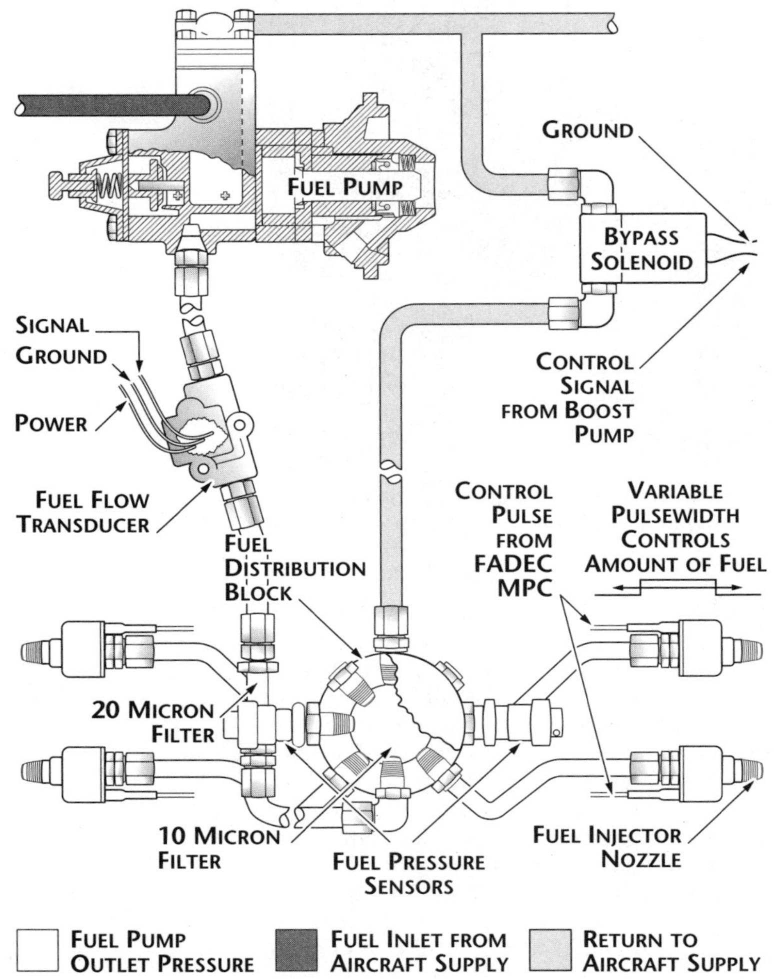


**Figure 1-7-3. Teledyne Continental PowerLink® FADEC System**

uses an array of engine data to calculate the correct engine timing and fuel mixture. Data includes cylinder head temperature, exhaust gas temperature, engine speed, camshaft position, fuel pressure, manifold air pressure and air temperature. Using this data, the unit is also able to compensate for density altitude.

The computer-controlled ignition system used with PowerLink® charges individual spark coils that provide hot, high-energy sparks throughout all modes of engine operation. Ignition timing is varied by the computers to ensure easy starting and that maximum-rated engine power is available in all flight conditions.

*Sequential-port fuel injection* provides one fuel injector per cylinder. Each fuel injector is solenoid operated under precise computer control to ensure the correct mixture for each ignition event. The amount of metered fuel will match the cylinder air charge for every given operating condition.

The pilot is provided with an Engine Performance Display that monitors and interprets all engine parameters, including the electronic propeller governor and electronic wastegate controls. This display keeps the pilot informed of trends and out-of-limit conditions. Additionally, it performs engine diagnostics and trend analysis, recording engine data compressed over the life of the engine.

PowerLink® FADEC also includes an engine diagnostic port where the technician can download vital engine data to a hand-held terminal or a laptop computer. Using the manufacturer's software, a detailed analysis of engine performance is available. This information may be used for troubleshooting, maintenance or determining performance trends.

The old question about what happens when the electrical power fails has also been addressed with this system. PowerLink® FADEC uses the aircraft power-distribution systems for its primary source, but requires a second, isolated power source. This can be a dedicated battery or a second alternator installed on the aircraft. The backup battery is a simple approach, where a small, sealed-cell battery, dedicated to the FADEC system, provides up to three hours of operation. The health of the backup battery is monitored, and the pilot can determine its condition prior to takeoff.

The PowerLink® FADEC system is comprised of two complete systems of highly reliable components, any one of which is capable of running the engine at full power. All critical components are, at the very least, doubly redundant.

## Maintenance and Installation of Engine Electronic Systems

Maintenance of engine electronic systems must be accomplished in accordance with the manufacturer's service information. Some systems include built-in troubleshooting systems and provide information regarding the repair or replacement of components. Often, field repairs to electronic equipment are limited to removal and replacement of major components.

Because this technology is new to the industry and is advancing at a tremendous speed, the aviation maintenance technician must be alert to stay out of trouble. New products are often introduced to the aviation market prior to their certification by the FAA. This can be accomplished as simply as selling them for use in home-built and experimental aircraft. Use in experimental aircraft is an excellent way for new technology to prove itself and build the history of reliability and accuracy necessary for FAA approval and installation on aircraft with standard certification.

Always look for the approval basis prior to installing any new equipment or certifying an aircraft airworthy following a 100-hour or annual inspection. Approvals may come from several sources, and you must be able to determine which is acceptable for the installation that you are working on. New technology is seldom included on the Type Certificate Data Sheet, so you will probably be looking for a Supplemental Type Certificate (STC), FAA Parts Manufacturing Approval (PMA) or a Technical Standard Order (TSO).

Installation must be made in accordance with acceptable certification documentation and the aircraft returned to service with the appropriate logbook entries and/or FAA Form 337.

Section 8

# Engine Fuel and Fuel Metering Systems

## Fuel System Requirements

Improvements in aircraft and engines have increased the demands on the fuel system, making it more complicated and increasing the installation, adjustment and maintenance problems. The fuel system must supply fuel to the carburetor or other metering device under all conditions of ground and air operation. It must function properly at constantly changing altitudes and in any climate. The system should be free of tendency to vapor lock, which can result from changes in ground and in-flight climatic conditions.

On small aircraft, a simple gravity-feed fuel system consisting of a tank to supply fuel to the engine is often installed. On multi-engine aircraft, complex systems are necessary so that fuel can be pumped from any combination of tanks to any combination of engines. Provisions for transferring fuel from one tank to another may also be included on large aircraft.

## Vapor Lock

Normally, the fuel remains in a liquid state until it is discharged into the airstream and then instantly changes to a vapor. Under certain conditions, however, the fuel may vaporize in the lines, pumps or other units. The vapor pockets formed by this premature vaporization restrict the fuel flow through units that are designed to handle liquids rather than gases. The resulting partial or complete interruption of the fuel flow is called *vapor lock*. The three general causes of vapor lock are the lowering of the pressure on the fuel, high fuel temperatures and excessive fuel turbulence. At high altitudes, the pressure on the fuel in the tank is low. This lowers the boiling point of the fuel and causes vapor bubbles to form. This vapor trapped in the fuel may cause vapor lock in the fuel system.

Transfer of heat from the engine tends to cause boiling of the fuel in the lines and the pump. This tendency is increased if the fuel in the tank is warm. High fuel temperatures often combine with low pressure to increase vapor formation. This is most likely to occur during a rapid climb on a hot day. As the aircraft climbs, the outside temperature drops, but the fuel does not lose temperature rapidly. If the fuel is warm enough at takeoff, it retains enough heat to boil easily at high altitude.

The chief causes of fuel turbulence are sloshing of the fuel in the tanks, the mechanical action of the engine-driven pump and sharp bends or rises in the fuel lines. Sloshing in the tank tends to mix air with the fuel. As this mixture passes through the lines, the trapped air separates from the fuel and forms vapor pockets at any point where there are abrupt changes in direction or steep rises. Turbulence in the fuel pump often combines with the low pressure at the pump inlet to form a vapor lock at this point.

Vapor lock can become serious enough to block the fuel flow completely and stop the engine. Even small amounts of vapor in the inlet line restricts the flow to the engine-driven pump and reduces its output pressure.

To reduce the possibility of vapor lock, fuel lines are kept away from sources of heat; also, sharp bends and steep rises are avoided. In addition, the volatility of the fuel is controlled in manufacture so that it does not vaporize too readily. The major improvement in reducing vapor lock, however, is the incorporation of *booster pumps* in the fuel system. These pumps keep the fuel in the lines to the engine-driven pump under pressure. The slight pressure on the fuel reduces vapor formation and aids in moving a vapor pocket along. The booster pump also releases vapor from the fuel as it passes through the pump. The vapor moves upward through the fuel in the tank and out the tank vents.

To prevent the small amount of vapor that remains in the fuel from upsetting its metering action, vapor eliminators are installed in some fuel systems ahead of the metering device or are built into this unit.

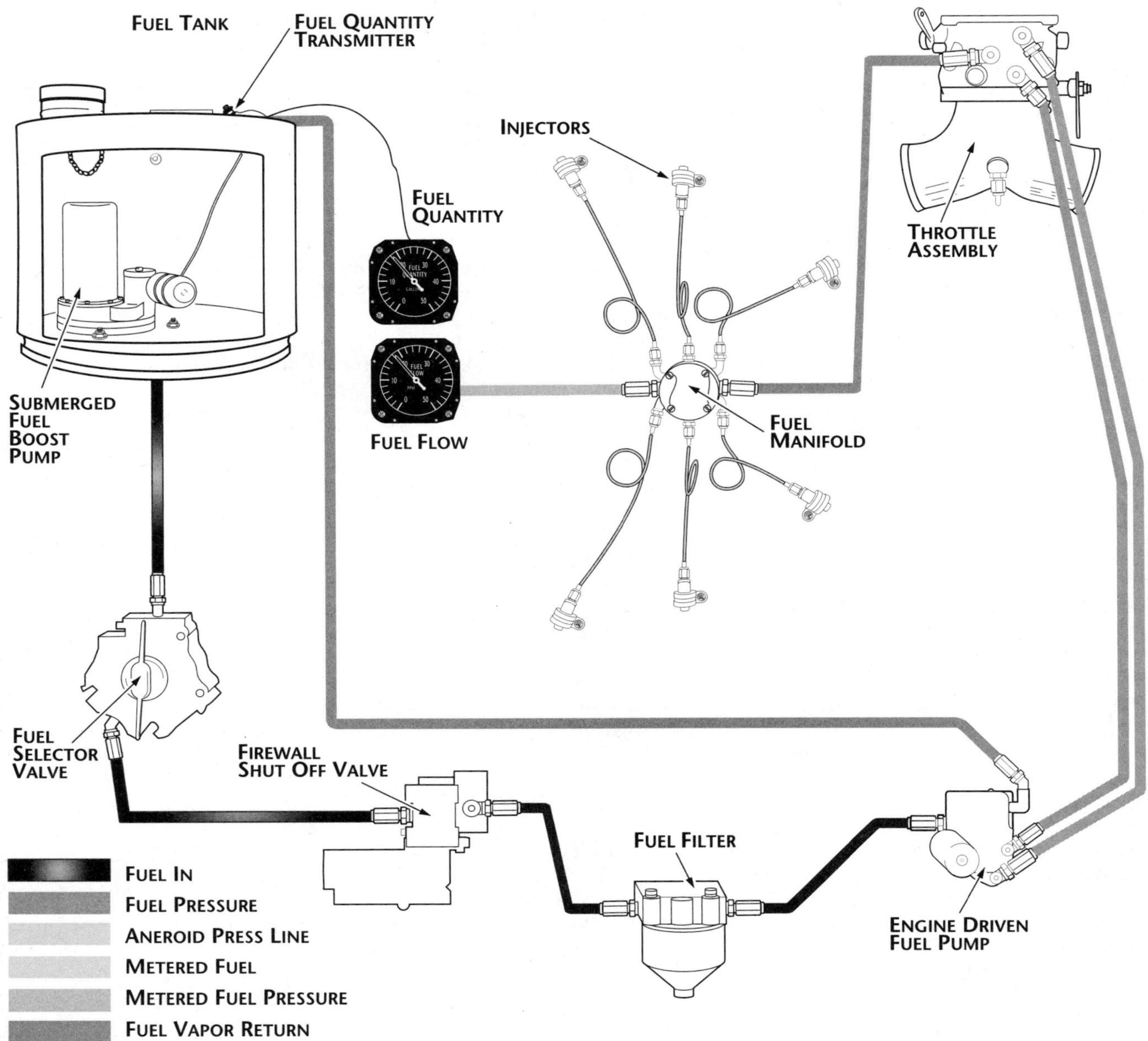


**Figure 1-8-1. A schematic of a basic aircraft fuel system**

## Basic Fuel System

The basic components of an aircraft fuel system (Figure 1-8-1) include:

- Fuel tanks
- Lines
- Selector valves
- Strainers
- Pumps
- Pressure gauges

Normally, an aircraft, regardless of its size, will have several fuel tanks. The design and location of these fuel tanks will vary with design and construction of the aircraft, as well as the design of the fuel system. Each fuel tank is connected to a *selector valve* with a fuel line. The selector valve controls the flow of fuel from the cockpit. Each fuel system must contain at least one *main strainer.* This main strainer is located at the lowest point of the fuel system and serves as a collection point for the moisture in the system. The system will also contain at least one strainer or filter to remove dirt and solid contaminate.

During starting, the booster pump forces fuel through a bypass in the engine-driven pump to the metering device. Once the engine-driven pump is rotating at sufficient speed, it takes over and delivers fuel to the metering device at the specified pressure.

The airframe fuel system begins with the fuel tank and ends at the engine fuel system. The engine fuel system usually includes the engine-driven pumps and the fuel-metering systems. In aircraft powered with a reciprocating engine, the fuel-metering system consists of the air- and fuel-control devices from the point where the fuel enters the first control unit until the fuel is injected into the supercharger section: intake pipe or cylinder. For example, the fuel metering system of the Continental IO-470L engine consists of the fuel/air-control unit, the injector pump, the fuel manifold valve and the fuel-discharge nozzles (see Figure 1-8-1).

The fuel metering system on reciprocating engines meters fuel at a predetermined ratio to airflow. The airflow to the engine is controlled by the carburetor or fuel/air-control unit.

The fuel-metering system of the gas turbine engine consists of a jet fuel control and may extend to and include the fuel nozzles. On some turboprop engines, a temperature datum valve is a part of the engine fuel system. The rate of fuel delivery is a function of air mass flow, compressor inlet temperature, compressor discharge pressure, r.p.m. and combustion chamber pressure.

The fuel-metering system must operate satisfactorily to ensure efficient engine operation as measured by power output, operating temperatures and range of the aircraft. Because of variations in design of different fuel metering systems, the expected performance of any one piece of equipment, as well as the difficulties it can cause, will vary.

## Fuel-metering Devices for Reciprocating Engines

This section explains the systems that deliver the correct mixture of fuel and air to an engine's combustion chambers. In the discussion of each system, the general purpose and operating principles are stressed, with particular emphasis on the basic principles of operation. No attempt is made to give detailed operating and maintenance instructions for specific types and makes of equipment. For the specific information needed to inspect or maintain a particular installation or unit, consult the manufacturer's instructions.

The basic requirement of a fuel-metering system is the same, regardless of the type of system used or the model engine on which the equipment is installed. It must meter fuel proportionately to air to establish the proper fuel/air mixture ratio for the engine at all speeds and altitudes at which the engine may be operated. In the fuel/air mixture curves shown in Figure 1-8-2, note that the basic best-power and best-economy fuel/air mixture requirements for all reciprocating engines are approximately the same.

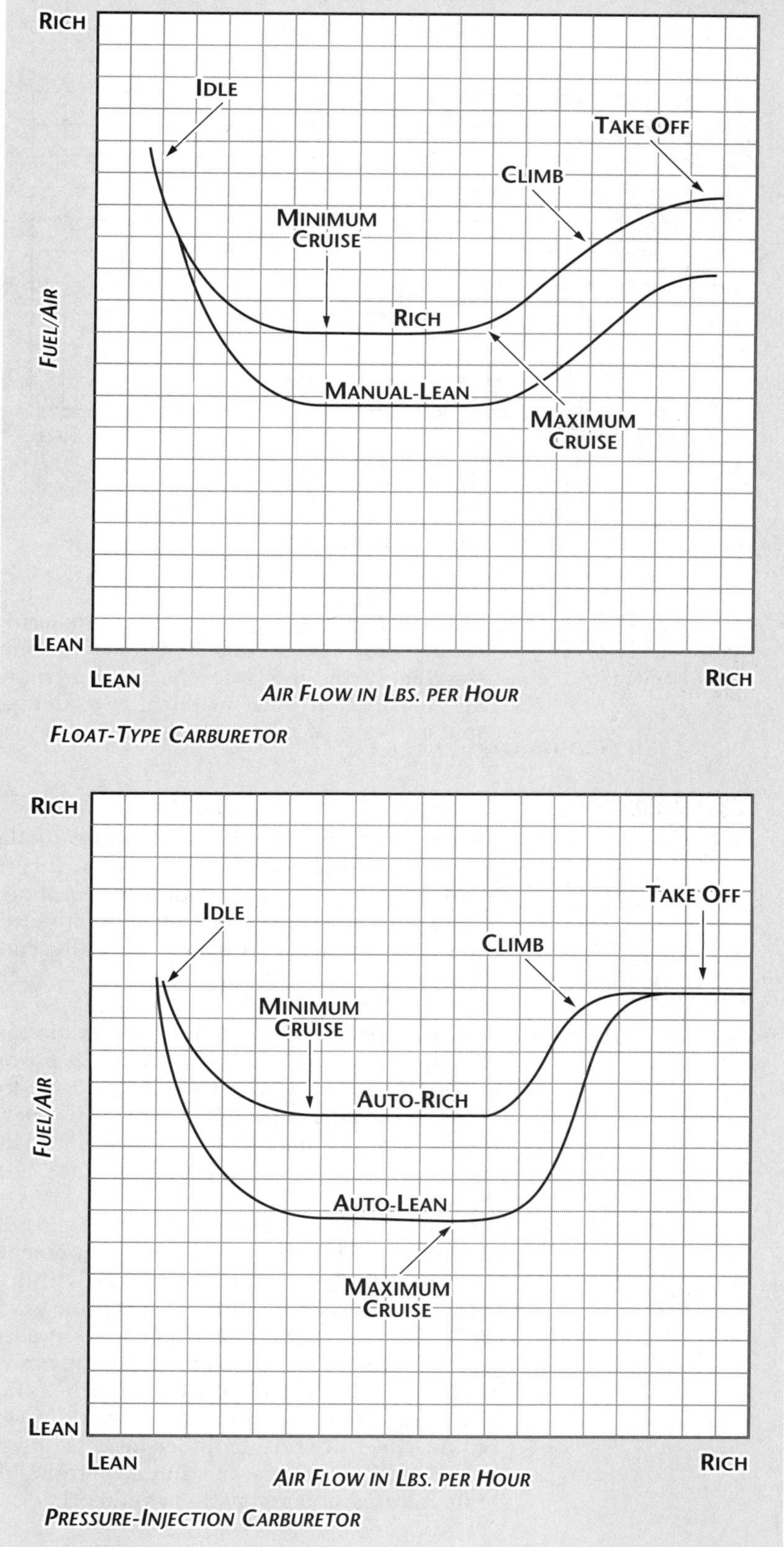


Figure 1-8-2. Fuel/air mixture curves

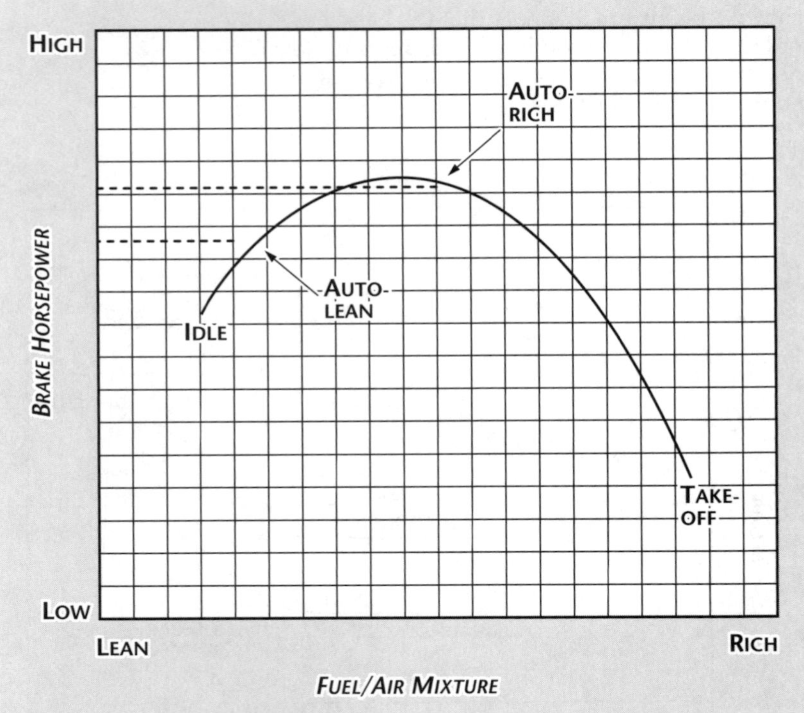


Figure 1-8-3. Power versus fuel/air mixture curve

The second requirement of a fuel-metering system is to atomize and distribute the fuel from the carburetor into the mass airflow. This distribution must be such as to deliver equal amounts of fuel in the fuel/air mixture to each cylinder.

A carburetor has a tendency to run richer at altitude than on the ground because of the change in the air density. Because of this, it is necessary for the carburetor to be equipped with a mixture-control device. This mixture-control device may be either manually controlled or automatically controlled.

The rich mixture requirements for an aircraft engine are established by running a power curve to determine the fuel/air mixture for obtaining maximum usable power. This curve (Figure 1-8-3) is plotted at 100-r.p.m. intervals from idle speed to takeoff speed.

Since it is necessary in the power range to add fuel to the basic fuel/air mixture requirements to keep cylinder-head temperatures within a safe range, the fuel mixture must become gradually richer as powers above cruise are used (see Figure 1-8-2, previous page). In the power range, the engine will run on a much leaner mixture, as indicated in the curves. However, on the leaner mixture, cylinder-head temperature would exceed the maximum permissible temperatures, and detonation would occur.

The best economy setting is established by running a series of curves through the cruise range, as shown in the graph in Figure 1-8-4, the low point (auto-lean) in the curve being the fuel/air mixture where the minimum fuel-per-horsepower is used. In this range, the engine will operate normally on slightly leaner mixtures and will also operate on richer mixtures than the low-point mixture. If a mixture leaner than that specified for the engine is used, the leanest cylinder of the engine is apt to backfire, because the slower burning rate of the lean mixture results in a continued burning in the cylinder when the next intake stroke starts.

## Fuel/Air Mixtures

Gasoline and other liquid fuels will not burn at all unless they are mixed with air. If the mixture is to burn properly within the engine cylinder, the ratio of air to fuel must be kept within a certain range.

It would be more accurate to state that the fuel is burned with the oxygen in the air. Seventy-eight percent of air by volume is nitrogen, which is inert and does not participate in the combustion process, and 21 percent is oxygen. Heat is generated by burning the mixture of gasoline and oxygen. Nitrogen and gaseous by-products of combustion absorb this heat energy and, by expansion, turn it into power. The mixture proportion of fuel and air by weight is of extreme importance to engine performance. The characteristics of a given mixture can be measured in terms of flame speed and combustion temperature.

The composition of the fuel/air mixture is described by the mixture ratio. For example, a mixture with a ratio of 12 to 1 (12:1) is made up of 12 lbs. of air and 1 lb. of fuel. The ratio is expressed in weight because the volume of air varies greatly with temperature and pressure. The mixture ratio can also be expressed as a decimal. Thus, a fuel/air ratio of 12:1 and a fuel/air ratio of 0.083 describe the same mixture ratio. Air and gasoline mixtures as rich as 8:1 and as lean as 16:1 will burn in an engine cylinder. The engine develops maximum power with a mixture of approximately 12 parts of air and 1 part of gasoline.

From the chemist's point of view, the perfect mixture for combustion of fuel and air would be 0.067 lb. of fuel to 1 lb. of air (a mixture ratio of 15:1). The scientist calls this chemically correct combination a *stoichiometric mixture* (pronounced stoy-key-o-metric). With this mixture — given sufficient time and turbulence — all the fuel and all the oxygen in the air will be completely used in the combustion process. The stoichiometric mixture produces the highest combustion temperatures because the proportion of heat released to a mass of charge (fuel and air) is the greatest. However, the mixture is seldom used because it does not result in either the greatest economy or the greatest power for the airflow or manifold pressure.

If more fuel is added to the same quantity of air charge than the amount giving a chemically perfect mixture, changes of power and temperature will occur. The combustion gas temperature will be lowered as the mixture is enriched, and the power will increase until the fuel/air ratio is approximately 0.0725. From 0.0725 fuel/air ratio to 0.080 fuel/air ratio, the power will remain essentially constant, even though the combustion temperature continues downward. Mixtures from 0.0725 fuel/air ratio to 0.080 fuel/air ratio are called *best-power mixtures,* since their use results in the greatest power for a given airflow or manifold pressure. In this fuel/air ratio range, there is no increase in the total heat released, but the weight of nitrogen and combustion products is augmented by the vapor formed with the excess fuel. Thus, the working mass of the charge is increased. In addition, the extra fuel in the charge (over the stoichiometric mixture) speeds up the combustion process, which provides a favorable time factor in converting fuel energy into power.

Besides reduction of temperature, enriching a fuel/air ratio above 0.080 results in a loss of power. The cooling effects of excess fuel overtake the favorable factor of increased mass. The reduced temperature and slower rate of burning lead to an increasing loss of combustion efficiency.

If, with constant airflow, the mixture is leaned to below a 0.067 fuel/air ratio, power and temperature will decrease together. This time, the loss of power is not a liability but an asset. The purpose in leaning is to save fuel. Air is free and available in limitless quantities. The object is to obtain the required power with the least fuel flow and to let the air consumption take care of itself.

A measure of the economical use of fuel is called *specific fuel consumption* (SFC), which is the pounds of fuel, per hour, per horsepower. Thus, SFC = lbs. fuel/hr./hp. By using this ratio, the engine's use of fuel at various power settings can be compared. When leaning below 0.067 fuel/air ratio with constant airflow, even though the power diminishes, the cost in fuel to support each horsepower hour (SFC) is also lowered temporarily. While the mixture charge is becoming weaker, this loss of strength occurs at a rate slower than that of the reduction of fuel flow. This favorable tendency continues until a mixture strength known as *best economy* is reached. With this fuel/air ratio, the required horsepower is developed with the least fuel flow or, to put it another way, a given fuel flow produces the most power.

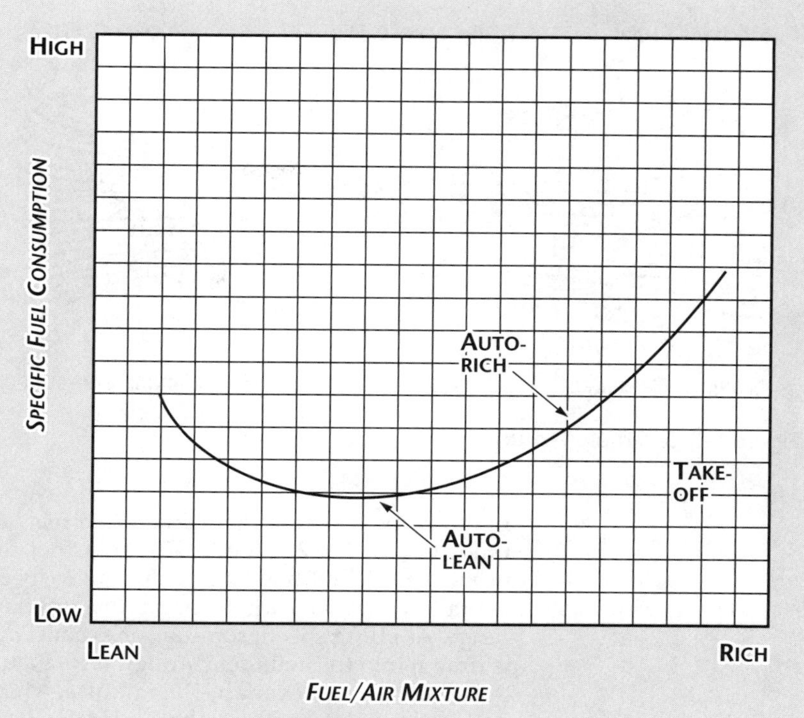


**Figure 1-8-4. Specific fuel consumption curve**

The best-economy fuel/air ratio varies somewhat with r.p.m. and other conditions but, for cruise powers on most reciprocating engines, it is sufficiently accurate to define this range of operation as being from 0.060 to 0.065 fuel/air ratios with retard spark, and from 0.055 to 0.061 fuel/air ratios with advance spark. These are the most commonly used fuel/air ratios on aircraft where manual leaning is practiced.

Below the best-economical mixture strength, power and temperature continue to fall with constant airflow while the SFC increases. As the fuel/air ratio is reduced further, combustion becomes so cool and slow that power for a given manifold pressure becomes so low as to be uneconomical. The cooling effect of rich or lean mixtures results from the excess fuel or air over that needed for combustion. Internal cylinder cooling is obtained from unused fuel when fuel/air ratios above 0.067 are used. The same function is performed by excess air when fuel/air ratios below 0.067 are used.

Varying the mixture strength of the charge produces changes in the engine operating condition, affecting power, temperature and spark-timing requirements. The best power fuel/air ratio is desirable when the greatest power from a given airflow is required. The best economy mixture results from obtaining the given power output with the least fuel flow. The fuel/air ratio which gives most efficient operation varies with engine speed and power output.

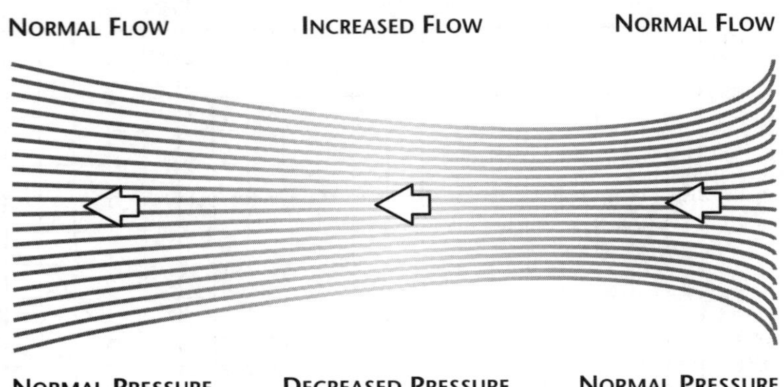


Figure 1-8-5. Simple venturi

In the graph showing this variation in fuel/air ratio (Figure 1-8-2, page 1-173), note that the mixture is rich at both idling and high-speed operation and is lean through the cruising range. At idling speed, some air or exhaust gas is drawn into the cylinder through the exhaust port during valve overlap. The mixture, which enters the cylinder through the intake port, must be rich enough to compensate for this gas or additional air. At cruising power, lean mixtures save fuel and increase the range of the airplane. An engine running near full power requires a rich mixture to prevent overheating and detonation. Since the engine is operated at full power for only short periods, the high fuel consumption is not a serious matter. If an engine is operating on too lean a mixture and adjustments are made to increase the amount of fuel, the power output of the engine increases rapidly at first, then gradually until maximum power is reached. With a further increase in the amount of fuel, the power output drops gradually at first, then more rapidly as the mixture is further enriched.

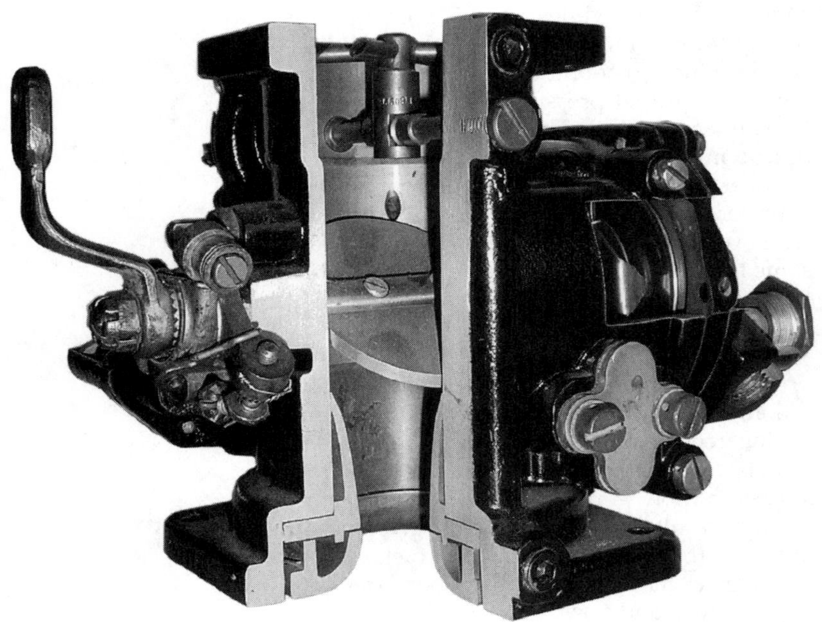

Figure 1-8-6. A carburetor cutaway to show the venturi

There are specific instructions concerning mixture ratios for each type of engine under various operating conditions. Failure to follow these instructions will result in poor performance and often in damage to the engine. Excessively rich mixtures result in loss of power and waste of fuel. With the engine operating near its maximum output, very lean mixtures will cause a loss of power and, under certain conditions, serious overheating. When the engine is operated on a lean mixture, the cylinder-head temperature gauge should be watched closely. If the mixture is excessively lean, the engine may backfire through the induction system or stop completely. Backfire results from slow burning of the lean mixture. If the charge is still burning when the intake valve opens, it ignites the fresh mixture and the flame travels back through the combustible mixture in the induction system.

## Carburetion Principles

### *Venturi Principles*

As previously discussed in this chapter, it is the function of a carburetor to regulate the proper amount of fuel to be mixed with the mass airflow. To accomplish this, a *venturi* is used in the throat of the carburetor to measure the airflow in the induction system. As the velocity of the air in the venturi increases, the pressure will decrease. It is this change in pressure that is used to regulate the fuel flow. Figure 1-8-5 shows the change in pressure as the air flows through the venturi. The operation of the carburetor is dependent on the pressure difference between the center of the venturi and the air inlet.

### *Application of Venturi Principle to Carburetors*

The carburetor is mounted on the engine so that air to the cylinders passes through the *barrel*, the part of the carburetor that contains the venturi. The size and shape of the venturi depends on the requirements of the engine for which the carburetor is designed. A carburetor for a high-powered engine may have one large venturi or several small ones. The air may flow either up or down the venturi, depending on the design of the engine and the carburetor (Figure 1-8-6). Those in which the air passes downward are known as *downdraft carburetors*, and those in which the air passes upward are called *updraft carburetors.*

Air can be drawn through a rubber tube by placing one end in the mouth and exerting a sucking action. The pressure inside the tube is lowered and atmospheric pressure pushes air into the open end. Air flows through the induction system in the same manner. When a piston moves toward the crankshaft on the intake stroke, the pressure in the cylinder is lowered. Air rushes through the carburetor and intake manifold to the cylinder due to the higher pressure at the carburetor intake. Even in a supercharged engine operating at high manifold pressure, there is still a low pressure at the engine side of the carburetor. Atmospheric pressure at the air intake pushes air through the carburetor to the supercharger inlet.

The throttle valve is located between the venturi and the engine. Mechanical linkage connects this valve with the throttle lever in the cockpit. By means of the throttle, airflow to the cylinders is regulated and controls the power output of the engine. It is the throttle valve in your automobile carburetor that opens when you step on the gas. More air is admitted to the engine, and the carburetor automatically supplies enough additional gasoline to maintain the correct fuel/air ratio. The throttle valve obstructs the passage of air very little when it is parallel with the flow. This is the wide-open position. Throttle action is illustrated in Figure 1-8-7. Note how it restricts the airflow more as it rotates toward the closed position.

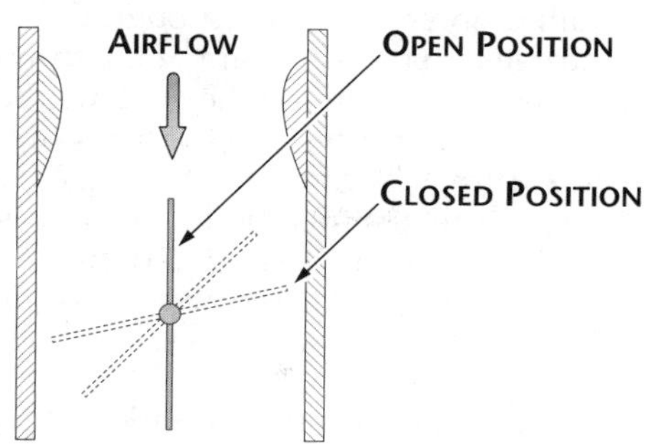


Figure 1-8-7. Throttle action

## *Metering and Discharge of Fuel*

In the illustration showing the discharge of fuel into the air stream (Figure 1-8-8), locate the inlet through which fuel enters the carburetor from the engine-driven pump. The float-operated needle valve regulates the flow through the inlet and, in this way, maintains the correct level in the fuel chamber. This level must be slightly below the outlet of the discharge nozzle to prevent overflow when the engine is not running.

The discharge nozzle is located in the throat of the venturi at the point where the lowest drop in pressure occurs as air passes through the carburetor to the engine cylinders. Thus, there are two different pressures acting on the fuel in the carburetor: a low pressure at the discharge nozzle and a higher (atmospheric) pressure in the float chamber. The higher pressure in the

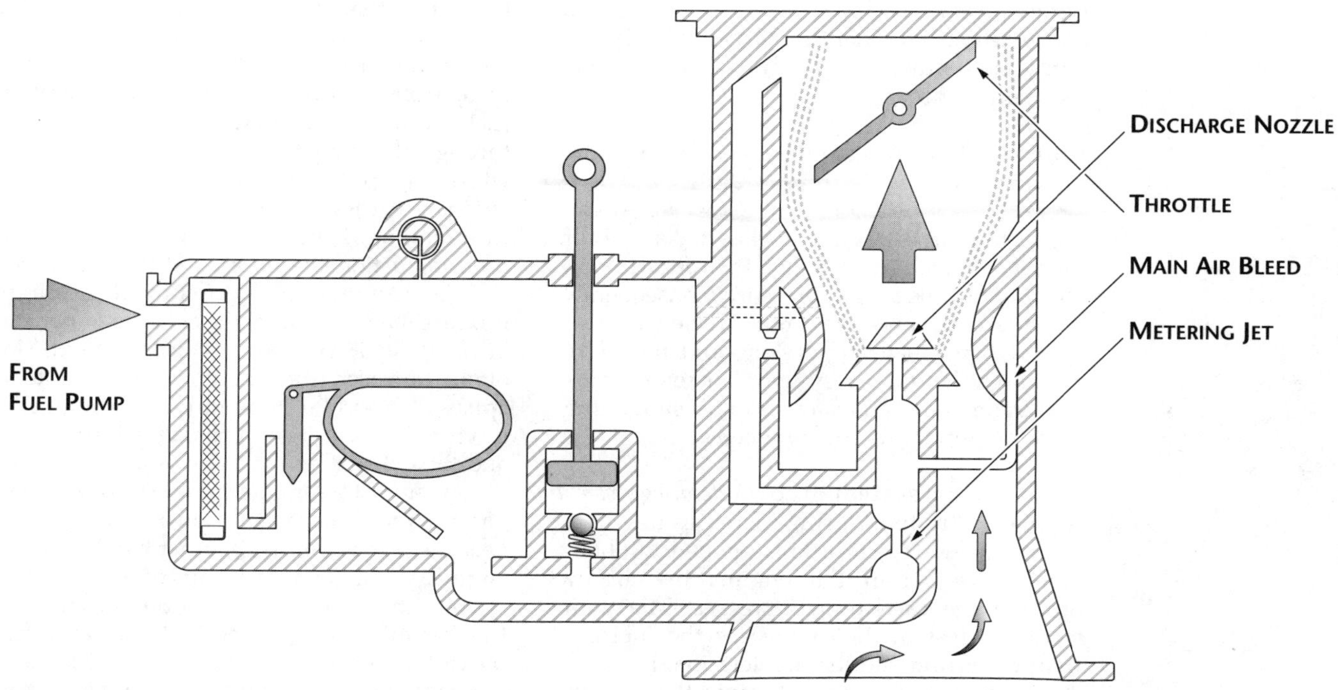


Figure 1-8-8. Fuel discharge

float chamber forces the fuel through the discharge nozzle into the airstream. If the throttle is opened wider to increase the airflow to the engine, there is a greater drop in pressure at the venturi throat. Because of the higher differential pressure, the fuel discharge increases in proportion to the increase in airflow. If the throttle is moved toward the closed position, the airflow and fuel flow decrease.

The fuel must pass through the metering jet (Figure 1-8-8) to reach the discharge nozzle. The size of this jet determines the rate of fuel discharge at each differential pressure. If the jet is replaced with a larger one, the fuel flow will increase, resulting in a richer mixture. If a smaller jet is installed, there will be a decrease in fuel flow and a leaner mixture.

## Carburetor Systems

To provide for engine operation under various loads and at different engine speeds, each carburetor has six systems:

1. Main metering
2. Idling
3. Accelerating
4. Mixture control
5. Idle cutoff
6. Power enrichment or economizer

Each of these systems has a definite function. It may act alone or with one or more of the others.

**Main metering system.** The *main metering system* supplies fuel to the engine at all speeds above idling. The fuel discharged by this system is determined by the drop in pressure in the venturi throat.

**Idling system.** A separate system is necessary for idling because the main metering system is unreliable at very low engine speeds. At low speed, the throttle is nearly closed. As a result, the velocity of the air through the venturi is low and there is little drop in pressure. Consequently, the differential pressure is not sufficient to operate the main metering system, and no fuel is discharged from the system. Therefore, most carburetors have an *idling system* to supply fuel to the engine at low engine speeds.

**Accelerating system.** The *accelerating system* supplies extra fuel during increases in engine power. When the throttle is opened to obtain more power from the engine, the airflow through the carburetor increases. The main metering system then increases the fuel discharge. During sudden acceleration, however, the increase in airflow is so rapid that there is a slight time lag before the increase in fuel discharge is sufficient to provide the correct mixture ratio with the new airflow. By supplying extra fuel during this period, the accelerating system prevents a temporary leaning-out of the mixture and gives smooth acceleration.

**Mixture-control system.** The *mixture-control system* determines the ratio of fuel to air in the mixture. By means of a cockpit control, the operator can select the mixture ratio to suit operating conditions. In addition to these manual controls, many carburetors have automatic mixture controls so that the fuel/air ratio, once it is selected, does not change with variations in air density. This is necessary because, as the airplane climbs and the atmospheric pressure decreases, there is a corresponding decrease in the weight of air passing through the induction system. The volume, however, remains constant and, since it is the volume of airflow which determines the pressure drop at the throat of the venturi, the carburetor tends to meter the same amount of fuel to this thin air as to the dense air at sea level. Thus, the natural tendency is for the mixture to become richer as the airplane gains altitude. The automatic mixture control prevents this by decreasing the rate of fuel discharge to compensate for the decrease in air density.

**Idle-cutoff system.** The carburetor has an *idle-cutoff system* so that the fuel can be shut off to stop the engine. This system, incorporated in the manual mixture control, stops the fuel discharge from the carburetor completely when the mixture-control lever is set to the idle-cutoff position. In any discussion of the idle-cutoff system, this question usually comes up: Why is an aircraft engine stopped by shutting off the fuel rather than by turning off the ignition? To answer this question, it is necessary to examine the results of both methods. If the ignition is turned off with the carburetor still supplying fuel, fresh fuel/air mixture continues to pass through the induction system to the cylinders while the engine is coasting to a stop. If the engine is excessively hot, this combustible mixture may be ignited by local hot spots within the combustion chambers, and the engine may keep on running or kick backward. Again, the mixture may pass out of the cylinders unburned but be ignited in the hot exhaust manifold. More often, however, the engine will come to an apparently normal stop but have a combustible mixture in the induction passages, the cylinders and the exhaust system. This is an unsafe condition, since the engine may kick over after it has been stopped and seriously injure anyone near the propeller. On the other hand, when the engine is shut down by means of the idle-cutoff system, the spark plugs continue to ignite the fuel/air mixture until the fuel discharge from the carburetor ceases. This alone should prevent the engine from coming to a stop with a combustible mixture in the cylinders.

Some engine manufacturers suggest that, just before the propeller stops turning, the throttle be opened wide so that the pistons can pump fresh air through the induction system, the cylinders and the exhaust system as an added precaution against accidental kick-over. After the engine has come to a complete stop, the ignition switch is turned to the OFF position.

**Power-enrichment system.** The *power-enrichment system* automatically increases the richness of the mixture during high-power operation. In this way, it enables the variation in fuel/air ratio necessary to fit different operating conditions. Remember that, at cruising speeds, a lean mixture is desirable for reasons of economy while, at high-power output, the mixture must be rich to obtain maximum power and to aid in cooling the engine. The power-enrichment system automatically brings about the necessary change in the fuel/air ratio. Essentially, it is a valve that is closed at cruising speeds and opens to supply extra fuel to the mixture during high-power operation. Although it increases the fuel flow at high power, the power-enrichment system is actually a fuel-saving device. Without this system, it would be necessary to operate the engine on a rich mixture over the complete power range. The mixture would then be richer than necessary at cruising speed to ensure safe operation at maximum power. The power-enrichment system is sometimes called an *economizer* or a *power compensator.*

Although the various systems have been discussed separately, the carburetor functions as a unit. The fact that one system is in operation does not necessarily prevent another from functioning simultaneously. At the same time that the main metering system is discharging fuel in proportion to the airflow, the mixture-control system determines whether the resultant mixture will be rich or lean. If the throttle is suddenly opened wide, the accelerating and power-enrichment systems act to add fuel to that already being discharged by the main metering system.

## *Carburetor Types*

In the discussion of the basic carburetor principles, the fuel was shown stored in a float chamber and discharged from a nozzle located in the venturi throat. With a few added features to make it workable, this becomes the main metering system of the *float-type carburetor* (Figure 1-8-9). This type of carburetor, complete with idling, accelerating, mixture-control, idle-cutoff and power-enrichment systems, is probably the most common of all carburetor types.

**Figure 1-8-9. A float-type carburetor**

A float-type carburetor has three major disadvantages:

1. The float bowl is incapable of maintaining a constant amount of fuel during abrupt maneuvers.
2. Complete vaporization of the fuel is difficult because of the low pressures at which the fuel is discharged into the airflow.
3. The float-type carburetor is very susceptible to icing, because the fuel is discharged at the point of lowest pressure in the induction system. The drop of pressure in the venturi also results in decreased temperature. This aids in the formation of ice in the venturi and on the throttle valve.

A *pressure-type carburetor* (Figure 1-8-10, next page) discharges fuel into the airstream at a pressure well above atmospheric. This results in better vaporization and permits the discharge of fuel into the airstream on the engine side of the throttle valve. With the discharge nozzle located at this point, the drop in temperature due to fuel vaporization takes place after the air has passed the throttle valve and at a point where engine heat tends to offset it. Thus, the danger of fuel vaporization icing is practically eliminated. The effects of rapid maneuvers and rough air on the pressure-type carburetors are negligible, since its fuel chambers remain filled under all operating conditions.

Figure 1-8-10. A pressure-type carburetor

## *Carburetor Icing*

There are three general classifications of carburetor icing common for all aircraft:

1. Fuel-evaporation ice
2. Throttle ice
3. Impact ice

*Fuel-evaporation ice*, or *refrigeration ice*, is formed because of the decrease in air temperature resulting from the evaporation of fuel after it is introduced into the airstream. It frequently occurs in those systems where the fuel is injected into the air upstream from the carburetor throttle, as in the case of float-type carburetors. It occurs less frequently in the systems in which the fuel is injected into the air downstream from the carburetor. Engines employing *spinner injection* or *impeller injection* of fuel are free of this type of ice, except those that have turning vanes (to change the direction of flow) at the entrance to the impeller. In this type, ice can be deposited on the turning vanes. Refrigeration ice can be formed at carburetor air temperatures as high as 100°F over a wide range of atmospheric humidity conditions, even at relative humidity well below 100 percent. Generally, fuel-evaporation ice will tend to accumulate on the fuel-distribution nozzle, the turning vanes and any protuberances in the carburetor. This type of ice can lower manifold pressure, interfere with fuel flow and affect mixture distribution.

*Throttle ice* is formed on the rear side of the throttle, usually when the throttle is in a partially closed position. The rush of air across and around the throttle valve causes a low pressure on the rear side, which sets up a pressure differential across the throttle and has a cooling effect on the fuel/air charge. Moisture freezes in this low-pressure area and collects as ice on the low-pressure side. Throttle ice tends to accumulate in a restricted passage. The occurrence of a small amount of ice may cause a relatively large reduction in airflow and manifold pressure. A large accumulation of ice may jam the throttles and cause them to become inoperable. Throttle ice seldom occurs at temperatures above 38°F.

*Impact ice* is formed either from water present in the atmosphere as snow, sleet or sub-cooled liquid water, or from liquid water which impinges on surfaces that are at temperatures below 30°F. Because of the effects of inertia, impact ice collects on or near a surface that changes the direction of the airflow. This type of ice may build up on the carburetor elbow, as well as the carburetor screen and metering elements. The most dangerous impact ice is that which collects on the carburetor screen and causes a very rapid throttling of airflow and power. In general, danger from impact ice exists only when ice forms on the leading edges of the aircraft structure.

Under some conditions, ice may enter the carburetor in a comparatively dry state and will not adhere to the screen or walls, therefore not affecting engine airflow or manifold pressure. This ice may enter the carburetor and gradually build up internally in the carburetor air metering passages and affect carburetor metering characteristics.

## *Float-type Carburetors*

A float-type carburetor consists essentially of a main air passage through which the engine draws its supply of air, a mechanism to control the quantity of fuel discharged in relation to the flow of air, and a means of regulating the quantity of fuel/air mixture delivered to the engine cylinders.

The essential parts of a float-type carburetor are illustrated in Figure 1-8-11. These parts are:

- Float mechanism and its chamber
- Main metering system
- Idling system
- Mixture-control system
- Accelerating system
- Economizer system

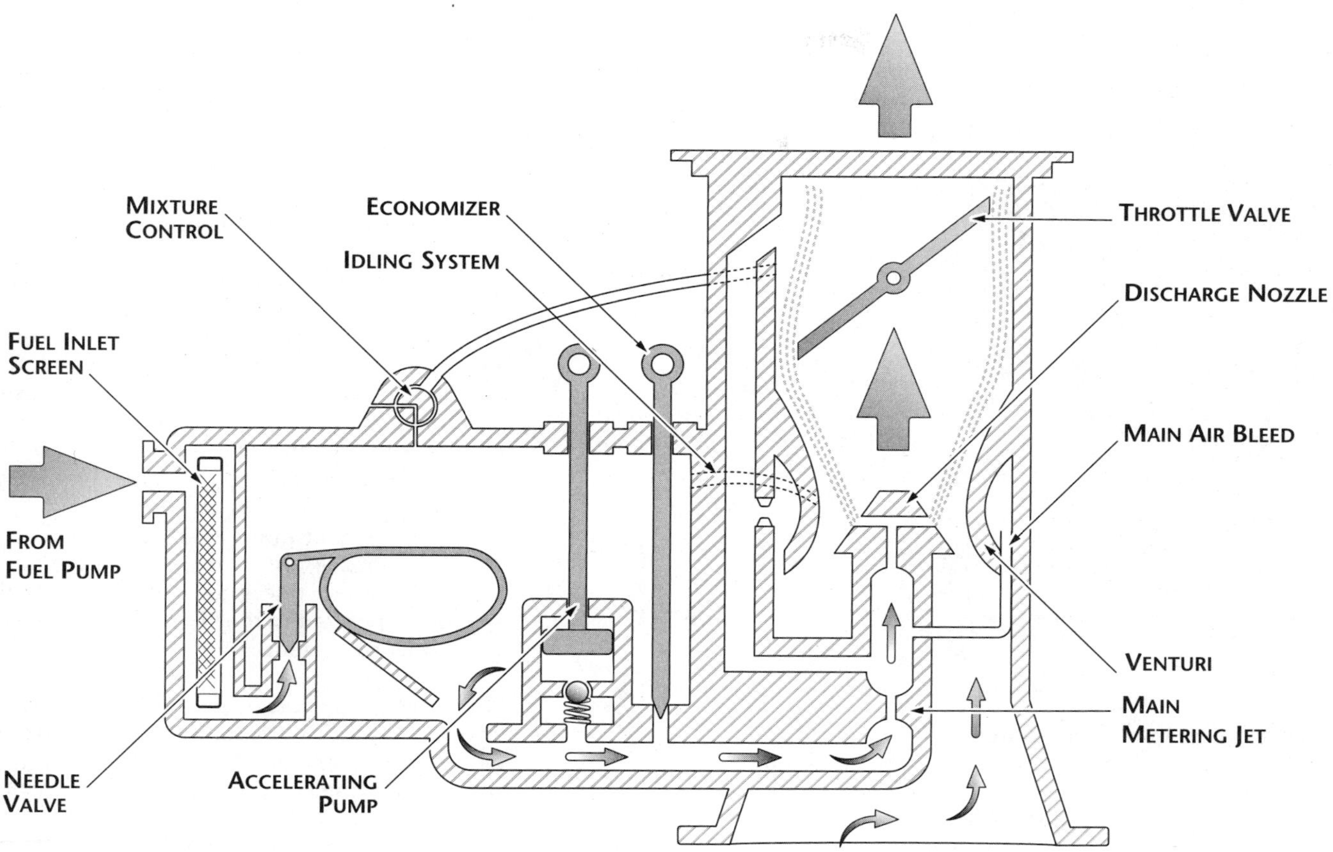

Mixture Control
Economizer
Idling System
Throttle Valve
Discharge Nozzle
Fuel Inlet Screen
Main Air Bleed
From Fuel Pump
Venturi
Main Metering Jet
Needle Valve
Accelerating Pump

**Figure 1-8-11. A float-type carburetor**

**Float mechanism.** The *float chamber* of the carburetor is located between the fuel supply and the fuel-metering system. The function of the float chamber is to supply nearly a constant level of fuel to the main metering nozzle. Normally, the fuel level is approximately 1/8 inch below the discharge nozzle. The fuel level is maintained slightly below the discharge nozzle to allow the proper amount of fuel flow and to prevent fuel from flowing out the nozzle when the engine is not running.

The constant level of fuel is maintained in the float chamber by the use of a *float* (Figure 1-8-12) and *needle valve* and *seat*. As the fuel level in the float chamber lowers, the float lowers. This pulls the hardened steel needle off of the bronze or brass seat, which allows more fuel to enter into the chamber. As the fuel level rises, the needle makes contact with the seat again, stopping fuel flow. During engine operation at a constant power setting, the metering jet will be drawing a constant amount of fuel, and the needle valve will be supplying a constant flow of fuel to the float chamber.

With the fuel at the correct level, the discharge rate is controlled accurately by the air velocity through the carburetor and the atmospheric pressure on top of the fuel in the float chamber.

A vent or small opening in the top of the float chamber allows air to enter or leave the chamber as the level of fuel rises or falls. This vent passage is open into the engine air intake; thus, the air pressure in the chamber is always the same as that existing in the air intake.

**Figure 1-8-12. A cutaway carburetor with exposed float mechanism**

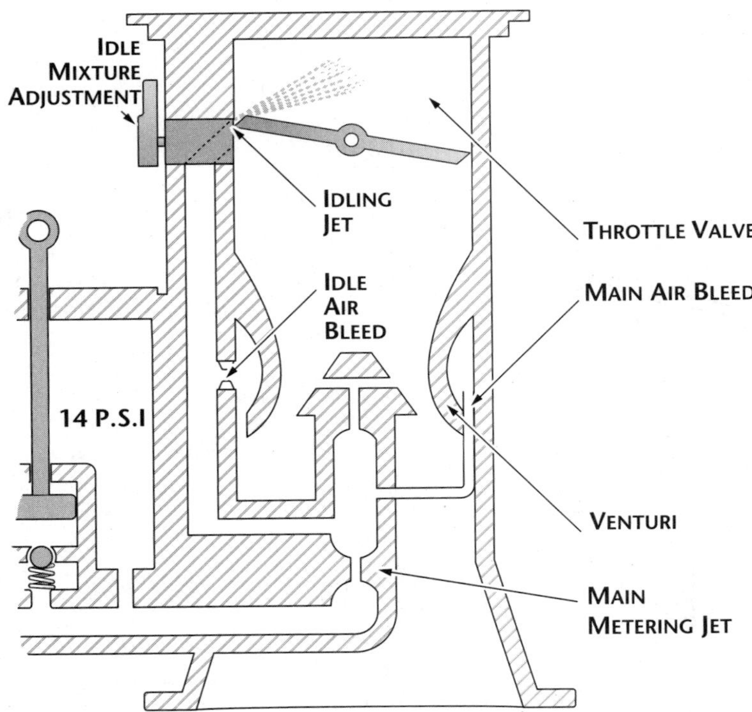


Figure 1-8-13. Main metering system

**Main Metering System.** The main metering system supplies fuel to the engine at all speeds above idling and consists of:

- Venturi
- Main metering jet
- Main discharge nozzle
- Passage leading to the idling system
- Throttle valve

Since the throttle valve controls the mass airflow through the carburetor venturi, it must be considered a major unit in the main metering system, as well as in other carburetor systems. A typical main metering system is illustrated in Figure 1-8-13.

The venturi performs three functions:

1. Proportions the fuel/air mixture
2. Decreases the pressure at the discharge nozzle
3. Limits the airflow at full throttle

The fuel-discharge nozzle is located in the carburetor barrel so that its open end is in the throat, or narrowest part, of the venturi.

A main metering orifice, or jet, is placed in the fuel passage between the float chamber and the discharge nozzle to limit the fuel flow when the throttle valve is wide open.

When the engine crankshaft is turning with the carburetor throttle open, the low pressure created in the intake manifold acts on the air passing through the carburetor barrel. Due to the difference in pressure between the atmosphere and the intake manifold, air will flow from the air intake through the carburetor barrel into the intake manifold. The volume of airflow depends upon the degree of throttle opening.

When air flows through the venturi, the velocity of the air increases. As the velocity of the air increases, its pressure decreases, creating low pressure in the throat of the venturi. The fuel discharge is located in the low-pressure area. Because the float chamber is vented to atmospheric or inlet pressure, a pressure differential is created. The pressure differential, or *metering force*, causes the fuel to flow from the float chamber to the discharge nozzle. The discharge nozzle discharges the fuel into the mass airflow as a fine spray or mist. This fuel is quickly vaporized in the air as it flows through the induction system.

The metering force in most carburetors increases as the throttle opening is increased. A pressure drop of at least 0.5 inch Hg is required to raise the fuel in the discharge nozzle to a level where it will discharge into the airstream. At low engine speeds where the metering force is considerably reduced, the fuel delivery from the discharge nozzle would decrease if an *air bleed* (air metering jet) were not incorporated in the carburetor.

The greater decrease of fuel flow in relation to the decrease in pressure differential is caused by two factors:

1. The decreased pressure differential in the carburetor means the fuel is discharged in larger droplets. These larger droplets are more resistant to flow than the atomized fuel, which hinders fuel flow.
2. The reduced pressure differential makes it harder to draw fuel out of the float chamber into the discharge nozzle.

The basic principle of the air bleed can be explained by simple diagrams, as shown in Figure 1-8-14. In each case, the same degree of suction is applied to a vertical tube placed in the container of liquid. As shown in illustration A of Figure 1-8-14, the suction applied on the upper end of the tube is sufficient to lift the liquid a distance of about 1 inch above the surface. If a small hole is made in the side of the tube above the surface of the liquid, as in illustration B of Figure 1-8-14, and suction is applied, bubbles of air will enter the tube and the liquid will be drawn up in a continuous series of small slugs or drops. Thus, air "bleeds" into the tube and partially reduces the forces tending to retard the flow of liquid through the tube. However, the large opening at the bottom of the tube effectively prevents any great amount of suction from being exerted on the air bleed hole or vent. Similarly, an air bleed hole that is too large in proportion to the size of the tube would reduce the suction available to lift the liquid. If the system is modified by placing a metering orifice in the bottom of the tube and air is taken in below the fuel level by means of an air bleed tube, a finely divided emulsion of air and liquid is formed in the tube, as shown in illustration C of Figure 1-8-14.

In a carburetor, a small air bleed is led into the fuel nozzle slightly below the fuel level. The open end of the air bleed is in the space behind the venturi wall, where the air is relatively motionless and approximately at atmospheric pressure. The low pressure at the tip of the nozzle not only draws fuel from the float chamber but also draws air from behind the venturi. Air bled into the main metering fuel system decreases the fuel density and destroys surface tension. This results in better vaporization and control of fuel discharge, especially at lower engine speeds.

The *throttle valve* is located on the engine side of the venturi. The throttle valve, or *butterfly valve*, controls the engine speed by regulating the amount of airflow to the engine. The valve is a round plate that rotates on a shaft and is turned to open or close the carburetor passage. In some applications, more than one throttle valve is used, and they may be attached to the same shaft or to separate shafts. If separate shafts are used, they must be connected with mechanical linkage. This linkage must be adjusted so that the valves operate in a synchronized manner.

**Idling system.** With the throttle valve closed at idling speeds, air velocity through the venturi is so low that it cannot draw enough fuel from the main discharge nozzle. In fact, the spray of fuel may stop altogether. However, low pressure (piston suction) exists on the engine side of the throttle valve. In order to allow the engine to idle, a fuel passageway is incorporated to discharge fuel from an opening in the low-pressure area near the edge of the throttle valve. This opening is called the *idling jet*. With the throttle open enough so that the main discharge nozzle is operating, fuel does not flow out of the idling jet. As soon as the throttle is closed far enough to stop the spray from the main discharge nozzle, fuel flows out the idling jet. A separate air bleed, known as the *idle-air bleed*, is included as part of the idling system. It functions the same as the main air bleed. An idle-mixture adjusting device is also incorporated. A typical idling system is illustrated in Figure 1-8-15.

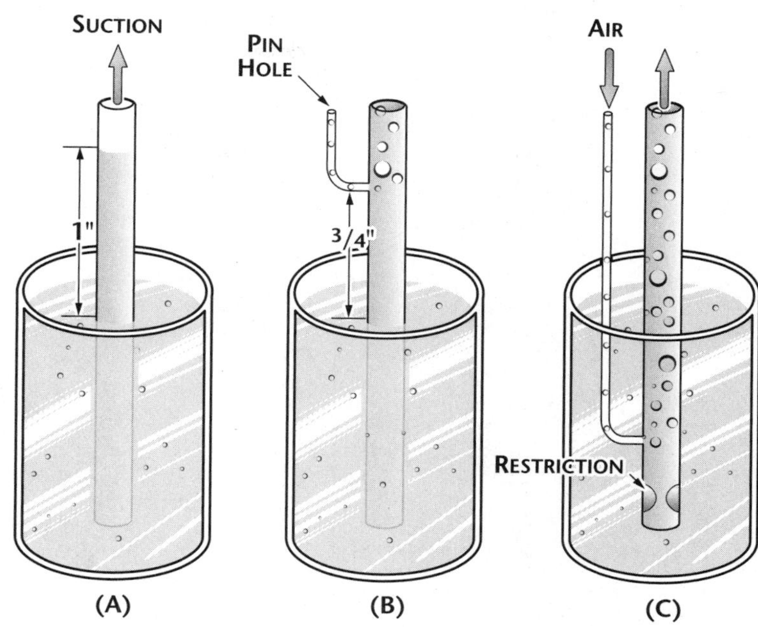


Figure 1-8-14. Air-bleed principle

**Mixture control system.** As altitude increases, the air becomes less dense. At an altitude of 18,000 ft., the air is only half as dense as it is at sea level, meaning that a cubic foot of space contains only half as much air at 18,000 ft. as at sea level. An engine cylinder full of air at 18,000 ft. contains only half as much oxygen as a cylinder full of air at sea level.

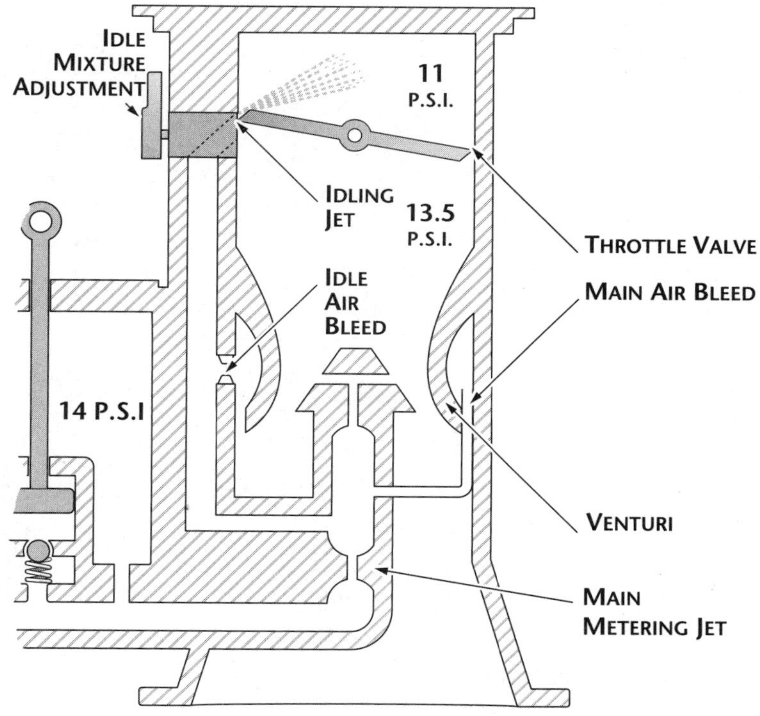


Figure 1-8-15. Idling system

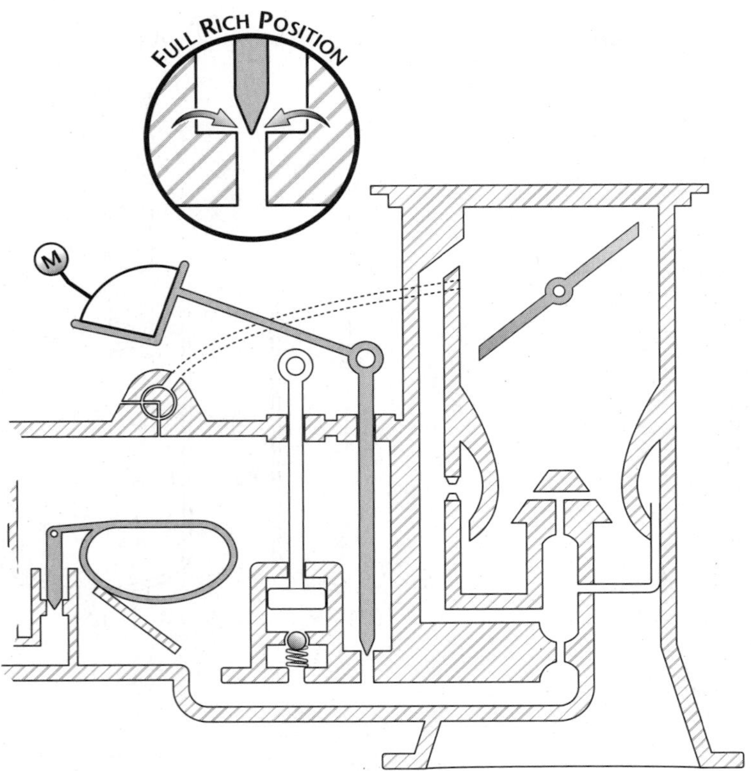


Figure 1-8-16. Needle-type mixture-control system

The low pressure in the venturi is affected only by the air velocity and not the air density. Therefore, the discharge nozzle discharges the same amount of fuel at high altitude as it does on the ground. Because of this, the fuel mixture increases as altitude increases. This must be compensated for with a mixture control.

On float-type carburetors, two types of purely manual or cockpit-controllable devices are in general use for controlling fuel/air mixtures: the needle type and the back-suction type.

With the *needle-type system,* manual control is provided by a needle valve in the base of the float chamber (Figure 1-8-16). This can be raised or lowered by adjusting a control in the cockpit. Moving the control to RICH opens the needle-valve wide, which permits the fuel to flow unrestricted to the nozzle. Moving the control to LEAN closes the valve partway and restricts the flow of fuel to the nozzle.

The *back-suction-type mixture-control system* may also be found on certain makes of carburetors. In this system (Figure 1-8-17), a certain amount of venturi low pressure acts upon the fuel in the float chamber so that it opposes the low pressure existing at the main discharge nozzle. An *atmospheric line,* incorporating an adjustable valve, opens into the float chamber. When the valve is completely closed, pressures on the fuel in the float chamber and at the discharge nozzle are almost equal, and fuel flow is reduced to its leanest. With the valve wide open, pressure on the fuel in the float chamber is greatest, and fuel mixture is richest. Adjusting the valve to positions between these two extremes controls the mixture.

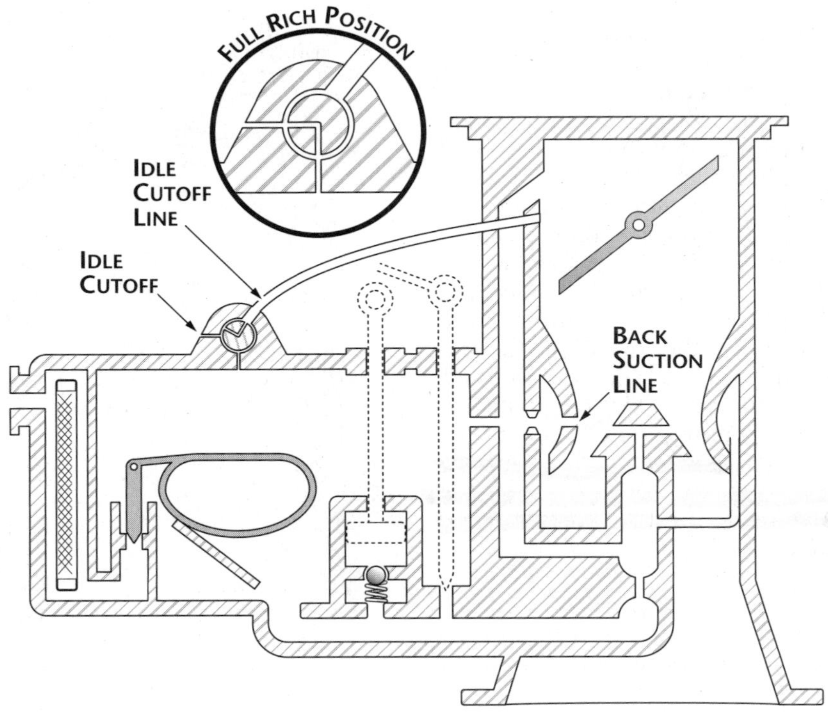


Figure 1-8-17. Back-suction-type mixture-control system

The cockpit quadrant is usually marked LEAN near the back and RICH at the forward end. The extreme back position is marked IDLE-CUT-OFF and is used when stopping the engine.

On float carburetors equipped with needle-type mixture control, placing the mixture control in idle-cutoff seats the needle valve, thus shutting off fuel flow completely. On carburetors equipped with back-suction mixture controls, a separate idle-cutoff line, leading to the extreme low pressure on the engine side of the throttle valve, is incorporated (see the dotted line in Figure 1-8-17). The mixture control is so linked that, when it is placed in the idle-cutoff position, it opens another passage that leads to piston suction. When placed in other positions, the valve opens a passage leading to the atmosphere. To stop the engine with such a system, close the throttle and place the mixture in the idle-cutoff position. Leave the throttle closed until the engine has stopped turning over, then open the throttle completely.

**Accelerating system.** When the throttle valve is opened quickly, a large volume of air rushes through the air passage of the carburetor, though the amount of fuel that is mixed with the air is less than normal. This is because of the slow response rate of the main metering system. As a result, after a quick opening of the throttle, the fuel/air mixture leans out momentarily.

To overcome this tendency, the carburetor is equipped with a small fuel pump called an *accelerating pump.* A common type of accelerating system used in float carburetors is illustrated in Figure 1-8-18. It consists of a simple piston pump operated through linkage by the throttle control, and a line opening into the main metering system or the carburetor barrel near the venturi. When the throttle is closed, the piston moves back, and fuel fills the cylinder. If the piston is pushed forward slowly, the fuel seeps past it back into the float chamber. If pushed rapidly, it will emit a charge of fuel and enrich the mixture in the venturi.

**Economizer System.** For an engine to develop maximum power at full throttle, the fuel mixture must be richer than for cruise. The additional fuel is used for cooling the engine,to prevent detonation. An economizer is essentially a valve which is closed at throttle settings below 60-70 percent of rated power. This system, like the accelerating system, is operated by the throttle control.

A typical *economizer system*, as shown in Figure 1-8-19, consists of a needle valve that begins to open when the throttle valve reaches a predetermined point near the wide-open position. As the throttle continues to open, the needle valve is opened further, and additional fuel flows through it. This additional fuel supplements the flow from the main metering jet directly to the main discharge nozzle.

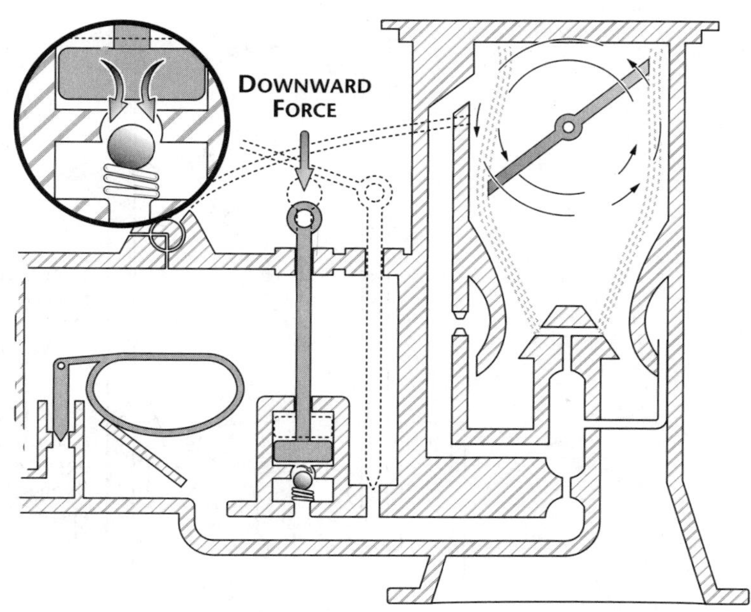


Figure 1-8-18. Accelerating system

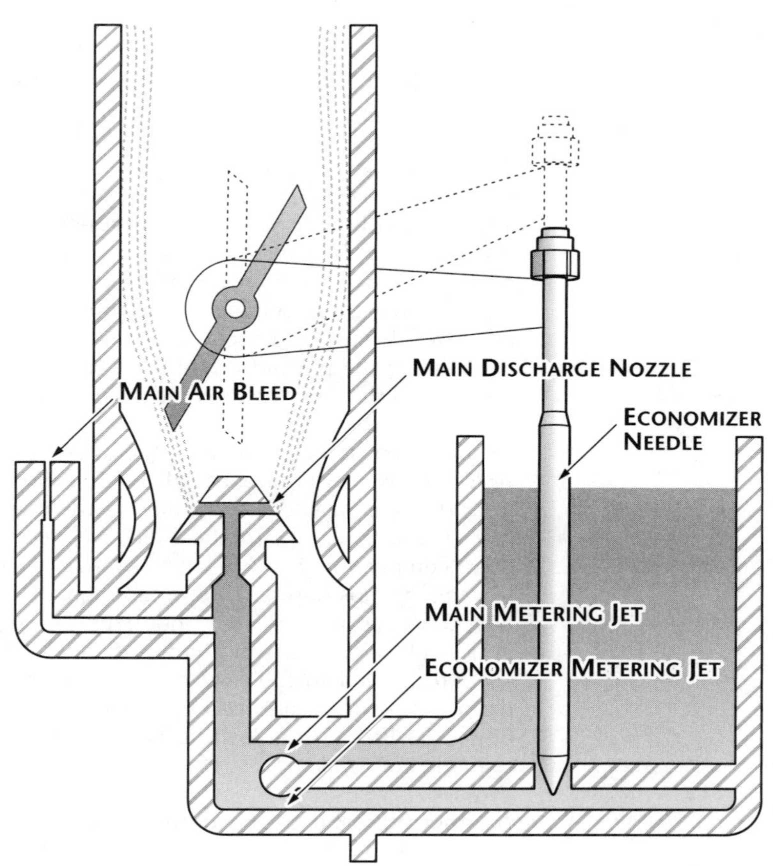


Figure 1-8-19. A needle-type economizer system

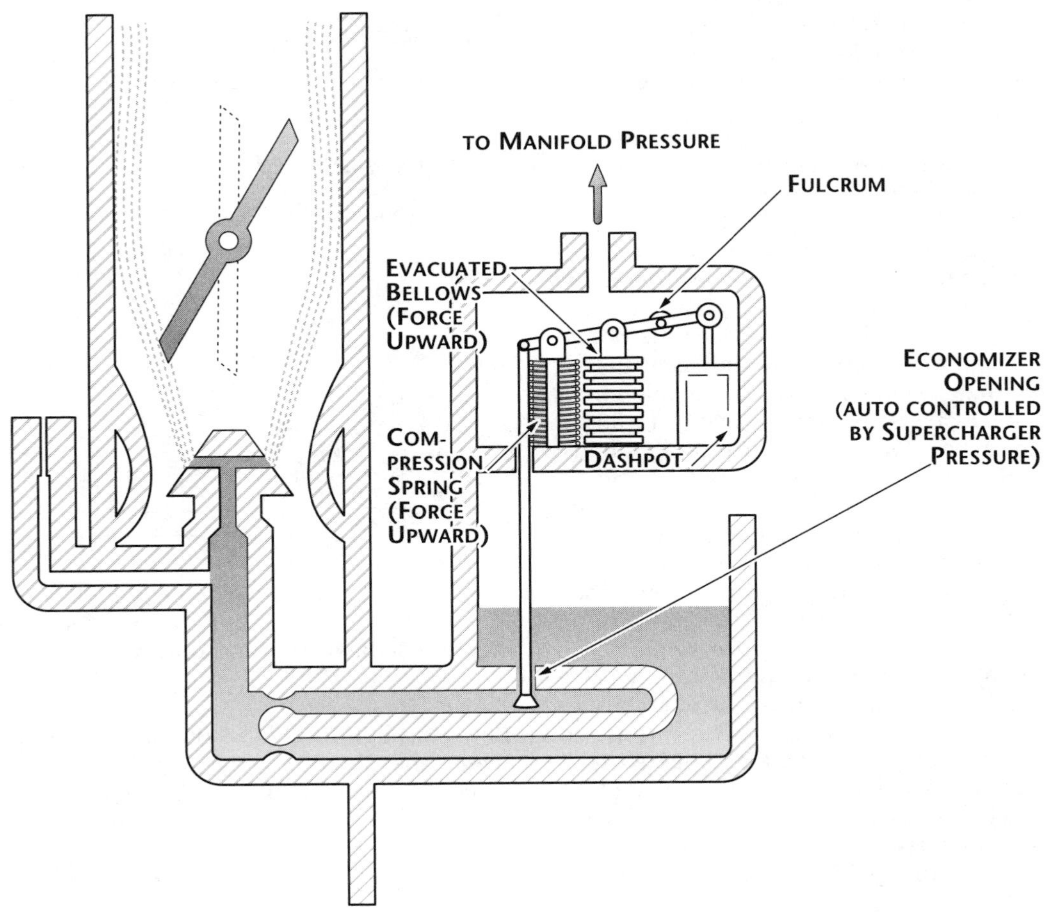


**Figure 1-8-20. A pressure-operated economizer system**

A *pressure-operated economizer system* is shown in Figure 1-8-20. This type has a sealed bellows located in an enclosed compartment. The compartment is vented to engine manifold pressure. When the manifold pressure reaches a certain level, the bellows is compressed and opens a valve in a carburetor fuel passage, supplementing the normal quantity of fuel being discharged through the main nozzle.

Another type of economizer is the *back-suction system* shown in Figure 1-8-21. Fuel economy in cruising is provided by reducing the effective pressure acting on the fuel level in the float compartment. With the throttle valve in cruising position, suction is applied to the float chamber through an economizer hole and back-suction economizer channel and jet. The suction applied to the float chamber opposes the nozzle suction applied by the venturi. Fuel flow is reduced, thus leaning the mixture for cruising economy.

Another type of mixture control system uses a metering valve that is free to rotate in a stationary metering sleeve. Fuel enters the main and idling systems through a slot cut in the mixture sleeve. Fuel metering is accomplished by the relative position between one edge of the slot in the hollow metering valve and one edge of the slot in the metering sleeve. Moving the mixture control to reduce the size of the slot provides a leaner mixture for altitude compensation.

## Pressure-injection Carburetors

*Pressure-injection carburetors* are distinctly different from float-type carburetors, as they do not incorporate a vented float chamber or suction pickup from a discharge nozzle located in the venturi tube. Instead, they provide a pressurized fuel system that is closed from the engine fuel pump to the discharge nozzle. The venturi serves only to create pressure differentials for controlling the quantity of fuel to the metering jet in proportion to airflow to the engine.

While not in common use today, the pressure carburetor represents an important step in the development of today's RSA-type fuel-injection system. Understanding the operating principles of the pressure carburetor will help you to understand the RSA fuel-injection system.

**PS-series Carburetor.** The *PS-series carburetor* is a low-pressure, single-barrel, injection-type carburetor. The carburetor consists essentially of the air section, the fuel section and the discharge nozzle, all mounted together to form a complete fuel-metering system. The carburetor shown in Figure 1-8-22 is a pressure-injection carburetor.

In this type of carburetor (Figure 1-8-23, next page), metering is accomplished on a mass airflow basis. Air flowing through the main venturi creates suction at the throat of the venturi, which is transmitted to the B chamber in the main regulating part of the carburetor and to the vent side of the fuel-discharge nozzle diaphragm. The incoming air pressure is transmitted to a chamber of the regulating part of the carburetor and to the main discharge bleed in the main fuel-discharge jet. The discharge nozzle consists of a spring-loaded diaphragm connected to the discharge-nozzle valve, which controls the flow of fuel injected into the main discharge jet. Here, it is mixed with air to accomplish distribution and atomization into the airstream entering the engine.

In the PS-series carburetor, as in the pressure-injection carburetor, the *regulator spring* has a fixed tension, which tends to hold the poppet valve open during idling speeds, or until the D chamber pressure equals approximately 4 p.s.i. The *discharge-nozzle spring*, on the other hand, has a variable adjustment which, when tailored to maintain 4 p.s.i., will result in a balanced pressure condition of 4 p.s.i. in chamber C of the discharge-nozzle assembly, and 4 p.s.i. in chamber D. This produces a zero drop across the main jets at zero fuel flow.

At a given airflow, if the suction created by the venturi is equivalent to 1/4-lb., this pressure decrease is transmitted to chamber B and to the vent side of the discharge nozzle. Since the area of the air diaphragm between chambers A and B is twice as great as that between chambers B and D, the 1/4-lb. decrease in pressure in chamber B will move the diaphragm assembly to the right to open the poppet valve. Meanwhile, the decreased pressure on the vent side of the discharge-nozzle assembly will cause a lowering of the total pressure from 4 p.s.i. to 3 3/4 p.s.i. The greater pressure of the metered fuel (4 1/4 -p.s.i.) results in a differential across the metering head of 1/4 lb. (for the 1/4-lb. pressure differential created by the venturi).

The same ratio of pressure drop across the jet-to-venturi suction will apply throughout the range. Any increase or decrease in fuel inlet pressure will tend to upset the balance in the various chambers in the manner already

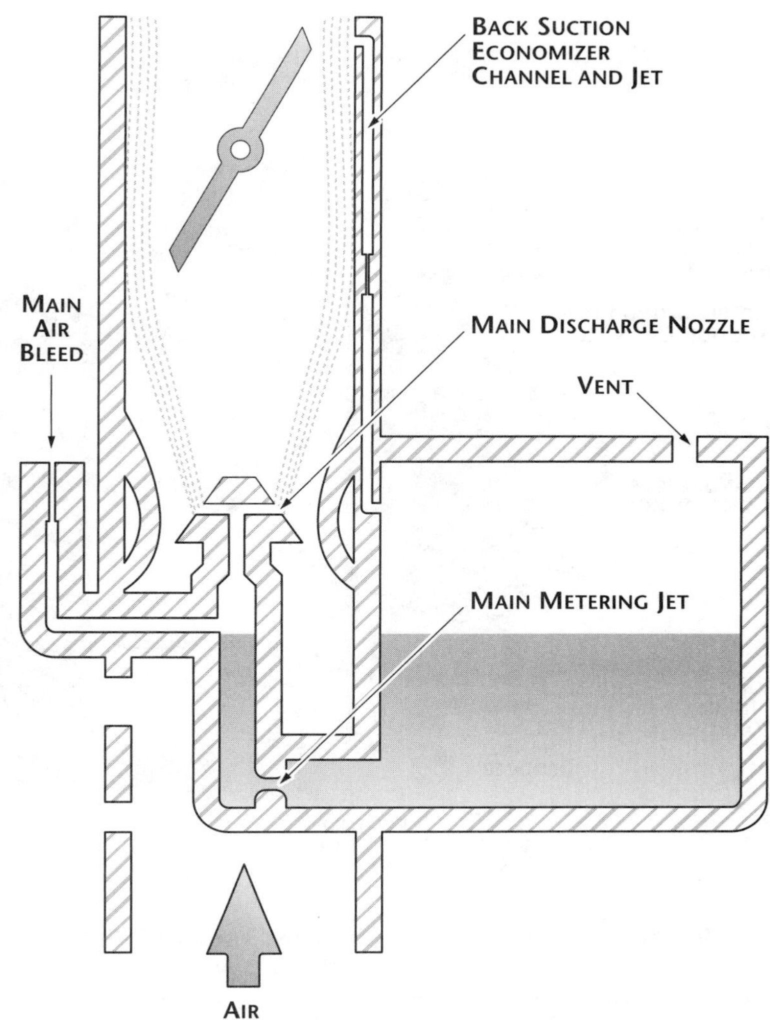


Figure 1-8-21. Back-suction economizer system

Figure 1-8-22. A PS-5 pressure injection carburetor

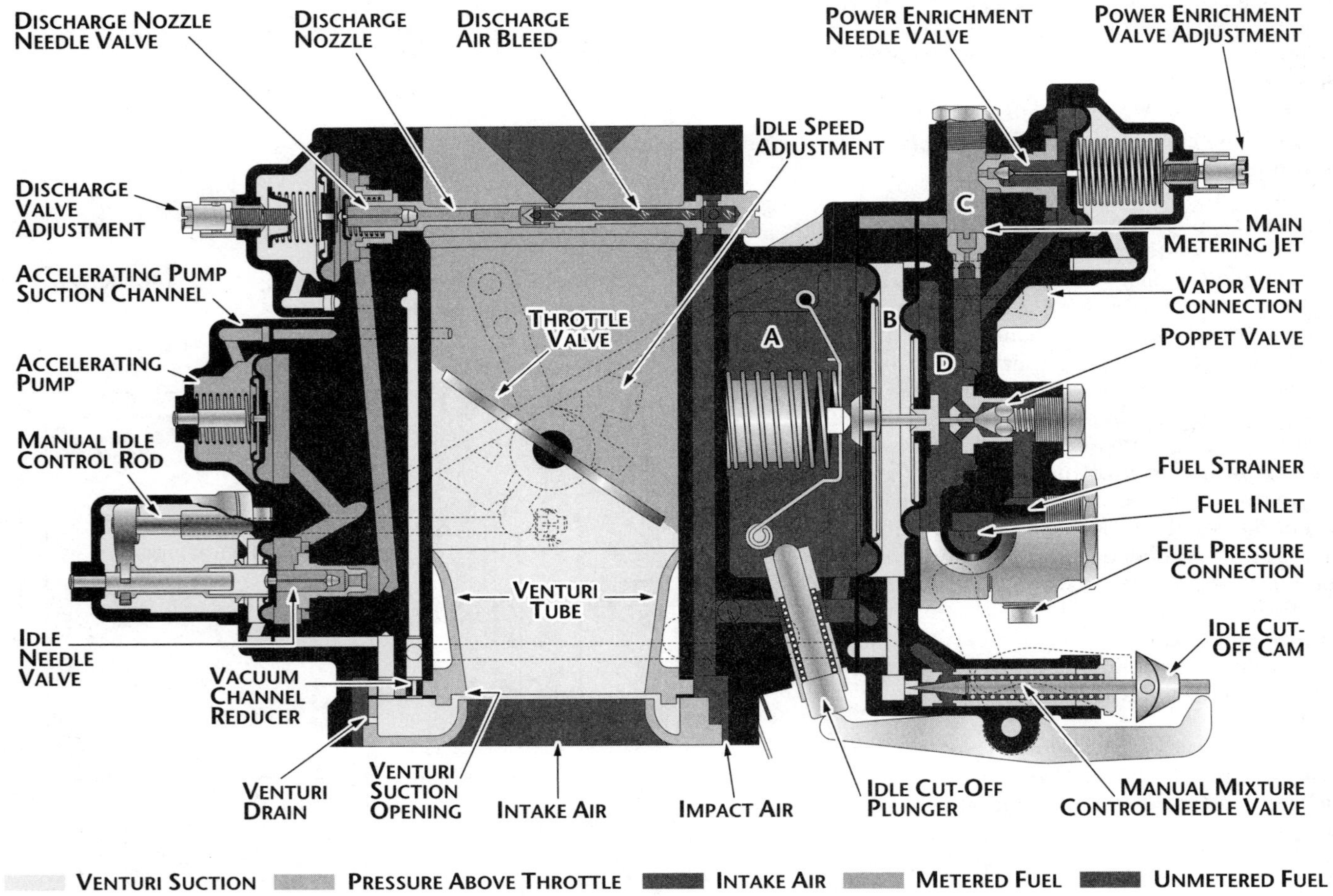

Figure 1-8-23. Schematic diagram of the PS-series carburetor

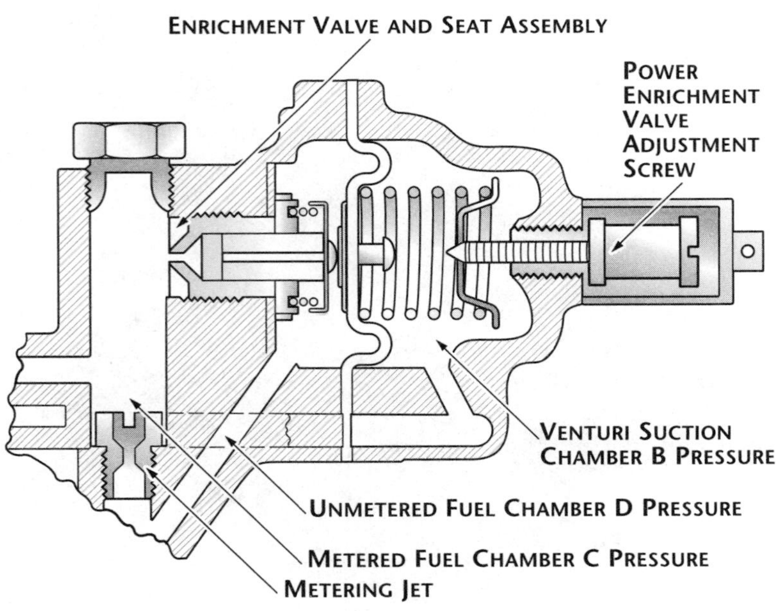


Figure 1-8-24. Airflow power enrichment valve

described. When this occurs, the main fuel regulator diaphragm assembly repositions to restore the balance.

The mixture control, whether operated manually or automatically, compensates for enrichment at altitude by bleeding impact-air pressure into B chamber, thereby increasing the pressure (decreasing the suction) in B chamber. Increasing the pressure in chamber B tends to move the diaphragm and poppet valve more toward the closed position, thus restricting fuel flow to correspond proportionately to the decrease in air density at altitude.

The idle valve and economizer jet can be combined in one assembly, controlled manually by the movement of the valve assembly. At low airflow positions, the tapered section of the valve becomes the predominant jet in the system, controlling the fuel flow for the idle range. As the valve moves to the cruise position, a straight section of the valve establishes a fixed-orifice effect that controls the cruise mixture. When the valve is pulled full-open by the throttle valve, the jet is pulled completely out of the seat, and the seat size becomes the controlling jet. This jet is calibrated for takeoff power mixtures.

An airflow-controlled power-enrichment valve can also be used with this carburetor. It consists of a spring-loaded, diaphragm-operated metering valve. Refer to Figure 1-8-24 for a schematic view of an airflow power-enrichment valve. One side of the diaphragm is exposed to un-metered fuel pressure, and the other side to venturi suction plus spring tension. When the pressure differential across the diaphragm establishes a force strong enough to compress the spring, the valve will open and supply an added amount of fuel to the metered fuel circuit, in addition to the fuel supplied by the main metering jet.

**Accelerating pump.** The *accelerating pump* is a spring-loaded diaphragm assembly located in the metered-fuel channel with the opposite side of the diaphragm vented to the engine side of the throttle valve. With this arrangement, opening the throttle results in a rapid decrease in suction. This decrease in suction permits the spring to extend and move the accelerating pump diaphragm. The diaphragm and spring action displace the fuel in the accelerating pump and force it out the discharge nozzle.

Vapor is eliminated from the top of main fuel chamber D through a bleed hole, then through a vent line back to the main fuel tank in the aircraft.

**Manual mixture control.** Manual mixture control provides a means of correcting for enrichment at altitude. It consists of a needle valve and seat that form an adjustable bleed between chamber A and chamber B. The valve can be adjusted to bleed off the venturi suction to maintain the correct fuel/air ratio as the aircraft gains altitude.

When the mixture control lever is moved to the IDLE-CUTOFF position, a cam on the linkage actuates a rocker arm, which moves the idle-cutoff plunger inward against the release lever in chamber A. The lever compresses the regulator diaphragm spring to relieve all tension on the diaphragm between A and B chambers. This permits fuel pressure — plus poppet valve spring force — to close the poppet valve, stopping the fuel flow. Placing the mixture-control lever in IDLE-CUTOFF also positions the mixture-control needle valve off its seat and allows metering suction within the carburetor to bleed off.

## Direct Fuel-injection Systems

The *direct fuel-injection system* has many advantages over a conventional carburetor system. There is less danger of induction system icing, since the drop in temperature due to fuel vaporization takes place in or near the cylinder. Acceleration is also improved because of the positive action of the injection system. In addition, direct fuel injection improves fuel distribution. This reduces the overheating of individual cylinders often caused by variation in mixture due to uneven distribution. The fuel-injection system also gives better fuel economy than a system in which the mixture to most cylinders must be richer than necessary to ensure that the cylinder with the leanest mixture will operate properly.

Fuel-injection systems vary in their details of construction, arrangement and operation. The Precision Airmotive RS/RSA fuel systems (formerly manufactured by Bendix) and Continental fuel-injection systems are the systems in most common use. They are described to provide an understanding of the operating principles involved. For the specific details of any one system, consult the manufacturer's instructions for the equipment involved.

### *RS Type Fuel-injection System*

The RSA fuel system (Figure 1-8-25) consists of an injector, flow divider and fuel-discharge nozzle. It is a continuous-flow system that measures engine-air consumption and uses airflow forces to control fuel flow to the engine.

Figure 1-8-25. An RSA fuel injector

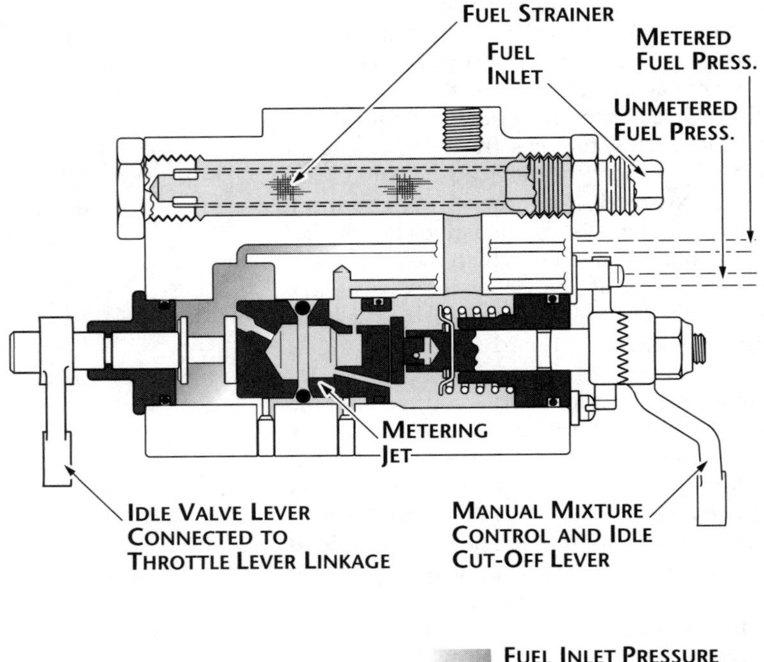


Figure 1-8-26A. Fuel injector, including airflow section

**Fuel injector.** Some fuel injectors are equipped with an automatic mixture-control unit. The fuel injector assembly consists of:

- Airflow section. The airflow consumption of the engine is measured by sensing impact pressure and venturi throat pressure in the throttle body. These pressures are vented to the two sides of an air diaphragm. Movement of the throttle valve causes a change in engine-air consumption. This results in a change in the air velocity in the venturi. When airflow through the engine increases, the pressure on the left of the diaphragm is lowered (Figure 1-8-26A) due to the drop in pressure at the venturi throat. As a result, the diaphragm moves to the left, opening the ball valve. This pressure differential is referred to as the air-metering force.

- Regulator section. The regulator section consists of a fuel diaphragm that opposes the air metering force. Fuel inlet pressure is applied to one side of the fuel diaphragm and metered fuel pressure is applied to the other side. The differential

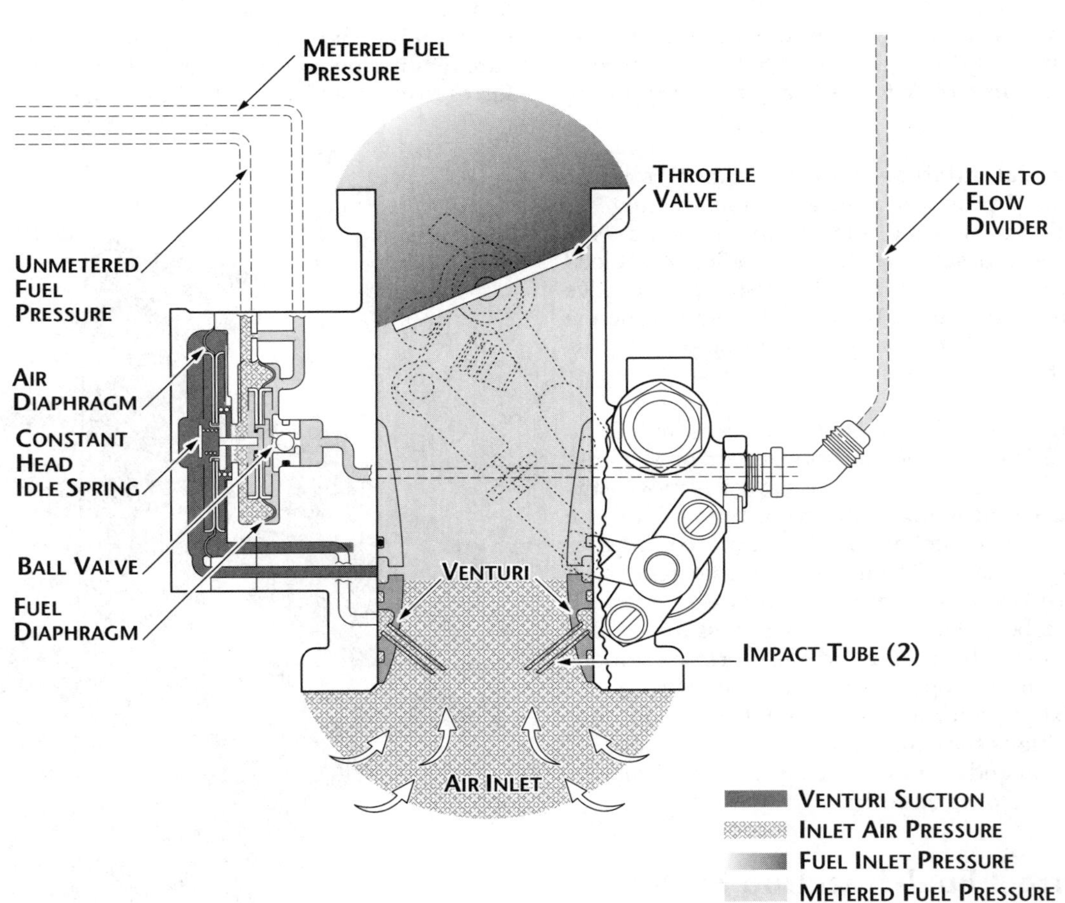


Figure 1-8-26B. Fuel metering section

pressure across the fuel diaphragm is called the fuel-metering force.

The distance the ball valve opens is determined by the difference between the pressures acting on the diaphragms. This difference in pressure is proportional to the airflow through the injector. Thus, the volume of airflow determines the rate of fuel flow.

Under low power settings, the difference in pressure created by the venturi is insufficient to accomplish consistent regulation of the fuel. A constant-head idle spring is incorporated to provide a constant fuel-differential pressure. This allows an adequate final flow in the idle range.

- Fuel metering section. The fuel-metering section, shown in Figure 1-8-26B, is attached to the air-metering section and contains an inlet fuel strainer, a manual mixture-control valve, an idle valve and the main metering jet. In some models of injectors, a power-enrichment jet is also located in this section. The purpose of the fuel-metering section is to meter and control the fuel flow to the flow divider.

**Flow divider.** The metered fuel is delivered from the fuel-control unit to a pressurized flow divider. This unit keeps metered fuel under pressure, divides fuel to the various cylinders at all engine speeds, and shuts off the individual nozzle lines when the control is placed in idle-cutoff.

Referring to the schematic diagram in Figure 1-8-26C, metered fuel pressure enters the flow divider through a channel that permits fuel to pass through the inside diameter of the flow-divider needle. At idle speed, the fuel pressure from the regulator must build up to overcome the spring force applied to the diaphragm and valve assembly. This moves the valve upward until fuel can pass out through the annulus of the valve to the fuel nozzle. Since the regulator meters and delivers a fixed amount of fuel to the flow divider, the valve will open only as far as necessary to pass this amount to the nozzles. At idle, the opening required is very small; thus, the fuel for the individual cylinders is divided at idle by the flow divider.

As fuel flow through the regulator is increased above idle requirements, fuel pressure builds up in the nozzle lines. This pressure fully opens the flow-divider valve, and fuel distribution to the engine becomes a function of the discharge nozzles.

A fuel-pressure gauge, calibrated in pounds-per-hour fuel flow, can be used as a fuel-flow meter with the RSA injection system. This

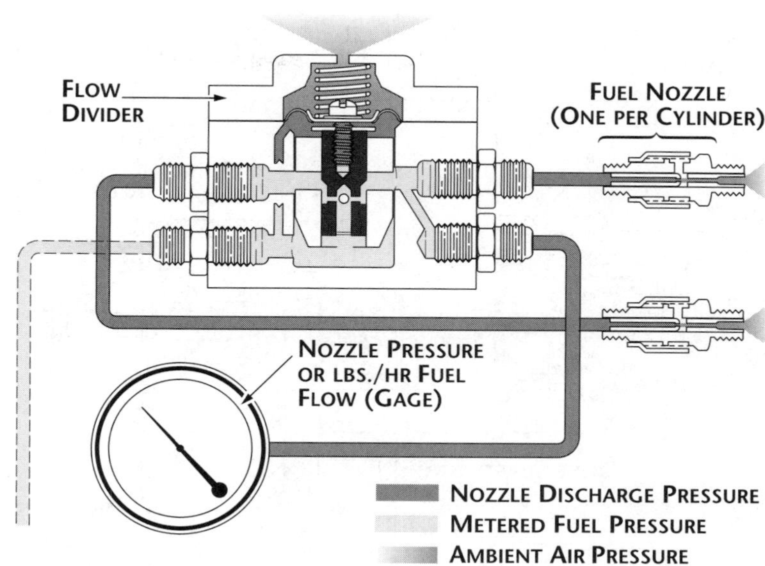


Figure 1-8-26C. Flow divider

gauge is connected to the flow divider and senses the pressure being applied to the discharge nozzle. This pressure is in direct proportion to fuel flow and indicates the engine power output and fuel consumption.

**Fuel-discharge nozzles.** The fuel-discharge nozzles (Figure 1-8-26C) are of the air-bleed configuration. There is one nozzle for each cylinder located in the cylinder head. The nozzle outlet is directed into the intake port. Each nozzle incorporates a calibrated jet. The jet size is determined by the available fuel-inlet pressure and the maximum fuel flow required by the engine. The fuel is discharged through this jet into an ambient air-pressure chamber within the nozzle assembly. Before entering the individual intake valve chambers, the fuel is mixed with air to aid in atomizing the fuel. A higher-than-normal indicated fuel flow is usually a plugged fuel nozzle.

## *TCM Fuel-injection System*

The Teledyne Continental fuel-injection system injects fuel into the intake valve port in each cylinder head. The system consists of a fuel injector pump, a control unit, a fuel manifold and a fuel-discharge nozzle. It is a continuous-flow type that controls fuel flow to match engine airflow. The continuous-flow system permits the use of a rotary-vane pump that does not require timing to the engine.

**Fuel-injection pump.** The fuel pump is a positive-displacement, rotary-vane type, with a splined shaft for connection to the accessory drive system of the engine. A spring-loaded, diaphragm-type relief valve is provided. The relief-valve diaphragm chamber is vented to

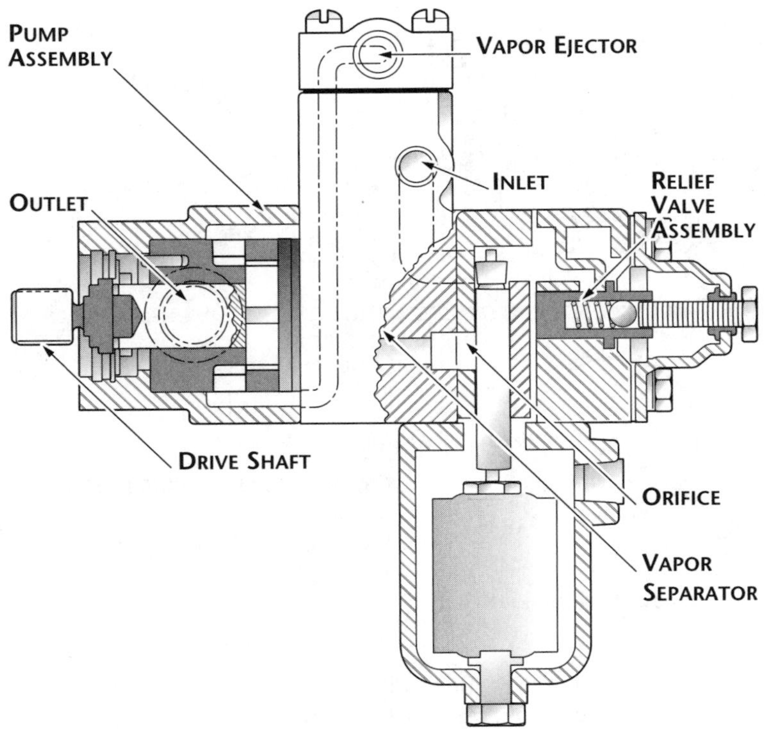

Figure 1-8-27. Fuel-injection pumps

atmospheric pressure. A cross-sectional view of a fuel-injection pump is shown in Figure 1-8-27 .

Fuel enters at the *swirl well* of the vapor separator. Here, vapor is separated by a swirling motion so that only liquid fuel is delivered to the pump. The vapor is drawn from the top center of the swirl well by a small pressure-jet of fuel and is directed into the *vapor return line.* This line carries the vapor back to the fuel tank.

Figure 1-8-28. Fuel/air control unit

Ignoring the effect of altitude or ambient air conditions, the use of a positive-displacement, engine-driven pump means that changes in engine speed affect total pump flow proportionally. Since the pump provides greater capacity than is required by the engine, a recirculation path is required. By arranging a calibrated orifice and relief valve in this path, the pump delivery pressure is also maintained in proportion to engine speed. These provisions assure proper pump pressure and fuel delivery for all engine operating speeds.

A check valve is provided so that boost-pump pressure to the system can bypass the engine-driven pump for starting. This feature also suppresses vapor formation under high ambient temperatures of the fuel. Furthermore, this permits use of the auxiliary pump as a source of fuel pressure in the event of engine-driven pump failure.

**Fuel/air-control unit.** The purpose of the fuel/air-control assembly is to control the engine-air intake and set the metered fuel pressure to the proper fuel/air ratio. The air throttle is located at the manifold inlet and is controlled by the throttle lever in the cockpit. The air-throttle valve controls the flow of air to the engine (Figure 1-8-28 and 1-8-29).

The air throttle assembly is an aluminum casting which contains the shaft and butterfly valve assembly. The casting bore size is tailored to the engine size, and no venturi or other restriction is used.

**Fuel control assembly.** The fuel control body is made of bronze for best bearing action with the stainless steel valves. Its central bore

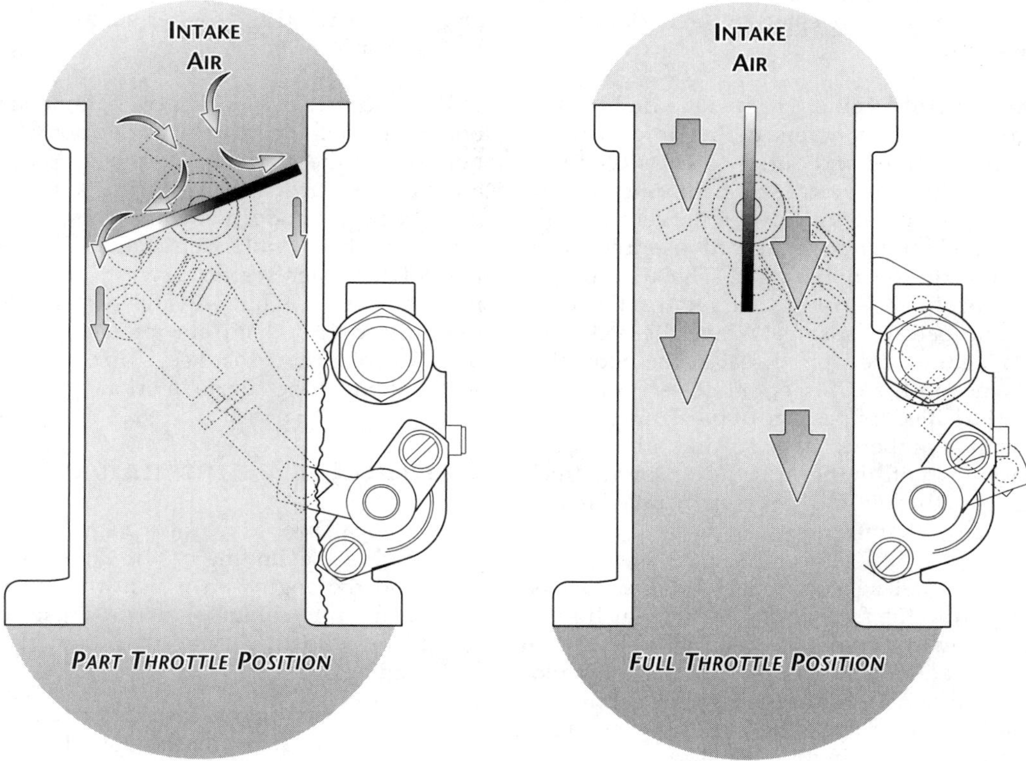


**Figure 1-8-29. Fuel/air-control unit**

contains a metering valve at one end and a mixture control valve at the other end. Each stainless steel rotary valve includes a groove that forms a fuel chamber.

Fuel enters the control unit through a strainer and passes to the metering valve (Figure 1-8-30). This rotary valve has a cam-shaped edge on the outer part of the end face. The position of the cam at the fuel delivery port controls the fuel passed to the manifold valve and the nozzles. The fuel return port connects to the return passage of the center metering plug. The alignment of the mixture control valve with this passage determines the amount of fuel returned to the fuel pump.

By connecting the metering valve to the air throttle, the fuel flow is properly proportioned to airflow for the correct fuel/air ratio. A con-

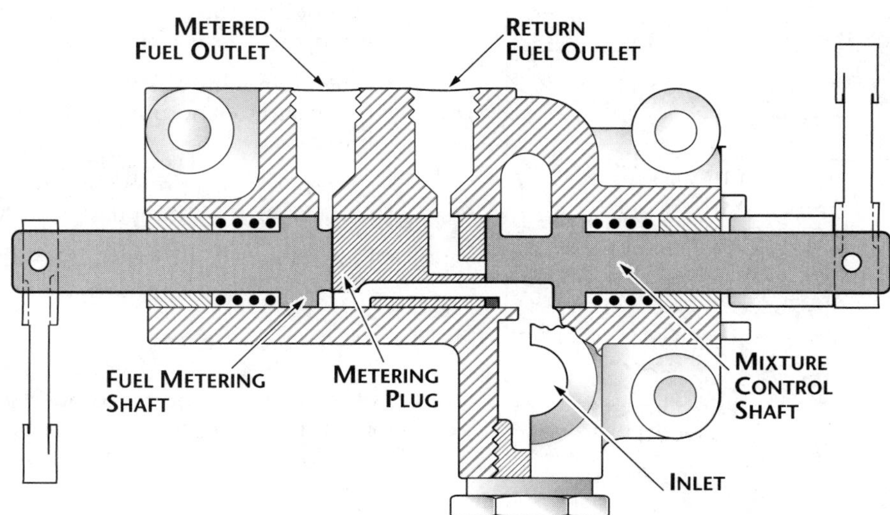


**Figure 1-8-30. Fuel-control assembly**

trol level is mounted on the mixture control valve shaft and connected to the cockpit mixture control.

**Fuel manifold valve.** The fuel manifold valve (Figure 1-8-31) contains a fuel inlet, a diaphragm chamber and outlet ports for the lines to the individual nozzles. The spring-loaded diaphragm operates a valve in the central bore of the body. Fuel pressure provides the force for moving the diaphragm. The diaphragm is enclosed by a cover that retains the diaphragm-loading spring. When the valve is down against the lapped seat in the body, the fuel lines to the cylinders are closed off. The valve is drilled for passage of fuel from the diaphragm chamber to its base, and a ball valve is installed within the valve. All incoming fuel must pass through a fine screen installed in the diaphragm chamber.

From the fuel-injection control valve, fuel is delivered to the *fuel-manifold valve*, which provides a central point for dividing fuel flow to the individual cylinders. In the fuel-manifold valve, a diaphragm raises or lowers a plunger valve to open or close the individual cylinder fuel supply ports simultaneously.

**Fuel-discharge nozzle.** The fuel-discharge nozzle is located in the cylinder head with its outlet directed into the intake port. The nozzle body, illustrated in Figure 1-8-32, contains a drilled central passage with a counterbore at each end. The lower end is used as a chamber for fuel/air mixing before the spray leaves the nozzle. The upper bore contains a removable orifice for calibrating the nozzles. Nozzles are calibrated in several ranges, and all nozzles furnished for one engine are of the same range and are identified by a letter stamped on the hex of the nozzle body.

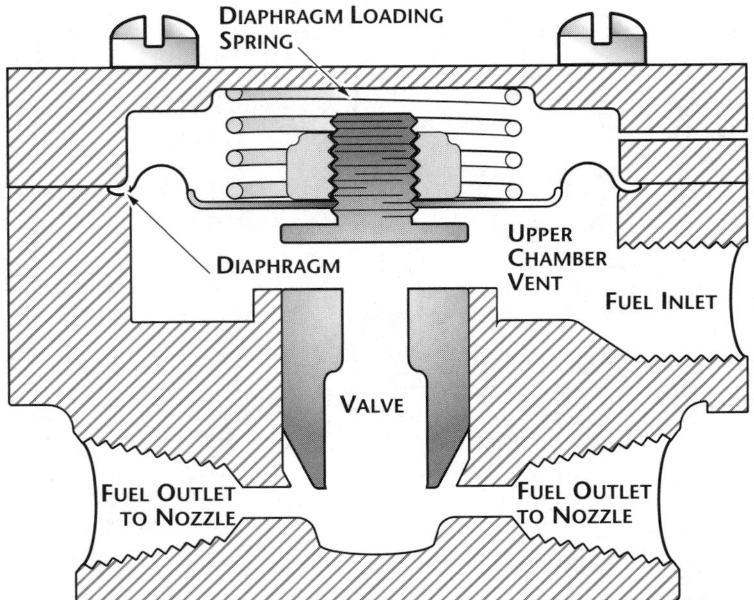


Figure 1-8-31. Fuel manifold valve assembly

Drilled radial holes connect the upper counterbore with the outside of the nozzle body. These holes enter the counterbore above the orifice and draw air through a cylindrical screen fitted over the nozzle body. A shield is press-fitted on the nozzle body and extends over the greater part of the filter screen, leaving an opening near the bottom. This provides both mechanical protection and an abrupt change in the direction of airflow that keeps dirt and foreign material out of the nozzle interior.

## Carburetor Maintenance

The removal procedure of a carburetor will vary slightly, depending on the type of carburetor and the engine on which it is installed. Although the manufacturer's instruction should be followed, some general guidelines are listed below.

- Before removing the carburetor, make certain that the fuel valve is shut off.
- Disconnect all control linkage.
- Lockwire the throttle valve in the closed position.
- Disconnect the fuel inlet line and the vapor return line.
- Disconnect the pressure-gauge line, the primer lines and the temperature-indicating lines.
- Remove the carburetor mounting hardware.
- Keep the carburetor upright after removing, because the float chamber will still contain fuel.
- Do not alter the rigging of the throttle or mixture-control cable.
- When removing a down-draft carburetor, be certain that nothing is allowed to drop into the engine.
- Always install a protective cover over the carburetor-mounting flange of the engine.
- Plug all lines to prevent contamination.

### *Installation of a Carburetor*

Check the carburetor for proper lockwiring before installation on an engine. Be sure that all shipping plugs have been removed from the carburetor openings.

Remove the protective cover from the carburetor-mounting flange on the engine. Place the carburetor-mounting flange gasket in position. On some engines, bleed passages are incorporated in the mounting pad. The gasket must be installed so that the bleed hole in the gasket is aligned with the passage in the mounting flange.

Inspect the induction passages for the presence of any foreign material before installing the carburetor.

As soon as the carburetor is placed in position on the engine, close and lockwire the throttle valves in the closed position until the remainder of the installation is completed.

Where it is feasible, place the carburetor deck screen in position to further eliminate the possibility of foreign objects entering the induction system.

When installing a carburetor that uses diaphragms for controlling fuel flow, connect the fuel lines and fill the carburetor with fuel. To do this, turn on the fuel booster pump and move the mixture-control setting from the idle-cutoff position. Continue the flow until oil-free fuel flows from the drain. This indicates that the preservative oil has been flushed from the carburetor.

Turn off the fuel flow, plug the fuel inlet and vapor vent outlet, and then allow the carburetor, filled with fuel, to stand for a minimum of 8 hours. This is necessary in order to soak the diaphragms and render them pliable to the degree they were when the unit was originally calibrated.

Tighten the carburetor-mounting bolts to the value specified in the table of torque limits in the applicable maintenance manual. Tighten and safety any other nuts and bolts incidental to the installation of the carburetor before connecting the throttle and mixture-control levers. After the carburetor has been bolted to the engine, check the throttle and mixture-control lever on the unit for freedom of movement before connecting the control cables or linkage. Check the vapor vent lines from the carburetor to the aircraft fuel tank for restriction by blowing through the line.

> **NOTE:** *When installing fuel fittings that have pipe threads, it is common practice to use thread sealer. It must, however, be used sparingly. Good practice is to insert the fitting one thread into the carburetor, then apply a drop or two of the sealant to the remaining threads. Then, when the fitting is tightened, there is no chance of the sealer getting inside the carburetor. It would be easy for excess sealer to plug up an orifice. It would then require that the carburetor be removed, disassembled, cleaned, reassembled and reinstalled.*

Teflon® plumber's tape is an example of a commercial product with wide usage that must not be used in aircraft maintenance. The chance of a stray particle of the tape entering a system is too great. To the best of the author's knowledge, no manufacturer's instructions in all of aviation maintenance recommend the use of commercial thread sealing tape.

## *Rigging Carburetor Controls*

Connect and adjust carburetor- or fuel-metering equipment throttle controls so that full movement of the throttle is obtained with corresponding full movement of the control in the cockpit. In addition, check and adjust the throttle-control linkages so that *springback* on the throttle quadrant in the aircraft is equal in both the full-open and full-closed positions. Correct any excess play or looseness of control linkage or cables.

When installing carburetors or fuel-metering equipment incorporating manual-type mixture controls that do not have marked positions, adjust the mixture-control mechanism to provide an equal amount of springback at both the RICH and LEAN ends of the cockpit control quadrant when the mixture control on the carburetor- or fuel-metering equipment is moved through the full range. Where mixture controls with detents are used, rig the control mechanism so that the designated positions on

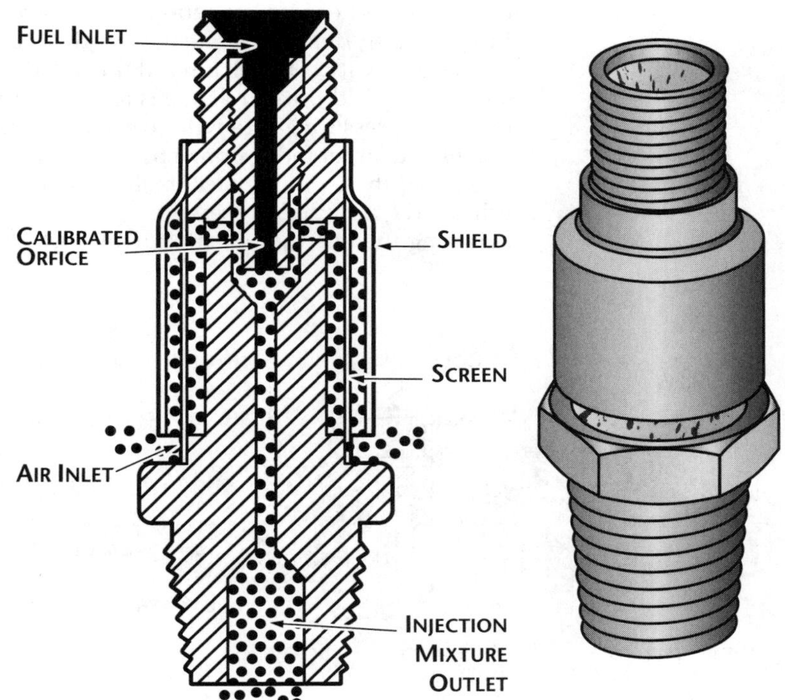


Figure 1-8-32. Fuel discharge nozzle

the control quadrant in the aircraft will agree with the corresponding positions on the fuel-metering unit.

In all cases, check the controls for proper positioning in both the advance and retard positions. Correct excess play or looseness of control linkage or cables. Safety all controls properly to eliminate the possibility of loosening from vibration during operation.

## Field Adjustments

### *Adjusting Idle Mixture*

Excessively rich or lean idle mixtures result in incomplete combustion within the engine cylinder, with resultant formation of carbon deposits on the spark plugs and subsequent spark plug fouling. In addition, excessively rich or lean idle mixtures make it necessary to taxi at high idle speeds with resultant fast taxi speeds and excessive brake wear. Each engine must have the carburetor idle mixture tailored for the particular engine and installation, if best operation is to be obtained.

Engines that are properly adjusted, insofar as valve operation, cylinder compression, ignition and carburetor idle-mixture are concerned, will idle at the prescribed r.p.m. for indefinite periods without loading up, overheating or spark plug fouling. If an engine will not respond to idle-mixture adjustment with the resultant stable idling characteristics previously outlined, it is an indication that some other phase of engine operation is not correct. In such cases, determine and correct the cause of the difficulty. On all aircraft installations where manifold pressure gauges are used, the manifold pressure gauge will give a more consistent and larger indication of power change at idle speed than will the tachometer. Therefore, utilize the manifold pressure gauge when adjusting the idle fuel/air mixture. Check and adjust the idle mixture and speed on all type reciprocating engines as discussed in the following paragraphs.

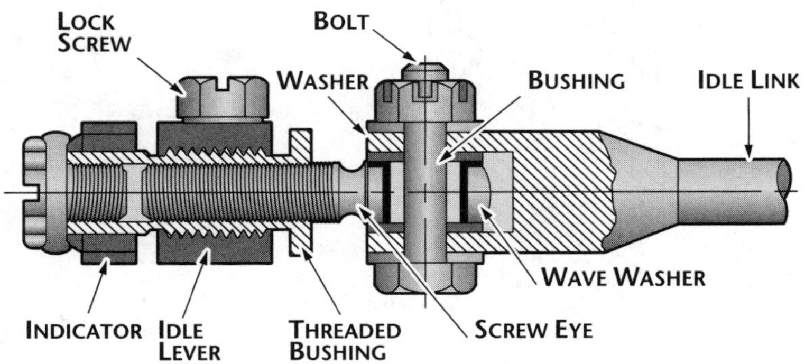


**Figure 1-8-33. Idle-mixture-adjusting mechanism for Stromberg injection carburetors**

Always make idle-mixture adjustments with cylinder head temperatures at normal values (about 150°-170°C) and never with temperatures approaching the maximum allowable.

The idle-mixture adjustment is made on the idle fuel-control valve. It should not be confused with the adjustment of the idle-speed stop. The importance of idle-mixture adjustment cannot be overstressed. Optimum engine operation at low speeds can be obtained only when proper fuel/air mixtures are delivered to every cylinder of the engine. Excessively rich idle mixtures and the resultant incomplete combustion are responsible for more spark plug fouling than any other single cause. Excessively lean idle mixtures result in faulty acceleration. Furthermore, the idle-mixture adjustment affects the fuel/air mixture and engine operation well up into the cruise range.

On an engine having a conventional carburetor, the idle mixture is checked by manually leaning the mixture with the cockpit mixture control. Move the carburetor mixture control slowly and smoothly toward the idle-cutoff position. At the same time, watch the manifold pressure gauge to determine whether the manifold pressure decreases prior to increasing as the engine ceases firing. The optimum mixture is obtained when a manifold pressure decrease immediately precedes the manifold pressure increase as the engine ceases firing. The amount of decrease will vary with the make and model of the engine, and the installation. As a general rule, the amount of manifold pressure decrease will be approximately 1/4 inch.

If the installation does not incorporate a manifold-pressure gauge, the tachometer must be used to adjust the mixture. Generally, as the mixture control is moved in to idle-cutoff position, the tachometer should indicate a 10-50 r.p.m. increase just prior to engine stoppage.

On direct fuel-injection engines, the mixture change during manual leaning with the mixture control is usually so rapid that it is impossible to note any momentary increase in r.p.m. or decrease in manifold pressure. Therefore, on these engines, the idle mixture is set slightly leaner than best-power and is checked by enriching the mixture with the primer. To check the idle mixture on a fuel-injection installation, first set the throttle to obtain the proper idle speed.

Next, momentarily depress the primer switch while observing the tachometer and manifold pressure gauge. If the idle mixture is correct, the fuel added by the primer will cause a momentary increase in engine speed and a momentary drop in manifold pressure. If the increase in engine speed or the decrease in manifold pressure exceeds the limits specified for the particular installation, the idle mixture is too lean (too far on the lean side of best-power). If the r.p.m. drops off when the mixture is enriched with the primer, the idle-mixture setting is too rich.

Before checking the idle mixture on any engine, warm up the engine until oil and cylinder head temperatures are normal. Keep the propeller control in the INCREASE R.P.M. setting throughout the entire process of warming up the engine, checking the idle mixture and making the idle adjustment. Keep the mixture control in AUTO-RICH except for the manual leaning required in checking the idle mixture on carburetor-equipped engines.

When using the primer to check the idle mixture on fuel-injection engines, merely flick the primer switch quickly on and off; otherwise, too much additional fuel will be introduced and a satisfactory indication will be obtained even though the idle mixture is set too lean.

If the check of the idle mixture reveals it to be too lean or too rich, increase or decrease the idle-fuel flow as required, then repeat the check. Continue checking and adjusting the idle mixture until it checks out properly. During this process, it may be desirable to move the idle-speed stop completely out of the way and to hold the engine speed at the desired r.p.m. by means of the throttle. This will eliminate the need for frequent readjustments of the idle stop as the idle mixture is improved and the idle speed picks up. After each adjustment, clear the engine by briefly running it at higher r.p.m. This prevents fouling of the plugs that might otherwise be caused by incorrect idle mixture.

After the idle mixture has been set, it should be checked to determine that it functions properly at all operating temperature and power settings.

On PS-series injection-type carburetors and RSA fuel-injection master control units, the idle-control link located between the idle valve stem in the fuel-control unit and the idle-control lever on the throttle shaft incorporates a bushing arrangement at each end (see Figure 1-8-33). Be sure that the bolt is tight and has the washers, wave washers and bushings assembled. In addition, there must be no play between the link and the lever. If there is any play at either end of the link, erratic mixtures will result.

If sufficient mixture change cannot be accomplished by the normal idle-mixture adjustment on the PS-series injection carburetor, disconnect the link at the idle-valve end by removing the bolt, washers and wave washers. Then, to further alter the mixture, turn out (to enrich) or in (to lean). One turn of the screw eye is equivalent to 13 notches, or clicks, on the normal idle adjustment.

### *Idle Speed Adjustment*

After adjusting the idle mixture, reset the idle stop to the idle r.p.m. specified in the aircraft's maintenance manual. The engine must be warmed up thoroughly and checked for ignition-system malfunctioning. Throughout any carburetor adjustment procedure, run the engine up periodically to approximately half of normal rated speed to clear the engine.

Some carburetors are equipped with an *eccentric screw* to adjust idle r.p.m. Others use a *spring-loaded screw* to limit the throttle valve closing. In either case, adjust the screw as required to increase or decrease r.p.m. with the throttle retarded against the stop. Open the throttle to clear the engine, then close the throttle and allow the r.p.m. to stabilize. Repeat this operation until the proper idling speed is obtained.

## Carburetor and Fuel-injection System Overhaul

While not often done today, the overhaul of a float-type carburetor may be performed in the field. The most common method of overhauling a float-type carburetor is to purchase an overhaul kit from the manufacturer. This kit will include all parts and instructions necessary to complete the overhaul (Figure 1-8-34, next page). As with any overhaul, it is very important to make certain that you are using current service information.

The overhaul of a pressure-type carburetor or fuel-injection system is much more complex and requires special tools and testing equipment. The overhaul of these units should normally be performed by an appropriately licensed FAA-certified repair station.

## Engine-driven Fuel Pumps

There are two styles of engine-driven fuel pumps in general use: the *gear-type* and the *vane-type*. The fuel comes in at one side of the pair of gears (the side on which they turn away from one another) and is carried around to the other side in the space between the teeth. At that point, it escapes through the outlet and will start to build pressure. The fuel flow, therefore, is around the outside of the gears, between the teeth and the casing. It will not work if it is driven in the reverse direction.

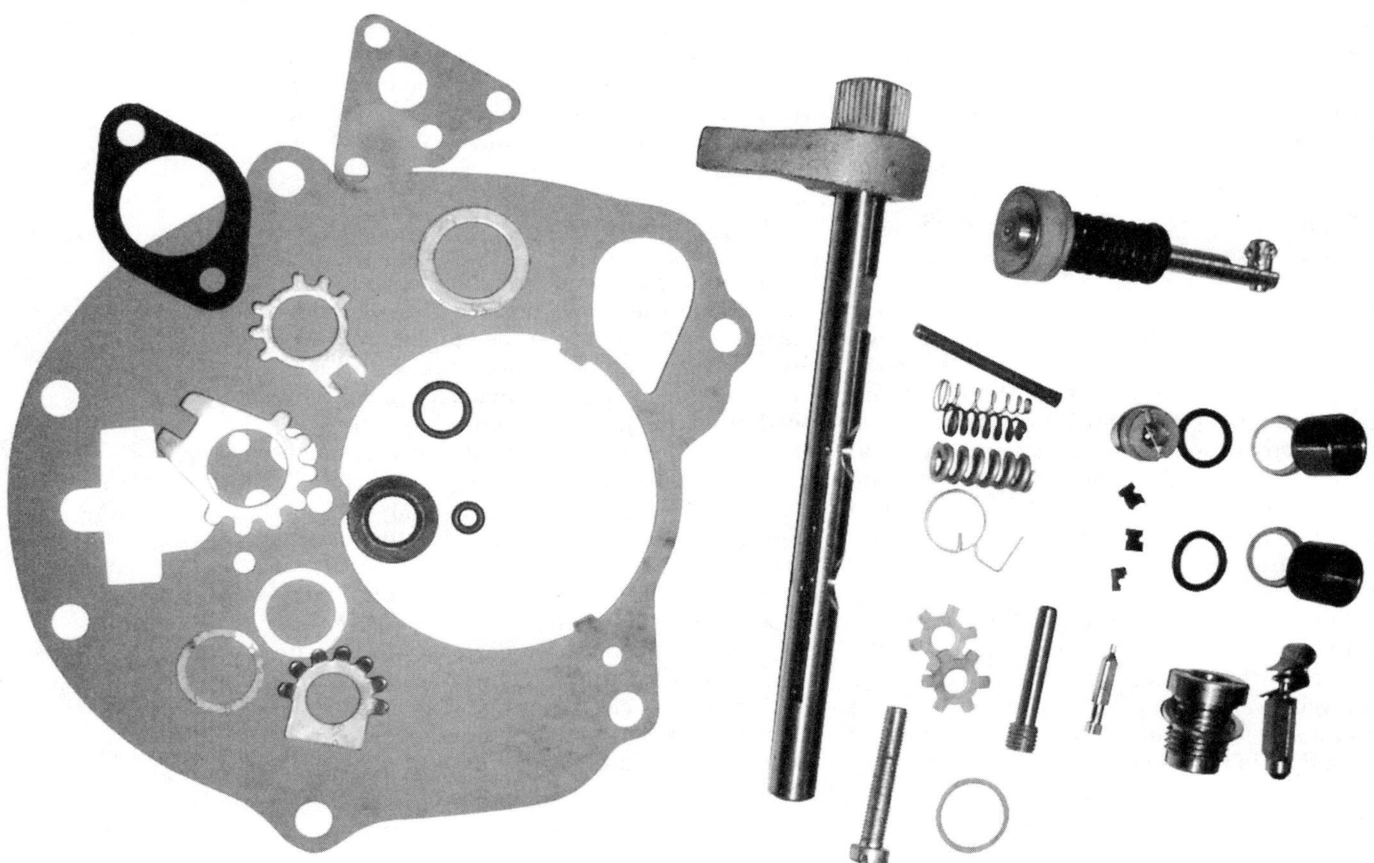

**Figure 1-8-34. A carburetor overhaul kit includes all required replacement parts. Never reuse old gaskets and seals.**

Because these gear pumps can build up considerable pressure, they are widely used for oil pumps as well as fuel pumps. They do not, however, give much suction, and if the liquid is not fed to them, may need priming to get them started. *Priming* means filling the pump full of liquid so that the teeth have something to bite on besides air.

The vane-type, shown in Figure 1-8-35, is provided with both a pressure-relief valve and a bypass valve. The working part of the pump consists of four vanes, which are merely flat-sided strips of metal. They are carried in a short, tubular member called the *rotor.* The whole thing turns in a circular housing, but this housing is off-center (eccentric).

When the vane assembly turns in this eccentric housing, some of the vanes will project farther than the others from the tubular members. In other words, while the vanes rotate, they also slide back and forth in their slots. The fuel, coming in at one side, is caught by the vanes, which act like paddle wheels, and held between each pair of vanes and the housing.

As each vane delivers its quota of fuel to the other side of the housing, it gradually recedes into the tubular member so that it does not pull the liquid back with it in the reverse direction.

On top of the pump, you will notice a screw. This is to regulate the pressure of a spring on the pressure-regulating valve. The top of the valve diaphragms, where you see the spring, has an outlet to the air. This outlet can be connected, if you wish, to the carburetor air scoop or to the supercharger, so as to maintain a uniform pump discharge at varying altitudes.

In this type of pump, the built-in relief valve lets some of the fuel pass back to the other side of the vanes whenever the pressure on the carburetor line becomes excessive. This does away with the necessity for an overflow line back to the tank.

Most engine-operated fuel pumps, such as the vane- and gear-types, are bolted directly to the engine crankcase.

## Fuel System Maintenance

The inspection of a fuel system consists of an examination of the system for conformity to design requirements, together with functional tests to prove correct operation.

Since there are considerable variations in the fuel systems used on different aircraft, no attempt has been made to describe any particular system in detail. It is important that the manufacturer's instructions for the aircraft concerned be followed when performing inspection or maintenance functions.

## Complete System Inspection

Inspect the entire system for wear, damage or leaks. Make sure that all units are securely attached and properly safetied.

The drain plugs or valves in the fuel system should be opened to check for the presence of sediment or water. The filter and sump should also be checked for sediment, water or slime. The filters or screens, including those provided for flow meters and auxiliary pumps, must be clean and free from corrosion.

The controls should be checked for freedom of movement, security of locking and freedom from damage due to chafing.

The fuel vents should be checked for correct positioning and freedom from obstruction; otherwise, fuel flow or pressure fueling may be affected. Filler neck drains should be checked for freedom from obstruction.

If booster pumps are installed, the system should be checked for leaks by operating the pumps. During this check, the ammeter or loadmeter should be read and the readings of all the pumps, where applicable, should be approximately the same.

## Fuel Tanks

All applicable panels in the aircraft skin or structure should be removed and the tanks inspected for corrosion on the external surfaces, for security of attachment and for correct adjustment of straps and slings. Check the fittings and connections for leaks or failures.

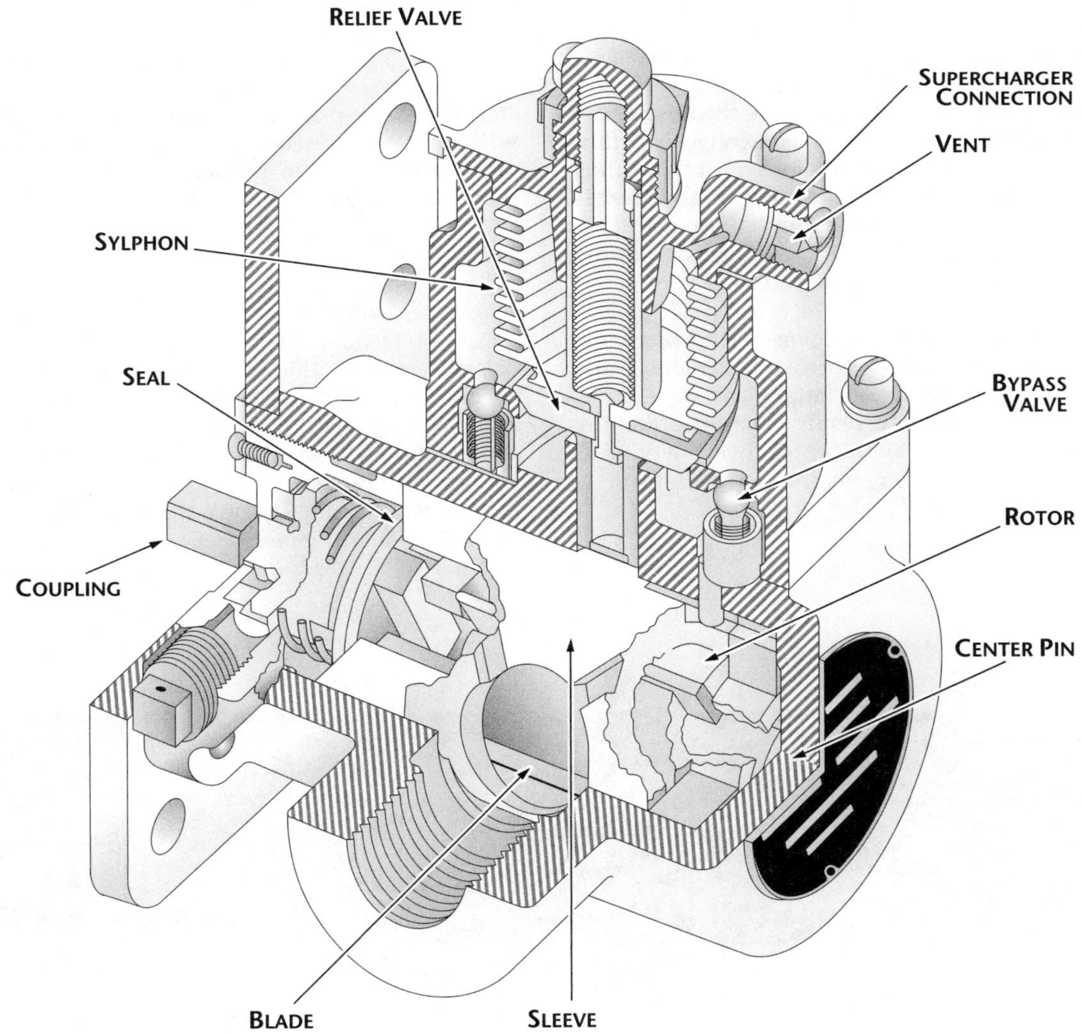


Figure 1-8-35. Cutaway view of a vane-type pump, showing both the relief and bypass valves

## Lines and Fittings

Be sure that the lines are properly supported and that the nuts and clamps are securely tightened. To tighten hose clamps to the proper torque, use a hose-clamp torque wrench. If this wrench is not available, tighten the clamp finger-tight plus the number of turns specified for the hose and clamp. If the clamps do not seal at the specified torque, replace the clamps, the hose or both. After installing new hose, check the clamps daily and tighten if necessary. When this daily check indicates that cold flow has ceased, inspect the clamps at less frequent intervals.

Replace the hose if the plies have separated (Figure 1-8-36), if there is excessive cold flow, or if the hose is hard and inflexible. Permanent impressions from the clamp and cracks in the tube or cover stock indicate excessive cold flow. Replace hose that has collapsed at the bends or as a result of misaligned fittings or lines. Some hose tends to flare at the ends beyond the clamps. This is not an unsatisfactory condition, unless leakage is also present.

Blisters may form on the outer synthetic rubber cover of hose. These blisters do not necessarily affect the serviceability of the hose. When a blister is discovered on a hose, remove the hose from the aircraft and puncture the blister with a pin. The blister should then collapse. If fluid (oil, fuel or hydraulic) emerges from the pinhole in the blister, reject the hose. If only air emerges, pressure test the hose at one-and-a-half times the working pressure. If no fluid leakage occurs, the hose can be regarded as serviceable.

Puncturing the outer cover of the hose may permit the entry of corrosive elements, such as water, which could attack the wire braiding and ultimately result in failure. For this reason, puncturing the outer covering of hoses exposed to the elements should be avoided.

The external surface of hose may develop fine cracks, usually short in length, which are caused by surface aging. The hose assembly may be regarded as serviceable, provided the cracks do not penetrate to the first braid.

**Engine primers.** Most carbureted reciprocating engines require an *engine primer* to supply initial fuel to the cylinders. When turned over with the starter, the prime will cause the engine to fire and start rotating. This initial rotation is enough for airflow to establish itself within the induction system. Once sufficient air is flowing through the carburetor, the engine can run on its own, without further priming.

Primers come in two types, *hand pump* and *electrical pump*. Each engine has a specific starting procedure that should be followed, but each engine is a bit different. Electric primers also have a definite procedure involved in their use.

Hand primers must always be pushed in and locked whenever they have been unlocked for any reason. Fuel can flow through an unlocked primer and fill the engine crankcase and cylinders full of liquid fuel. It then becomes quite a project to get the engine drained, re-oiled and pre-oiled again prior to starting. If an engine is started and the primer is not locked, the engine will run very rich until the trouble is found and corrected. Open primers are so common that they should be on any technician's list of things to check for every time.

## Selector Valves

Selector valves (Figure 1-8-37) should be checked to determine that they rotate freely and without excessive backlash. The valve should have a clear detent when it is in position and be marked clearly. Check the valve for loose nuts or pins. Check worn control cables or pulleys or faulty bearings.

Figure 1-8-36. The most common hose damage is caused by abrasion.

## Pumps

During an inspection of booster pumps, check for the following conditions:

- Proper operation
- Leaks and condition of fuel and electrical connections
- Wear of motor brushes

Be sure the drain lines are free of traps, bends or restrictions.

Check the engine-driven pump for leaks and security of mounting. Check the vent and drain lines for obstructions.

**Figure 1-8-37. A five-port main fuel valve**

## Main Line Strainers

All fuel strainers (Figure 1-8-38, next page) should be removed, checked for contamination and cleaned at regular intervals. These intervals are listed in the aircraft maintenance manual. Check the strainers for contamination that might indicate a problem elsewhere in the fuel system. Pieces of rubber are often indications of faulty hoses.

## Fuel Quantity Gauges

If a *sight gauge* is used, be sure that the glass is clear and that there are no leaks at the connections. Check the lines leading to it for leaks and security of attachment.

Check the mechanical gauges for free movement of the float arm and for proper synchronization of the pointer with the position of the float.

On the electrical and electronic gauges, be sure that both the indicator and the tank units are securely mounted and that the electrical connections are tight.

## Fuel-pressure Gauge

The fuel-pressure gauge should be checked for leaks around the fittings, oscillation of the needle and proper markings. It is also important to make certain that needles zero when the engine and boost pumps are shut off.

## Pressure Warning Signal

Inspect the entire installation for security of mounting and condition of the electrical, fuel and air connections. Check the lamp by pressing the test switch. Check the operation by turning the battery switch on, building up pressure with the booster pump and observing the pressure at which the light goes out. If necessary, adjust the contact mechanism.

## Water-injection System

There are few of these now being used, but an anti-detonation injection system enables more power to be obtained from the engine at takeoff than is otherwise possible. These systems were used on some high-horsepower reciprocating engines and may be referred to as *Anti-Detonation Injection* (ADI) systems.

The carburetor (operating at high-power settings) delivers more fuel to the engine than it actually needs. A leaner mixture would produce more power, but the additional fuel is necessary to prevent overheating and detonation. With the injection of the *anti-detonant fluid* (a mixture of alcohol and water), the mixture can be leaned-out to that which produces maximum power, and the vaporization of the water-alcohol mixture then provides the cooling formerly supplied by the excess fuel.

Operating on this best-power mixture, the engine develops more power, even though the manifold pressure and r.p.m. settings remain unchanged. In addition, the manifold pressure can be increased to a point that would cause detonation without injection of the water-alcohol mixture. Thus, the increase in power with the anti-detonation injection is two-fold:

1. The engine can be operated on the best-power mixture
2. The maximum manifold pressure can be increased

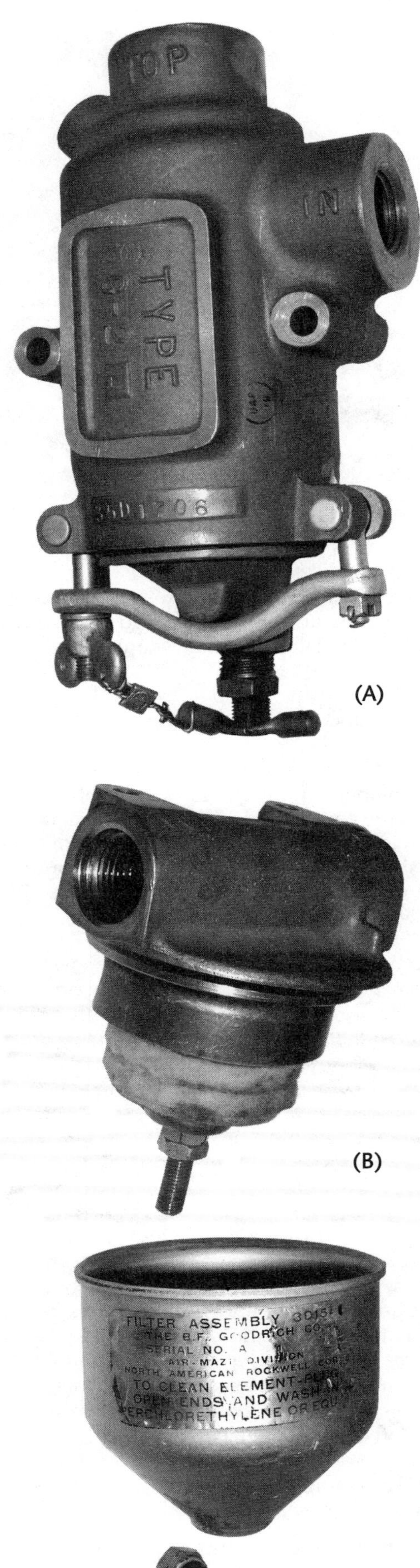


**Figure 1-8-38. Two types of fuel strainers: (A) is an AN-type C-2 strainer that uses a conical screen, while (B) uses a felt-type material to filter fuel for contaminants.**

*Section 9*

# *Induction and Airflow Systems*

A reciprocating engine induction system consists of a carburetor, an air-scoop or ducting that delivers air to the carburetor and the intake manifold. The *intake manifold* consists of a series of long, curved tubes that deliver the fuel/air mixture to the cylinders.

Some induction systems contain a temperature-indicating system and a means by which warm air can be supplied to the induction system. This warm air is used to prevent the formation of ice in the induction system. A turbocharger or supercharger may be incorporated in the intake system to compress the fuel/air mixture. If an engine has a turbocharger or supercharger installed, it is classified as a *supercharged* engine. If not installed, the engine is classified as *naturally aspirated* or *non-supercharged*.

## Non-supercharged Induction Systems

Most light aircraft use naturally aspirated engines. The induction system of these engines may contain either a carburetor or a fuel-injection system. The carburetor may be either of the float-type or the pressure-type, while a fuel-injection system may be either of the constant-flow or pulsating-flow type.

Figure 1-9-1 is a diagram of an induction system used in a non-supercharged (naturally aspirated) engine equipped with a carburetor. In this induction system, carburetor cold air is admitted at the leading edge of the nose cowling below the propeller spinner and is passed through an air filter into air ducts leading to the carburetor. An air valve is located at the carburetor for selecting an alternate warm air source to prevent carburetor icing.

The cold-air valve admits air from the outside air scoop for normal operation and is controlled by a control knob in the cockpit. The warm-air valve admits warm air from the engine compartment for operation during icing conditions and is spring loaded to the CLOSED position. When the cold air door is closed, engine suction opens the spring-loaded warm-air valve. If the engine should backfire with the warm-air valve open, spring tension automatically closes the warm-air valve to keep flames out of the engine compartment.

The carburetor air filter is installed in the air scoop in front of the carburetor air duct. Its purpose is to stop dust and other foreign matter from entering the engine through the carburetor. The screen consists of an aluminum alloy frame and a deeply crimped screen, arranged to present maximum screen area to the air stream.

The carburetor air ducts consist of a fixed duct, riveted to the nose cowling, and a flexible duct between the fixed duct and the carburetor air valve. The carburetor air ducts provide a passage for cold, outside air to the carburetor.

Air enters the system through the ram-air intake. The intake opening is located in the slipstream, so the air is forced into the induction system, giving a ram effect.

The air passes through the ducts to the carburetor. The carburetor meters the fuel in proportion to the air and mixes the air with the correct amount of fuel. The carburetor can be controlled from the cockpit to regulate the flow of air and, in this way, power output of the engine can be controlled.

The carburetor air temperature-indicating system shows the temperature of the air at the carburetor inlet. If the bulb is located at the engine side of the carburetor, the system measures the temperature of the fuel/air mixture.

## Supercharged Induction Systems

Supercharging systems used in reciprocating engine induction systems are normally classified as either *internally driven* or *externally driven* (turbo-supercharged).

Internally driven superchargers compress the fuel/air mixture after it leaves the carburetor, while externally driven superchargers (turbochargers) compress the air before it is mixed with the metered fuel from the carburetor. Each increase in the pressure of the air or fuel/air mixture in an induction system is called a *stage*. Superchargers can be classified as single-stage, two-stage or multi-stage, depending on the number of times compression occurs. Superchargers may also operate at different speeds. Thus, they can be referred to as single-speed, two-speed or variable-speed superchargers.

Combining the methods of classification provides the nomenclature normally used to describe supercharger systems. Thus, from a simple, single-stage system that operates at one fixed-speed ratio, it is possible to progress to a single-stage, two-speed, mechanically clutched system or a single-stage, hydraulically clutched supercharger. Even though two-speed or multi-speed systems permit varying the output pressure, the system is still classified as a single-stage of compression if only a single impeller is used, since only one increase (or decrease) in compression can be obtained at a time.

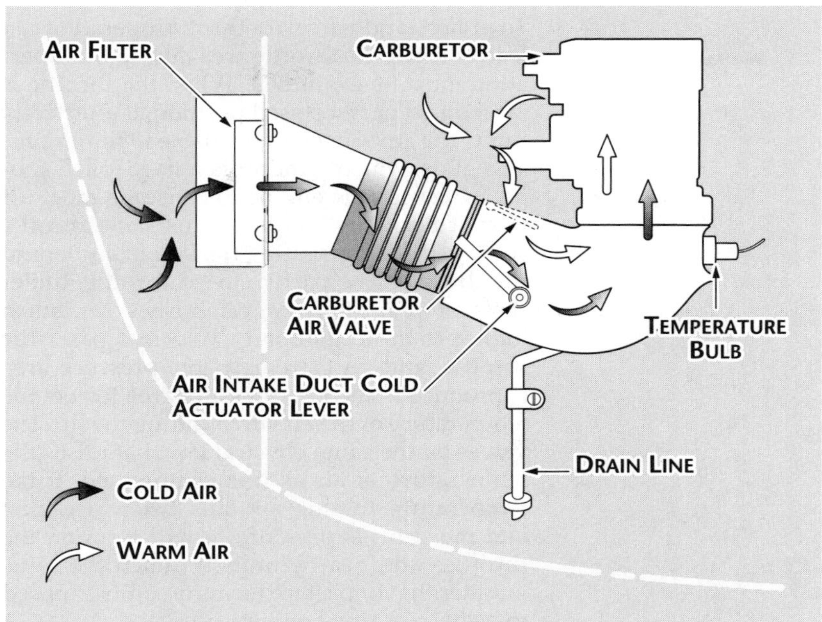


Figure 1-9-1. Non-supercharged induction system using a carburetor

## Induction-system Icing

A short discussion concerning the formation and the place of formation of induction system ice is helpful, even though the mechanics are not normally concerned with operations that occur only when the aircraft is in flight. But the technician should know something about induction-system icing because of its effect on engine performance. Even when inspection shows that everything is in proper working order, induction-system ice can cause an engine to act erratically and lose power in the air, yet the engine will perform perfectly on the ground. Many engine troubles commonly attributed to other sources are actually caused by induction-system icing.

Induction-system icing is an operating hazard because it can cut off the flow of the fuel/air charge or vary the fuel/air ratio. Ice can form in the induction system while an aircraft is flying in clouds, fog, rain, sleet, snow or even clear air that has a high moisture content (high humidity). Induction-system icing is generally classified in three types:

1. Impact ice
2. Fuel-evaporation ice
3. Throttle ice

To understand why part-throttle operation can lead to icing, the throttle area during this operation must be examined. When the throttle is placed in a partly closed position, it effectively limits the amount of air available to the engine. The glide, which windmills a fixed-pitch propeller, causes the engine to consume more air than it normally would at this same throttle setting, thus aggravating the lack of air behind the throttle. The partly closed throttle, under these circumstances, establishes a much higher-than-normal air velocity past the throttle, and an extremely low-pressure area is produced. The low-pressure area lowers the temperature of the air surrounding the throttle valves by the same physical law that raises the temperature of air as it is compressed. If the temperature in this air falls below freezing and moisture is present, ice will form on the throttles and nearby units in much the same manner that impact ice forms on units exposed to below freezing temperatures.

Induction-system ice can be prevented or eliminated by raising the temperature of the air that passes through the system, using a pre-heater located upstream near the induction-system inlet and well ahead of the dangerous icing zones. Heat is usually obtained through a control valve that opens the induction system to the warm air circulating in the engine compartment. When there is danger of induction-system icing, move the cockpit control toward the HOT position until a carburetor-air temperature is obtained that will provide the necessary protection.

Throttle ice, or any ice that restricts airflow or reduces manifold pressure, can best be removed by using full carburetor heat. If the heat from the engine compartment is sufficient and the application has not been delayed, it is only a matter of a few minutes until the ice is cleared. If the air temperature in the engine compartment is not high enough to be effective against icing, the preheat capacity can be increased by closing the cowl flaps and increasing engine power. However, this may prove ineffective if the ice formation has progressed so far that the loss of power makes it impossible to generate sufficient heat to clear the ice.

Improper or careless use of carburetor heat can be just as dangerous as the most advanced stage of induction-system ice. Increasing the temperature of the air causes it to expand in volume and decrease in density. This action reduces the weight of the charge delivered to the cylinder and causes a noticeable loss in power because of decreased volumetric efficiency. In addition, high intake air temperature may cause detonation and engine failure, especially during takeoff and high-power operation. Therefore, during all phases of engine operation, the carburetor temperature must afford the greatest protection against icing and detonation. When there is no danger of icing, the heat control is normally kept in the COLD position. It is best to leave the control in this position if there are particles of dry snow or ice in the air. The use of heat may melt the ice or snow, and the resulting moisture may collect and freeze on the walls of the induction system.

To prevent damage to the heater valves in the case of backfire, carburetor heaters should not be used while starting the engine. During ground operation, only enough carburetor heat should be used to give smooth engine operation. The carburetor-air inlet temperature gauge must be monitored to be sure the temperature does not exceed the maximum value specified by the engine manufacturer.

On some aircraft, the basic de-icing system is supplemented by a fluid de-icing system. This auxiliary system consists of a tank, a pump, suitable spray nozzles in the induction system and a cockpit-located control unit. This system is intended to clear ice whenever the heat from the engine compartment is not high enough to prevent or remove ice. The use of alcohol as a de-icing agent tends to enrich the fuel mixture; however, at a high-power output, such slight enrichment is desired. At low throttle settings, however, the use of alcohol may over-enrich the mixture. For this reason, alcohol should be applied with great care.

## Induction-system Filtering

Contaminants that are always present in the air passing through the engine must be removed to provide for proper engine operation and to prevent engine damage.

While dust is merely an annoyance to most individuals, it is a serious source of trouble to an aircraft engine. Dust consists of small particles of hard, abrasive material that can be carried into the engine cylinders by the very air the engine breathes. It can also collect on the fuel-metering elements of the carburetor, upsetting the proper relation between airflow and fuel flow at all powers. It acts on the cylinder walls by grinding down these surfaces and the piston rings. It then contaminates the oil and is carried through the engine, causing further wear on the bearings and gears. In extreme cases, an accumulation may clog an oil passage and cause oil starvation.

Although dust conditions are most critical at ground level, dust of sufficient quantity to obscure a pilot's vision has been reported in flight. In some parts of the world, dust can be carried to extremely high altitudes. Continued operation under such conditions without engine protection will result in extreme engine wear and excessive oil consumption.

When operation in a dusty atmosphere is necessary, the engine can be protected by an alternate induction-system air inlet that incorporates a dust filter. This type of air filter system normally consists of a *filter element*, a *door* and an *electrically operated actuator.* When the filter system is operating, air is drawn through a louvered access panel that does not face directly into the airstream. With this entrance location, considerable dust is removed as the air is forced to turn and enter the duct. Since the dust particles are solid, they tend to continue in a straight line, and most of them are separated at this point. Those that are drawn into the louvers are easily removed by the filter.

In flight, with air filters operating, consideration must be given to possible icing conditions which may occur from actual surface icing or from freezing of the filter element after it becomes rain-soaked. Some installations have a spring-loaded filter door that automatically opens when the filter is excessively restricted. This prevents the airflow from being cut off when the filter is clogged with ice or dirt. Other systems use an *ice guard* in the filtered-air entrance.

The ice guard consists of a coarse-mesh screen located a short distance from the filtered-air entrance. In this location, the screen is directly in the path of incoming air, so the air must pass through or around the screen. When ice forms on the screen, the air, which has lost its heavy moisture particles, will pass around the iced screen and into the filter element.

The efficiency of any filter system depends upon proper maintenance and servicing. Periodic removal and cleaning of the filter element is essential to satisfactory engine protection.

## Induction-system Inspection and Maintenance

During all regularly scheduled engine inspections, the induction system should be checked for leaks, cracks and security. Leaks of the induction system can sometimes be detected by a fuel stain on the leaking component. A rough-idling engine may also be an indication of an induction leak. Cracks can be detected visually or with a dye-penetrant type of inspection. All hose clamps and mounting bolts should be checked for tightness.

The carburetor air filter should be checked regularly. Paper element filters should be replaced. Reusable-type foam filters should be thoroughly washed in Stoddard solvent, allowed to dry and treated with a thin coat of oil before reinstallation.

**Induction-system troubleshooting.** Table 1-9-1 provides a general guide to the most common induction-system troubles.

## Internally Driven Superchargers

Internally driven superchargers are used almost exclusively in high-horsepower reciprocating engines. Internal superchargers have been replaced by exhaust driven turbochargers but some may still be found on radials and older opposed engines. Except for the construction and arrangement of the various types of superchargers, all induction systems with internally driven superchargers are almost identical. The reason for this similarity is that all aircraft engines require the same

| INDUCTION-SYSTEM TROUBLESHOOTING | | |
|---|---|---|
| PROBABLE CAUSE | ISOLATION PROCEDURE | REMEDY |
| *ENGINE FAILS TO START* | | |
| Induction system obstructed | Inspect airscoop and air ducts | Remove obstruction(s) |
| Air leaks | Inspect carburetor mounting and intake pipes | Tighten carburetor and repair/replace intake pipe |
| *ENGINE RUNS ROUGH* | | |
| Loose air ducts | Inspect air ducts | Tighten air ducts |
| Leaking intake pipes | Inspect intake pipe packing nuts | Tighten nuts |
| Engine valves sticking | Remove rocker arm cover and check valve action | Lubricate and free sticking valves |
| Bent or worn valve push rods | Inspect push rods | Replace worn or damaged push rods |
| *LOW POWER* | | |
| Restricted intake duct | Examine intake duct | Remove restrictions |
| Broken door in carburetor air valve | Inspect air valve | Replace air valve |
| Dirty air filter | Inspect air filter | Clean air filter |
| *ENGINE IDLES IMPROPERLY* | | |
| Shrunken intake packing | Inspect packing for proper fit | Replace packing |
| Hole in intake pipe | Inspect intake pipes | Replace defective intake pipe(s) |
| Loose carburetor mounting | Inspect mount bolts | Tighten mount bolts |

**Table 1-9-1. Induction-system troubleshooting chart**

air temperature control to produce good combustion in the engine cylinders. For example, the temperature of the charge must be warm enough to ensure complete fuel vaporization and, thus, even distribution; but at the same time it must not be so hot that it reduces volumetric efficiency or causes detonation. With these requirements, all induction systems that use internally driven superchargers must include pressure- and temperature-sensing devices and the necessary units required to warm or cool the air.

**Single-stage, single-speed supercharger system.** The simple induction system in Figure 1-9-2 is used to explain the location of units and the path of the air and fuel/air mixture.

Air enters the system through the *ram air intake*. The intake opening is located so that the air is forced into the induction system, giving a ram effect.

The air passes through ducts to the carburetor. The carburetor meters the fuel in proportion to the air and mixes the air with the correct amount of fuel. The carburetor can be controlled from the cockpit to regulate the flow of air. In this way, the power output of the engine can be controlled.

The *manifold pressure gauge* measures the pressure of the fuel/air mixture before it enters the cylinders. It is an indication of the performance that can be expected of the engine.

The *carburetor air-temperature indicator* measures the temperature of either the inlet air or the fuel/air mixture. Either the air inlet or the mixture temperature indicator serves as a guide so that the temperature of the incoming charge may be kept within safe limits.

If the temperature of the incoming air at the entrance to the carburetor scoop is 100°F, there will be approximately a 50°F drop in temperature because of the partial vaporization of the fuel at the carburetor discharge nozzle. Partial vaporization takes place, and the air temperature falls due to absorption of the heat by vaporization. The final vaporization takes place as the mixture enters the cylinders, where higher temperatures exist.

The fuel, as atomized into the airstream flowing in the induction system, is in a globular form. The problem, then, becomes one of uniformly breaking up and distributing the fuel remaining in globular form to the various cylinders. On engines equipped with a large number of cylinders, the uniform distribution of the mixture becomes a greater problem, especially at high engine speeds, when full advantage is taken of large air capacity.

One method of improving fuel distribution is shown in Figure 1-9-3. This device is known as a *distribution impeller.* The impeller is attached directly to the end of the rear shank of the crankshaft by bolts or studs. Since the impeller is attached to the end of the crankshaft

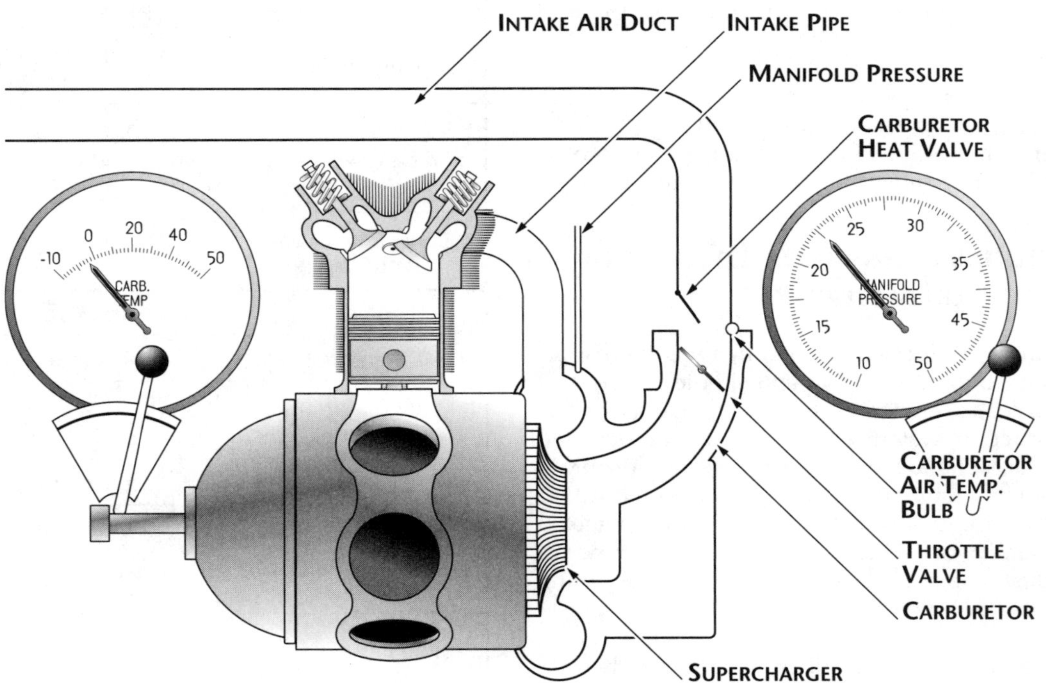


**Figure 1-9-2. Simple induction system**

and operates at the same speed, it does not materially boost or increase the pressure on the mixture flowing into the cylinders. But the fuel remaining in the globular form will be broken up into finer particles as it strikes the impeller, thereby coming in contact with more air. This will create a more homogeneous mixture, with a consequent improvement in distribution to the various cylinders, especially on acceleration of the engine or when low temperatures prevail.

When greater pressure is desired on the fuel/air fixture in the induction system to charge the cylinders more fully, the diffuser or blower section contains a high-speed impeller. Unlike the distribution impeller, which is connected directly to the crankshaft, the *supercharger,* or *blower impeller,* is driven through a gear train from the crankshaft.

The impeller is located centrally within the diffuser chamber. The diffuser chamber surface may be any one of two general designs (Figure 1-9-4, next page):

- Venturi-type
- Vane-type
- Airfoil-type

The *venturi-type diffuser* is equipped with plain surfaces, sometimes more or less restricted sectionally to form the general shape of a venturi between the impeller tips and the manifold ring. This type has been most widely used on medium-powered, supercharged engines or those in which lower volumes of mixtures are to be handled and where turbulence of the mixture between the impeller tips and the manifold chamber is not critical.

On large-volume engines ranging from 450 hp. upward, in which the volume of mixture to be handled at higher velocities and turbulence is a more important factor, either a *vane-type* or *airfoil-type diffuser* is widely used. The vanes or airfoil section straighten the airflow within the diffuser chamber to obtain an efficient flow of gases.

The intake pipes on early radial engines extended in a direct path from the manifold ring to the intake port on the cylinder. In the later designs, however, the intake pipes extend from the manifold ring on a tangent and the pipe is curved as it extends toward the intake port, which has also been streamlined or shaped to promote efficient flow of gases into the cylinder. This reduces turbulence to a minimum. This has been one of the important methods of increasing the *breathing capacity,* or volume of air, that a given design of engine might handle. Increases in supercharger efficiency have been one of the major factors in increasing the power output of modern engines.

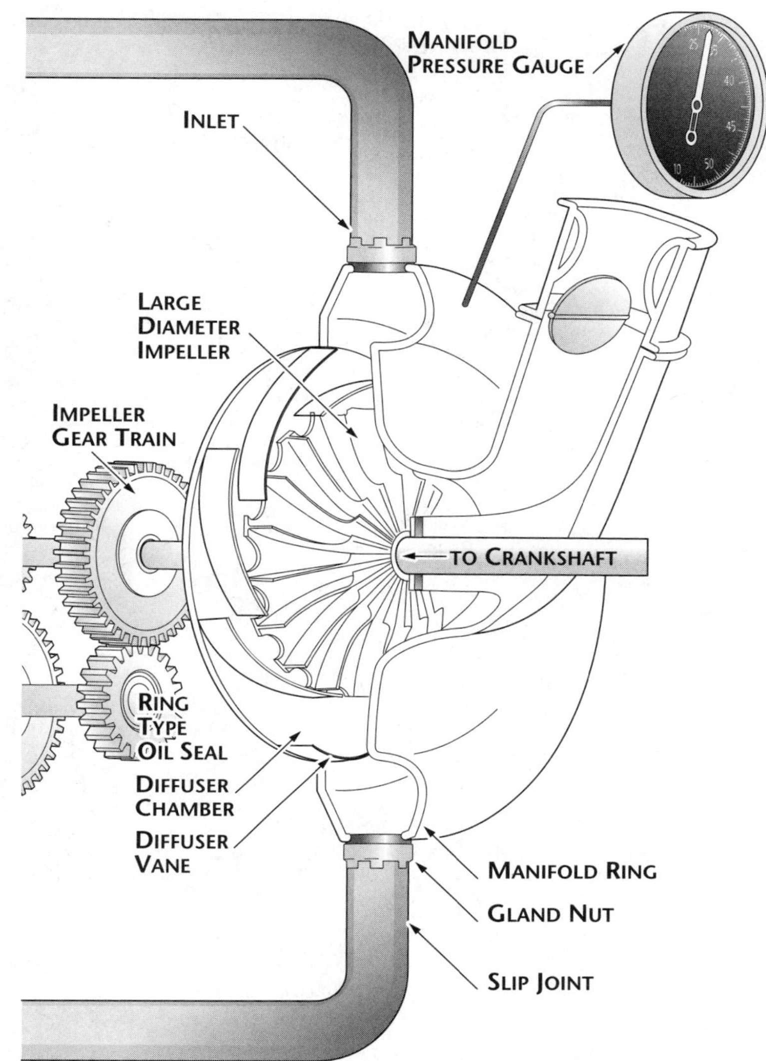


**Figure 1-9-3. Distribution impeller arrangement used on a radial engine**

The gear ratio of the impeller gear train varies from approximately 6:1 to 12:1. Impeller speed on an engine equipped with a 10:1 impeller gear ratio operating at 2,600 r.p.m. will be 26,000 r.p.m. This requires that the impeller unit be a high-grade forging, usually of aluminum alloy, carefully designed and constructed. Because of the high ratio of all supercharger gear trains, considerable acceleration and deceleration forces are created when the engine speed is increased or decreased rapidly. This necessitates that the impeller be splined on the shaft. In addition, some sort of spring-loaded or antishock device must be incorporated in the gear train between the crankshaft and impeller.

An *oil seal* is usually provided around the impeller shaft just forward of the impeller unit. The functions of the seal on this unit are to minimize the passage of lubricating oil and vapors from the crankcase into the diffuser chamber when the engine is idling and to

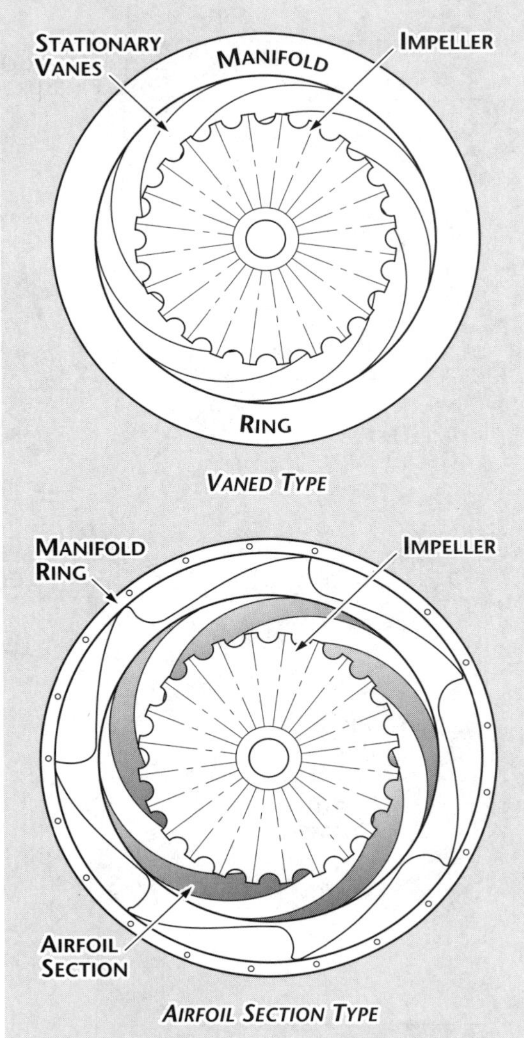


**Figure 1-9-4. Supercharger diffuser designs**

minimize the leakage of the fuel/air mixture into the crankcase when the pressure on the mixture is high at open throttle.

The clearance between the diffuser section and the impeller is obtained by varying the length of the oil seal or the thickness of spacers, commonly called *shims*. Close clearance is necessary to give the greatest possible compression of the mixture and to eliminate, insofar as is possible, leakage around the fore and aft surfaces of the impeller. The impeller shaft and intermediate drive shaft assemblies may be mounted on anti-friction balls or roller bearings or on friction-type bushings.

The impeller shaft and gear are usually forged integrally of very high-grade steel. The impeller end of the shaft is splined to give as much driving surface as possible. The intermediate shaft and large and small gears also are one piece. Both of these units are held within very close running balance or dynamic limits due to the high speeds and stresses involved.

## Externally Driven Superchargers

Externally driven superchargers, or *turbochargers*, as they are commonly called, are essentially compressors driven by the exhaust gases of the engine. The compressed air obtained from these compressors is delivered to the inlet of the carburetor or the fuel/air control unit of the engine.

## Turbo-supercharger System for Large Reciprocating Engines

During WWII, airplane designers couldn't get engines to develop enough horsepower for airplanes to reach the higher altitudes. Even if they could get them up there, the airplanes couldn't carry a worthwhile load. The answer was *compound supercharging*. In essence, the internal supercharger is supplemented by an external turbo-supercharger driven by a portion of the exhaust gas from the aircraft engine. This type of supercharger is mounted ahead of the carburetor, as shown in Figure 1-9-5, to pressurize the air at the carburetor inlet. If the air pressure entering the carburetor is maintained at approximately sea level density throughout the aircraft's climb to altitude, there will be none of the power loss experienced in aircraft not equipped with compound superchargers.

However, this type of supercharger imposes an induction-system requirement not needed in other supercharger installations. As air moves through the turbo, its temperature is raised because of compression. If the hot air charge is not properly cooled before it reaches the internal supercharger, the second stage of supercharging will produce a final charge temperature that is too great.

**Intercooling**. The air in turbo-equipped induction systems is cooled by an *intercooler* (Figure 1-9-5), so named because it cools the charge between compression stages rather than after the last stage. The hot air flows through tubes in the intercooler in much the same manner that water flows in the radiator of an automobile. Fresh outside air, separate from the charge, is collected and piped to the intercooler so that it flows over and cools the tubes. As the induction air charge flows through the tubes, heat is removed and the charge is cooled to a degree that the engine can tolerate without detonation occurring. Control for the cooling air is provided by intercooler shutters that regulate the amount of air that passes over and around the tubes of the intercooler.

**Turbocompounding.** Trying to get the last bit of power from a large engine installation, the

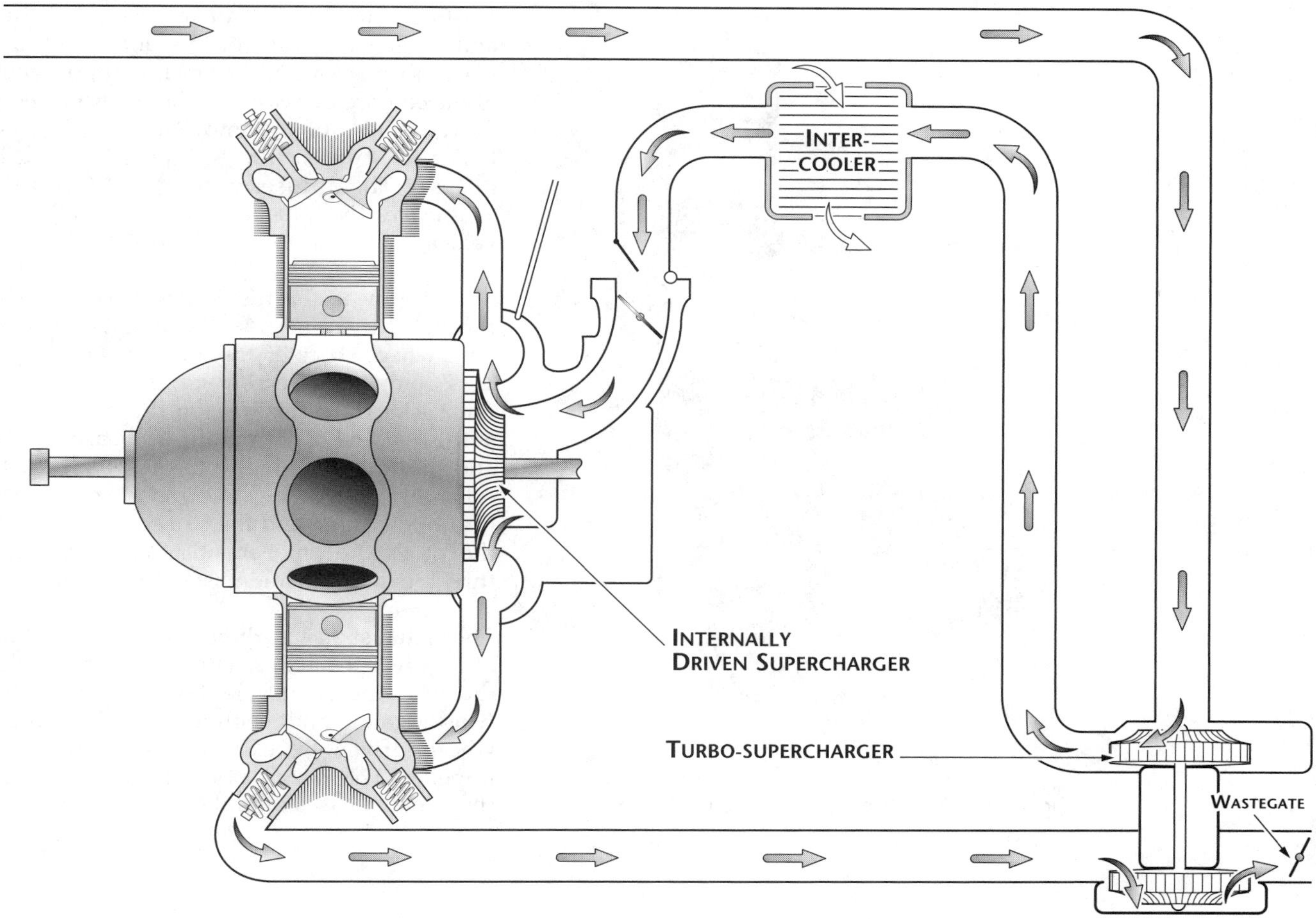

Figure 1-9-5. Induction system with turbo-supercharger

Curtiss-Wright Corporation developed a system called *turbocompounding*. Primarily used on early passenger airliners, it consisted of a turbine wheel connected to the internal gear train in the engine. Instead of compressing something, the turbocompound impeller was driven by exhaust gases. Thus, the power derived from the impeller was added to the total horsepower of the engine. Before turbocompounding became universal, the jet age arrived and large reciprocating engine production ceased.

## Turbo-supercharger Construction

The typical turbo-supercharger is composed of three main parts:

1. Compressor assembly
2. Exhaust gas turbine assembly
3. Pump and bearing casing

These major sections are shown in Figure 1-9-6. In addition to the major assemblies, there is a baffle between the compressor casing and the exhaust-gas turbine that directs cooling air to the pump and bearing casing, and shields the compressor from the heat radiated by the turbine. In installations where cooling air is limited, the baffle is replaced by a regular cooling shroud that receives its air directly from the induction system.

Figure 1-9-6. Main sections of a typical turbo-supercharger

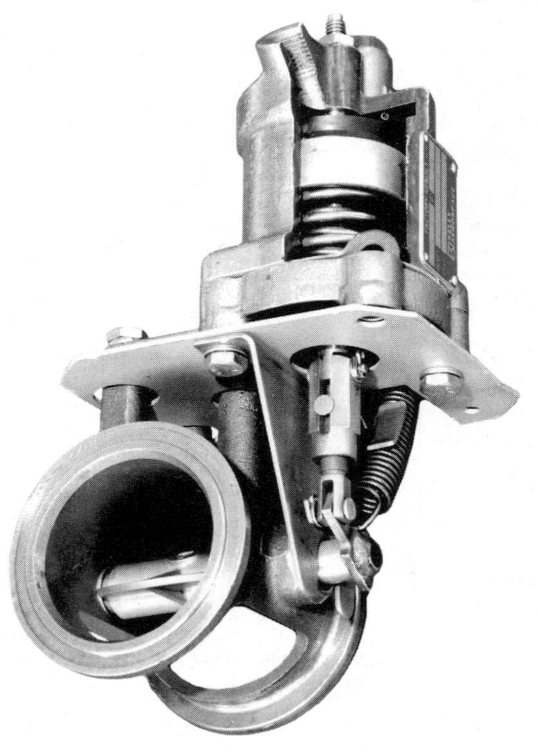

Figure 1-9-7. Waste gate assembly

The compressor assembly (right side of Figure 1-9-6) is made up of an impeller, a diffuser and a casing. The air for the induction system enters through a circular opening in the center of the compressor casing, where it is picked up by the blades of the impeller, which gives it high velocity as it travels outward toward the diffuser. The diffuser vanes direct the airflow as it leaves the impeller and converts the high velocity of the air to high pressure.

Motive power for the impeller is furnished through the impeller's attachment to the turbine wheel shaft of the exhaust-gas turbine. This complete assembly is referred to as the rotor. The rotor revolves on the ball bearings at the rear end of the pump and bearing casing and the roller bearing at the turbine end. The roller bearing carries the radial (centrifugal) load of the rotor, and the ball bearing supports the rotor at the impeller end and bears the entire thrust (axial) load and part of the radial load.

The exhaust gas turbine assembly (left side of Figure 1-9-6) consists of the turbine wheel (bucket wheel), nozzle box, butterfly valve (waste gate), and cooling cap. The turbine wheel, driven by exhaust gases, drives the impeller. The nozzle box collects and directs the exhaust gases onto the turbine wheel, and

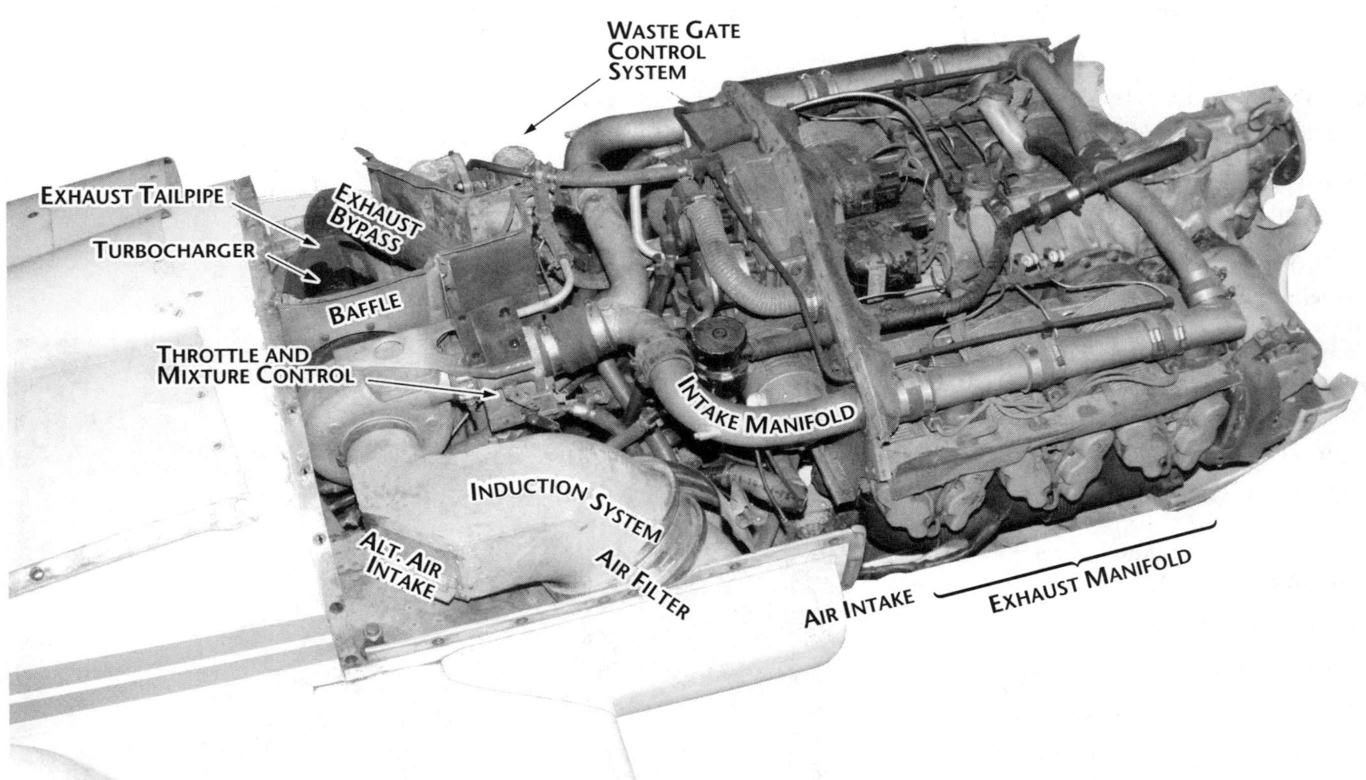


Figure 1-9-8. Turbocharger induction and exhaust system

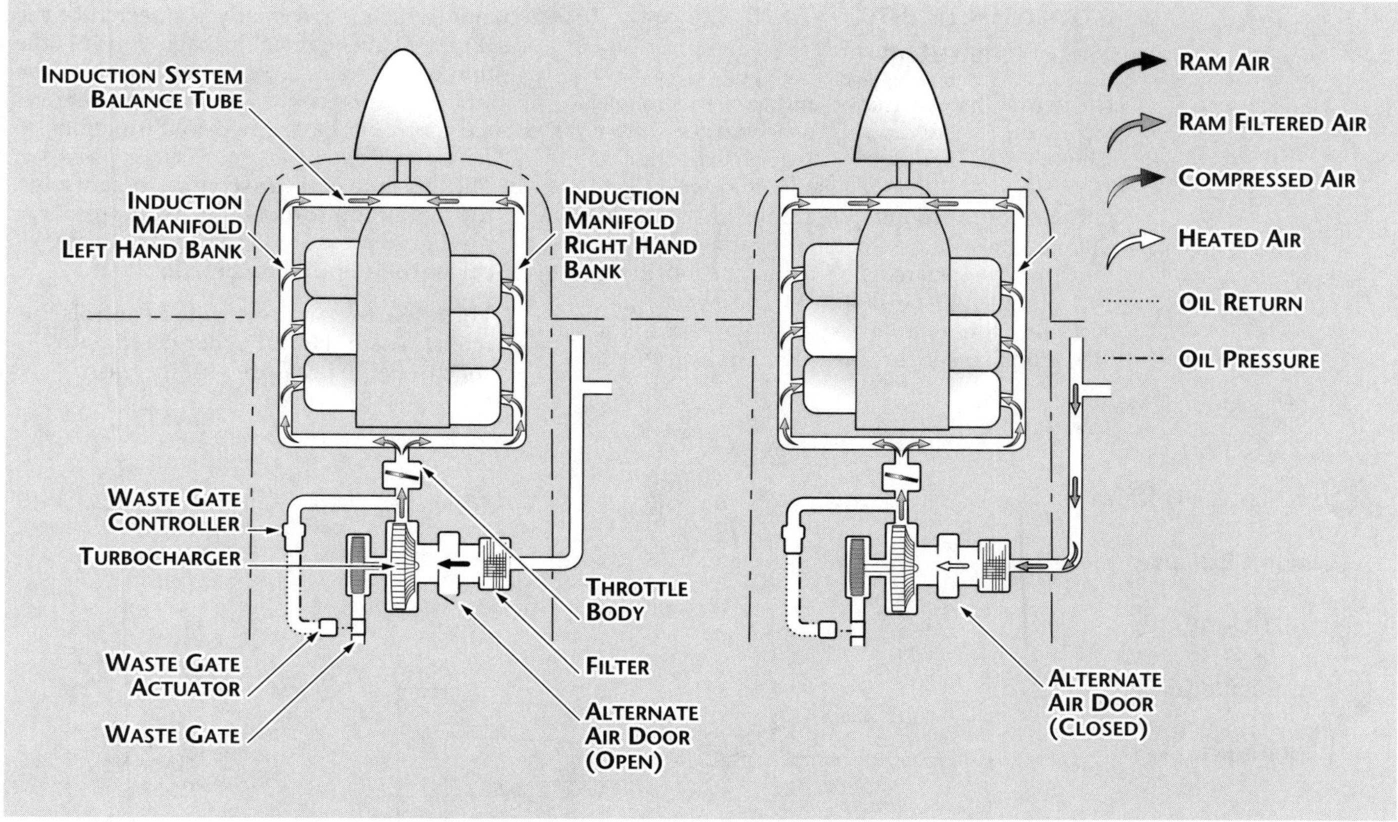


**Figure 1-9-9. Induction air system schematic**

the waste gate regulates the amount of exhaust gases directed to the turbine by the nozzle box. The cooling cap controls a flow of air for turbine cooling.

The waste gate (Figure 1-9-7) controls the volume of the exhaust gas that is directed onto the turbine and thereby regulates the speed of the rotor (turbine and impeller).

If the waste gate is completely closed, all the exhaust gases are "backed up" and forced through the nozzle box and turbine wheel. If the waste gate is partially closed, a corresponding amount of exhaust gas is directed to the turbine. The exhaust gases, thus directed, strike the outer edge of the turbine, and cause the rotor (turbine and impeller) to rotate. The gases are then exhausted overboard through the spaces between the buckets. When the waste gate is fully open, nearly all of the exhaust gases pass overboard through the tailpipe.

## Turbocharger Systems

An increasing number of engines used in light aircraft are equipped with externally driven supercharger systems. These superchargers are powered by the energy of exhaust gases and are usually referred to as *turbocharger systems* rather than *turbo-superchargers.*

On many small aircraft engines, the turbocharger system is designed to be operated only above a certain altitude, for example, 5,000 ft., since maximum power without supercharging is available below that altitude.

The location of the air induction and exhaust systems of a typical turbocharger system for a small aircraft is shown in Figure 1-9-8 .

## Induction Air System

The induction air system shown in Figure 1-9-9 consists of a filtered ram-air intake located on the side of the nacelle. An alternate air door within the nacelle permits compressor suction to automatically admit alternate air (heated engine compartment air) if the induction filter becomes clogged. The alternate air door can be operated manually in the event of filter clogging. A separately mounted exhaust-driven turbocharger is included in each air induction system. The turbocharger is automatically controlled by a pressure controller, to maintain manifold pressure at approximately 34.5 inches Hg from sea level to the critical altitude (typically 16,000 ft.) regardless of temperature. The turbocharger is completely automatic, requiring no pilot action up to the critical altitude.

## Controllers and Waste Gate Actuator

The waste gate actuator and controllers use engine oil for power (see turbocharger system schematic in Figure 1-9-10). The turbocharger is controlled by the waste gate and waste gate actuator, an absolute pressure and a rate-of-change controller. A pressure ratio controller controls the waste gate actuator above the critical altitude (16,000 ft.). The waste gate bypasses the engine exhaust gases around the turbocharger turbine inlet. The waste gate actuator, which is physically connected to the waste gate by mechanical linkage, controls the position of the waste gate butterfly valve. The absolute pressure controller and the rate-of-change controller have a two-fold function:

1. The absolute pressure controller controls the maximum turbocharger compressor discharge pressure (34 ±5 in. Hg to critical altitude, approximately 16,000 ft.)
2. The rate-of-change controller controls the rate at which the turbocharger compressor discharge pressure will increase.

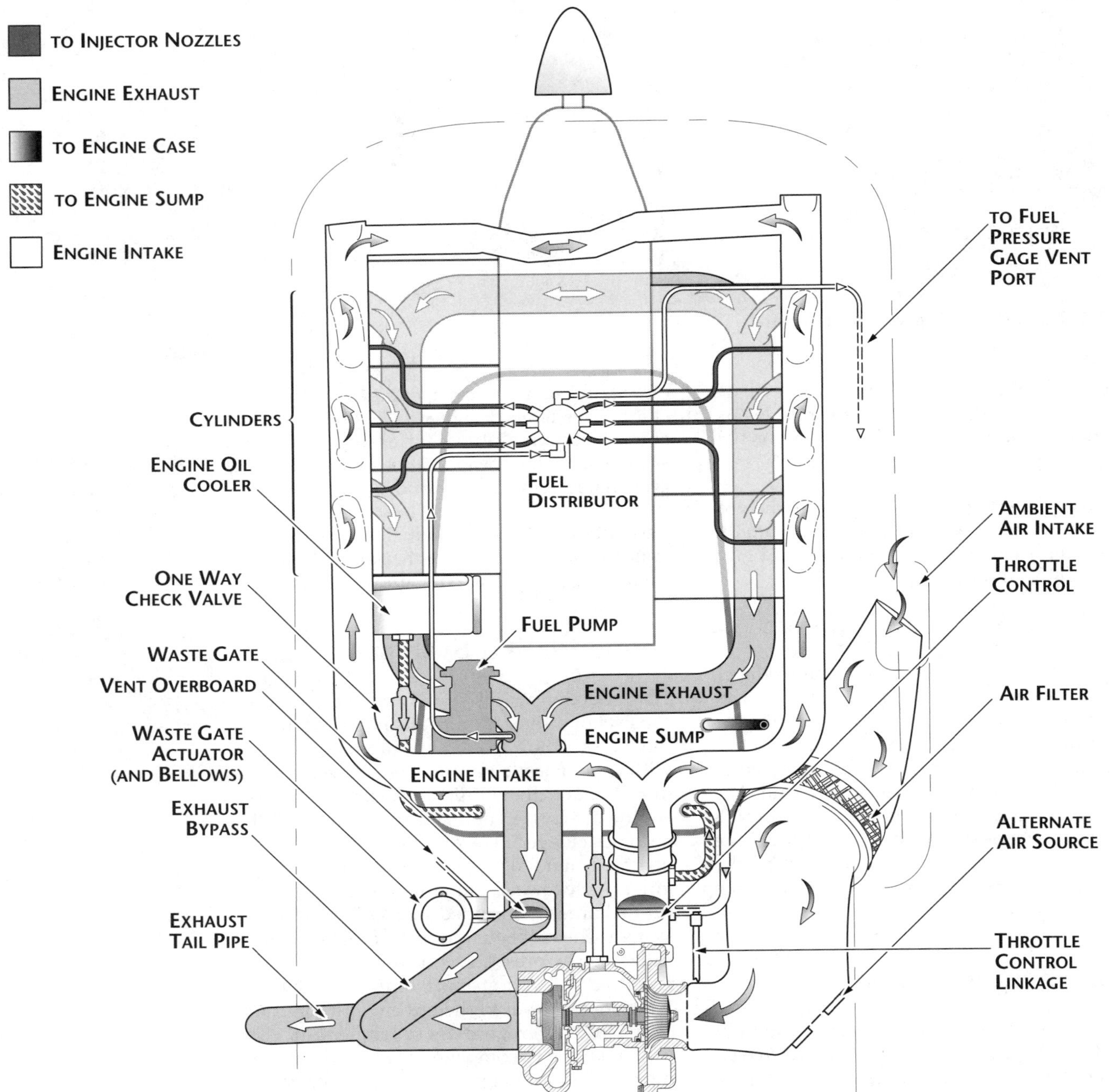


Figure 1-9-10. Schematic of a typical turbocharger system

## Sea-level Boosted Turbocharger System

Some turbocharger systems are designed to operate from sea level up to their critical altitude. These engines, sometimes referred to as sea-level boosted engines, can develop more power at sea level than an engine without turbocharging.

Figure 1-9-11 (next page) is a schematic of a sea-level boosted turbocharger system. This system is automatically regulated by three components shown in the schematic:

1. The exhaust bypass valve assembly
2. The density controller
3. The differential pressure controller

It should be noted that some turbocharger systems are not equipped with automatic control devices. They are similar in design and operation to the system shown in Figure 1-9-11 (next page), except that the turbocharger output is manually controlled.

By regulating the waste gate position and the fully open and closed positions (Figure 1-9-11, next page), a constant power output can be maintained. When the waste gate is fully open, all the exhaust gases are directed overboard to the atmosphere, and no air is compressed and delivered to the engine air inlet.

Conversely, when the waste gate is fully closed, a maximum volume of exhaust gases flow into the turbocharger turbine, and maximum supercharging is accomplished. Between these two extremes of waste gate position, constant power output can be achieved below the maximum altitude at which the system is designed to operate.

A critical altitude exists for every possible power setting below the maximum operating ceiling, and if the aircraft is flown above this altitude without a corresponding change in the power setting, the waste gate will be automatically driven to the fully closed position in an effort to maintain a constant power output. Thus, the waste gate will be almost fully open at sea level and will continue to move toward the closed position as the aircraft climbs, in order to maintain the pre-selected manifold pressure setting.

When the waste gate is fully closed (leaving only a small clearance to prevent sticking) the manifold pressure will begin to drop if the aircraft continues to climb. If a higher power setting cannot be selected, the turbocharger's critical altitude has been reached. Beyond this altitude, the power output will continue to decrease.

The position of the waste gate valve, which determines power output, is controlled by oil pressure. Engine oil pressure acts on a piston in the waste gate assembly that is connected by linkage to the waste gate valve. When oil pressure is increased on the piston, the waste gate valve moves toward the closed position, and engine output power increases.

Conversely, when the oil pressure is decreased, the waste gate valve moves toward the open position, and output power is decreased.

The position of the piston attached to the waste gate valve is dependent on bleed oil that controls the engine oil pressure applied to the top of the piston. Oil is returned to the engine crankcase through two control devices, the density controller and the differential pressure controller. These two controllers, acting independently, determine how much oil is bled back to the crankcase, and thus establish the oil pressure on the piston.

The density controller is designed to limit the manifold pressure below the turbocharger's critical altitude, and regulates bleed oil only at the full throttle position. The pressure and temperature sensing bellows of the density controller react to pressure and temperature changes between the fuel injector inlet and the turbocharger compressor. The bellows, filled with dry nitrogen, maintains a constant density by allowing the pressure to increase as the temperature increases. Movement of the bellows re-positions the bleed valve, causing a change in the quantity of bleed oil, which changes the oil pressure on top of the waste gate piston (see Figure 1-9-11, next page).

The differential pressure controller functions during all positions of the waste gate valve other than the fully open position, which is controlled by the density controller. One side of the diaphragm in the differential pressure controller senses air pressure upstream from the throttle; the other side samples pressure on the cylinder side of the throttle valve (Figure 1-9-11, next page). At the wide open throttle position, when the density controller controls the waste gate, the pressure across the differential pressure controller diaphragm is at a minimum and the controller spring holds the bleed valve closed. At part throttle position, the air differential is increased, opening the bleed valve to bleed oil to the engine crankcase and re-position the waste gate piston.

Thus, the two controllers operate independently to control turbocharger operation at all positions of the throttle. Without the overriding function of the differential pressure controller during part-throttle operation, the density controller would position the waste gate valve for maximum power. The differential pressure

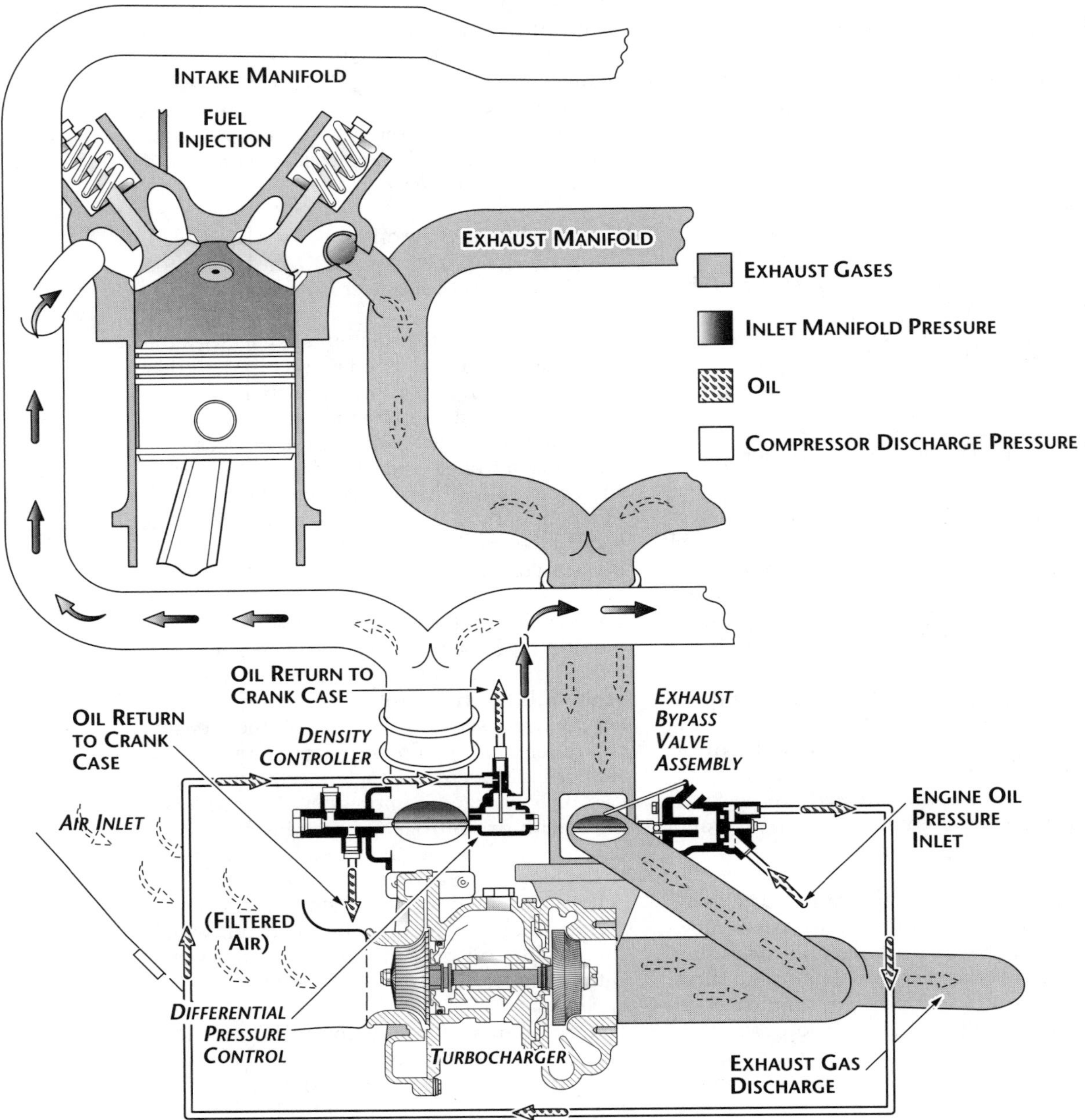


**Figure 1-9-11. Turbocharger controllers and actuator system schematic**

controller reduces injector entrance pressure and continually re-positions the valve over the whole operating range of the engine.

The differential pressure controller reduces the unstable condition known as *bootstrapping* during part-throttle operation. Bootstrapping is an indication of unregulated power change that results in the continual drift of manifold pressure. This condition can be illustrated by considering the operation of a system when the waste gate is fully closed. During this time, the differential pressure controller is not modulating the waste gate valve position. Any slight change in power caused by a change in temperature or r.p.m. fluctuation will be magnified and will result in manifold pressure change since the slight change will cause a change in the amount of exhaust gas flowing to the turbine. Any change in exhaust gas flow to the turbine will cause a change in power output that will be reflected in manifold pressure indications. Bootstrapping, then, is an undesirable cycle of turbocharging events, causing the manifold pressure to drift in an attempt to reach a state of equilibrium.

*Bootstrapping* is sometimes confused with the condition known as *overboost,* but bootstrapping is not a condition that is detrimental to engine life. An overboost condition is one in which manifold pressure exceeds the limits prescribed for a particular engine and can cause serious damage.

Thus, the differential pressure controller is essential to smooth functioning of the automatically controlled turbocharger, since it reduces bootstrapping by reducing the time required to bring a system into equilibrium. There is still a great deal more throttle sensitivity with a turbocharged engine than with a naturally aspirated engine. Rapid movement of the throttle can cause a certain amount of manifold pressure drift in a turbocharged engine. This condition, less severe than bootstrapping, is called *overshoot*. While overshoot is not a dangerous condition, it can be a source of concern to the pilot or operator who selects a particular manifold pressure setting only to find it has changed in a few seconds and must be reset. Since the automatic controls cannot respond rapidly enough to abrupt changes in throttle settings to eliminate the inertia of turbocharger speed changes, overshoot must be controlled by the operator. This can best be accomplished by slowly making changes in throttle setting, accompanied by a few seconds wait for the system to reach a new equilibrium. Such a procedure is effective with turbocharged engines, regardless of the degree of throttle sensitivity.

## Turbocharger System Troubleshooting

Table 1-9-2 includes some of the most common turbocharger system malfunctions, together with their cause and repair. These troubleshooting procedures are presented as a guide only and should not be substituted for applicable manufacturer's instructions or troubleshooting procedures.

| TROUBLESHOOTING TURBOCHARGER SYSTEMS | | |
|---|---|---|
| SYMPTOM/TROUBLE | PROBABLE CAUSE | REMEDY |
| *Aircraft fails to reach critical altitude* | Damaged compressor or turbine wheel | Replace turbocharger |
| | Exhaust system leaks | Repair leaks |
| | Faulty turbocharger bearings | Replace turbocharger |
| | Waste gate will not close fully | Refer to *Waste gate* entries in the SYMPTOM/TROUBLE column |
| | Malfunctioning controller | Refer to *Differential controller malfunctions* in the SYMPTOM/TROUBLE column |
| *Engine surges* | Bootstrapping | Ensure engine is operated in proper range |
| | Waste gate malfunction | Refer to *Waste gate* entries in the SYMPTOM/TROUBLE column |
| | Controller malfunction | Refer to *Differential controller malfunctions* in the SYMPTOM/TROUBLE column |
| *Waste gate will not close fully* | Waste gate bypass valve bearings tight | Replace bypass valve |
| | Oil inlet orifice blocked | Clean orifice |
| | Controller malfunction | Refer to *Differential controller malfunctions* and *Density controller malfunctions* in the SYMPTOM/TROUBLE column |
| *Differential controller malfunctions* | Seals leaking | Replace controller |
| | Diaphragm broken | Replace controller |
| | Controller valve stuck | Replace controller |
| *Density controller malfunctions* | Seals leaking | Replace controller |
| | Bellows damaged | Replace controller |
| | Valve stuck | Replace controller |

**Table 1-9-2. Turbocharger-system troubleshooting chart**

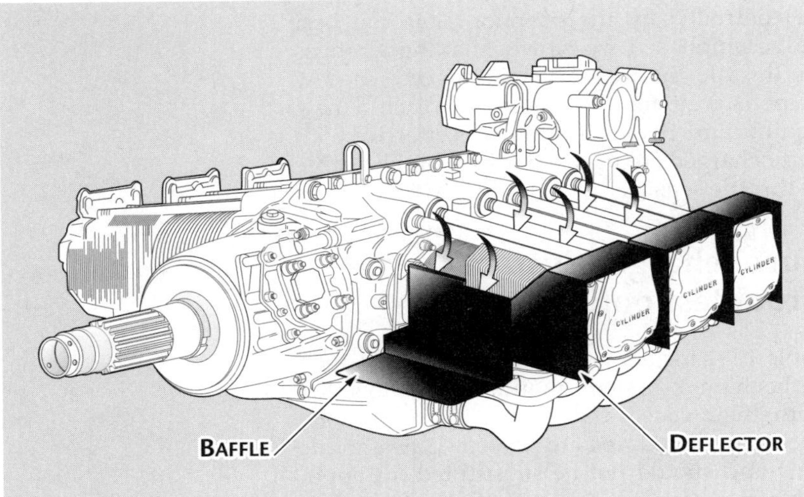


Figure 1-10-1. Cylinder baffle and deflector system

## The Induction Manifold

The induction manifold provides the means of distributing air, or the fuel/air mixture, to the cylinders. Whether the manifold handles a fuel/air mixture or air alone depends on the type of fuel metering system used. On an engine equipped with a carburetor, the induction manifold distributes a fuel/air mixture from the carburetor to the cylinders. On a fuel-injection engine, the fuel is delivered to injection nozzles, one in each cylinder, which provides the proper spray pattern for efficient burning. Thus, the mixing of fuel and air takes place in the cylinders or at the inlet port to the cylinder. On a fuel-injected engine the induction manifold handles only air.

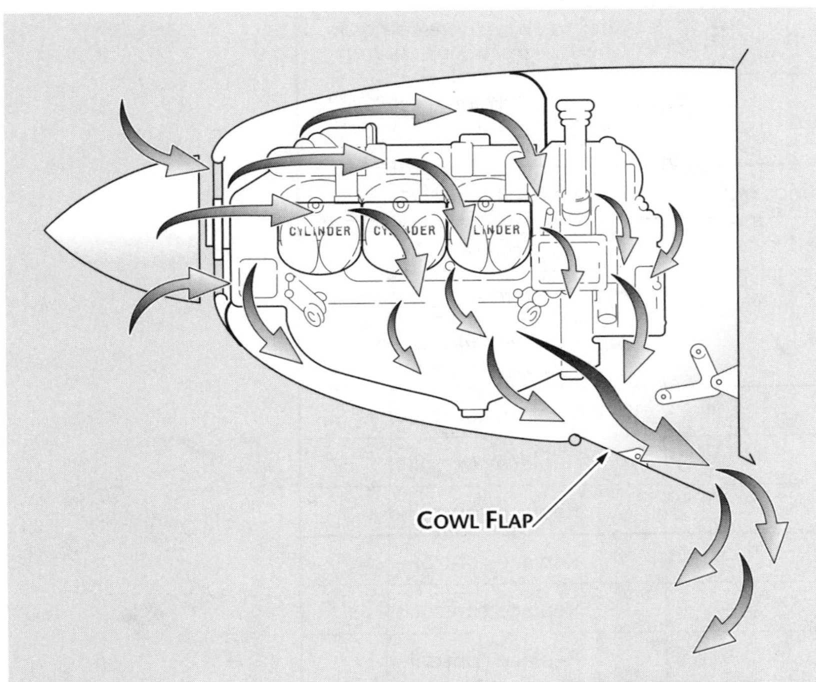


Figure 1-10-2. Regulating the cooling airflow

The induction manifold is an important item because of the effect it can have on the fuel/air mixture which finally reaches the cylinder. Fuel is introduced into the air stream by the carburetor in a liquid form. To become combustible, the fuel must be vaporized in the air. This vaporization takes place in the induction manifold, which includes the internal supercharger if one is used. Any fuel that does not vaporize will cling to the walls of the intake pipes. Obviously, this affects the effective fuel/air ratio of the mixture that finally reaches the cylinder in vapor form.

This explains the reason for the apparently rich mixture required to start a cold engine. In a cold engine, some of the fuel in the air stream condenses out and clings to the walls of the manifold. This is in addition to that fuel that never vaporized in the first place. As the engine warms up, less fuel is required because less fuel is condensed out of the air stream and more of the fuel is vaporized, thus giving the cylinder the required fuel/air mixture for normal combustion.

Any leak in the induction system has an effect on the mixture reaching the cylinders. This is particularly true of a leak at the cylinder end of an intake pipe. At manifold pressures below atmospheric pressure, such a leak will lean out the mixture. This occurs because additional air is drawn in from the atmosphere at the leaky point. The affected cylinder may overheat, fire intermittently or even cut out altogether.

*Section 10*

# *Reciprocating Engine Cooling Systems*

An internal-combustion engine is a heat machine that converts chemical energy in the fuel into mechanical energy at the crankshaft. It does not do this without some loss of energy, however, and even the most efficient aircraft engines may waste 60-70 percent of the original energy in the fuel. Unless most of this waste heat is rapidly removed, the cylinders may become hot enough to cause complete engine failure.

Excessive heat is undesirable in any internal-combustion engine for three principal reasons:

1. It affects the behavior of the combustion of the fuel/air charge
2. It weakens and shortens the life of engine parts
3. It impairs lubrication

If the temperature inside the engine cylinder is too great, the fuel/air mixture will be preheated, and combustion will occur before the desired time. Since premature combustion causes detonation, knocking, and other undesirable conditions, there must be a way to eliminate heat before it causes damage.

One gallon of aviation gasoline has enough heat value to boil 75 gallons of water. It is easy to see that an engine that burns one gallon of fuel per minute releases a tremendous amount of heat. About one-fourth of the heat released is changed into useful power. The remainder of the heat must be dissipated so that it will not be destructive to the engine. In a typical aircraft powerplant, half of the heat goes out with the exhaust, and the other half is absorbed by the engine. Circulating oil picks up part of this soaked-in heat and transfers it to the air stream through the oil cooler. The engine cooling system takes care of the rest.

Cooling is a matter of transferring the excess heat from the cylinders to the air, but there is more to such a job than just placing the cylinders in the air stream.

**Cylinder fins.** The surface area of the cylinder is increased by use of cooling fins. This increased surface area increases the efficiency of heat transfer by radiation and convention. If an excessive amount of the cooling fin is broken off, the cylinder will not properly cool and a hot spot will occur. Cylinders should be replaced when an excessive amount of cylinder fin damage occurs.

**Deflectors and baffles.** Deflectors and baffles are used to force air over the cylinder fins. This action results in the air carrying heat away from the cooling fins. Blast tubes are attached to the baffles to direct streams of cooling air to hot spots such as ignition leads and engine accessories (see Figure 1-10-1).

To provide for proper operating temperatures of the engine under all conditions of flight ,devices must be used regulate the temperature.

**Cowl flaps.** One of the most common methods of regulating engine temperature is the use of cowl flaps as illustrated in Figure 1-10-2.

Cowl flaps are used to control temperature by regulating the amount of air that passes over the engine. If the engine temperature is approaching red line, the cowl flaps should be opened. This will increase the amount of airflow over the engine, which will decrease its temperature. If the engine temperature is too low, the opening of the cowl flaps is reduced to decrease the amount of air flowing over the engine. The cowl flaps are controlled from inside the cockpit by either mechanical linkage, an electrical motor or hydraulic power.

**Augmentors.** Some aircraft use augmentors (Figure 1-10-3) to provide additional cooling airflow. Each nacelle has two pairs of tubes running from the engine compartment to the rear of the nacelle. The exhaust collectors feed exhaust gas into the inner augmenter tubes. The exhaust gas mixes with air that has passed over the engine and heats it to form a high-temperature, low-pressure, jet-like exhaust. This low-pressure area in the augmentors draws additional cooling air over the engine. Air entering the outer shells of the augmentors is heated through contact with the augmenter tubes but is not contaminated

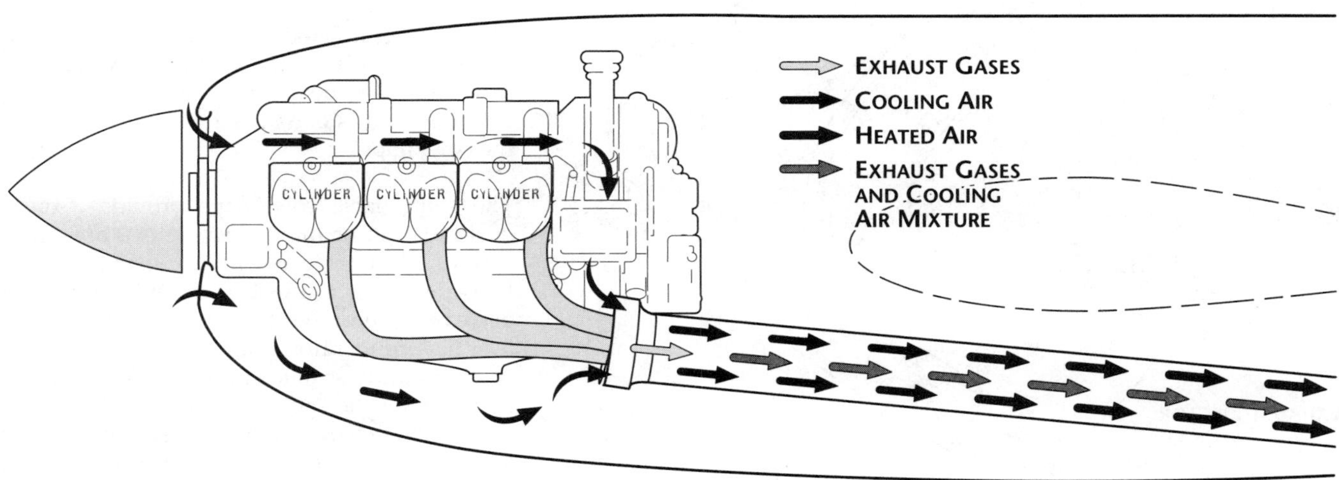


Figure 1-10-3. Augmentor

with exhaust gases. The heated air from the shell goes to the cabin heating, defrosting, and anti-icing system.

Augmentors use exhaust gas velocity to cause an airflow over the engine so that cooling is not entirely dependent on the prop wash. Vanes installed in the augmentors control the volume of air. These vanes usually are left in the trail position to permit maximum flow. They can be closed to increase the heat for cabin or anti-icing use or to prevent the engine from cooling too much during descent from altitude. In addition to augmentors, some aircraft have residual heat doors or nacelle flaps that are used mainly to let the retained heat escape after engine shutdown. The nacelle flaps can be opened for more cooling than that provided by the augmentors.

A modified form of the previously described augmenter cooling system is used on some light aircraft. Figure 1-10-4 is an outline diagram of such a system.

As shown in Figure 1-10-4, the engine is pressure-cooled by air taken in through two openings in the nose cowling, one on each side of the propeller spinner. A pressure chamber is sealed off on the top side of the engine with baffles properly directing the flow of cooling air to all parts of the engine compartment. Warm air is drawn from the lower part of the engine compartment by the pumping action of the exhaust gases through the exhaust ejectors. This type of cooling system eliminates the use of controllable cowl flaps and assures adequate engine cooling at all operating speeds.

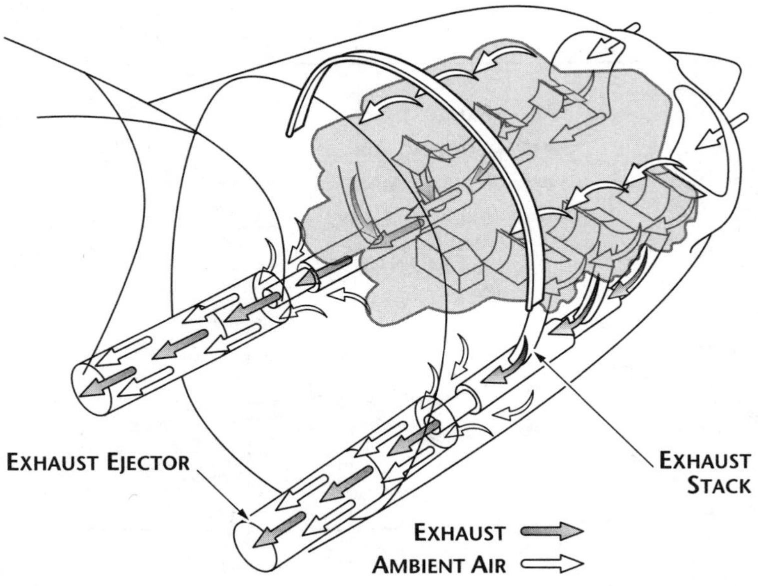


Figure 1-10-4. Engine cooling and exhaust system

Many light aircraft use only one or two engine cowl flaps to control engine temperature. As shown in Figure 1-10-5, two cowl flaps, operated by a single control in the cabin, are located at the lower aft end of the engine nacelle. Cutouts in the flaps permit extension of engine exhaust stacks through the nacelle. The flaps are operated by a manual control in the cockpit to control the flow of air directed by baffles around the cylinders and other engine components.

Some small aircraft that have horizontally opposed engines use this type of cowling. The cowl flaps are controlled by an electrically operated gill-type flaps on the trailing edge of each cowling.

## Cooling-system Maintenance

The engine cooling system of most reciprocating engines usually consists of the engine cowling, cylinder baffles, cylinder fins, and some type of cowl flaps. In addition to these major units, there is also some type of temperature-indicating system (cylinder head temperature).

The cowling performs two functions:

1. It streamlines the bulky engine to reduce drag.
2. It forms an envelope around the engine, which forces air to pass around and between the cylinders, absorbing the heat dissipated by the cylinder fins.

The cylinder baffles are metal shields, designed and arranged to direct the flow of air evenly around all cylinders. This even distribution of air aids in preventing one or more cylinders from being excessively hotter than the rest.

The cylinder fins radiate heat from the cylinder walls and heads. As the air passes over the fins, it absorbs this heat, carries it away from the cylinder, and is exhausted overboard through the cowl flaps.

The controllable cowl flaps provide a means of decreasing or increasing the exit area at the rear of the engine cowling. Closing the cowl flaps decreases the exit area, which effectively decreases the amount of air that can circulate over the cylinder fins. The decreased airflow cannot carry away as much heat; therefore, there is a tendency for the engine temperature to increase. Opening the cowl flaps makes the exit area larger. The flow of cooling air over the cylinders increases, absorbing more heat, and the tendency is then for the engine temperature to decrease. Good inspection and maintenance in the care of the engine cooling system will aid in overall efficient and economical engine operation.

Figure 1-10-5. Small aircraft cowl flaps

## Inspection of Cowling

The cowling support structure and skin panels should be inspected during each engine and airframe inspection. This should include inspecting for scratches, dents, cracks, missing or loose fasteners as well as tears on the skin. The cowling latches should also be inspected for proper operation and loose fasteners.

When inspecting the cowling the cowl flaps should also be inspected for proper operation and adjustment. Check the hinge pins for tightness and lubricate according to the aircraft manufacturer's maintenance instructions.

Repair and replacement of cowling components should be made in accordance with the aircraft manufacturer's repair instructions.

## Engine Cylinder Cooling Fin Inspection

The cooling fins are of the utmost importance to the cooling system, since they provide a means of transferring the cylinder heat to the air. Their condition can mean the difference between adequate or inadequate cylinder cooling. The fins are inspected at each regular inspection.

Fin area is the total area (both sides of the fin) exposed to the air. During the inspection, the fins should be examined for cracks and breaks (Figure 1-10-6). Small cracks are not a reason for cylinder removal. These cracks can be filed, or even sometimes stop-drilled to prevent any further cracking. Rough or sharp corners on fins can be smoothed out by filing, and this action will eliminate a possible source of new cracks. However, before re-profiling cylinder cooling fins, consult the manufacturer's service or overhaul manual.

Figure 1-10-6. Cracked cylinder fins

The definition of fin area becomes important in the examination of fins for broken areas. It is a determining factor for cylinder acceptance or removal. For example, on a certain engine, if more than 12 inches in length of any one fin, as measured at its base, is completely broken off, or if the total fins broken on any one cylinder head exceed 83 sq. inches of area, the cylinder is removed and replaced. The reason for removal in this case is that an area of that size would cause a hot spot on the cylinder, since very little heat transfer could occur.

Where adjacent fins are broken in the same area, the total length of breakage permissible is 6 inches on any two adjacent fins, 4 inches on any three adjacent fins, 2 inches on any four adjacent fins, and 1 inch on any five adjacent fins. If the breakage length in adjacent fins exceeds this prescribed amount, the cylinder should be removed and replaced. These breakage specifications are applicable only to the engine used in this discussion as a typical example. In each specific case, applicable manufacturer's instructions should be consulted.

## Cylinder Baffle and Deflector System Inspection

Reciprocating engines use some type of intercylinder and cylinder head baffles to force the cooling air into close contact with all parts of the cylinders. Figure 1-10-1 shows a baffle and deflector system around a cylinder. The air baffle blocks the flow of air and forces it to circulate between the cylinder and the deflectors.

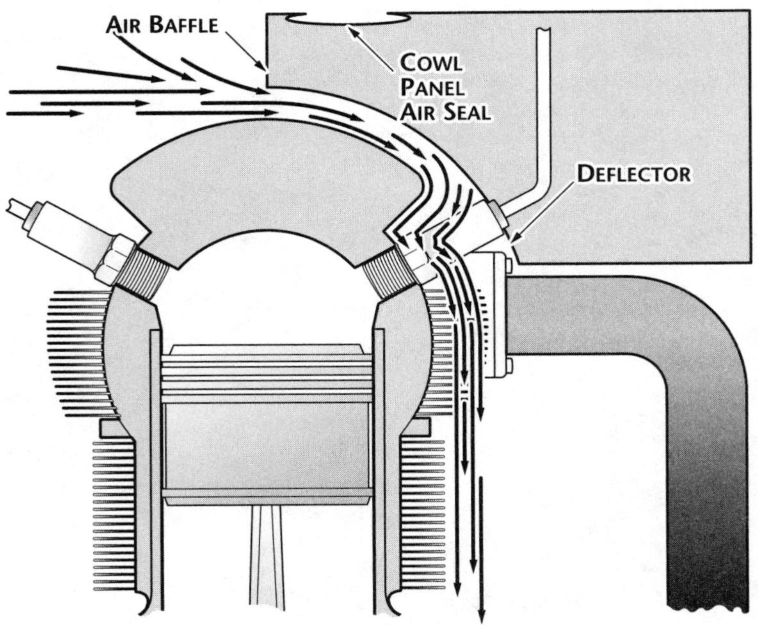


Figure 1-10-7. Cylinder head baffle and deflector system

Figure 1-10-7 illustrates a baffle and deflector arrangement designed to cool the cylinder head. The air baffle prevents the air from passing away from the cylinder head and forces it to go between the head and deflector.

Although the resistance offered by baffles to the passage of the cooling air demands that an appreciable pressure differential be maintained across the engine to obtain the necessary airflow, the volume of cooling air required is greatly reduced by employing properly designed and located cylinder deflectors. The airflow approaches the nacelle and piles up at the face of the engine, creating a high pressure in front of the cylinders. This piling up of the air reduces the air velocity. The cowl flaps, of course, produce the low-pressure area. As the air nears the cowl flap exit, it is speeded up again and merges smoothly with the air stream. The pressure differential between the front and the rear of the engine forces the air past the cylinders through the passages formed by the deflectors.

The baffles and deflectors should not only be inspected during regular engine inspections but also at anytime the engine cowling is removed. They should be inspected for cracks, dents, and loose mounting fasteners. If a crack or dent is found, the piece should either be repaired or replaced.

## Cowl Flap Installation and Adjustment

Since the cowl flaps are the most important components that effects engine cooling, the technician must use extreme care in their installation, adjustment and inspection. The technician must make certain that the open and closed tolerances of the cowl flaps are correct. If the cowl flaps are not allowed to close tightly, the engine temperature will be too low when the engine is warming up. If the cowl flaps are not allowed to open fully the engine may run to high power settings.

It is important to install the cowl flaps correctly, adjust the jackscrews, adjust the open and close limit switches and inspect the system.

The following checks and inspections are typical of those made to maintain an efficient cowl flap system:

- Check the cowl flaps for response by actuating the cockpit control from the open to the closed, and back to the open position. The flaps should respond rapidly and smoothly. If a cowl flap indicator is installed, observe the indications received for synchronization with the flaps in the open and closed positions.

- With the cowl flaps open, check for cracks, distortion or security of mounting. Grasp the flap at the trailing edge, and shake laterally and up and down to determine the condition of the bushings, bearings or the turnbuckles. Looseness of the flaps during this check indicates worn bushings or bearings that should be replaced. Inspect the hinges and hinge terminals for wear, breaks or cracks; check the hinges for security of mounting on the cowl support.

Measure the open and closed positions of the cowl flaps to check for specified tolerances and adjust as necessary.

## Cylinder Temperature Indicating System

This system usually consists of an indicator, electrical wiring, and a thermocouple. The wiring is between the instrument and the nacelle firewall. At the firewall, one end of the thermocouple leads connects to the electrical wiring, and the other end of the thermocouple leads connects to the cylinder.

The thermocouple consists of two dissimilar metals, generally constantan and iron, connected by wiring to an indicating system. If the temperature of the junction is different from the temperature where the dissimilar metals are connected to wires, a voltage is produced. This voltage sends a current through wires to the indicator, a current-measuring instrument graduated in degrees.

The thermocouple end that connects to the cylinder is either the bayonet or gasket type. To install the bayonet type, the knurled nut is pushed down and turned clockwise until it is snug. In removing this type, the nut is pushed down and turned counterclockwise until released. The gasket type fits under the spark plug and replaces the normal spark plug gasket (see Figure 1-10-8).

When installing a thermocouple lead, remember not to cut off the lead because it is too long, but coil and tie up the excess length. The thermocouple is designed to produce a given amount of resistance. If the length of the lead is reduced, an incorrect temperature reading will result. The bayonet, or gasket, of the thermocouple is inserted or installed on the hottest block-tested cylinder of the engine.

When the thermocouple is installed and the wiring connected to the instrument, the indicated reading is the cylinder temperature. On a cold engine, the cylinder bead temperature indicator will indicate the free outside air temperature. That is one test for determining that the instrument is working correctly.

The cover glass of the cylinder head temperature indicator should be checked regularly to see that it has not slipped or that it is not cracked. The cover glass should be checked for indications of missing or damaged decals that indicate temperature limitations. If the thermocouple leads were excessive in length and need to be coiled and tied down, the tie should be inspected for security or chafing of the wire. The bayonet or gasket should be inspected for cleanness and security of mounting. When operating the engine, if the cylinder head temperature pointer fluctuates, all the electrical connections could be checked.

*Section 11*

## *Reciprocating Engine Exhaust Systems*

The reciprocating engine exhaust system is fundamentally a scavenging system that collects and disposes of the high-temperature noxious gases as they are discharged by the engine. Its basic requirement is to dispose of the gases with complete safety to the airframe and the occupants of the aircraft. The exhaust system can perform many useful functions, but its first duty is to provide protection against the potentially destructive action of the exhaust gases. Modern exhaust systems, though comparatively light, adequately resist high temperatures, corrosion, and vibrations. They generally provide long, trouble-free operation with a minimum of maintenance.

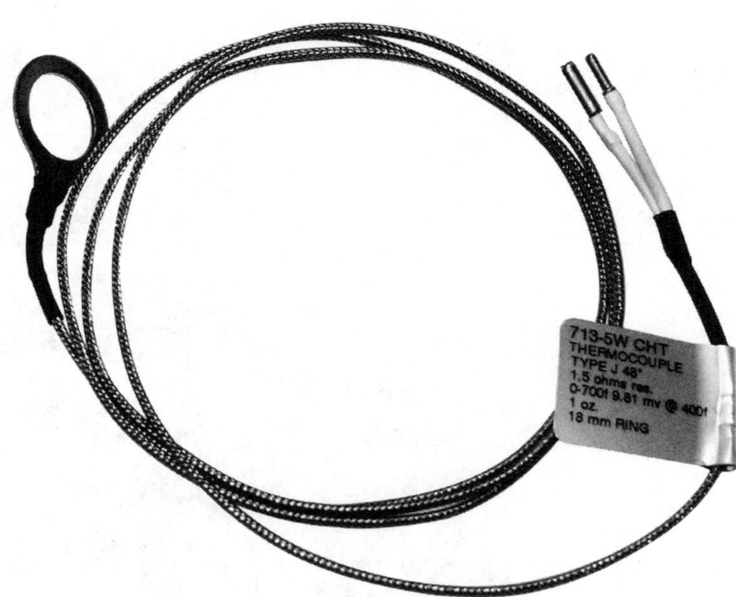


**Figure 1-10-8. Cylinder head temperature thermocouple replaces the spark plug gasket**

There are two general types of exhaust systems in use on reciprocating aircraft engines: the *straight stack* (open) system and the *collector system*. The straight stack system is generally used on non-supercharged engines and low-powered engines where noise level is not too objectionable. The collector system is used on most large non-supercharged engines and on all turbo-supercharged engines and installations where it would improve nacelle streamlining or provide easier maintenance in the nacelle area. On turbo-supercharged engines, the exhaust gases must be collected to drive the turbine compressor of the supercharger. Such systems have individual exhaust headers that empty into a common collector ring with only one outlet. From this outlet, the hot exhaust gas is routed via a tailpipe to the nozzle box of the turbo-supercharger to drive the turbine.

**Figure 1-11-1. A classic example of straight stacks on an Allison engine**

**Figure 1-11-2. An example of a radial engine exhaust collector ring**

Although the collector system raises the back pressure of the exhaust system, the gain in horsepower from turbo-supercharging more than offsets the loss in horsepower that results from increased back pressure.

**Straight stack.** The straight stack system is relatively simple. It consists of a short length of exhaust tubing (header pipe) connected to the exhaust port of each cylinder by high temperature lock nuts (see Figure 1-11-1). Each of these stacks dumps exhaust gas overboard without being connected to a common manifold. The removal and installation of this type of system consists essentially of removing and installing the hold down nuts and clamps.

**Exhaust collector.** While the radial engines normally use an exhaust collector, commonly called an exhaust collector ring (see Figure 1-11-2), opposed-type reciprocating engines often use a manifold-type exhaust system. The manifold exhaust system can be considered a collector type of system because the exhaust gases are collected in a common manifold.

The manifold system uses a short length of exhaust pipe connected to the exhaust port of each cylinder with high temperature lock nuts. These individual stacks empty exhaust gases into a manifold, which carries the exhaust gases overboard. Most opposed-type engines have a manifold for each side of the engine (or for each cylinder bank), but there are opposed engines that empty the exhaust gases from all of the cylinders into a common manifold.

Figure 1-11-3 shows the location of the exhaust system components of a typical horizontally opposed engine. The exhaust system in this installation consists of a down-stack from each cylinder, an exhaust collector tube on each side of the engine, and an exhaust ejector assembly protruding aft and down from the firewall. The down-stacks are connected to the cylinders with high-temperature locknuts and secured to the exhaust collector tube by ring clamps.

A cabin heater exhaust shroud is installed around each collector tube (see Figure 1-11-4).

**Exhaust augmentors.** Some reciprocating exhaust systems include exhaust augmentors. An exhaust augmenter is a length of large diameter heat- and corrosion- resistant tubing constructed with a bell mouth. The diameter of the augmenter is large enough to have the exhaust system tubing empty into it and still have sufficient room for ram air to be carried through it. The exhaust gases from the engine are carried to the augmenter, either by the individual stacks or a common manifold. The exhaust gases are emptied into the bellmouth, carried through the augmenter and emptied overboard by the ram air flowing through the augmenter.

The bellmouth design of the augmenter creates a venturi effect that increases the flow of air through the augmenter. Because most augmentors are located inside of the engine cowling, this venturi effect provides a more efficient method of engine cooling.

Augmenter tubes may contain a controllable vane (butterfly valve) that is controlled by the pilot in the cockpit. By controlling the size of the opening of the controllable vane, the pilot can regulate the operating temperature of the engine. The controllable vane can only be closed to obstruct approximately 45 percent of the augmenter tube. This prevents excessive exhaust back-pressure, which would cause the engine to lose power.

**Augmenter inspection.** On exhaust systems equipped with augmenter tubes, the augmenter tubes should be inspected at regular intervals for proper alignment, security of attachment, and general overall condition. Even where augmenter tubes do not contain heat exchanger surfaces, they should be inspected for cracks along with the remainder of the exhaust system. Cracks in augmenter tubes can present a fire or carbon monoxide hazard by allowing exhaust gases to enter the nacelle, wing or cabin areas.

**Mufflers.** Many reciprocating engine exhaust systems contain one or more mufflers to reduce the noise level of the engine. The mufflers used in reciprocating engines are normally a large diameter section of heat- and corrosion- resistant steel with baffles attached to the inside (see Figure 1-11-5).

The baffles on the inside of the muffler are subjected to a large amount of heat as the exhaust gases pass over them. Because this great amount of heat may cause erosion of the baffles, it is very important to inspect the internal condition of the muffler periodically. If the baffles should fail and be broken off, they may cause an exhaust system blockage, which could cause the engine to fail.

**Heat exchangers.** The great amount of heat that leaves the engine through the exhaust system is often used to provide heat for the cabin of the aircraft (see Figure 1-11-6, next page).

Heat exchangers are used as a method of gathering the heat from the exhaust system. Remember that the engine exhaust carries noxious gases that must not be allowed to enter the cockpit. The heat exchangers only harness the heat of the exhaust system, not the exhaust gases.

The heat exchanger is a shroud that covers a portion of the exhaust system tubing, normally around the muffler or a Y in the exhaust manifold. The shroud has fresh ram air, from

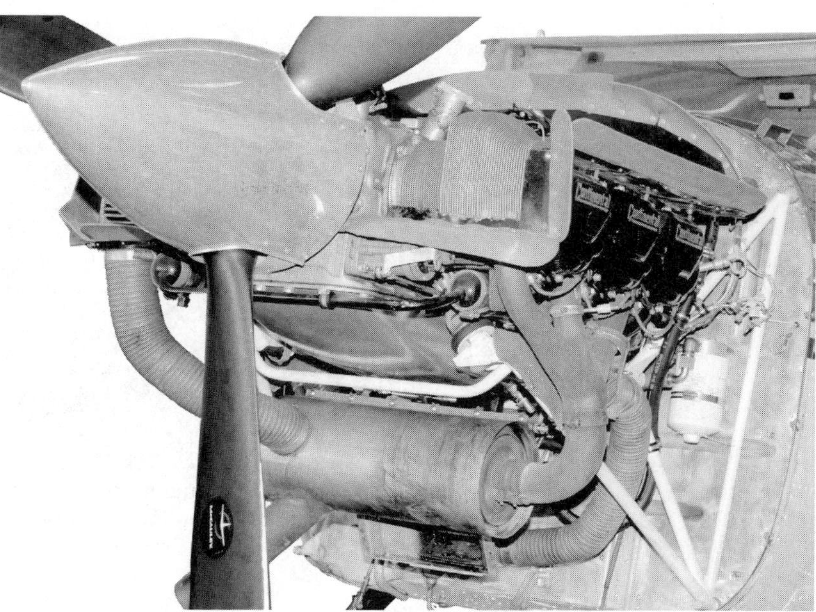

Figure 1-11-3. Exhaust system of a horizontally opposed engine

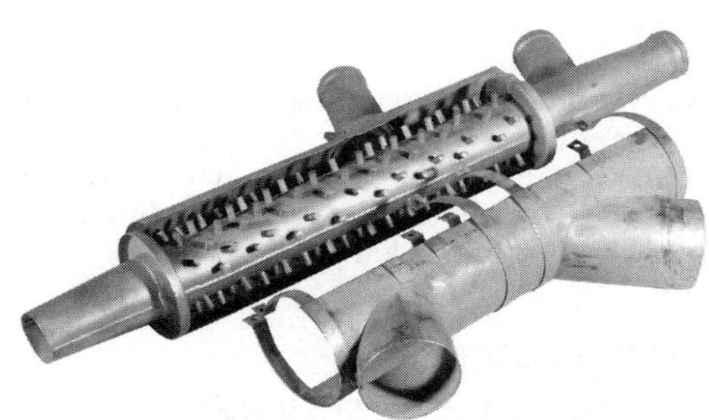

Figure 1-11-4. A disassembled heater exhaust shroud

Figure 1-11-5. A muffler without the heat exchanger wrap

**Figure 1-11-6. The heat exchanger shroud, or heat muff, shows the flexible ducting that goes to the cabin heater valve on the firewall.**

a source outside the cowling, flowing through it. As the air flows through the heat exchanger, it is heated by the hot exhaust system tubing. This hot air can then be routed into the cabin to provide *cabin heat*.

Heat exchangers are also used in many reciprocating engine aircraft to provide a source or hot air to be routed to the engine induction system. This hot air can be used to warm the intake air to prevent carburetor icing.

## Exhaust System Maintenance Practices

An exhaust system failure is considered a severe hazard because of the possibility of an engine fire or carbon monoxide poisoning. An exhaust system failure may also result in partial or complete engine power loss. Most engine exhaust system failures occur within 400 hours of engine operation, and reach a maximum rate of occurrence at between 100 and 200 hours of operation.

**Exhaust system inspection.** While the type and location of exhaust system components vary somewhat with the type of aircraft, the inspection requirements for most reciprocating engine exhaust systems are very similar. The following paragraphs include a discussion of the most common exhaust system inspection items and procedures for all reciprocating engines. Figure 1-11-7 shows the primary inspection areas of three types of exhaust systems.

Before the removal and installation of representative exhaust systems are discussed, a precaution to be observed when performing maintenance on any exhaust system should be mentioned. Galvanized or zinc-plated tools should never be used on the exhaust system, and exhaust system parts should never be marked with a lead pencil. The lead, zinc or galvanized mark is absorbed by the metal of the exhaust system when heated, creating a distinct change in its molecular structure. This change softens the metal in the area of the mark, causing cracks and eventual failure.

After the exhaust system and all components have been installed, the engine cowling should be securely installed. The engine should be run up and allowed to operate at normal operating temperature for a short period of time. After the engine has cooled down, the cowling should be removed and the exhaust system inspected for visual damage.

All connections of the exhaust system should be inspected for evidence of exhaust leakage. Exhaust leaks are normally indicated by streaks streaming from the connection. These streaks will show up as flat gray, brown or black in color. These streaks will sometimes have a sooty appearance. Most exhaust system leaks are caused by misalignment of system components, improper or worn gaskets or loose or damaged clamps. When an exhaust leak is found the clamps, or retaining fasteners should be loosened and the members repositioned. After repositioning, the clamps or retaining fasteners should be re-torqued. Run the engine and then repeat the inspection procedure. If the connection again shows signs of leakage, the clamps, bolts and gasket should be replaced with new ones. After the exhaust system leak has been stopped all fasteners should be safetied.

With the cowling removed, all necessary cleaning operations can be performed. Some exhaust units are manufactured with a plain sandblast finish. Others may have a ceramic-coated finish. Ceramic-coated stacks should be cleaned by degreasing only. They should never be cleaned with sandblast or alkali cleaners.

During the inspection of an exhaust system, close attention should be given to all external surfaces of the exhaust system for cracks, dents or missing parts. This also applies to welds, clamps, supports, and support attachment lugs, bracing, slip joints, stack flanges, gaskets and flexible couplings. Each bend should be examined, as well as areas adjacent to welds; any dented areas or low spots in the system should be inspected for thinning and pitting due to internal erosion by combustion products or accumulated moisture. An ice pick or similar pointed instrument is useful in probing suspected areas. The system should be disassembled as necessary to inspect internal baffles or diffusers.

If a component of the exhaust system is inaccessible for a thorough visual inspection or is hidden by non-removable parts, it should be removed and checked for possible leaks. This can often best be accomplished by plugging the openings of the component, applying a suitable internal pressure (approximately 2 p.s.i.), and submerging it in water. Any leaks will cause bubbles that can be readily detected.

The procedures required for an installation inspection are also performed during most regular inspections. Daily inspection of the exhaust system usually consists of checking the exposed exhaust system for cracks, scaling, excessive leakage and loose clamps.

## Exhaust Systems with Turbocharger

When a turbocharger or a turbo-supercharger system is included, the engine exhaust system operates under greatly increased pressure and temperature conditions. Extra precautions should be taken in exhaust system care and maintenance. During high altitude operation, the exhaust system pressure is maintained at or near sea level values. Due to the pressure differential, any leaks in the system will allow the exhaust gases to escape with torch-like intensity that can severely damage adjacent structures.

A common cause of malfunction is *coke deposits* (carbon buildup) in the waste gate unit, causing erratic system operation. Excessive deposit buildups may cause the waste gate valve to stick in the closed position, causing an overboost condition. Coke deposit buildup in the turbo itself will cause a gradual loss of power in flight and a low manifold pressure reading prior to takeoff. Experience has shown that periodic de-coking, or removal of carbon deposits, is necessary to maintain peak efficiency. Clean, repair, overhaul and adjust the system components and controls in accordance with the applicable manufacturer's instructions. Some manufacturers have a specific cool-down period included in their maintenance checklist. It is to keep the oil flowing through the turbocharger bearings, allowing them too cool somewhat before shutting off the engine.

## Exhaust System Repairs

It is normally recommended that field repairs are not made to the components of the exhaust system. Generally, it is considered a good practice to replace the component with a new one, or one that has been repaired by a certified repair station. Welded repairs are avoided by the technician because of the difficulty of accurately identifying the base metal and matching that alloy with the filler rod.

However, when welded repairs are necessary, the original contours should be retained; the exhaust system alignment must not be warped or otherwise affected. Repairs or sloppy weld beads that protrude internally are not acceptable, as they cause local hot-spots and may restrict exhaust gas flow. When repairing or replacing exhaust system components, the proper hardware and clamps should always be used. Steel or low-temperature self-locking nuts should not be substituted for brass or special high-temperature locknuts used by the manufacturer. Old gaskets should never be reused. When disassembly is necessary, gaskets should be replaced with new ones of the same type provided by the manufacturer.

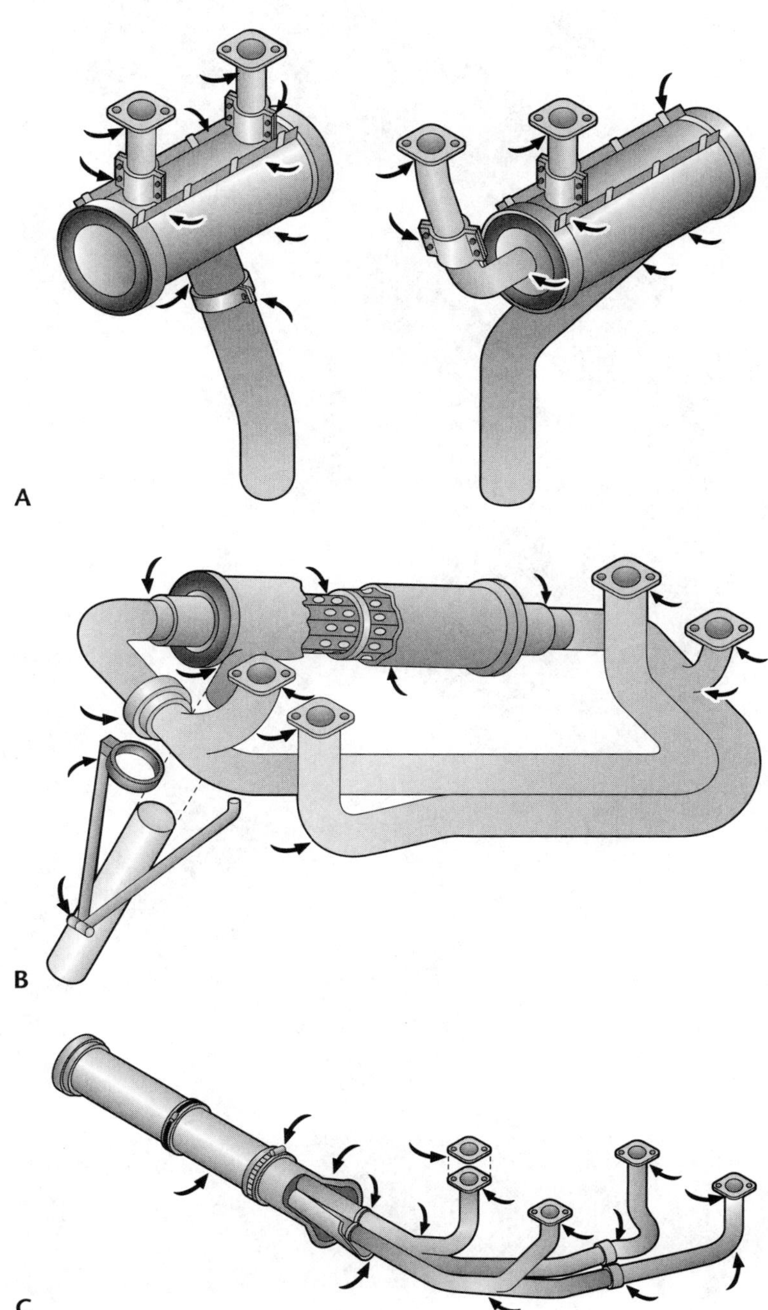


Figure 1-11-7. Primary inspection (A) separate system; (B) crossover-type system; (C) exhaust/augmenter system

Chapter 2

# TURBINE engines

## Section 1

## Basic Turbine Engines

**Learning Objectives:**

- *Basic Turbine Engines*
- *Engine Operating Principles*
- *Nomenclature and Construction*
- *Engine Sections*
- *Engine Fuel Systems*
- *Engine Oil Systems*
- *Engine Exhaust Systems*
- *Engine Cowling and Nacelles*
- *Engine Ignition, Starter, and Generator Drive Systems*
- *Auxiliary Systems*
- *Instrument Systems*
- *Engine Familiarization and Differences*
- *Maintenance and Troubleshooting*

*Left.* **The development of the gas turbine engine is the biggest factor in the evolution of modern transport.**

Though jet propulsion powerplants have been in practical use in aviation for just 50 years, the application of jet reaction has been around for more than 2,000 years. History books credit a Greek mathematician and engineer named Hero with the first positive example of jet propulsion. Hero's machine, called an Aeolipile, is illustrated in Figure 2-1-1. A sealed sphere was filled with water, and then heated on a wood fire and brought to a boil. Steam escaped from orifices on the connected tubes. The energy from the escaping steam spun the ball at high speed. What the Aeolipile actually illustrated was jet reaction.

It remained for Sir Isaac Newton, in 1678, to actually develop a law of physics to explain it. Jet reaction is the principle behind the successful use of the turbine (or jet) engine in aviation. It is a practical application of Newton's third law, which states that for every action there is an equal and opposite reaction.

The first real example of a turbine is credited to an Italian engineer named Giovanni Branca. A turbine is a mechanical device that transforms heat energy into a rotating mechanical force. In 1629, he developed a steam turbine and applied the principal of reduction gearing to obtain useful power. Branca is credited with two applications of his powerplant. One of designs turned a reciprocating mechanism that the drove a stamp mill for crushing gravel. The other turbine-powered machine drove a horizontal drill (Figure 2-1-2). Neither were fantastic successes, but they did move the discovery process forward.

Figure 2-1-1. Hero's Aeolipile was the earliest demonstration of the principles of jet reaction.

During the next 350 years, various pioneers conducted a series of experiments with jet propulsion, gas turbines, ramjets, thrust augmentors and thrust spoilers. Because of a lack of suitable compressors and a method to drive them, most never got past the experimental stage. It was only after the Industrial Revolution and the introduction of modern materials science in the 20th century that practical turbine power has become a reality.

During the early 1920s Dr. Stanford Moss, an American, developed a workable exhaust driven turbosupercharger for the piston engine. This was the first time exhaust gases were used to drive a compressor. Moss' improved impeller designs made possible the centrifugal compressors that could be used for the development of successful jet engines.

On January 16, 1930, Englishman Frank Whittle, filed an application for his first jet propulsion powerplant patent. Whittle's design used exhaust gases generated by the engine to drive a centrifugal compressor. The remaining energy was available to provide forward thrust. The drawing that was submitted to the patent office for his engine is shown in Figure 2-1-3.

During the same time period, a German engineer, Hans von Ohain, was developing an engine of the same type. Because neither man knew of the other's work, jet engine development took a parallel track. Both Whittle and von Ohain started serious development work on their engines in 1936.

Hans von Ohain's work came to the attention of the Heinkel company, a successful German aircraft manufacturer. They provided much needed funding for his initial development work. These early engines greatly resembled the turbo-chargers used in modern reciprocating aircraft.

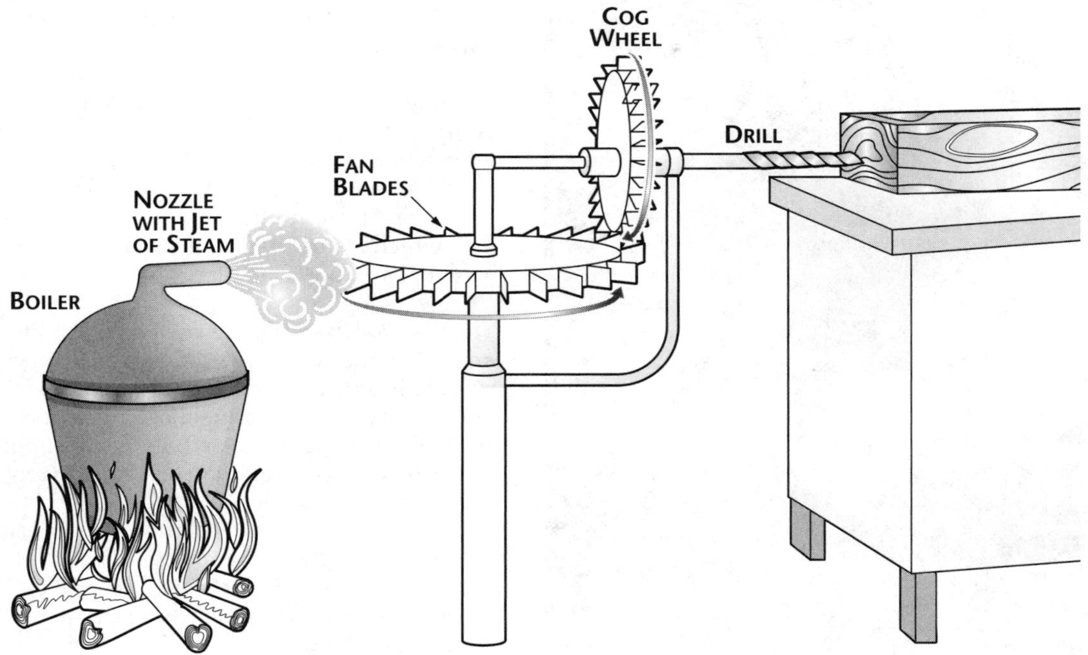


Figure 2-1-2. Giovanni Branca's steam-driven turbine was used to drive a drill for boring holes in wood. It was the earliest combination of a turbine and gear reduction.

Whittle's initial efforts were met with significant doubt in the aviation community. The British War Department initially refused to provide any developmental funding, yet Whittle persisted in working towards a successful prototype engine.

Germany was much quicker to realize the benefits of practical turbine technology. In August of 1939, the German Heinkel HE-178, powered by a Heinkel HeS3 centrifugal flow jet engine designed by von Ohain, was the first jet-powered aircraft to fly successfully. Whittle's engine did not fly until two years later. In May 1941, the Gloster Pioneer jet fighter prototype, the W1, made its first flight. It was powered by a Whittle engine.

Germany produced several jet-powered aircraft during WWII, the most notable being the ME-262. More than 1,600 ME-262s were produced during the last stages of the war. The first Allied jet fighter was the British built Gloster Meteor. Produced in limited numbers, it saw no actual combat, but was used to pursue and shoot down German V-1 Buzz Bombs.

American jet engine development began after a prototype of Whittle's engine was shipped to the General Electric Company in the United States. GE was placed under contract to improve and develop the engine for production. Simultaneously, the Bell Aircraft Corporation started design of the XP-59A airplane, the first American jet fighter. The XP-59A, Figure 2-1-4, with its GE engines, first flew in October 1942.

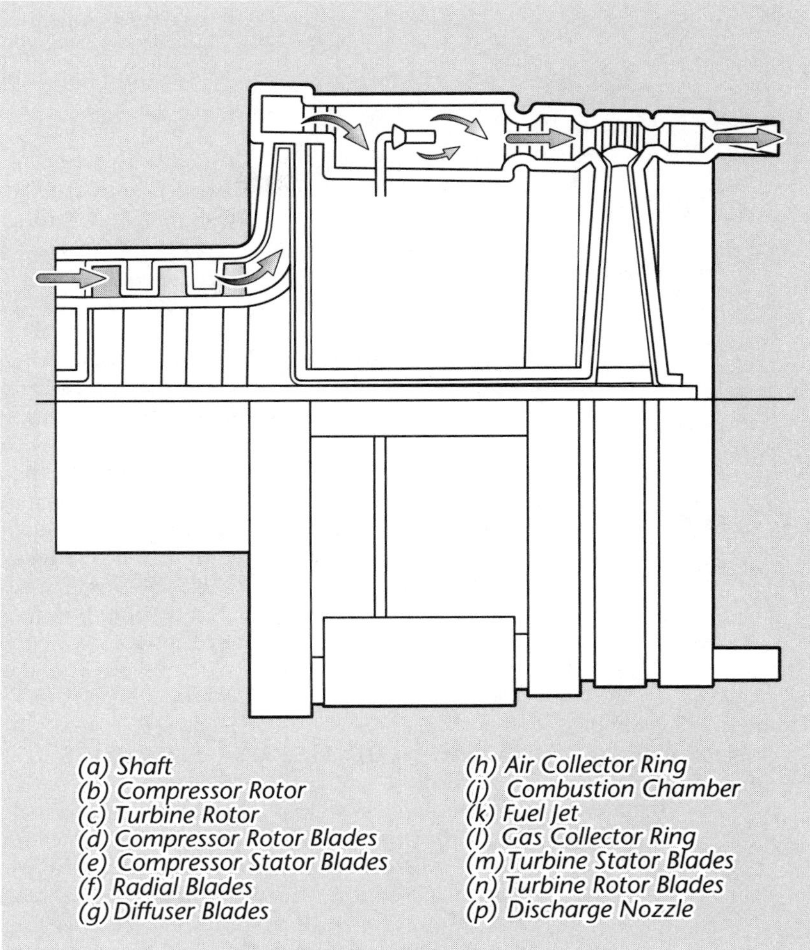


**Figure 2-1-3. The powerplant in Frank Whittle's first patent used a combined axial-centrifugal compressor and multiple combustion chambers.**

**Figure 2-1-4. The XP-59A, with its GE engines, first flew in October 1942.**

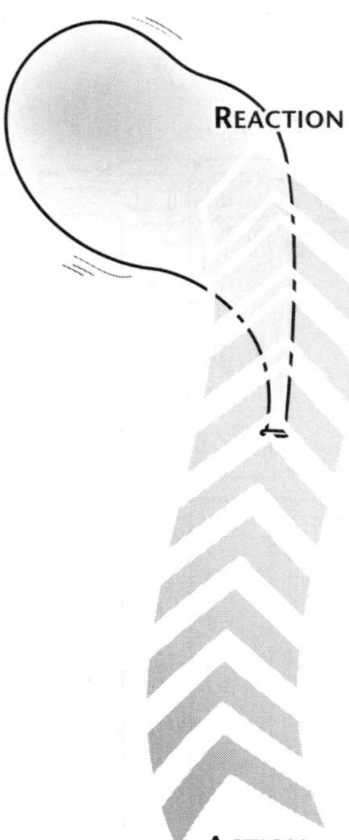


Figure 2-1-5. A simple balloon shows the laws of action and reaction.

The XP-59, later the P-59, was built in limited numbers and was primarily used to develop and test early jet aerodynamics and handling characteristics.

The first production contract for a jet fighter was awarded to the Lockheed Company. The Lockheed XP-80 was designed and built in six months. It first flew in January 1944.

From these beginnings, development of the jet engine powerplant has progressed at an alarming rate. After World War II, the Allies obtained copies of much of the German engine development information. Additionally, many German engineers, including Hans von Ohain, came to the United States and continued their work in turbine engine development. This new resource and already acquired knowledge combined to develop some of the first successful commercial jet engines. It was only slightly more than a decade later that turbine powered aircraft started to take over the world of commercial aviation.

## Basic Propulsion Principles

A propulsion system is a machine that produces thrust to push an object forward. On airplanes, thrust is usually generated through some application of Newton's third law of action and reaction. A gas, or working fluid, is accelerated by the engine, and the reaction to this acceleration produces a force on the engine. The rocket is one example of a jet-propulsion device.

An ordinary toy balloon can be used to demonstrate jet propulsion. Blow up the balloon and then hold it closed with your fingers. The balloon now contains air, which is really a gas under pressure.

The pressure is exerted equally in all directions. It presses the same amount of force against the top, the bottom and the sides of the balloon. Since the pressure is equal in all directions, the total propulsive effect is zero.

Remember that the balloon possesses the energy of the compressed air inside it, just as gasoline in an automobile possesses energy that is released when it is ignited by a spark, and in the same manner that a rifle cartridge possesses the energy of the powder in the cartridge case.

To see that energy put into action, release your grasp at the opening of the balloon. The balloon leaves your hand and flies across the room, collapses and falls to the floor. During its short flight, it has demonstrated the principle of jet propulsion that drives a rocket into the air. Figure 2-1-5 illustrates the application of the law of action and reaction within the balloon.

The pressure at the top of the balloon is equal to the pressure against the bottom of the balloon until it is released. When released, the compressed air escapes through the opening, thus upsetting the balance of pressure between the top and the bottom.

The pressure against the top was greater than the pressure against the bottom, hence the balloon, now subjected to unbalanced pressure, has moved in the direction of the greater force and continued to move until the pressure inside the balloon has left. Recoil from a gun or a fire hose is a result of this same principle.

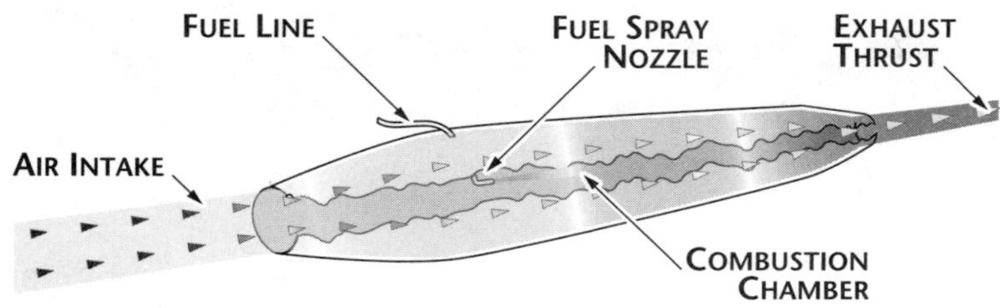


Figure 2-1-6. A simplified drawing of a ramjet engine

## Types and Descriptions of Propulsion Engines

**Ramjet propulsion engine**. In the ramjet engine, thrust is produced by passing the hot exhaust from the combustion of a fuel through a nozzle. The nozzle accelerates the flow, and the reaction to this acceleration produces thrust. To maintain the flow through the nozzle, the combustion must occur at a pressure that is higher than the pressure at the nozzle exit.

In a ramjet, ramming external air into the combustor, using the forward speed of the vehicle, produces high pressure. The external air that is brought into the propulsion system becomes the working fluid. Ramjets produce thrust only when the vehicle is already moving; ramjets cannot produce thrust when the engine is stationary or static. Since a ramjet cannot produce static thrust, some other propulsion system must be used to accelerate the vehicle to a speed where the ramjet begins to produce thrust.

Ramjets are ideally suited for very high-speed flight within the atmosphere. For high supersonic or hypersonic flight, the ideal propulsion system is a ramjet.

Figure 2-1-6 shows a drawing of a typical ramjet engine. On the left is the inlet, which brings outside air into the engine. The inlet is shaped in a manner such that it compresses the high-speed air entering it. This shape is only effective in compressing the air when the engine is moving forward at a high rate of speed. For ground testing a ramjet can be operated by adding an external source of high-speed air just in front of the inlet.

After the air is compressed, fuel is injected and mixed for combustion. This occurs just downstream of the inlet. The resulting flame is stabilized in the engine by the flame holder ring. The combustion process greatly increases the pressure within the engine. The hot exhaust then passes through the nozzle, which is shaped to accelerate the flow and produce thrust. The constant stream of air entering the inlet prevents the combustion pressure from flowing backwards out of the front of the engine. Ramjets are best suited for aircraft operating at constant high speeds. They do not operate as well under varying speed conditions. They have been most often used on high-speed missiles.

**Pulsejet propulsion engine.** The pulsejet uses an intermittent-firing duct. As the airplane travels swiftly through the atmosphere, compression waves are forced into the combustion chamber. The waves are controlled by a series of shutters or flaps in the duct inlet. These shutters are forced open by the ramming action of the incoming air and closed when the pressure builds up inside. They open and close intermittently. The fuel is continuous, but the combustion is intermittently fired by an electric spark at a rate of 40 times per second in some models. This type of engine is not used for commercial aircraft.

**Rocket propulsion engine.** In a rocket engine, stored fuel and stored oxidizer are mixed and exploded in a combustion chamber. The hot exhaust is then passed through a nozzle, which accelerates the flow. For a rocket, the accelerated gas, or working fluid, is the hot exhaust; the surrounding atmosphere is not used. That's why a rocket will work in space, where there is no surrounding air, and a jet engine or propeller will not work. Jets and propellers rely on the atmosphere to provide the working fluid. In other words, a rocket carries its working fluid with it, rather than obtaining it from its surroundings. Carrying the working fluid on the aircraft adds a significant weight penalty and is one of the greatest limitations to developing successful commercial space travel.

There are two main categories of rocket engines; liquid rockets and solid rockets. In a liquid rocket, the fuel and the source of oxygen (oxidizer) necessary for combustion are stored separately and pumped into the combustion chamber of the nozzle where burning occurs. In a solid rocket, the fuel and oxidizer are mixed together and packed into a solid cylinder. The space shuttle in Figure 2-1-7 uses liquid fueled main engines and solid rocket boosters. The liquid fueled main engines are located at the rear of the shuttle fuselage and their fuel is carried in separate tanks within the large cylindrical tank under the belly. The solid rocket boosters are the large missile shaped objects attached to either side.

**Figure 2-1-7. A space shuttle uses liquid fueled main engines and solid rocket boosters.**

Under normal temperature conditions, the fuel and oxidizer mixture of a solid rocket will not self ignite; but will burn when exposed to a source of heat. Some type of igniter is used to initiate the burning of a solid rocket motor at the end of the propellant facing the nozzle.

Once the fuel starts to burn, hot exhaust gas is produced, which is used to propel the rocket, and a flame front is produced, which moves into the propellant. Once the burning starts, it will proceed until all the propellant is burned.

Most liquid rocket fuels are highly volatile and many will self-ignite when exposed to the atmosphere. There are significant advantages and disadvantages to both types of rockets.

With a liquid rocket, the thrust can be stopped by turning off the flow of fuel, but special

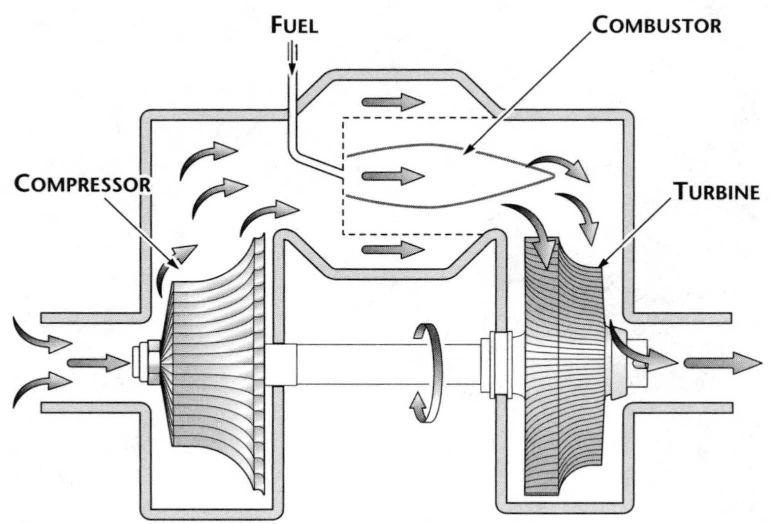


Figure 2-1-8. A turbine engine in its most basic form

precautions must be used when handling the fuel. With a solid rocket, once it ignition has occurred, you would have to destroy the casing to stop the engine. Liquid rockets tend to be heavier and more complex because the fuel must be pumped, and as a safety precaution, you usually wait to fuel the rocket until just before launch. A solid rocket is much easier to handle, as they can sit, fully fueled, for years before firing.

**Gas turbine propulsion engine.** The earliest turbine engines resembled our modern turbochargers in their basic construction. (See Figure 2-1-8.) The turbine in its simplest form, just like the reciprocating engine, is an air pump. Air is introduced into a compressor. The compressor performs the same function as the compression stroke of a reciprocating engine. Fuel is added to the compressed air mixture. This compressed fuel/air mixture is then ignited. The combustion creates heat and pressure. This hot, high-pressure air is ducted over a turbine. The turbine is connected back to the compressor causing it to spin. The remaining energy is exhausted out of the jet tailpipe providing forward thrust.

Turbine propelled forward thrust can be accomplished through many different forms and the energy created can be used in a number of different ways. Figure 2-1-9 shows different variations of the gas turbine engine. While each of the engines is different, they share some parts in common. Each of these engines has an inlet, a compressor section, combustion section, a turbine section and an exhaust nozzle. The compressor, burner and turbine are called the core of the engine, since all gas turbine engines have these components. The core is also referred to as the gas generator, since the output of the core is hot exhaust gas. The gas is passed through a turbine that in turn drives the compressor. The remaining combustion energy can be used in a variety of ways.

Each of these methods provides distinct benefits in certain situations. In the simplest application, the turbojet, the exhaust gas is directed through a nozzle and directly produces forward thrust. All the other turbine variations direct the exhaust gas through at least one additional set of turbines. When these turbines

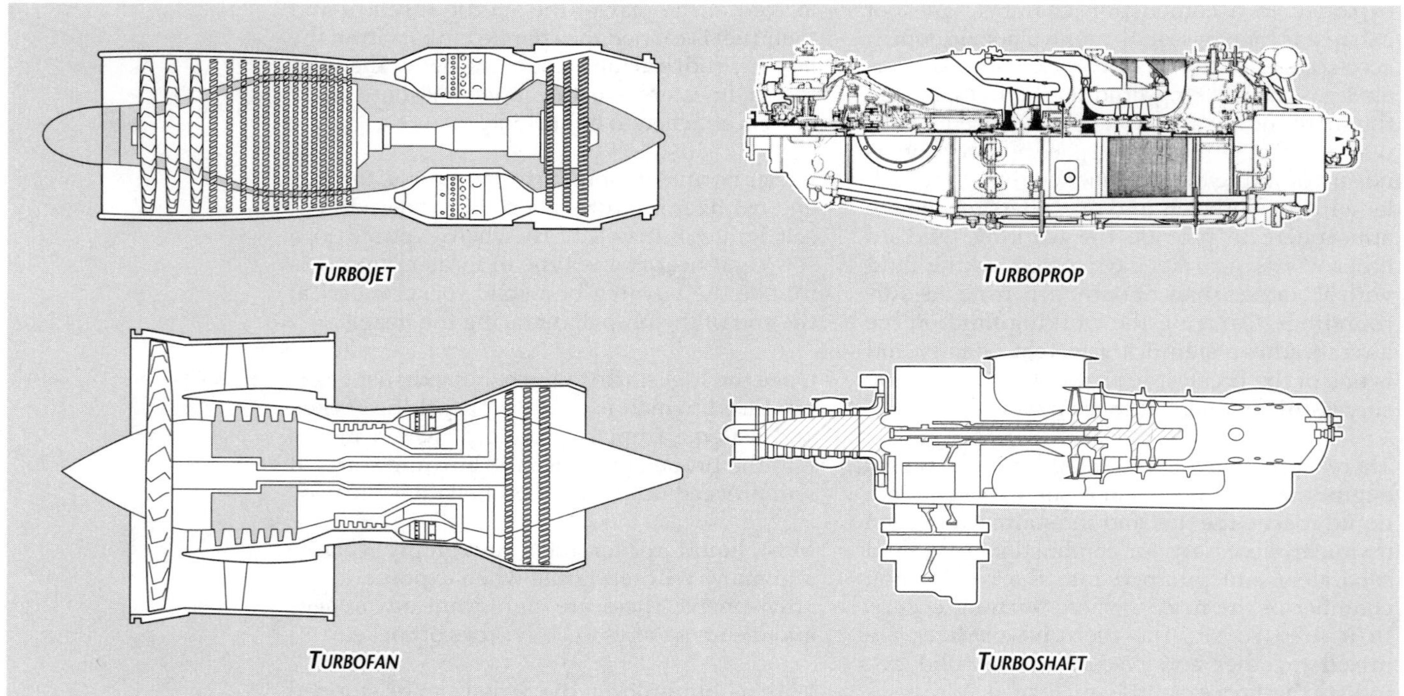


Figure 2-1-9. Different types of gas turbine arrangements

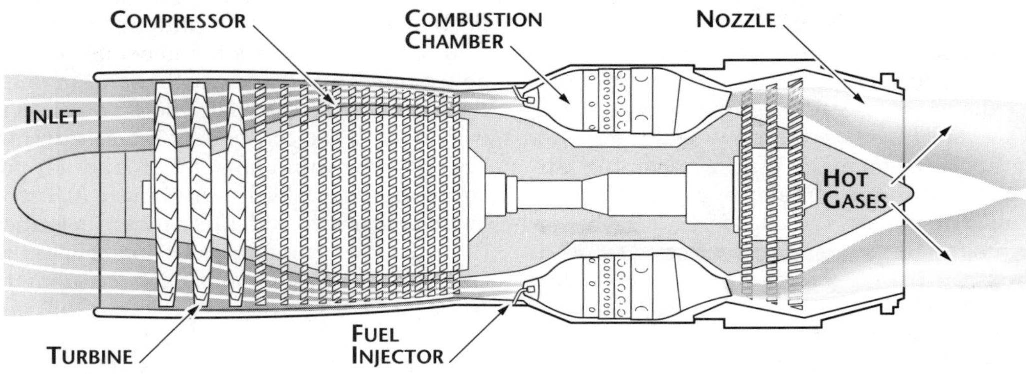


Figure 2-1-10. A drawing of a simplified turbojet engine

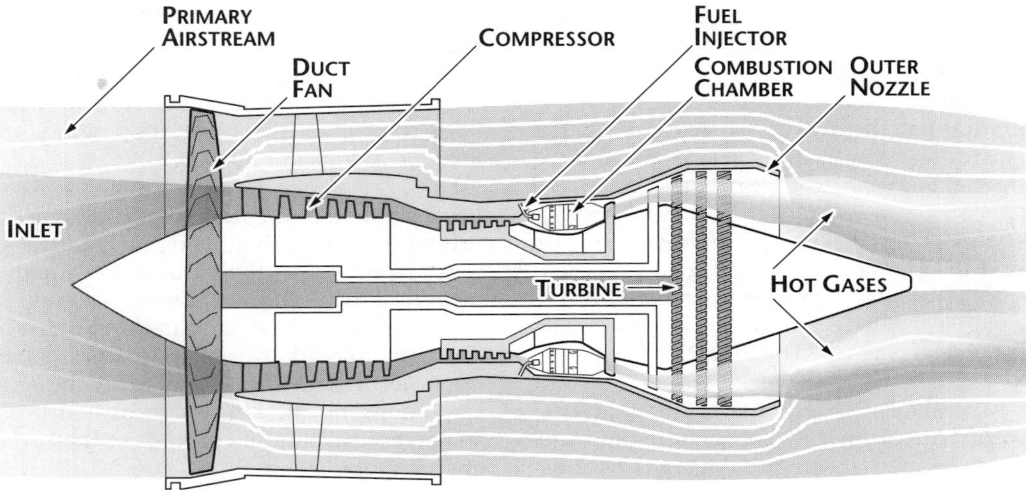


Figure 2-1-11. A drawing of a simplified turbofan engine

are attached to a ducted fan, we call the engine a turbofan. When they are connected to a reduction gearbox to drive a propeller or a rotor, we call it a turboprop or turboshaft engine. Each of these types is described in detail.

The compressor and turbine are linked by the central shaft and rotate together. This group of parts is called the turbo machinery. It can also be referred to as the rotating group or a rotor.

## Types and Uses of Aircraft Gas Turbine Engines

**Turbojet engine**. In a *turbojet* engine, shown in Figure 2-1-10, the high pressure in the combustor is generated by a piece of machinery called a compressor. The air at the end of the combustor flows into the turbine mounted on the same shaft as the compressor. The turbine absorbs only enough of the energy of the rearward rushing gases of combustion to operate the compressor and the accessories; the remaining energy, which retains its high velocity, exits through the exhaust nozzle and provides thrust. The sole function of the turbine is to turn the compressor; it does not contribute in any other manner to the production of thrust. This engine contains no other turbine sections.

**Turbofan engine**. A *turbofan* engine, shown in Figure 2-1-11, is the most modern variation of the basic gas turbine engine. The core engine is surrounded by a fan, which is attached to an additional turbine at the rear. The fan shaft passes through the core shaft to the turbine that drives the fan. This type of arrangement is called a two-spool engine (one spool for the fan, one spool for the core). A spool consists of a compressor, shaft and turbine. Some Rolls Royce engines have an additional spool, known as the intermediate pressure spool. It adds an additional compressor and an additional turbine section in between the typical low and high pressure arrangement.

The incoming air passes through the engine inlet. Some of the incoming air passes through the fan and continues on into the core or gas generator. The core contains the combustor that uses fuel to increase the energy in the gas path. The exhaust gases pass through a set of cores and the fan turbines, then out of the exhaust nozzle. The remainder of the incoming air passes through the fan and bypasses, or goes around, the core of the engine. So a turbofan gets some of its thrust from the core, but most of its thrust comes from the fan.

Two different duct designs are used with fan engines. The air leaving the fan can be ducted through exit vanes and a nozzle, then overboard. In a separate or two-exhaust nozzle system, the air passes directly to the atmosphere without mixing with the core air. This system uses one nozzle for core air and one for fan air. The air can also be ducted along the outer case of the basic engine, to be discharged through a common exhaust nozzle. This type of arrangement is called a ducted turbofan engine. In this configuration, the fan air is mixed with the exhaust core gas path air before it is discharged through the mixed-exhaust nozzle.

The ratio of the airflow (in lbs/sec) that goes around the engine to the air that goes through the core is called the *bypass ratio.* In high bypass-ratio engines, up to 80 percent of the total engine's thrust will be developed by the fan.

The general principle of the fan engine is to convert more of the fuel energy into pressure. With more of the energy converted to pressure, a greater product of pressure area can be achieved. Because the fuel flow rate for the core is changed only a small amount by the addition of the fan, a turbofan generates considerably more thrust for nearly the same amount of fuel used by the core. The end result is savings in fuel, with the consequent increase in range. This is why turbofans are found on high-speed transport (airliner) aircraft (0.8 mach). Turbofans are considerably more fuel-efficient than turbojet engines. Even low bypass-ratio turbofans are still more fuel- efficient than turbojets.

Many modern fighter planes actually use low bypass-ratio turbofans equipped with afterburners or thrust augmenters (afterburners can use extra fuel at the exhaust nozzle area to increase thrust). They can then cruise efficiently but still have high thrust for higher speeds. Even though the fighter plane can fly much faster than the speed of sound, the air going into the engine must do so at less than the speed of sound. Therefore, the airplane inlet controls airflow into the engine, which must be at subsonic speeds.

Because more of the fuel energy is turned into pressure in the turbofan engine, another stage must be added in the turbine to provide the power to drive the fan. These turbine stages absorb some of the energy in the exhaust path. The fan more than makes up for the drop-off in jet thrust from the core of the engine. Depending on the fan design, it will produce somewhere around 80 percent of the turbofan engine's total thrust.

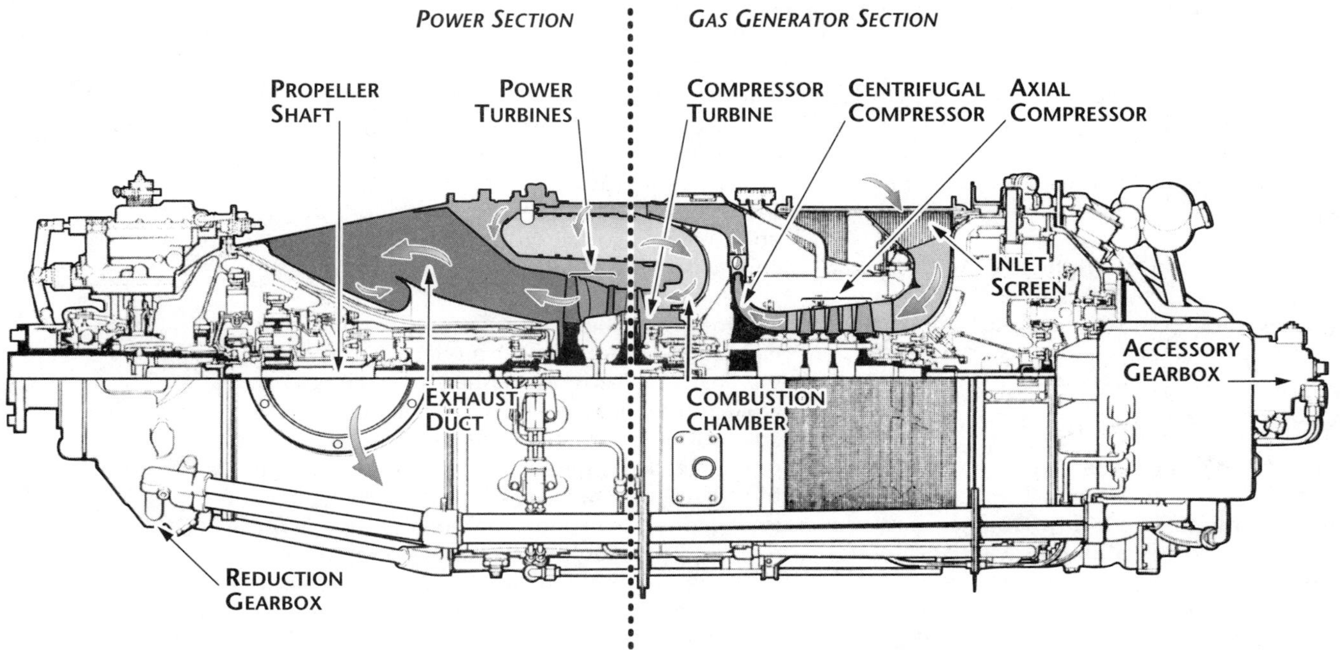


**Figure 2-1-12. A cutaway view of a Pratt & Whitney PT6A engine. Notice the division between the power section and the gas generator section.**

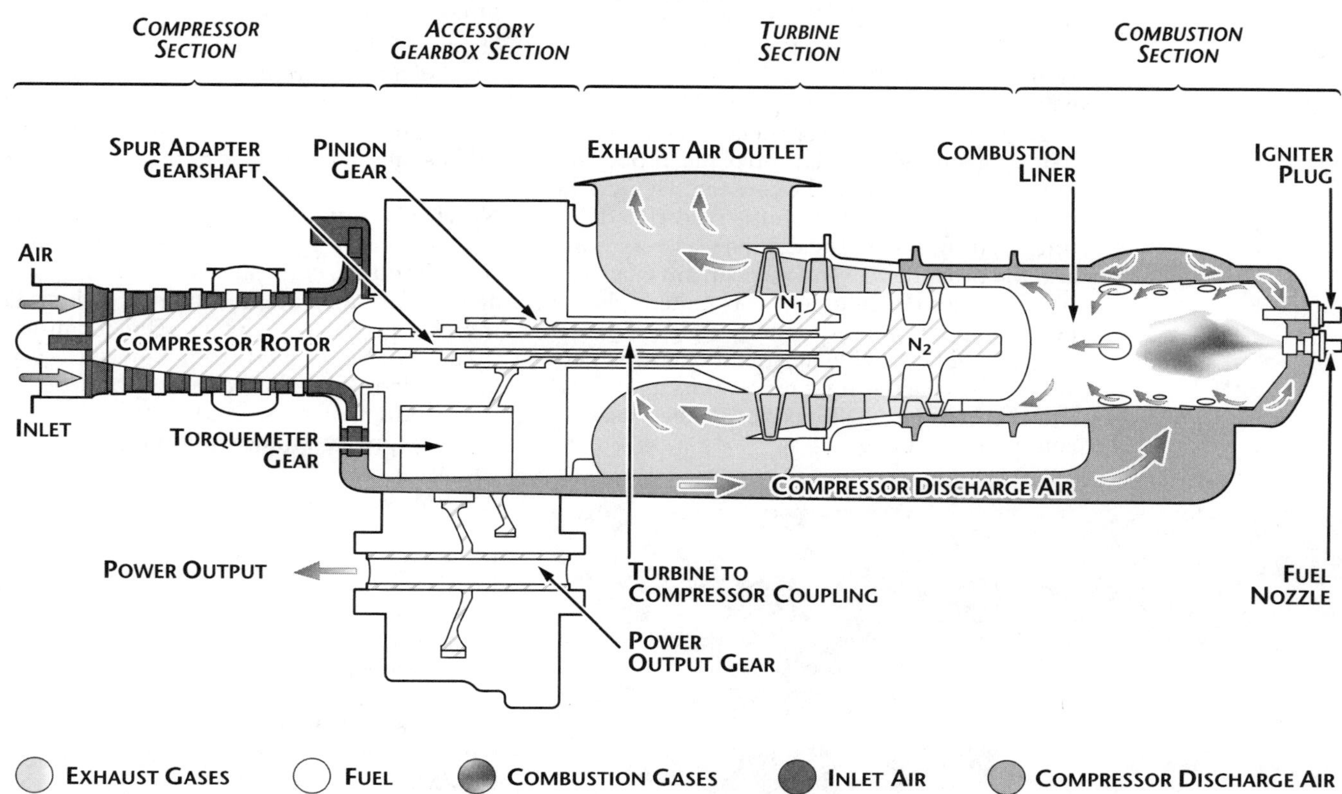


**Figure 2-1-13. A Rolls Royce turboshaft engine diagram showing the main sections of the engine.**

Turbofans, sometimes call *fanjets*, are the most widely used gas turbine engine. The turbofan is a compromise between the good operating efficiency and high-thrust capability of a turboprop and the high-speed, high-altitude capability of a turbojet.

**Turboprop engine**. There are two main parts to a *turboprop* propulsion system: the core engine (gas generator), and the speed reduction gearbox and propeller, as can be seen in Figure 2-1-12. The core is a basic gas turbine engine, except that instead of expending all the hot exhaust through the nozzle to produce thrust, the energy of the exhaust not only turns the compressor turbine but also turns an additional turbine. The increased power is generated by the exhaust gases passing through these additional turbine stages.

If this added turbine is connected to the compressor directly (mechanically), it is called a fixed-turbine engine. If the added turbine is not mechanically connected to the gas generator or core of the engine, it is called a free-turbine (power turbine) engine. This power turbine drive shaft is connected to a reduction gearbox. The gearbox reduces the shaft's speed because of the high speeds used in the engine. This is needed because propellers must turn at a relatively slow speed compared to the engine.

The gearbox is then connected to a propeller that produces approximately 85 percent of the thrust. The exhaust gases also contribute to engine power output through jet reaction, although the amount of energy available for jet thrust is considerably reduced. This is because most of the energy of the core exhaust has gone into turning the compressor and the reduction gear box/propeller. Turboprop engines are mostly used on military, commuter or business aircraft.

**Turboshaft engine**. A gas turbine engine that delivers power through a shaft to operate something other than a propeller is referred to as a *turboshaft* engine. The turboshaft engine, shown in Figure 2-1-13, uses the gas generator core of the engine to turn an output shaft. The power takeoff may be coupled directly to the engine turbine, or the shaft may be driven by a turbine of its own (free turbine) located in the exhaust stream. The free turbine rotates independently. This principle is used extensively in current-production turboshaft engines. The largest use of this type of engine in aviation is in the helicopter industry and for some types of auxiliary power units (APU).

**Unducted fan engine concept (UDF).** During the 1970s and 1980s the price of fuel took an alarming climb. Major design projects were undertaken to design and produce more fuel-efficient engines. One of the most touted departures from conventional engines was the *unducted fan* concept (Figure 2-1-14). Test engines were actually produced and one was mounted on an airplane for flight trials. The engines had large propellers, shaped more like boat propellers than airplane propellers, mounted at the rear of the cowling. Though the design showed great promise, it also had significant drawbacks. The engine was very noisy and the propeller control mechanism was complex. Further development was halted when it became apparent that modern high bypass jet engines could provide most of the desired benefits of the UDF without many of its associated drawbacks. The UDF has never been used in commercial service.

**Figure 2-1-14. An experimental General Electric unducted fan engine**

## Section 2

# Turbine Engine Operating Principles

The principle used by a turbojet engine as it provides force to move an airplane is based on Newton's law of momentum. This law shows that a force is required to accelerate a mass; therefore, if the engine accelerates a mass of air, it will apply a force on the aircraft. The turbopropeller engine generates thrust by giving a relatively small acceleration to a large quantity of air. The turbojet engine achieves thrust by imparting greater acceleration to a smaller quantity of air. The turbofan engine achieves thrust by using more of an equal-parts amount of air mass and acceleration.

Ambient air enters the inlet of the engine where it is subjected to changes in temperature, pressure and velocity. The compressor then increases the pressure and temperature of the air mechanically. The air continues at constant pressure to the burner section, where its temperature is increased by the combustion of fuel. The energy is taken from the hot gas by expanding through a turbine that drives the compressor, and by expanding through an exhaust nozzle designed to discharge the exhaust gas at high velocity to produce thrust.

**Newton's law**. The high-velocity exhaust gases from a gas turbine engine may be considered a continuous force, imparting against the aircraft in which it is installed, thereby producing thrust. The formula for thrust can be derived from Newton's second law, which states that force is proportional to the product of mass and acceleration. This law is expressed in the following formula:

$$F = M \times A$$

where:

F = Force in pounds
M = Mass in slugs
A = Acceleration in ft. per sec$^2$.

In the above formula, mass is similar to weight, but it is actually a different quantity. Mass refers to the quantity of matter, while weight refers to the pull of gravity on that mass. At sea level, under standard conditions, one pound of mass will have a weight of one pound.

To calculate the acceleration of a given mass, the gravitational constant is used as a unit of comparison. The force of gravity is 32.2 ft./sec$^2$ (feet per second, squared). This means that a free-falling one-pound object will accelerate at the rate of 32.2 feet per second for each second that gravity acts upon it. Since the object mass weighs 1 lb., which is also the actual force imparted to it by gravity, we can assume that a force of 1 lb. will accelerate a one-pound object at the rate of 32.2 ft./sec$^2$.

Also, a force of 10 lbs. will accelerate a mass of 10 lbs. at the rate of 32.2 ft./sec$^2$. This is assuming there is no friction or other resistance to overcome. It is now apparent that the ratio of the force (in pounds) is to the mass (in pounds) as the acceleration in ft./sec$^2$ is to 32.2. Using M to represent the mass in pounds, the formula may be expressed like this:

$$F = \frac{M \times A}{G}$$

where:

F = Force
M = Mass
A = Acceleration
G = Gravity

In any formula involving work, the time factor must be considered. It is convenient to have all time variables in equivalent units: seconds, minutes or hours. In calculating jet thrust, the phrase "pounds of air per second" is convenient, since the time factor for the force of gravity is the same; its time factor is in seconds.

## Airflow Principles

Bernoulli's principle states that whenever a stream of any fluid has its velocity increased at a given point, the pressure of the stream at that point is less than the rest of the stream. Bernoulli's principle has its practical application to the gas turbine engine through the design of the air ducts.

The two types of ducts, shown in Figure 2-2-1, are the convergent and the divergent. The convergent duct increases velocity and decreases pressure. The divergent duct decreases velocity and increases pressure. The convergent principle is the one usually used for the exhaust nozzle. The divergent principle is the one most frequently used in the compressor, where the air is slowing and pressurizing.

The Brayton cycle is the name given to the thermodynamic cycle of a gas turbine engine to produce thrust. This is a varying-volume constant-pressure cycle of events and is commonly called the constant-pressure cycle. A more recent term is continuous-combustion cycle. The four continuous and constant events are the intake, compression, expansion (includes power) and exhaust.

In the intake cycle, air enters at ambient pressure and a constant volume. It leaves the intake at an increased pressure and a decrease in volume. At the compressor section, air is received from the intake at an increased pressure, slightly above ambient temperature and at a slight decrease in volume. Air enters the compressor, where it is compressed. It leaves the compressor with a large increase in pressure and a decrease in volume. This is caused by the mechanical action of the compressor. The next step, the expansion, takes place in the combustion chamber. Expansion is achieved by burning fuel, which expands the air by adding heat. The pressure remains relatively constant, but a marked increase in volume takes place. The expanding gases move rearward through the turbine inlet guide vanes and turbine rotor, where the energy of the gases is converted to mechanical energy to drive the turbine. The turbine is connected to the compressor, in turn, driving it to compress more incoming air.

The exhaust section, which is a convergent duct, converts the expanding volume and decreasing pressure of the gases to a final high velocity. The force created inside the jet engine to keep this cycle continuous has an equal and opposite reaction (thrust) to move the aircraft forward. In turbofan and turboprop engines, additional turbines absorb most of the gases' energy to drive a fan or propeller.

## Thrust Calculations

Using the formula from the previous page, compute the force necessary to accelerate a mass of 50 lbs. to an acceleration of 100 ft./sec$^2$, as follows:

$$F = \frac{50 \text{ lbs.} \times 100 \text{ ft./sec}^2}{32.2 \text{ ft./sec}^2}$$

$$F = \frac{50 \times 100}{32.2}$$

$$F = 155 \text{ lbs.}$$

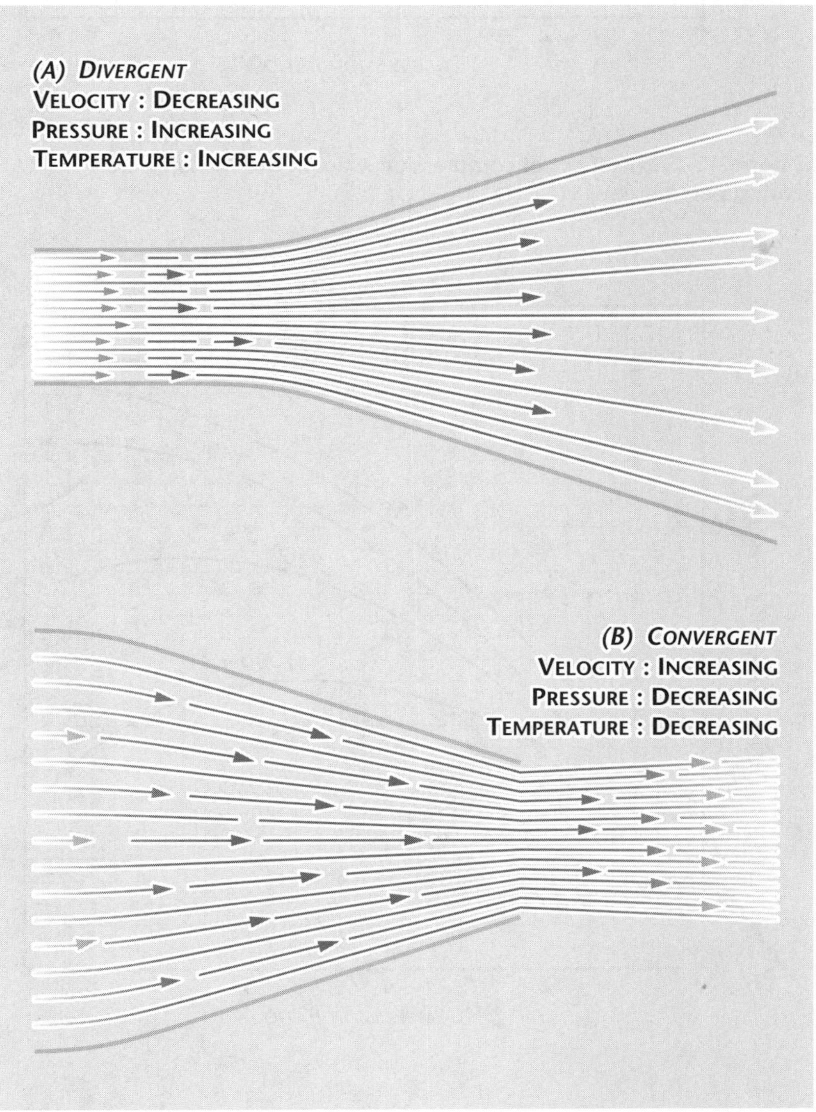


**Figure 2-2-1. Converging and diverging ducts provide the means to change the velocity and pressure of the air as it moves through the engine.**

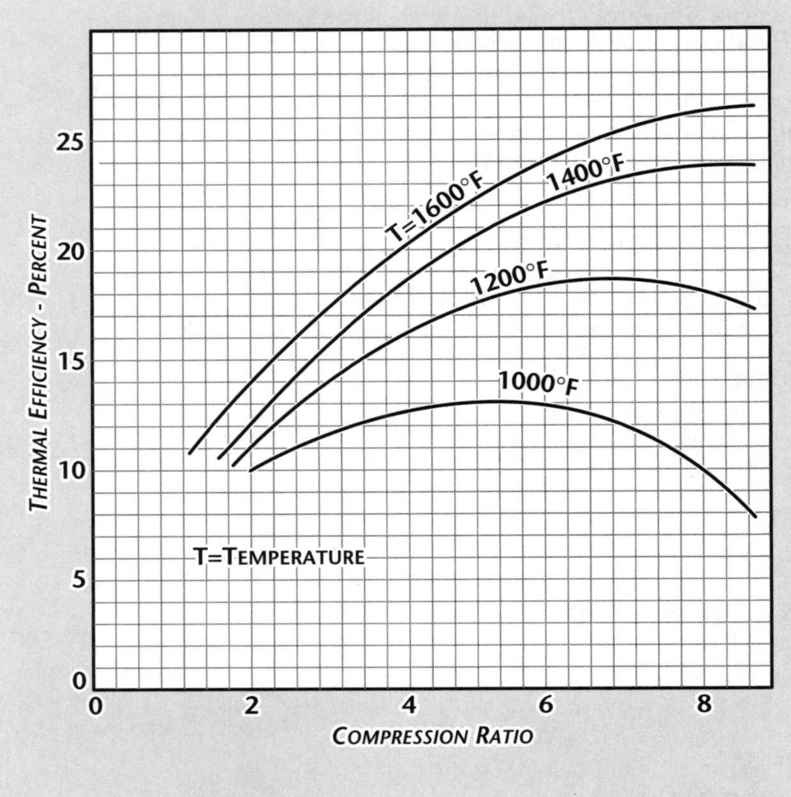


**Figure 2-2-2. The effect of compression ratio on thermal efficiency is a critical parameter.**

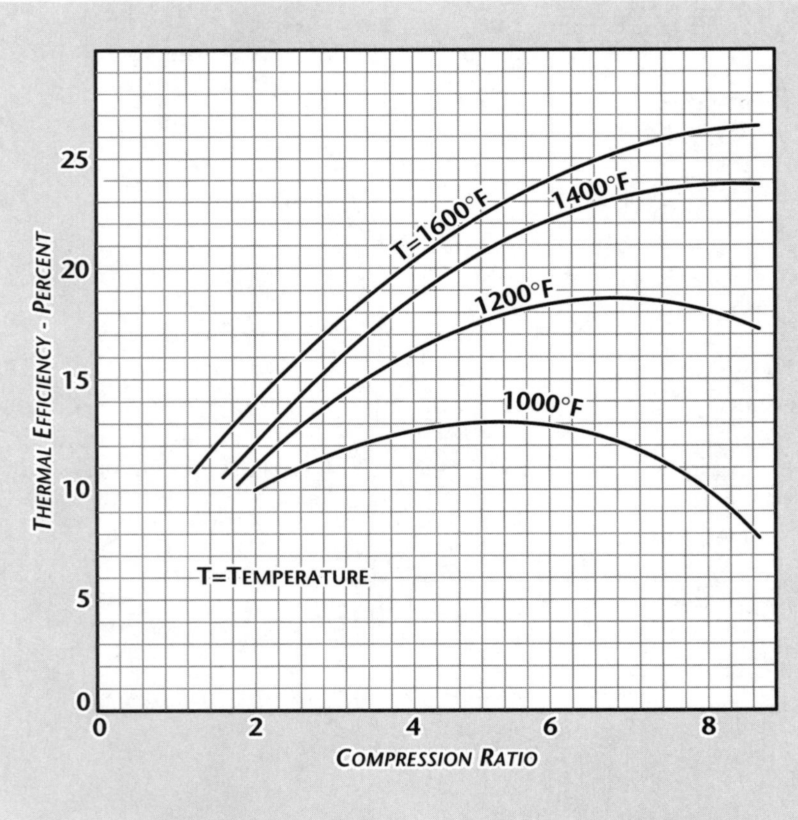


**Figure 2-2-3. The effect of outside air temperature on thrust output: as temperature drops, the thrust increases.**

This illustrates that if the velocity of 50 lbs. of mass per second is increased by 100 ft./sec$^2$, the resulting thrust is 155 lbs. Since the turbojet engine accelerates a mass of air, the following formula can be used to determine jet thrust:

$$F = \frac{M_s (V_2 - V_1)}{G}$$

where:

F = Force in lbs.
$M_a$ = Mass flow in lbs./sec
$V_1$ = Inlet velocity
$V_2$ = Jet velocity

$V_2 - V_1$ = Change in velocity; difference between inlet velocity and jet velocity

G = Acceleration of gravity, or 32.2 ft./sec$^2$

As an example, the formula for changing the velocity of 100 lbs. of mass airflow per second should be applied as follows:

$$F = \frac{100\ (800 - 600)}{32.2}$$

$$F = \frac{20{,}000}{32.2}$$

F = 621 pounds

It is easy to see from this formula that the thrust of a gas turbine engine can be increased by two methods: first, by increasing the mass flow of air through the engine, and second, by increasing the jet velocity.

**Performance and engine parameters.** Thermal efficiency is a prime factor in gas turbine performance. It is the ratio of total work produced by the engine to the amount of chemical energy supplied in the form of fuel.

The three most important factors affecting the thermal efficiency are turbine inlet temperature, compression ratio, and the component efficiencies of the compressor and turbine. Other factors that affect thermal efficiency are compressor inlet temperature and burner efficiency.

Figure 2-2-2 shows the effect that changing compression ratio has on thermal efficiency when compressor inlet temperature and the component efficiencies of the compressor and turbine remain constant. In actual operation, the turbine engine exhaust gas temperature varies directly with turbine inlet temperature at a constant compression ratio. R.p.m. is a direct measure of compression ratio; therefore, maximum thermal efficiency can be obtained by maintaining the highest possible exhaust gas temperature, at constant r.p.m. Since the engine life (the time it takes engine components to wear out) is greatly reduced at a high turbine-inlet temperature, the operator should not exceed the exhaust gas temperatures specified for continuous operation.

Figure 2-2-3 shows that the thrust output improves rapidly with a reduction in OAT (outside air temperature) at constant altitude, r.p.m. and airspeed. This increase occurs partly because the energy required per pound of airflow to drive the compressor varies directly with the temperature, thus leaving more energy to develop thrust. In addition, the thrust output will increase, since the air at reduced temperature has an increased density. The increase in density causes the mass flow through the engine to increase.

The altitude effect on thrust, as shown in Figure 2-2-4, can also be discussed as a density and temperature effect. In this case, an increase in altitude causes a decrease in pressure and temperature. Since the temperature lapse rate is less than the pressure lapse rate as altitude is increased, the density is decreased. Although the decreased temperature increases thrust, the effect of decreased density more than offsets the effect of the colder temperature. The net result of increased altitude is a reduction in the thrust output.

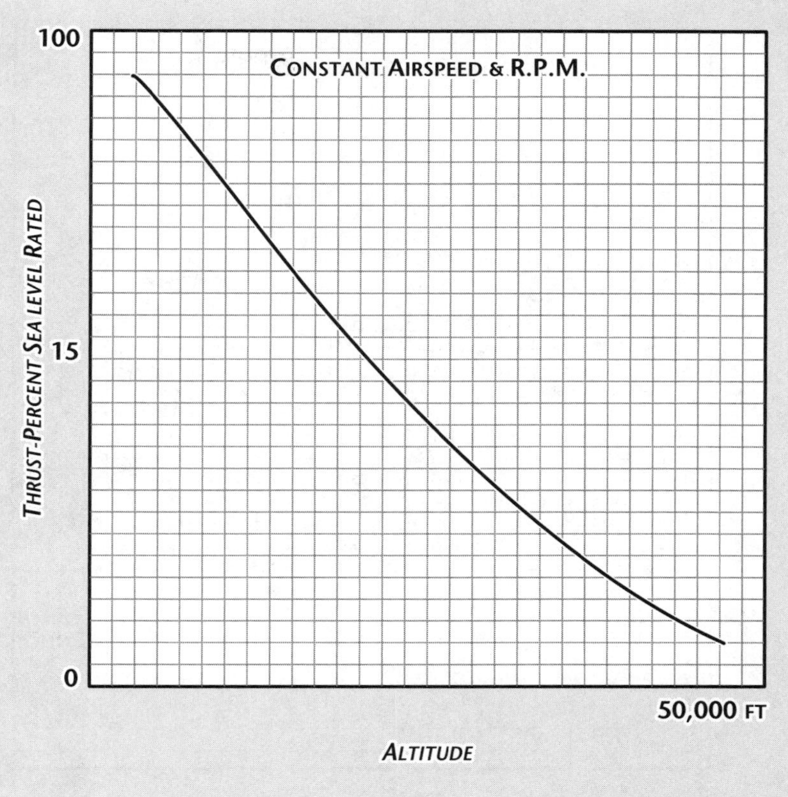


Figure 2-2-4. The effect of altitude on thrust output: as altitude increases the thrust tends to decrease.

The effect of airspeed on the thrust of a gas turbine engine is shown in Figure 2-2-5. To explain the airspeed effect, it is first necessary to understand the effect of airspeed on the factors that combine to produce net thrust. These factors are: specific thrust and engine airflow. Specific thrust is defined as the pounds of net thrust developed per pound of airflow per second.

As airspeed is increased, the ram drag increases rapidly. The exhaust jet velocity remains relatively constant; therefore, the effect of the increase in airspeed results in decreased specific thrust, as shown in Figure 2-2-5. In the low-speed range, the specific thrust decreases faster than the airflow increases and causes the net thrust to increase until sonic velocity is reached. The effect of the combination on net thrust is illustrated in Figure 2-2-6 (see the following page).

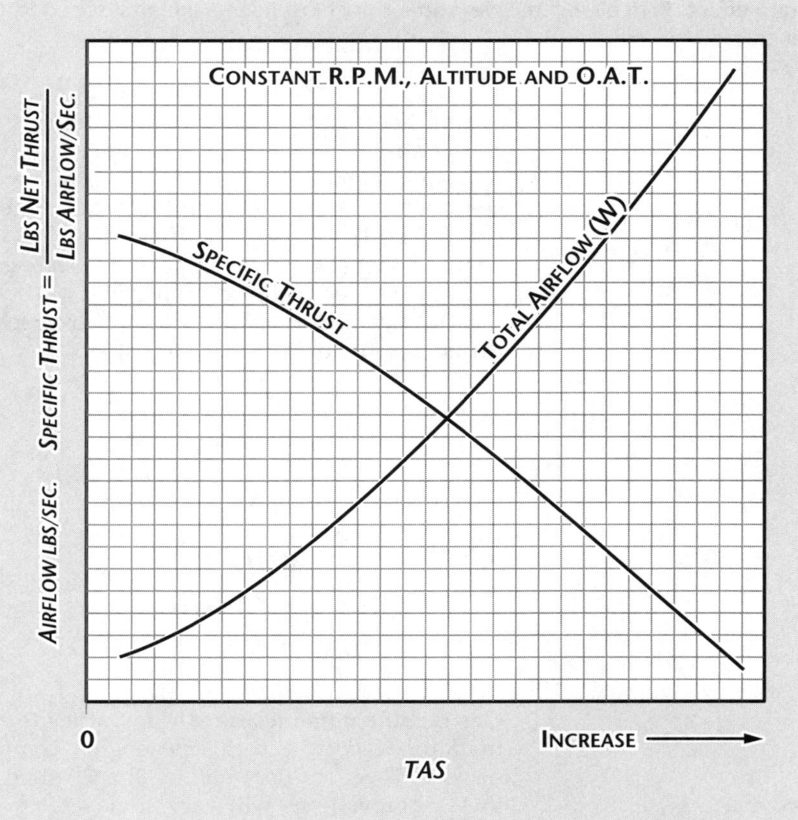


Figure 2-2-5. The effect of airspeed on net thrust will decrease without ram effect. Ram air will recover some thrust lost as speed is increased.

A rise in pressure above existing outside atmospheric pressure at the engine inlet, as a result of the forward velocity of an aircraft, is referred to as ram. Since any ram effect will cause an increase in compressor entrance pressure over atmospheric pressure, the resulting pressure rise will cause an increase in the mass airflow and jet velocity, both of which tend to increase thrust.

Although ram effect increases the engine thrust, the thrust being produced by the

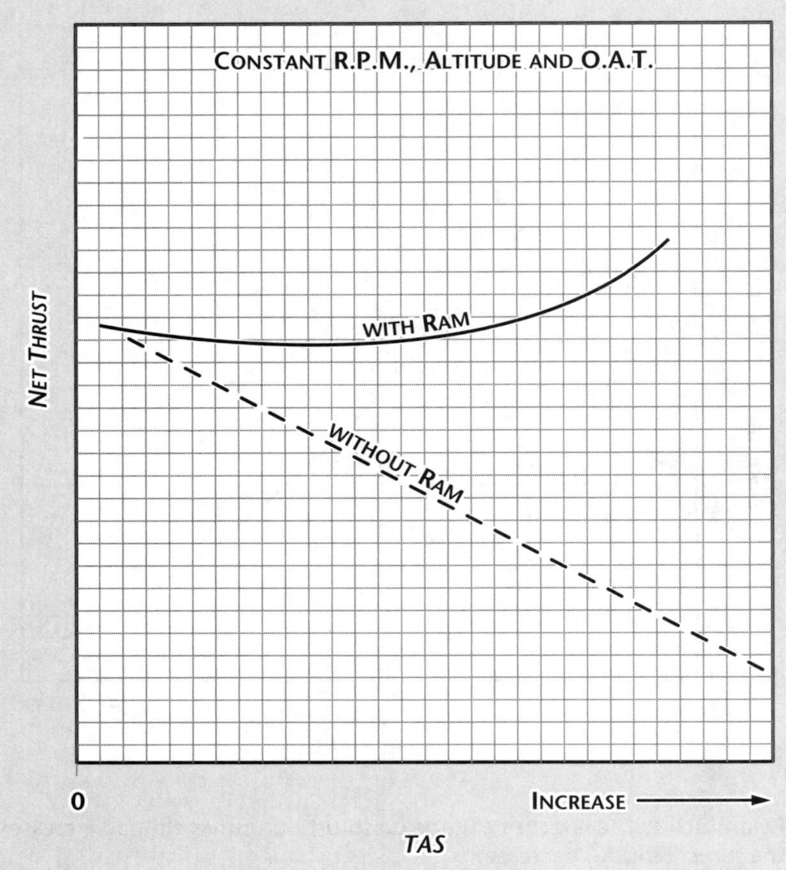


**Figure 2-2-6. The effect of airspeed on net thrust will decrease without ram effect. Ram air will recover some thrust lost as the aircrafts' speed is increased.**

engine decreases for a given throttle setting as the aircraft gains airspeed. Therefore, two opposing trends occur when an aircraft's speed is increased. What actually takes place is the net result of these two different effects. An engine's thrust output temporarily decreases as aircraft speed increases from static, but soon ceases to decrease; toward the higher speeds, thrust output begins to increase again.

## *Section 3*

# *Turbine Engine Nomenclature and Construction*

**Gas turbine nomenclature.** Many of the terms used to describe gas turbine engine components and sections use abbreviations. Although these abbreviations will vary somewhat from one engine manufacturer to another, many are common. Table 2-2-1 contains a list of abbreviations and their most common meanings.

**Table 2-2-1. Index of turbine engine abbreviations, terms, and symbols.**

| | -A- |
|---|---|
| ABBR | Abbreviations |
| ABMM | Airbus Maintenance Manual Absolute |
| ABS | Absolute |
| AC | Alternating Current |
| ACARS | ARINC Communications and Reporting System |
| ACCEL | Acceleration |
| AGB | Angle Gearbox |
| AOX | Air/Oil Heat Exchanger |
| AP | Access Port |
| APU | Auxiliary Power Unit |
| ARINC | Aeronautical Radio Incorporated |
| ATA | Air Transport Association |
| ATRCCS | Automatic Turbine Rotor Clearance |
| ATS | Autothrottle System |
| AUTO | Automatic |
| AVM | Airborne Vibration Monitor |
| | -B- |
| BAL | Balance |
| BPR | Bypass Ratio |
| BRT | Bright |
| BVA | Bleed Valve Actuator |
| | -C- |
| C | Celsius |
| CONT | Continouos |
| CPR | Compressor Pressure Ratio |
| CPU | Central Processing Unit |
| CRT | Cathode Ray Tube |
| | -D- |
| DADC | Digital Air Data Computer |
| DECEL | Deceleration |
| °C | Degrees Celcius |
| °F | Degrees Fahrenheit |
| DEP | Data Entry Plug |
| DMU | Data Management Unit |
| DOT3 | Differential Oil Temperature (No. 3 Bearing Scavenge Oil Temperature MinusEngine Scavenge Oil Temperature) |
| | -E- |
| EBU | Engine Buildup Unit (components which con-Figure 2-an engine for an aircraft installation; usually supplied by other than P&W) |
| ECAAM | Electronic Centralized Aircrfat Monitoring |
| ECS | Environmental Control System |
| EEC | Electronic Engine Control |
| EEROM | Electrically Erasable Read Only Memory |
| EGT | Exhaust Gas Temperature (Also referred to as Tt 4.95 or Tt5) |
| eng | Engine |
| EPR | Engine Pressure Ratio |
| EPR ACT | Actual EPR |
| EPR COM | Commanded EPR |
| EPR MAX | Maximum EPR |
| | -F- |
| F | Thrust |
| FADEC | Full Authority Digital Electronic Control |
| FD | Fuel Drain |
| FLT | Flight |
| FLXTOTEMP | Flexible Takeoff Temperature |
| FMU | Fuel Metering Unit |
| Fn | Net Thrust |

| | |
|---|---|
| Fnf | Fan Net Thrust |
| Fnp | Primary Net Thrust |
| Fnt | Total Net Thrust |
| FP | Fuel Pump |
| ft | Feet |
| | -G- |
| gal | Gallon |
| GND | Ground |
| | -H- |
| Hg | Mercury |
| HPC | High Pressure Compressor |
| HPT | High Pressure Turbine |
| hr | Hour |
| HX | Heat Exchanger |
| | -I- |
| ID | Inside Diameter |
| IDG | Integrated Drive Generator |
| IDGS | Integrated Generator System |
| Ign | Ignition |
| IGV | Inlet Guide Vane |
| in | Inch |
| | -K- |
| K | Constant |
| Kg | Kilograms |
| KPa | Kilo Pascal |
| | -L- |
| L | Left |
| lb. | Pound |
| lb./hr. | Pounds Per Hour |
| LPC | Low Pressure Compressor |
| LPT | Low Pressure Turbine |
| LRU | Line Replacable Unit |
| LVDT | Linear Variable Differential Transformer |
| | -M- |
| MAX | Maximum |
| Mb | Millibars |
| MGB | Main Gearbox |
| mils | Thousands of an Inch |
| Mn | Mach Number |
| MSG | Message |
| | -N- |
| N | Rotational Speed In rpm Or Percent rpm |
| NAC | Nacelle |
| $N_1$ | Low Pressure Compressor Rotational Speed |
| $N_2$ | High Pressure Compressor Rotational Speed |
| $N_2c_2$ | Corrected High Pressure Rotor Speed (derived from Tt2) |
| $N_2c_{25}$ | Corrected High Pressure Rotor Speed (derived from Tt2.5) |
| $N_2$dot | Derivative of High Pressure Rotor Speed With Respect to Time |
| ND | Navigation Display |
| No. | Number |
| | -O- |
| OD | Outside Diameter |
| | -P- |
| P | Pressure |
| Pamb | Ambient Pressure |
| Pb | Burner Pressure |
| $P_{br}$ | Breather Pressure |
| PDU | Pneumatic Drive Unit |
| $P_f$ | Fine Filtered Supply Pressure |

| | |
|---|---|
| PFD | Primary Flight Display |
| $P_{fr}$ | Fuel Pump Interstage Pressure |
| PLA | Power Level Angle |
| PMA | Permanent Magnet Alternator |
| PNAC | Nacelle Pressure |
| PNEU | Pneumatic |
| PO | International Standard Day Sea Level Pressure |
| pph | Pounds Per Hour |
| PRI | Primary |
| $P_s$ | Static Pressure |
| psi | Pounds Per Square Inch |
| psia | Pounds Per Square Inch Absolute |
| psid | Pounds Per Square Inch Differential (Differential Pressure) |
| psig | Pounds Per Square Inch Gauge |
| $P_s3$ | High Pressure Compressor Discharge Static Pressure |
| $P_t$ | Total Pressure |
| $P_t2$ | Compressor Inlet Total Air Pressure |
| $P_t2.5$ | Low Pressure Compressor Exit Total Air Pressure (also referred to as intercompressor pressure) |
| $P_t4.95$ | Exhaust Gas Pressure (also referred to as $P_t4.9$ or $P_t5$) |
| P&W | Pratt & Whitney |
| | -Q- |
| QEC | Quick Engine Change |
| qt | Quart |
| Qty | Quanity |
| | -R- |
| R | Right |
| REF | Reference |
| REV | Reverse |
| REV UNLK | Reverse Unlock |
| rpm | Revolutions Per Minute |
| RVDT | Rotational Variable Differential Transformer |
| | -S- |
| SEC | Secondary |
| SVA | Stator Vane Actuator |
| | -T- |
| Tamb | Ambient Temperature |
| TAT | Total Air Temperature |
| TBD | To Be Determined |
| TBV | Thrust Balance Vent |
| TCC | Thrust Control Computer |
| TEC | Turbine Exhaust Case |
| TLA | Thrust Lever Angle |
| TNGV | Turbine Nozzle Guide Vane |
| Toil | Temperature Engine Oil |
| T/R | Thrust Reverser |
| TRA | Thrust Lever Resolver Angle |
| TRC | Thermatic Rotor Control |
| TRP | Thrust Rating Panel |
| $T_s$ | Temperature Static |
| TSFC | Thrust Specific Fuel Consumption |
| $T_t$ | Total Temperature |
| $T_{t2}$ | Compressor Inlet Total Air |
| $T_{t2.5}$ | Low Pressure Compressor Exit Total Air Temperature |
| $T_{t3}$ | High Pressure Compressor Exit Total Air Temperature |
| $T_t4.95$ | Exhaust Gas Total Temperature (also referred to as $T_t4.9$ and $T_t5$) |

**Table 2-2-1. Cont'd.**

| | |
|---|---|
| TVBCA | Turbine Vane and Blade Cooling Air |
| | -U- |
| US | United States |
| | -V- |
| VAC | Volts, Alternating Current |
| VDC | Volts, Direct Current |
| VIB | Vibration |
| | -W- |
| W | Rate of Flow (Expressed as Weight of Flow) |
| $W_a$ | Rate of Airflow |
| $W_{af}$ | Rate of Fan Airflow |
| $W_{ap}$ | Rate of Primary Airflow |
| $W_{at}$ | Rate of Total Airflow (Wat + Waf) |
| $W_f$ | Rate of Fuel Flow |
| $W_{ft}$ | Fuel Flow Temperature |
| | -SYMBOLS- |
| < | Less Than |
| > | Greater Than |
| Δ | Delta, Differential, Finite Difference |
| ∂ | Theta: RelativeTemperature Ratio |
| δ | Delta: Relative Pressure Ratio |
| = | Equal, Equivalent |
| ≈ | Approximately |
| % | Percent |
| ↑ | Above or Higher |
| ↓ | Below or Lower |
| : | Proportional To |
| → | No Change |

**Engine stations**. Engine stations are used to mark points in the engine's gas path. Stations are marked with the use of numbers, as can be seen in the lower half of Figure 2-3-1. Engine stations can vary from one manufacturer to another, but many of the stations are the same from engine to engine. The core, or gas generator, of the engine generally uses single-digit numbers, while the fan path (in the case of a turbofan) uses two-digit numbers to locate positions in the gas path.

**Flanges.** Flanges, see top of Figure 2-3-1, are identified by the use of letters. Starting with the front and moving to the back of the engine, 'A' is the first flange, 'B' is the second flange and so on. Flanges are just as the name implies, they are the flanges that separate sections of the engine and are, in most cases, the points where the engine is joined together. Most removable flanges are held together with a series of bolts and nuts extending around the outside of the engine radially. Refer to Figure 2-3-2.

**Bearings**. The type and size of the antifriction bearings used in gas turbine engines vary considerably. The two most common are the ball bearings and roller bearings. Examples of these types of bearings are shown in Figure 2-3-3. The bearings are made up of an inner and outer race, with the cage holding the rollers or balls. This cage keeps the roller or balls

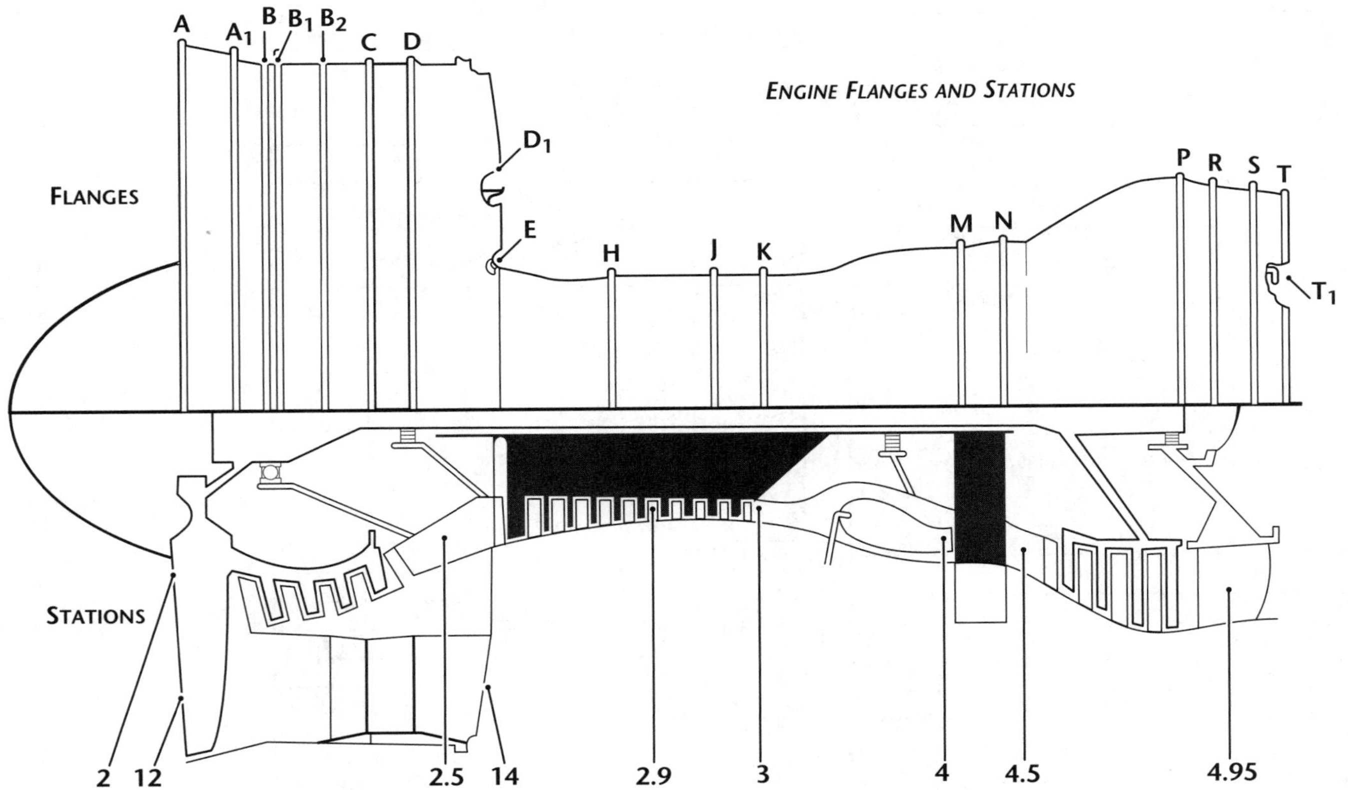


Figure 2-3-1. This diagram points out the engine stations (bottom of diagram) and the flanges (top of diagram).

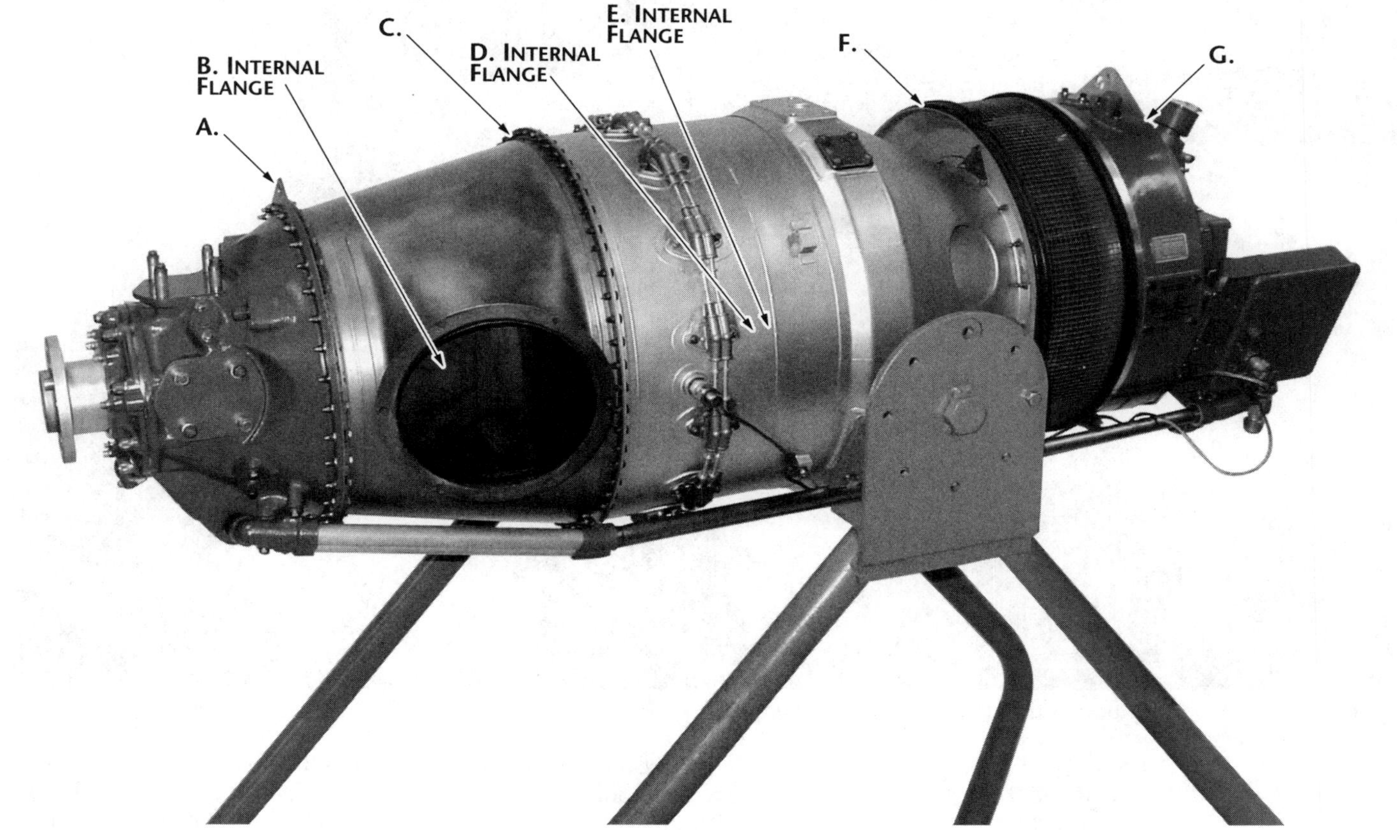

**Figure 2-3-2. The flanges in this PT6 engine are held together with a series of bolts and nuts extending around the outside of the engine radially.**

aligned between the two races, which support the turning shaft. Straight roller bearings can accept only radial loads and will not support the shaft under thrust (axial loads). The thrust load is carried by the ball bearing, which can carry radial and thrust loads.

Each engine configuration differs, but a common method is using one ball bearing and one or more roller bearings per spool shaft. Remember that a spool consists of a turbine wheel, shaft, and compressor. In order to support the spool, a series of bearings are generally needed. An example would be a spool supported by one ball bearing and at least one roller bearing (see Figure 2-3-4, on the following page).

A minimum of one ball bearing is required in each assembly that generates or is subjected to a thrust load. Roller bearings offer a larger bearing contact surface than ball bearings and are the preferred bearing where thrust loads are not present. The use of one ball bearing to handle the thrust load and roller bearings as additional supports where needed is the most common design. This configuration of bearings allows the engine flexibility when it expands and contracts due to temperature changes while operating.

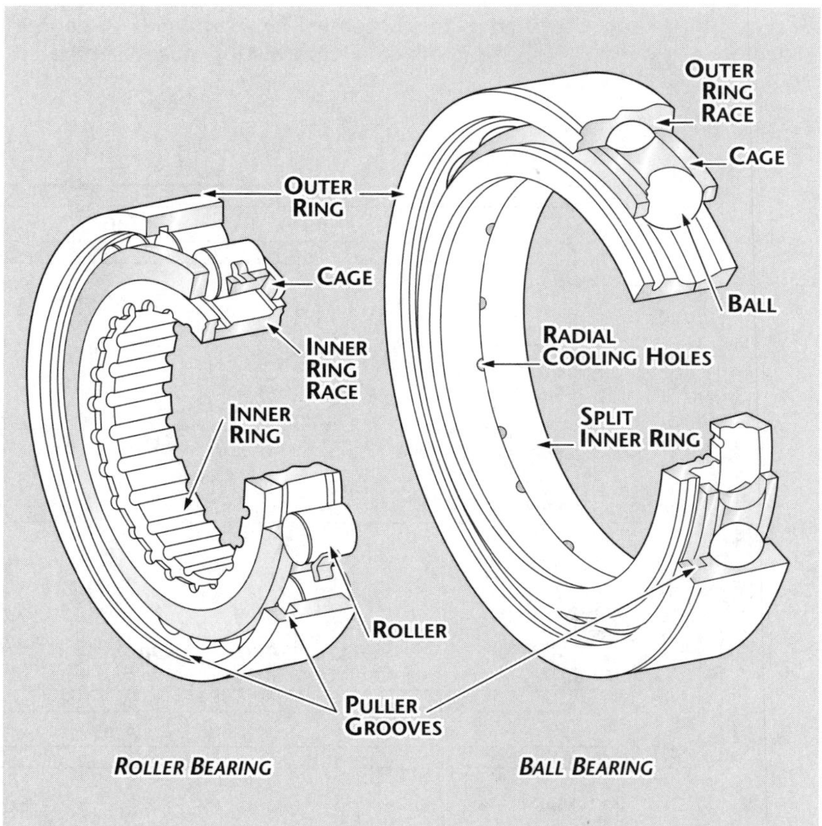


**Figure 2-3-3. This drawing shows the typical antifriction bearings used in turbine engines: ball and roller.**

**Figure 2-3-4. A bearing supports the shaft of a PT6 engine.**

**Seals**. Several types of seals are used to prevent leakage from the engine bearing cavities, cooling airflows and to prevent gas path air from entering into the internal cavities of the engine. Key points in the engine will have hot combustion gases and clean lubricating oil in close proximity. Engine life is seriously degraded if the oil becomes contaminated. Keeping the air, oil, and combustion gases separated within an engine with many moving parts is a major design challenge. The type of seal used is dependent upon several factors, including temperature, pressure, wearability, and ease of installation and removal.

Labyrinth seals, shown in Figure 2-3-5, are widely used to retain oil in bearing cavities and as a method to control internal airflows. A laby-

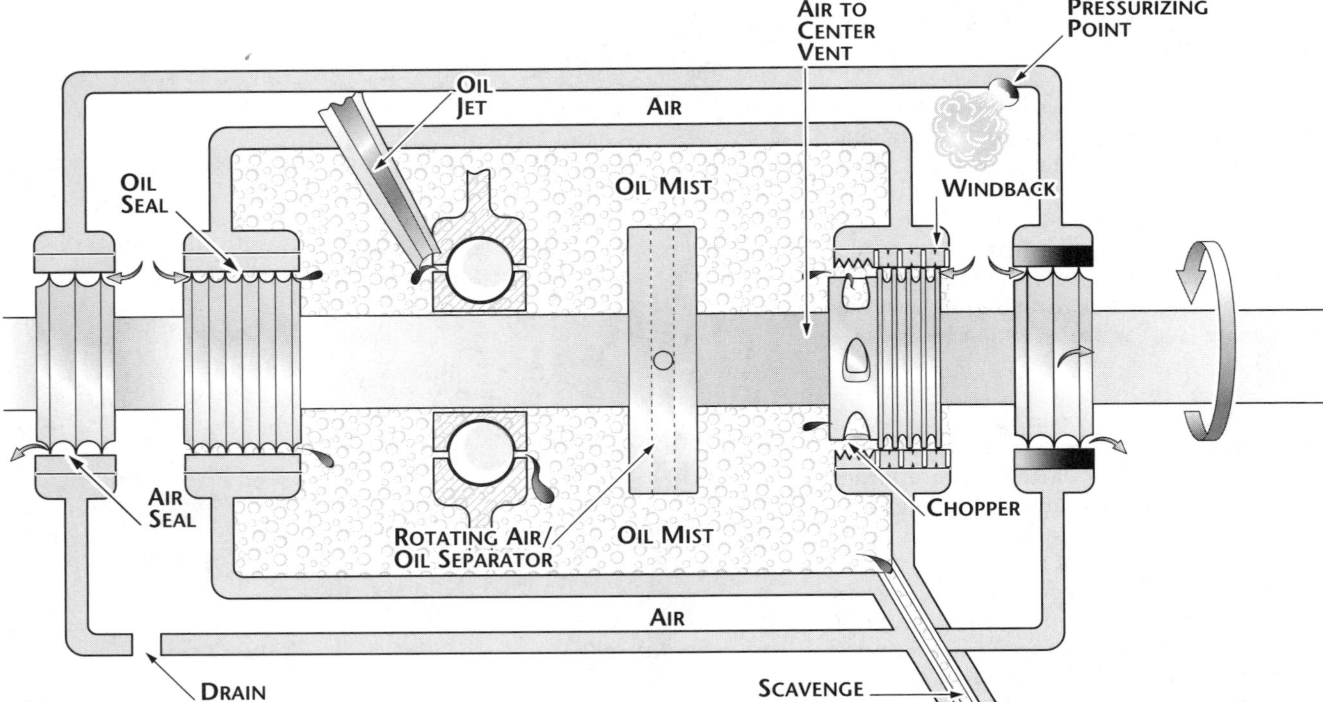


**Figure 2-3-5. This drawing shows the operation of a labyrinth seal, which prevents the oil from leaking out of a bearing cavity.**

Figure 2-3-6. Bearing cavity seal

rinth seal comprises a finned rotating member with a non-rotating bore, which is lined with a soft abrading material, or a high-temperature honeycomb structure. This type of seal does not come into physical contact with the shaft. Instead it operates in very close proximity. The multiple fins provide for a predictable pressure drop within or across the seal.

The clearance varies throughout the flight cycle, dependent upon the thermal growth of the parts and the natural flexing of the rotating members. Across each seal fin, there is a predictable pressure drop. An engineer can calculate the exact number of seal fins required to balance the air pressure on one side with the oil pressure on the other. Designing the seal with the right number of fins results in a restricted flow of sealing air from one side of the seal to the other. When this seal is used for bearing cavity sealing (oil seal), it prevents oil leakage by allowing the air to flow from the outside of the cavity into the cavity (see Figure 2-3-6). This outside airflow also induces a positive pressure in the bearing cavity, which assists in scavenging the oil back to the oil tank. The seal can also be designed to prevent oil or air from moving either way through the seal. The pressure drop across the seal fins creates an area within the seal where the pressures on either side are equalized. At this point, the air and oil pressures are equal and the air is contained on one side, while the oil is contained on the other.

A ring seal, shown in Figure 2-3-7, consists of a ring housed in a groove in a non-rotating housing. The clearance between the ring and rotating shaft is much smaller than with the labyrinth seal. The ring is allowed to move within its groove, allowing the seal ring to move as the shaft flexes. Ring seals can be used for bearing cavity sealing, except in areas of the engine where high temperatures would cause the oil to coke up. The term "coke up" refers to the process where the oil solvents are vaporized, leaving a carbon buildup on the seal components. This action would result in the ring seizing in its groove. Gaps would be created between the shaft and the seal ring as the shaft flexes, which would result in leaks.

A carbon seal, refers to a non-rotating carbon ring, which is in contact with a rotor on the rotating shaft. The rotor has a highly polished face, which contacts the carbon seal surface. Springs are generally used to maintain contact between the carbon seal and the rotor face. This type of seal relies upon a high degree of contact between the face of the rotor and the carbon seal. This type of seal should not allow oil or air leakage across it. The heat caused by friction

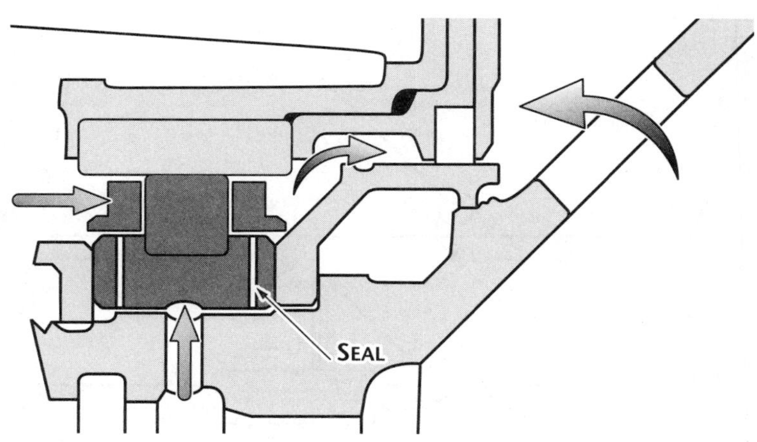


Figure 2-3-7. A diagram of a ring seal, used to seal internal engine parts

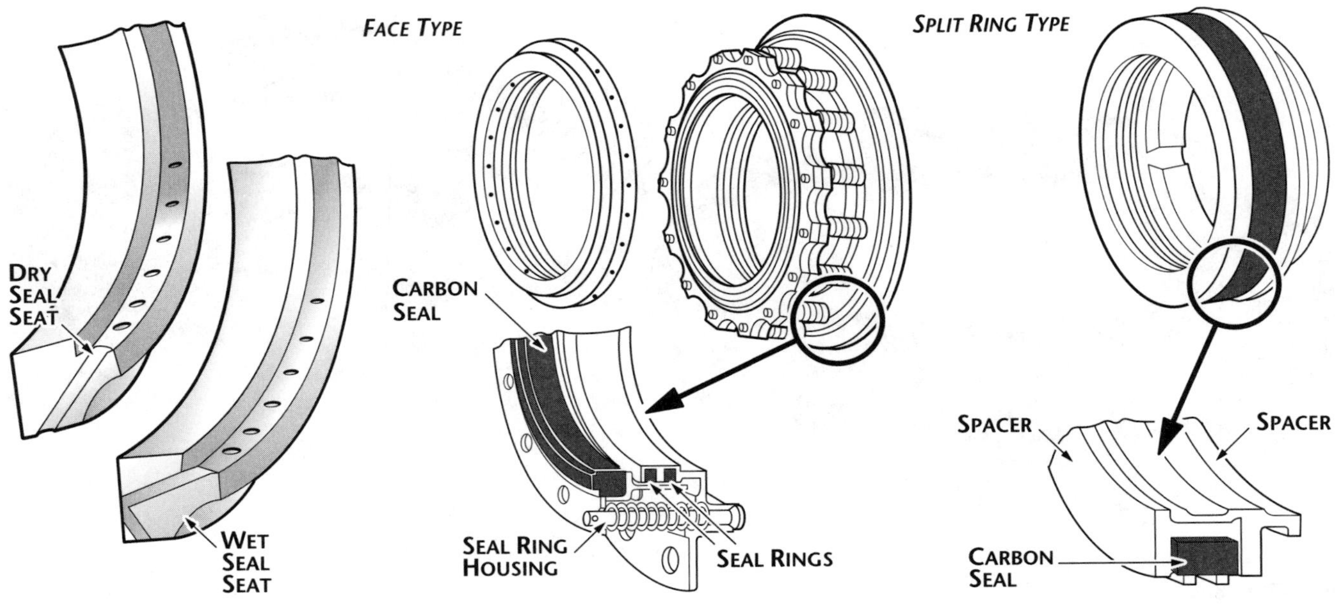


Figure 2-3-8. Applications of a typical turbine engine carbon seal and rotor.

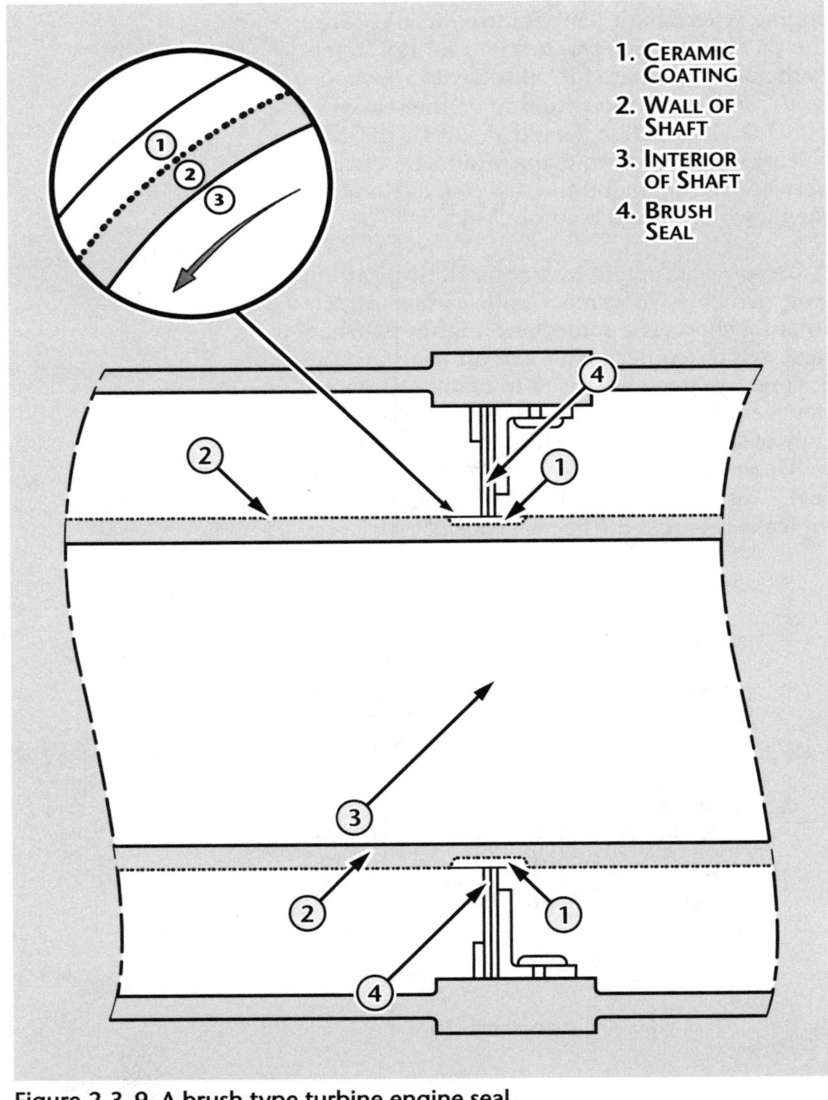


Figure 2-3-9. A brush type turbine engine seal.

is dissipated mainly because of the properties of the carbon material. An example of a carbon seal can be seen in Figure 2-3-8.

Brush seals, shown in Figure 2-3-9, are made up of a non-rotating ring of fine wire bristles. They are in continuous contact with a rotating shaft, rubbing against a hard ceramic coating. This type of seal has the advantage of withstanding radial rubs without increasing leakage. Prevention of hot gas path air ingestion into areas inside the engine - such as around turbine disk, bearing cavities and other components in the hot section of the engine - is achieved by continuously supplying the required quantity of cooling (secondary) air into the engine cavities. This action opposes the inward flow of hot gas. The flow and pressure of the secondary airflow is controlled by some of the aforementioned seals.

**O-rings & packings**. Turbine engines also use traditional o-rings and packings to seal many of the areas that do not require special seals. These are described in detail in the chapter on hardware and seals. The aircraft maintenance technician must use extreme care when replacing o-rings in a turbine engine. Most turbines are lubricated with synthetic oils. These oils will cause rapid breakdown of conventional rubber o-rings. This can quickly lead to engine failure.

Just because an o-ring is the same size and shape does not mean it can be used. Always locate the correct part number o-ring. O-rings made for turbine engines use special rubber compounds that are compatible with turbine oils and/or fuels.

## Section 4

# Turbine Engine Sections

A turbine engine consists of the following major sections and systems:

- Air inlet section
- Compressor section
- Combustion section
- Turbine section
- Exhaust section (including thrust reversers and afterburners)
- Accessory section
- Systems necessary for starting, lubrication and fuel supply and those necessary for auxiliary purposes such as anti-icing and cooling

**Inlet sections**. The air inlet is designed to direct the incoming air into the compressor or fan section smoothly and with minimal energy loss. Most of this energy loss is a result of drag (the friction of the air flowing over the inlet). Nacelles and inlets are carefully designed to achieve optimum flow. This can be a critical factor in aircraft performance. On the Beechcraft King Air series aircraft, a modified inlet design increases cruise speed by as much as 20 knots (a 10% improvement) (See Figure 2-4-1). The flow of air into the compressor should also be free of turbulence to achieve maximum airflow into the engine. The inlet design must also take into account airflow changes that occur as an aircraft maneuvers and the inlet air enters from different angles. The design must maintain a smooth even airflow in all flight conditions.

There are many different types of inlets. The most common inlet used on turbofan and turbojet transport is a simple ring surrounding the front of the engine. A typical inlet is shown in Figure 2-4-2.

One concern with this inlet is the possible formation of ice on the inlet ring. This type of inlet is

**Figure 2-4-1(a). Original King Air Inlet. (b) This redesigned inlet provides significant speed and fuel efficiency gains.**

**Figure 2-4-2. A typical high bypass turbofan inlet nacelle**

**Figure 2-4-3. A bell mouth with an inlet screen is used to test engines in a test cell.**

generally anti-iced by passing hot bleed air from the compressor around the front face of the inlet.

Another type of inlet is used for engine testing in a test cell, a bell mouth inlet, shown in Figure 2-4-3. The round inlet allows the air to flow into the diameter of the bell mouth, making the diameter 100-percent efficient, or as close as possible. This type of inlet uses a protective screen to prevent large objects from being draw into the inlet by the force of the air moving through the engine.

Inlets for supersonic aircraft are designed to provide air to the engine that is subsonic. An example of this inlet type can be seen in Figure 2-4-4. Airflow at the compressor entrance and airflow through the engine must remain below the speed of sound. The spike is designed in a manner that allows it to be repositioned. As such, the pilot can control where the supersonic pressure wave will form in relation to the

Figure 2-4-4. The spike in the inlet duct of the SR71 is part of its inlet air control system.

air inlet, thus controlling the subsonic airflow behind the pressure wave. In addition, drop-down doors to let subsonic air enter the inlet are also part of the nacelle structure of some supersonic aircraft. During certain conditions they provide additional airflow needed.

**Compressor section**. The compressor section of the gas turbine engine has one main function: to supply air in sufficient quantity to satisfy the requirements of the engine. The compressor must increase the pressure of the air received from the air inlet duct and then discharge it to the burners in the correct quantity and at the pressures required.

A secondary function of the compressor is to supply bleed air for various purposes in the engine and aircraft (see Figure 2-4-5). The bleed air is taken from any of the various pressure stages of the compressor. The exact locations of the bleed air ports are, of course, dependent on the pressure or temperature required. The ports are openings in the compressor case adjacent to the particular stage from which the air is to be bled.

Figure 2-4-5. Pressure regulating compressor bleed valve

**Figure 2-4-6. Compressor bleed valve with piping and intercooler**

The bleed air valve (Figure 2-4-6) fulfills two functions. It regulates the compressor pressure and maintains it at the level required for proper combustion. Any excess pressure is released from the compressor before it enters the combustion chamber. In some cases, it is vented overboard, but it can also be used to drive aircraft systems. Most modern turbines are designed to provide an excess of bleed air during all phases of engine operation. Aircraft designers to use the bleed air as a source of hot, high-pressure air to power many vital systems.

Air is often bled from the intermediate- or high-pressure stage, since at this point pressure and air temperature are at a maximum. At times, it may be necessary to cool this high-pressure air before it is used for operation of aircraft systems. In most cases, the air is passed through a precooler, which is a heat exchanger that has fan air passing through it. This cools the hot-bleed air so it is suitable for use in the aircraft's systems.

User- or customer-bleed air is utilized in many different aircraft systems. Some of these systems are:

- Cabin pressurization, heating and cooling
- Anti-icing
- Pneumatic starting of engines
- Auxiliary drive units (ADU)
- Air turbine hydraulic units

**Types of compressors**. The two principle types of compressors in gas turbine engines are the centrifugal flow, shown in Figure 2-4-7, and the axial flow. In the centrifugal-flow engine, the compressor achieves its purpose by picking up the entering air and accelerating it outwardly by centrifugal action. Then the air's velocity is given up and the pressure of the air is increased by the use of a divergent duct (diffuser).

In the axial-flow engine, the air is compressed while flowing straight through the engine. This action avoids the energy loss caused by turning the airflow. From inlet to exit, the air flows along an axial path, hence the name axial-flow compressor. This compressor uses a row of rotor blades, which increase the velocity of the airflow, and a set of stator blades following the rotor blades that increase the air pressure.

There are three types of blade actions that can be used in an axial flow compressor; *impulse*, *reaction*, and *reaction-impulse*. With an impulse blade there is no change in pressure between the rotor inlet and exit. The blades' discharge velocity is the same as its inlet velocity. Change in the state of the airflow is accomplished by the nozzle guide vanes. The nozzle guide vanes are shaped such that they reduce the pressure but increase the velocity.

With the reaction type blade the nozzle guide vanes do little to the airflow but change its direction. Pressure decreases and velocity increases are the result of the convergent shape of the passage between the rotor blades.

A reaction-impulse blade is a combination of the two processes. Both the blades and the guide vanes work together to change pressure and velocity.

The compression ratio of each stage of compression (increase in pressure) is a ratio of approximately 1.25:1 per stage. The compression ratio of a centrifugal compressor can be as high as 8-10:1. Although each type of compressor operates differently, their function is the same: to provide compressed air for the combustion section.

Figure 2-4-7. A centrifugal compressor impeller, like those used on many turbine engines.

**Centrifugal-flow compressors**. The centrifugal-flow compressor consists basically of an impeller (rotor), a diffuser (stator) and a compressor manifold, as illustrated in Figure 2-4-8. The two main functional elements are the impeller and the diffuser. Although the diffuser is a separate unit and is placed inside and bolted to the manifold, the entire assembly (diffuser and manifold) is often referred to as the diffuser. The impeller's function is to bring

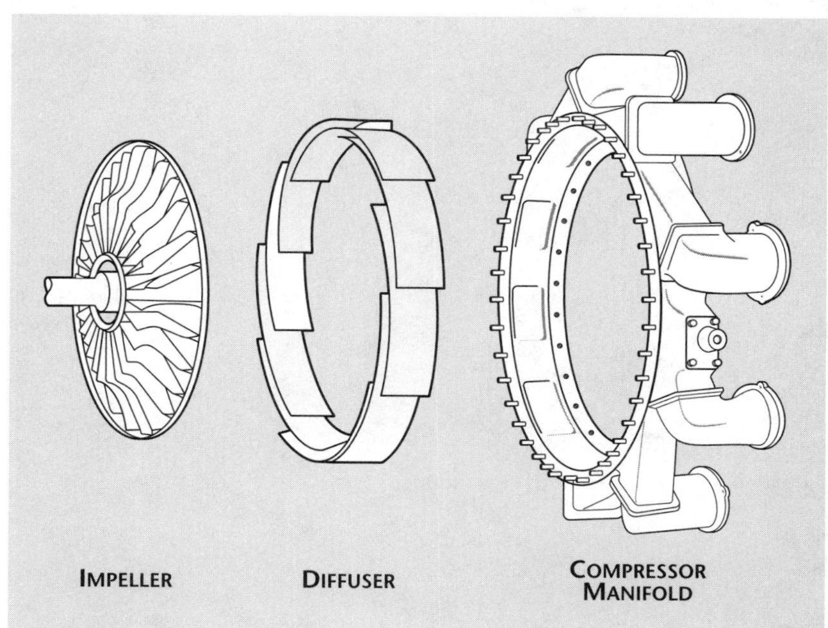


Figure 2-4-8. This drawing shows the components that make up a centrifugal compressor.

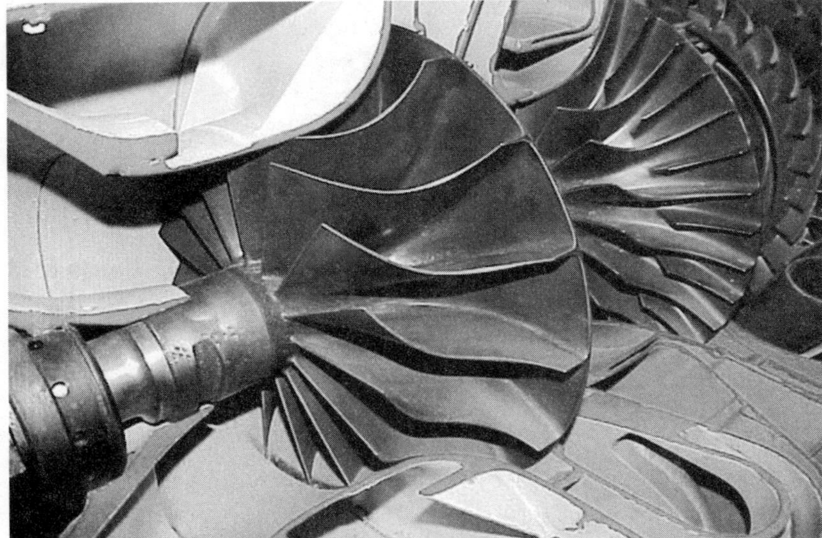

Figure 2-4-9. Two centrifugal compressors inline. The second compressor's intake is the outlet of the first compressor.

in and accelerate the air outwardly to the diffuser. There are two types of centrifugal compressors: single entry and double entry. The principle difference between the two types of impellers is the inlet ducting arrangement. In the double-entry type of impeller, air can flow in from both sides. The single-entry impeller allows for air to enter only from one side of the impeller. These are shown in Figure 2-4-9

To help perform this function in an efficient manner, vanes (cascade vanes or turning vanes) are fitted inside the engine air passages to reduce air pressure losses by presenting a smooth turning surface.

Figure 2-4-10. This picture shows an axial-flow compressor with its two main components (rotor and stator).

**Axial-flow compressor**. The axial-flow compressor, Figure 2-4-10, have two main elements, a *rotor* and a *stator.* Rotors rotate, and stators are stationary. There is always a series of rotors before a series of stators. The rotor has blades fixed radially on a drum. The rotor blades, turning at high speed, draw in air at the compressor inlet and the air is forced rearward into the stator vanes. The action of the rotor increases the velocity of the air and the stator increases the pressure of the air at each stage. As the air passes through each combination of rotor and stator, it is accelerated rearward through several stages. With this increased velocity (imparted by the rotor), energy is transferred through the air in the form of velocity energy. The stator blades act as diffusers at each stage, partially converting high velocity to pressure. Each consecutive pair of rotor and stator blades constitutes a pressure stage. The number of rows of blades (stages) is determined by the amount of airflow and total pressure rise required.

With a greater the number of stages, there is a higher compression ratio. The compression ratio is the outlet pressure divided by the inlet pressure. If the outlet pressure is 300 p.s.i. and the inlet pressure is 14.7 p.s.i., then the compression ratio of this engine would be approximately 20.4:1. Most present-day engines utilize 10-16 stages, or combinations of rotor and stators.

The stator has rows of vanes, as shown in Figure 2-4-11, which project radially toward the rotor axis and fit closely on either side of each stage of the rotor (see Figure 2-4-12). The compressor case, into which the stator vanes are fitted, is horizontally divided into halves. Either the upper or lower half may be removed for inspection or maintenance of rotor and stator blades. The function of the stator vanes is twofold. First, they are designed to receive air from the rotor. They also control the direction of air to the next rotor stage in order that they may achieve the maximum possible compressor blade efficiency.

At the first stage of the compressor, an inlet guide vane assembly sometimes precedes the rotor blades. The guide vanes direct the airflow into the first-stage rotor blades at the proper angle and impart a swirling motion to the air entering the compressor. This pre-swirl, in the direction of engine rotation, improves the aerodynamic characteristics of the compressor by reducing the drag on the first-stage rotor blades. The inlet guide vanes are curved vanes, usually welded to inner and outer shrouds. If the inlet guide vanes are moveable, they are called variable inlet guide vanes (VIG). These can be used to control the amount of air that can be drawn into the compressor. At the discharge end of the compressor, the stator vanes

**Figure 2-4-11. This cutaway view of an axial-flow compressor shows the relationship and location of the rotors and stators.**

are constructed to straighten the airflow using the straightening vanes (outlet vane assembly).

The casings of axial-flow compressors not only support the stator vanes and provide the outer wall of the axial airflow path, but they also provide the means for extracting compressor air for customer- or user-bleed air. This also provides access for bleed valves that are used to prevent compressor stall.

The rotor blades are usually made of stainless steel or titanium. Methods of attaching the blades in the rotor disk rims vary in different designs, but they are commonly fitted into disks by either a bulb-type or dovetail joint

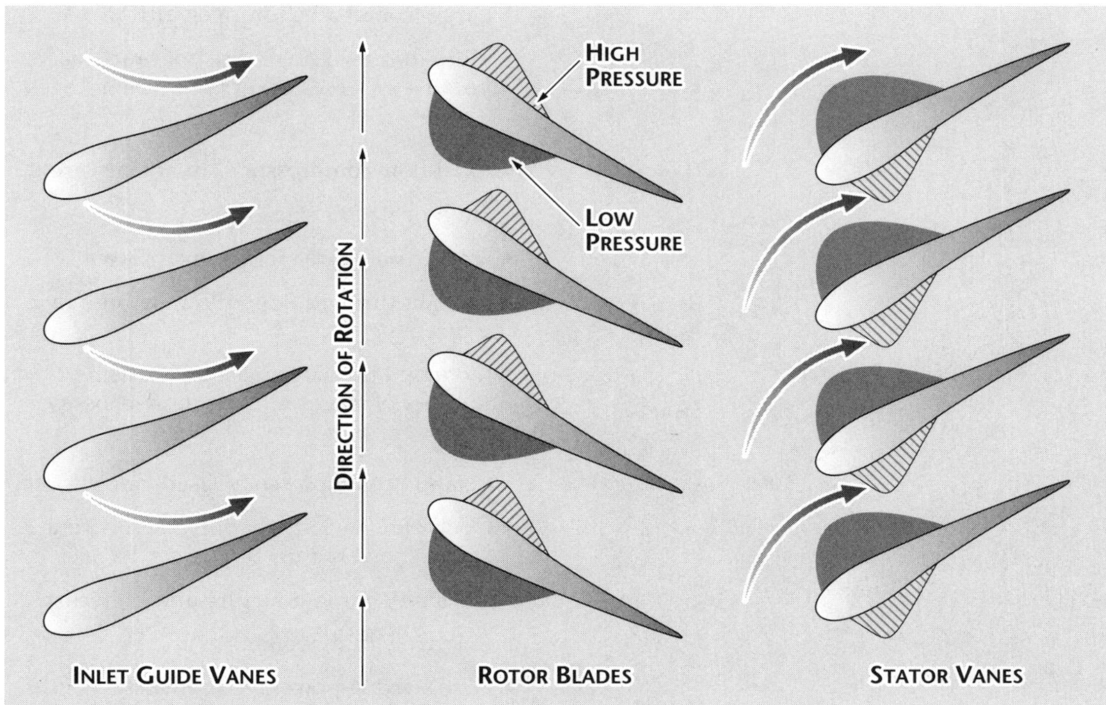


**Figure 2-4-12. Cascade arrangement of stators and rotors**

**Figure 2-4-13. Compressor rotor blades are retained by the dovetail method in the rotor disk.**

assembly (see Figure 2-4-13). The blades are then locked into place.

Compressor blade tips are reduced in thickness by cutouts, referred to as blade profiles. These profiles prevent serious damage to the blade or housing should the blades contact the compressor housing. This condition can occur if rotor blades become excessively loose or if rotor support is reduced by a malfunctioning bearing. Even though blade profiles greatly reduce such possibilities, occasionally a blade may break or crack under stress of rubbing and cause considerable damage to compressor blades and stator vane assemblies.

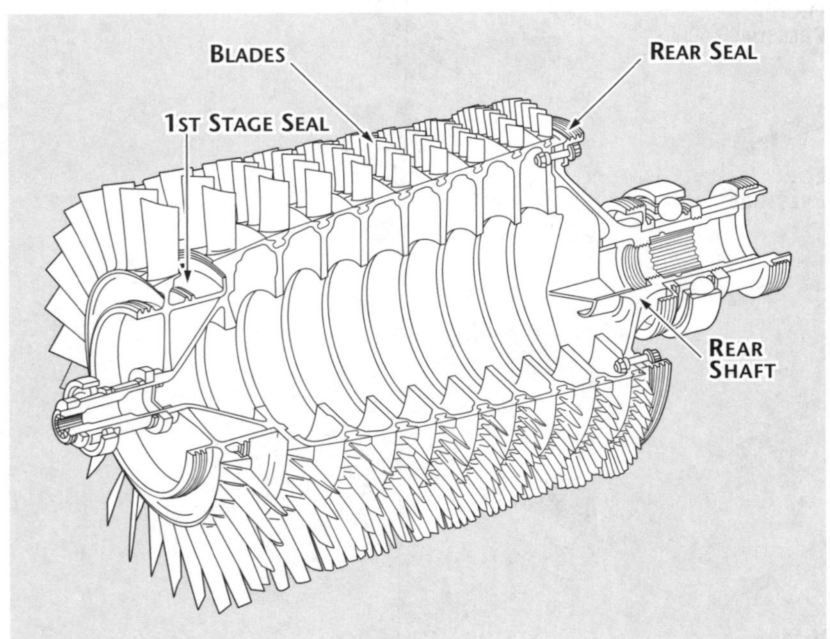


**Figure 2-4-14. In an axial flow compressor, there is a reduction of area from front to rear.**

The blades vary in length from entry to discharge, because the annular working space (drum-to-casing) is reduced progressively toward the rear by the decrease in the casing diameter (see Figure 2-4-14). This reduction will allow the air to flow through the compressor at a fairly constant velocity.

There are two configurations of the axial compressor currently in use, the single-spool compressor and the dual-spool compressor. A spool consists of a compressor shaft and turbine, as can be seen in Figure 2-4-15. The dual-spool compressor is found in most modern turbofan engines. It incorporates two compressors with their respective turbines and interconnecting shafts, which form two physically independent rotating systems. Air exiting the low-pressure compressor feeds into the inlet of the high-pressure compressor. The advantages and disadvantages of both types of compressors are included in the following list. Even though each compressor type has its merits and limitations, performance potential and application are the deciding factor in their use. The centrifugal-flow compressor's advantages are:

- High-pressure rise per stage
- Good efficiencies over wide rotational speed range
- Simplicity of manufacture, thus low cost
- Low weight
- Low starting-power requirements

The centrifugal-flow compressor's disadvantages are:

- Large frontal area for given airflow
- More than two stages are not practical because of losses in turns between stages

The axial-flow compressor's advantages are:

- High peak efficiencies
- Small frontal area for given airflow
- Straight-through flow, allowing high ram efficiency
- Increased pressure-rise by increasing number of stages with negligible losses

The axial-flow compressor's disadvantages are:

- Good efficiencies over only narrow rotational speed range
- Difficulty of manufacture and high cost
- Relatively high weight
- High starting-power requirements. (This is partially overcome by split compressors)

**Figure 2-4-15. A turbine engine spool with two high stages and two low stages**

**Diffuser section**. The diffuser is the divergent section of the engine that is between the compressor and the combustion chamber(s). Typically, it is the point of highest pressure inside the engine. It has the all-important function of changing high-velocity compressor discharge air to a lower velocity. This is accomplished by the shape of the diffuser. The diffuser has a larger diameter at the combustion chamber entrance than it does at the compressor exit.

A decrease in velocity produces an increase in pressure. This prepares the air for entry into the combustion section at a lower velocity and with a decrease in ram pressure. When fuel is added to the higher pressure, lower velocity, air in the combustion chamber it will burn with a flame that will not extinguish.

**Combustion section**. The basic combustion section, shown in Figures 2-4-16 and 2-4-17, houses the combustion process, which raises the temperature of the air passing through the engine. This process releases energy contained in the air/fuel mixture. The major part of this energy, about two-thirds, is required at the turbine to drive the compressor. This is the primary reason that turbine engine fuel consumption at idle is high. Even at idle, a large amount of energy is required to turn the compressor. The remaining energy creates the reaction, or propulsion, and passes out the rear of the engine in the form of high-velocity air. This energy is used, in the case of the turbofan, to turn a set of turbines that, in turn, drive the fan spool. A far smaller amount of energy is then utilized for exhaust thrust. Turboshaft and turboprop engines use this energy to drive the output shaft or propeller, with very little used for exhaust thrust.

The primary function of the combustion section is to burn the fuel/air mixture, thereby adding heat energy to the air. To do this efficiently, the combustion chamber must:

- Provide the means for mixing fuel and air in proper amounts, to assure good combustion

**Figure 2-4-16. A picture of a turboshaft single-burner can (combustion chamber).**

- Burn this mixture efficiently
- Cool the hot combustion products to a temperature that the turbine blades can withstand under operating conditions
- Deliver the hot gases to the turbine section

The location of the combustion section is directly between the compressor and the turbine sections. All combustion chambers contain the same basic elements:

- A fuel drainage system to drain off unburned fuel after system shutdown

**Figure 2-4-17. An example of an annular combustion chamber (combustor).**

**Figure 2-4-18. A picture of the can and annular types of combustion chambers. The can-annular is a combination of these two types.**

- A casing
- A perforated inner liner
- A fuel-injection system
- Some means for initial ignition

There are three basic types of combustion chambers that can be seen in Figure 2-4-18:

- The multiple-chamber or can type
- The annular type
- The can-annular type

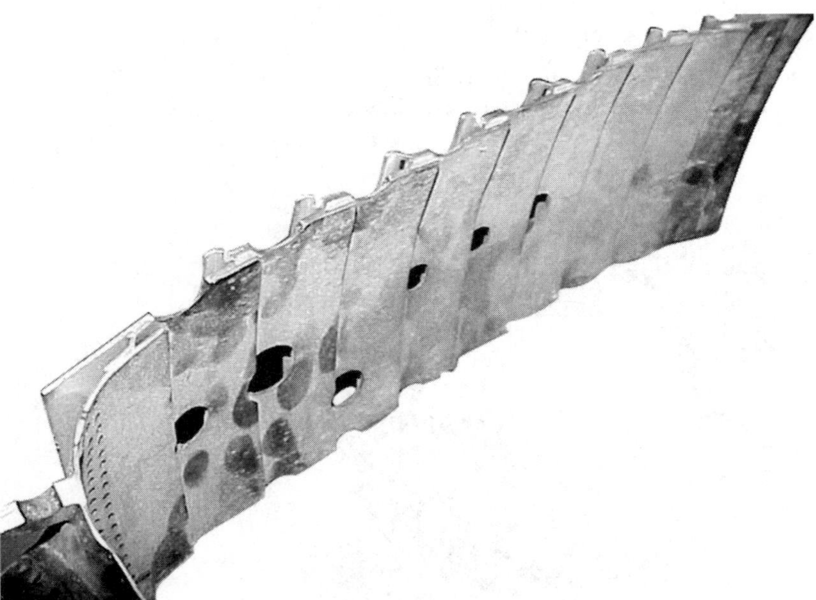

**Figure 2-4-19. A cutaway of an inter-combustion chamber showing the louvers and cooling holes in the section wall.**

Each of the can-type combustion chambers consists of an outer case or housing, within which there is a perforated stainless steel (highly heat-resistant) combustion chamber liner, or inner liner (see Figure 2-4-19). The outer case is divided to facilitate liner replacement. The larger section, or chamber body, encases the liner at the exit end, and the smaller chamber cover encases the front or inlet end of the liner.

The interconnector (flame propagation) tubes are a necessary part of the can-type combustion chambers. Since each can is a separate, independently operating burner, there has to be a way to spread combustion during the initial starting operation. This is accomplished by interconnecting all the chambers.

Another very important requirement in the construction of combustion chambers is providing the means for draining unburned fuel. This drainage prevents gum deposits in the fuel manifold, nozzles and combustion chambers. These deposits are caused by the residue left when fuel evaporates. The danger of afterfire exists if the fuel is allowed to accumulate. If the fuel is not drained, the excess fuel in the combustion chamber might burn at the next startup, allowing the exhaust gas temperatures to go beyond safe operating limits.

Draining the excess fuel is accomplished by a spring-loaded valve that opens when the combustion pressure is low. When the engine is operating, the valve is held closed by the pressure in the combustion chamber (see Figure 2-4-20).

The air entering the combustion chamber is divided into two separate streams of air. These steams of air are called primary and secondary air flows. The primary or combustion air is directed inside the liner at the front end, where it mixes with the fuel and is burned. Secondary or cooling air passes through the outer casing via holes in the liner. This cooling air prevents the combustion gases from coming into contact with the liner's inner diameter.

Louvers are provided along the axial length of the liners to direct a cooling layer of air along the inside wall of the liner. This layer of air also tends to control the flame pattern by keeping it centered in the liner, thereby preventing burning of the liner walls. Farther downstream in the gas path, cooling air joins the combustion gases through larger holes toward the rear of the liner, diluting the combustion gases with cooling air.

To aid in atomization of the fuel, holes are provided around the inlet end of the fuel nozzle. There are vanes that swirl the air, thereby slowing its linear velocity. By slowing the gas flow in this area, these pre-swirl vanes impart

Figure 2-4-20. A combustion chamber drain valve, which allows fuel to exit the chamber at shutdown.

Figure 2-4-21. An annular combustion chamber showing the inner and outer liners. The white ceramic coating is used to prevent burning of the chamber wall.

a swirling motion to the fuel spray itself, which results in better atomization of the fuel. The improved atomization leads to better burning and increased efficiency. This better mixing of the fuel and air will also reduce smoke and exhaust emissions by allowing the fuel-air mixture to burn more completely.

The annular combustion chamber, see Figure 2-4-21, consists basically of a housing and a liner. The liner is made up of an undivided circular shroud extending all the way around the outside of the turbine shaft housing. The liner may be constructed of heat resistant steel. This type of liner can be of the straight-flow or reverse-flow type. In the reverse-flow type, the combustion gases are burned, flowing 180° to the normal gas path direction.

The can-annular combustion chambers, Figure 2-4-22, are arranged radially around the axis of the engine (the axis in this instance being the rotor shaft housing). The combustion chambers are enclosed in a removable steel shroud, which covers the entire burner section. This feature makes the burners readily available for any required maintenance. The burners are interconnected by flame tubes, which facilitate the engine-starting process.

**Turbine section**. The turbine transforms a portion of the kinetic (velocity) energy of the exhaust gases into mechanical energy, used to drive the compressor and accessories. This is the sole purpose of the compressor turbine stages, and this function absorbs approximately 60-80 percent of the total energy in the exhaust gases. The exact amount of energy absorption at the turbine is somewhat dependent upon the load applied to the turbine.

In the case where additional turbine stages drive a propeller (turboprop) or a fan (turbofan) more energy is extracted from the gas path. The

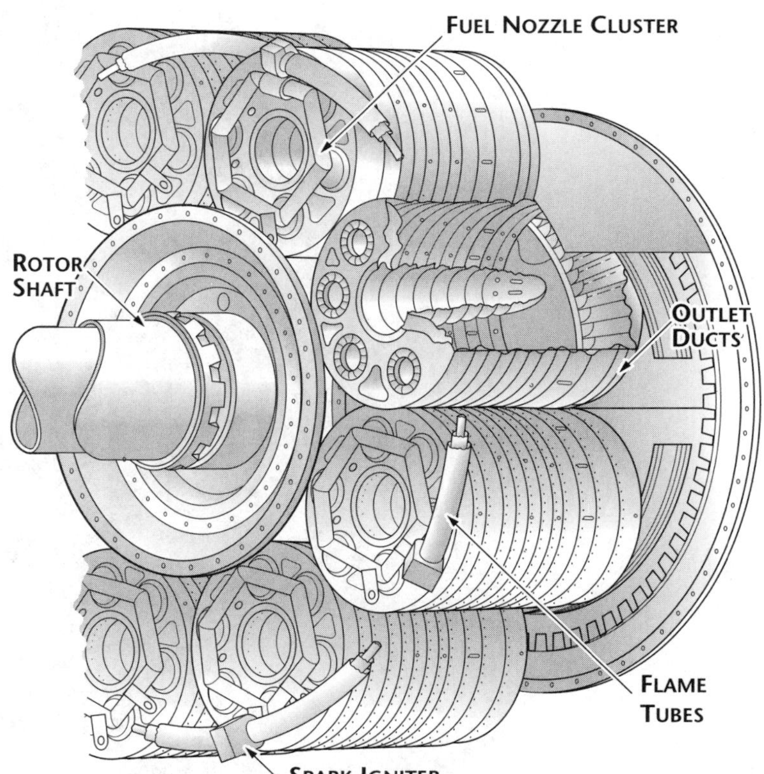


Figure 2-4-22. A can-annular combustion chamber, showing the cans arranged annularly around the engine centerline

**Figure 2-4-23. Three stages of turbine engine inlet guide vanes and blades**

turboshaft engine will also contain additional turbine stages that will drive an output shaft. When additional energy is used to drive a prop, fan or output shaft, the amount of exhaust thrust through the nozzle is reduced.

The turbine section of a gas turbine engine is located downstream of the combustion chamber section. Specifically, it is directly behind the combustion chamber outlet and the first stage turbine inlet guide vanes. This location is the point where the greatest energy is available. The turbine absorbs both heat and pressure energy here and converts some of it into rotating motion.

The turbine assembly consists of two basic elements, the turbine inlet guide vanes and the turbine rotor blades. These two turbine components are shown in Figure 2-4-23.

The turbine stator vane element is known by a variety of names. Turbine nozzle vanes, turbine guide vanes and nozzle diaphragm are the three terms most commonly used. These phrases can often be used interchangeably. The turbine nozzle vanes are located directly aft of the combustion chambers and immediately forward of the turbine wheel. See Figure 2-4-24.

The function of the turbine nozzles is twofold. First, after the combustion chamber has introduced the heat energy into the mass airflow and delivered it evenly to the turbine nozzles, it becomes the job of the nozzles to prepare the mass airflow for driving the turbine rotor (See Figure 2-4-25). The stationary vanes of the tur-

**Figure 2-4-24. Guide vanes and temperature probes on a Pratt & Whitney PT6**

bine nozzles are contoured and set at such an angle that they form a number of small nozzles discharging the gas at extremely high speed; thus, the nozzle converts a varying portion of the heat and pressure energy to velocity energy which can then be converted to mechanical energy through the rotor blades.

The second purpose of the turbine nozzle is to deflect the gases to a specific angle in the direction of turbine wheel rotation. Since the gas flow from the nozzle must enter the turbine blade passageway while it is still rotating, it is essential to aim the gas in the general direction of turbine rotation.

The turbine nozzle assembly consists of an inner shroud and an outer shroud between which are fixed turbine nozzle vanes. The number of vanes employed varies with different types and sizes of engines. Some engines have all the vanes mounted in one ring, while many engines mount only two or three vanes in a cluster. The latter design, accommodates replacement of as few as two vanes at a time.

The rotor element of the turbine section consists essentially of a shaft, a disk and blades (see Figure 2-4-26). The turbine wheel is a dynamically balanced unit consisting of blades attached to a rotating disk. The disk, in turn, is attached to the main power-transmitting shaft of the engine. The gases leaving the turbine nozzle vanes act on the blades or the turbine wheel, causing the assembly to rotate at a very high rate of speed. The high rotational speed imposes severe centrifugal loads on the turbine wheel and at the same time the elevated temperatures result in a lowering of the strength of the material. Consequently, the engine speed and temperature must be controlled to keep turbine operation within safe limits.

**Blade creep**. A turbine blade always increases its length during operation. It is part of the heat expansion process. When the blade cools again, it shrinks to its original dimension, almost! Because of heat and centrifugal force the blade actually grows by an infinitesimal amount, hence the name *blade creep*. The change over time is noticeable to the point that a hot section may have to be performed to stop severe blade rub.

The turbine disk is the rotating disk that the turbine blades are mounted in. When the turbine blades are installed, the disk then becomes the turbine wheel. The disk acts as an anchoring component for the turbine blades. Since the disk is connected to the turbine shaft, the blades are able to transmit to the rotor shaft the energy they extract from the exhaust gases.

The turbine shaft must have some means for attachment to the compressor rotor hub. This

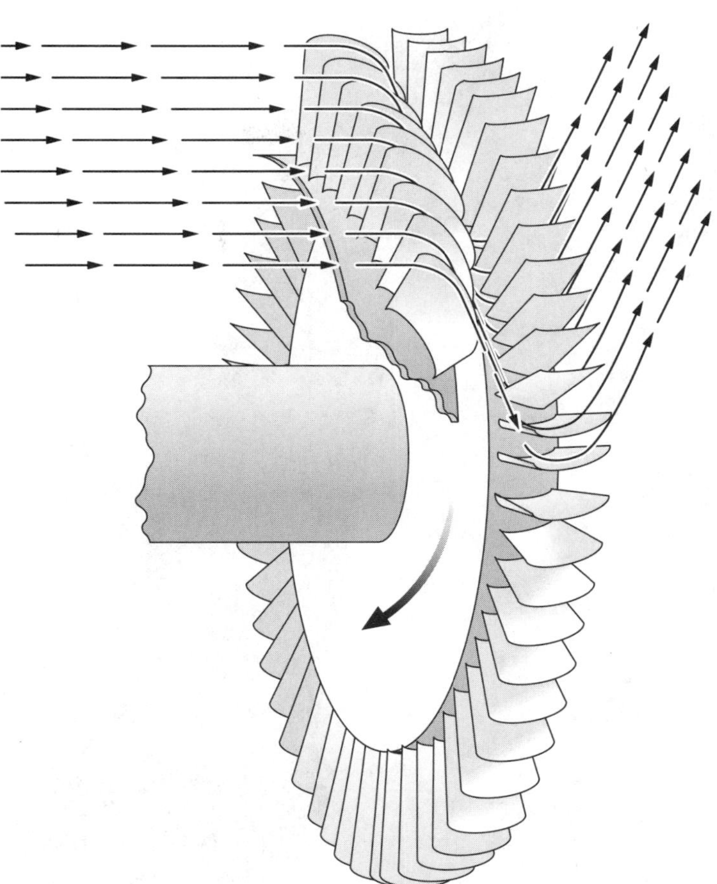

Figure 2-4-25. Guide vanes redirect the combustion gases onto the turbine for the most efficient airflow

Figure 2-4-26. Turbine rotor blades are mounted on the turbine disk, which is mounted on the turbine wheel.

Figure 2-4-27. The fir tree design is used for retention of the blades in the disk.

is usually accomplished by a spline cut on the forward end of the shaft. The spline fits into a coupling device between the compressor and turbine shafts.

There are various ways of attaching turbine blades, some similar to compressor blade attachment. The most satisfactory method used is the fir-tree design shown in Figure 2-4-27. The fir-tree design is so-named because it resembles an upside down pine tree. This design allows for a large contact area and provides a highly secure mounting method, even when subjected to extremely high rotational speeds. The blades are retained in their respective grooves by a variety of methods; some of the more common ones are peening, welding, locktabs, and riveting.

Most turbines are open at the outer perimeter of the blades; however, a second type called the shrouded turbine is sometimes used. The shrouded turbine blades, in effect, form a band around the outer perimeter of the turbine wheel. This improves efficiency and vibration characteristics and permits lighter stage weights; on the other hand, it limits turbine speed and requires more blades (see Figure 2-4-28).

In turbine rotor construction, it occasionally becomes necessary to utilize turbines of more than one stage. A single turbine wheel often cannot absorb enough power from the exhaust gases to drive the components dependent on the turbine for their power and thus, it is necessary to add additional turbine stages.

A turbine stage consists of a row of stationary vanes or nozzles, followed by a row of rotating blades (turbine wheel). In some models

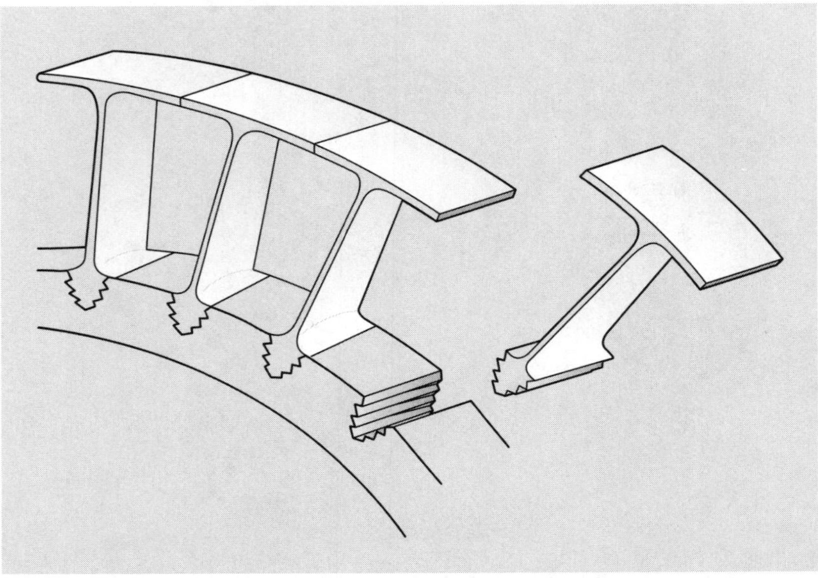

Figure 2-4-28. Shrouded turbine blades make a complete outer ring at the tips of the blades when installed on the disk.

Figure 2-4-29. The high- and low-pressure turbines, which turn the high-pressure compressor and the low-pressure compressor (fan).

of the turboprop engine, as many as five turbine stages have been successfully utilized. Regardless of the number of wheels necessary for driving engine components, there is always a turbine nozzle preceding each wheel.

The turbine rotor arrangement for a dual-rotor turbine, such as is required for a split-spool compressor, is similar to the arrangement in Figure 2-4-29, except that the additional turbine stages drive the fan or low-pressure compressor.

This covers the primary sections of the turbine engine. The remaining sections are the exhaust section, the accessory section and the auxiliary systems. These parts of the engine are discussed in detail in the following sections of this chapter.

*Section 5*

# *Turbine Engine Fuel Systems*

## Fuel Systems

The basic function of the fuel system is to provide the engine with the proper amount of fuel under all operating conditions. Several components are needed to perform this function, such as: pumps, filters, fuel controls, flow dividers and fuel nozzles. In turbine-powered aircraft, varying the flow of fuel to the combustion chambers provides engine speed and power control. The quantity of fuel supplied must be adjusted automatically to correct for changes in ambient temperature or pressure.

The fuel system must deliver fuel to the combustion chambers not only in the right quantity, but also in the right condition for satisfactory combustion. The fuel must be injected into the combustion chambers in a combustible condition during all expected operating conditions. The fuel system must also supply fuel so that the engine can be easily started on the ground and in the air. When the engine is being started or accelerating to its normal running speed, combustion must be sustained. It must also be possible to increase or decrease the power at will, to obtain the thrust required for any operating condition. Turboprop and turboshaft engines also use variable-pitch propellers or rotors; thus, two controllable variables — fuel flow and propeller/rotor blade angle — share the selection of thrust.

**Gas turbine fuels.** Engine fuel must be tailored to the engine and vice versa. There must be enough quantities of fuel available to the engine. Some significant properties of aviation fuels are discussed below.

**Heat energy content.** The energy content or heating value of a fuel is expressed in heat units (British thermal units - BTUs). Fuel satisfactory for aircraft engines must have high heat energy content per unit of weight. Fuel with high heat energy content weighs less than a fuel with low heat energy content, making more of the load-carrying capacity of the aircraft available for the payload. The heat energy content for aviation gasoline is about 18,700 BTUs/pound, and for jet fuels about 18,200 BTUs/pound.

A volatile liquid is one capable of readily changing from a liquid to a vapor when heated or when contacting a gas into which it can evaporate. Since liquid fuels must be in a vaporous state to burn volatility is an important property to consider when choosing a suitable fuel for an aircraft engine. Jet fuels are very satisfactory because they can be blended during the refining process to give the desired characteristics. Because of the nature of constant pressure combustion in gas turbine engines a highly volatile fuel is not necessary. Jet fuels are of rather low volatility, while aviation gasoline is highly volatile. By comparing a highly volatile fuel like aviation gasoline to a less volatile one like Jet 'A' fuel the following effects become apparent. The highly volatile fuel:

- Starts easier in cold temperatures.
- Has a slightly better combustion efficiency.
- Leaves less deposit in the combustion chamber and on the turbine blades.
- Is a greater fire hazard.
- Creates a greater danger of vapor lock of the fuel system.
- Has high evaporation losses through the breather of the fuel tank at high altitudes.
- The last two difficulties are practically nonexistent with fuels having low volatility.

**Stability.** The fuels used in aircraft engines must be stable. Because aviation fuels are sometimes stored for long periods, they must not deposit sediment.

Aviation fuel must be free from water, dirt, and sulfur. Small amounts of water will not usually cause any difficulty because water can be removed from the fuel system by draining. Large amounts of water, however, can cause complete engine failure. It is very important that corrosive sulfur be eliminated from fuel. The sulfur content of fuel may form corrosive acids when brought in contact with the water vapor formed in the combustion process.

The *flash point* is the lowest temperature at which fuel will vaporize enough to form a combustible mixture of fuel vapor and air above the fuel. It is found by heating a quantity of fuel in a special container while passing a flame above the liquid to ignite the vapor. A distinct flash of flame occurs when the flash point temperature has been reached.

The *fire point* is the temperature that must be reached before enough vapors will rise to produce a continuous flame above the liquid fuel. It is obtained in much the same manner as the flash point.

**Specific gravity**. Specific gravity is the ratio of the density (weight) of a substance (fuel) compared to that of an equal amount of water at 60°F. Specific gravity is expressed in terms of degrees API. Pure water has a specific gravity of 10. Liquids heavier than water have a number less than 10. Liquids lighter than water have a number greater than 10. An example is Jet A, whose specific gravity in degrees API is 57. The American Petroleum Institute (API) has chosen pure water by which to measure the specific gravity of fuels.

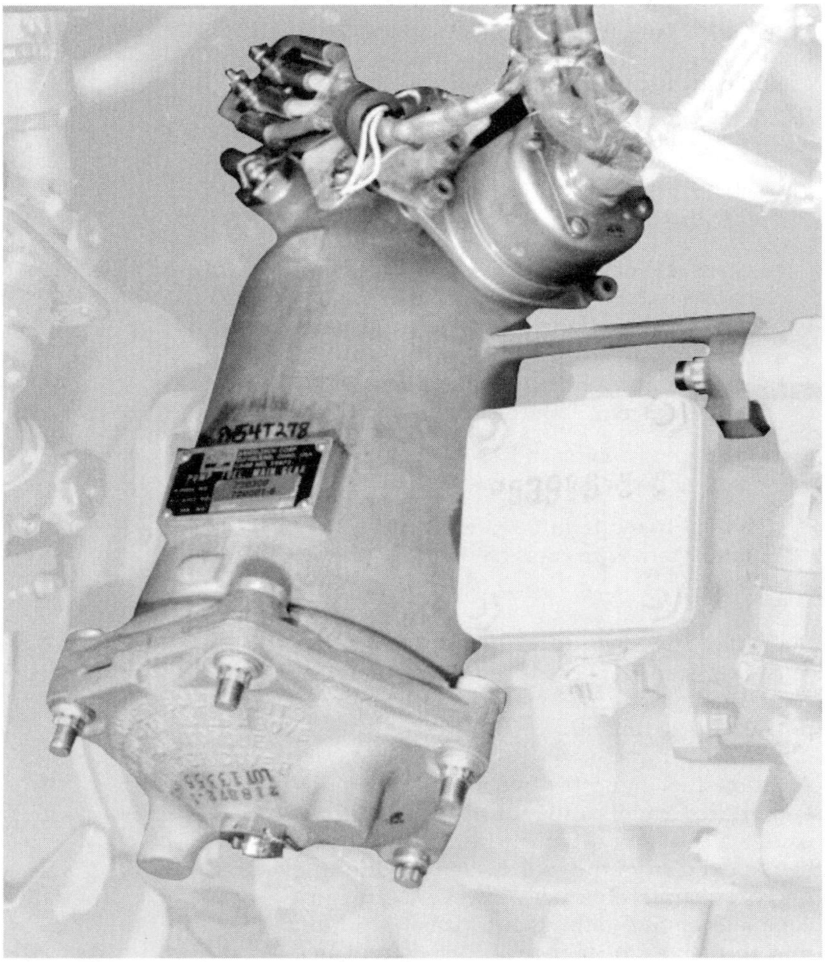

> **NOTE:** *Both flash and fire points give a relative measure of the safety properties of fuel. A high flash point denotes that a high temperature must be reached before dangerous handling conditions are encountered. The minimum flash point permitted in a fuel is usually written into the specifications.*

One of the major differences between the wide-boiling and kerosene types of fuel is the fuel volatility. Kerosene-type fuels have Reid vapor pressures of less than 0.5 pound and flash points higher than 100°C. Wide-boiling-range fuels generally have lower freezing points than kerosene fuels.

Jet A, Jet A1, and Jet B are commercial fuels that conform to the American Society for Testing Materials specification ASTM-D-1655.

Wide-cut gasoline (Jet B) has an advantage of a low freezing point (-76°F). ASTM Jet A and A-1 differ primarily in their fuel freezing points. Jet A is considered suitable down to fuel temperatures of -36° C; Jet A-1, to -54°C.

Jet fuels vary from water white to light yellow; color-coding, however, does not apply to these fuels. Unlike AvGas, they contain no dye.

Additives in jet fuels include oxidation and corrosion inhibitors, metal deactivators, and icing inhibitors. Icing inhibitors also function as biocides to kill microbes in aircraft fuel systems.

Kerosene-type fuels consist essentially of heavier hydrocarbon fractions that are denser than the wide-cut gasoline hydrocarbon factions. Because of this greater density of the hydrocarbons, it has a higher calorific or heating value per gallon.

**Gas turbine fuel contamination**. Precautions are generally taken to ensure that aircraft fuel contains as little water as possible, though it is virtually impossible to have jet fuel that contains no water. The level of water in the fuel is measured in parts per million (ppm) by volume. Until this water collects in the bottom of the tank so it can be drained from the system through the sumps, it will be carried in the fuel. When the fuel temperature drops below freezing, the water droplets can freeze solid. Gas turbine fuel systems are very susceptible to the formation of ice in the fuel filters. When the fuel in the aircraft fuel tanks cools to 32°F or below, water droplets can freeze and plug the fuel system filters. To prevent this from occurring, the system can be equipped with a system to warm the fuel. Fuel heaters generally use bleed air or oil through a heat exchanger to warm the fuel.

**Figure 2-5-1. A turbine engine fuel pump**

An aircraft, after a long flight at high altitudes, will have fuel temperatures and tank surfaces that are colder than the air being drawn into the tank during descent. When moisture-laden air enters the tank space, condensation may occur in the tank. Because of the higher viscosity of cold fuel, this water will not settle out readily and will be carried in the fuel. All turbine fuels can contain some dissolved water. Large quantities of free water (over 30 parts per million) can cause engine performance loss or even flameout.

Keeping the fuel free of water will prevent icing problems and reduce microbiological growth and tank corrosion. Excess water in the fuel can accumulate in the bottom of the tank, where microbiological growth can occur. This can cause blockage of filters and tank corrosion.

**Fuel pumps**. Turbine fuel pumps (engine-driven), as shown in Figure 2-5-1, deliver a continuous supply of fuel at the proper pressure and at all times during operation of the aircraft engine. The engine-driven fuel pumps must be capable of delivering the maximum needed flow at high pressure, in order to obtain satisfactory nozzle spray and accurate fuel regulation.

Turbine fuel pumps may be divided into two distinct categories:

- Constant displacement
- Variable displacement

Their use depends on the placement in the system and on the type of fuel control system used. Variable-displacement pumps (centrifugal pumps) are often used as boost pumps located in the fuel tanks. These are the pumps that provide fuel to the engine-driven pumps. They may also be used as the primary engine driven fuel pump. Pump displacement is changed to meet varying fuel flow requirements; that is, the amount of fuel that is discharged from the pump can be made to vary at any one speed. With a pump of variable flow the applicable fuel control unit can automatically and accurately regulate the pump pressure and fuel delivery to the engine.

Fuel pumps for turbine engines are often positive-displacement gear-type pumps. The term positive displacement means that the gear cavity will supply a fixed quantity of fuel to the engine for every revolution of the pump gears. Gear-type pumps have approximately straight-line flow characteristics (there is a direct relationship between pump speed and fuel flow), whereas fuel requirements fluctuate with flight or ambient air conditions. Hence, a pump of adequate capacity at all engine operating conditions will have excess capacity over most of the range of operation. This characteristic requires the use of a pressure-relief valve for disposing of excess fuel. A typical constant-displacement gear-type pump is illustrated in Figure 2-5-1.

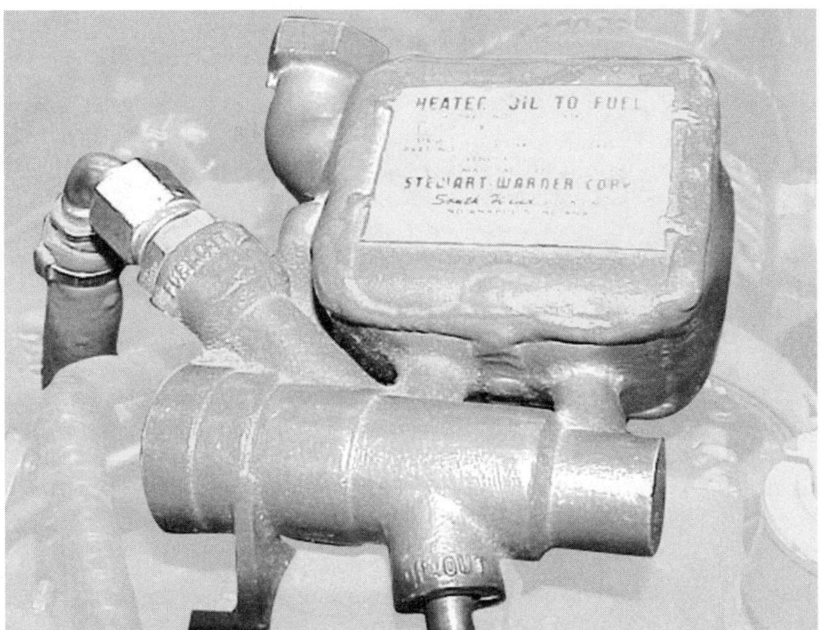

**Figure 2-5-2. A fuel heater warms the fuel for better vaporization.**

**Fuel heater**. A fuel heater, like the one shown in Figure 2-5-2, operates as a heat exchanger to warm the fuel. The heater can use engine bleed air or engine-lubricating oil as a source of heat. The function of a fuel heater is to protect the engine's fuel system from ice formation. However, should ice form, the heater can also be used to thaw ice on the fuel screen. In some installations, the fuel filter is fitted with a pressure-differential warning switch, which illuminates a warning light on the cockpit instrument panel when sufficient ice has accumulated to cause a pressure difference between the inlet and outlet sides of the filter.

**Fuel filters**. A low-pressure filter is installed between the supply tanks and the engine fuel system to protect the engine-driven fuel pump and various control devices. An additional high-pressure fuel filter is installed between the fuel pump and the fuel control. The three most common types of filter in use are the micron filter, the wafer screen filter and the plain screen mesh filter.

**Micron filter**. The micron filter has the greatest filtering action and, as the name implies, is rated in microns. A micron is the thousandth part of one millimeter. The porous cellulose material frequently used in construction of the filter cartridges is capable of removing foreign matter measuring from 10-25 microns. The minute openings make this type of filter sus-

Figure 2-5-3. A turbine engine micron filter is used to filter the fuel.

Figure 2-5-4. A wafer screen filter uses a replaceable wafer disc element to filter the fuel.

ceptible to clogging; therefore, a bypass valve is a necessary safety factor. An example of a micron filter is shown in Figure 2-5-3.

**Wafer screen filter**. The wafer screen type of filter, shown in Figure 2-5-4, has a replacement element that is made of layers of screen discs of bronze, brass, steel, or similar material. This type of filter is capable of removing minute particles. It also has the strength to withstand high pressure.

**Plain mesh filter**. The plain screen mesh filter is the most common type. It has long been used in internal combustion engines of all types for fuel and oil strainers. It is simply a wire screen, often formed into a cone, placed in the fuel flow path. Any particles larger than the holes in the screens are prevented from moving further downstream. In present-day turbine engines, it is used in units where filtering action is not so critical, such as in fuel lines before the high-pressure pump filters. The mesh size of this type of filter varies greatly according to the purpose for which it is used.

Gas turbine engines are equipped with a fuel-control unit that automatically satisfies the requirements of the engine. Fuel controls can be divided into three basic groups:

- Hydromechanical
- Electronic-hydromechanical
- Full authority digital electronic control (FADEC)

Early fuel controls were hydromechanical only. While effective in controlling engine operation, the mechanical linkages within them limit their capabilities. To properly control the fuel flow in a wide variety of operating environments, the mechanical linkages become quite complex.

In an effort to improve the functioning of the fuel control unit, the electronic-hydromechanical FCU was developed. The electronics in these units primarily act in a supervisory capacity. They monitor the engine operation and, through the use of a torque motor linked to the hydromechanical FCU, prevent the engine from operating in an overspeed or overthrust condition.

The current state of the art is a unit known as a FADEC, or Full Authority Digital Engine Control. There are no mechanical linkages used in this system, rather the entire operation is controlled electronically. The aircraft throttle is connected to a *precision position transmitter.* FADECs receive an electric signal from this transmitter. The FADEC then controls the operation of the engine provide the most efficient operation for the desired thrust setting. Electronics were initially developed to compute the fuel flow more accurately. The electronic controls have increased efficiency and prolonged engine life. This increased efficiency saves fuel and reduces maintenance costs.

Regardless of the type, all fuel controls accomplish essentially the same basic functions. The two sections of a fuel control, computing and metering, are used to determine the amount of fuel flow and then meter the proper amount of flow. The fuel control senses power lever position, engine r.p.m., either compressor inlet pressure or temperature, and burner pressure or compressor discharge pressure. These variables affect the amount of thrust that an engine will produce for a given fuel flow.

## Hydromechanical Fuel Controls

Hydromechanical turbine engine fuel controls can be extremely complicated devices, as shown in Figure 2-5-5. The hydromechanical types are composed of speed governors, servo systems, sleeve and pilot valves, feedback or follow-up devices, and metering systems. This type of control is mechanically linked to the engine through the accessory section, and it senses engine gas generator speed. The fuel control must control r.p.m., prevent overheating, provide a suitable air-fuel mixture and prevent surging (compressor stall). Some fuel controls have two operating levers while other only have one. On the units with two levers, the power lever controls engine r.p.m. in all engine modes, while the second lever turns fuel metering on or off. Single lever units have all functions controlled by one lever.

A fuel control in its simplest form is generally used on an APU, an example of which can be seen in Figure 2-5-6. Following through the operation of a FCU (fuel control unit), the fuel entering the FCU first passes through a positive-displacement gear-type pump. Upon discharge from the pump, the fuel passes through a filter element. The filter is installed at this point to filter any wear debris that might be discharged from the pump element. Filtered fuel then enters the fuel control unit.

Engine acceleration, exhaust temperature and speed are automatically controlled within established limits by the metering of the fuel flow by the fuel control unit. Engine acceleration is controlled by the action of the acceleration-limiter valve in the fuel-control unit. This valve references compressor air discharge pressure to fuel pressure. This ratio is used to determine the position of a bypass valve. The valve allows some fuel to bypass in order to maintain the required ratio of fuel flow to compressor air pressure.

**Figure 2-5-5. A JT8D hydromechanical fuel control mounted on the engine**

**Figure 2-5-6. An APU fuel control mounted on the engines accessory case.**

During startup and the first stages of acceleration, when compressor air pressure is very low, the valve does not bypass any fuel. The fuel pressure at which the valve permits normal bypass of fuel is the acceleration-limiter valve cracking-pressure setting of the fuel-control unit. Engine speed is controlled by action of the governor in the fuel-control unit. An example is shown in Figure 2-5-7. Normally a flyweight-type governor, it is driven by the main fuel pump drive shaft, which, in turn, is driven through the accessory gear train.

To better understand the functioning of a typical fuel control, a representative JFC (Jet Fuel Control) series hydromechanical unit will be explained in detail. The details of this process vary from one fuel control to another, even within the JFC series units, but all accomplish the same basic functions in a similar fashion. Some fuel controls use a single lever to control all of the functions, others have a separate lever for the fuel shut-off function. The JFC unit described is a single lever type. The basic internal operation of this unit is shown in Figure 2-5-8.

The fuel control unit is a lightweight, high-capacity, fuel-flow-metering unit that is designed to permit selection of a desired engine thrust for the ambient operating conditions encountered during flight. Engine thrust during ground operation and under various flight conditions is controlled by a single power lever, which is also used to regulate fuel for engine starting and shutdown. The variables sensed by the fuel control are power-lever angle, burner pressure (PB), high-pressure compressor speed ($N_2$) and compressor inlet temperature ($Tt_2$). By utilizing these variables, the fuel control accurately governs the engine steady-state selected through a speed-governing system of the proportional or droop type. The fuel control also utilizes these same variables to control fuel flow for acceleration and deceleration.

These same variables control fuel flow for acceleration and deceleration. When engine speed exceeds the preset governed speed, the governor flyweights move outward to open a slide valve to bypass fuel, which decreases engine speed. When engine speed drops, the governor flyweights move inward to close the slide valve to permit greater fuel flow, which increases engine speed to normal governed conditions.

Figure 2-5-7. A PT6 fuel control with fuel pump and filter

The fuel control consists of a fuel-metering system and a computing system.

**Fuel Metering System.** The metering system regulates fuel supplied to the engine by the engine-driven fuel pump to provide the engine thrust demanded by the pilot. Fuel regulation is also controlled by engine operating limitations, as sensed and scheduled by the fuel control computing system. The computing systems senses and combines various operational parameters to govern the output of the metering system of the fuel control under all engine operating conditions.

High-pressure fuel is supplied to the control inlet from the engine-driven pump. Fuel is filtered by both a coarse (80-mesh) screen and a fine (40-mesh) screen at the inlet of the control. The course screen protects the metering system from large particles of fuel contaminants. If this screen becomes clogged, a filter relief valve will open, permitting continued operation with unstrained fuel. The fine screen protects the computing system against solid contaminants.

Next, the fuel encounters the pressure-regulating valve, which is designed to maintain a constant pressure differential across the throttle valve. All high-pressure fuel in excess of that required to maintain this pressure differential is bypassed to the pump interstage by the pressure-regulating valve. This valve is servo controlled, whereby the actual pressure drop across the throttle valve orifice is compared, by the sensor, with a selected pressure drop, and any error is hydraulically amplified. The amplified error positions the pressure-regulating-valve spring, altering the force balance of this valve so that sufficient high-pressure fuel is bypassed to maintain the selected pressure drop. The pressure-regulating-valve sensor also incorporates a bimetallic disc to compensate for any variation in the specific gravity of the fuel, which results from fuel temperature change.

The high-pressure fuel, as regulated by the pressure-regulating valve, then passes through the throttle valve. This valve consists of a contoured plunger that is positioned by the computing system of the control within a sharp-edged orifice. By virtue of the constant pressure drop maintained across the throttle valve, fuel flow is a function of the plunger position. An adjustable stop is provided to

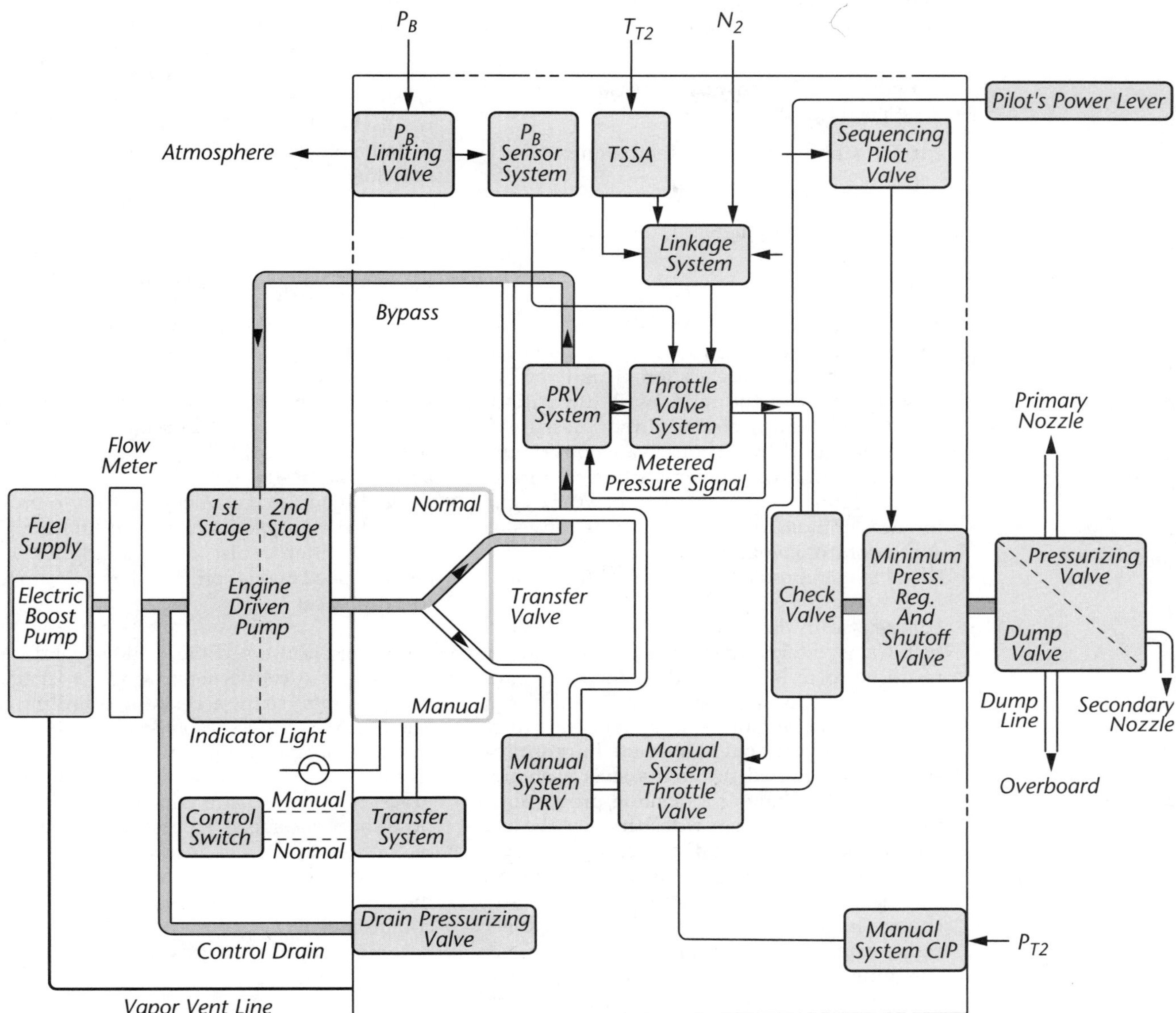


**Figure 2-5-8. JFC fuel control flow schematic**

limit the motion of this plunger in the decrease fuel direction to permit selection of the proper minimum fuel flow.

The final component to act upon the metered flow prior to its exit from the control is the minimum pressure and shut-off valve. This valve is designed to shut off the flow of metered fuel to the engine when the power lever on the control is moved to the OFF position. This causes the power-lever-operated sequencing valve to transmit a high-pressure signal to the spring side of the shut-off valve, forcing the latter against the seat, thus shutting off the flow of fuel to the engine. When the power lever is moved out of the OFF position, the high-pressure signal is replaced by pump interstage pressure, and when metered fuel pressure has increased sufficiently to overcome the spring force, the valve opens and fuel flow to the engine is initiated. Thereafter, the valve will provide a minimum operating pressure within the fuel control, ensuring that adequate pressure is always available for operation of the servos and valves at minimum flow conditions.

The power-lever-operated sequencing valve also incorporates a windmill bypass feature, which functions when the shutoff valve is closed. This feature bleeds throttle valve discharge flow to the fuel pump interstage to increase the throttle valve pressure drop and thereby cause the pressure-regulating valve to open. Damage to the fuel pump from excessive pressure is thus prevented during engine windmilling. The sequence valve functions in both the normal and the manual operating systems.

**Fuel Computing System.** The computing system measures and compares a number of variables. These variables are:

- $N_2$ High Pressure Compressor Rotor Speed (RPM)
- $T_{T2}$ Compressor Inlet Temperature
- $P_B$ Burner Can Pressure
- $W_F/P_B$ Ratio of metered fuel flow to burner can pressure

The computing system positions the throttle valve to control steady-state engine speed, acceleration, and deceleration. This is accomplished by using the ratio $W_F/P_B$ (the ratio of metered fuel flow to engine burner pressure) as a control parameter. Throttle valve positioning by means of this parameter is achieved through a multiplying system whereby the $W_F/P_B$ signal for acceleration, deceleration, or steady-state speed control is multiplied by a signal proportional to $P_B$ to provide the required fuel flow.

$P_B$ is sensed in the following manner. A motor bellows is internally exposed to $P_B$ and the resulting force is increased by the force of an evacuated bellows of equal size directly connected to the motor bellows. The net force, which is proportional to absolute burner pressure, is transmitted through a lever system to a set of rollers having a position proportional to $W_F/P_B$. These rollers ride between the bellows-actuated lever and a multiplying lever. The force proportional to $P_B$ is thus transmitted through rollers to the multiplying lever. Any change in the roller position ($W_F/P_B$) or the $P_B$ signals upsets the equilibrium of this lever, changing the position of a flapper-type servo valve, which is supplied with regulated high-pressure fuel through a fixed bleed orifice. The resulting change in servo pressure between the two orifices is used to control the position of a piston attached to the throttle valve plunger. The motion of this piston compresses or relaxes a spring that will return the multiplying lever to the equilibrium position. An adjustable minimum-ratio stop on the $W_F/P_B$ signal controls engine deceleration. This arrangement provides a linear relationship between decreasing $W_F$ and $P_B$, which results in blow-out free decelerations.

An adjustable maximum-ratio stop on the $W_F/P_B$ signal controls engine acceleration. This stop is positioned by an acceleration-limiting cam that is rotated by a speed-sensing servo system and translated by a compressor inlet temperature ($T_{T2}$) sensing servo system. The cam is so contoured as to define a schedule of $W_F/P_B$ versus engine speed for each value of $T_{T2}$ that will permit engine accelerations that avoid engine overtemperature and surge limits, without compromising engine acceleration time.

A burner pressure limiter incorporated in the fuel control senses burner pressure with respect to ambient pressure. When this differential exceeds a preset maximum, the pressure signal to the burner pressure motor bellows is reduced by bleeding through the limiter valve to ambient pressure. This causes a limitation in fuel flow, which prevents burner pressure from exceeding a maximum safe value.

A flyweight-type, engine-driven, speed-sensing governor controls movement of the speed servo piston through a pilot valve. When $N_2$ speed changes, the flyweight force varies and the pilot valve is positioned to meter either low-or high-pressure fuel to the speed servo piston. The motion of the piston repositions the pilot valve until the speed-sensing system returns to equilibrium. The piston incorporates a rack that meshes with a gear segment on the three-dimensional acceleration cam to provide the speed signal for acceleration limiting. This piston position is also used to indicate actual engine speed and is connected by a droop lever to a droop cam.

The temperature-sensing bellows and servo assembly are connected through a lever and yoke assembly to the acceleration-limiting cam. The position of this servo piston is indicative of compressor inlet temperature ($T_{T2}$) and is used to translate the acceleration cam, which integrates the temperature and speed signals. The position of the speed-set cam is also translated by the servo piston by means of a cross-link to the acceleration cam. Engine steady-state condition is therefore a function of high-pressure compressor ($N_2$) speed, compressor inlet temperature ($T_{T2}$), burner pressure ($P_B$), and power lever position.

In the event that the primary control system malfunctions, the FCU is also provided with a manual mode. A cockpit switch activates the manual mode. The computing system is bypassed and fuel control is directly controlled by the pilot. Manual mode is only used in emergencies because it is very easy to overtemp, stall out, or blow out during acceleration or deceleration.

This example illustrates how mechanically complex the typical hydromechanical fuel control is. Each of the internal mechanisms is designed to control the specific engine model that the FCU is intended to operate with. Different engines, and even different models of the same engine, will have FCUs that operate on different internal ratios.

In the field, most maintenance is limited to external adjustments and trimming. When major repair or overhaul is required, the unit is usually sent to an authorized repair station and overhauled there.

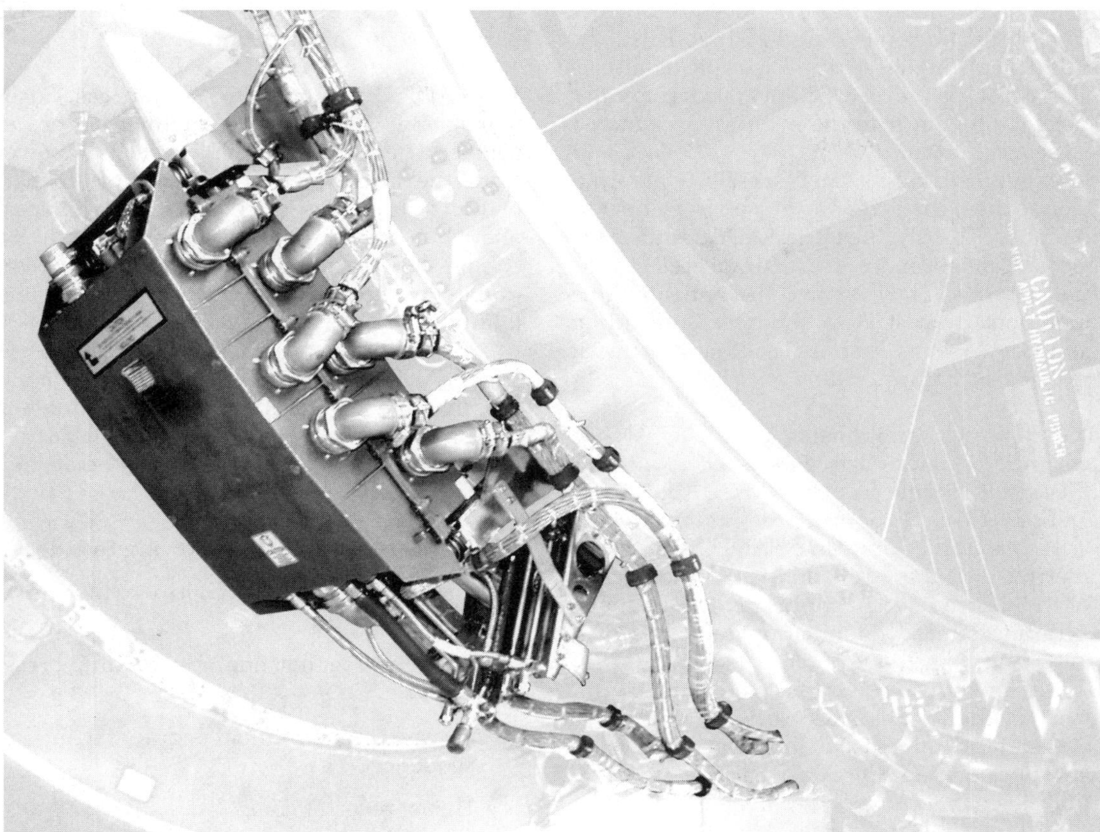

Figure 2-5-9. Full-authority digital electronic control unit mounted on the engine fan case.

## Electronic-Hydromechanical Fuel Controls

Hydromechanical FCUs are often limited to controlling fuel flow by linear relationships because of their mechanical linkages. Many engine parameters do not follow strict linear curves. Newer engine designs have added various levels of electronics to the fuel metering process in order to better match fuel flow to operating characteristics. The next step in the development of the fuel control is the electronic-hydromechanical fuel control. These systems add an Electronic Engine Control (EEC) unit to the system. The EEC in these systems acts primarily in a supervisory capacity. The pilot still controls the fuel control through a mechanical linkage, but an additional electronic unit is interconnected.

The EEC monitors certain engine parameters and takes over some of the pilot's workload. In straight hydromechanical systems, the pilot must continually monitor turbine temperatures and power settings (EPR for turbofans and turbojets, Torque for turboprops and turboshafts). When operating close to the power limits, such as during takeoff, the flight crew must pay particular attention to these settings to prevent overstressing the engine. In Electronic-Hydromechanical units, the EEC takes over the task of monitoring these limits. When these limits are met, the EEC uses a torque motor to adjust the hydromechanical fuel control to a slightly lower power setting.

These electronic fuel-control system units incorporate amplifiers, thermocouples, relays, electrical servo systems, torque motors, switches and solenoids. Field maintenance is limited to troubleshooting and removal and replacement of the electronics boxes. The boxes are then sent to a qualified repair facility to be tested and repaired. Their internal electronics are quite complex and only qualified personnel should open the boxes.

## Full Authority Digital Engine Controls (FADEC)

The current state-of-the-art fuel control design is the Full Authority Digital Engine Control or FADEC. FADEC units have no mechanical linkages to the cockpit. The throttle is connected to a precision potentiometer. This sends an electrical signal to the FADEC indicating the power level commanded by the pilot. The FADEC computer then calculates the correct fuel flow and other engine settings necessary to achieve this power level in the most economical fashion. A typical unit can be seen in Figure 2-5-9.

The FADEC has three distinct differences from earlier fuel metering systems. It has no mechanical linkages to the cockpit. It also is able, through the use of advanced computer systems, to eliminate the compromises forced upon mechanical systems. The fuel flow and other engine controls can be precisely manipulated to meet the demands of any given operating mode. Thirdly, the latest FADEC units have become much more than simply electronic fuel controls. They control the entire engine operation, from start to shutdown, including auxiliary systems such as ignition and thrust reverse (see Figure 2-5-10).

The computer components in a FADEC system may be referred to as a FADEC, an EEC (Electronic Engine Control), EFCU (Electronic Fuel Control Unit), or other similar term. Many of the electronics units in a FADEC system are referred to using terms similar to those used for the electronic components in the Electronic-Hydromechanical combination units.

The following text describes the FADEC unit used on the Airbus A320 series aircraft. It is an advanced unit that demonstrates the wide variety of engine functions that a late generation FADEC can control (Figure 2-5-11).

Both engines are equipped with a FADEC system. This system is also commonly referred to as the Electronic Control Unit (ECU). It is a digital control system that performs complete engine management. The airframe Engine Interface Unit (EIU) transmits to the FADEC data for engine management.

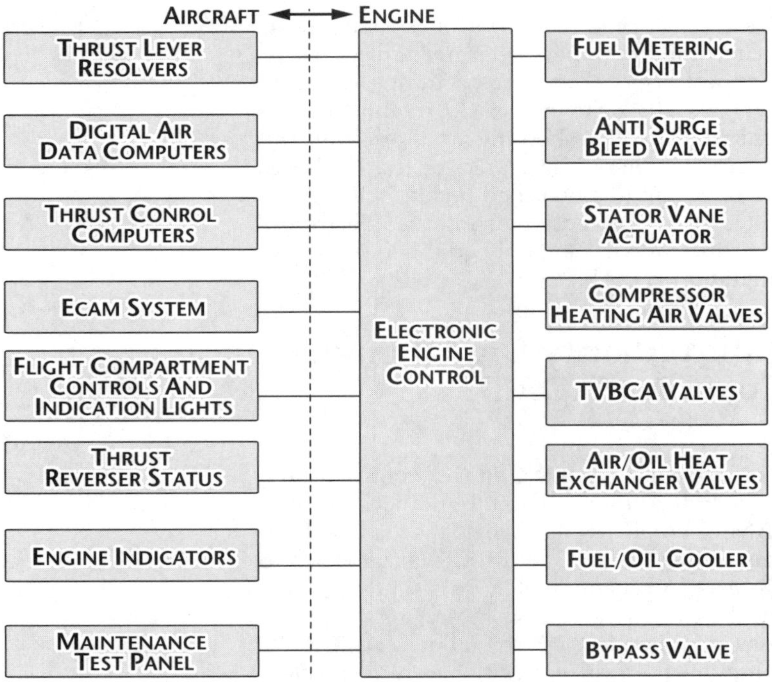

The FADEC, mounted on the fan case, uses a magnetic alternator as an internal power source. The FADEC initially receives power from the aircraft electrical system. As part of the failsafe design, the FADEC's internal alternator provides needed power once the $N_2$ compressor reaches 12%, allowing for operation in the event of an aircraft electrical system failure.

Each FADEC has dual channel redundancy. One channel controls the engine while the other remains in standby. Should the active channel fail, the standby channel automatically takes control.

The FADEC performs the following functions:

- Power management / control of the gas generator
- Protection against engine exceeding preset limits
- Automatic and Manual engine starting sequence
- Thrust reverser control
- Providing engine information to the EICAS
- Detection, isolation, and recording of failures

Descriptions of these functions follow.

**Power management and control of the gas generator.** The FADEC controls fuel flow to provide the desired engine thrust output. The pilot moves the throttle (referred to as a thrust lever) to set the thrust. The throttle is connected to a precision position sensor. This sensor tells the FADEC the thrust level chosen. The FADEC controls the fuel flow to supply the desired thrust in all aircraft flight modes.

Fuel from the aircraft fuel tanks is supplied to the engine drive pump. The high-pressure compressor drives this pump. Fuel first enters the low-pressure portion of the pump. From there, it passes through fuel/oil heat exchanger before entering the high-pressure side of the fuel pump.

Fuel supply to the engine is controlled by the FADEC operated fuel metering valve located within the engine HMU (hydromechanical unit). The fuel metering valve position is relayed back to the FADEC by the FMV resolver. The HMU also has a fuel bypass valve that maintains a constant pressure drop across the fuel-metering valve, ensuring the metered fuel is proportional to the valve position. Fuel

AIRCRAFT ↔ ENGINE
THRUST LEVER RESOLVERS
DIGITAL AIR DATA COMPUTERS
THRUST CONROL COMPUTERS
ECAM SYSTEM
FLIGHT COMPARTMENT CONTROLS AND INDICATION LIGHTS
THRUST REVERSER STATUS
ENGINE INDICATORS
MAINTENANCE TEST PANEL
ELECTRONIC ENGINE CONTROL
FUEL METERING UNIT
ANTI SURGE BLEED VALVES
STATOR VANE ACTUATOR
COMPRESSOR HEATING AIR VALVES
TVBCA VALVES
AIR/OIL HEAT EXCHANGER VALVES
FUEL/OIL COOLER
BYPASS VALVE

Figure 2-5-10 FADEC data inputs and outputs

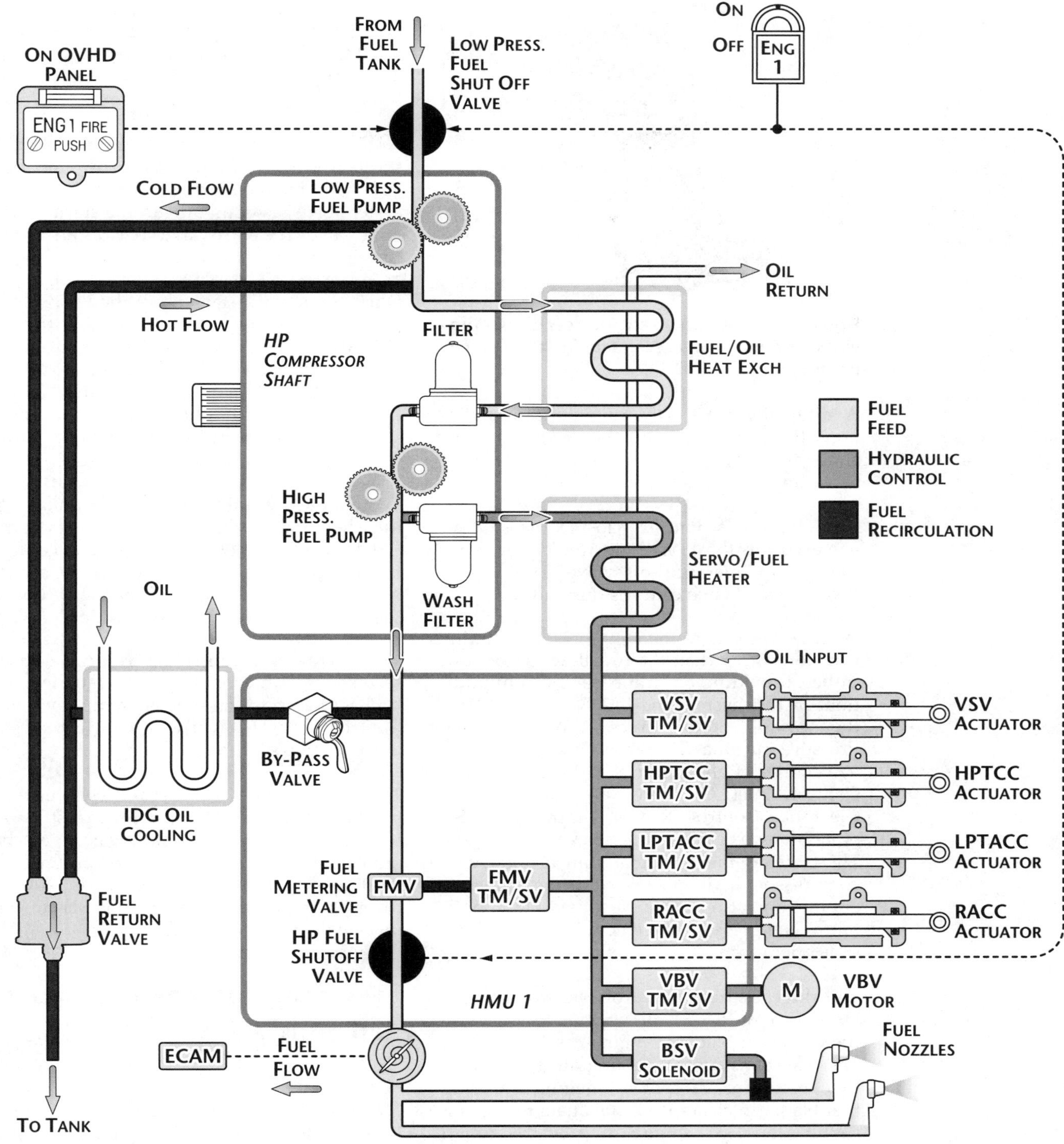


**Figure 2-5-11 FADEC fuel flow schematic**

is supplied to the combustion chamber at the required volume pressure and temperature based on FADEC commands.

Fuel can be supplied to the combustor by either 10 or 20 fuel nozzles. Ten of these nozzles are available permanently. The remaining ten are available when the burner stage valve (BSV) is open. The FADEC opens the BSV when the engine requires a high fuel-air ratio. The BSV is normally closed when low idle has been selected or during engine deceleration.

The FADEC computes the fuel flow that will maintain the desired $N_1$ speed. While $N_1$ is primary, the FADEC allows the $N_2$ speed to vary, within a predetermined minimum and maximum range. The FADEC also controls fuel flow to manage engine acceleration and deceleration, prevent engine stall or flameout.

**Figure 2-5-12. A turbine engine flow divider provides for primary and secondary fuel flow to the nozzles.**

Additionally, the FADEC automatically maintains the desired engine thrust levels while, simultaneously, providing sufficient bleed air to the airframe systems.

Fuel shut-off is done by closing the shut-off valve within the HMU and the low-pressure fuel pump inlet shut-off valve. The cockpit ENG master switch controls this valve.

With a FADEC equipped engine, many engine functions other than the fuel flow can be controlled providing optimum engine configuration. In this system the FADEC also controls the Variable Stator Vanes (VSV), the Low Pressure Turbine Clearance Control (LPTCC) valves, the High Pressure Turbine Clearance Control (HPTCC) valves and Rotor Active Clearance Control (RACS) system. The VSV system moves the compressor variable vanes. These vanes move to maintain an adequate compressor stall margin during transient engine operation. They also maintain optimum compressor efficiency during steady-state operations.

The three clearance control systems (LPTCC, HPTCC and RACS) all operate to maintain the optimum blade to engine case clearances. These systems control cooling air provided to their respective area of the engine, ensuring that blade tip clearance is adequate to prevent rubbing, but small enough to provide for high engine efficiency. By integrating control of all of these functions, significantly better engine fuel efficiency can be achieved.

**Protection against engine exceeding preset limits.** The FADEC continually monitors $N_1$ and $N_2$ speeds, preventing overspeed conditions from occurring. $N_1$ and $N_2$ speeds are measured by their corresponding tach generators. These signals are transmitted to the FADEC computer.

The system also operates to prevent overtemperature conditions throughout the entire operating range. The engine's interturbine temperature thermocouples are connected to the FADEC. It monitors turbine temperature and takes appropriate action when temperature limits are approached.

**Automatic and manual engine starting sequence.** FADEC systems allow for fully automatic engine starting. The A320 starting system consists of an air turbine starter and a start valve. The start valve controls air flow to the air turbine starter.

During automatic start mode, the FADEC controls the start valve, the igniters, and the fuel HP valves. The FADEC can detect a hot start, a hung start, a stall, and a no light-up condition. The FADEC automatically aborts the start in the event of a start failure, even continuing to spool the engine until temperatures are within safe limits during hot starts.

The engines can also be started manually by the flight crew. During manual starts, the FADEC controls most engine fuel and configuration functions, but does not have the authority to abort the start. It displays engine operating data on the EICAS in both modes.

**Thrust reverser control.** The engine thrust reverser system consists of four actuators and latches, door position switches and a Hydraulic Control Unit (HCU). The system actuates in less than two seconds.

The FADEC controls thrust reverser function. It deploys the reversers when the thrust lever is in reverse and both main landing gear switches indicate that the aircraft is on the ground. The FADEC also controls engine thrust during reverser operation, moving from idle to max reverse thrust as required.

## Providing engine information to the EICAS

The FADEC system receives all of the engine information transmitted to analog instruments in conventional systems. This data is monitored by the FADEC computer and used by the system in monitoring and controlling the engine. The data is also transmitted to the cockpit EICAS panel. These systems will be discussed in more detail in the instrumentation section.

## Flow dividers

A flow divider, shown in Figure 2-5-12, creates primary and secondary fuel supplies which are discharged through separate, concentric spray tips, providing proper spray angle at all fuel flows. Fuel enters the inlet of the nozzle and passes through a screen. When fuel pressure reaches a predetermined pressure, the pressure opens the flow divider, and fuel is directed into a second drilled passage in the stem of the fuel nozzle. Fuel from the secondary passage is directed into the secondary spin chamber. The fuel then passes through a secondary spray tip into the combustion liners.

Fuel pressurizing valves will usually trap fuel forward of the manifold, giving a positive cutoff. This cutoff prevents fuel from dribbling into the manifold and through the fuel nozzles, eliminating most after-fires and carbonization of the fuel nozzles. Carbonization occurs because combustion chamber temperatures are lowered and the fuel is not completely burned. A flow divider performs essentially the same function as a pressurizing valve. It is used, as the name implies, to divide flow to the duplex fuel nozzles. It is not unusual for units performing identical functions to have a different nomenclature between engines.

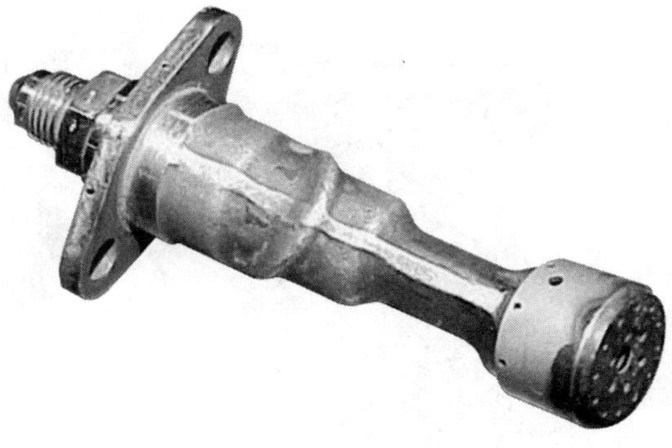

**Figure 2-5-13. A simplex air blast fuel nozzle with mounting flange and fuel connection.**

## Fuel spray nozzles and fuel manifolds

Fuel spray nozzles form part of the fuel system and atomize or vaporize the fuel so that it will ignite and burn efficiently. Although fuel spray nozzles are an integral part of the fuel system, their design is closely related to the type of combustion chamber in which they are installed.

The fuel nozzles inject fuel into the combustion area in a highly atomized, precisely patterned spray so that burning is completed evenly and in the shortest possible time and in the smallest possible space. It is very important that the fuel be evenly distributed and well centered in the flame area within the liners. This is to preclude the formation of any hot spots in the combustion chambers and to prevent the flame from burning through the liner.

Fuel nozzle types vary considerably between engines, although for the most part, fuel is sprayed into the combustion area under pressure through small orifices in the nozzles. The two types of fuel nozzles generally used are the simplex and the duplex configurations. The duplex nozzle usually requires a dual manifold and a pressurizing valve or flow divider for dividing primary and main fuel flow, but the simplex nozzle requires only a single manifold for proper fuel delivery.

The simplex fuel nozzle was the first type of nozzle used in turbojet engines and was replaced in most installations with the duplex nozzle, which gave better atomization at starting and idling speeds. The simplex nozzle (Figure 2-5-13) is still being used in certain configurations. Each of the simplex nozzles consists of a nozzle tip, an insert and a strainer made up of fine-mesh screen and a support.

The duplex fuel nozzle is the nozzle most widely used in most gas turbine engines. As mentioned previously, its use requires a flow divider, but at the same time it offers a desirable spray pattern for combustion over a wide range of operating pressures. A nozzle typical of this type is illustrated in Figure 2-5-14.

The air-spray nozzle takes a portion of the primary combustion air and mixes it with the fuel. By this mixing action, the spray tip is aerated. This action prevents the local fuel-rich tip concentrations produced by other types of spray nozzles, thus giving a reduction in both carbon formation and exhaust smoke. An advantage

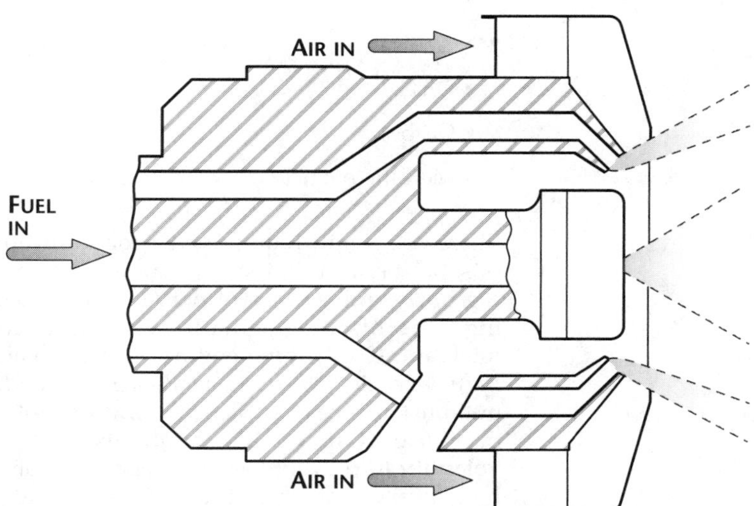


**Figure 2-5-14. A duplex air-blast fuel-spray nozzle with primary and secondary flow paths.**

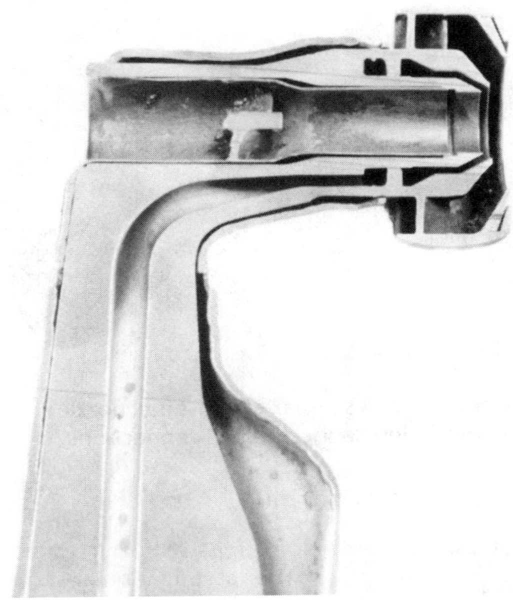

**Figure 2-5-15. A cutaway air-blast fuel nozzle tip, used to mix fuel and combustion air.**

of the air-spray nozzle is that the pressure required for atomization of the fuel is lower, allowing the use of comparatively smaller and lighter fuel pumps. An example of an air-spray nozzle is shown in Figure 2-5-15.

## Section 6

# Turbine Engine Oil Systems

Turbine oil is constantly exposed to many harmful substances that reduce its ability to protect moving parts. Some of the main contaminants can be:

- Moisture
- Acids
- Dirt
- Carbon
- Metallic particles

There are many requirements for lubricating oils, but because of the small number of moving parts and the use of ball and roller bearings, the turbine engine uses a less viscous lubricant. Gas turbine engine oil must have a high viscosity for good load-carrying ability but must also be of sufficiently low viscosity to provide good flowability. It must also be of low volatility to prevent loss by evaporation at the high altitudes at which the engines operate.

In addition, the oil should not foam and should be essentially nondestructive to natural or synthetic rubber seals in the lubricating system. Also, with high-speed antifriction bearings, the formation of carbons or varnishes must be held to a minimum. The many requirements for lubricating oils are met in the synthetic oils developed specifically for turbine engines. Synthetic oil has two principle advantages over petroleum oil. It has less tendency to deposit lacquer and coke and less tendency to evaporate at high temperature. Turbine oils are classified by type 1 (Mil-L-7808) and type 2 (Mil-L-23699).

Lubrication systems can be classified as pressure-relief, full-flow or total-loss systems. The pressure-relief system is very common and will be described in detail later in this section. The full-flow system is used when the bearing-cavity pressures are high, with an increase in engine speed. This system overcomes high bearing-cavity pressures by dispensing with the pressure-relief valve.

As engine speed increases (bearing-cavity pressure increase), so does the oil pump's output pressure. This higher pump pressure allows for lubricating oil to reach the bearings. This system is generally used on larger turbofan engines. The total-loss system is used when the engine will operate for short periods. As the name implies, the oil is generally lost out of the exhaust or other drains on the engine.

Both wet- and dry-sump lubrication systems are used in gas turbine engines. Most turbine engines use a dry-sump lubrication system. Wet-sump engines store the lubricating oil in the engine proper, while dry-sump engines utilize an external tank mounted, usually, on the engine or somewhere in the aircraft structure near the engine. To ensure proper temperature, oil is routed through either an air-cooled oil cooler, a fuel-cooled oil cooler, or both.

In a gas turbine dry-sump lubrication system, the oil supply is carried in the oil tank. With this type of system, a larger oil supply can be carried and the temperature of the oil can readily be controlled by the use of an oil cooler. Some manufacturers desire to have the oil quantity checked soon after shutdown to prevent over servicing.

The vent in an oil tank keeps the tank pressure from rising above or falling below that of the outside atmosphere. In the accessory case, the vent (or breather) is a screen-protected opening, which allows accumulated air pressure within the accessory case to escape to the atmosphere. The scavenged oil carries air into the accessory case, and this air must be vented; otherwise, the pressure buildup within the accessory case, or bearing cavities, could stop the flow of oil. The vent system inside the tank (Figure 2-6-1) is arranged so that the airspace is vented at all

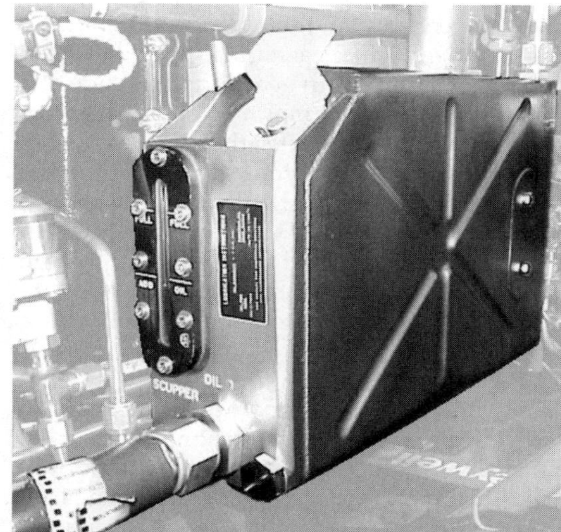

**Figure 2-6-1. Turbine engine oil tank, used to store the oil outside of the engine**

times, even though oil may be forced to the top of the tank by deceleration of the aircraft.

All oil tanks have extra space, allowing for expansion of the oil after heat is absorbed from the bearings and gears, and after the oil foams as a result of circulating through the system. Some tanks also incorporate a de-aerator tray for separating air from the oil returned by the scavenger system. Usually these de-aerators are the can type, in which oil enters at a tangent. The air released is carried out through the vent system in the top of the tank.

The oil pump is designed to supply oil, under pressure, to the parts of the engine that require lubrication. Many oil pumps consist not only of a pressure supply element, but also of scavenge elements as well. In all types of pumps, the scavenge elements have a greater pumping capacity than the pressure element. This is to prevent oil from collecting in the bearing sumps. The pumps may be one of several types, each type having certain advantages and limitations.

The most common oil pump is the gear-type pump. The gear-type oil pump, illustrated in Figure 2-6-2, has only two elements, one for pressure oil and one for scavenging. However, some pumps may have two or more elements for scavenging and one or more for pressure.

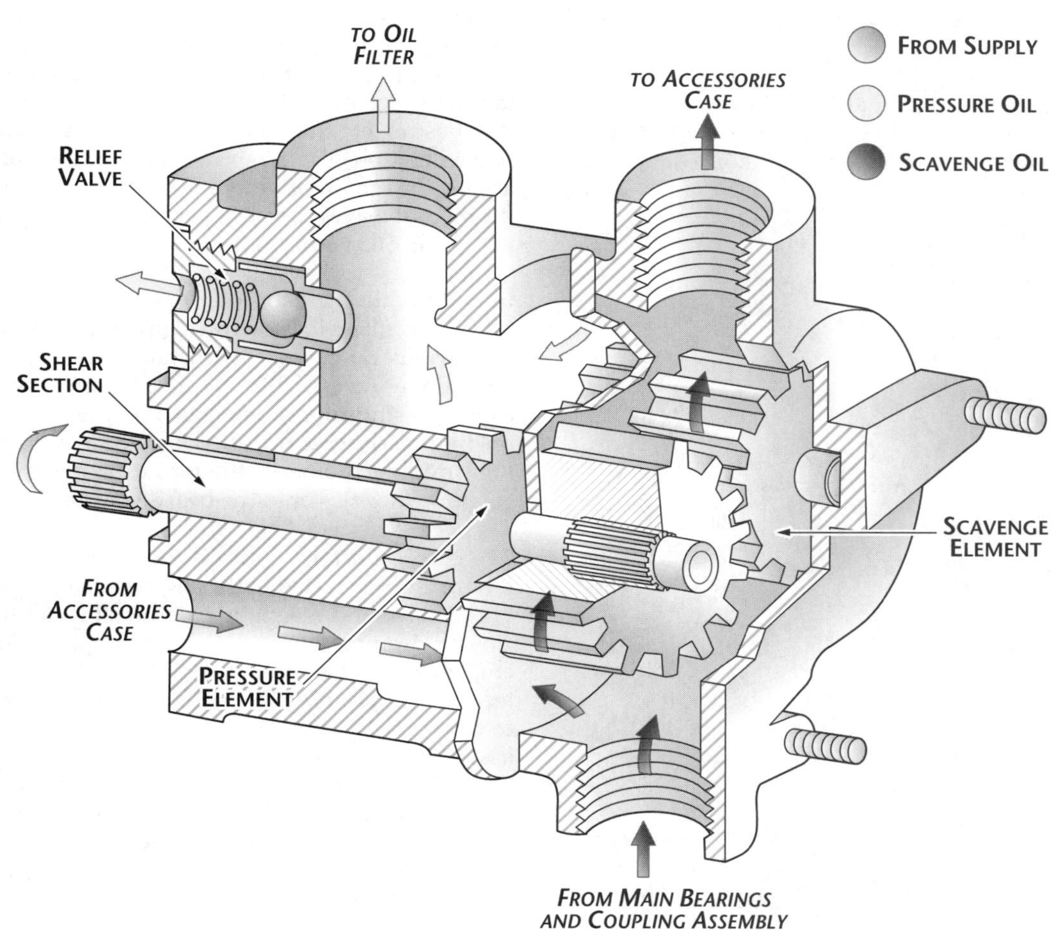


**Figure 2-6-2. Typical turbine engine oil pump, with pressure and scavenge sections.**

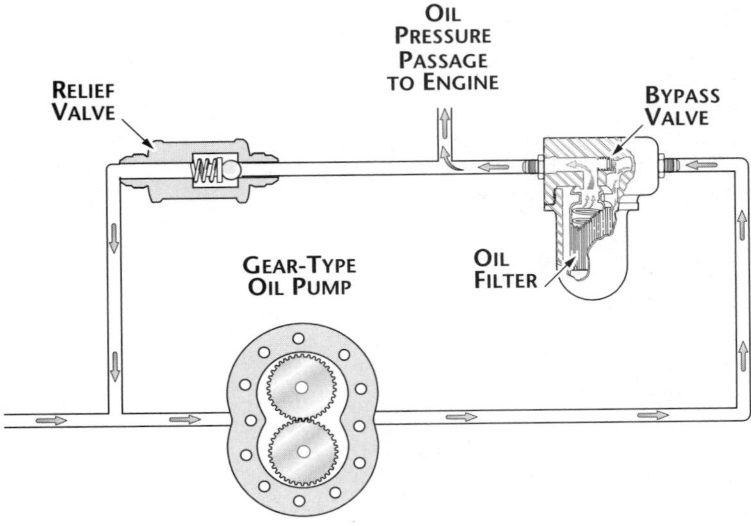


Figure 2-6-3. This diagram shows the arrangement of the oil pressure pump, filter, relief valve, and bypass valve.

A pressure-regulating (relief) valve in the discharge side of the pump (Figure 2-6-3) limits the output pressure of the pump. This valve will bypass oil to the pump inlet when the outlet pressure exceeds a predetermined limit. Also shown is the shaft shear section, which causes shaft to shear if the pump gears should seize.

**Gerotor pumps**. Frequently *gerotor* pumps are used to supply large amounts of liquid at the correct pressure. A gerotor pump is another type of gear pump that works more like a Wankel engine than a regular gear pump. It consists of a four-tooth gear turning inside of a five-tooth rotor that is mounted on an eccentric. As the gear turns it rotates the eccentric across the intake port. As the rotor turns further, the fluid contained in the fifth space is moved to the exhaust, while the eccentric opens up another space over the intake. Although more frequently found on older engines, gerotors are still in use. The operation is shown in figure 2-6-4.

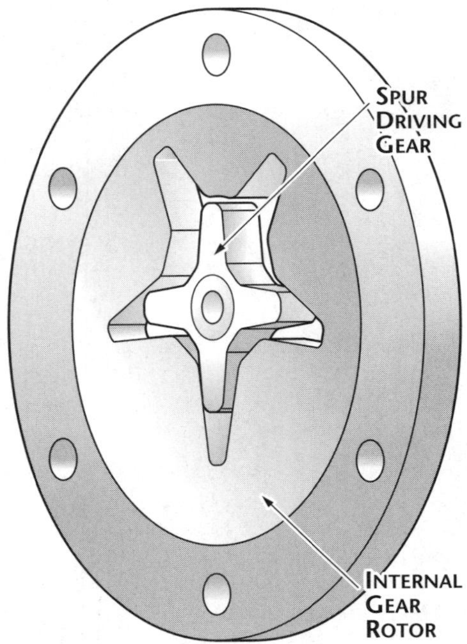


Figure 2-6-4. Gerotor pump operation

Filters are an important part of the lubrication system, since they remove foreign particles that may be in the oil. This is particularly important in gas turbines, as very high engine speeds are attained and the antifriction types of ball and roller bearings would become damaged quite rapidly if lubricated with contaminated oil.

Also, there are usually a number of drilled or core passages leading to various points of lubrication. Since these passages are usually small, they are easily clogged. One common type of oil filter uses a replaceable laminated paper element, such as those used in hydraulic systems.

Another type of filter, used as a main oil strainer, is shown in Figure 2-6-5. The filtering element interior is made of stainless steel. Each of the oil filters mentioned has certain advantages. In each case, the filter selected is the one that best meets the individual needs of a particular engine. The filters discussed generally are used as main oil filters; that is, they strain the oil as it leaves the pump before being piped to the various points of lubrication.

In addition to the main oil filter, there are also secondary filters located throughout the system for various purposes. For instance, there may be a finger-screen filter, which is sometimes used for straining scavenged oil. Also, there are fine-mesh screens, called last-chance filters, for straining the oil just before it passes from the spray nozzles onto the bearing surfaces.

The chip collector or chip detector, as seen in Figure 2-6-6, is bolted to the engine and consists of the valve and magnetic plug housing, a self-closing drain valve, and a magnetic plug. The chip detector has a magnetic plug, which collects ferrite particles in the scavenged oil prior to the oil being returned to the engine oil tank assembly. The drain valve provides a positive shutoff to prevent oil leakage when the magnetic plug is removed. The drain plug is removed and inspected for any magnetic particles that might be in the oil. These particles could be an early warning of engine bearing wear.

Some chip detectors also contain an electrical sensor. This type has two magnetic plugs next to each other. An example is shown in Figure 2-6-7. If enough metal accumulates, it will close the electrical circuit and activate a warning light in the cockpit. This allows the flight crew to shut-down the engine before further damage is caused by metal contamination. Many of these chip detector systems also have a ground test function capability.

The components of a typical main oil filter include a housing, which has an integral relief (or bypass) valve and, of course, the filtering

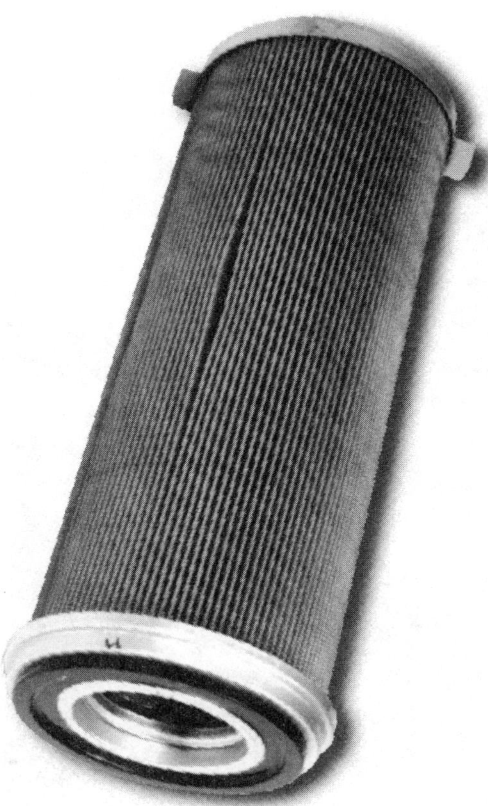

Figure 2-6-5. An oil filter used to filter all of the engine's oil

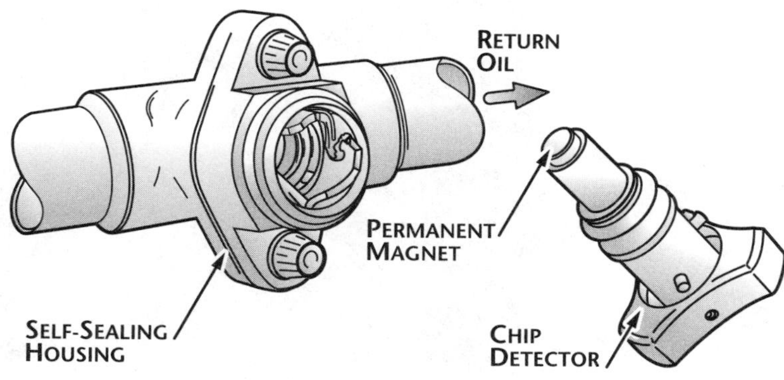


Figure 2-6-6. A chip detector with a magnetic tip is used to attract any loose particles in the oil system.

Figure 2-6-7. Magnetic chip detector with electrical contacts and ground test switch

Figure 2-6-8. Bearing cavity with bearing removed to show oil nozzle

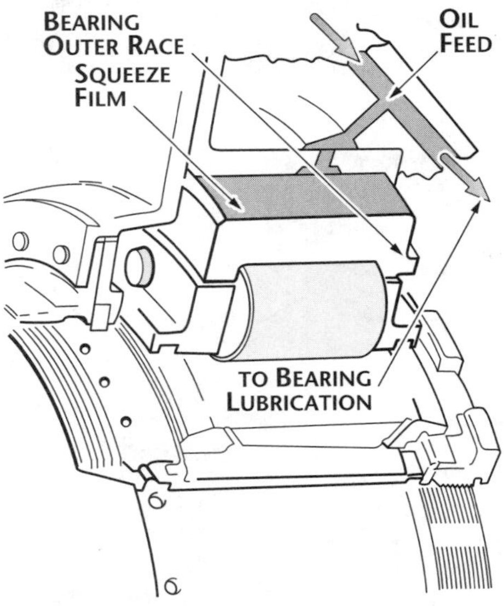


Figure 2-6-9. An oil dampened bearing that uses an oil film between the outer race and the structure to reduce vibration transmitted to the rest of the engine and airframe.

element. The filter bypass valve prevents the oil flow from being stopped if the filter element becomes clogged. The bypass valve opens whenever a certain pressure is reached. If this occurs, the filtering action is lost, allowing unfiltered oil to be pumped to the bearings.

Oil jets (or nozzles) are located in the pressure lines adjacent to, or within, the bearing compartments and rotor-shaft couplings (See Figure 2-6-8). The oil from these nozzles is delivered in the form of an atomized spray. This spray maintains a constant flow of oil directly onto the bearing, thus insuring consistent lubrication. The oil jets are easily clogged because of the small orifice in their tips; consequently, the oil must be free of any foreign particles.

If the last-chance filters in the oil jets should become clogged, bearing failure usually results, since nozzles are not accessible for cleaning except during engine overhaul. To prevent damage from clogged oil jets, main oil filters are checked frequently for contaminates.

Many engines use a special type of bearing, called a squeeze-film, hydraulic, or oil-dampened bearing, to reduce the effect of dynamic loads, which are transferred from the rotating assemblies through the bearing housing and into the airframe as vibration. The outer race of the bearing has a very small clearance that is filled with pressure oil, as can be seen in Figure 2-6-9. The oil film dampens the radial motion of the rotating assembly that would otherwise be transferred through the bearing housing. This action reduces the vibration level in the engine and airframe, and can help prevent damage resulting from vibration.

The instrumentation for the oil system is usually an oil pressure and temperature gauge. The oil pressure gauge connection is located in the pressure line between the pump and the various points of lubrication. The oil temperature gauge connection is usually located in the pressure inlet to the engine.

Check valves are sometimes installed in the oil supply lines of dry-sump oil systems to prevent reservoir oil from seeping (by gravity) through the oil pump elements and high-pressure lines into the engine after shutdown. Check valves, by stopping flow in an opposite direction, prevent accumulations of undue amounts of oil in the accessory gearbox and bearing cavities.

Oil coolers are used in the lubricating systems of some turbine engines to reduce the temperature of the oil to a degree suitable for circulation through the system. Two basic types of oil coolers in general use are the air-cooled oil cooler and the fuel-cooled oil cooler.

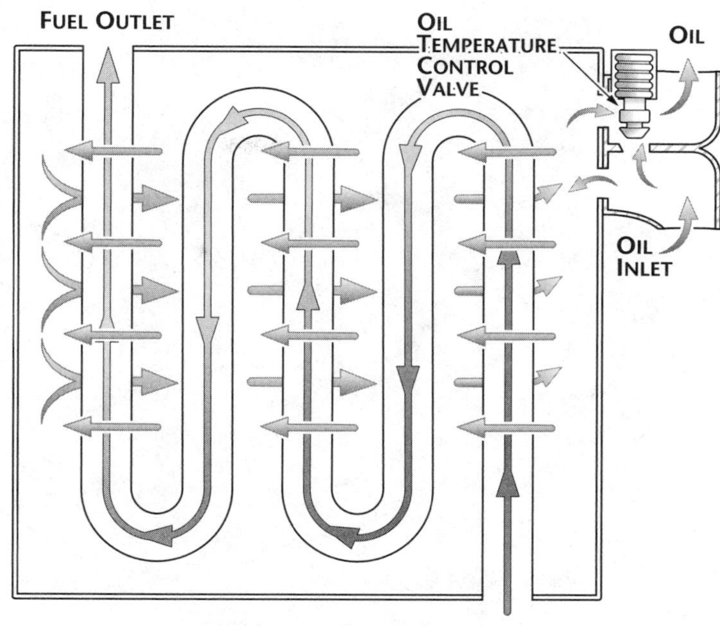


Figure 2-6-10. A fuel-cooled oil cooler that uses fuel to cool the engine oil

The fuel-cooled oil cooler, illustrated in Figure 2-6-10, acts as a fuel/oil heat exchanger in that the fuel cools the oil and the oil heats the fuel.

The air-cooled oil cooler (Figure 2-4-11) normally is installed at the forward end of the engine or in the fan duct. It is similar in construction and operation to the air-cooled cooler used on reciprocating engines.

The fuel/oil heat exchanger is designed to cool the hot oil and to preheat the fuel for combus-

Figure 2-4-11. Air-cooled oil cooler

**Figure 2-7-1. This is a simple exhaust duct with a W type thrust reverser attached.**

tion. Though there is a thermostatic valve that controls the oil flow as needed, fuel flowing to the engine must pass through the heat exchanger. The fuel/oil heat exchanger consists of a series of joined tubes with an inlet and outlet port. The oil enters the inlet port, moves around the fuel tubes and goes out the oil outlet port.

## Section 7

# Turbine Engine Exhaust Systems

**Exhaust section.** The term *exhaust duct* applies to the engine exhaust pipe or tail pipe including the jet nozzle of a non-afterburning engine. Although an *afterburner* might also be considered a type of exhaust duct, afterburning is, in itself, a subject.

If the engine exhaust gases could be discharged directly to the outside air in an exact axial direction at the turbine exit, an exhaust duct might not be necessary. A larger total thrust can be obtained from the engine if the gases are discharged from the aircraft at a higher velocity than that permissible at the turbine outlet. An exhaust duct is added to collect and straighten the gas flow as it comes from the turbine. It also increases the velocity of the gases before they are discharged from the exhaust nozzle at the rear of the duct. Increasing gas velocity increases its momentum and the thrust produced.

An engine exhaust duct is often referred to as the *engine tail pipe*. The duct is essentially a simple, stainless steel conical or cylindrical pipe (figure 2-7-1.) The engine tail cone and struts are usually included at the rear of the turbine. The struts support the rear bearing and impart an axial direction to the gas flow, the tail cone helps smooth the flow. Immediately aft of the turbine outlet and usually just forward of the flange to which the exhaust duct is attached, the engine has a sensor for turbine discharge pressure.

In large engines, it is frequently not practical to measure internal temperature at the turbine inlet. Therefore, the engine is usually also instrumented for exhaust gas temperature at the turbine outlet. One or more thermocouples in the exhaust case can provide adequate sampling of exhaust gases. Pressure probes are also inserted in this case to measure pressure of gases coming from the turbine. The gradually diminishing cross-sectional area of a conventional convergent type of exhaust duct is capable of keeping the flow through the duct constant at velocities not exceeding Mach 1.0 at the exhaust nozzle.

Frequently a bad fuel nozzle can be found by looking in the tail pipe for hot streaking, or over temperature spots.

## Exhaust Ducts

Turboshaft engines in helicopters do not develop thrust using the exhaust duct. If thrust were developed by the engine exhaust gas, it would be impossible to maintain a stationary hover; therefore, helicopters use divergent ducts. These ducts reduce gas velocity and dissipate any thrust remaining in the exhaust gases. On many fixed-wing aircraft, the exhaust duct used is the convergent type, which accelerates the remaining gases to produce thrust. An example is shown in Figure 2-7-2. It has the effect of increasing aircraft speed and fuel efficiency by harnessing the remaining jet thrust left after turning the propeller drive turbine. This adds additional Shaft Horsepower (SHP) to the engine rating. Equivalent shaft horsepower (ESHP) is the combination of thrust and SHP.

Conventional convergent exhaust nozzle. The rear opening of the exhaust duct is the jet nozzle, or exhaust nozzle as it is often called. The nozzle acts as an orifice, the size of which determines velocity of gases as they emerge from the engine. In most non-after-burning engines, this area is critical; for this reason, it is fixed at the time of manufacture. The exhaust (jet) nozzle area should not be altered in the field because any change in the area will change both the engine performance and the exhaust gas temperature. See Figure 2-7-3.

**Figure 2-7-2. The Fairchild Merlin uses residual exhaust thrust to add to the power developed by the propeller.**

**Figure 2-7-3. The exhaust nozzle creates a converging nozzle, increasing the velocity of the exhaust air.**

Some early engines, however, were trimmed to their correct RPM or exhaust gas temperature by altering the exhaust-nozzle area. When this is done, small tabs that may be bent as required are provided on the exhaust duct at the nozzle opening. On other early engines, small adjustable pieces called mice are fastened as needed around the perimeter of the nozzle to change the area. Occasionally, engines are equipped with variable area nozzles that are opened or closed, usually automatically, with an increase or decrease in fuel flow. The velocity of the gases within a convergent exhaust duct is usually held to a subsonic speed. The velocity at the nozzle approaches Mach 1.0 (the velocity at which the nozzle will choke) on turbojets and low-bypass-ratio turbofans during most operating conditions.

**Convergent-divergent exhaust nozzle.** When a divergent duct is employed in combination with a conventional exhaust duct, it is called a *convergent-divergent* exhaust duct (Figure 2-2-1 on page 2-11). In the C-D nozzle, the convergent section is designed to handle the gases while they remain subsonic and to deliver them to the throat of the nozzle just as they attain sonic velocity. The divergent section handles the gases after they emerge from the throat and become supersonic further increasing their velocity.

Pressure generated within an engine cannot be converted to velocity, particularly when a convergent nozzle is used. The additional pressure results in additional thrust that must be added when the total thrust developed by the engine is computed. The additional thrust is developed inefficiently. It would be much better to convert all of the pressure within the engine to velocity and develop all of the engine thrust by means of changes in momentum. In theory, a C-D nozzle does this. Because it develops this additional part of the total thrust more efficiently, it enables an engine to produce more total net thrust than the same basic engine would generate if it were equipped with a conventional convergent duct and nozzle.

The C-D nozzle would be nearly ideal if it could always be operated under the exact conditions for which it was designed. However, if the rate of change in the duct area is either too gradual or too rapid for the calculated increase in weight of the gases, unsteady flow downstream of the throat will occur with an accompanying loss of energy. This ultimately means loss of thrust. If the rate of increase in the area of the duct is too little, the maximum gas velocity that can be reached will be limited. If the rate of increase is too great, the gas flow will break away from the surface of the nozzle, and the desired increase in velocity will not be obtained.

As exhaust gases accelerate or decelerate with changing engine and flight conditions, their pressure fluctuates above or below the pressure ratio for which the nozzle was designed. When this occurs, the nozzle no longer converts all of the pressure to velocity, and the nozzle begins to lose efficiency.

The solution to this dilemma is a C-D nozzle with a variable configuration that can adjust itself to changing pressure conditions. Several types of C-D nozzles are in use, mostly on military aircraft.

## Afterburners

In order for fighter planes to fly faster than sound (supersonic), they have to overcome a sharp rise in drag near the speed of sound. A simple way to get the necessary thrust is to add an afterburner to a basic turbojet. In a basic turbojet, some of the energy of the exhaust from the burner is used to turn the turbine. The afterburner is used to put back some energy by injecting fuel directly into the hot exhaust. An afterburner takes care of this problem. Basically an afterburner is a ramjet attached to the exhaust of a turbojet engine. An afterburner uses the fact that approximately 25 percent of the air passing through a turbojet engine is consumed by combustion, and the remaining 75 percent is capable of supporting additional combustion if it had more fuel. Most modern fighter aircraft employ an afterburner on either a low bypass turbofan or a turbojet.

The high velocity air necessary for a ramjet to operate is supplied by attaching the afterburner to the turbine discharge section of a conventional jet engine. The gases leaving the engine have sufficient velocity to satisfy a ramjets requirements. The remarkable thing about an afterburner is its simplicity. It consists of only four fundamental parts: the afterburner duct, the fuel nozzles (spraybars), the flame holders, and a two position or variable-area exhaust nozzle. When the afterburner is not operating, the afterburner duct also serves as the basic engine tailpipe.

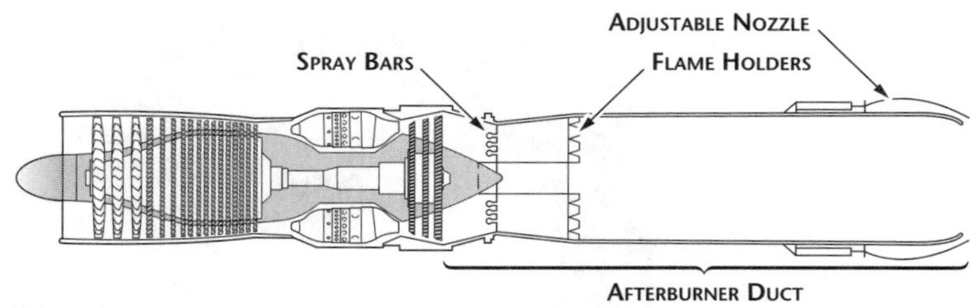


**Figure 2-7-4. An afterburner is an addition to the tailpipe of an engine.**

In the diagram (Figure 2-7-4) the nozzle of the basic turbojet engine has been extended and a ring of flame holders has been added in the nozzle. Additional fuel is injected into the hot exhaust stream. The fuel burns and produces additional thrust (See Figure 2-7-5), but it does not burn as efficiently as it does in the combustion section of the basic engine. More thrust is generated; more much fuel is also consumed. With the increased temperature of the exhaust, the flow area of the nozzle has to be increased to pass the same mass flow. Therefore, afterburning nozzles must be designed with variable geometry, and are heavier and more complex than simple turbojet nozzles. When the afterburner is turned off, the engine performs like a basic turbojet.

The nozzle of a turbojet is usually designed to take the exhaust pressure back to free stream pressure. When the afterburner is in operation, fuel is introduced through the fuel nozzles and is burned within the duct, causing considerable additional expansion of the gases after they leave the basic engine. This increases their momentum and provides more thrust. When an afterburner is used, the increase in total fuel consumption is approximately 300%.

**Figure 2-7-5. Shock diamonds coming from engine in afterburner**

Figure 2-7-6. Adjustable Afterburner Nozzle

**Afterburner exhaust nozzle.** Because of the increase in expansion ratio of the gases, the afterburner exhaust nozzle must be capable of being increased in area during afterburning in order to assure that basic engine operation will continue to be normal. The exhaust nozzle must also be capable of being closed to its original small area during normal operation. Early systems used an adjustable clamshell type nozzle. Nozzles having variable vanes throughout the entire range of operation are used in modern military airplanes. They are adjusted automatically to coincide with the amount of thrust called for by the cockpit throttle position. The adjustable nozzle, shown in Figure 2-7-6, also shows the coating of ceramic heat resistant material that has been applied to the inside of the adjustable nozzle vanes.

Afterburners are only used on fighter planes. The one supersonic airliner that used afterburners was the Concorde. The Concorde turned the afterburners off once it gets into cruise. Otherwise, it would run out of fuel before reaching its destination. It was retired from commercial service in 2003. No other supersonics transports are currently in service. Afterburners offer a mechanically simple way to augment thrust and are used on both turbojets and turbofans.

## Section 8

# Turbine Engine Cowling and Nacelles

As airplanes get bigger, more powerful, and faster, cowling becomes much more than just a cover for the engine. Not only must an aerodynamic shape be maintained, but also there must be enough underlying structure to maintain the necessary strength. At the same time it must be removable, not only for engine changes, but for almost any engine or accessory service. Gone are the days when a dozen Dzus fasteners can hold it all together.

**Turboprop cowls**. Of all the turbine cowling installations, those for turboprop engines look the most like piston engine installations. Typically consisting of a nose bowl, fixed lower cowl, and removable top cowls, they are fairly simple in construction with a minimum of underlying stiffeners.

The installation on the PT-6 run-up stand, shown in Figure 2-8-1A, is a good example of a normal turboprop cowling installation. The engine mount is clearly visible, as are the oval formers attached to it. The formers are fabricated into channels for strength. The lower cowl in solidly attached, while the upper cowl can be opened for maintenance and service. The cowl latches hold both halves solidly together during operation. In 2-8-1B, the formed internal cowl stiffeners are riveted to both upper and lower cowlings. The entire assembly is more than strong enough to withstand the propeller blast. Inspection and repair process is the same as for any normal aircraft structure.

**Turbojet cowling**. As engines get bigger and more powerful, cowling remains the same, only bigger and stronger with more secure mountings. In Figure 2-8-2 the mount pylon for a fuselage mounted engine pod can be seen. Also visible are the forged assemblies that are part of the aft fuselage structure. The actual engine mounts attach to the forgings.

Engine mounts for these types of aft side mounted engines generally consist of two half-circle forgings that mount on the engine proper. These forgings are then mounted to the fuselage engine attach points.

Figure 2-8-3 (on the next page), shows an engine change on a side-mounted powerplant being completed. Notice that the shop crew is using the correct engine change fitting on the crane. DO NOT FREELANCE when it comes to engine slings. Use the manufacturers approved tool. It was designed to hold the engine at the proper angle(s) so as not to put any part of

Figure 2-8-1. In this PT-6 cowl installation notice the formers that reinforce the cowl when it is closed (A), as well as the stiffeners on the individual cowling sections (B).

2-8-2. The mount pylon for a fuselage mounted engine pod

**Figure 2-8-3. An engine change on a side mounted jet engine. Note the proper engine-hoisting adapter is being used to lift the engine.**

*Photo courtesy of Duncan Aviation.*

the mount or structure in a bind. Binding, or a forced fit, can and will lead to cracks in the structure. Those cracks may lead to eventual separation of the engine assembly.

**Large airplane cowling.** Cowling is normally broken down into sections. First is the nose ring. Not only does the nose ring set the form for the rest of the cowl, it also houses the anti-ice system for the cowling. Hot bleed air is used to keep ice from forming, rather than being used for deicing.

In the case of a fanjet, the low-pressure fan is the next piece of cowl in line. It not only ducts the airflow, but also provides a protected tip plane path for the rotors. Remember, 75% or more of the engines thrust passes through this piece of cowling. Right behind the fan shroud comes the main cowl. Figure 2-8-4 shows the open cowl on a Boeing 747. Notice there is quite a bit of access to the engine with the cowl open. On engines with fan cascades, this section includes a translating mechanism to allow it to slide aft exposing the reverser section.

On engines with W or bucket reversers, the last cowl section is where the thrusters reside. In actuality, only some of the thrust reverser is attached to the cowling, with the balance being attached to the engine structure. Nevertheless, the entire assembly must be able to hold up to quite a bit of wear and tear just from the subsonic airflow.

On cold stream reverser engines, the last cowl section streamlines the fan air flow over the combustion section of the engine. It may also act as a mixer section, combining cold bypass air with hot exhaust air. On some engines, the fan air is contained within the cowling and mixed with the exhaust inside the aft cowl sections. On other engines, the

**Figure 2-8-4. JT9D on a Boeing 747 showing the many cowling sections.**

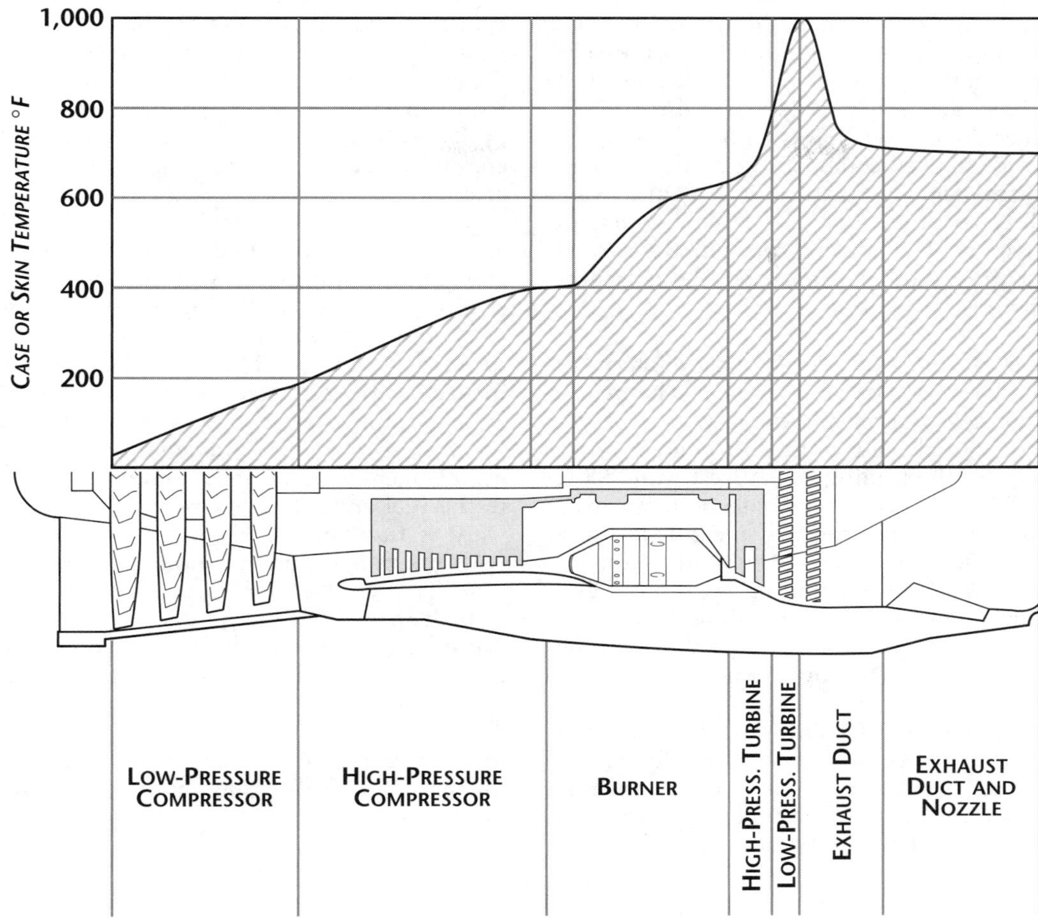


**Figure 2-8-5. Typical outer-case temperature for dual-axial compressor turbojet engine.**

fan air only passes through the inlet ring and forwardmost cowl section.

## Cooling System

The intense heat generated when fuel and air are burned mandates that some means of cooling be provided for all internal combustion engines. Reciprocating engines are cooled either by passing air over fins on the cylinders or by passing a liquid coolant through jackets surrounding the cylinders. Cooling is made easier because combustion occurs only during every fourth stroke of a four-stroke-cycle engine.

In contrast, the burning process in a gas turbine engine is continuous, and nearly all of the cooling air must pass through the inside of the engine. If enough air were admitted to the engine to provide an ideal air-fuel ratio of 15:1, internal temperatures would increase to more than 4000°F. In practice, a larger amount of air than is needed for combustion is admitted to the engine. This large surplus of air cools the hot sections of the engine to acceptable temperatures ranging from 1100° to 1500°F. Operating a turbine engine at too low an idle speed can cause serious overheating problems. This is because the airflow within the engine is not sufficient to carry away the excess heat. Always observe recommended minimum idle speeds.

**Cooling of engine outer case.** Figure 2-8-5 illustrates the approximate engine outer case (skin) temperatures encountered in a properly cooled dual-axial compressor turbojet engine. Because of the effect of cooling, the temperatures of the outside of the case are considerably less than those encountered inside the engine. The hottest spot occurs opposite the entrance to the first stage of the turbine. Although the gases have begun to cool a little at this point, the conductivity of the metal in the case carries the heat directly to the outside skin.

**Cooling of combustion chamber and gas producer.** The combustion-chamber burner cans or liners are constructed to induce a thin, fast-moving film of air over both the inner and outer surfaces of the can or liner. Can-annular-type burners are frequently provided with a center tube to lead cooling air into the

**Figure 2-8-6. Turbine blade with cooling vents**

center of the burner to promote high combustion efficiency and rapid dilution of the hot combustion gases, while minimizing pressure losses. Combustion liners often have a series of vent holes along their length, allowing cool air to enter the combustion chamber throughout. This aids in even combustion and helps maintain consistent temperatures in the burner can. In all types of gas turbines, large amounts of relatively cool air join and mix with the burned gases aft of the burners to cool the hot gases just before they enter the turbines.

Almost all turbine blade airfoils are internally cooled by means of compressor discharge air. The nozzles are typically convection-and film-cooled with the air exiting through a series of holes in the leading edge. See Figure 2-8-6. Aft of the leading edge, film-cooling air exits through convex and concave side gill holes and trailing edge slots. It cools by internal convection, exiting both by trailing edge holes and by inner band holes. The turbine blades are convection-cooled through radial holes in the dovetail. Air exits through tip holes.

**External cooling.** Cooling air inlets are frequently provided around the exterior of the engine to permit entrance of air to cool the turbine case, the bearings, and the turbine nozzle. In some instances, internal air is bled from the engine compressor section and is vented to the bearings and other parts of the engine. Air vented into or from the engine is ejected into the exhaust stream. When an accessory case is mounted at the front of an engine, it is cooled by inlet air. When located on the side of the engine, the case is cooled by outside air flowing around it.

The engine exterior and the engine nacelle are cooled by passing air between the case and the shell of the nacelle. The engine compartment frequently is divided into two sections. The forward section is built around the engine air inlet duct; the aft section is built around the engine. A fume-proof seal is provided between the two sections. The advantage of such an arrangement is that fumes from possible leaks in the fuel and oil lines contained in the forward section cannot become ignited by contact with the hot sections of the engine. In flight ram air provides ample cooling of the two compartments. On the ground air circulation is provided by the effect of reduced pressure at the rear of the engine compartment produced by gases flowing from the exhaust nozzle.

Fan jet engines are basically cooled in the same manner, with the air coming from the airflow produced by the bypass air.

**Figure 2-8-7. This tailpipe view shows the corrugated nozzle built into the exit duct.**

## Noise Suppression

All gas turbine engines have one thing in common; they are extremely noisy! So noisy in fact that noise control measures are a major design element. An attempt to control noise by legislation has resulted in increasingly more restrictive noise laws. Today, all jet aircraft must meet Stage 3 noise pollution standards. Most new aircraft have met the standard with design and operational changes to the point that the ear-splitting takeoff and climbout are more or less a thing of the past. Aircraft engine manufacturers are now busy trying to meet the new Stage 4 noise standards. In the near future, these new standards will be phased in.

**Sources of noise.** Jet engines produce three types of noise:

- Inlet duct and fan blade noise
- Internal Compressor noise
- Exhaust noise

## Exhaust Noise

Jet exhaust noise is caused by the violent and turbulent mixing of exhaust gasses with the atmosphere. There is a shearing action between the high exit speed of the exhaust (added to the aircraft's speed) and the relative speed of the airstream. Small eddies of disturbed air just aft of the exhaust nozzle produce high frequency noise, while further downstream larger eddies produce low frequency noise. In many installations the jet exhaust will exceed the local speed of sound and create a regular shock wave. The combination of these three things produces noise that is extreme.

The answer to controlling exhaust noise is simple. Reduce the velocity relative to the atmosphere and accelerate the mixing rate, preferably inside the exhaust nozzle.

**Corrugated type nozzle**. To accomplish this task a corrugated exhaust nozzle was developed. Basically, additional air is exhausted through the corrugations and mixed with the exhaust in such a manner as to slow the exhaust/atmosphere shearing effect. This is accomplished by actually increasing the contact area of the total exhaust stream by surrounding the propelling nozzle with the exhaust from the corrugated nozzle. Where the extra air comes from depends on the engine type. Normal gas turbines use low drag inlets on the engine cowl to gain the extra air for mixing. Low bypass turbofans were modified to use fan exhaust as the source. Internal corrugations are still used on some newer corporate jets. An example of a corrugated type nozzle is shown in Figure 2-8-7.

**Lobe type nozzles**. The lobe type exit ducts were actually an aft part of the nozzle, forming several separate ducts. The exhaust is divided and flowed through each nozzle, along with a small central nozzle. The result is that the exhaust can more rapidly mix with the atmosphere without the extreme shearing effect, thus less noise. The total nozzle area remains the same as if it were a single nozzle. Because of extra weight and maintenance complications the lobe type nozzle is not used on newer airplane designs.

**High bypass nozzles**. From the outset high bypass engines have a lower exhaust velocity than either low bypass or conventional jet engines. Additionally, they have a large quantity of secondary airflow from the fan exhaust. This allows the hot and cold stream air flows to be mixed within the engine itself, thus not requiring complicated exhaust nozzle configurations.

**Compressor noise.** One of the other noisiest components is the compressor or fan. As each rotor turns its airflow produces a wake, similar to the wake behind a projecting rock in a stream. The noise comes from pressure fields and turbulence caused by the wakes as they interact with other rotors and stators within the engine.

**Figure 2-8-8. These PT6 nose case gears are one example of components that contribute to internal engine noise.**

The intensity of the wake is a function of the distance between the rows of blades and vanes. Shorter distances produce more noise, while longer distances reduce noise at the expense of compressor efficiency.

In high bypass engines the fan also produces the same type of noise, but of much lesser intensity. The vanes and struts in the secondary airflow are much farther away from the rotor.

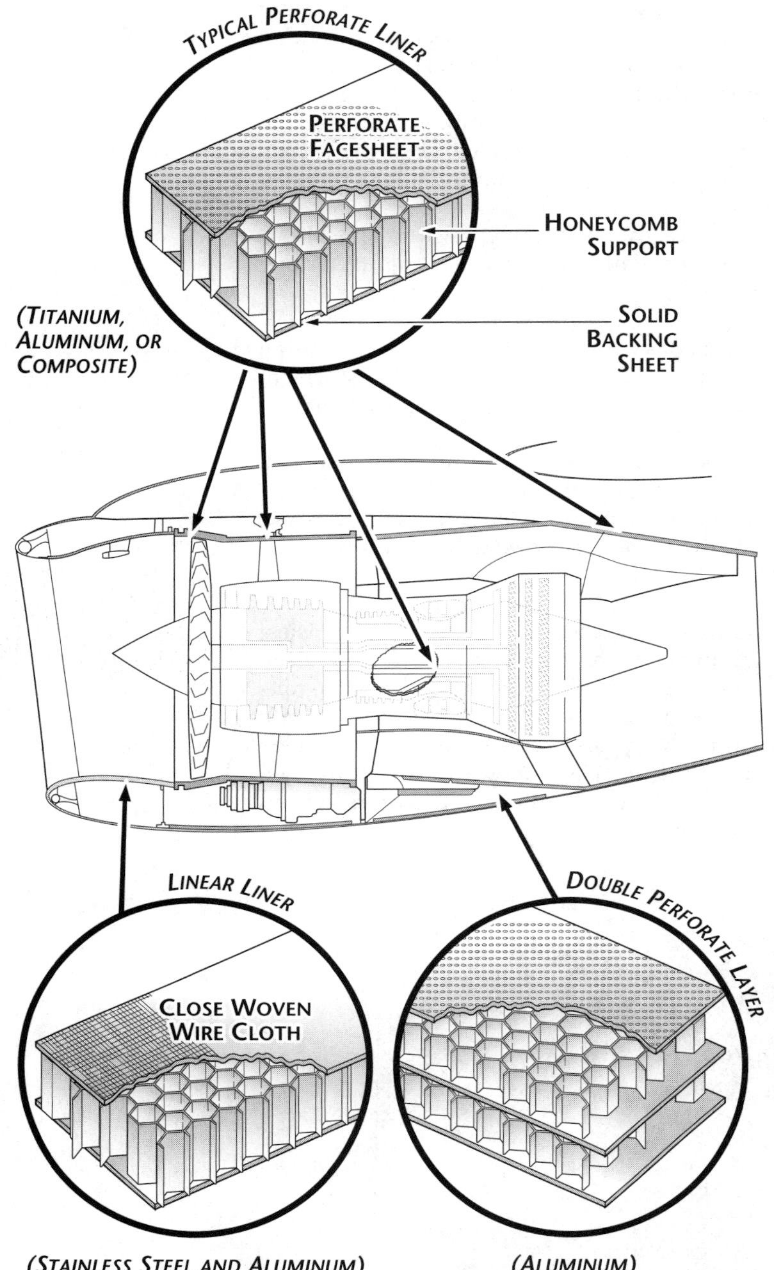


**Figure 2-8-9. Sound absorbing panels are part of the cowling in most installations.**

Most turbine blades are air cooled from the inside. Compressed air exits through cooling holes drilled in the blade. By drilling holes in the trailing edge extra air can be used to fill the wake behind the blade and also help reduce the wake and its turbulence. Reduced turbulence equals less noise.

**Internal engine noise.** The interior of a running jet engine is a violent place and most violence produces noise. Combustion chambers, gear trains, bearings, pumps, and flame fronts all contribute their share of noise (Figure 2-8-8.).

The most effective method of dealing with this collection of noise is to put barriers between it and the outside of the cowling. Thus, each different design of cowling has some method of dealing with the internal noise problem. The most common is sound absorbing material; metal-faced honeycomb panels are an example. Most sound insulation material is added to the cowling so as to have the cooling airflow pass between it and the engine. See Figure 2-8-9.

## Hush Kits

As noise regulations came into being there were a lot of older airplanes that did not meet the requirements. The past several decades have seen the introduction of increasingly more restrictive noise regulations. Many airports actively monitor noise levels generated by arriving and departing aircraft, and levy large fines against aircraft operators that exceed specified maximum noise levels.

Noise reduction can be accomplished through a number of methods. One is to replace the engines with quieter ones. This has been done on a few aircraft models, but is typically cost prohibitive. The engineering and certification work is extremely expensive, sometimes exceeding the basic value of the non-re-engined aircraft. There are, however, some potential benefits. Modern, quieter engines are also more fuel efficient, providing operational savings. Additional range or payload capabilities are other possible benefits.

A second method is to operate with reduced power settings. This reduces noise, but also reduces the payload that an aircraft can carry. It may also significantly reduce the safety margins in the event of an engine failure. Reduced power settings are not typically a practical solution.

The most common noise reduction option is the installation of a hush kit. An example is shown in Figure 2-8-10. Several specialty maintenance shops engineered modifications that would allow these older, noisier aircraft to continue to operate legally. The original engines are used, but have modifications made to reduce the noise they create. Most conversions used either corrugated or lobe type exhaust nozzles, along with soundproofing in the nacelles. Sometimes they also involve minor changes in power settings or operating methods. The goal is to greatly reduce the audible noise levels, without the expense of re-engining or suffering the safety consequences of operating with reduced power. In other cases these modifications were used in conjunction with newer, less noisy ver-

Figure 2-8-10. Early Rolls-Royce Spey turbofan on a Gulfstream G-III equipped with a hush kit

sions of the original engines, thus reducing engineering costs for the upgrades. While not inexpensive, hush kits are less expensive than a newer airplane.

## Thrust Reversers

Stopping an aircraft after landing is a problem that increases with higher airspeeds, greater gross weights, higher wing loadings, and increased landing speeds. Many times wheel brakes can no longer be entirely relied upon to slow the aircraft within a reasonable distance. The reversible pitch propeller has solved the problem for reciprocating engine aircraft and turboprop-powered airplanes. This method is described in detail in the chapter on propeller systems. Turbojet aircraft rely on reversing the thrust of the engines (*thrust reversers*).

Thrust reversers are standard equipment on commercial, airline, and many corporate aircraft. Some small corporate aircraft use a device called a thrust attenuator (an example is shown in Figure 2-8-11). It is simply a flap that moves out into the jet exhaust. While it does not provide any braking capability, it fully blocks the residual thrust that the turbine generates at idle, thus reducing the wear on the brakes during landing.

The most successful thrust reversers are the *mechanical blockage* type. Mechanical blockage is accomplished by placing a removable obstruction in an air stream, usually the exhaust gas path.

An acceptable thrust reverser must not affect engine operation either when the reverser is operating or when it is not. It must be able to

Figure 2-8-11. Cessna CitationJet thrust attenuator

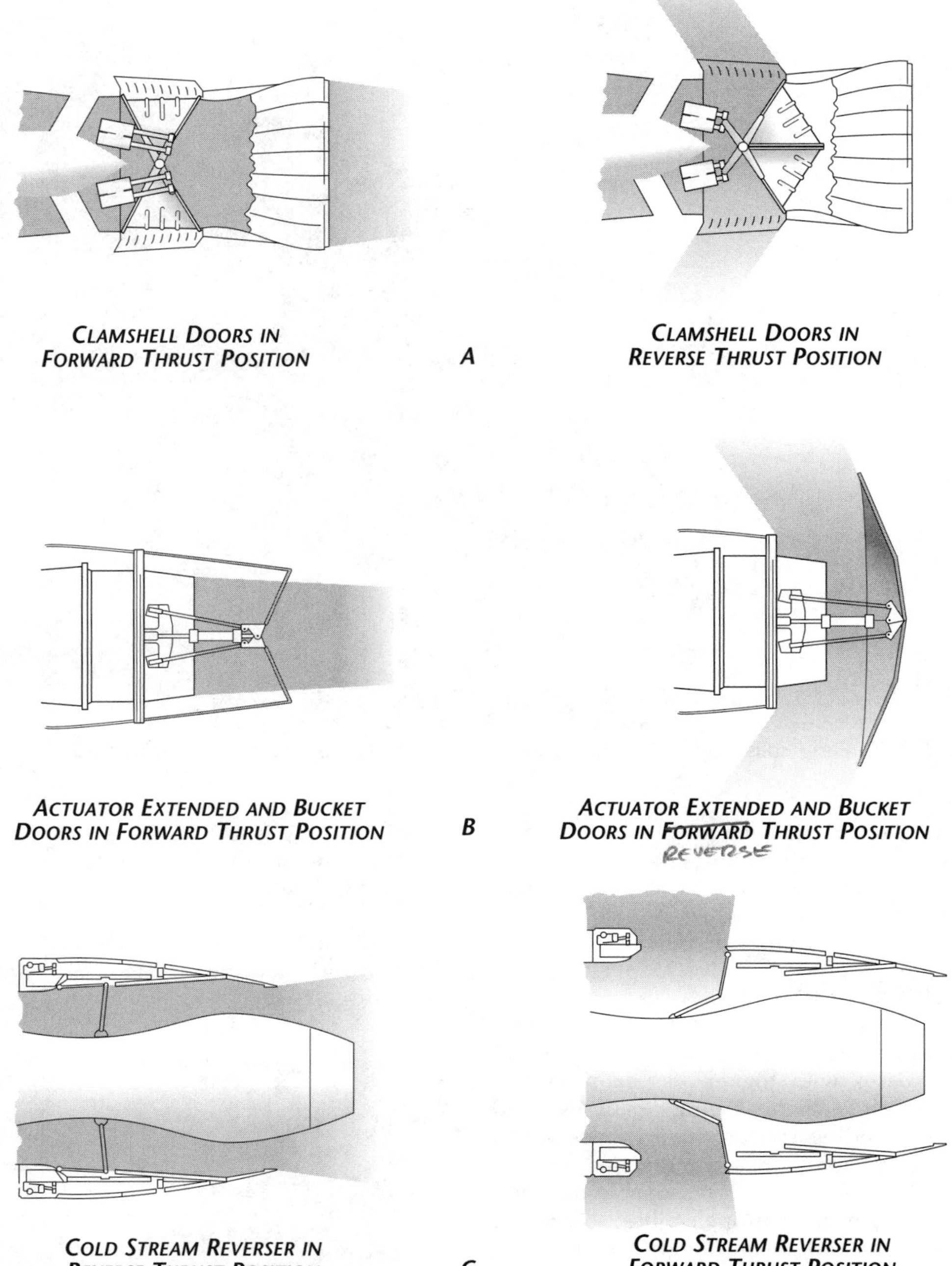


**Figure 2-8-12. The three basic types of thrust reversers; W type (A), hot stream (B) and cold stream (C).**

withstand high temperatures, be mechanically strong, relatively light weight, reliable, and *fail-safe*.

Thrust reverser operation is typically controlled by the engine throttle position. It is critical that thrust reversers have safety mechanisms preventing their deployment in the air. Air activation of a thrust reverser can cause the airplane to lose control and crash. Preventing unsafe deployment is typically accomplished through a connection with a landing gear safety switch. The thrust reverser will only activate when there is weight on the wheels. Most systems also have a safety switch preventing thrust reverse operation after the engine fire extinguisher system has been activated.

When not in use, the thrust reverser must be streamlined into the configuration of the engine nacelle. Figure 2-8-12 shows the three common types of thrust reversers in use.

**Hot stream reversers**. Reverse thrust is an operation that must be selected by the cockpit crew. Firstly, the throttles must be fully aft, the airplane must be on the ground, and the engines must be running. In a typical setup, the pilot then pulls the throttles over a detent, and back into reverse range. The reversers will then deploy and the engines will spool up to approximately 40% power. Figure 2-8-13 shows a hot stream reverser on a Cessna Citation being deployed. Deployment is very rapid and typically occurs in less than a second. The photo sequence was taken with a high-speed camera, and is typical of all hot stream thrust reversers.

**Cold stream reversers.** The major difference with cold stream reverse thrust is that only the cold fan air is reversed. Because the hot combustion exhaust stream does not reverse, there is always some forward thrust during reverse thrust application, however, the fan thrust braking action through the reverser exceeds the forward thrust output of the turbine exhaust section.

**W thrust reverser (clamshell).** The W-clamshell shown in 2-8-12A has operated very satisfactorily. When open, the reverser is approximately one nozzle diameter to the rear of the engine exhaust nozzle. When folded and not in use, it nests neatly around the engine exhaust duct, forming the rear section of the engine nacelle cowling. The W-clamshell reverser has the added advantage of splitting the exhaust gas discharge, which tends to prevent flow instability and severe buffeting. The clamshell is opened and closed by pneumatic pistons operated by high-pressure bleed air from the engine compressor. The speed at which the reverser operates is sufficient to meet most emergencies. Most W type reverser systems are pneumatically operated.

**Bucket thrust reverser.** The bucket type of thrust reverser is another popular system for reversing thrust. Unlike the W system, the buckets are mounted on the aft end of the exhaust nozzle. Basically, the buckets unlock and slide aft on rails approximately their own length, then the two parts open, as shown in 2-8-12B.

**Cascade reversing system.** Reverse thrust on fanjets is accomplished by a system called a *cascade reversing system*. Most cascade reversing systems are of the cold stream type, however, some are combined with a W or bucket type thrust reverser to reverse all of the forward thrust of the engine. If you refer to Figure 2-8-12C once again, you will notice that the system basically reverses the secondary airflow produced by the fan, instead of reversing the hot stream gasses. The turning (*cascade*) vanes are mounted circumferentially in the engine nacelle. When at rest (in the battery position) the cascades are covered by the *blocking doors* that are attached to the moveable portion of

Figure 2-8-13. A typical hot stream reverser in sequence. In practice this process happens very quickly.

**Figure 2-8-14. Deployed cascade thrust reversers on a military C-17A. Cascade reversers on most modern turbofans are not visible until they are deployed.**

*Photo courtesy of Burkhard Domke*

the cowling. When a reverse thrust command is received, the cowling slides aft, the blocker doors drop down to divert the cold stream airflow, and the cascade vanes redirect the airflow to reverse the thrust. See Figure 2-8-14.

*Section 9*

# *Turbine Engine Ignition, Starter, and Generator Drive Systems*

## Ignition Systems

Most turbojet engines are equipped with a high-energy, capacitor-type ignition system. The typical gas turbine engine is equipped with two identical independent ignition units. Thus, as a safety factor, the ignition system is actually a dual system, designed to operate two igniter plugs. The ignition system, shown in Figure 2-9-1, consists of three main components, an ignition unit (exciter box) with high-tension output, an igniter lead and an igniter plug. The ignition system can have continuous-duty capability, although its main use is during starting. A cockpit selector switch is used for selecting normal operation or continuous ignition modes.

Since turbine ignition systems are normally operated for a brief period during the engine-starting cycle, they require little maintenance. Continuous ignition is used during certain flight configurations, such as take-off and landing. The ignition system is designed to operate using power supplied by the permanent magnet generator (PMG) or from the aircraft power buss. The PMG is an AC generator that is driven directly by the engine. When the engine is turning, the PMG is developing power. The PMG is also used to provide power for the FADEC control units.

**Figure 2-9-2. Igniter plug**

The capacitor-discharge high-voltage ignition unit (exciter) is normally a solid-state, high-tension capacitor-discharge unit that provides a spark to the igniter plug. A rate of one to six sparks per second, depending upon input voltage, will provide an output of 18,000-24,000 volts.

Most units have one output circuit, but the engine is generally equipped with two systems. The circuit and igniter plug release sufficient energy for all ground and air starting requirements. The igniter lead provides a path for electrical energy to travel from the ignition unit to the igniter plug. The igniter lead is encased in metal sheathing and is attached to the igniter plug (Figure 2-9-2).

The igniter plug can be of two types, the air-gap type and the self-ionizing shunted-gap type. The shunted-gap type uses a semi-conductor between the firing electrodes, which reduces the voltage needed in the system. This type of system is sometimes referred to as a low-voltage system. When energized, the igniter plug releases a high-energy arc to ignite the atomized fuel/air mixture in the combustion chamber. Turbine engines can be ignited readily in ideal atmospheric conditions, but since they often operate in low temperatures and at high altitudes, it is imperative that the system be capable of supplying a high-heat-intensity spark. Thus, a high voltage is supplied to arc across a wide igniter-spark gap, providing the

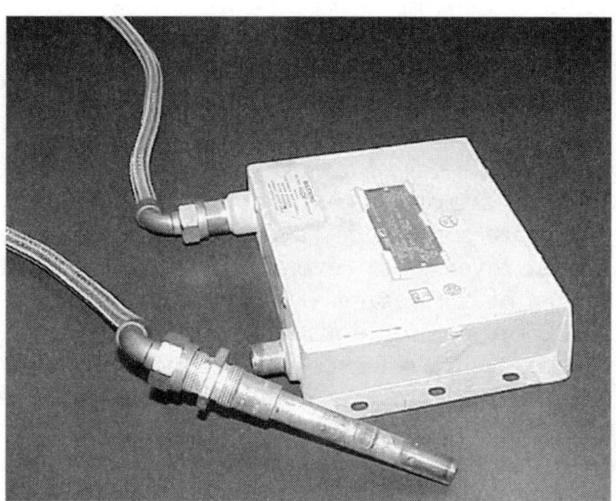

**Figure 2-9-1. A typical turbine engine ignition system that shows the exciter, lead, and igniter**

ignition system with a high degree of reliability under widely varying conditions of altitude, atmospheric pressure, temperature, fuel vaporization and input voltage.

## Starter Systems

Gas turbine engines are started by first rotating the compressor. On dual-axial-compressor engines, the high-pressure compressor, $N_2$, is the only one rotated by the starter. To start a gas turbine engine, it is necessary to accelerate the compressor to provide sufficient air to support combustion in the burners. Once fuel has been introduced and the engine has lit off, the starter must continue to assist the engine to reach a self-accelerating speed that will allow the engine to accelerate by itself. The torque supplied by the starter must be in excess of the torque required to overcome compressor inertia and the friction loads of the engine. In larger engines, these loads can require in several hundred horsepower to turn at high enough speed. This requires the use of an air turbine starting system.

Figure 2-9-3. A starter generator unit mounted on a PT-6

The basic types of starting systems, which are used on gas turbine engines, are electric motor and air turbine. The electric motor type of starter usually uses DC power to turn the engine. This type of starter is limited by the amount of horsepower they can provide, so they are used on smaller turbine engines.

Other types of starting systems have been used, but their use is very limited. These include impingement starting systems, combustion starters and even hydraulic systems. An impingement starter consists of jets of compressed air piped to the inside of the compressor, or turbine case, so that the jet air-blast is directed onto the compressor or turbine rotor blades, causing them to rotate. Combustion starters were very small turbine engines that were started and then used to turn the engine's compressor. Hydraulic motors can also be used to turn the compressor during starting.

Figure 2-9-4. Starter-generator unit

### *Types of Starters*

**Starter generators**. Many smaller gas turbine aircraft are equipped with starter-generator systems. These starting systems use a combination starter-generator that operates as a starter motor (DC electric motor) to drive the engine during starting. It has an additional set of fields that allow it to also operate as a generator. This system has the advantage of needing only one unit, instead of two, in the installation (Figure 2-9-3). It cannot, however, act as both a starter and a generator at the same time. Besides simplifying the electrical system there is a large weight savings.

Starting the engine consists of engaging the start portion of the system by turning on the master switch, setting the necessary fuel controls, and pushing the start switch. After the engine has reached a self-sustaining speed and the starter is disengaged, the unit can be switched to operate as a generator to supply the electrical system power. Starter-generator units are desirable from an economical standpoint, since one unit performs the functions of both starter and generator. Additionally, the total weight of electrical system is reduced by the weight of one unit. See Figure 2-9-4.

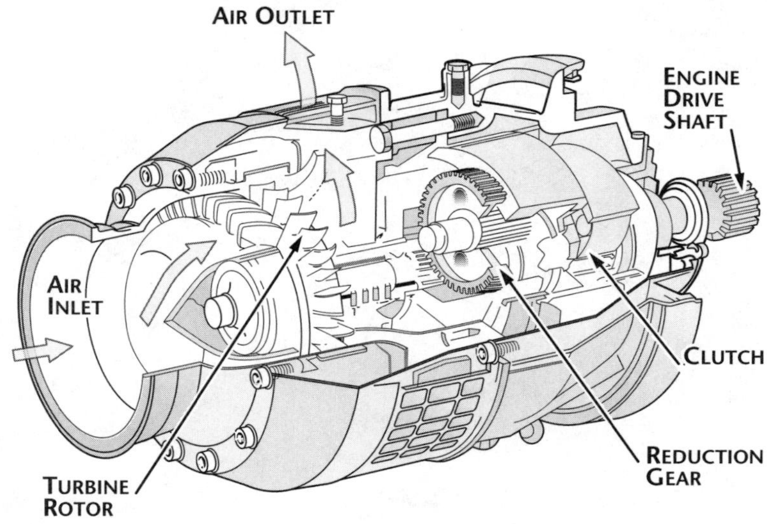


Figure 2-9-5. A cutaway view of an air turbine starter

Most starter generators are rated at 28 volts, 300 amps maximum. Generally, the units draw electrical power from the main bus bar; the bus bar can be supplied by the aircraft battery or a ground power unit GPU. The starter really does not care as long as the capacity is enough to turn the engine to self-accelerating speed. Most aircraft electrical systems have an undercurrent sense circuit that will not allow the starter to power up if there is not enough capacity in the battery. To power up the starter with an under capacity battery would both overheat the starter and more than likely produce a hung start, or a no start, on the engine. At worst the engine could produce a hot start and exceed the engine temperature limits. This is often referred to as "overtemping".

Once self-accelerating speed is reached and the starter is deselected, the generator is selected. This puts the generator online to the generator bus bar and produces regulated electricity. The units have a reverse current system that will not allow the generator section of the unit to absorb power. Anytime the power output drops below the system voltage, the unit is electrically disconnected.

**Air turbine starters.** The air turbine starters are designed to provide high starting torque from a small, lightweight source. The typical air turbine starter weighs from one-fourth to one-half as much as an electric starter capable of starting the same engine. It is capable of developing twice as much torque as the electric starter. The typical air turbine starter consists of an axial-flow turbine that turns a drive coupling through a reduction gear train and a starter-clutch mechanism.

The air to operate an air turbine starter is supplied from either a ground-operated compressor, the bleed air from another engine or an Auxiliary Power Unit (APU). Figure 2-9-5 is a cutaway view of an air turbine starter. Introducing air of sufficient volume and pressure into the starter inlet operates the starter; the air passes into the starter turbine housing, where it is directed against the rotor blades by the nozzle vanes, causing the turbine rotor to turn. As the rotor turns, it drives the reduction gear train and clutch arrangement.

The air turbine starter is used to start large gas turbine engines. The starter is mounted on an engine pad and its drive shaft is splined by mechanical linkage to the engine compressor.

There is little maintenance for an AMT to perform on an air starter. The major check is to make sure there is no FOD damage to the rotor.

**Direct current motor starters.** Direct current motor starters are essentially the same as direct current starters used on reciprocating engines. These are discussed in detail in the starter section of Chapter One. The biggest difference between turbine starters and reciprocating

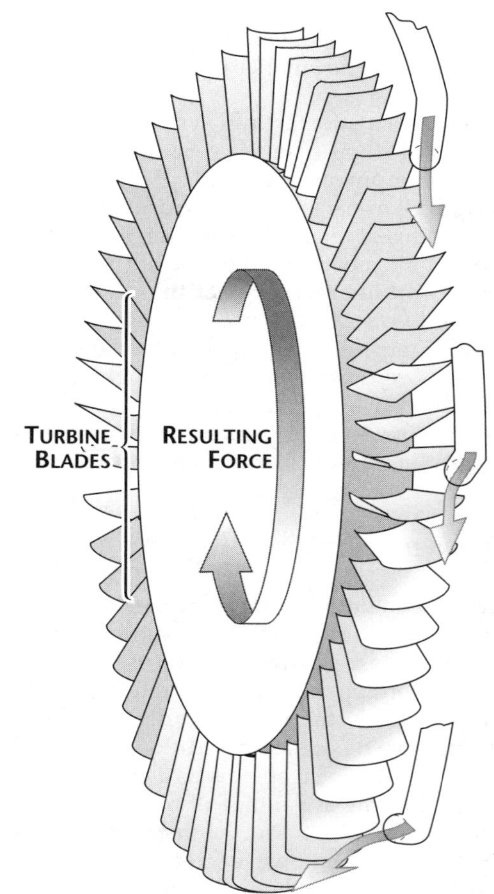


Fig 2-9-6. Air impingement system

starters is their operating r.p.m. and their size. The r.p.m. requirements for starting turbine engines are often higher than the typical reciprocating engines. Repairs for this unit are the same as for any other DC starter.

The development of the starter-generator, and the corresponding weight savings, has all but eliminated the use of the dedicated direct current starter in commercial service.

**Hydraulic Motor Starters.** Hydraulic motors can also be used to start turbine engines. They are typically mounted on an engine accessory gearbox pad just like other starter types. Hydraulic starters are most prevalent on small APU turbines. Maintenance procedures are the same as for any other hydraulic motor.

**Air impingement starters.** In turbine impingement starting systems, high-pressure air is directed onto the turbine rotor blades (see Figure 2-9-6). An external starting unit supplies starting air. The air is then ducted through the impingement manifold and directed onto the turbine blades. A check valve in the starting manifold prevents loss of gases after the engine has been started. Air impingement systems are no longer in widespread commercial use.

**Cartridge starters.** Cartridge starters are constructed much like air turbine starters. The primary difference between the two is their method of power. An air turbine starter is powered by a compressed air source. A cartridge starter is powered by a cartridge. This is typically a black powder cartridge. The cartridge is placed in a chamber at the entrance to the starter. When the cartridge is triggered, the by-product of the combustion is pressure – just like firing a gun. This pressure is directed across the turbine in the cartridge starter, spinning it just like an air turbine starter. Some cartridge type starters can also be powered by an external compressed air source, thus acting as an air turbine starter. Repairs for the cartridge starter are essentially the same as for an air turbine starter.

While cartridge starters require the use of a new cartridge before each start, they are totally independent of external ground power units. As such, they have been used extensively by military aircraft. Some early commercial turbine aircraft used cartridge starters, but they are largely a thing of the past.

## Generator Constant Speed Unit (CSD)

### *Description and Operation*

Alternating current motors are typically lighter for a given power than an equivalent direct current motor. Significant weight savings can be obtained through the use of AC motors, however, this causes some engineering challenges. AC motors need a constant cycle (or frequency) to the current supplied to them. The frequency of a generator is a function of the speed at which it is turned. This becomes a problem when the turbine changes speeds to meet the flight demands of the aircraft. The constant speed drive was developed to solve this problem.

Accordingly, the constant speed drive converts the varying speed of the jet engine to a constant speed so that the generator it drives will produce AC current at 400 hertz, within narrow limits. There is one generator drive and its associated generator for each engine on the aircraft. The generator CSU is mounted on the engine gearbox. Each of the constant speed drive installations consists essentially of a hydro-mechanical transmission with electromechanical controls governing the output rotation speed (Figure 2-9-7).

### *Transmission*

The transmission is capable of either adding to, or subtracting from, the speed received from the engine gearbox in order to provide constant output speed to keep the generator on frequency. Mechanical (flyweight) governor action keeps each generator's output close to 400 hertz by controlling the generators drive speed.

**Overdrive.** If the input speed supplied to the transmission is lower than that needed to produce the required output speed, the transmission hydraulically adds the necessary speed to the speed of the engine gearbox through the

**Figure 2-9-7. A CSD can drive a very large ac generator and is capable of absorbing 200 horsepower.**

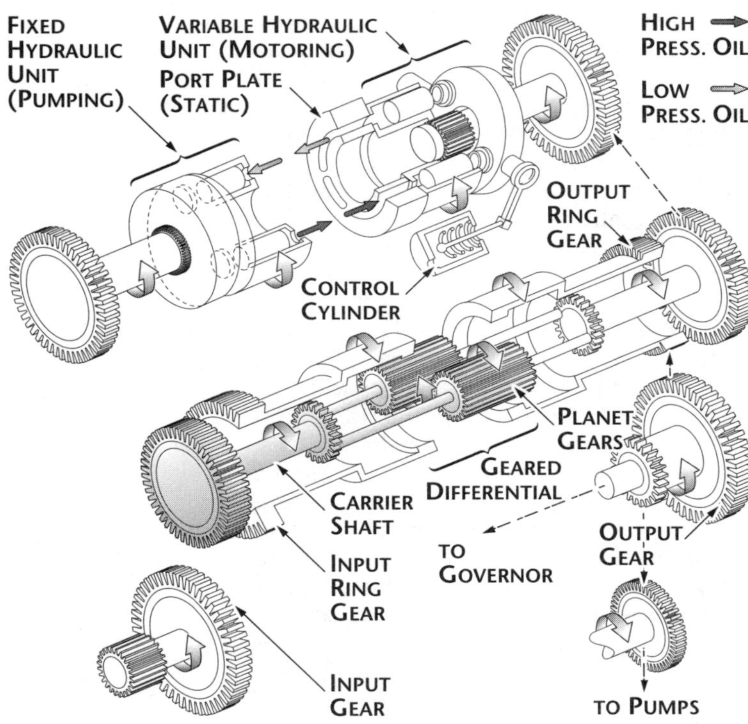


Figure 2-9-8. Operational schematic of a CSD in overdrive phase

differential, thus maintaining constant output speed. When the transmission is adding speed hydraulically, it is operating in *overdrive.*

**Straight through drive.** If the input speed supplied to the transmission is sufficient to produce the required output speed, the transmission drives the generator directly through the differential. When the transmission is neither adding nor subtracting speed hydraulically, it is operating in *straight through drive.*

Figure 2-9-9. Variable hydraulic unit

**Underdrive.** If the input speed supplied to the transmission exceeds that needed to produce the required output speed, the transmission hydraulically subtracts the necessary speed from the speed of the engine gearbox through the differential, thus maintaining constant output speed. When the transmission is subtracting speed hydraulically, it is operating in *underdrive.*

## *The Power Unit*

The power used to drive the generator is controlled and transmitted from the aircraft engine through the combined effects of the *differential*, the *variable hydraulic unit*, and the *fixed hydraulic unit*. See Figure 2-9-8.

**Differential.** The differential consists of a carrier shaft, two planetary gears, and two ring gears (the input and output ring gear). The ratio between the ring gears and the carrier shaft is 2:1.

At any speed and load condition, a torque load is imposed on the output ring gear by the output gear. The input gear that turns the carrier shaft supplies input torque. (see Figure 2-9-8). If there were no torque on the input ring gear, it would run freely at whatever speed would allow the output ring gear to stop. Because the ring gear to carrier shaft ratio is 2:1, the speed of the input ring gear at this condition would be double that of the carrier shaft. Since a given output speed is desired, the input ring gear must be constrained.

If the input ring gear is constrained to zero speed, the output ring gear will run at double the carrier shaft speed. If the input ring gear is forced to rotate in a direction opposite to that of the carrier shaft, the output ring gear will run at a speed more than double that of the carrier shaft. If the input ring gear is allowed to rotate in the same direction as the carrier shaft, the output ring gear will run at a speed less than twice that of the carrier shaft. Thus, the differential is an adding device that is controlled through the input ring gear to add to or subtract from the speed of the engine gearbox to achieve the desired output.

**Variable hydraulic unit.** The variable hydraulic unit consists of a cylinder block, reciprocating pistons, a variable angle wobbler, and a control piston. The variable unit is connected to the aircraft engine by direct gearing; consequently, the speed of the cylinder block is always proportional to the input speed and the direction of rotation is always in the same direction. See Figure 2-9-9.

When the transmission is operating in overdrive, the variable hydraulic unit will function as a hydraulic pump. Refer back to Figure 2-9-8. To enable the variable unit to pump oil, the governor ports control oil to the control piston, which in turn positions the wobbler so oil will be compressed as the pistons are forced into the rotating cylinder block. This high pressure (working pressure) oil is ported to the fixed hydraulic unit.

As the input speed increases and the need to add speed decreases, the governor will port less oil to the control cylinder until the variable wobbler is in a position approximately normal to the pistons. When the face of the variable wobbler is approximately perpendicular to the pistons, no oil (except that required to provide for power losses due to leakage) is pumped or received by the variable unit. At this time the transmission is operating in straight through drive.

When the transmission is operating in underdrive, the variable hydraulic unit will function as a motor. To enable the variable unit to operate as a motor (receive oil from the pumping unit), the governor ports oil away from the control cylinder causing the wobbler to be positioned so the volume for accommodating oil in the piston bores on the high pressure side is increased; consequently, oil flows from the fixed hydraulic unit to the variable unit. See Figure 2-9-10.

**Fixed hydraulic unit.** The fixed hydraulic unit consists of a cylinder block, reciprocating pistons, and a fixed angle wobbler. The direction of rotation and speed of the fixed hydraulic unit is determined by the volume of oil pumped or received by the variable hydraulic unit. This volume of oil is determined by the angular position of the variable wobbler and the speed of the variable block.

When the transmission is operating in overdrive, the fixed hydraulic unit functions as a hydraulic motor. High pressure oil pumped from the variable unit forces the fixed unit pistons to slide down the inclined wobbler face, thus causing the cylinder block to rotate. The block's rotation forces the input ring gear to turn in a direction opposite to the carrier shaft rotation and adds to the speed of the engine gearbox through the differential, thus maintaining constant output speed.

As the input speed increases and the need to add speed to the output decreases, the variable hydraulic unit pumps less oil to the fixed hydraulic unit until the cylinder block stops

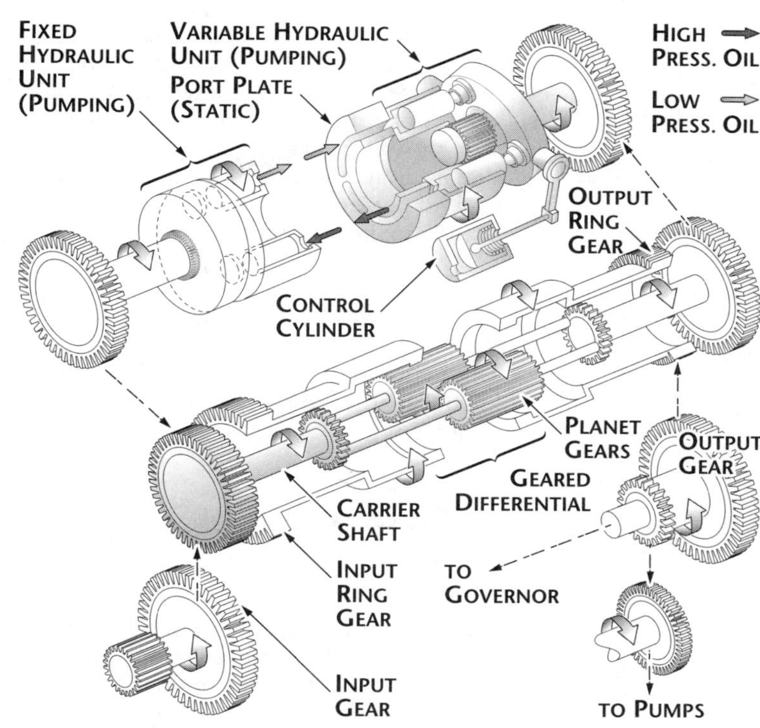


Figure 2-9-10. Operational schematic of a CSD in underdrive phase

rotating. At this time, the transmission is operating in straight through drive.

When the transmission is operating in underdrive, the fixed hydraulic unit functions as a pump. The variable wobbler in the variable hydraulic unit is positioned so the variable hydraulic unit can receive oil from the fixed hydraulic unit. The fixed hydraulic unit's pistons are forced into the cylinder block as they slide up the inclined wobbler face, thus pumping high pressure oil to the variable hydraulic unit and allowing the cylinder block to rotate in the direction opposite that of overdrive operation. The opposite block rotation allows the input ring gear to turn in the same direction as the carrier shaft rotation and subtracts speed from the speed of the engine gearbox through the differential, thus maintaining constant output speed.

## *Governing System*

The basic governor is a spring-biased, flyweight operated hydraulic control valve. It functions to control porting of transmission charge oil to the control cylinder. The rotating sleeve in the governor is driven by the output gear and, hence, is responsive to transmission output speed. Flyweights pivoted on this sleeve move a valve stem located within the sleeve against the bias of a spring. During steady state operation, supply pressure is reduced to the required control pressure by an orifice action at the edges of the stem control groove. Depending on valve stem position, charge oil is ported to the control piston or control oil is drained to the transmission case. A magnetic trim device is included with the basic governor to apply the corrective signals from the frequency and load controller without any additional parts above those already in the governor.

### *The Hydraulic System*

The hydraulic system is described in the following text. It consists of the following components:

- Charge pump
- The scavenge pump
- The charge relief valve

**Charge pump.** The charge pump is located in the hydraulic circuit between the all-attitude reservoir and the transmission. The charge pump supplies oil to the cylinder blocks, governor, control piston, and the lubricating system.

**Scavenge pump.** The scavenge pump is located in the hydraulic circuit between the transmission sump and the aircraft cooler. The scavenge pump picks up lube oil and internal leakage andpumps it through the aircraft cooler into the all-attitude reservoir.

**Charge relief valve.** The charge relief valve is located in the hydraulic circuit between the charge filter and the transmission sump. The purpose of the charge relief valve is to regulate the operating pressure of the charge oil system. The valve accomplishes this function by metering the discharge of oil from the charge oil system to maintain the pre-set charge pressure.

### *All-attitude Reservoir*

The all-attitude reservoir is designed to perform the following functions:

- Separate entrained air from the system oil
- Provide a supply of de-aerated oil to the transmission through a wide range of acceleration loadings and aircraft attitudes
- The reservoir contains no moving parts and performs its functions automatically, utilizing the energy of the transmission scavenge oil

Operation of the reservoir. Return oil is always de-aerated and ported to the lower reservoir chamber, regardless of transmission attitude. The inlet suction port is located in approximately the middle of the lower chamber, and the volume of oil delivered to the chamber is such that, regardless of transmission attitude, the inlet port will always be surrounded by oil. Static pressure in this chamber is relatively low due to the constant removal of oil through the inlet suction port.

Regardless of transmission attitude, the upper reservoir chamber will always receive a mixture of air and oil, foam, or mist. As the air settles out of this mixture, the oil will leak back into the lower reservoir. A vent between the upper chamber and the transmission case ensures that reservoir and case pressure are the same.

Filter-pressure differential indicator. The filter has a pop-out button that will actuate when the pressure drop across the filter is 44 pounds per square inch, indicating that the filter is becoming clogged. The filter is also equipped with a bypass around the filter element to ensure a flow of oil in case the element becomes completely clogged.

The disconnect. The disconnect is an electrically actuated device which decouples the input splined coupling from the input spline shaft and is activated from the cockpit.

When the driving dogs have been separated, the input spline shaft, which is still being driven by the aircraft engine, spins freely in the transmission without causing transmission rotation. Reset may be accomplished, following engine shutdown, by pulling out on the reset handle until the solenoid nose pin snaps into position.

# *Section 10*

# *Turbine Engine Auxiliary Systems*

## Secondary Airflow Systems

Secondary airflow can be divided by engine internal air use and aircraft pneumatic system use. Secondary airflow systems do not directly contribute to the engine's thrust. Therefore, the more air that is drawn from the engine's gas path, the more fuel that will be required to operate the engine. To reduce engine performance losses, the air is taken from the compressor only as needed to operate aircraft systems. Internal bleed air is drawn away from the gas

**Figure 2-10-1. A turbine blade with cooling holes to prevent the heat of combustion from coming into contact with the surface of the blade.**

path and is used for several different engine functions. The engine internal air system is used for increasing engine efficiency, increasing engine life, providing internal cooling, controlling disk loads, sealing, anti-icing and controlling engine internal clearances. To increase the engine's efficiency, the operating temperatures must be increased.

The temperature the turbine blades and inlet guide vanes can withstand limits the maximum turbine temperature. By using internal airflow through hollow turbine vanes and blades, the engine's overall exhaust temperatures can be raised. This air flows over the surface of the blades and vanes through holes, shown in Figure 2-10-1, that protect the surface from the higher turbine temperatures.

Internal air can be used to cool bearing areas around the combustion area and stabilize temperatures on the engine's interior components. By using internal airflow against the area of the turbine disk, the axial loads applied to the bearings can be reduced. The internal airflow is expelled overboard or into the engine main gas stream.

Another important use of the secondary air systems is to supply compressed air (bleed air) for aircraft systems operation. This air is generally bled from the compressor at the high-pressure stage and at an intermediate stage. At high engine speeds, the high-pressure compressor stages provide the air needed by the aircraft pneumatic systems. At lower engine speeds, both air sources - high-pressure and intermediate - are used for operation of the aircraft systems. Secondary bleed air is used in a wide variety of ways, including some of the following applications:

- Cabin pressurization, heating and cooling
- Anti-icing
- Starting engines
- Auxiliary drive units

## Compressor Stall and Surge Protection Bleed Air Systems

One of the characteristics of a gas turbine engine is its tendency to stall under certain operating conditions. Compressor stall occurs in many different types of gas turbine engines. Depending on the operating conditions, stall or surge can occur in various forms and intensities. Compressor stall or surge, in its most violent stage, can cause engine damage and a loud audible noise. Surge is the airflow velocity in the rear of the compressor slowing down to the point of restricting airflow into the compressor. Compressor stalls can be initiated at both the entrance to and the exit from the compressor. Careful inlet designs minimize the chance of an intake induced stall.

Until the fuel flow is proper for the engine condition, the cycle will repeat itself many times over. A disruption of airflow will cause the air velocity over the blades to slow, which will cause the blades to stall because of a high angle of attack.

The compressor blades and vanes are airfoils. If the angle of attack becomes too great or if the velocity of air flowing over an airfoil is too low, airflow separation occurs and the airfoil stalls. This results in a loss in compressor efficiency, *compressor pressure ratio* (CPR), and a reduction in airflow velocity. CPR is the pressure differential between the inlet pressure and the compressor outlet pressure. In order to produce engines with superior fuel consumption, it is necessary to operate with as high a compressor pressure ratio as possible.

The ability of the compressor to pump air is a function of r.p.m. At low r.p.m., the compressor does not have the same ability to pump air as it does at higher r.p.m. In order to keep the angle of attack and air velocity within desired limits, it is necessary to unload the compressor in some manner during starting and low-power operation. The compressor has less restriction to the flow of air through the use of a compressor surge/stall bleed air valve system. This air is not used for aircraft systems and is dumped directly back to the atmosphere.

**Figure 2-10-2. Compressor bleed valve**

The compressor bleed air system consists of an (open-closed) air bleed control valve, or a variable air bleed control valve (open-modulating-closed). Engines can be equipped with both or only one type of bleed valve.

A slot, located at the appropriate stage, allows an equal bleed of compressor air into a manifold (annulus) that is an integral part of the compressor case. See Figure 2-10-2. The manifold or annulus forms the mounting flange for the air bleed control valve.

Compressor discharge air pressure sensing, for air bleed control valve operation, is obtained at a sensing port on the diffuser or outlet of

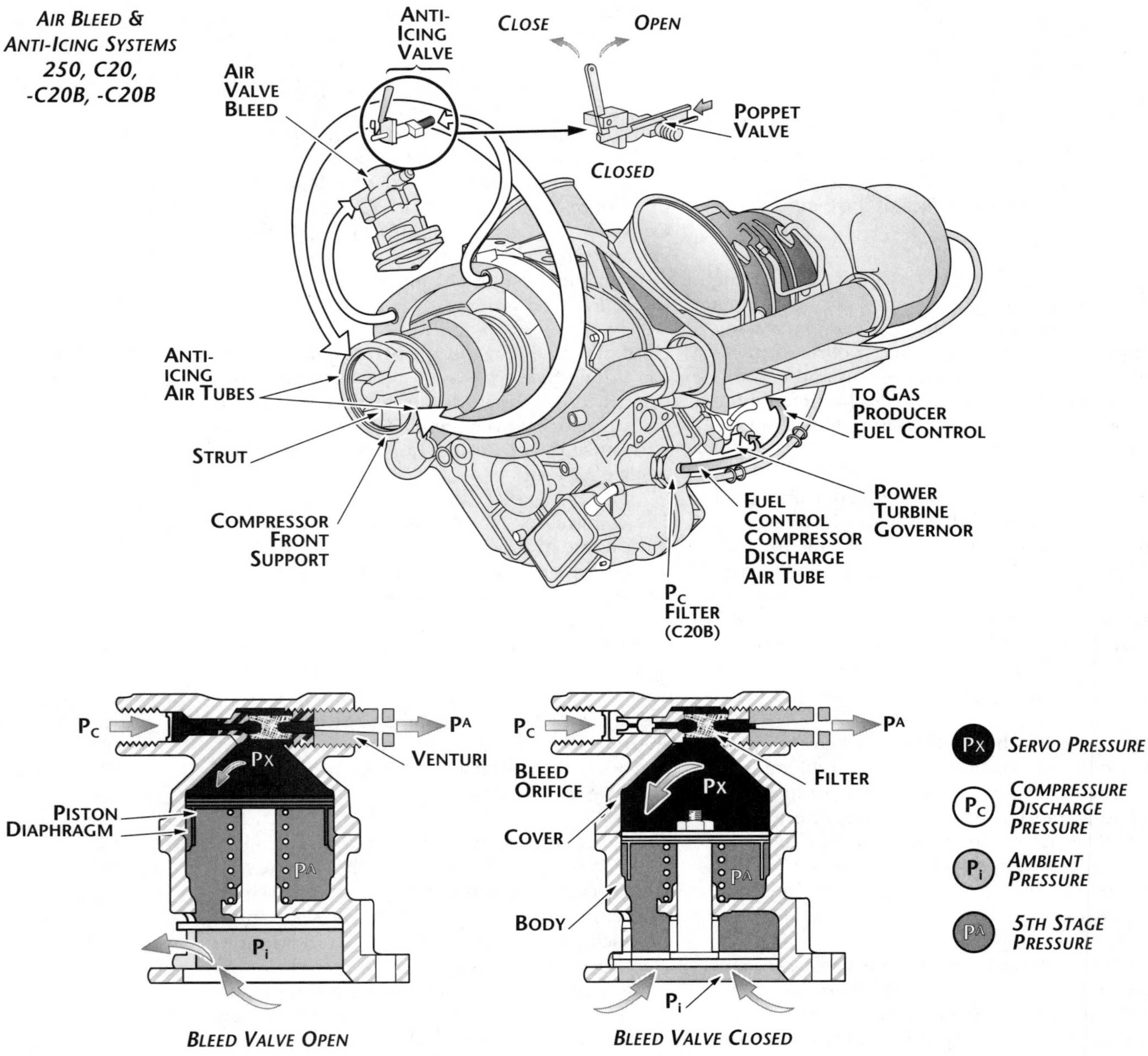

**Figure 2-10-3. An illustration showing the operation of the anti-icing system and the surge bleed valve.**

the compressor. The air bleed control valve is generally open during starting and ground idle operation and it remains open until a predetermined pressure ratio is obtained across the bleed valve diaphragm. The valve begins to move from the open to the closed position.

Some air bleed control valves, as shown in Figure 2-10-3, consist of a piston, a diaphragm, a retaining plate, a spring, a filter, a bleed nozzle, a bleed valve, a restrictor orifice (jet), a bleed valve body, a bleed valve cover, and miscellaneous small parts. The following pressures are associated with the typical operation of an example air bleed control valve:

- $P_c$ - Compressor discharge pressure
- $P_x$ - Modulated pressure
- $P_i$ - Compressor stage pressure
- $P_a$ - Ambient pressure

When the engine is not in operation, the bleed valve is positioned fully open by a spring located inside of the vented piston chamber. The spring, along with $P_i$ pressure directed onto the bleed valve end, is used during engine starting and acceleration to position the bleed valve fully open.

During engine operation, $P_c$ pressure is directed through an inlet filter and a restrictor (jet) into the $P_x$ chamber. $P_x$ air is then directed to $P_a$ through a nozzle (venturi). The rate of airflow from $P_c$ to $P_x$ to Pa determines the value of $P_x$ pressure for any given $N_1$ r.p.m. $P_x$ pressure is separated from the $P_a$ pressure and spring chamber by means of the rolling diaphragm.

Operation of the valve is a function of preselected ratios of $P_i$ to $P_a$ and $P_c$ to $P_x$ to $P_a$ pressures. When $P_x$ is less than $P_i$ plus the spring force, the bleed valve is positioned open. When $P_x$ is greater than $P_i$ plus the spring force, the bleed valve is positioned open during engine starting and acceleration until the $P_x$ pressure increases sufficiently to overcome the combined value of the spring and $P_a$ pressure. The bleed valve then closes and remains closed at all speeds above the closing r.p.m.

On a two-spool engine, the fan or low-pressure compressor (LPC) stall can occur when the high-pressure compressor (HPC) speed slows in relation to the LPC speed. This is due to the LPC developing more airflow than the HPC can accept. This condition occurs mostly during deceleration, when the HPC spools slow more quickly than the heavier LPC spool.

Inlet guide vanes, shown in Figure 2-10-4, are used to smooth or straighten the airflow into the engine. These vanes can be fixed or variable. Variable inlet guide vanes can also control the amount of air allowed to flow into the intake of the engine. In the first few compressor stages, variable stator vanes can also be used to reduce the airflow at slow speeds. The variable inlet guides vanes (VIGV) and variable stator vanes (VSV) will tend to be closed at lower engine speeds, preventing excess air from entering the compressor. Generally, the front stages of the compressor, at low r.p.m. (engine speed), can bring in more air than the compressor back stages can handle. So the VIGV and VSV are closed until the compressor r.p.m. becomes high enough to handle the airflow, then they will move toward the open position.

**Figure 2-10-4. Stationary inlet guide vanes on the inlet of a JT3D.**

**Anti-icing system**. The inlet of turbine engines must be free of ice because of the possibility it could break off after it has formed on the intake components and cause damage to the engine. Most engines are equipped with an anti-icing system that provides hot air to the inlet cowl, the inlet guide vanes, fan dome and the areas that are subject to the formation of ice during icing conditions. This system is entirely separate and independent of any other bleed air system. Operation of the engine anti-icing can be selected, when required, by the pilot.

As air passes through the compressor, it is compressed and the air is heated considerably; thus, compressor discharge air, which is extracted from the compressor, is an excellent

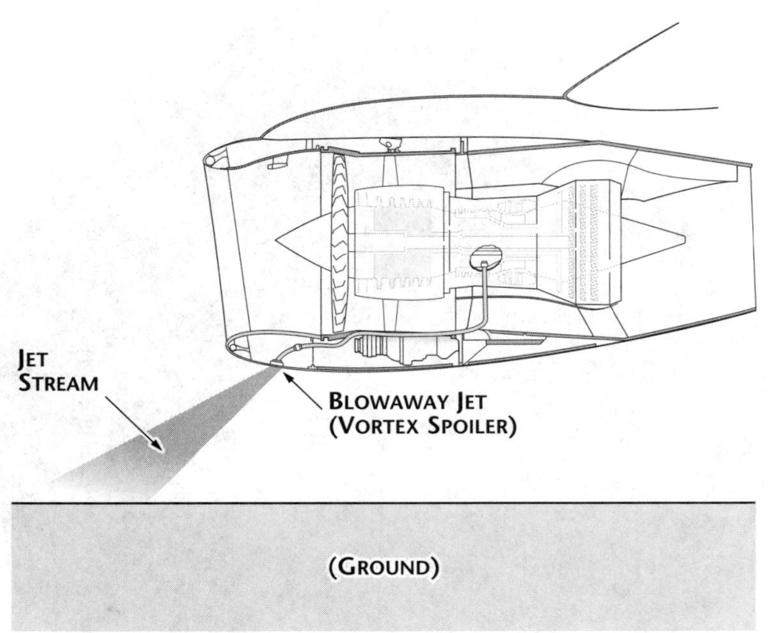


Figure 2-10-5. Vortex interrupter system action

source of the hot air required by the engine anti-icing system.

A typical anti-icing system, shown in Figure 2-10-3, consists of an anti-icing valve, two anti-icing tubes and passages within the compressor front support. Anti-icing tubes bring hot compressor air to the poppet valve, which can be opened to allow hot air to flow into the engine's inlet guide vanes. The actuating lever controls the position of the poppet valve. When anti-icing operation is required, the pilot actuates the linkage to move the actuating lever away from the poppet valve. Compressor hot air pressure flows through the valve and tubing into the inlet vanes and other inlet components as needed.

## Vortex Interrupter System

The large amount of air flow required by jet engines causes a problem when the airplane is taxied or run up. The inrush of air into the inlet duct creates whirlwind type vortices on the ground in front of the nacelles. In essence, the engines become huge vacuum cleaners, sucking up small rocks, sand, and any FOD from the ramp or taxiway. They will even suck up water from the runways. The closer to the ground an inlet is, the easier it can draw in material from the surface in front of it. Turbine rotors and vanes are very expensive to replace. Even if they are not damaged by debris, the dirt alone will reduce the efficiency of the engine and necessitate an increase in the frequency of compressor washes.

Airframe manufacturers solved the problem using bleed air. An air jet is placed at the bottom of the inlet cowl. Bleed air is blown from the jet forward at an angle sufficient to break up any spiral airflow and basically sweep a path in front of the engines. The airflow reduces the engine's ability to draw in material from the ground in front of it and it also pushes any dirt away from the inlet path. The system is called a *Vortex Interrupter System* (see Figure 2-10-5).

*Section 11*

# *Engine Instruments*

Safe, economical and reliable operation of turbine engines depends on accurately measuring engine operation. This is accomplished through the science of instrumentation. In hydromechanically controlled systems, the pilot has primary responsibility for monitoring the instruments and maintaining the engine power settings within preset limits. Modern electronic control systems (primarily the FADEC and EEC) have moved the task of keeping the engine within safe operating parameters to these electronic systems. Nonetheless, the pilot must still maintain an awareness of the operating condition and settings of the engine. Figure 2-11-1 graphically shows those parts of the turbine that are monitored by the typical instrument configuration.

## Types of Engine Instruments

### *Tachometers*

A turbine engine tachometer (tach) is a bit different than those used in reciprocating engine systems. To begin with, it measures in percent (% r.p.m.) of engine r.p.m. Also r.p.m. is not normally used to set engine power. Engine power is set by measuring the EPR. The tachometer is used mainly for engine start and to indicate an overspeed condition. Because turbine engine output can vary so much with a change in atmospheric conditions, it is possible to have a 90% engine one day and a 110% engine the next day. If r.p.m. was used to set engine power, it would be quite common to not have full power one day and overtemp the engine the next day.

**High pressure tachometer ($N_2$).** On a single-spool engine the HP tach is the only one available. It provides the $N_2$ indication. On a split-spool engine the *high-pressure* (HP) *tachometer* indicator (Figure 2-11-2) measures the high-pressure compressor r.p.m. There is an $N_2$ tach for each engine. The indicators show HP compressor rpm in percent. The units are

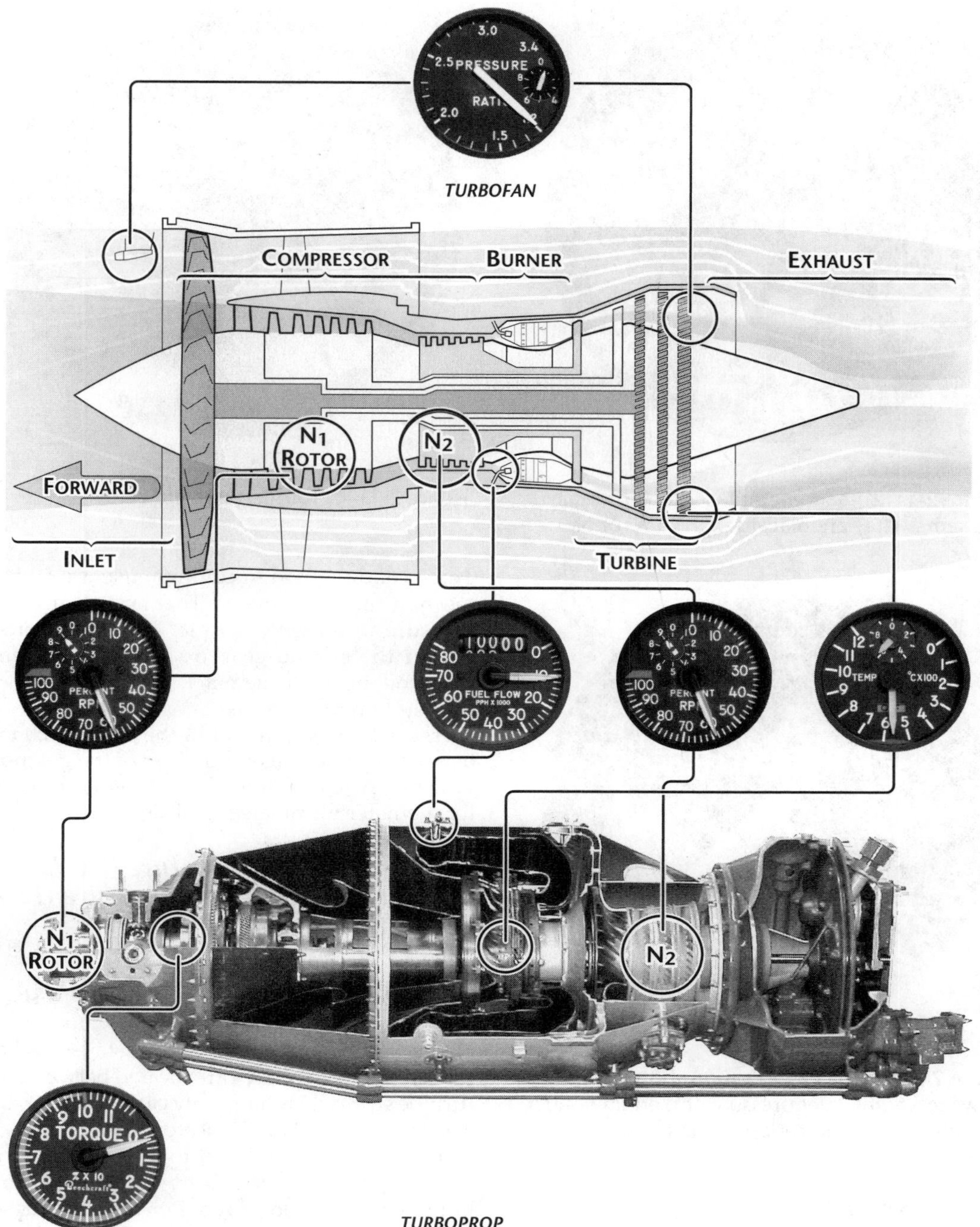


**Figure 2-11-1. Turbine engine instruments and engine sensor locations**

self-powered by a tach generator located on the gearbox of each engine.

**Low pressure tachometer ($N_1$).** A *low pressure* (LP) *tachometer* indicator for each engine is used on all single spool engines and fan jets, either high or low bypass. The indicators show LP compressor rpm in percent.

## Engine Pressure Ratio (EPR) Indicator

An *engine pressure ratio* (EPR) indicator for each engine measures the engine pressure ratio as a measure of the thrust being developed by the engine. The indicator provides an indication of engine power in the form of the ratio of exhaust total pressure ($P_7$) to intake total pressure ($P_1$). EPR also requires the monitoring of several other parameters to be correct. Before the digital age, EPR required the use of charts to actually compute the correct reading for the particular conditions. With digital fuel controls, all the variations are compared and computed automatically. Thus EPR is an actual presentation of engine power. See Figure 2-11-3.

Figure 2-11-2. Tachometer faces look basically the same. They are placarded as to $N_1$ or $N_2$.

Figure 2-11-3. An EPR indication shows the ratio between engine inlet pressure and engine outlet pressure. It is an indication of the thrust of an engine.

## Torquemeter Indicator

Turboprop aircraft measure engine power output through the use of a *torquemeter*. The electric torquemeter system in turboprop aircraft measures the horsepower produced by the engine. Two methods are in use. The first measures torque at an extension shaft. Each system consists of a transmitter (part of the engine extension shaft), a phase detector, and an indicator. The system measures the torsional deflection (twist) of the extension shaft as it sends power from the engine to the propeller. The more power the engine produces, the greater the twist in the extension shaft. Magnetic pickups detect and measure this deflection electronically. The indicator registers the amount of deflection in shaft horsepower.

Figure 2-11-4. Torquemeter indicator

The second method in common use is a hydromechanical torquemeter. This system uses oil pressure to measure torque. Rotational forces within the engine gearbox are used to compress oil in a torquemeter sensing chamber. The system then measures the difference in pressure between the oil in the sensing chamber and the internal oil pressure of the gearbox. This pressure differential is transmitted to the torquemeter via an electrical signal.

## Exhaust Gas Temperature (EGT) and Inter-turbine Temperature (ITT) Indicators

**Exhaust gas temperature probes.** *Exhaust gas temperature* (EGT) probes can be located either between or behind the turbine stages. When the temperature probes are located between the turbine stages it is commonly called an *inter turbine temperature* (ITT) indicator. These temperature probes are connected to an instrument in the cockpit. EGT and ITT are engine operating limits and are used to monitor the integrity of the turbine and check engine operating conditions. They are also the first indication that an engine has started to run during a start.

**Turbine inlet temperature (TIT).** *Turbine inlet temperature* (TIT) is the most important consideration in engine instrumentation. TIT determines how much fuel can be burned before critical temperature is reached. Before digital instrumentation, it was very difficult to read TIT. Additionally, it would require an armload of charts to compute the proper fuel burn. Thus, turbine outlet temperature (TOT) was almost always instrumented instead. In the digital age, it is not difficult to measure TIT and it is used as a modifying parameter for TOT. Basically, it is another item that a FADEC system takes care of. See Figure 2-11-5.

## Fuel Flow Indicator

*Fuel flow indicators* for each engine are also a necessity. The indicators indicate the fuel flow rate for each engine at the entry to the HP fuel pump. Typically fuel flow is indicated in pounds per hour. When troubleshooting the relationship between abnormal fuel flow and other engine parameters can be a valuable aid. As another positive factor of the digital age, an ECAS system figures that all out for you. A typical steam gauge style fuel flow indicator is shown in Figure 2-11-6.

(A)

## Turbine Vibration Indicator

The *Turbine Vibration Indicators* (TVI) provide a continuous monitoring of the balance of the rotating assemblies in the engine in order to detect a possible internal failure that could result in an engine failure. The indication is normally picked up from the low pressure compressor or the low pressure turbine. Vibration indications for both engines are normally presented on a single indicator with a switch to allow either compressor or turbine vibration to be displayed. Onboard vibration monitoring systems are not new. However, in the digital age they are even more useful. Not only can vibrations be monitored, but a vibration can be cross referenced to a specific set of instrument readouts, producing a better picture of what is going on at a moment in time.

(B)

Figure 2-11-5. (A) EGT and (B) TIT are, for all intents and purposes, the same thing. EGT is the most critical temperature during engine operation.

## Engine Auto Synchronizer

An *engine auto synchronizer* receives signals from both engine tachometers. The amplifier produces a signal that operates an actuator on the master engine throttle linkage. Thus, it varies the slave engine r.p.m. to match the master engine r.p.m. The left throttle lever does not move while the throttle linkage is being positioned by the synchronizer. The range of control of the synchronizer is between six and ten percent. The system is fail-safe in that the pilot, by operating the throttles manually, will have full control of both engines.

Each different airplane system will have a specific set of instructions on how to move the throttles with the ENGINE SYNC engaged. Engine Auto Sync is not available on all aircraft.

Figure 2-11-6. Fuel flow is normally indicated in pounds per hour. An abnormal fuel flow is always an indication that requires troubleshooting.

## Engine Top Temperature Control

On engines with FADEC controls, this is monitored and controlled automatically. It senses the temperature from thermocouples in the turbine primary exhaust gas stream, preventing over

Figure 2-11-7. As with any combustion engine, oil pressure is critical. A loss of pressure will rapidly bring on a complete engine failure.

Figure 2-11-8. Excessively high oil temperature is a sign of an impending problem. With high temperatures bearings and gears do not get adequate lubrication.

temperature of the turbine gas. Normally, one element in each thermocouple is used for cockpit *Turbine Gas Temperature* (TGT) indication. The second element passes signals to a temperature control signal amplifier. When the TGT exceeds the maximum, the amplifier signals the temperature control actuator on the engine fuel flow regulator. The actuator operates, resulting in a reduced fuel flow, thus preventing the TGT from exceeding the limits.

These systems do not operate during engine ground starting. Control of the TGT during start is only by the pilot's manual operation of the throttle lever. This system can reduce the fuel flow by as much as six to ten percent, which could slightly reduce HP compressor speed.

### Engine Oil System

Most engine oil systems are a conventional dry sump system incorporating pressure and scavenge systems. The oil is contained in an engine-mounted tank, which is fitted with a pressure filling adapter. The system includes one pressure pump, which draws oil from the tank and provides for pressure lubrication of all bearings and gears. Positive scavenging provides for pressure lubrication of all bearings and gears. Positive scavenging of all bearings and gearboxes is normally done with multiple scavenge pumps.

**Engine oil pressure indicator.** An *engine oil pressure* indicator (Figure 2-11-7) is necessary for each engine. Oil tank pressure is sensed from a transmitter mounted on the rear of the tank. When engine oil pressure is below normal operating range, warning indicators on the *Master Warning Panel* will illuminate.

**Engine oil temperature indicator.** Engine oil temperature indicators (Figure 2-11-8) are used for each system. The indicator is typically scaled from 50°C to 150°C. Temperature is sensed by a probe in the oil tank.

**Oil cooling.** A *fuel-oil cooler* is provided to maintain oil temperature within design limits. Typically the oil cooler is designed to circulate the oil through a matrix that is cooled by the fuel supply. The oil cooler is protected by a bypass valve to limit pressure buildup due to clogging or cold oil. Under most operating conditions, the heat transfer to the fuel provides sufficient fuel heating. An oil temperature transmitter is fitted to the outlet from the oil cooler.

## Instrument Markings

Turbine engine instrument marking are just slightly different than the markings for piston engines. Refer to Table 2-11-1 for these markings.

**Engine Indicating and Crew Alerting System (EICAS).** Most new aircraft use digital electronic instrumentation that is presented on a CRT (Cathode Ray Tube) or LCD (Liquid Crystal Display). Glass cockpit technology has all but eliminated the traditional analog gauge in new transport aircraft construction. In this system, known as the *Engine Indicating and Crew Alerting System or EICAS*, all of the engine sensors transmit their information to the EICAS computer (Figure 2-11-9). The EICAS computer then sends digital data to the cockpit

| TURBINE ENGINE INSTRUMENT MARKINGS | |
|---|---|
| Green arc | Normal operating Range |
| Yellow Arc | Caution Range |
| White Arc | Special Operations Range |
| Red Arc | Prohibited Range |
| Red Radial Line | Do Not Exceed |
| Blue Radial Line | Special Operating Condition |
| Red Triangle, Diamond, or Dot | Max Limit for High Transit Indications |

**Table 2-11-1. Turbine engine instrument markings**

screen telling it what information to display. In most cases, this information is presented on a computer monitor in a form that graphically resembles the face of an analog instrument.

All of the information presented on conventional gauges is displayed on the cockpit display. Typically there is only one cockpit display unit. This unit may replace as many as 36 conventional analog instruments. A significant benefit is greatly reduced maintenance in the cockpit and enhanced aircraft reliability. The presentation can easily be tailored to meet specific operational needs. In the event of a screen failure, the EICAS information can be displayed on a flight display screen or vice versa.

One benefit of the EICAS system is the ability to alert the flight crew of failures or abnormal conditions. The screen shows system status messages, in addition to basic engine data, during all phases of flight. It also can display color-coded alert messages communicating both the type and severity of a failure. The crew can also set the display to only show the primary parameters full time and allow the EICAS to monitor secondary systems, presenting information on those systems only when certain parameters are exceeded. This simplifies the information the crew must absorb and reduces crew workload and fatigue.

A typical EICAS system will provide the following primary information to the flight crew:

- Actual and commanded $N_1$ speed
- Transient $N_1$ (the difference between actual and commanded $N_1$)
- Max potential and permissible $N_1$

**Figure 2-11-9 Typical EICAS display in a modern corporate jet glass cockpit**

*Photo courtesy of Raytheon Aircraft*

**Figure 2-11-10. King Air starting sequence showing the propeller unfeathering automatically on startup**

Additional indications occur if $N_1$ limits are exceeded.

- Current thrust lever position
- Thrust Reverser system status
- Exhaust Gas Temperature (EGT)
- Max permissible EGT

Additional indications occur if EGT limits are exceeded.

- $N_2$ rotor speed
- Fuel Flow

Additional secondary data is also presented. These include fuel used, oil quantity, oil pressure, oil temperature, vibration sensor data, oil filter status, fuel filter status, ignition system status, start valve position, engine bleed pressure, and nacelle temperature.

The EICAS system also provides a number of advantages for the AMT. The system can automatically record trend monitoring information, eliminating hand recording (and the potential for mistakes). After an engine or system fault, all of the data listed previously can be called up and evaluated during the troubleshooting process. The AMT has much more information available than with conventional systems. Maintenance personnel can also use the EICAS on the ground to monitor many systems and access the computer's Built-In-Test-Equipment (BITE). Many systems on modern digitally controlled aircraft have the ability to perform extensive self-testing. The computer can analyze circuits and system status and relay the information back to the EICAS panel. BITE systems are covered in more detail in a later section of the chapter.

In addition to displaying engine related information, the EICAS also presents vital information regarding pressurization, aircraft configuration and other key systems to the crew. This part of the EICAS is covered in detail in the instrumentation chapter.

## Engine Operation, Inspection, Troubleshooting & Testing

Starting the turbine engine requires that the compressor be rotated to a speed at which self-acceleration is initiated. The power to perform this duty is significant and may be provided by a number of different means, generally dictated by the size of the engine. Smaller turbine engines, typically under 6,000 lbs. thrust, may utilize an electric motor, which may be either a DC electric or a starter-generator type.

Air turbine starters are comprised of an axial-flow turbine (which drives a coupling through a gearbox) and starter-clutch assembly. Air to operate this type of starter is generally supplied from either a ground source or another engine.

In some cases, an on-board compressed-air bottle supplies the required air under pressure. A cartridge-type starter is actually a variety of the air turbine starter that uses hot gases generated by a solid-fuel cartridge, rather than compressed air from another source.

For starting, large turbofan engines generally use compressed air and electrical power provided by the on-board auxiliary power unit (APU) or from another running engine. The compressed air normally turns an air turbine starter. Maintenance and inspection of starting systems are discussed in the appropriate section.

An air impingement system has been used on turbojet engines. This system utilizes air from either an operating engine or another external source to turn the turbine. Such a system uses non-return valves to control the flow and direction of the high-pressure air, and nozzles to change pressure energy into kinetic energy to turn the turbine.

In a two-spool engine, only the high-pressure compressor is turned in the process of starting. The low-pressure compressor, or the power turbine, in the case of a turboprop, will begin to rotate when the combustion process reaches a self-accelerating level. Once the compressor begins to rotate and has reached a specified r.p.m., ignition is turned on, followed by fuel, which is accomplished by either moving the *throttle/power lever* or a *condition lever* to the appropriate position. In some cases, actuating a switch moves a solenoid valve to introduce fuel. Following ignition of the fuel and air, indicated by a rise in EGT, the starter continues to assist in compressor rotation until the r.p.m. has reached a speed at which the combustion process provides power for self-acceleration.

When self-acceleration is achieved, the ignition is turned off, followed by cut-out of the starter. Throughout this process, fuel flow and EGT must be monitored closely for expected values. The specific r.p.m. values for application of ignition, fuel and cut-out of the starter and the limiting values for EGT are very much type dependent.

Consider, for example, more specific details for starting the Pratt & Whitney PT6 turboprop engine. Although the following procedures are typical for this engine, operators should make reference to the applicable instructions for starting details. Controls for this engine generally include a power lever, a propeller lever and a condition lever that controls fuel. For starting, the power lever is placed in the idle position, the propeller lever is moved full forward to High r.p.m. and the condition lever remains in the fuel cut-off position until the starting sequence calls for its movement. The start procedure is as follows:

1. Starter switch on. Minimum gas generator speed for starting is 4,500 r.p.m. or 12 percent Ng.
2. Check that engine oil pressure is climbing to a normal value.
3. Ignition switch on.
4. After Ng r.p.m. has stabilized at a satisfactory r.p.m., move the fuel control lever to the Lo-idle position.
5. Observe that the engine accelerates to the specified low-idle r.p.m. and that maximum ITT (also known as the $T_5$ temperature on the PT-6) is not exceeded.

If a light-off does not occur within 10 seconds after moving the condition lever to Lo-idle, abort the start by shutting off the fuel, starter and ignition. A 30-second drain period should then be observed. Then a 15-second dry motor run should be performed prior to subsequent start attempts. Caution should be taken to avoid exceeding the starter limits during all start procedures. Starter limits are specified in the aircraft or engine manuals.

The above is given only to illustrate a typical turboprop starting procedure. With the free turbine engine, of which the PT-6 is a representative example, the propeller shuts down in the feather position and during starting operations is driven out of feather by oil pressure supplied by the propeller governor. This is a fail-safe design that automatically feathers the propeller in the event of engine failure (Figure 2-11-10).

For a fixed-shaft turboprop engine, start procedures will be similar, but the propeller must be in a low pitch or latched position prior to starting. The reason for this is that the starter on a fixed-shaft engine must also turn the propeller, which, if in the feather position, will create sufficient drag to prevent a successful start procedure. Therefore, a fixed-shaft engine must be in a low-pitch position to minimize the drag of the propeller for starting purposes.

If the operator finds that the propeller on a fixed-shaft turboprop engine is in the feather position at start-up, the propeller must be driven to low pitch, utilizing the electric unfeathering pump on board the airplane, before a start may be attempted.

Failure to follow the procedures outlined in the airplane operating manual, or malfunctions of equipment, may lead to a hot start, a hung start, or a false start.

**Hot start**. A hot start is indicated by EGT rising above the starting limit, generally resulting from an excessively rich fuel/air mixture ratio. An early indication of this problem may be a rapid rise in fuel flow above that generally expected, particularly in the event that compressor r.p.m. does not continue to rise in a typical fashion. Excessive fuel, combined with inadequate air flow, leads to a rich fuel/air mixture ratio that ultimately results in a hot start. In the event of a hot start, prompt intervention is required to avoid or limit damage to the hot section of the engine. Fuel should be cut off while continuing to motor the engine, both for cooling purposes and to reduce the fuel/air mixture ratio that led to the hot start.

Following shutdown of the engine, an inspection and further maintenance action may be necessary if a hot start has occurred. Two factors dictate the required procedures: the temperatures observed and the duration of the event. Each engine has limits based on these two factors that will direct the required inspection and subsequent maintenance procedures. In the case of a short duration hot start at a minimal over temperature value, only an inspection of certain hot section components may be necessary. In the worse case scenario, hot section maintenance, including removal and replacement of turbine stages and other components, may be the end result.

**Hung start**. Hung starts occur when the engine lights off but fails to self-accelerate. A hung start will prevent the engine from reaching idle r.p.m. This type of start should be responded to promptly. When the compressor r.p.m. is observed to remain at a low value (below idle), the operator should shut off the fuel and continue to motor the engine as called for by the airplane operating instructions. Subsequent troubleshooting may focus on the EGT value during the start procedure, if it remained low. The problem may be fuel-related. If so, fuel flow to the fuel control should be verified, then operation of the fuel control bypass valve and enrichment pressure switch should also be confirmed.

**False (no) start**. A false start, or no start occurs when light off fails to take place. Generally speaking, this scenario has its roots in either a fuel delivery or ignition problem and it is important to determine which of these is the cause prior to subsequent start attempts. If no fuel delivery is the issue, then correction of that matter can be followed by a normal start procedure. However, when an ignition problem causes a false start, the presence of fuel in the combustion chamber could lead to a hot start in the follow up attempt to start the engine. Under these circumstances, it is important to motor the engine prior to starting. In every case, compliance with the operating instructions for the engine in question is critical.

Note that problems with starting are generally less likely to occur for engines with automatic starting systems or electronic controls. With this type of advanced equipment, equipment malfunctions are automatically noted before they become an issue that might otherwise lead to a hot, hung, or no start condition. Having said this, operators must nonetheless exercise vigilance during the start procedure to prevent the unlikely occurrence of a hot start or other starting problem. Generally speaking, this requires that EGT, fuel flow, and spool speeds be carefully monitored for expected values during start up.

Operational testing procedures are performed for new engines prior to delivery in order to validate performance, security of the engine, and to check for leaks. Other operational checks performed in the field include idle checks and power assurance tests, whereby hot day take off thrust may be confirmed. A dry motor check may be required to assure that the starting system is operating nominally, the engine rotates freely and the instrumentation is indicating properly. During such checks, neither the ignition nor the fuel is activated and the starting system is energized only for that period of time necessary to observe r.p.m. and oil pressure instrumentation indications.

As mentioned previously, new engines are subject to performance validation runs typically performed in a fully instrumented test cell with a bell mouth air inlet (see Figure 2-4-3) in place to eliminate turbulence as the air enters the engine. These test operations require that engine parameters be corrected for non-standard conditions such that meaningful performance analysis can be conducted regardless of the ambient temperature and pressure. Greek letters DELTA ($\delta$) and THETA ($\theta$) symbolize the correction factors for pressure and temperature, respectively. Refer to the following material for correction factors.

| NOMENCLATURE FOR CORRECTING PARAMETERS TO SEA LEVEL VALUES |
|---|
| Delta ($\delta$): P / Po = P / 14.7 |
| Theta ($\theta$): T / To = (59°F + 460) = T ambient R/519R |
| P: Ambient pressure in PSI (lbs per square inch) |
| Po: Standard-day barometric pressure (14.7 or 29.92) |
| T: Ambient temperature in degrees Rankine |
| To: Standard-day temperature in degrees Rankine |

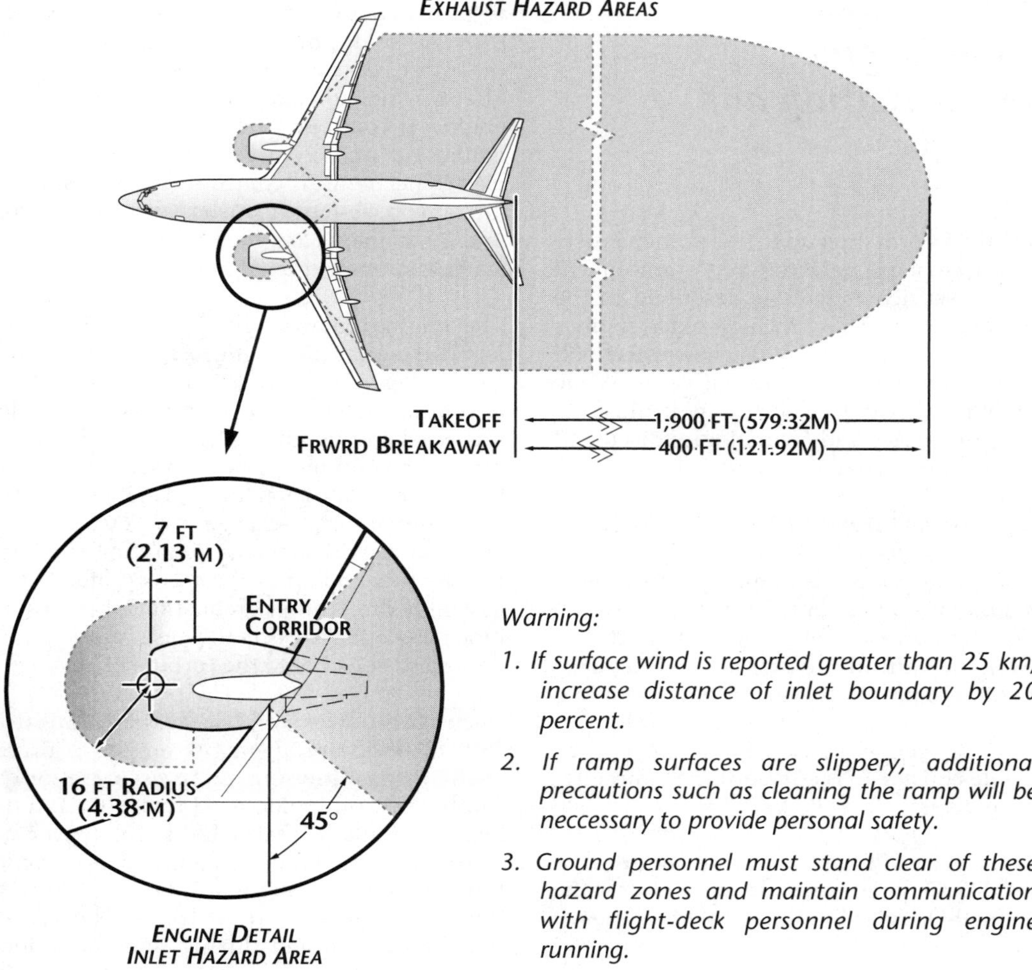


**Figure 2-11-11. Takeoff thrust power hazard areas with regard to turbine engine operation**

Prior to conducting test operations, attention should be given to the engine air inlet and the ramp area immediately in front of the engine for the presence of foreign objects that could lead to foreign engine damage (FOD). Given the high velocity gas produced by both the fan and the exhaust, every effort must be made to assure a clear area for a distance dictated by the thrust of the engine (see Figure 2-11-11).

High-bypass ratio turbine engines also present a danger to people and objects in front of the engine, and this is particularly the case with wing-mounted turbofan engines having minimal ground clearance. Specific clear-distance information is available for each powerplant. During test operations or maintenance engine runs, personnel stationed at the nose of the airplane and both wing tips should monitor the surrounding area and notify flight deck personnel of any potential problem.

Starting any engine always produces a fire hazard, and jet engines are no exception. If a tailpipe fire starts from excess fuel, shut off the fuel flow and continue to motor the engine, trying to blow out the fire. If the fire is too large for that process, $CO_2$ extinguishers should be used. They will not contaminate the engine interior or damage any auxiliary systems. Remember that $CO_2$ is also a cooling agent. If a fire were to start during shutdown it is possible to cause the case to shrink quickly, causing rotor drag.

Some engine installations have specific shutdown procedures. Many times the manufacturer will specify a short cool-off period before actual shutdown. This procedure requires the pilot to monitor engine temperatures after landing. When the engine has idled for a short period of time, the internal temperature will fall below a preset minimum. Once this minimum temperature has been met, the pilot may shut-down the engine. The purpose is to allow the turbine blades to cool by the same amount as the engine case, reducing the possibility of severe tip rub.

Section 12

# Turbine Engine Familiarization and Differences

Most of the information up to this point has necessarily been of a general nature. A sampling of specific in-service engines is described in the following section. These include older engines still found in service as well as newer technology engines introduced in recent years. While not comprehensive, this section will illustrate how the principles and technologies discussed earlier are applied in commercial service.

**Honeywell gas turbine APU.** GTCP85 two-bearing series gas turbine engine. Most commercial airliners use a small gas turbine as an auxiliary power unit. Its purpose is to provide high pressure bleed air to drive the main engine's air turbine starters and to turn a generator to provide electrical power while on the ground. Often mounted in the tail cone (see figure 2-12-1), the APU exhaust is sometimes mistaken for an additional aircraft engine exhaust.

The GTCP85 APU is used on the Boeing 727 and 737 series aircraft as well as the McDonnell-Douglas DC-9 and MD-88 series. It is a self-contained power source and requires only fuel electrical power for operation, which is provided by aircraft systems. The engine provides shaft power for driving generators or other devices connected to the accessory drive. The starting, acceleration and operation of the engine is controlled by an integral system of automatic and coordinated pneumatic and electromechanical controls. A minimum of airframe customer-furnished external controls and instruments are used for initiation engine starts and observing engine operation.

Figure 2-12-1. A typical APU engine used on large transport aircraft.

The engine cutaway in Figure 2-12-2 shows a two-stage centrifugal compressor and a single-stage radial inflow turbine rotating on a common shaft. The compressor impellers are attached pneumatically through interstage ducts. The turbine plenum assembly encloses the turbine components and provides a receiver for compressor discharge air. A combustion chamber provides a place for mixing and burning the fuel and air at the correct rate. A torus assembly directs the combustion gases to a turbine nozzle that directs the gases at the proper angle and speed onto the turbine blades.

Pneumatic power supplied by the APU is used for starting aircraft main engines, aircraft cabin air conditioning and pressurization, air for the APU oil cooler, APU enclosure and fuel heaters (some aircraft). Electrical power can also be provided independently of - or in combination with - pneumatic power. Electrical power has priority at all times. The APU is temperature-limited during pneumatic loading by the use of a fuel control-connected thermostat. During a cooling period before shutdown, the bleed air valve is automatically closed. This lowers the air load, the fuel control reduces fuel flow, and the engine cools. Some aircraft can use the APU generator as an emergency backup for powering the aircraft electrical systems in flight if a main engine has to be shut down.

Ambient air drawn into the hub of the first-stage impeller is discharged through the diffuser ducting to the hub of the second-stage impeller. At the second-stage impeller, the air velocity is further increased and discharged through the second-stage diffuser, where velocity is decreased and total pressure is increased. The pressure rise from ambient to second-stage discharge is approximately 3:1. The pressure rise represents the contribution of pneumatic energy provided by the compressor section to the engine power cycle. Combustion takes place within the combustion liner. An atomizer injects fuel into the center of the liner, where it is mixed with air and lit by an igniter plug.

First, sufficient air is supplied through specially designed holes in the combustion liner to

Figure 2-12-2. Cutaway view of the Honeywell GTCP85 series APU.

aid the combustion process. Then, as the flame progresses down its length, additional air is added to dilute this burning mixture in order to reduce the gas temperature, as overheating can damage the APU turbine inlet guide vanes. The holes in the liner are designed to produce efficient burning while ensuring that the gases cool properly before reaching the inlet guide vanes and turbine. The turbine assembly consists of two major pieces: the turbine wheel and turbine nozzle assemblies.

The nozzles are convergent ducts; gases flowing from a larger to a smaller area, causing an increase in air velocity. Air entering the turbine nozzle is directed against the outer blade tips of the turbine wheel. Since the air is moving at high velocity, it imparts a force against the blade tips, which produces a torque to rotate the wheel. The air then flows toward the hub, where it must turn again and push against the exducer (radial outflow) blades, thus yielding more of its energy, before escaping through the exhaust duct.

A mechanically driven generator and other accessories necessary for APU operation (fuel control, oil pump, etc.) are mounted on an accessory gearbox and are connected to the compressor/turbine shaft through a gear-reduction system.

During operation of the APU, a condition known as *IDLE* will exist, when the APU operates at full speed (100 percent r.p.m.) and no load is applied to the APU. No load means no pneumatic or electrical power is extracted from the APU. When power is extracted from the generator, a load is imposed on the shaft that causes the compressor and turbine r.p.m. to slow down. Additional increases in load will produce further decreases in shaft r.p.m. In order to keep the unit on speed, the turbine will have to produce more power to match the load requirements. This is accomplished by increasing fuel flow to the combustion chamber.

The APU can also supply heated, compressed air. As this energy is extracted from the compressor section, the available pneumatic energy to the turbine section is reduced. This will also increase the exhaust temperatures because of reduced airflow. Figure 2-12-3 illustrates this operation.

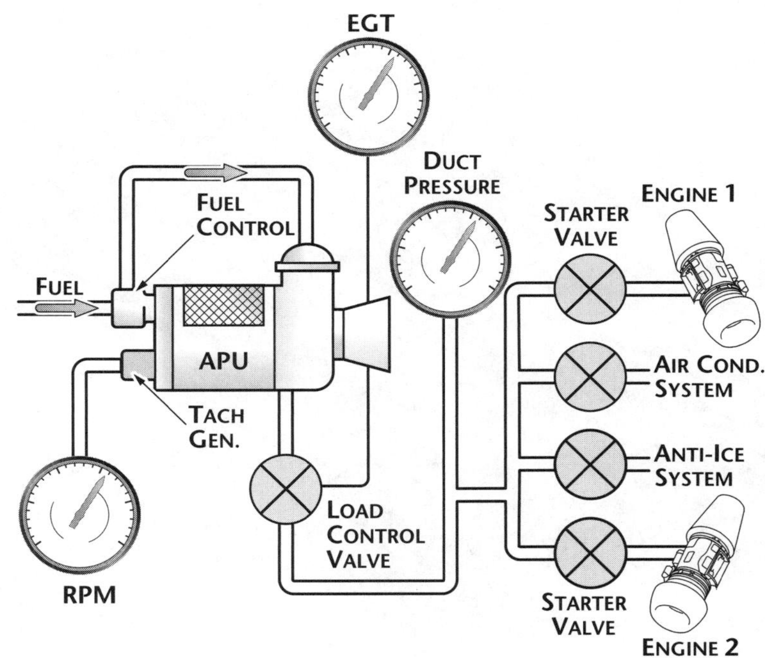


Figure 2-12-3. APU plumbing schematic

Figure 2-12-4. A Rolls Royce Allison 250 series engine

APU air used to provide for aircraft systems passes through a pneumatically actuated bleed air valve. The load-control thermostat controls the opening of this valve. As more bleed air is demanded, more air is taken from the compressor section, causing the APU to slow down. The governing section in the fuel control senses a reduction in speed and increases fuel flow to keep speed essentially consistent. As more fuel is added, EGT rises; at a certain point, the load-control thermostat will prevent further opening of the load-control valve. This limits the amount of air being bled and ensures that the maximum EGT is not exceeded.

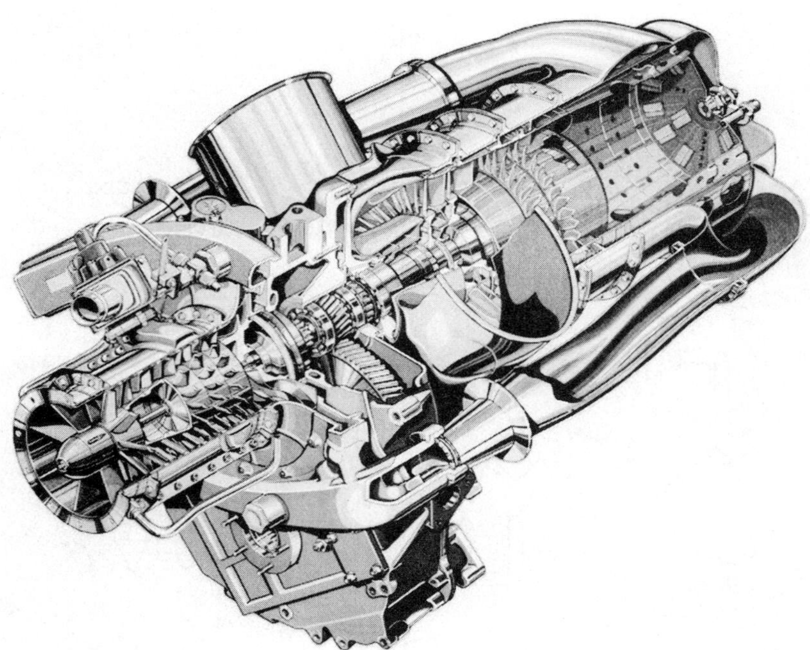

Figure 2-12-5. A cutaway view of a Rolls Royce Allison 250 turboshaft series engine.

In order to initiate the APU's start cycle, all pertinent switches and controls must be in their proper positions as specified in the aircraft checklist; including fuel shutoff valves, electrical controls, doors, etc. The start switch is moved to the start position momentarily, the switch is spring-loaded to return to the run position when released. This action energizes relays that open fuel valves and doors, relays that hold the APU circuits, and the starter relay that carries heavy current to the starter.

When the starter has cranked the engine to sufficient speed, oil pressure from the lubrication pump will build (3±5 p.s.i.) to close the oil pressure sequence switch. This completes a circuit and energizes the ignition and the fuel solenoid, which opens to permit fuel into the combustor. The fuel ignites from the spark provided by the igniter plug. Light-off (combustion) normally occurs from 10-20 percent r.p.m. When combustion has started, the EGT gauge in the cockpit will show an increase in exhaust gas temperature. The pressure rise produced by the heat of combustion increases the velocity of airflow through the turbine, which begins producing power to assist the starter in accelerating the engine.

Airflow is still too low, however, for the turbine to do the job alone. Acceleration will not continue without the assistance of the starter. Therefore, during the early stages of light-off and acceleration, the operator must pay very close attention to the exhaust gas temperature and r.p.m. to ensure the APU is accelerating smoothly.

## Gas Turbine Flight Engines

**Rolls-Royce Allison 250 turboshaft series.** The Rolls-Royce Allison 250 series turboshaft engine is one of the most popular helicopter powerplants in the world. Over 23,000 engines have been manufactured since production started in 1958. It has been modernized and updated over the years and powers the Bell JetRanger and 222 series, the MD Helicopters 500 & 600 series, the Aerospatiale AS350 as well as the Sikorsky S-76. It is almost impossible to be involved in helicopter maintenance without working with this powerplant. With the addition of a prop drive gearbox, the engine has also been used to power a number of small turbo-prop aircraft. The 250 is available with power output ratings ranging from 318 shaft horsepower to 650 shaft horsepower.

The Rolls-Royce Allison 250 series turboshaft engine, shown in Figure 2-12-4, features a "free" power turbine. A free turbine engine has no mechanical connection between the gas generator turbine and the power turbine. This connection is formed by an air link that will rotate the power turbine due to the gas generator gas flow. The engine consists of a combination axial-centrifugal compressor, a single "can" type combustor, a turbine assembly, which incorporates a two-stage power turbine and an exhaust collector, and an accessory gearbox, which incorporates a gas producer gear train and a power turbine gear train. A cutaway view of the engine can be seen in Figure 2-12-5.

The compressor assembly consists of a compressor front-support assembly, compressor rotor assembly, compressor case assembly and compressor diffuser assembly. See Figure 2-12-6.

The combustion assembly is comprised of the combustion outer case and a combustion liner. The combustion outer case is a fabricated stainless steel part with two tapped bosses for mounting the burner drain valve and plug, a fuel nozzle and a spark igniter. The burner drain valve threads into the boss nearest the gravitational bottom, and the other boss is plugged. The fuel nozzle and spark-igniter boss are on the rear side of the combustion outer casing. The fuel nozzle and spark igniter thread into their respective bosses and extend into the combustion liner dome. The fuel nozzle positions and supports the aft end of the combustion liner, and the spark igniter locates the combustion liner in a circumferential position.

**Figure 2-12-6. Cutaway view detailing the compressor front-support assembly, compressor rotor assembly, compressor case and compressor diffuser**

The combustion outer casing is flanged on the front for mounting of the combustion assembly to the gas producer turbine support. See Figure 2-12-7. The combustion liner must provide for rapid mixing of fuel and air and it must control the flame length and position such that the flame does not contact any metallic surface.

**Figure 2-12-7. Rolls Royce Allison 250 installed in a Bell JetRanger**

**Figure 2-12-8. A typical Pratt & Whitney PT6A turboprop engine, showing the profile of the engine.**

The turbine assembly of the engine is that part of the turbine which incorporates the components necessary for the development of power and the exhausting of gases. The turbine assembly has a two-stage gas-producer turbine and a two-stage power turbine. Power to drive the compressor rotor is furnished by the gas-producer turbine rotor through a direct drive. The power turbine rotor converts the remaining gas energy into power that is delivered to the power pads of the engine. Exhaust gases from the power turbine are directed into the exhaust collector, which provides for exhaust flow through two elliptical ducts at the top of the engine.

**Pratt & Whitney PT-6.** The PT6 powerplant, an example of which is shown in Figure 2-12-8, is a lightweight, free-turbine engine. Arguably the most popular light turboprop ever manufactured, the PT6 powers more than fifty different aircraft types. These include the ever popular Beechcraft King Air, 99 and 1900 series, Cessna Caravan and Conquest, DeHavilland Twin Otter and Dash 7, Piper Cheyenne, Pilatus PC-9 and PC-12, Shorts 330 and 360 as well as a number of agricultural aircraft. More than 30,000 examples power over 17,000 aircraft worldwide.

Horsepower for the PT6-series ranges from approximately 400 to 1,970 SHP (shaft horsepower). The engine utilizes two independent turbine sections: a compressor turbine, driving the compressor in the gas generator section; and a power turbine (two-stage in later-model engines), driving the output shaft through a reduction gearbox. As can be seen in Figure

**Figure 2-12-9. A cutaway view of the Pratt & Whitney PT6A turboprop engine.**

2-12-9, inlet air enters at the rear of the engine through an annular plenum chamber formed by the compressor inlet case. From there, it is directed forward to the compressor. The compressor consists of three axial stages, combined with a single centrifugal stage, assembled as an integral unit. The rotating compressor blades add energy to the air passing through them by increasing its velocity.

A row of stator vanes located between each stage of compressor rotor blades diffuses the air, converting velocity into pressure, and directs it at the proper angle to the next stage of compression. For the tips of the centrifugal impeller, the air passes through diffuser tubes (diffuser vanes in older-model engines), which turn the air 90° in one direction and also convert velocity to pressure. The diffused air then passes through straightening vanes to the annulus surrounding the combustion chamber liner.

The combustion chamber liner consists of an annular weldment having perforations of various sizes that allow entry of compressor delivery air. The flow of air changes direction 180° as it enters and mixes with fuel. The location of the liner outside the turbines eliminates the need for a long shaft between the compressor and the compressor turbine, thus reducing the overall length and weight of the engine. Fuel is injected into the combustion chamber liner through 14 simplex nozzles. The fuel is supplied by a dual manifold (single on early engine models, see Figure 2-12-10), consisting of primary and secondary transfer tubes and adapters. An example is shown in Figure 2-12-11.

The fuel/air mixture is ignited by two spark igniters, or glow plugs, that protrude into the combustion liner. The spark igniters are used normally only during starting of the engine. The resultant hot gases expand from the liner, reverse direction (180°) in the exit duct zone and pass through the compressor turbine inlet guide vanes to the single-stage compressor turbine. The guide vanes ensure that the expanding gases impinge on the turbine blades at the correct angle, with minimum loss of energy. The still-expanding gases are then directed forward to drive the power turbine section, which in turn drives the prop shaft through a planetary reduction gearbox. The reduction gearbox embodies an integral torque meter device, which is instrumented to provide an accurate indication of engine power (ft-lb. of torque).

The compressor and power turbines are located in the approximate center of the engine, with their respective shafts extending in opposite directions. This feature provides for simplified installation and inspection procedures. The exhaust gases from the power turbine are directed through an annular plenum into the atmosphere via twin opposed ports provided in the exhaust duct.

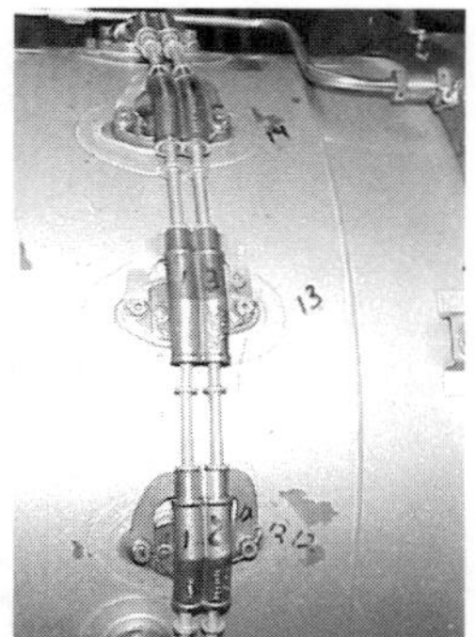

**Figure 2-12-11. Primary and secondary fuel manifolds on the Pratt & Whitney PT-6A engine.**

**Figure 2-12-10. Single fuel manifold on an uncowled PT6 engine**

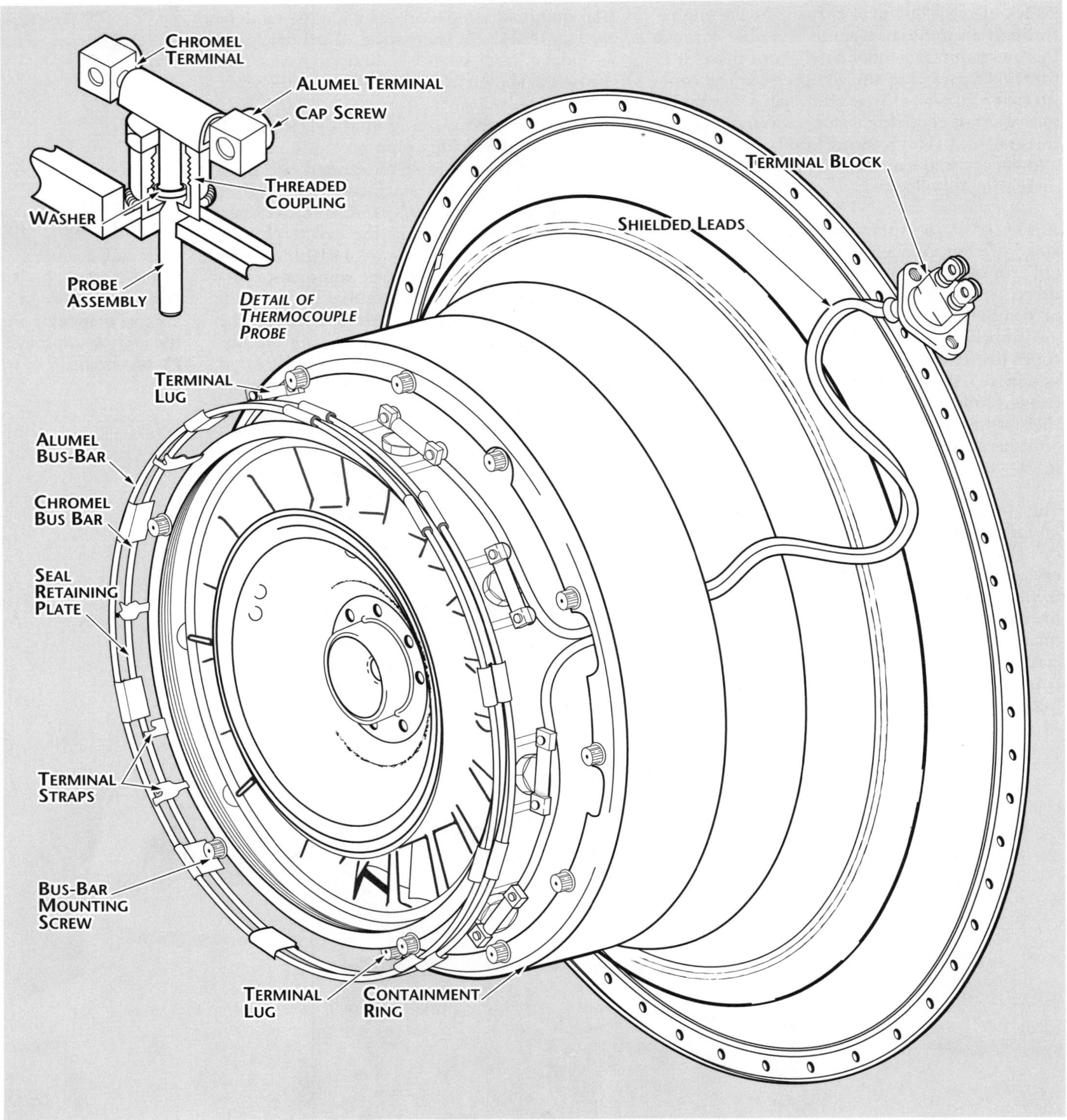

**Figure 2-12-12. The PT-6A turboprop engine's $T_5$ interturbine temperature thermocouple assembly**

Interturbine temperature ($T_5$), see Figure 2-12-12, is monitored by an integral set of thermocouple probes, a bus-bar and a harness assembly installed between the compressor and power turbines, with the probes projecting into the gas path. A terminal block, mounted on the gas generator case (exhaust case in late-model engines), provides a connection point to cockpit instrumentation.

All engine-driven accessories, with the exception of the propeller governor, power turbine over-speed governor and power turbine tachometer-generator, are mounted on the accessory gearbox at the rear of the engine (see Figure 2-12-13). These components are driven off the compressor by means of a coupling shaft that extends the drive through a conical tube in the center section of the oil tank (see Figure 2-12-14). The rear location of accessories provides

**Figure 2-12-13. A typical PT-6A turboprop engines accessory section, showing the starter, tach generator, fuel control, and igniter assembly.**

for a clean engine and simplifies any subsequent maintenance procedures. The engine is self-sufficient, since its gas generator-driven oil system provides lubrication for all areas of the engine, including pressure for the torque meter and power for propeller pitch control.

The engine oil supply is contained in an integrated oil tank that forms the rear section of compressor inlet case. The tank has a total capacity of 2.3 U.S. gallons and is provided with a dipstick and drain plug. Fuel supplied to the engine from an external source is further pressurized by an engine-driven fuel pump. The rate of flow to the fuel manifold is then controlled by the fuel control unit (FCU). These components can be seen in the detail photo shown in Figure 2-12-15.

**Figure 2-12-14. The oil tank of a typical PT-6A engine, just forward of the accessory case**

**Figure 2-12-15. A hydromechanical fuel control used on a PT6 engine**

**JT9D series engine**. The Pratt & Whitney JT9D was the first successful large high bypass turbofan in commercial service. Initially designed to power the Boeing 747 series jet transports, it can also be found on the Boeing 767, McDonnell-Douglas DC-10, and the Airbus A300 and A310. While the JT9D is an older technology engine and is no longer available on new construction aircraft, it is still in service worldwide. An example is shown in Figure 2-12-16. The JT9D is a good example of an early high bypass turbofan. It is mechanically controlled with only limited electronics. The JT9D has largely been replaced by the third generation PW4000 (described in a later section) in new production aircraft.

The JT9D was first run in December of 1966. It powered the first 747 flown in 1969. Early versions used water injection to provide cooling at high power settings. Advanced cooling technology and better metallurgy has eliminated the need for water injection on later model engines, even on the high thrust versions.

The JT9D has a fan diameter of 93 or 94 inches. Engine thrust ratings for the series range from 43,600 lbs. to 56,000 lbs. The engine, less tailpipe, weighs from 8,450 lbs. to 9,295 lbs. depending on the version. The basic construction and nomenclature of the engine is explained, starting from the front of the engine. Even though each engine type has some differences, the basic systems and construction of the engines is the same.

The JT9D is an example of the conventional two-spool axial flow front fan high bypass turbofan. It has an overall pressure ratio of 26 to 1. The first stage consists of the front fan acting as the first stage of the low pressure compressor. The fan has a nose cone to smooth airflow into the engine. This nose cone will often have a stripe or similar marking painted across it.

This nose cap stripe is present on many high bypass engines. On a busy airport ramp, the noise level can be quite high. A modern, high bypass engine is relatively quiet in this environment. It can be quite difficult under these conditions to determine if an engine is operating. As explained earlier, there is a large danger zone if front of an operating engine. Maintenance and ground personnel must take extreme care when working around an operating engine. This stripe provides a visual cue and makes it easier to work safely in a noisy environment.

The fan is made up of 46 (40 on some versions) titanium alloy blades. One noticeable feature of the JT9D fan is its use of two mid-span fan

Fig 2-12-16. Pratt & Whitney JT9D high bypass turbofan on a Boeing 747-200

**Figure 2-12-17. JT9D fan showing the distinctive dual mid-span blade supports**

blade supports. The fan is connected to the low pressure compressor. The JT9D is a high bypass type engine with a bypass ratio of 4.8 to 1. An example is shown in 2-12-17. There is a vibration pickup sensor mounted on the top of the fan housing (see Figure 2-12-18). It generates an electrical signal proportional to acceleration caused by engine vibration. This signal is processed and displayed in the cockpit on the airborne vibration monitoring system, allowing early indication of engine problems. Abnormal vibration can be an indication of compressor or turbine blade damage, bearing distress, rotor imbalance, or failure of an engine accessory or drive gear.

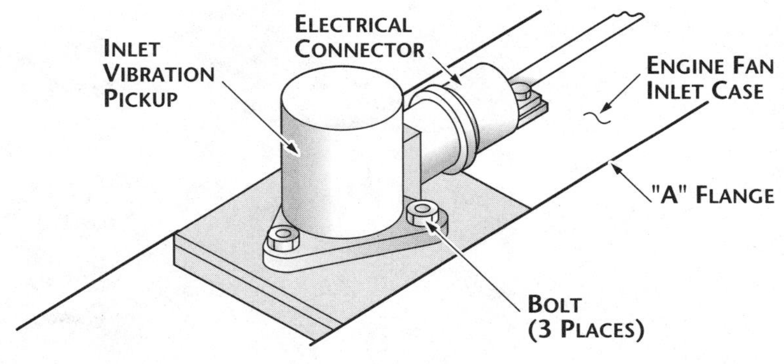


**Fig 2-12-18. Fan vibration pickup**

**Fig 2-12-19. JT9D Accessory gearbox with CSD and air turbine starter visible**

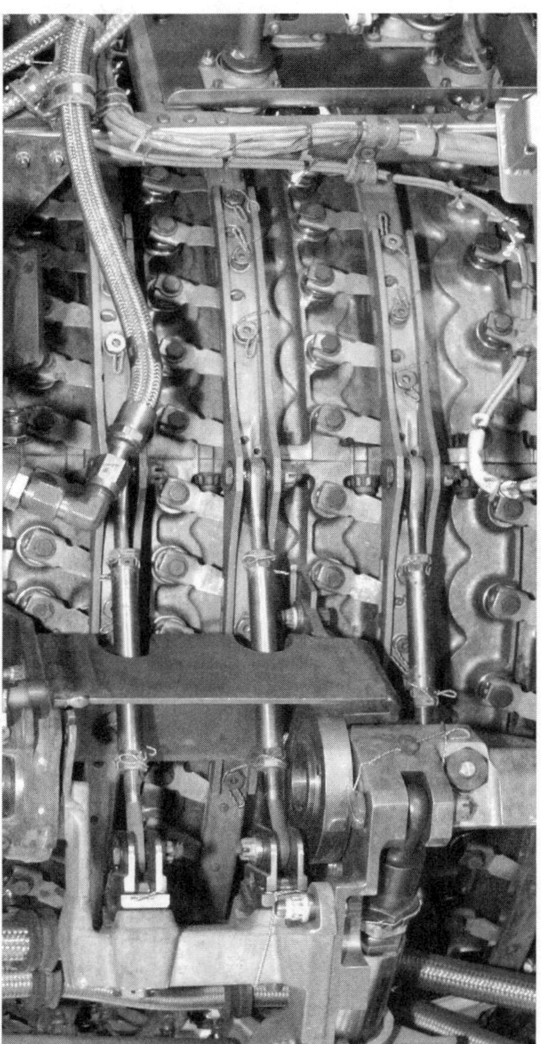

**Fig 2-12-20. Stator vane controls**

The low pressure compressor has three stages in addition to the fan. There are, depending on the stage and the engine variant, between 100 and 132 dovetailed titanium alloy blades in each compressor disc. The fan and low pressure compressor have a maximum speed of 3,650 rpm. The No. 1 bearing is located in the forward bearing housing. It is a ball bearing and acts as the $N_1$ rotor forward support and thrust bearing.

The main accessory gearbox is mounted under the central diffuser (see Figure 2-12-19). An angle gear box drives it from the front of the high pressure compressor. The gearbox typically mounts a CSD, fuel pump, fuel control, air turbine starter, hydraulic pump, alternator and $N_2$ tach. On some aircraft it also has the primary reverser motor or backup hydraulic and fuel pumps.

An oil tank provides a continuous supply of oil to the engine oil pressure pump. The oil tank capacity varies with each model. Tanks from 5 to 10 gallons are used. The oil pump feeds four main bearing cavities. Oil is scavenged from the main bearing compartments into the main gearbox sump. Oil is pumped from the sump to the oil tank by a scavenge pump. Chip detectors are located in the main sump, the gearbox scavenge lines and the No. 3 and 4 bearing chambers.

The high pressure compressor has eleven stages. There are, depending on the stage and the engine variant, from 60 to 108 dovetailed titanium or nickel alloy blades in each compressor disc. The first three stators have variable blades (Figure 2-12-20). The HP compressor also has an intermediate IGV stage. The HP compressor has a maximum speed of 7,850 to 8,080 rpm depending on the engine variant. The No. 2 bearing is also contained in the forward bearing housing. This bearing is a ball bearing and it acts as the $N_2$ rotor forward support as well as the $N_2$ thrust bearing.

The high pressure compressor incorporates variable stator vanes at the entrance of the 5th through the 8th stages. The engine vane control automatically positions these to improve stall margin for engine starting, acceleration and during partial power operation. Figure 2-12-20 shows stator vane controls. Four mach (or total pressure) probes located around the compressor are used to calculate the optimum stator vane position. These probes are positioned between the low and high-pressure compressors (Figure 2-12-21).

The combustor consists of the short bulkhead type annular burner cans. These cans are constructed of nickel alloy. They are coated to withstand higher heat levels. Two igniter plugs with a dual AC capacitor system providing power initiate the ignition. Engines equipped with water injection use a different igniter plug than those not equipped.

A Hamilton Standard hydromechanical fuel control unit supplies fuel. This unit regulates the fuel supply to the engine to control engine r.p.m., prevent overheating, surging, rich blowout and lean flameout. Some later versions of the JT9D have a supervisory type electronic-hydromechanical fuel control system. The engine fuel systems includes a fuel filter with a differential pressure switch (to detect fuel filter icing), a fuel temperature sensor, and a fuel control idle select solenoid.

The JT9D has a two stage high-pressure turbine. This turbine drives the high pressure compressor. The first stage of the turbine has 116 aircooled blades. These are of the single-crystal type. The second stage has 138 blades. The second stage blades are either solid or air-cooled, depending on the variant, with the higher thrust engines having air-cooled blades. In engines with 50,000 lbs of thrust or higher, the turbine blades are also of the single-crystal type. The No. 3 bearing is located in the mid-bearing housing. It is a roller bearing and is the $N_2$ rotor rear support bearing.

The low-pressure turbine has four stages. The turbine blades are of solid, nickel alloy construction. They are mounted with a fir tree arrangement. The No. 4 bearing is located in the rear bearing housing. It is a roller bearing and is the $N_1$ rotor rear support bearing.

The JT9D has a plug nozzle in the exhaust section.

The fan reverser is equipped with a translating sleeve. This sleeve is mechanically linked to the blocker doors. As the sleeve moves aft, it closes the blocker doors and exposes the fixed cascade vanes. Airflow from the fan is then directed across the cascades, providing reverse thrust. See Figure 2-12-22.

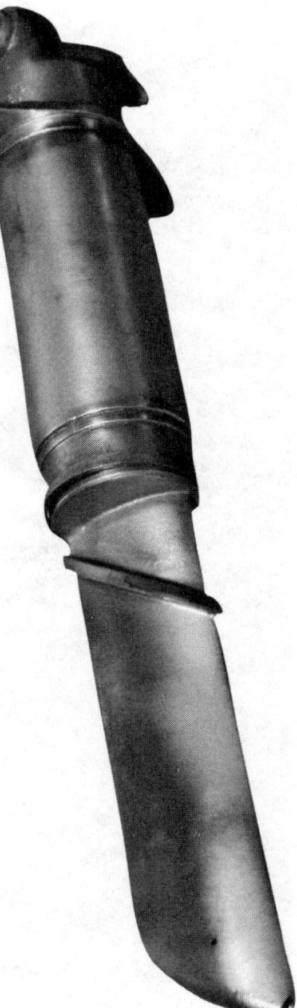

**Fig 2-12-21. Mach probe**

**Fig 2-12-22. Thrust Reverser Cascades**

**Figure 2-12-23. Front quartering view of uncowled RB.211 installed on a Boeing 757**

**RB.211 series engine**. The RB.211 was initially developed to power the Lockheed L-1011 TriStar. It experienced significant development problems and was delivered later than initially planned. Once the original design problems were resolved, it has provided stellar service to a variety of commercial airliners. The high time engine was in service in excess of 24,000 hours prior to removal for overhaul. The RB.211 can be found on the Lockheed L-1011, the Boeing 747, 757 and 767.

The RB.211 has a fan diameter of 84 to 86 inches. Engine thrust ratings for the series range from 37,400 lbs. to 95,000 lbs. The engine, less tailpipe, weighs from 7,294 lbs. to 9,874 lbs. depending on the version. The basic construction and nomenclature of the engine is explained starting from the front. An example is shown in Figure 2-12-23. The engine has seven basic modules. Each module can be changed easily. With the exception of the low-pressure fan case module, all of the modules can be replaced on-wing. Each module has its own log card and serial number. Even though each engine type has some differences, the basic systems and construction of the engines is the same.

The RB.211 is an example of the traditional Rolls-Royce three-spool axial flow front fan high bypass turbofan (see Figure 2-12-24). It has an overall pressure ratio of 33 to 1. The first stage consists of the front fan acting as the low pressure compressor. The fan has a composite nose cone. Twenty-four titanium alloy hollow wide-chord blades start the airflow into the engine. A steel shaft coupled to the low pressure turbine drives the fan. It is a high bypass type engine with a bypass ratio of 4.3 to 1. Just behind the fan is a supercritical outlet guide vane ring.

The fan blades are designed for easy replacement in the event of FOD damage. The blades can be individually changed out without the use of special tooling. The nose cone and a fan blade retaining ring must be removed. The fan blades can then be slide forward out of the fan assembly. A replacement blade is then slid into place. The low pressure fan case incorpo-

**Figure 2-12-24. Close-up of front of RB.211 fan showing the wide chord titanium fan blades**

rates a blade containment ring in the event of a blade failure.

The low-pressure compressor feeds into a seven-stage intermediate pressure (IP) compressor. The IP compressor is a seven stage compressor. Two drums, one of welded titanium discs and another of welded steel discs, are bolted together to create one rotor assembly. The compressor uses titanium compressor blades. A titanium shaft connected to the intermediate turbine drives this assembly. Steel stator blades are mounted in an aluminum and steel compressor case. The intermediate pressure compressor also utilizes single stage titanium inlet guide vanes. They are of the variable type. Eight of these blades are enlarged to permit an internal oil tube to pass through them. These lines provide lubricating oil to the low pressure and intermediate compressor roller bearings.

Engine accessories (Figure 2-12-25) are driven by a radial drive from the HP shaft to a gearbox on the fan casing. This drive assembly is mounted in the intermediate case just ahead of the high pressure compressor. An integrated drive generator, hydraulic pumps and engine related systems are powered by the gearbox. The RB.211 is started using an air turbine starter. It can be powered by air from the APU, an external source, or another running engine. The starter spins the high pressure spool.

Oil is provided by a 27 liter oil tank integral to the engine gearbox. The oil fill cap and pressure oil filter are located on the left-hand side of the engine.

**Figure 2-12-25. Close-up of engine accessories**

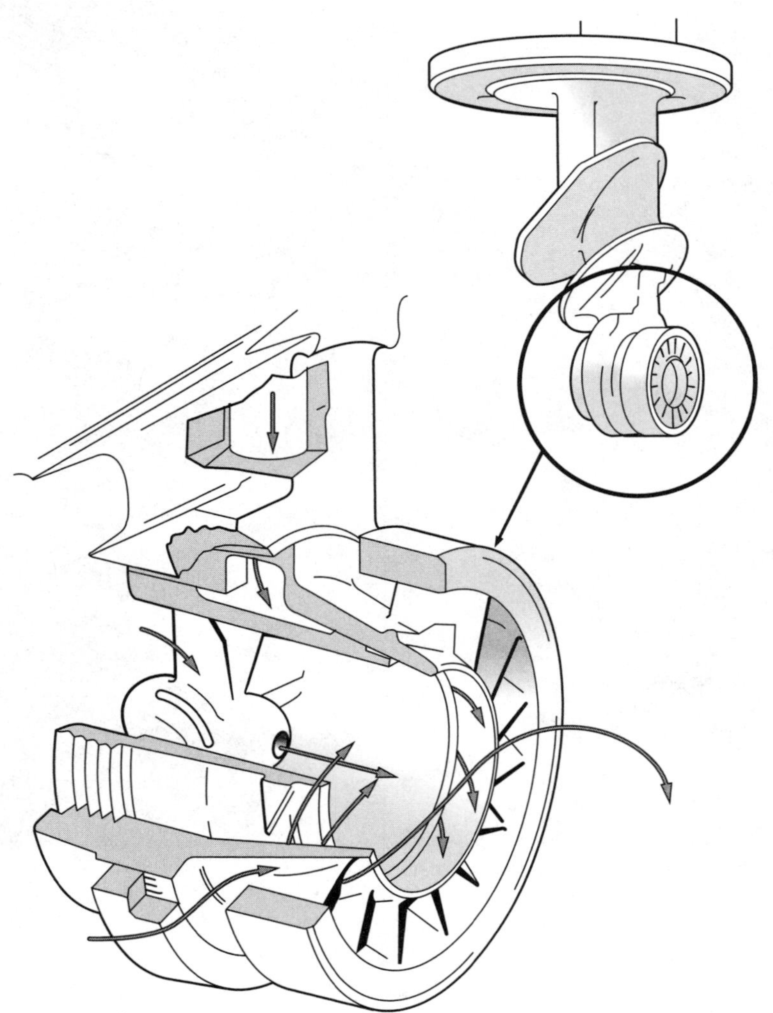

Figure 2-12-26. Air spray annular fuel injection atomizers

Lubrication is provided by a continuous circulation dry sump oil system. The system provides lubricating oil to four bearing chambers and the gearbox. Oil is proved under pressure by a single high pressure oil pump. Five scavenge pumps return oil from throughout the engine. There is a separate scavenge pump for the front bearings, thrust bearings, HP/IP turbine bearings, tail bearing and the external gearbox. As with most gas-turbines, the engine includes both roller and ball bearings.

The master chip detector is located in the return line between the scavenge pumps and the oil cooler. Additional chip detectors are located in each of the individual scavenge lines ahead of the scavenge pumps.

The high pressure compressor is the final section of the compressor. The HP compressor rotor is a bolted assembly consisting of a single steel disc plus multiple titanium and nickel alloy discs. On these discs are mounted a combination of steel, titanium, and nickel alloy compressor blades. The HP compressor case is made of steel and carries Nimonic stator blades. There is a vibration pickup sensor mounted at the aft end of the HP compressor. This sensor provides information to the flight crew about out-of-balance conditions within the engine, permitting inspection or shut-down prior to failure.

The RB.211 combustion chamber is of the fully annular type. It features a steel outer casing and a nickel alloy combustor. It also has an inner combustion case which serves to contain high pressure compressor delivery air. Some of the high pressure compressor delivery air is fed into the liner at various points along its length. This type of delivery aids in efficient and stable combustion and provides for proper cooling at the liner outlet.

Eighteen through-flow annular combustion chambers are used. Fuel is injected through fuel spray nozzles containing annular atomizers (Figure 2-12-26). These fuel nozzles project into the front of the front combustion liner. Ignition is provided by high energy igniter plugs in the number 8 and number 12 burners. The ignition unit has a high and a low energy position. The high energy position is used for ground and air starting, whereas the low energy output is used when automatic and continuous operation modes have been selected.

Fuel flow is controlled by a mechanical fuel flow governor and an electronic engine controls system. The electronic engine control (EEC) prevents overspeed or overthrust conditions by sending a down-trim signal to the fuel flow governor. It normally operates in a supervisory role, preventing the engine from exceeding preset limits. In the event of an EEC failure, the system reverts to full hydromechanical control.

The high pressure turbine is the first stage in the turbine section. It drives the high pressure compressor. The HP turbine features directionally solidified nickel alloy blades and a nickel alloy turbine disc. It is mounted on roller bearings. Blade cooling is by both convection and film methods. The film air is supplied both through holes in the disc and the blade shank. The blades use a conventional fir tree mounting. The HP turbine has nickel guide vanes. Film cooling air is fed into hollow vanes at both inboard and outboard ends. A cutaway RB.211 is shown in 2-12-27.

The second turbine stage drives the intermediate pressure compressor. Like the high pressure turbine, it is a single stage with directionally solidified nickel alloy blades and nickel alloy turbine disc. Fir tree blade mounting is again used. Circumferential lockplates hold the blades in position. Only the IP nozzle

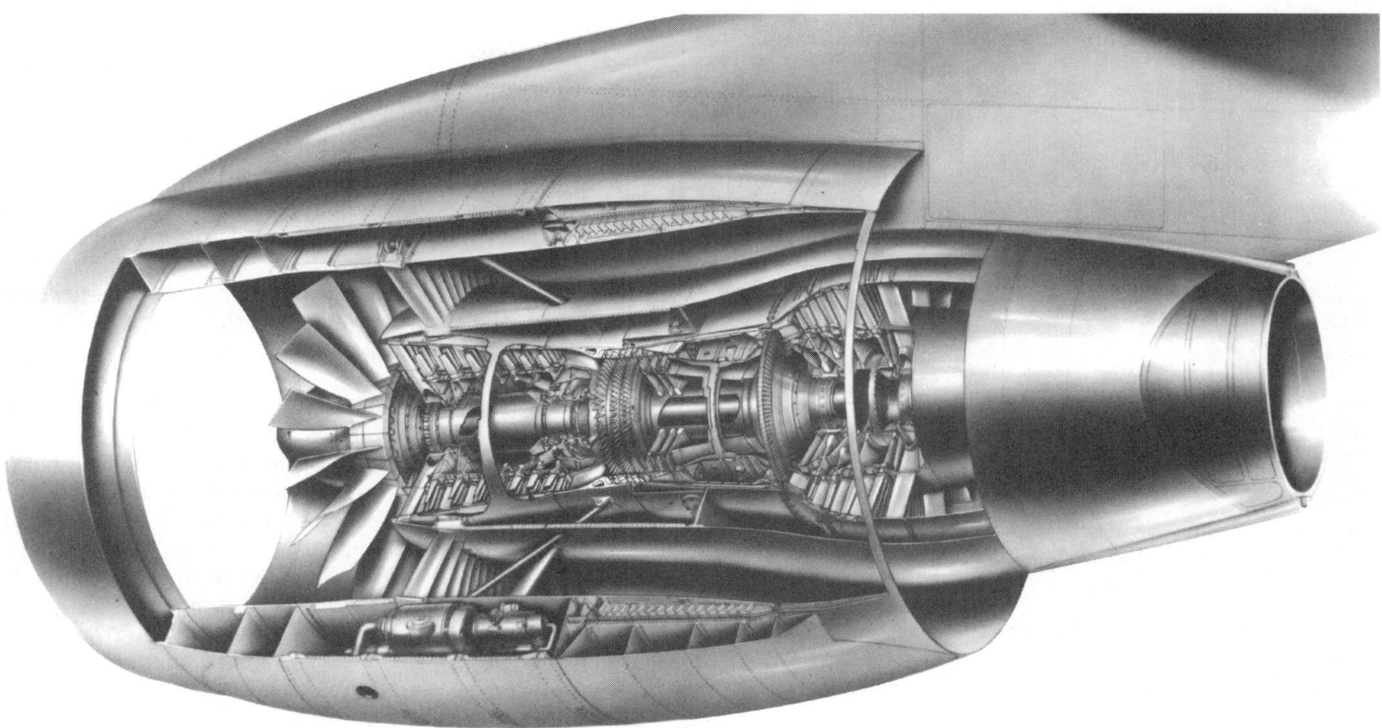

Figure 2-12-27. Cutaway RB.211

Image Courtesy of Lockheed

guide vanes are cooled. The vanes are hollow and take air from the third stage of the HP compressor.

The RB.211's second vibration monitoring pickup is located in the IP turbine. This sensor, in conjunction with the one in the compressor, can be tuned to various vibration frequencies to help isolate which of the three turbines is vibrating.

The low pressure turbine is the third and final turbine section. It drives the fan (acting also as the low pressure compressor). This stage again has nickel alloy blades, but the lower temperatures at this point permit the use of steel turbine discs. The LP casing is cooled by fan stream air to control expansion and maintain consistent rotor blade tip clearances.

The exhaust nozzle is attached to the rear of the LP turbine. The RB.211 has an integrated exhaust nozzle with a deep-chute forced mixer (see Figure 2-12-28).

For most engine installations, Rolls-Royce provides a complete powerplant system including the core engine, thrust reverser, cowling, noise suppression, fan cowling, and exhaust duct. The reverser system is a cold-stream system with no reversing of the core engine gases. The hydraulically actuated reverser cowl moves aft, exposing the reverser cascades. At the same time, blocker doors move into place diverting the fan air through the cascades. The EICAS system also gives a visual indication by displaying the letters "REV" in yellow while the cowls are translating aft. The letters change to a green when fully deployed.

Figure 2-12-28. Exhaust nozzle of RB.211

Figure 2-12-29. The PW 4000 series engine, shown from the aft lower side with the cowlings open.

Figure 2-12-30. The PW 4000 series engine wide-chord fan blades with midspan supports

**PW-4000 series engine**. A third generation turbofan, the PW-4000 series of engines, seen in Figure 2-12-29, consists of several fan diameters: the 94-inch, 100-inch and the 112-inch fan engines. It is in service on the Airbus A310, A330, Boeing 747, 767 and 777 and McDonnell-Douglas DC-10 and MD-11. The PW4000 is a new technology replacement for the JT9D. While similar in size and appearance, it is not a development of the JT9D, but rather an all new engine. One of the design goals was to reduce the parts count by 50%, thus simplified maintenance and spares support.

Engine thrust rating for the series can vary from 56,000 lbs. to in excess of 100,000 lbs. of thrust. PW4000 is the basic series designation. Specific variants change the last two digits to indicate the thrust rating. For example, the PW4084 is an 84,000 lb. thrust engine whereas the PW4098 is a 98,000 lb. thrust variant. The basic construction and nomenclature of the engine is explained, starting from the front of the engine. Even though each engine type has some differences, the basic systems and construction of the engines is the same.

The compressor inlet cone, also referred to as the nose cone or spinner, is mounted to the very front of the engine and serves as an aerodynamic fairing. The nose cone provides for smooth airflow into the engine. Its construction is of kevlar and epoxy resin with a polyurethane coating. Two cone segments are used: the rear segment, which is bolted to the front of the low-pressure compressor (LPC) hub; and the front segment, bolted to the rear cone segment. The complete assembly is pre-balanced to reduce vibration. There are 12 vent holes for anti-icing air, equally spaced around the rear of the cone segment, allowing the anti-icing air to vent back to the fan duct.

The next component is the fan that is the first stage of the LPC rotor. It compresses the air that flows into the engine. The LPC first stage (see Figure 2-12-30) has 38 wide-chord fan blades and a single-part span shroud. Each blade is moment-weighted for balance and individually marked. The fan blades are installed into the LPC hub in dovetail slots, which hold the blades radially, while a split-ring blade lock holds them axially. The fan blades can be replaced as moment-weighted pairs or individually, if the replacement blade is within the required moment-weight. Each fan blade has a rubber seal below the blade platform to prevent air leakage.

At the entrance to the primary gas path, the first stage of the stator assembly is located aft of the fan blades. The fan exit fairing divides the primary and secondary airstreams. The remaining stages further increase the pressure of the primary airstream. The LPC build group includes a five-stage rotor and stator assembly (stages 1, 1.6, 2, 3 and 4 ) The LPC rotor is supported by the No. 1 ball bearing, which is on the front compressor hub and turned by the low-pressure turbine. The fan exit fairing is attached to the fan exit inner case assembly, which supports the LPC stator assembly.

The fan case makes a flow path for the fan discharge air and is the part of the engine structure that supports the nacelle inlet cowl. The inlet cowl contains the fan blade tip rubstrips that prevent blade tip wear. The fan blade containment ring prevents the fan blades from exiting out of the engine radially should they break off. Fan exit guide vanes are used to straighten the discharge air before it goes into the thrust-reverser fan air duct. The 2.5 bleed valve is attached to the fan exit inner case at the LPC exit. The 2.5 bleed air goes out of the fan exit inner case through 14 slots, to the fan airstream.

The intermediate case is the primary structural component of the engine. An example is shown in Figure 2-12-31. It has the supports for three main engine bearings. These are the No. 1 bearing (LPC), No. 1.5 (LPC/LPT shaft) and the No. 2 bearing (HPC). The intermediate case also includes the fourth-stage compressor stator assembly.

The high-pressure compressor (HPC) increases the pressure of the primary air from the low compressor and sends it into the diffuser. The HPC has an 11-stage rotor and stator assembly. The first set of airfoils is the inlet guide vane (IGV) assembly. Each of the first four stator stages has variable vanes moved by a unison ring assembly. The HPC rotor is supported at the front by the No. 2 bearing and at the rear by the No. 3 bearing. The HPC is turned by the HPT. The HPC turns the tower shaft that drives the angle and main gearbox. The HPC supplies eighth-stage bleed air for aircraft use. The ninth-stage bleed air is used for engine stability, rotor cooling and No. 1.5 bearing seal pressurization. The twelfth-stage bleed air is used to cool the No. 3 bearing, cool parts of the turbine and provide for the No. 3 bearing seal pressurization. The fifteenth-stage bleed air is used to balance the thrust load on the No. 2 bearing, for muscle pressure, for airflow sensing, for aircraft use, for the No. 3 bearing seal pressurization and to cool parts of the turbine.

The diffuser straightens the airflow from the compressor exit, and increases the pressure and reduces the velocity of the primary air, which is sent into and around the combustor, where the fuel is mixed with air. The fuel/air is burned to add energy to the primary gas path air-flow. The diffuser case is attached with bolts to the high-pressure compressor rear case.

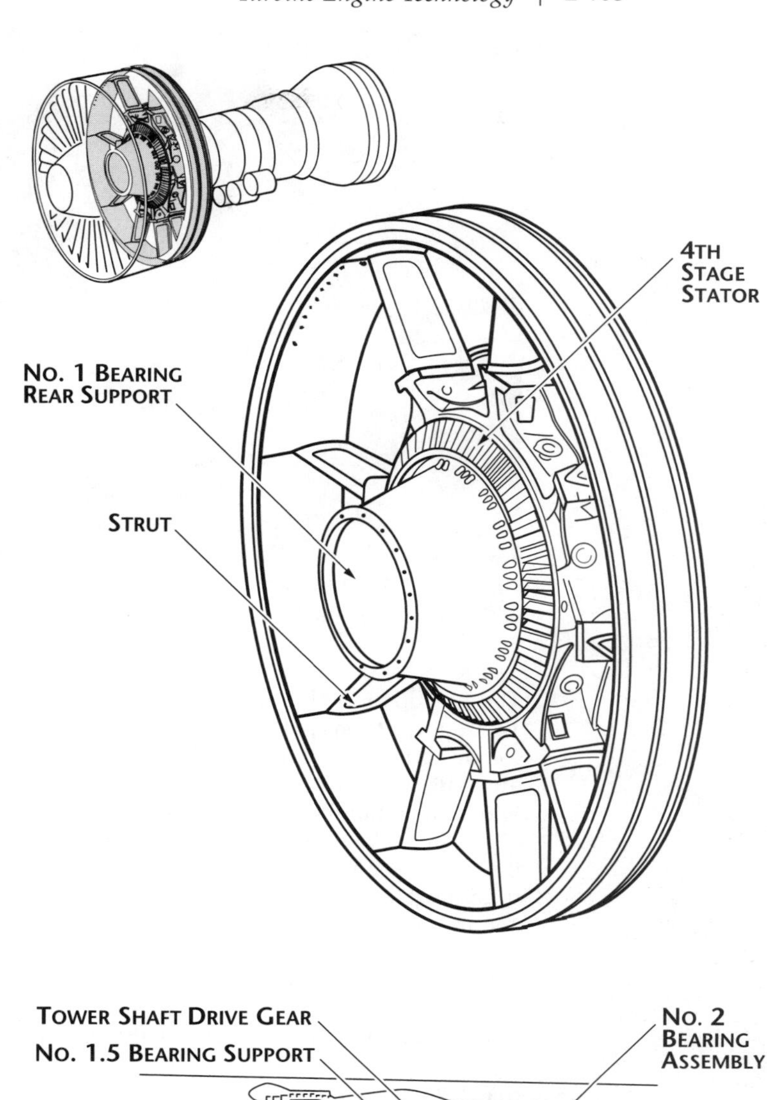


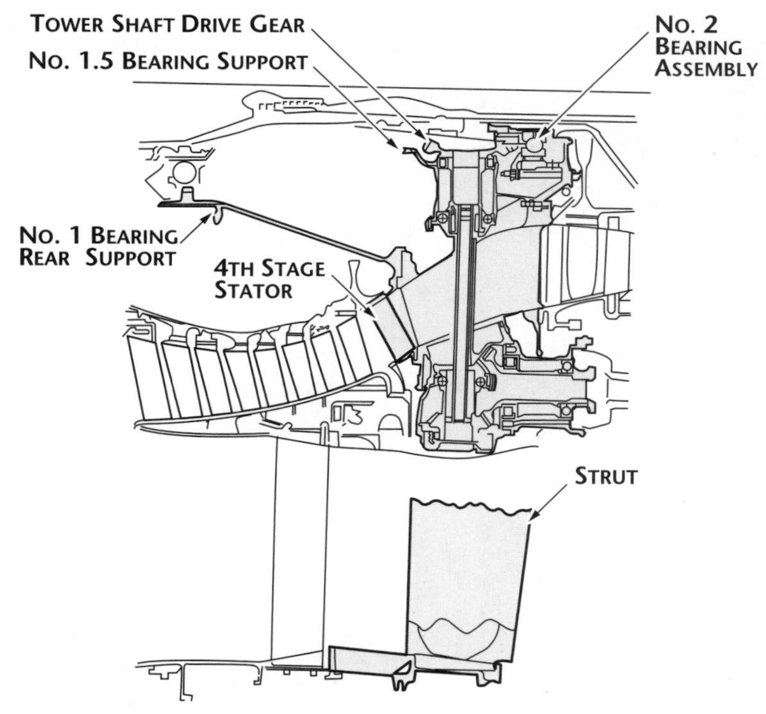


**Figure 2-12-31. The PW 4000 engine's intermediate case houses several engine components and forms the main support for the engine's connection to the airframe.**

**Figure 2-12-32. The PW 4000 combustion chamber showing the fuel nozzles and cooling holes, along with the turbine inlet guide vanes, for the first and second stage turbines**

The combustion chamber, shown in Figure 2-12-32, permits gas path air to enter the chamber for combustion, dilution and cooling. Cooling air from the diffuser flows around the inner and outer chamber walls. The gas path flow also enters the chamber, through small holes, into each segment. Then it flows against the inner surface of the combustion chamber as a cooling film. This action prevents the combustion chamber from being subjected to temperatures that would be harmful to its surface.

The combustion chamber has eight fuel injector manifolds that transmit fuel to the 24 fuel injectors, which are attached around the diffuser case. The combustor also includes two igniter plug bosses, which provide ignition for combustion. The combustion chamber is an annular design with double-pass cooled louvers formed by two parts: the outer combustion chamber that is part of this build group, and the inner combustion chamber that is part of the turbine nozzle build group. Air from the gas path flow also enters the internal passages of each turbine inlet guide vane. As it passes out through a pattern of holes in the surface of the vane, it causes a protective air film on the surface of the vane, preventing it from being subjected to high temperatures.

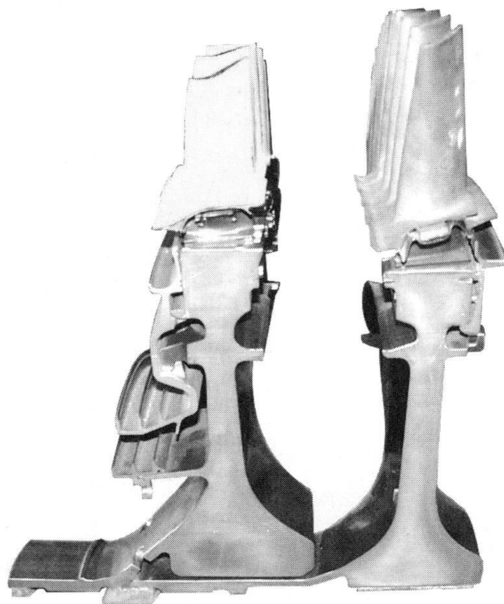

**Figure 2-12-33. A cutaway view of the first and second stage high-pressure turbine disk with blades attached.**

The gas path air goes into the annular cooling duct that operates as a metering nozzle, then goes out of the nozzle and flows against the turbine rotor. There are honeycomb seals that interface with the first-stage turbine rotor, keeping leakage of the cooling air used for the first-stage turbine rotor and blades to a minimum.

The high-pressure turbine, see Figure 2-12-33, supplies the force to turn the high-pressure compressor. The high-pressure turbine includes one case and vane assembly, two disk and blade rotor assemblies and one rotating inner air seal. The components that are air cooled in the HPT are the: first-stage disk and blade assembly, second-stage disk and blade assembly, second-stage vanes and inner air seal. The 60 first-stage turbine blades are made via a single crystal-casting process. The 82 second-stage turbine blades are made via a directional-solidification casting process. Both of these casting processes are used to make the blades more heat resistant and stronger.

Abradable ceramic outer air seal segments form a shroud around the blade tips. The second-stage turbine vanes (21 vane cluster assemblies, two vanes per cluster) are cooled internally by twelfth-stage HPC air that enters the cooling air annulus through cooling air ports. The turbine case cooling (TCC) components include cooling air manifolds, which are attached to brackets on the outside of the HPT case. They control the flow of fan air through those manifolds to cool the HPT case to reduce its diameter during cruise flight. This action reduces the HPT blade tip clearances. This increases the engine's efficiency at cruise flight.

Figure 2-12-34. A segment of the low pressure turbine stages, showing the rotor blades and turbine inlet guide vanes

The Low-Pressure Turbine (LPT) section (see Figure 2-12-34) contains the turbine nozzle guide vanes that send hot gases from the combustion chamber to the first-stage turbine blades at the correct angle and speed. The cooling duct sends cooling air to the first-stage turbine rotor and blades. The turbine nozzle build group includes: the inner combustion chamber, the first-stage HPT cooling duct, and 17 first-stage HPT nozzle guide vane cluster assemblies (two vanes per cluster). The low-pressure turbine supplies the force to turn the low-pressure compressor through a drive shaft. The low-pressure turbine has four stages. Starting with the third stage: 39 nozzle guide vane cluster (three vanes per cluster), 128 blades. The fourth-stage LPT turbines consist of 44 vane clusters (three vanes per cluster) and 130 blades. The fifth-stage of the LPT consists of 38 vane clusters (three vanes per cluster) and 118 blades. The sixth-stage of the LPT consists of 36 vane clusters (three vanes per cluster) and 128 blades. The fifth-stage disk supports the other three sides as follows: the third- and fourth-stage disks are cantilevered from the front of the disk, while the sixth-stage disk is attached to the rear. Internal cooling air from the HPC ninth stage is supplied to the LPT to reduce the temperature at the inner wall of the transition duct.

The turbine exhaust case (TEC) supports the No. 4 bearing and holds the exhaust nozzle and plug. The TEC has attachment points for the rear-engine mount and for ground-handling tools. The struts make the primary airflow straight before it enters the area of the exhaust plug and nozzle, and has opening and attachment points for four probes: two $T_t$4.95 probes and two combination $P_t$4.95/$T_t$4.95 probes.

The accessory gear box (AGB) is driven by the tower shaft, which is turned by the HPC.

The angle gearbox drives the main gearbox. The AGB is installed at the rear of the intermediate case at the 6 o'clock position and is supported by two mount lugs. The lay shaft transfers power from the AGB to the main gear box. As shown in Figure 2-12-35, all of the main

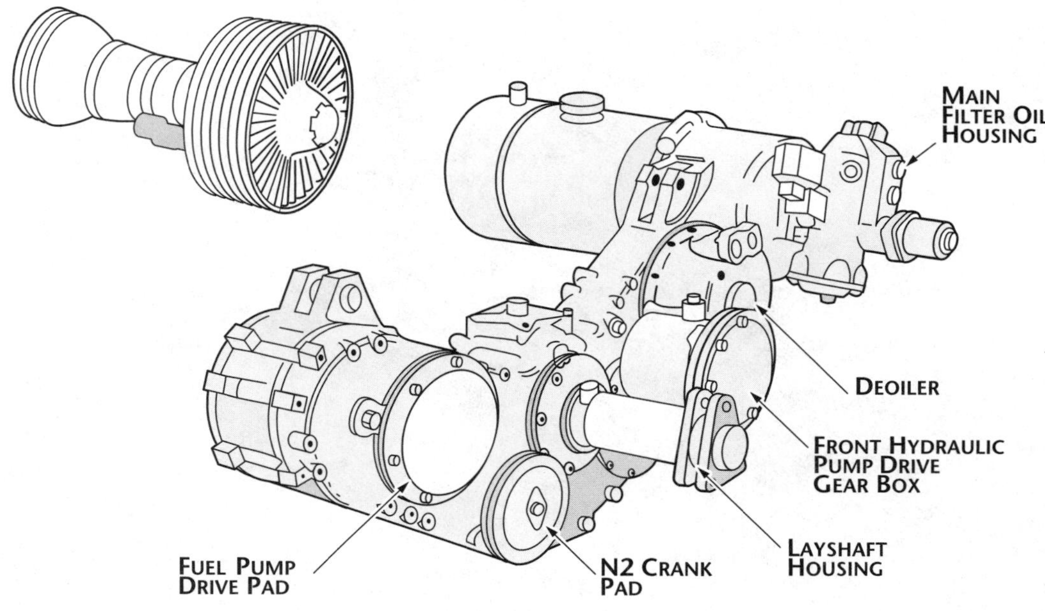


Figure 2-12-35. A drawing of the main gearbox on the PW 4000 series engine.

gearbox accessory drives are modules that can be easily put into, or pulled out of, the MGB. The main gearbox drives the accessories for the engine and aircraft. The MGB housing is an aluminum casting that contains a chip detector and is a line-replaceable unit.

## Section 13

# Turbine Engine Inspection, Maintenance and Troubleshooting

Maintenance and inspection requirements are based on both flight hours and also on operating cycles. An operating cycle is defined as one engine start. It is assumed that the engine must stop at the end of the cycle before the next cycle begins. It is for this reason that some high-cycle operations will only stop the engines on one side of the aircraft when they land for a short time. This practice reduces the total cycle count of the other engines, thus allowing them to operate longer.

Some parts will also have a life-limit, whereby their removal and replacement is mandated after a certain number of hours-in-service or cycles-in-service, regardless of condition. In view of the critical nature of these components, care is taken to track these times and record replacements as the life-limit time is reached.

Each inspection will include record keeping relative to such components. Airplanes operating on long routes have lower flight cycles per hour than do short-haul airplanes, and the maintenance and inspection requirements reflect these differences. Engines on airplanes with higher flight-cycle duty undergo more thermal cycles, leading to additional wear and, therefore, shorter inspection and maintenance intervals. Commuter airlines in short haul service can have operating cycles as short as fifteen minutes. Long-haul overseas airline routes may have cycles in excess of 14 hours.

Overhaul or replacement times are specified by the manufacturer in accordance with the approved maintenance program specified by FAR part 125.247. Program changes are based on service experiences and subject to FAA approval and, in the case of airline equipment, recommendations by the MRB (Maintenance Review Board).

Many types of engine maintenance and inspection can be performed with the engine still attached to the aircraft. This is often referred to as "on-wing" maintenance. However, overhauls and many heavy maintenance procedures require that the engine be removed from the aircraft and taken to a repair station. These repair stations will have the required tools and inspection equipment necessary to perform

**2-13-1. Pratt & Whitney JT9D in engine lift stand**

the needed inspections, repairs or overhaul. Turbine engines are not overhauled in the field. Detailed overhaul instructions are beyond the scope of this text.

Specialized equipment, including engine stands and hoists, is necessary to remove a turbine engine without damaging it. See Figure 2-13-1. Only trained, qualified personnel should remove the engine. All hoses, lines, controls and other attachments should be properly capped to prevent dirt or other foreign material from entering the aircraft systems or engine. They should also be properly marked to facilitate reconnection of a replacement engine.

Once removed, the engine should be prepared for shipment by wrapping it in a protective material. All turbine engines need to be properly secured before they are transported. See Figure 2-13-2. Most engine manufacturers have specialized shipping stands that are designed to make this process relatively straightforward. At this point the engine is shipped to a repair station for overhaul. In some cases, airlines will have an in-house repair station and overhaul capability, although it may not be located at the same place as the aircraft needing an overhaul.

Overhauled engines are returned on the same types of shipping stands. The reinstallation process again requires the use of specialized tooling, lifts and hoists to position the engine for mounting. After mounting the engine, all of the lines, hoses and controls should be reconnected properly. Most aircraft manufacturers provide detailed written instructions for this process.

## Turbine Engine Inspection

There are a number of types of inspections that the Aircraft Maintenance Technician will be involved in on a regular basis. Due to the wide variation in engines and manufacturer inspection requirements, specific requirements and procedures for inspections will not be provided here. An overview of inspection types and procedures follows.

For purposes of inspection, the engine can be divided into cold and hot sections. The cold section consists of the inlet, compressor and diffuser, while the hot section is comprised of the combustion chamber, turbine, exhaust and thrust reverser.

## Inspection Classifications

Inspections can be also classified as routine and non-routine. Periodic inspections are of the routine type and will be scheduled based on either number of flight cycles or number of flight hours. Airline procedures may call out inspections following each flight, often referred to as line checks or No. 1 service. These checks often include fueling of the airplane and a walk-

2-13-2. Pratt & Whitney JT9D wrapped and mounted on transport stand

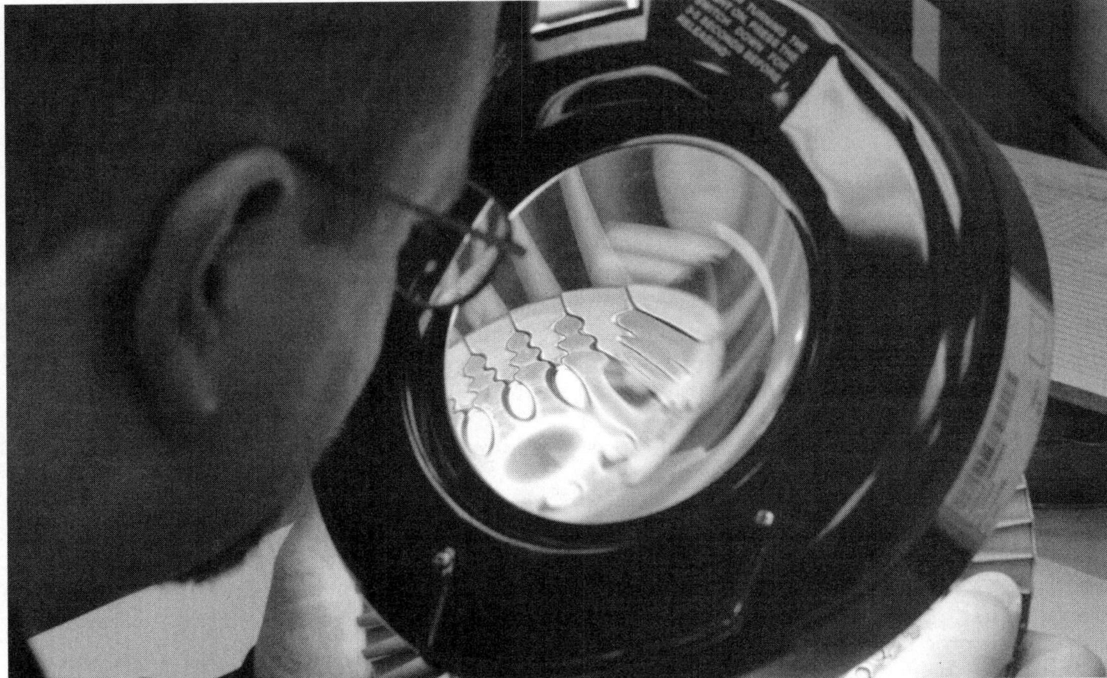

**Figure 2-13-3. An AMT inspects turbine blades for signs of damage**

around inspection of the engine inlet and fan, as well as the exhaust and thrust reverser. No. 2 service may consist of engine fluid quantity check and service, a review of the flight maintenance logbook and a more detailed visual inspection of the engine inside the nacelle.

"A" checks are generally performed after a specified number of hours, often at the 100-hour interval, and generally include the following duties:

- Visually inspect the engine exhaust and rear turbine stages for damage (See Figure 2-13-3 and 2-13-4).
- Visually inspect the thrust reverser for damage, and check for sound operation of the translating cowls and other moving components of the system.
- Check and fill the engine oil tanks as required. Record the amount of oil added in the inspection log.
- Inspect and service the constant-speed drive as necessary.

"B" checks are more extensive and differ more from airplane to airplane, but may generally include the following:

- Visually inspect all cowling, fan and first-stage compressor blades
- Apply approved lubrication to cowl latches
- Service oil system, including oil filter and screen. Visually check for appearance of metal contamination in both, as an indicator of abnormal wear.
- Check oil quantity, carefully following installation directions for time to perform following shutdown of the engine.
- Check the CSD and starter oil, servicing as required and recording quantities added in the inspection log.
- Check the ignition system for operation, observing all precautions to avoid potentially lethal voltage.
- Inspect thrust reversers, paying close attention to the deflector doors or cascade vanes for the presence of cracks.
- Check for sound operation and security of the translating cowls.

"C" and "D" check are generally quite extensive in scope and very much type-specific in procedure. These checks are routine in nature, as they are scheduled based on flight hours and cycles, and they require many man-hours to perform at facilities appropriately equipped for heavy maintenance activities.

Non-routine maintenance may become necessary as a result of foreign object damage (FOD), over-temperatures and over-speeds, chemical contamination, or extreme "G" loads due to turbulence or maneuvering. In the event of temperature or r.p.m. excursions above the limits, maintenance and repair procedures will generally be specified based on the degree and duration of the events. In some cases, only a visual or borescope examination may be mandated; in extreme cases, removal and replacement of rotating assemblies may be necessary. In every case, the specific procedures will be called for in the appropriate maintenance publication.

Other events that may lead to non-routine maintenance include operation without oil pressure, excessive maneuvering loads or turbulence, or engine stall and surge. The required inspection methods will be dictated by the severity and duration of the event.

**Borescope inspections**. Turbine engines are designed with ports for borescoping at a number of different locations, allowing for inspection of critical components without disassembly. The borescope consists of a tube with one or more mirrors that reflect the light to an eyepiece, where it is focused to form an image for viewing. Fiberscopes are an evolution of the borescope that use fiber optics and a remotely controllable section at the tip to allow inspection around blind corners. See Figure 2-13-5.

These instruments, the borescope and fiberscope, have vastly improved on-wing maintenance and have directly impacted the number of disassembly inspections required for routine and non-routine maintenance. Electronic imaging has helped to improve the definition available with this type of inspection and allows for archiving of images that represent the internal condition of the engine at each inspection interval. This equipment includes a video processor that digitizes the image, a monitor, an inspection probe and a light source.

Borescoping is only one visual means of inspection. Optical viewing, which includes the use of hand scopes, microscopes and mirrors, is still the dominant means of identifying problems in an engine. However, some faults and defects are difficult or impossible to locate using visual methods alone. Dye-penetrant or fluorescent-penetrant inspection procedures are commonly used to identify cracks in combustion liners, for instance, and may be applied to various components of the exhaust and thrust reverser. This will reveal cracks on the surface of materials only; other methods must be used if it is necessary to examine parts for subsurface cracks.

**NDT inspection**. Ultrasonic, eddy current and x-ray equipment provide the means to identify and locate sub-surface cracks in different materials. Magnetic particle inspection is used to test ferrous materials for cracks on or immediately below the surface. For all of these inspection methods, cleanliness of the parts in question is critical.

Some terms that describe damage and typical causes follow below.

The cold section is subject to accumulations of dust and dirt that, over time, impact the aerodynamic efficiency of the blades. Foreign object damage (FOD) is another cumulative issue that particularly affects the fan. Through a centrifuge effect, the fan tends to expel foreign objects in a radial direction. Consequently, the remainder of the compressor may be spared the worst damage.

**Figure 2-13-4. Turbine blades can usually be visually inspected while still on the aircraft.**

Increased EGT, changes in relative spool speeds and engine pressure ratios, and increased acceleration time may all indicate a damaged or contaminated compressor. Inspections in these areas include visual methods, including application of the borescope, as required. The fan and fan case must be carefully inspected for foreign object damage or cracks, so that the proper decision is made concerning repairs.

**Figure 2-13-5. Borescope inspection allows the combustion chamber and turbine blades to be inspected without disassembling the engine.**

During borescope inspections turbine blades should be examined carefully. Of particular interest is an examination for stress rupture cracks. These are hairline cracks that run at right angles to the length of the blade. They are caused by an over temperature.

## Bearing Inspection

### *Determining Causes*

Defective mounting, improper operating conditions and similar causes can usually be detected by visual inspection of the bearing.

**Normal bearing under satisfactory operating conditions.** Absence of trouble in a bearing that has operated under radial load can easily be recognized. When bearings which were properly mounted, operated under good load conditions, kept clean and properly lubricated, the path of the balls in the highly polished races shows as a dulled surface similar to a lapped surface, wherein the microscopically fine grinding scratches have been smoothed out. There has been no appreciable removal of material from the surface, as indicated by the fact that there has been no measurable decrease in the diameter of the balls, though their entire surface has been dulled.

Other indications that operating conditions have been satisfactory are the uniformity of the ball paths, their exact parallelism with the side of the races, indicating correct alignment, and the centering of the ball track in the race, which indicates that the bearing carried a purely radial load. Normally, the outer race, if fixed, should carry load for less than half the circumference.

Angular-contact bearings that have operated a considerable length of time under ideal conditions, carrying a partial thrust load, show the ball path riding the sides of the race approaching the edge of the shoulder, but not reaching it.

**Angular-contact bearing under satisfactory operating conditions.** If operating conditions are correct, the balls in angular-contact bearings will have their entire surfaces dulled, appearing just like the ball paths, but with one important exception. If a thrust load has been present continuously for the entire period of operation, and if it was large enough relative to the radial load to keep the balls in contact at all times, the balls will show a circumferential band, indicating that:

1. The balls did not normally spin or change their axis of rotation
2. The fatigue life of the balls has been reduced because of the small area of ball surface used
3. If the bands have appreciable depth there has been material wear caused by improper operating conditions

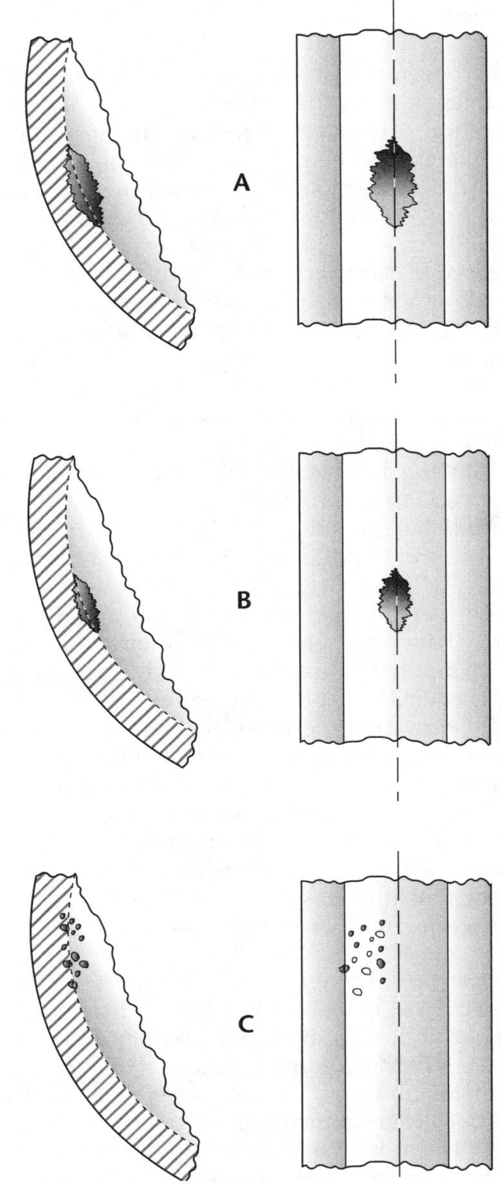


**Figure 2-13-6. Three types of race surface deterioration: (A) Hard coarse foreign matter causes rounded depressions; (B) Overloads cause fatigue failure characterized by sharp jagged edges; (C) Etching causes hemispherical depressions.**

If when operating under conditions that cause circumferential bands, the bearing setting was disturbed or the loads were removed or shifted, the axis of spin of the balls would shift, causing immediate roughness or noisy operation if the balls have been worn appreciably.

**Damage caused by foreign matter.** Ball bearings are particularly sensitive to dirt or foreign matter, which is always more or less abrasive, because the very high unit pressure between the ball and race and the rolling motion that tends to entrap the pieces, particularly if they are small.

This is most common cause of damage to ball bearings. The foreign matter may get into the bearing during the initial assembly, during repairs, by seepage from the atmosphere into the bearing housing during operation, or it might even get in as adulterants in the lubricant.

The character of the damage caused by different types of foreign matter getting into ball bearings varies considerably with the nature of the foreign matter. Fine foreign matter or foreign matter that is soft enough to be ground fine by the rolling action of the balls will have an effect the same as that which results from the presence of a fine abrasive or lapping material. The races become worn in the ball paths, the balls wear and the bearings become loose and noisy. The lapping action increases rapidly as the steel removed from the bearing surfaces contributes to more lapping wear.

Hard coarse foreign matter such as iron, steel, metallic particles introduced when assembling the machine produce small depressions of a character considerably different than those produced by overload failure. Acid etching or corrosion jamming of the hard particles between the balls and the races may cause the inner race to turn on the shaft or the outer to turn in the housing.

Water, acid, other corrosive materials, or corrosives from deterioration of the lubricant, produce a type of failure that is indicated by a reddish brown coating and very etched holes over the entire exposed surface of the races. Frequently, the etching does not show on the ball path because the rolling action of the balls pushes the lubricant, loaded corrosive, away from the ball path. The corrosive oxides formed act as lapping agents that cause wear and produce dull gray color on the balls and the ball paths, as contrasted with the reddish brown color of the remainder of the surface. Figure 2-13-6 shows three types of race deterioration.

**Effects of corrosion.** Corrosion makes itself evident over the entire exposed surface of the race by developing small etched holes.

**Damage caused by overload.** Overload causes fatigue failure of the material as evidenced by the breaking out of the surface layer of steel. Such a failure starts in a small area and spreads rapidly and would eventually spread over both the races and the ball surfaces. Most frequently failure starts on the inner race. A wide ball path is added indication of excessive loading.

**Thrust load wrong direction**. A thrust load in the wrong direction on an angular-contact ball bearing will make the balls ride on the edge of the race groove where it joins the shallow shouldered counterbore and, thus, breaks up the race surfaces. Under these conditions the balls will frequently break up or split, even though their material and heat-treatment was perfect.

**Misalignment and off-square**. On a conventional single-row bearing that had been subjected to off-square operating conditions, the outer race ball path clearly moves from the right hand side over to the left and back again (Figure 2-13-7). The ball path can be a full 1/8 in. off-square in about a three inch diameter raceway. The ring itself could tilt only about

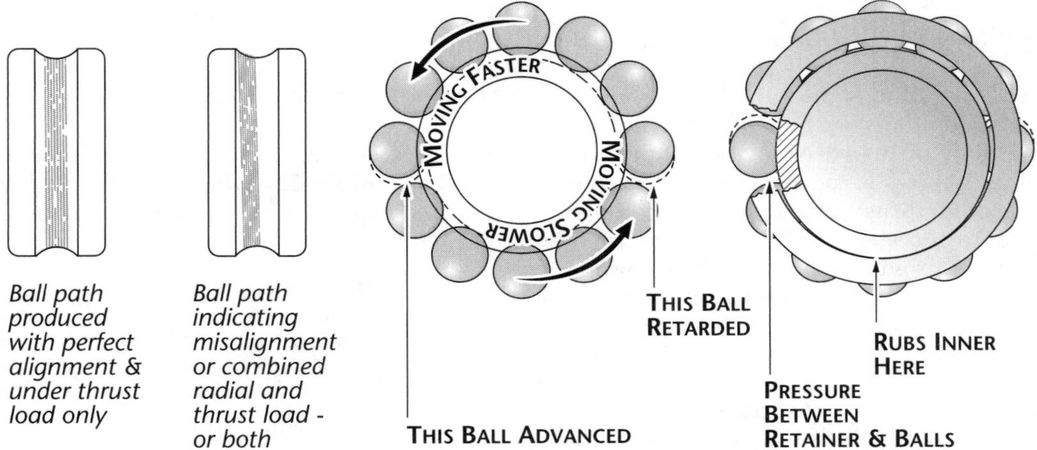


**Figure 2-13-7. Conventional ball bearings that were installed misaligned will show these wear patterns.**

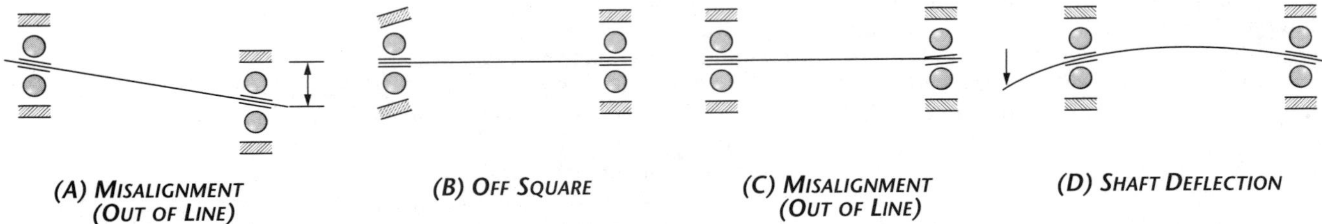


**Figure 2-13-8. The most common types of misalignment**

1/64 in., but produce a greater change in the ball path because of the change in contact angle. The inner race would have a very wide ball path since the ball wanders back and forth across nearly its full width.

Almost all separator failures in a standard single-row type of ball bearing can be traced back to an off-square condition of either the inner or outer race. The failure is caused by variations in ball speed jamming the balls into the separator pockets.

Ball speed variation caused by off-square operation produce the greatest effect when the balls are never free to readjust themselves, as in a tight bearing or in a bearing carrying mainly thrust loads. The variation in ball speed is then accumulative and the two side balls force the retainer downward until it strikes the inner or outer race. Outer race tilting can be easily determined in operation, being evidenced by the eccentrically operating retainer. With inner race cocking, the eccentricity of the retainer rotates with the inner race.

Four types of misalignment to which ball bearings may be subjected are shown in Figure 2-13-8. Out-of-line conditions rarely cause trouble. The exceptions occur only on long shafts where diameters are reduced to an absolute minimum, as on automobile rear wheels or airplane engine crankshafts.

## *Brinelling and False Brinelling*

A few ball bearings are brinelled when putting them on the shaft by being struck with a severe blow or by extremely heavy pressure. But this is rare, because most ball bearings have enough thrust capacity to permit forcing them onto the shaft by pressure against the outer race, provided the pressure is applied squarely. The exceptions are the smallest bearings and bearings with flatter races such as propeller shaft center support ball bearings and self aligning types. The rare occasions when regular single-row ball bearings are brinelled usually result from off-center blows or pressure when only three or four balls carry the entire load.

**True Brinelling.** Brinelling of a bearing caused by off-center blow in mounting is typical, though rare. Such bearings frequently go unnoticed in operation because the load does not cause the ball path to reach the brinell marks on the races. However, the balls acquire similar depressions and will make the bearing sound noisy or catchy, as though it had several small pieces of foreign matter in it. Failure may result at the flat spots on the balls or will occur on the races if carrying heavy enough thrust load for the ball path to reach the brinell marks.

**False Brinelling.** False Brinelling occurs when bearings do not rotate for extensive periods. Loads may be relatively light but slight changes in the surfaces of the raceways result from even minute axial or rotational movements and these only appear under each ball.

**False brinelling due to vibration without rotation.** False brinelling pockets occur in the presence of vibration without rotation and can be very greatly deepened by abrasive foreign matter. Exposing the bearing surfaces to air will oxidize the microscopic metal particles freed by the vibrating movement and these oxides in turn provide the abrasive action as in the extreme example shown here. False brinelling produces surface depressions far more rapidly when balls actually skid or slide on the race surfaces instead of just rolling. Examples of this are large axial vibrations in ball bearings or centrifugal forces on loose balls in controllable pitch propeller bearings.

## *Damage Caused by Heat*

Heat failures occur only at medium and high speed operations. The initial cause is frequently obscured especially at the highest speeds and may be:

1. Failure of lubricant by lubricant source being cut off, lubricant deterioration or contamination
2. Excessive load
3. Cramped bearings, either radially tight bearings caused by expansion of inner race when pressing on shaft or expansion of outer race when pressing on housing,

or axially tight by squeezing one bearing against another

4. Off-square producing heat at retainer
5. Heat from an external source.

Sometimes the initial cause may be diagnosed by noting which parts showed the initial heating, the most heating or by other similar indications.

**Effect of heat on balls.** Heat will not only badly oxidize separators but will also soften the balls and races, and particularly the balls because the heat cannot be conducted away from them as rapidly as from the races. Liberal lubrication may make continued operation possible for some time in spite of the partly softened balls and races

## Compressor Washes

The turbine compressor ingests large volumes of air. Because of the high airflow demands, it is not practical to filter this air. Airborne contaminants, the same things that cause the haze and smog we see in the sky around us, are drawn into the compressor. Some of this dust and dirt accumulates on the blades of the compressor. Compressor efficiency and overall engine efficiency is reduced by this dirt build-up. The end result is hotter starts, lower fuel efficiency, and lower thrust.

Contaminated compressors can be cleaned through compressor washes, which utilize potable water sprayed under pressure at the first stages of the compressor. Detergent washes involve the application of a cleaning agent to the water and are used when warranted. Specific procedures differ from airplane to airplane, but generally all require that certain air and fluid lines be capped off to prevent contamination during the process. The engine must be motored while the wash preparation is applied under pressure to the compressor.

Comparison of engine parameters, including spool speeds, EGT, and fuel flow recorded prior to and following the service, will demonstrate the degradation in performance that occurs as a result of dirt and contamination.

**Hot section inspections**. Hot-section inspection intervals are scheduled on either an "on-condition" or "time-in-service" basis. The use of borescope and other ongoing inspection procedures in many cases allows operators to utilize an "on-condition" schedule for hot-section inspection, which results in longer on-wing service and significant savings. The hot section includes (Figure 2-13-9) the combustion chamber and liners and the turbine section, all of which are typically disassembled from the engine for this inspection. During disassembly, care must be taken to identify location of components critical for reassembly.

Due to the specific nature of disassembly and inspection methods, only a brief description of the inspection procedures follows below. Upon disassembly (following manufacturer's instructions), visual inspection is applied to all of the components in the hot section. Cracks, distortions, corrosion, and evidence of

**Figure 2-13-9. This PT6 turbine wheel is being removed to check the combustion chamber during an on-wing hot section inspection.**

Figure 2-13-10. Determining turbine blade tip clearance during a hot section inspection.

Figure 2-13-11. Vanes are often numbered to better illustrate and document the variances in blade tip clearances, as well as for correct replacement.

Figure 2-13-12. A turbine wheel, uninstalled

excessive temperatures are the subject of these inspections.

In many cases, cracks in the combustion chamber may not be reason enough for rejection, but should be noted and examined for growth. Erosion of coatings on the compressor-turbine vane rings and shroud segments, for instance, is to be evaluated, with particular emphasis on the underlying parent material. Where parent material has been lost, component replacement is most likely necessary. Fuel nozzles are checked for leaks and operation, and the ignition system is checked for resistance. Any marking on hot section parts must not be done using a pencil or ink marker that contains lead. Although commercial felt tip markers are generally used, only markers specified by the manufacturer may be safely used.

**Shroud tip clearance**. One of the most critical dimensions within the hot section is the turbine tip shroud clearance (Figure 2-13-10). As total operating time accumulates on a turbine engine, this dimension increases, leading to both high EGT temperatures and lower efficiency. Correcting this problem requires the selection of new shroud segments (Figure 2-13-11), which are classed based on thickness. Following selection and temporary installation of the segments and the turbine wheel (Figure 2-13-12), the turbine tip clearance is again checked. Grinding of the shroud segments is performed on the inside circumference of the segments to achieve the precise tip clearance. The grinding process is shown in Figures 2-13-13 and 2-13-14.

**Turbine blade blending**. If minor blade damage is found, it can be repaired in the field. Blending out minor nicks and abrasions is normally part of a hot section inspection. Operations must be carried out from tip to root and not crossways. Major damage may still be repairable, but not field repairable.

## Maintenance Classifications

Maintenance can also be defined as being either heavy or line maintenance. Line maintenance typically involves removal and replacement of accessories and components considered to be line replaceable units (LRUs). Hot-section inspection may fall under the heading of line maintenance. Heavy maintenance is dictated when the procedures call for facilities, equipment, and methods not available under line maintenance. These activities are generally performed at an overhaul facility.

**Modular maintenance**. Modular maintenance is applicable to most modern, high-bypass-ratio turbofans and is generally a part of heavy maintenance procedures. Engines of this

**Figure 2-13-13. The grinding wheel is attached via the axial adjuster only after blanking material has been installed over the vanes and other moving parts, protecting them from dust and debris.**

type are assembled in build groups or modules, which enables more rapid and efficient removal, repair or replacement. This method of assembly reduces down-time, leading to increased dispatch reliability and decreased cost of operation.

In the case where no problems have been indicated, maintenance of the fuel system may consist simply of changing fuel filters and cleaning screens. Filters in the fuel system generally have a bypass function, which allows for continued fuel flow to the engine in the event that the filter becomes blocked with contami-

Figure 2-13-14. Grinding the shroud segments

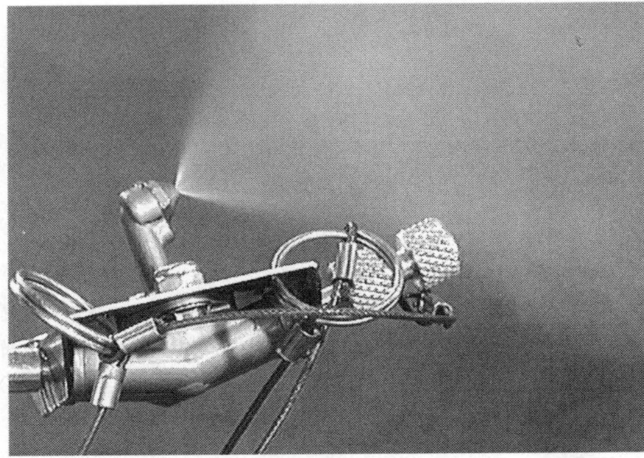

**Figure 2-13-15. Along with tests of the ignition system and visual examination of surfaces for cracking, warping, corrosion, and heat distress, fuel nozzles are checked for leaks during a hot section inspection.**

nants. When this occurs, a warning message is displayed on the flight deck, advising the flight crew of the situation. During maintenance procedures, both the bypass function and the warning light should be checked for proper operation.

**Fuel nozzle flow checks**. Fuel nozzle maintenance requires that checks be performed for leaks and proper flow under specified pressures. When pressure is applied to the nozzles, the pattern of spray is observed for correct shape, often described as *onion-bulb* at low-flow rates, to *tulip-shaped* under higher pressures. Only a specified percentage of streakiness observed in the spray pattern is allowed, and drops of unatomized fuel may represent a problem in fuel nozzle operation. Specific engine overhaul procedures will dictate the required action when problems of this nature occur. See Figure 2-13-15.

**Engine trimming**. For the process of fuel-control trimming, or engine trimming, as it sometimes referred to, engine r.p.m. is checked against fuel flow, EPR, and power lever positions. For engines equipped with variable stator vanes in the compressor, additional information on the position of the vanes at specified power settings is considered as well. Over accumulated hours of operation and the condition of the engine, including compressor and turbine efficiency, are affected by contamination and expected wear. EGT, over many hours of operation, will tend to increase relative to a specified EPR value and will eventually indicate that trimming of the fuel control is necessary.

Some manufacturers also prescribe trimming after specific maintenance, or when the fuel control or fuel control module is replaced. See Figure 2-13-16.

**Figure 2-13-16. A maintenance technician safeties the linkages on the FCU on a JT9D after performing an engine trimming procedure.**

Generally speaking, these steps include the following. Note that applicable trim speeds, EPR, EGT, and other values will vary from engine to engine. Successful fuel-control trimming requires careful observation of these specifications and close adherence to the procedures.

- Position the airplane into the wind and clear the area of all hazards and unnecessary personnel. Note that procedures for each airplane may dictate a maximum wind condition for these procedures.
- Install the necessary, calibrated equipment. Critical parameters that must be calibrated or confirmed for trimming include spool speeds and EPR.
- Locate, as required, the fuel-control trim stops.
- Using the published airplane trim curve, determine the target EPR for trimming. The ambient temperature and pressure is required for this step.
- Following a normal start, the engine should be operated at idle r.p.m. to allow temperatures and pressures to stabilize. Confirm that bleed valves and variable stator vanes are operating correctly, then bring the engine to the target trim speed.
- Record EPR and other thrust parameters. Compare with the nominal values on the trim curve and adjust the fuel control as necessary to provide the required EPR value.
- Record all spool speeds, EGT and EPR information during the subsequent confirmation test runs.

Obviously, specific procedures for different airplanes and engines vary, as do target speeds, temperature and thrust, but the objective remains the same - to maintain rated engine performance at required spool speeds and EGT. Fuel-control trimming is not necessary with Full Authority Digital Electronic Controls (FADEC), or similar systems that utilize computer or electronic controls. In these cases, the adjustments for engine condition and performance are both automatic and continuous.

**Ignition systems**. Turbine engine ignition systems are generally comprised of an exciter box, high-tension ignition lead and an igniter or glow plug located in the combustion chamber. Although most turbine ignition systems provide ignition only for starting, under some atmospheric and operating conditions, airplane operating procedures may call for continuous ignition. Nonetheless, due to the limited operation of these systems, they are generally not an operational issue. Maintenance of the ignition system most often involves servicing, cleaning and replacement, as needed, of the igniter. Replacement will be dictated when erosion and carbon deposits become an operational issue.

The operational check requires a technician be positioned close to the engine nacelle, while qualified personnel runs the ignition system. A steady snapping noise indicates that the igniter is functioning properly. If removal of an igniter becomes necessary, extreme care must be taken to avoid the potentially lethal charge stored in the condenser of the system. (Figure 2-13-17). Technicians should take every precaution outlined in the appropriate maintenance instructions.

Igniters may be reconditioned for service by following the manufacturer's recommendations for cleaning, drying and electrical testing. In some cases, the igniter is pressurized to simulate combustion chamber conditions, and then high-voltage is applied. Some test procedures allow for open-air testing, with the igniter connected to the aircraft ignition harness. Glow plugs are checked in this latter fashion and should glow a bright yellow after a specified period of time. Igniter boxes, although generally trouble-free, are often considered to be a line-replaceable unit. Troubleshooting an igniter that fails to snap or a glow plug that does not heat may be simply a process of temporarily installing another igniter to identify where the problem lies.

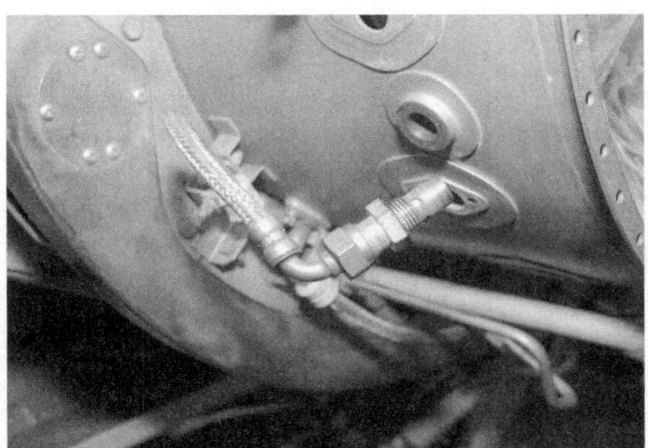

**Figure 2-13-17. Igniter with lead**

**Oil systems**. Modern turbine engine oil systems are most often of the dry-sump variety, all of which include pressure, scavenge, and breather subsystems. In a dry-sump system, oil is stored in an external tank from which it is supplied, under pressure, to the engine bearings and accessory drives. A scavenge system generally uses gear-type pumps to return the scavenged oil through the fuel-oil cooler; if oil temperatures are sufficiently high, the oil is sent back to the tank. When oil temperature is not at a prescribed limit that dictates cooling,

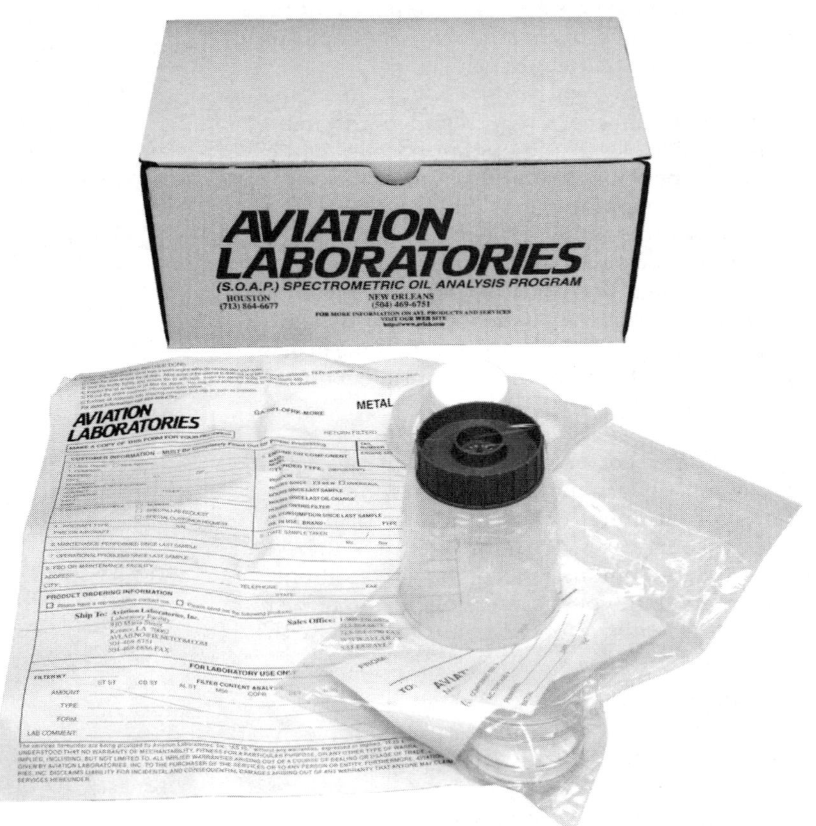


**Figure 2-13-18. Oil analysis kit**

then the oil is bypassed around the fuel-oil cooler and routed directly to the oil tank.

Maintenance of the lubrication system may require that the oil temperature-control valve in the fuel-oil cooler be checked for operation. Examination of maintenance and operational logs should be conducted for trends in oil pressure. If an oil pressure adjustment is required, the pressure-relief valve located downstream from the main pressure pump may be adjusted accordingly. It should be noted, however, that a change in oil pressure is a possible indicator of other problems, some of which may lead to catastrophic bearing failure, for instance. Therefore, careful inspection of the entire system should be undertaken prior to adjustment of the oil pressure-relief valve.

Oil filters in turbine engine service include the screen-disk, screen and cartridge types. The latter must be replaced on an hours-in-service basis, while the screen and screen-disk type may be cleaned and reused. Each of these filter installations will include a bypass function, built into the filter itself or in the oil passages, to allow for oil flow in the event of a blocked filter. The bypass system should be checked at inspection for proper operation, as should the flight-indicating light that warns of a blocked filter. Magnetic chip detectors installed in scavenge oil lines will warn of the presence of metal contamination in the oil. Metallic particles are attracted to the chip detector and, upon contact with the center plug, complete an electrical circuit that then powers the indicator light. At inspection chip detectors should be checked for operation and cleaned.

**Oil analysis**. All engines, transmissions, gear boxes, and such are subject to continual wear. As they wear the byproducts of that wear is generally suspended in the oil. By examining the byproducts of wear it is possible to determine something about the internal wear of the machine. Maintenance procedures often include *oil analysis*, whereby *optical emission spectrometry* or *atomic absorption* tests are performed to determine the contaminants residing in the oil. Figure 2-13-8 shows an example of an oil analysis kit. The *Spectrometric Oil Analysis Program* (S.O.A.P.) has become standard practice. S.O.A.P. analysis programs normally measure metal content in oil samples down to 1 part per million.

Because most parts in an engine are manufactured from different alloys, it is possible to judge the wear rates of individual parts. By graphing the test results obtained from different inspections, a visual wear rate of an engines internal parts can be observed. Figure 2-13-19 shows a typical oil analysis report. Should any part start wearing excessively for any reason, it will show up in the next sample as an increased content of a specific metal. It is frequently possible to find an impending failure before it fails, or while it can be fixed with maintenance instead of overhaul.

Such tests done consistently at specified intervals may indicate bearing wear and other issues, which then enable maintenance to be performed as a preventative act, rather than on the basis of repair and replacement.

In addition to these tests, the technician should also, while cleaning or replacing filters and screens, perform a visual inspection for the presences of metallic particles. Such findings may not always indicate that a true problem exists, but their discovery helps guard against the worse-case scenario.

## Vibration Trend Monitoring

All machines that have parts that rotate produce vibrations. Each different rotating part vibrates at a different frequency. With turbine engines operating at consistent r.p.m.'s, it is possible to measure the various vibration frequencies. It is a relatively simple process to derive normal frequency ranges in a graph and to compare them, either from prior charts or in real time. Any time a vibration frequency changes it is fairly simple to figure out what part is causing the change. As with oil analysis, it can frequently stop a failure from happening by performing a maintenance action. Several such analyzers are available, some built into the airplanes own inspection system.

## Troubleshooting Charts

The problem with using charts is that most of them were designed for specific engines and installations. Some are applicable to most, but not all, normal maintenance problems. The chart following is applicable to many different engines, and studying it will help you learn the interdependency of the different parts of an engine, based on instrument readings. As usual, when working on a live engine, use only the manufacturers current information. See Table 2-13-1.

**Starter systems.** Turbine engine starter maintenance includes checks for leaks, security and condition of the mount bracket, wiring, and air ducts. On smaller turbine engines using electrical starter motors, replacement of brushes may be a maintenance procedure. Given the high temperatures to which the starter is exposed, careful attention must be given to proper lubrication and to the condition of blast tubes installed for cooling of the unit. Other, more specific inspection procedures, applicable for each starting system, should be followed closely.

**BITE systems**. Gas turbine troubleshooting has improved dramatically with the advent of fault-detection and recording devices. Built-in troubleshooting equipment (BITE) systems installed record EGT information, spool speeds, fuel flow, EPR, and other engine parameters, which can then be displayed for both flight deck and maintenance personnel.

These data collection methods are often referred to as *engine condition monitoring*, or *trend-condition monitoring*, whereby maintenance personnel may observe changes and degradation of performance over many flight hours, often preventing small problems from becoming major maintenance issues. Precise and consistent patterns for recording the required information are critical to this process and are often done automatically with new-generation equipment.

When flight deck personnel are responsible for the manual collection of this information, the procedures may state that the airplane must be in level flight at a specified altitude and power setting for a prescribed period of time. Failure to follow the exact procedures may result in a great deal of parameter data scatter that invalidates the effort.

**AVIATION LABORATORIES**

910 MARIA STREET
KENNER, LOUISIANA 70062
(504) 469-6751

ANALYSIS RESULTS

DATE: 10/31/02

OIL ANALYSIS CUSTOMER
ATTN: CHIEF MECHANIC
PO BOX 100

ANYTOWN, USA

ENGINE S/N:PCE81175M
ENGINE MODEL:PT6A-41
AIRCRAFT:BE 200
S/N:BC12
TAIL NO.:N88A-LH

** CURRENT RESULT ** SAMPLE APPEARS NORMAL, SEND NEXT SAMPLE AT THE NORMAL INTERVAL.

SAMPLE DATE: 9/21/02
ANALYSIS DATE:10/31/02
SAMPLE NUMBER: 31
TSN:
TSO: 3957
OIL HRS: 620
FILTER HRS: 620
OIL ADDED: 4.0
FILTER MGS: 9
FLASHPOINT:

*** FILTER ANALYSIS RESULTS ***

| MATL: | ST ST | CB ST | AL ST | M50 | COPPER | SILVER | MAGNS | ALUM | GRIT | MISC |
|---|---|---|---|---|---|---|---|---|---|---|
| AMOUNT: | | TR | | | | | | TR | MA | |
| TYPE: | | | | | | | | | | |
| FORM: | | | | | | | | | | |

*** OIL ANALYSIS RESULTS ***

| Iron | Copper | Nickel | Chromium | Silver | Magnesium | Aluminum | Lead | Silicon | Titanium | Tin | Moly | Tan |
|---|---|---|---|---|---|---|---|---|---|---|---|---|
| 0.3 | 0.3 | 0.1 | 0.2 | 0.1 | 0.1 | 0.1 | 0.0 | 8.2 | 0.1 | 0.0 | 0.1 | |

COMMENTS: PACTEC. MORE

** PREVIOUS 1 RESULT ** SAMPLE APPEARS NORMAL, SEND NEXT SAMPLE AT THE NORMAL INTERVAL.

SAMPLE DATE:10/18/02
ANALYSIS DATE:10/31/02
SAMPLE NUMBER: 32
TSN:
TSO: 4034
OIL HRS: 698
FILTER HRS: 698
OIL ADDED:
FILTER MGS: 14
FLASHPOINT:

*** FILTER ANALYSIS RESULTS ***

| MATL: | ST ST | CB ST | AL ST | M50 | COPPER | SILVER | MAGNS | ALUM | GRIT | MISC |
|---|---|---|---|---|---|---|---|---|---|---|
| AMOUNT: | | TR | | | | | TR | TR | MA | |
| TYPE: | | | | | | | | | | |
| FORM: | | | | | | | | | | |

*** OIL ANALYSIS RESULTS ***

| Iron | Copper | Nickel | Chromium | Silver | Magnesium | Aluminum | Lead | Silicon | Titanium | Tin | Moly | Tan |
|---|---|---|---|---|---|---|---|---|---|---|---|---|
| 0.5 | 0.3 | 0.1 | 0.2 | 0.1 | 0.1 | 0.2 | 0.0 | 7.9 | 0.1 | 0.0 | 0.0 | |

COMMENTS: PACTEC. MORE

** PREVIOUS 2 RESULT ** SAMPLE APPEARS NORMAL, SEND NEXT SAMPLE AT THE NORMAL INTERVAL.

SAMPLE DATE: 7/09/02
ANALYSIS DATE:10/01/02
SAMPLE NUMBER: 04
TSN:
TSO: 3841
OIL HRS: 505
FILTER HRS: 505
OIL ADDED: 2.0
FILTER MGS: 5
FLASHPOINT:

*** FILTER ANALYSIS RESULTS ***

| MATL: | ST ST | CB ST | AL ST | M50 | COPPER | SILVER | MAGNS | ALUM | GRIT | MISC |
|---|---|---|---|---|---|---|---|---|---|---|
| AMOUNT: | | TR | | | | | | TR | MA | |
| TYPE: | | | | | | | | | | |
| FORM: | | | | | | | | | | |

*** OIL ANALYSIS RESULTS ***

| Iron | Copper | Nickel | Chromium | Silver | Magnesium | Aluminum | Lead | Silicon | Titanium | Tin | Moly | Tan |
|---|---|---|---|---|---|---|---|---|---|---|---|---|
| 0.2 | 0.1 | 0.0 | 0.2 | 0.1 | 0.2 | 0.2 | .2 | 0.5 | 0.0 | 0.0 | 0.1 | |

COMMENTS: PACTEC - MORE

FOR MATERIAL CODE SEE REVERSE SIDE

**Figure 2-13-19. Oil analysis report**

Hot-section problems are generally characterized by high EGT and fuel flow, while the gas generator ($N_2$, in most cases) usually stays the same or goes down. This would be the case, for instance, if tip clearance became excessive.

Under this scenario, excessive amounts of hot gas would pass around the tips of the turbine blades, which would result in high EGT and low gas-generator speed. Cold-section problems, on the other hand, may result in an increase in all these parameters: EGT, fuel flow, and gas generator speed. Typical of these problems would be compressor $P_3$ leaks, damaged compressor blades, or a contaminated compressor.

Other problems often have to do with faulty operation of bleed valves. When bleed valves remain closed at start up, compressor stalls may occur. Bleed valves stuck wide open may impact on the EPR at cruise r.p.m. Similar problems may occur if the scheduling of variable stator vanes is incorrect.

**Table 2-13-1. Troubleshooting charts**

| Condition | Probable Cause | Action |
|---|---|---|
| **TAKE-OFF ENGINE CHECK** | | |
| $P_{t7}$ or ERP at full throttle higher than computed. EGT, $N_2$ rpm and fuel flow high when compared with other engines.<br>EGT — High<br>N2 Tach — High<br>$P_{T7}$ or EPR — High<br>Fuel Flow — High<br>Fuel Inlet Press. — Normal<br>Oil Press. — Not Applicable<br>Oil Temp. — Not Applicable | • Possible miscalculation, or use of incorrect temperature when computing take-off EPR (or use of incorrect temperature and/or barometric pressure when computing $P_{t7}$)<br>• Engine trim too high<br>• Possible, but not probable, fuel control malfunction. | • Recheck take-off $P_{t7}$ or EPR computation for prevailing ambient conditions.<br>• Continue with the take-off at the pilot's discretion, throttling back, if necessary, to avoid exceeding maximum allowable values for EGT, $N_2$ rpm and that computed from curves or tables in the aircraft Flight Handbook for the maximum allowable values $P_{t7}$ or EPR at the prevailing ambient conditions.[a] |
| $Pt_7$ or EPR at full throttle or higher than computed for existing ambient conditions.<br>EGT — Possibly Slightly High<br>N2 Tach — Normal or Near Normal<br>$P_{T7}$ or EPR — High<br>Fuel Flow — Slightly High<br>Fuel Inlet Press. — Normal<br>Oil Press. — Not Applicable<br>Oil Temp. — Not Applicable | • Thrust overshoot. This condition, particularly in the case of a cold engine, is considered normal.<br>• Possible $Pt_7$ or EPR indicating system malfunction[b] | • Although thrust over shoot may be used for take-off if desired, it is unwise to rely on this engine characteristic for added thrust because the pilot has no assurance that the additional thrust due to overshot will continue for the duration of the take-off. The maximum allowable value for $P_{t7}$ or EPR computed for the prevailing ambient conditions from curves or tables in the aircraft Flight Handbook should not be exceeded.<br>• Throttle back if necessary.<br>• For an accurate check of engine thrust, the engine must be allowed to stabilize. In case of a cold engine, this may require from 3 to 5 minutes' operation at full throttle. |
| **ENGINE AT IDLE ON THE GROUND** | | |
| Abnormal idle rpm. $N_2$ rpm high or low when compared with the prescribed range of rpm at idle.<br>EGT — Near Normal<br>N2 Tach — Too Low or Too High<br>$P_{T7}$ or EPR — Not Applicable<br>Fuel Flow — Too Low or Too High<br>Fuel Inlet Press. — Normal<br>Oil Press. — Not Applicable<br>Oil Temp. — Not Applicable | • Loose throttle rigging and/or linkage<br>• Incorrect fuel control idle rpm adjustment<br>• Possible tachometer system malfunction<br>• Fuel control servo bleeds possibly contaminated. A malfunction of this nature may be recognized by very high rpm at idle (in the vicinity of 85%) and failure of the engine to decelerate properly. | • No action normally necessary, except at the discretion of the pilot.<br>• Report the circumstances in the aircraft flight log. As an aid to ground maintenance personnel with conditions of this nature, it is suggested that the pilot or flight engineer check and record the minimum fuel flow following a sharp deceleration from approximately 75% rpm.[c] |
| Hot start. Engine "lights up," but EGT goes overboard, exceeding the starting temperature limit. Acceleration will usually be erratic or slower than normal. $N_2$ rpm may or may not hang at some value below the normal rpm for idle[d]<br>EGT — Too High<br>N2 Tach — Rise, Then Erratic or Low<br>$P_{T7}$ or EPR — No Rise or Low<br>Fuel Flow — Normal for Starting or High<br>Fuel Inlet Press. — Normal for Starting<br>Oil Press. — Not Applicable<br>Oil Temp. — Not Applicable | • Over rich fuel/air ration in the combustion chamber. This condition might be the result of any one of the several possible causes indicated under Action | • Discontinue the starting attempt and investigate.<br>• Suspect possible malfunction of the starter power source, which would result in the insufficient power being supplied to the starter.<br>• Suspect starter cutout rpm set too low.<br>• Suspect possible fuel control malfunction ONLY if the fuel flow meter indicates an abnormal reading.<br>• Inspect engine air inlet duct for obstruction During cold weather, suspect possible ice in the fuel control, which will require the application of heat before a successful start can be made. |

| ENGINE AT IDLE ON THE GROUND (CONT'D) | | |
|---|---|---|
| CONDITION | PROBABLE CAUSE | ACTION |
| Hung or false start. Engine "lights up" but does not accelerate to normal idle rpm, the rpm hanging at some lower value, instead. EGT will probably (but not necessarily always) be higher than normal and may, or may not, exceed the starting temperature limit.<br>EGT: TOO HIGH; N2 TACH: BELOW IDLE RPM; $P_{T7}$ OR EPR: TOO LOW; FUEL FLOW: TOO LOW; FUEL INLET PRESS.: NORMAL FOR STARTING; OIL PRESS.: NOT APPLICABLE; OIL TEMP.: NOT APPLICABLE | • Insufficient power to the starter<br>• Binding rotor assemblies, damaged compressor or turbine blades, damaged accessory gearbox<br>• Starter improperly adjusted. Cuts out below engine self-accelerating speed.<br>• In a few cases, one malfunctioning igniter has resulted in almost identical conditions.[e] | • Discontinue the starting attempt and investigate.<br>• Suspect possible malfunction of the starter power source, which would result in insufficient power being supplied to the starter.<br>• Check compressor for free rotation. Inspect compressor and turbine blades and the accessory gearbox for possible damage.<br>• Suspect starter cutout rpm set too low.<br>• When combustion starters are used, check starter air and / or fuel supply for possible depletion before engine-starting cycle was completed.<br>• Suspect possible fuel control or fuel pump malfunction. |
| ENGINE STARTING (GROUND)[f] | | |
| No start after engaging the starter. Compressors rotate but no fuel flow observed.<br>EGT: NO RISE; N2 TACH: NORMAL TO STARTER CUTOUT SPEED; $P_{T7}$ OR EPR: NO RISE; FUEL FLOW: ZERO OR RISE TO NORMAL FOR STARTING, THEN DROP; FUEL INLET PRESS.: NORMAL OR LOW; OIL PRESS.: NOT APPLICABLE; OIL TEMP.: NOT APPLICABLE | • No fuel to engine | • Discontinue the starting attempt and investigate.<br>• Check aircraft fuel boost pump for "on."<br>• Check fuel in tank. Switch tanks, if necessary.<br>• Check emergency fuel shutoff valve for "open."<br>• During cold weather, suspect possible ice in fuel lines or sumps.<br>• Suspect a fuel bypass or relief valve hung open, which might prevent the engine from starting by causing the engine pump stage of fuel pump to bypass excessively.<br>• If fuel tank has been run dry or if maintenance has been preformed on the fuel system, which will require bleeding the system. |
| No start after engaging the starter. Compressors rotate and normal fuel flow is observed.<br>EGT: NO RISE; N2 TACH: NORMAL TO STARTER CUTOUT SPEED; $P_{T7}$ OR EPR: NO RISE; FUEL FLOW: NORMAL FOR STARTING; FUEL INLET PRESS.: NORMAL BOOST PUMP PRESSURE; OIL PRESS.: NOT APPLICABLE; OIL TEMP.: NOT APPLICABLE | • Malfunctioning engine ignition or aircraft relays.<br>• Malfunctioning p&d valve, or weak signal from fuel control to p&d valve (unlikely but possible) | • Discontinue the starting attempt and investigate.<br>• Check for low battery<br>• Check for open circuit in electrical power leads from power source to starter and ignition system.<br>• In a fuel puddle is observed in the engine tailpipe and no drainage is observed from the p&d valve, suspect ignition systems malfunction and check the ignition circuit breaker.<br>• Check for igniter plugs firing properly<br>• If excessive fuel drainage is observed, suspect a malfunctioning p&d valve or p&d valve signal. |

a = *Caution: With engines whose fuel controls sense inlet temperature and when operating an engine known to be trimmed too high, care must be exercised to monitor EGT very closely to avoid over temperatures during the "climb out" in the altitude range of 30,000 to 40,000 feet. Due to a slight lag in the fuel control sensing of compressor inlet air temperature (Tt2), the most critical operation for an over-trimmed engine has been shown by experience to be when the aircraft is passing through the range of altitude at which the outside air temperature no longer becomes colder with increasing altitude*

b = *Note: Although this condition could conceivably be cause by a malfunctioning fuel control, such a possibility is very unlikely and should be assumed only as a last resort.*

c = *Caution: If high idle rpm results from contaminated fuel control servo bleeds, engine might possibly not respond to throttle movement later in flight, which could be serious. Also, high idle rpm could result in accelerated wear of the aircraft brakes, with possible serious consequences. In case of doubt, abort the take-off.*

d = *Caution: At any time that the starting EGT reaches 500° to 550° C and is still climbing during the ground starting cycle, a hot start will almost invariably result.*

e = *Note: A delayed or otherwise abnormal start has been known to be the result of using JP-5 fuel in engines equipped with "short-reach" igniters.*

f = *Caution: Whenever a starting attempt is discontinued, the entire starting sequence must be repeated from the beginning after allowing the prescribed fuel draining or starter cooling period, whichever is longer. When necessary, the starter may be reengaged as soon as the engine has decelerated below 40% rpm.*

Chapter 3

# PROPELLER systems

**Learning Objectives:**

- *Basic Principles of Propellers*
- *Types of Propellers*
- *Fixed-pitch Propellers*
- *Constant-speed Propellers for Light Aircraft*
- *Turboprop Propellers*
- *Propeller Ice-control Systems*
- *Propeller Inspection and Maintenance*

The propeller has passed through many stages of development. Great increases in power output have resulted in the development of four- and six-bladed propellers of large diameter. However, there is a limit to the r.p.m. at which these large propellers can be turned. The centrifugal force at high r.p.m. tends to pull the blades out of the hub, and excessive blade-tip speed results in poor blade efficiency and in fluttering and vibration.

As an outgrowth of the problems of large propellers, a variable-pitch, constant-speed propeller system was developed. This system makes it necessary to vary the engine r.p.m. only slightly during flight and, therefore, increases efficiency. Roughly, such a system consists of a flyweight-equipped governor unit, which controls the pitch angle of the blades so that the engine speed remains constant. The governor can be regulated by controls in the cockpit so that any desired blade angle setting and engine operating speed can be obtained. A low-pitch, high-r.p.m. setting can be utilized for takeoff; after the aircraft is airborne, a higher-pitch and lower-r.p.m. setting can be used for cruise.

**_Left._ As far as we know the Wright Brothers were the first to actually apply aerodynamics to propeller design. Based on Bernoulli's principal, the Coanda effect, and plain old-fashioned air deflection, modern propellers are remarkably efficient. For fast acceleration, short takeoff and landings, and rapid climb performance propellers beat jets horsepower for horsepower.**

## Section 1

## Basic Propeller Principles

The aircraft propeller consists of two or more blades and a central hub to which the blades are attached. Each blade of an aircraft propeller is essentially a rotating wing. As a result of their construction, the propeller blades produce forces that create thrust to push the air

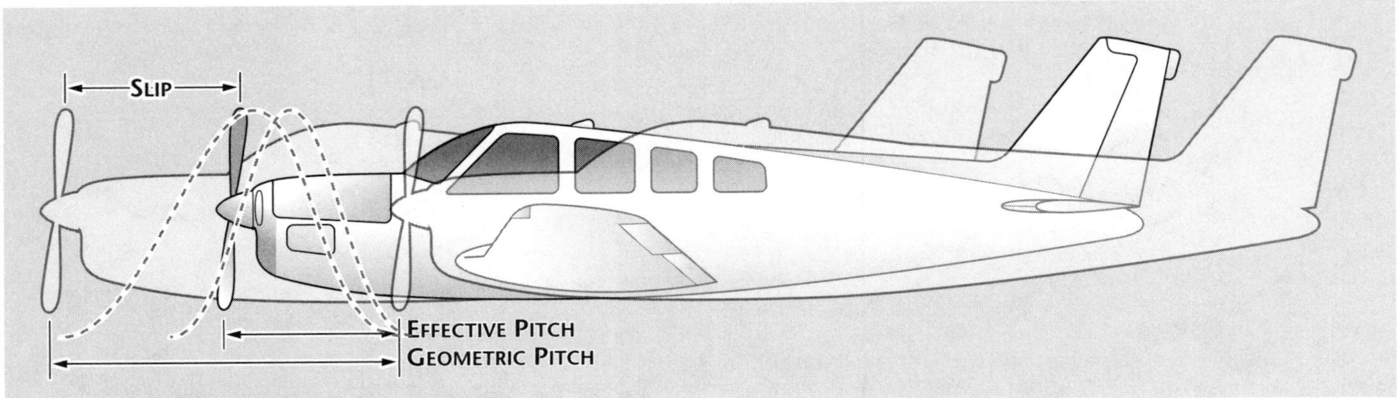


**Figure 3-1-1. Effective and geometric pitch**

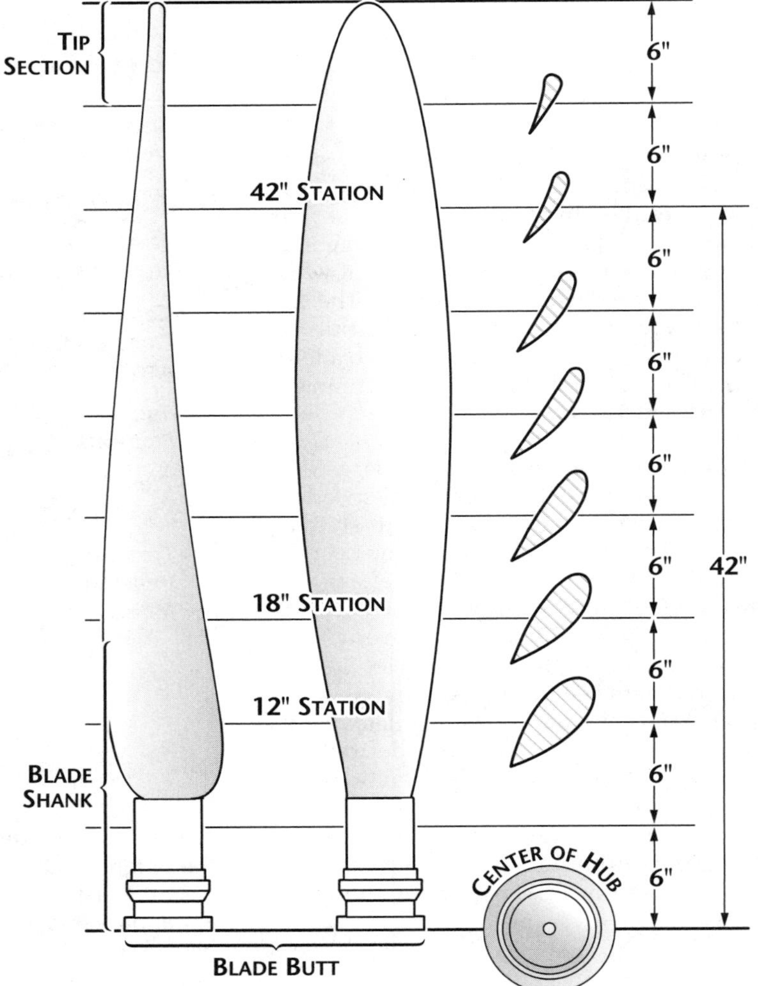


**Figure 3-1-2. Typical propeller blade elements**

rearward and pull the airplane through the air by its crankshaft.

The engine furnishes the power needed to rotate the propeller blades. The propeller is mounted on a shaft, which may be an extension of the crankshaft on low-horsepower engines. On high-horsepower engines, the propeller is mounted on a propeller shaft that is geared to the engine crankshaft. In either case, the engine rotates the airfoils of the blades through the air at high speeds and transforms the rotary power of the engine into thrust.

## Aerodynamic Factors

An airplane moving through the air creates a drag force opposing its forward motion. If an airplane is to fly on a level path, there must be a forward-acting force applied to it that is equal to the drag. This force is called *thrust*. The work done by the thrust is equal to the thrust times the distance it moves the airplane:

Work = Thrust x Distance

The power expended by the thrust is equal to the thrust times the velocity at which it moves the airplane:

Power = Thrust x Velocity

If the power is measured in horsepower units, the power expended by the thrust is termed *thrust horsepower* (t hp).

The engine supplies *brake horsepower* (b hp) through a rotating shaft, and the propeller converts it into thrust horsepower. In this conversion, some power is wasted. For maximum efficiency, the propeller must be designed to keep the amount of this waste as small as possible. Since the efficiency of any machine is the ratio of the useful power output to the power input, propeller efficiency is the ratio of thrust horsepower to brake horsepower. The usual symbol for propeller efficiency is the Greek letter $\eta$ (eta). Propeller efficiency varies from 50 to 87 percent, depending on how much the propeller *slips*.

Propeller slip is the difference between the *geometric pitch* of the propeller and its *effective pitch* (see Figure 3-1-1). Geometric pitch is the distance a propeller should advance in one rev-

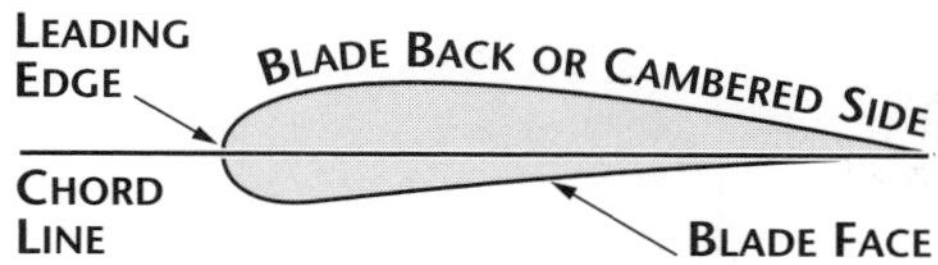


Figure 3-1-3. Cross-section of a propeller blade

olution; effective pitch is the distance it actually advances. Thus, geometric (or theoretical) pitch is based on no slippage; effective (or actual) pitch recognizes propeller slippage in the air.

## Nomenclature

The typical propeller blade can be described as a twisted airfoil of irregular planform. Two views of a propeller blade are shown in Figure 3-1-2. For purposes of analysis, a blade can be divided into segments that are located by station numbers in inches from the center of the blade hub. The cross sections of each 6-inch blade segment are shown as airfoils on the right side of Figure 3-1-2. Also identified in this figure are the *blade shank* and *blade butt.*

The blade shank is the thick, rounded portion of the propeller blade near the hub. It is designed to give strength to the blade. The blade butt, also called the *blade base* or *blade root,* is the end of the blade that fits in the propeller hub. The *blade tip* is that part of the propeller blade that is farthest from the hub. This generally applies to the last 6 inches of the blade.

A cross section of a typical propeller blade is shown in Figure 3-1-3. This section, or blade element, is an airfoil comparable to the cross section of an aircraft wing. The *blade back* is the cambered or curved side of the blade and is similar to the upper surface of an aircraft wing. The *blade face* is the flat side of the propeller blade. The *chord line* is an imaginary line drawn through the blade from the leading edge to the trailing edge. The *leading edge* is the thick edge of the blade that meets the air as the propeller rotates.

*Blade angle,* usually measured in degrees, is the angle between the chord of the blade and the plane of rotation (Figure 3-1-4). The chord of a propeller blade is determined in about the same manner as the chord of a wing. Because most propellers have a flat blade face, the chord line is often drawn along the face of the propeller blade.

*Pitch* is not the same as blade angle but, because pitch is largely determined by blade angle, the two terms are often used interchangeably. An increase or decrease in one is usually associated with an increase or decrease in the other.

## Forces Acting On a Propeller

As a propeller rotates, there are many forces that interact together to cause stresses on the propeller (see Figure 3-1-5, next page).

**Centrifugal force.** Centrifugal force is the force caused by the rotation of the propeller. Centrifugal force tries to pull the blades out of the hub. This force causes the greatest stress on the propeller. Centrifugal force may be as much as 7,500 times greater than the weight of the propeller.

**Thrust bending force.** The lift of the propeller is greatest at the tip, and the tip is the thinnest part of the blade. Thrust bending force bends the tip of the blade forward. From a straight side view of a propeller during takeoff, the bend can actually be seen.

**Torque bending force.** The torque bending force tries to bend the propeller blade in the opposite direction as the direction of rotation.

**Aerodynamic twisting moment.** Aerodynamic twisting moment (ATM) tries to twist the blade to the higher blade angle. Aerodynamic twisting moment is produced because the axis of rotation is at the chord line, but the center of lift is ahead of the chord line.

**Centrifugal twisting moment.** Centrifugal twisting moment (CTM) tries to decrease blade angle. This force opposes ATM. CTM is produced because all parts of the blade try to

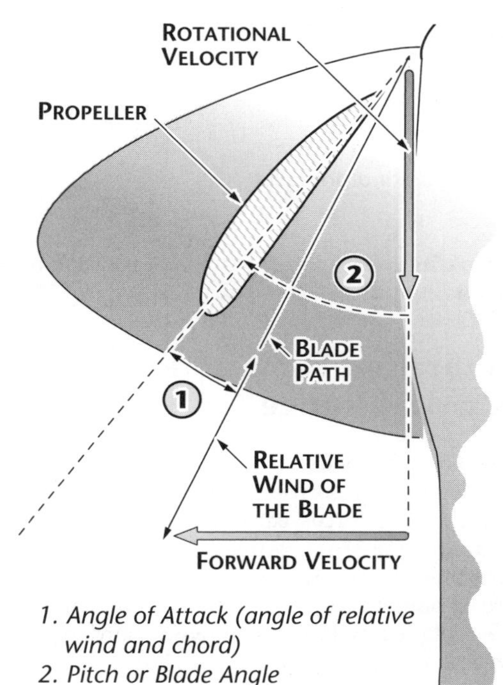


Figure 3-1-4. Propeller aerodynamic factors

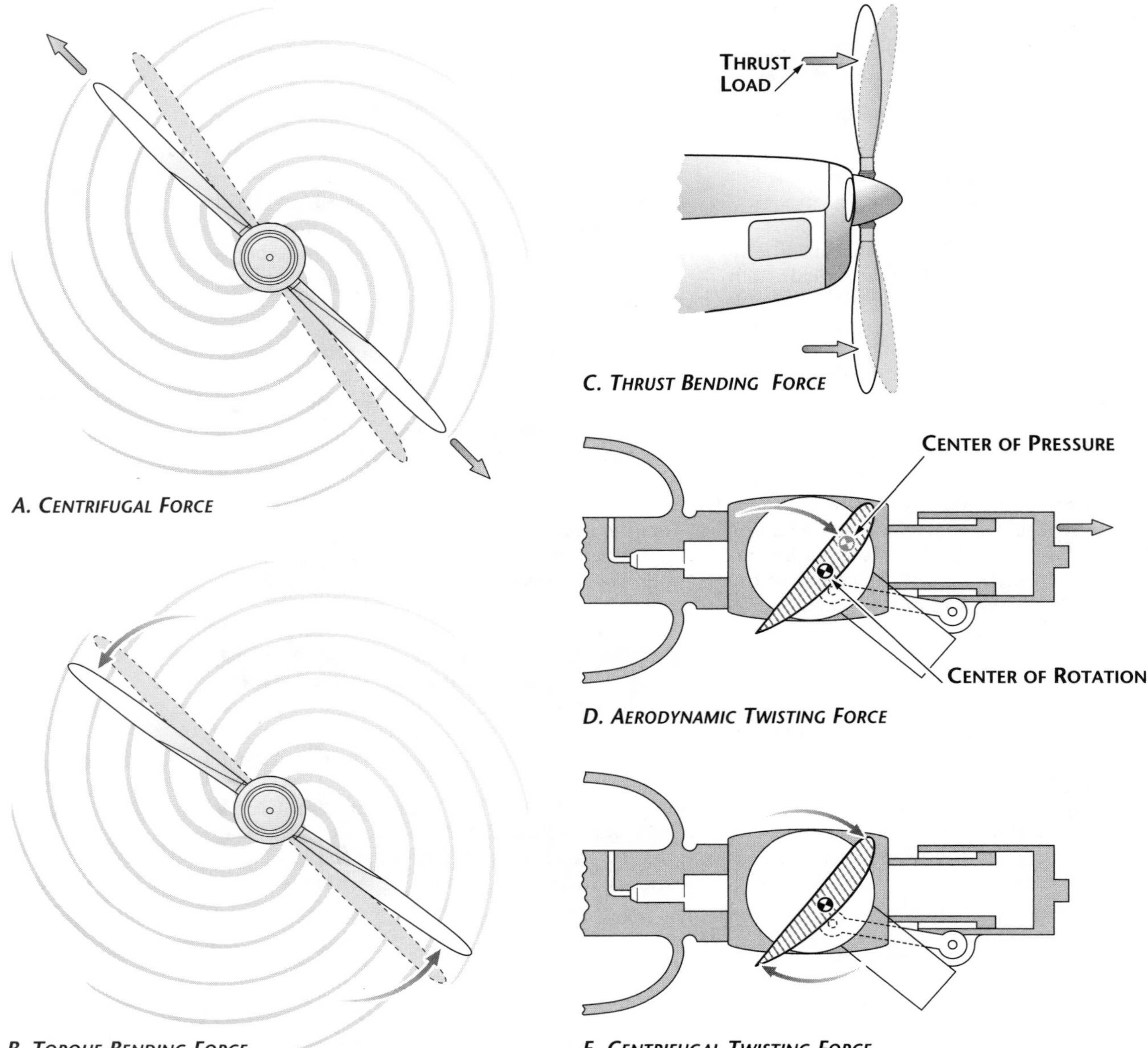


**Figure 3-1-5. Forces acting on a rotating propeller**

move in the exact same plane of rotation as the blade centerline.

## Vibration Force and Critical Range

When a propeller is connected to a running engine and produces thrust, aerodynamic and mechanical forces cause the blade to vibrate. The design of the propeller must compensate for these forces. If these forces were allowed to react on the propeller without being controlled, the blades will become work-hardened and break.

Aerodynamic forces are caused by the vibrations at the tip, where the effects of the transonic speeds cause buffeting and vibration. Most tip speeds during takeoff exceed the speed of sound. Therefore, controlling tip speed is of major importance to noise reduction.

Mechanical vibrations are caused by the power pulses of the engine.

Even though the propeller is designed to eliminate most of these forces, some aircraft-propeller-engine combinations have a *critical operating range*. This is an r.p.m. range in which vibrational forces are too great to allow for constant operation. At certain combinations of airspeed and engine r.p.m., these vibrations may create *harmonic stresses* that could lead to metal fatigue and eventually to propeller failure. This range is indicated on the engine tachometer by a red arc and should be operated in only long enough to pass through that r.p.m. range to a higher or lower r.p.m. setting. The Type Certificate Data Sheet for the engine/propeller combination will identify any critical r.p.m. ranges that are to be avoided.

## Propeller Theory

To understand the action of a propeller, consider first its motion, which is both rotational and forward. Thus, as shown by the vectors of propeller forces in Figure 3-1-6, a section of a propeller blade moves downward and forward. As far as the forces are concerned, the result is the same as if the blades were stationary and the air were coming at it from a direction opposite its path. The angle at which this air, or *relative wind,* strikes the propeller blade is called *angle of attack.* The air deflection produced by this angle causes the dynamic pressure at the engine side of the propeller blade to be greater than atmospheric, thus creating thrust.

As the aircraft speed increases, the angle at which the relative wind strikes the propeller blade changes. This, in turn, changes the angle of attack.

The shape of the blade also creates thrust, because it is like the shape of a wing. Consequently, as the air flows past the propeller, the pressure on one side is less than that on the other. As in a wing, this produces a reaction force in the direction of the lesser pressure. In the case of a wing, the area over the wing has less pressure, and the force (lift) is upward. In the case of the propeller, which is mounted in a vertical instead of a horizontal position, the area of decreased pressure is in front of the propeller, and the force (thrust) is in a forward direction. Aerodynamically, thrust is the result of the propeller shape and the angle of attack of the blade, and of the downwash produced.

Another way to consider thrust is in terms of the mass of air handled. In these terms, thrust is equal to the mass of air handled times the slipstream velocity minus the velocity of the airplane.

Thrust = Air Mass Handled x
(Slipstream Velocity – Airplane Velocity)

Thus, the power expended in producing thrust depends on the mass of air moved per second. On the average, thrust constitutes approximately 80 percent of the torque (total horsepower absorbed by the propeller). The other 20 percent is lost in friction and slippage. For any speed of rotation, the horsepower absorbed by the propeller balances the horsepower delivered by the engine. For any single revolution of the propeller, the amount of air handled depends on the blade angle, which determines how big a bite of air the propeller takes. Thus, the blade angle is an excellent means of adjust-

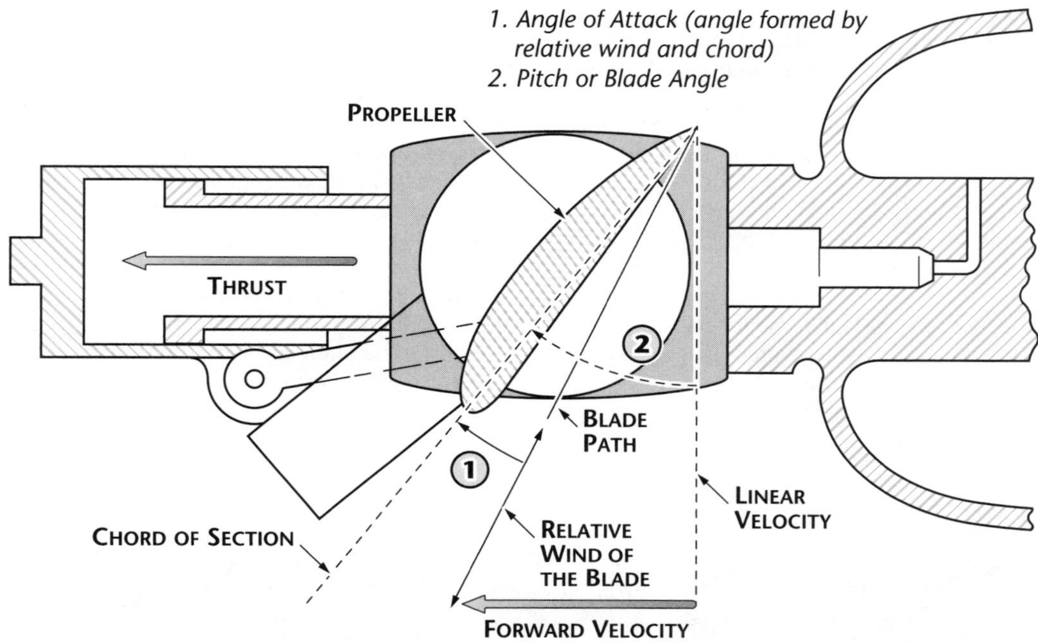


Figure 3-1-6. Propeller forces

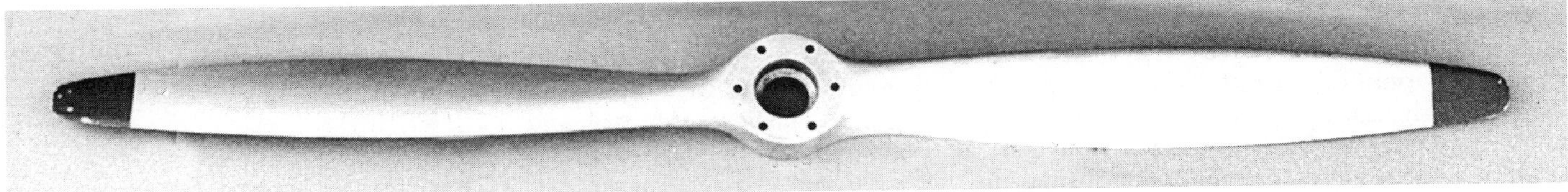

Figure 3-2-1. A fixed-pitch propeller

ing the load on the propeller to control the engine r.p.m.

Blade angle is also an excellent method of adjusting the angle of attack of the propeller. The blade angle must be adjusted to provide the most efficient angle of attack for any given engine and airplane speed. Lift versus drag curves, which are drawn for propellers as well as wings, indicate that the most efficient angle of attack is a small one, varying from 2° to 4° positive. The actual blade angle necessary to maintain this small angle of attack varies with the forward speed of the airplane.

## Section 2

# Types of Propellers

There are many types of propellers: the simplest are the fixed-pitch and ground-adjustable propellers. More complex systems include controllable-pitch, constant-speed, feathering and reversing-type propellers.

Fixed-pitch and ground-adjustable propellers are designed for best efficiency at one rotation and forward speed. In other words, they are designed to fit a given airplane-and-engine combination. A propeller may be used that provides the maximum propeller efficiency for takeoff, climb, cruising, or high speeds. Any change in these conditions lowers the efficiency of both the propeller and the engine. Controllable and automatic propellers allow for blade angle changes in flight.

**Fixed-pitch propeller.** The fixed-pitch propeller is normally constructed from a single piece of aluminum or wood (see Figure 3-2-1). The fixed-pitch propeller has a constant pitch, or blade angle, built into it and cannot be changed. This propeller is designed to operate efficiently at one power setting and one airspeed. With any change in aircraft speed or power setting, efficiency of the propeller is severely reduced.

The fixed-pitch propeller is normally found on small aircraft with low horsepower engines and relatively low flight speeds and altitudes.

**Ground-adjustable propeller.** The ground-adjustable propeller (Figure 3-2-2) operates like a fixed-pitch propeller, because its pitch cannot be changed in-flight. However, the pitch of a ground-adjustable propeller can be changed on the ground. This is accomplished by loosening the blade-mounting clamps and rotating the blades to the desired position and retightening the mounting clamp.

**Automatic propeller.** While not common, there are propellers dating back to the 1930s that provide the benefits of a variable-pitch or controllable-pitch propeller, automatically. These propellers need no governor, cockpit control or hollow crankshaft. The natural, physical forces acting on the blades and counterweights are utilized to accomplish the desired pitch change.

The *Aeromatic®* propeller is a two-bladed, variable-pitch unit that is entirely self-contained. It incorporates a single-piece hub of chrome-nickel-molybdenum steel and retains the blade flanges on large ball thrust bearings. A synchronizer gear between the blade flanges coordinates their movements. Adjustments for pitch range, balance and lubrication are all accessible externally.

**Controllable-pitch propeller.** A controllable-pitch propeller system permits the pilot to change the blade angle while the propeller is rotating. This allows the propeller to assume a blade angle that will give the best performance for particular flight conditions. The number

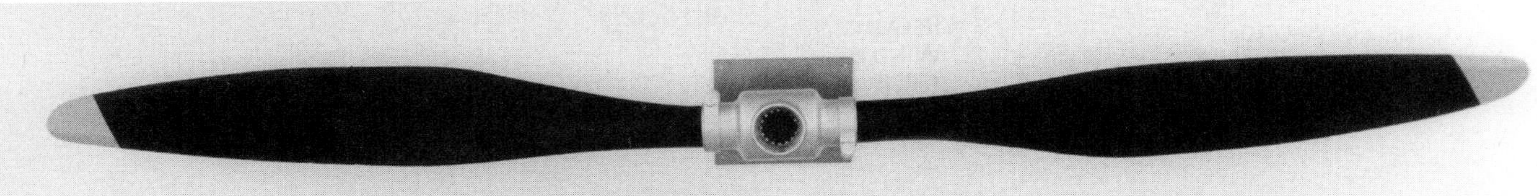

Figure 3-2-2. A ground-adjustable propeller

Figure 3-2-3. A Hamilton Standard two-position propeller

of pitch positions may be limited — as with a two-position, controllable propeller — or the pitch may be adjusted to any angle between the minimum and maximum pitch settings of a given propeller.

Two-position propellers are generally controlled using a *three-way propeller valve*. This selector valve directs oil from the engine lubrication system to the propeller to control the pitch angle (Figure 3-2-3).

The use of controllable-pitch propellers also makes it possible to attain the desired engine r.p.m. for a particular flight condition. As an airfoil is moved through the air, it produces two forces: *lift* and *drag*. Increasing propeller blade angle increases the angle of attack and produces more lift and drag. It also increases the horsepower required to turn the propeller at a given r.p.m. Since the engine is still producing the same horsepower, the propeller slows down. If the blade angle is decreased, the propeller speeds up. Thus, the engine r.p.m. can be controlled by increasing or decreasing the blade angle.

**Constant-speed propeller.** A constant-speed propeller, however, keeps the blade angle adjusted for maximum efficiency for most conditions encountered in flight. During takeoff, when maximum power and thrust are required, the constant-speed propeller is at a low blade angle or pitch. The low blade angle keeps the angle-of-attack small and efficient with respect to the relative wind. At the same time, it allows the propeller to handle a smaller mass of air per revolution.

This light load allows the engine to turn at high r.p.m. and convert the maximum amount of fuel into heat energy in a given time. The high r.p.m. also creates maximum thrust. Although the mass of air handled per revolution is small, the number of revolutions per minute is great, the slipstream velocity is high and, with the low airplane speed, the thrust is maximum.

After the initial takeoff, the engine r.p.m. starts to increase. When this happens, the constant-speed propeller changes to a higher blade angle. This keeps the angle of attack small and efficient. This higher blade angle increases the amount of air handled per revolution of the propeller, thereby decreasing the r.p.m. of the engine, which reduces engine wear and fuel consumption.

For climb after takeoff, the power output of the engine is reduced to *climb power* by decreasing the manifold pressure and increasing the blade angle to lower the r.p.m. Thus, the torque (horsepower absorbed by the propeller) is reduced to match the reduced power of the engine. The angle of attack is again kept small by the increase in blade angle. The greater mass of air handled per second in this case is more than offset by the lower slipstream velocity and the increase in airspeed.

At cruising altitude, when the airplane is in level flight and less power is required than is

**Figure 3-2-4. A constant-speed propeller with the propeller governor visible on the engine case**

used in takeoff or climb, engine power is again reduced by lowering the manifold pressure and increasing the blade angle to decrease the r.p.m. Again, this reduces torque to match the reduced engine power. Although the mass of air handled per revolution is greater, it is more than offset by a decrease in slipstream velocity and an increase in airspeed. The angle of attack is still small because the blade angle has been increased with an increase in airspeed.

The use of a *propeller governor* to increase or decrease propeller pitch is common practice (Figure 3-2-4). When the airplane goes into a climb, the blade angle of the propeller decreases just enough to prevent the engine speed from decreasing. Therefore, the engine can maintain its power output, provided the throttle setting is not changed. When the airplane goes into a dive, the blade angle increases sufficiently to prevent overspeeding. With the same throttle setting, the power output remains unchanged. If the throttle setting is changed instead of changing the speed of the airplane by climbing or diving, the blade angle will increase or decrease, as required, to maintain a constant engine r.p.m. The power output (and not r.p.m.) will therefore change in accordance with changes in the throttle setting. The governor-controlled, constant-speed propeller changes the blade angle automatically, keeping engine r.p.m. constant.

Most propeller pitch-changing mechanisms are operated by oil pressure (hydraulically), and use some type of piston-and-cylinder arrangement. The piston may move in the cylinder, or the cylinder may move over a stationary piston. The linear motion of the piston is converted by several different types of mechanical linkage into the rotary motion necessary to change the blade angle. The mechanical connection may be through gears, the pitch-changing mechanism turning a drive gear or power gear that meshes with a gear attached to the butt of each blade.

In most cases, the oil pressure for operating these various types of hydraulic pitch-changing mechanisms comes directly from the engine-lubricating system. When the engine-lubricating system is used, the engine oil pressure is usually boosted by a pump that is integral with the governor to operate the propeller. The higher oil pressure provides a quicker blade-angle change.

The governors used to control the hydraulic propeller pitch-changing mechanisms are geared to the engine crankshaft and, thus, are sensitive to changes in r.p.m. The governors direct the pressurized oil for operation of the propeller hydraulic pitch-changing mechanisms. When r.p.m. increases above the value for which a governor is set, the governor causes the propeller pitch-changing mechanism to turn the blades to a higher angle. This angle increases the load on the engine, and r.p.m. decreases. When r.p.m. decreases below the value for which a governor is set, the governor causes the pitch-changing mechanism to turn the blades to a lower angle, the load on the engine is decreased, and r.p.m. increases. Thus, a propeller governor keeps engine r.p.m. constant.

**Feathering propellers.** A feathering propeller is a controllable propeller that has the ability to change the pitch to an angle so that the blade angle is 90° to the apparent wind and will have no windmilling effect on an engine that has been shut down in flight. Feathering propellers must be used on multi-engine aircraft to reduce propeller drag to a minimum under engine-failure conditions.

**Reverse-pitch propellers.** A reverse-pitch propeller is a controllable propeller in which the blade angles can be changed to a negative value during operation. The purpose of the reversible-pitch feature is to produce a high negative thrust at low speed by using engine power. Reverse pitch is used principally as an aerodynamic brake to reduce ground roll after landing.

**Turboprop propellers.** The turboprop propeller systems get much more complex. They are operated by a gas turbine engine through a reduction-gear assembly. The turboprop fuel control and the propeller governor are connected and operate in coordination with each other. The power lever directs a signal from the cockpit to the fuel control for a specific amount of power from the engine. The fuel control and the propeller governor together establish the correct combination of r.p.m., fuel flow and propeller blade angle to create sufficient propeller thrust to provide the desired power.

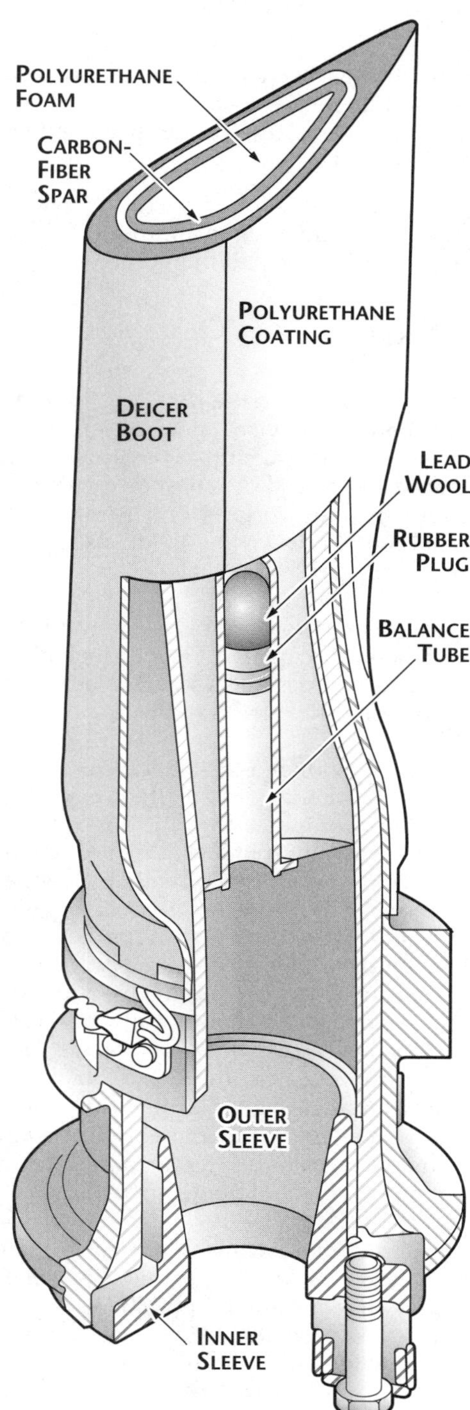


Figure 3-2-5. A cutaway illustration of a Dowty Rotol® composite propeller blade

## Blade Developments

Not all propellers are wood or aluminum. During the 1940s, steel propeller blades were developed that were extremely satisfactory for the times. Some of these blade styles are still in use on older piston and large turboprop airplanes.

**Steel blades.** Steel blades were a wartime development and were not easily repairable. They were, for their time, almost as valuable for the aluminum they saved as for the job they did.

**Micarta® blades.** Several civilian propellers were developed in the 1950s that use Micarta® blades. Micarta® is a linen/paper/resin-laminated material similar to what control pulleys, cable guides, wear plates and electrical parts are manufactured from. These blades are still around today, but used only on special airplanes that are mostly for display. Most Aeromatic® propellers used Micarta® blades.

**Composite blades.** Advanced composite materials have been used to manufacture some very excellent new generation propeller blades. Though composite blades first showed up as helicopter rotors, the migration to regular propeller blades has been rapid.

Figure 3-2-5 shows a drawing of a Dowty Rotol® composite blade in current use on medium commuter-type airplanes. Note the built-in lightning protection. Without the lightning protection, a strike could cause a considerable amount of damage. It also makes dissipation of static buildup possible.

Hartzell Propeller Inc. is producing an excellent series of composite-bladed propellers for the turboprop market. Built with foam cores and Kevlar® coverings (Figure 3-2-6), the composite propellers have been used on commuter airplanes for several years, where they regularly operate for 3,000-5,000 hours between overhauls. This is almost double the life of standard aluminum propeller blades.

There are many benefits from using composite blades that are not immediately apparent:

- More accurate manufacturing tolerances, allowing individual blades to be replaced, instead of complete sets
- Lighter weight as an installed unit, reducing the centrifugal load on all parts of the airframe and greatly lengthening component life and reducing stresses
- Flexibility even exceeding a wooden pro-

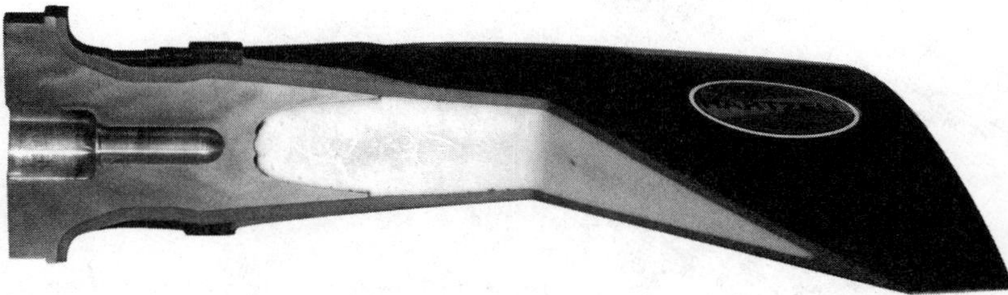

Figure 3-2-6. A portion of a Hartzell composite propeller blade shows the blade shank, Kevlar® outer covering and foam core.

Figure 3-2-7. A Fairchild Merlin with rounded tips

Figure 3-2-8. Square-tip propellers run more smoothly and quietly than their rounded counterparts.

peller eliminates fatigue-cycle constraints and gives a blade a nearly unlimited life

- Repairs have been developed that will allow material to be put back on the blade, instead of the blade constantly being filed and ground down
- With the elimination of fatigue, better airfoils and tip designs improve efficiency and reduce noise pollution problems

## Tip Design

When you think about the part that tip design plays in a propeller, it isn't much different from a wing-tip design, except that the propeller tip approaches the speed of sound on every take-off. The same tip vortices and aerodynamic drag that complicate a wing design also exist in a prop-tip choice.

For many years, only two tip designs were common: rounded and squared. Only during the last decade or so have designers seriously attempted to improve on the old tried-and-true styles.

**Round tips.** Rounded-tip blades (Figure 3-2-7) go back to the days of all-wood propellers. Rounded tips, provided they have proper capping, will generally split less than a square tip. The degree of roundness depends a great deal on the width of the propeller blade. A very wide blade would have a tip that approaches a half circle, with the diameter the same as the blade width. Narrower blades would have a rounded tip that seemed to be about a third of a circle faired into the leading and trailing edge. None offer any improvement in airflow off the tip. Many early radial-engine airplanes had long, narrow propellers with semi-elliptical round tips. This gave them a large area or rotation, but no improvement in tip disturbances.

**Square tips.** Squared tips (Figure 3-2-8) have an advantage, inasmuch as they have more blade area than a rounded tip. This is a major advantage on propeller blades that are very large and wide. A 5- to 10-percent increase in blade area means that the same blade can run with less pitch, or can be made somewhat shorter and run with a lower r.p.m. Lowering the r.p.m. helps lower tip speed, which causes the propeller to run more quietly and with less disturbance. Sometimes, a small difference in speed can make a large difference in noise. Square tips also will not be damaged as much should the prop pick up a piece of gravel and get a nick.

**Scimitar tips.** A fairly recent development in propeller tips is the scimitar tip (Figure 3-2-9), found on some McCauley propellers. These props have the tips swept back, from leading edge to trailing edge, at a very steep angle.

Figure 3-2-9. A recent innovation, scimitar tips feature a steep sweepback from leading to trailing edge.

They are designed very much like a swept wing, with the sweep occurring only at the tip. The idea is to decrease the tip disturbances, and induced drag, by delaying separation until the airflow has actually reached the point of the trailing edge. A swept tip should also allow for higher tip speed without approaching the speed of sound. If you recall your aerodynamics, swept wings delay the onset of separation.

**Q-tip® propellers.** Q-tip® propeller blades, manufactured by Hartzell, are formed by bending the tip section of the blade 90° toward the face side (Figure 3-2-10). Aerodynamic improvements include a reduced diameter and decreased tip speeds. This results in quieter operation and reduced tip vortices. The 90° bend reduces the vortices that, on traditional blades, pick up debris that can contact the blades and cause nicks, gouges and scratches.

In essence, it works like a winglet.

## Classification of Propellers

**Tractor propeller.** Tractor propellers are those mounted on the front of the drive shaft of a conventional airplane (Figure 3-2-11, next page). Most airplanes are tractor types. A major advantage of the tractor propeller is that lower stresses are induced in the propeller as it rotates in relatively undisturbed air.

Figure 3-2-10. A Hartzell Q-tip® propeller

**Figure 3-2-11. A tractor-type propeller installation is a conventional airplane with the propeller pulling it.**

**Pusher propellers.** Pusher propellers are those mounted on the downstream end of a drive shaft, behind the supporting structure. Pusher propellers are constructed as fixed- or variable-pitch propellers. Seaplanes and amphibious aircraft use a greater percentage of pusher propellers than other kinds of aircraft (Figure 3-2-12).

On land planes, where propeller-to-ground clearance usually is less than the propeller-to-water clearance of watercraft, pusher propellers are subject to more damage than tractor propellers. Rocks, gravel and small objects, dislodged by the wheels, quite often are thrown or drawn into a pusher propeller. Similarly, planes with pusher propellers are likely to encounter propeller damage from water spray thrown up by the hull during landing or takeoff from water. Consequently, the pusher propeller is often mounted above and behind the wings to prevent such damage.

**Figure 3-2-12. A Lake Amphibian is a classic pusher-propelled airplane.**

## *Section 3*

# *Fixed-pitch Propellers*

A significant segment of general aviation aircraft is still equipped with fixed-pitch propellers. There are two types: wooden and metal.

**Wooden propellers.** The construction of a fixed-pitch wooden propeller (Figure 3-3-1) is such that its blade pitch cannot be changed after manufacture. The choice of the blade angle is decided by the normal use of the propeller on an aircraft during level flight, when the engine will perform at maximum efficiency.

The impossibility of changing the blade pitch on the fixed-pitch propeller restricts its use to small aircraft with low-horsepower engines, in which maximum engine efficiency during all flight conditions is of lesser importance than in larger aircraft. The wooden fixed-pitch propeller, because of its low weight, rigidity, economy of production, simplicity of construction and ease of replacement, is well-suited for such small aircraft.

Before dismissing wooden propellers as a thing of the past, we need to look at how many are still being manufactured today. The Wright Flyer used wooden propellers, and a century later, Sensenich Propeller Manufacturing Company still makes about 4,000 wooden

Figure 3-3-1. Fixed-pitch wooden propeller assembly

propellers each year for certificated aircraft (Figure 3-3-2). Additionally, they and other manufacturers carve propellers for non-certified aircraft and other applications.

Figure 3-3-2. A Sensenich propeller on a Piper J3 Cub

Constructing propellers from wood helps avoid many of the vibration problems associated with metal propellers. Wood props dampen engine-induced vibrations by several magnitudes better than metal. Additionally, wood is not subject to the cumulative stresses from vibration and flexing that may cause failure in a metal propeller. As always, proper maintenance of the propeller is a major factor in its long-term serviceability. The advantages exhibited by wooden propellers have been the starting point for the design of composite propellers being manufactured today.

Wooden propellers are constructed from many thin sheets of hardwood, laminated together. Birch is the most common wood used in propeller construction, but woods such as oak, black walnut, cherry and mahogany are also used. The layers of wood are joined together with waterproof, resinous glue before the wood is shaped.

After the propeller has been roughed out, it is allowed to cure for about one week. This curing prevents warping and cracking of the finished product. After the propeller has cured, it is carefully shaped using templates and protractors to obtain the proper form.

After the propeller blades have been shaped, a covering of fabric is cemented to the outer portion of the blades, and metal tipping (Figure 3-3-3) is fastened to the leading edge and tips of the blades. This tipping is installed to provide protection to the blade from flying debris during ground operation.

Metal tipping may be made of terneplate, brass, Monel® or stainless steel. It is fastened to the blade by countersunk screws. The heads of the screws are fastened to the metal tipping by soldering to prevent them from coming out during operation. After soldering the screws, the solder must be filed smooth. The bottom of the tipping should have small holes drilled near the tip to allow moisture to escape.

To prevent the wood from shrinking, swelling and warping due to changing moisture content, the propeller must be sealed.

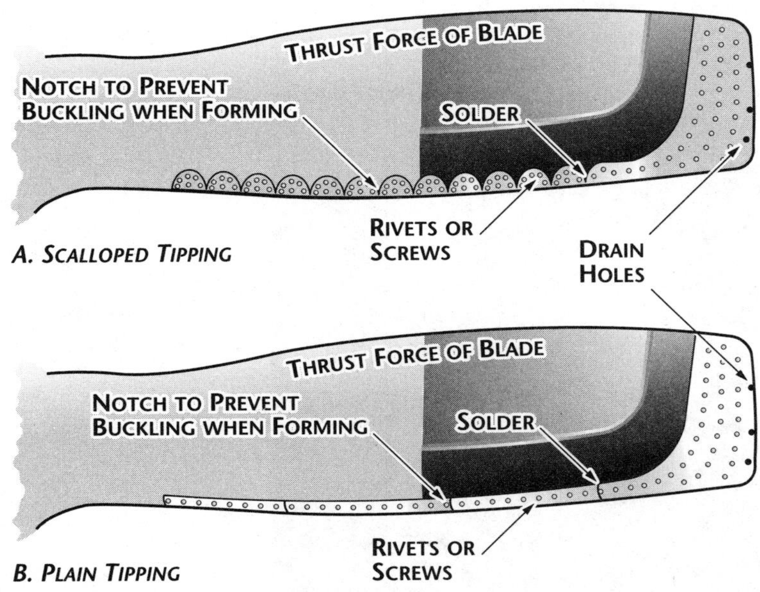


Figure 3-3-3. Installation of metal sheath and tipping

There are many types of sealers that are available today. The one most frequently used is a clear, waterproof varnish that is applied in multiple coats. After the propeller is varnished, it must be balanced before mounting on the aircraft.

Several types of hubs are used to mount wooden propellers on the engine crankshaft. The propeller may have a forged steel hub that fits a splined crankshaft; it may be connected to a tapered crankshaft by a tapered, forged steel hub; or it may be bolted to a steel flange forged on the crankshaft. In any case, several attaching parts are required to mount the propeller on the shaft properly.

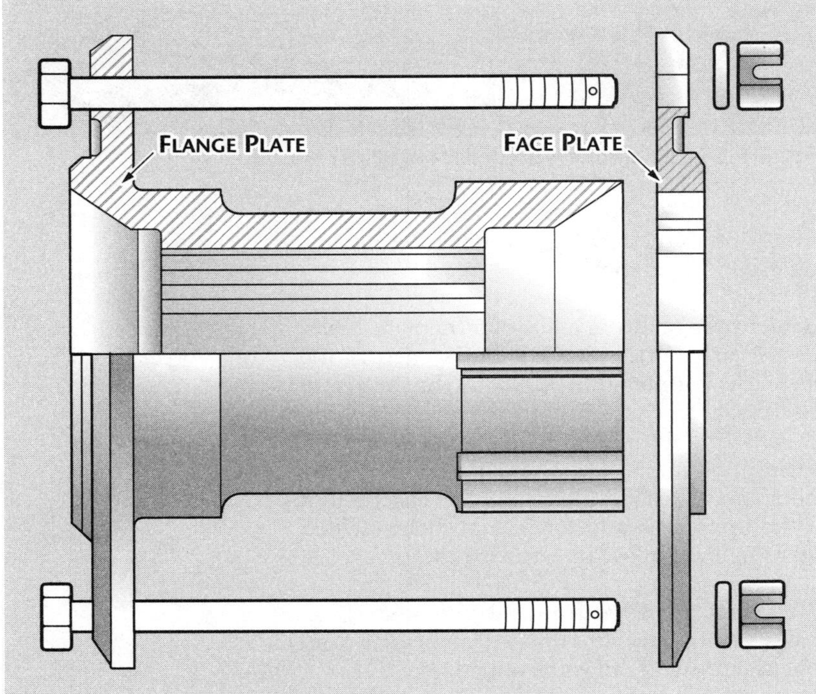


Figure 3-3-4. Typical hub assembly for a fixed-pitch propeller

Hubs fitting a tapered shaft usually are held in place by a *retaining nut* that screws onto the end of the shaft. On one model, a *lockring* is used to safety the retaining nut and to provide a puller for removing the propeller from the shaft. This nut screws into the hub and against the retaining nut. The lockring and the retaining nut are safetied together with lock wire or a *cotter pin*.

A front and rear cone may be used to seat the propeller properly on a splined shaft. The rear cone is a one-piece bronze cone that fits around the shaft and against the *thrust nut* (or spacer) and seats in the rear-cone seat of the hub. The front cone is a two-piece, split-type steel cone that has a groove around its inner circumference so that it can be fitted over a flange of the propeller's retaining nut. When the retaining nut is threaded into place, the front cone seats in the front-cone seat of the hub. A *snap ring* is fitted into a groove in the hub in front of the front cone, so that when the retaining nut is unscrewed from the propeller shaft, the front cone will act against the snap ring and pull the propeller from the shaft.

One type of hub assembly for the fixed-pitch wooden propeller is a steel fitting inserted in the propeller to mount it on the propeller shaft. It has two main parts, the *faceplate* and the *flange plate* (Figure 3-3-4). The faceplate is a steel disk that forms the forward face of the hub. The flange plate is a steel flange with an internal bore splined to receive the propeller shaft. The end of the flange plate opposite the flange disk is externally splined to receive the faceplate; the faceplate bore has splines to match these external splines. Both faceplates and flange plates have a corresponding series of holes drilled on the disk surface concentric with the hub center. The bore of the flange plate has a 15° cone seat on the rear end and a 30° cone seat on the forward end to center the hub accurately on the propeller shaft.

**Fixed-pitch metal propellers.** Metal fixed-pitch propellers are similar in general appearance to a wooden propeller, except that the sections are usually thinner. The metal fixed-pitch propeller is widely used on many models of light aircraft.

Many of the earliest metal propellers were manufactured in one piece of forged Duralumin®. Compared to wooden propellers, they were lighter in weight because of elimination of blade-clamping devices; they also offered a lower maintenance cost because they were made in one piece (Figure 3-3-5). They provided more efficient cooling because of

Figure 3-3-5. Fixed-pitch metal propellers are one-piece forgings of aluminum alloy

the effective pitch nearer the hub, and because there was no joint between the blades and the hub, the propeller pitch could be changed (within limits) by twisting the blade slightly.

Propellers of this type are now manufactured of one-piece anodized aluminum alloy. They are identified by stamping the propeller hub with the serial number, model number, FAA type certificate number, production certificate number and the number of times the propeller has been reconditioned. The complete model number of the propeller is a combination of the basic model number and suffix numbers to indicate the propeller diameter and pitch. An explanation of a complete model number, using the McCauley 1B90/CM propeller, is provided in Figure 3-3-6.

## Section 4

# *Constant-speed Propellers for Light Aircraft*

Hartzell and McCauley propellers for light aircraft are similar in operation. Information presented here is only to help illustrate the construction and operation of the various types of propellers. A basic understanding of the operation of these propeller systems will help the AMT diagnose and troubleshoot problems encountered with the propeller system. In all cases, the appropriate manufacturer's specifications and instructions must be consulted for information on specific models.

## Hartzell Constant-speed Propellers

Hartzell produces two styles of constant speed propellers, the *steel hub-type* and the *compact model*. The steel hub-type has its pitch-changing mechanism exposed, whereas the compact

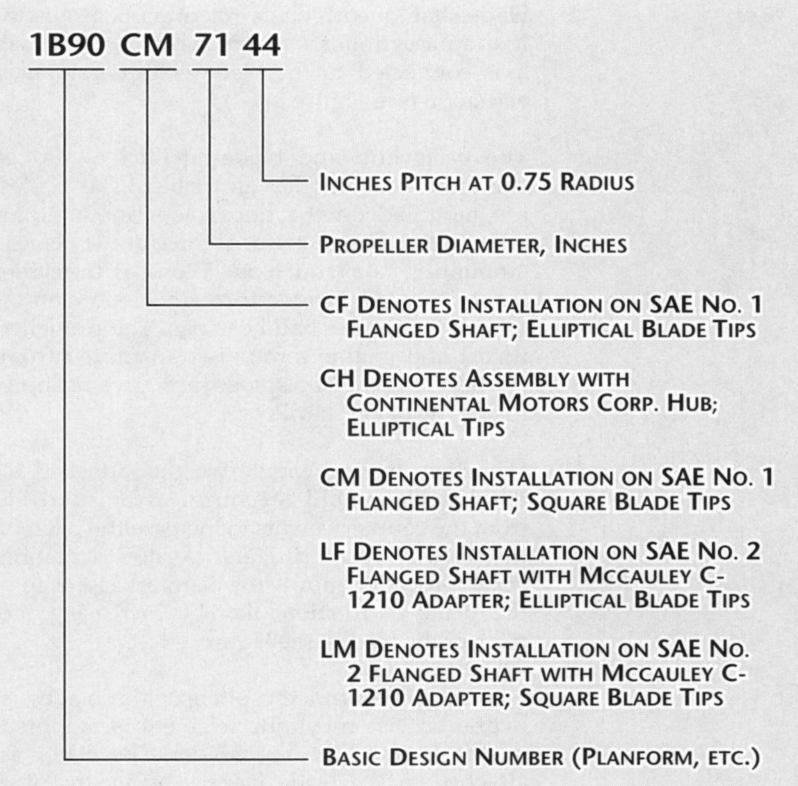


Figure 3-3-6. Complete propeller model number

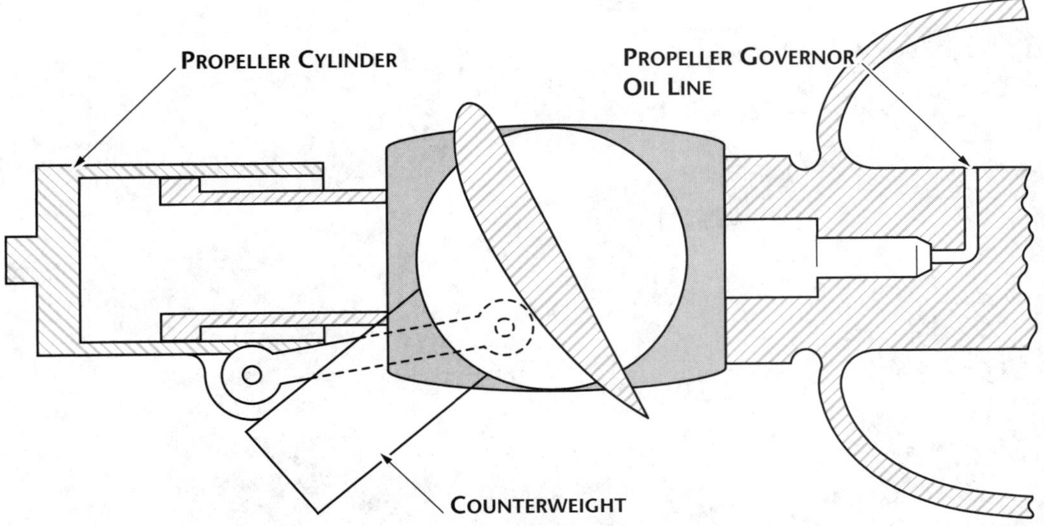


**Figure 3-4-1. Pitch-change mechanism for a counterweight propeller**

models have the pitch-changing mechanism inside the hub.

**Steel hub models.** All of the propellers in this series are similar in basic design and have many parts in common. Design variations generally relate to the horsepower of the engine for which the propeller was built. The steel hub consists of a central spider, which supports aluminum blades with a tube extending inside the blade roots. Blade clamps connect the blade shanks with blade-retention bearings. A hydraulic cylinder is mounted on the rotational axis connected to the blade clamps for pitch actuation (see Figure 3-4-1).

The basic hub and blade-retention setup is common to all models described. The blades are mounted on the hub spider for angular adjustment. The centrifugal force of the blades, amounting to as much as 25 tons, is transmitted to the hub spider through blade clamps and then through ball bearings. The propeller thrust and engine torque is transmitted from the blades to the hub spider through a bushing inside the blade shank.

Propellers having counterweights attached to the blade clamps utilize centrifugal force derived from the counterweights to increase the pitch of the blades. The centrifugal force, due to rotation of the propeller, moves the counterweights into the plane of rotation, thereby increasing the pitch of the blades (see Figure 3-4-2).

In order to control the pitch of the blades, a hydraulic piston-cylinder element is mounted on the front of the hub spider. The piston is attached to the blade clamps by means of a *sliding rod-and-fork system* for non-feathering models and a *link system* for the feathering models. The piston is actuated in the forward direction by means of oil pressure supplied by a governor, which overcomes the opposing force created by the counterweights.

## Constant-speed Models

When the engine is exactly at the r.p.m. set by the governor, the centrifugal reaction of the flyweights balances the force of the speeder spring, positioning the pilot valve so that oil is neither supplied to or drained from the propeller. With this condition, the propeller blade angle does not change. Note that the r.p.m. setting is made by varying the amount of compression in the *speeder spring.* The positioning of the *speeder rack* is the only action controlled manually. All others are controlled automatically within the governor.

If the engine speed drops below the r.p.m. for which the governor is set (see Figure 3-4-3), the rotational force on the engine-driven governor flyweights becomes less. This allows the speeder spring to move the pilot valve downward. With the pilot valve in the down position, oil from the gear-type pump flows through passages to the propeller and moves the cylinder outward. This, in turn, decreases the blade angle and permits the engine to return to the on-speed setting.

If the engine speed increases above the r.p.m. for which the governor is set, the flyweights move against the force of the speeder spring and raise the pilot valve. This permits oil in the propeller to drain out through the governor drive shaft. As the oil leaves the propeller, the centrifugal force acting on the counterweights

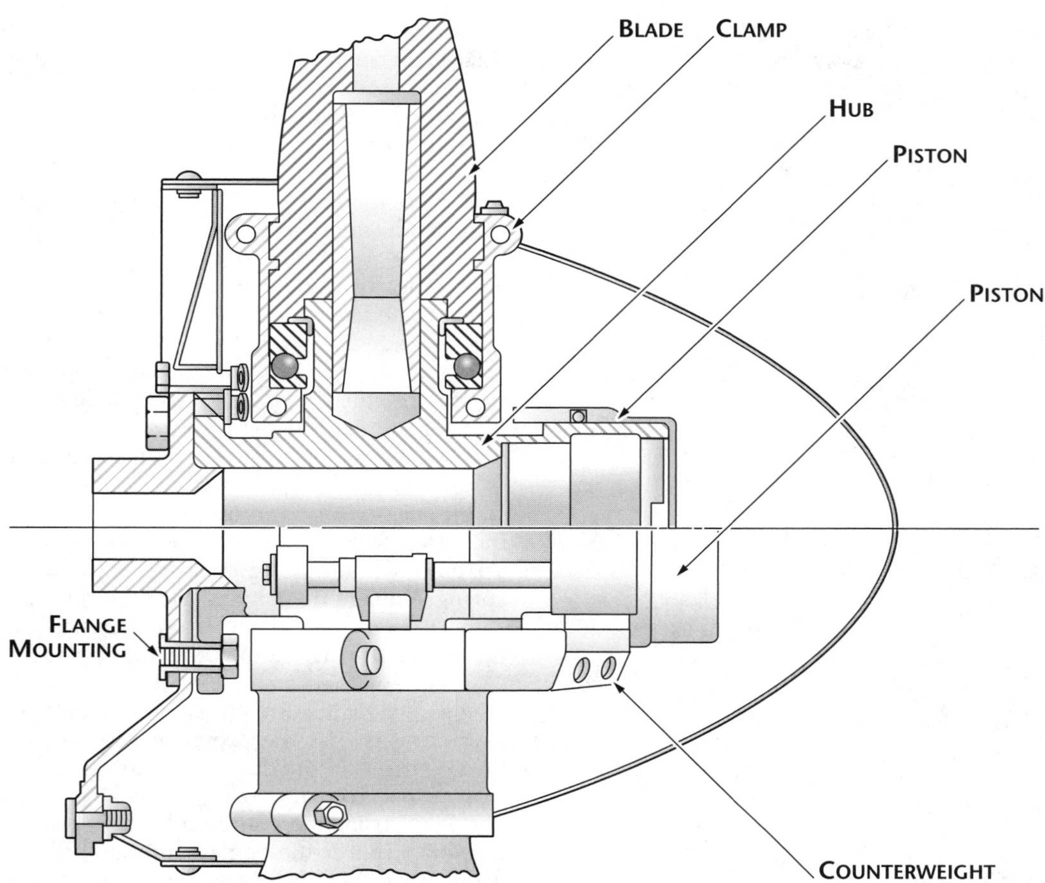


Figure 3-4-2. Constant-speed prop that uses counterweights

turns the blade to a higher angle, which decreases engine r.p.m.

Some Hartzell steel-hub propellers do not have counterweights. This reduces the overall weight of the propeller installation. On these systems, the pitch-changing forces are reversed. Blade centrifugal twisting force reduces pitch, and governor oil pressure acting on the cylinder increases pitch.

## Feathering Propellers

Feathering propellers operate similarly to the non-feathering ones, except that the *feathering spring* assists the counterweights to increase the pitch (see Figure 3-4-4, next page).

Feathering is accomplished by releasing the governor oil pressure, allowing the counter-weights and feathering spring to feather the blades. This is done by pulling the governor pitch-control back to the limit of its travel, which opens up a port in the governor and allows the oil from the propeller to drain back into the engine. The time necessary to feather depends upon the size of the oil passage from the propeller to the engine, and the force

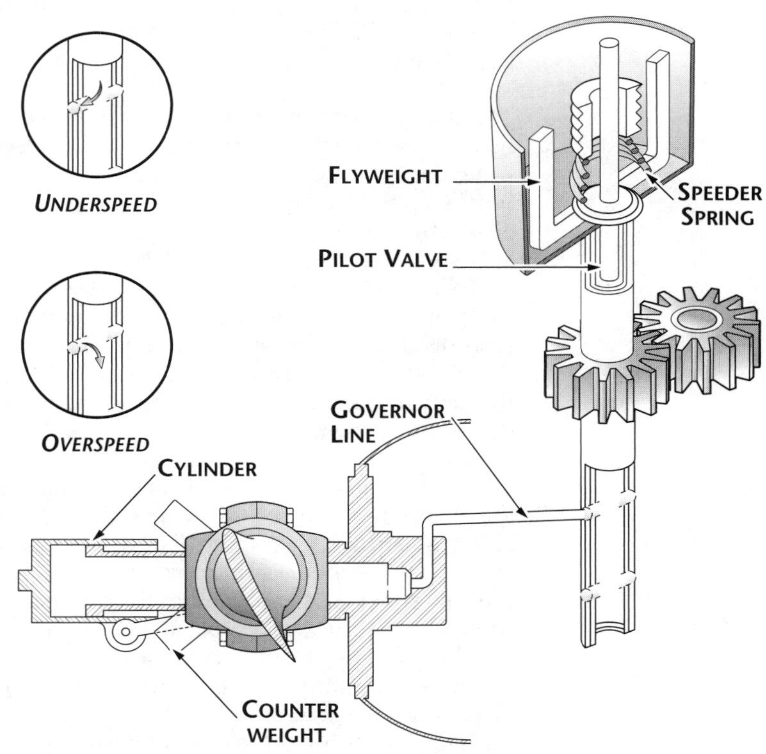


Figure 3-4-3. On-speed, basic operation

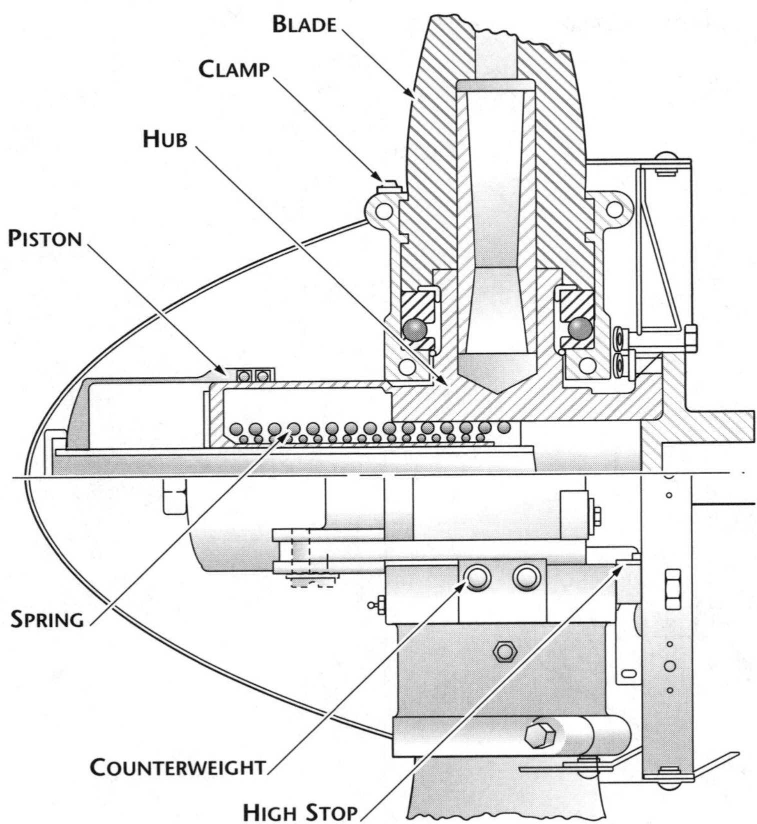


**Figure 3-4-4. Constant-speed feathering**

exerted by the spring and counterweights. The larger the passages through the governor and the heavier the spring, the quicker the feathering action. Elapsed time for feathering with this system is usually 3-10 seconds.

The ability to unfeather the blades, or re-establish normal pitch, within the same elapsed time is not considered important for the light, twin-engine airplane. The possibility of feathering the wrong propeller in an emergency is remote, as the wrong action will become apparent in ample time to be corrected. Furthermore, the requirement to restart the dead engine for landing does not exist, as the light twin can easily be landed with only one engine. About the only requirement for unfeathering is for demonstration purposes.

*Unfeathering* is accomplished by repositioning the governor control to the normal flight range and restarting the engine (Figure 3-4-5). As soon as the engine cranks over a few turns, the governor starts to unfeather the blades, and soon windmilling takes place, which speeds up the process of unfeathering. In order to facilitate cranking of the engine, feathering blade angle is set at 80°-85° at the 3/4 point on the blade, allowing the air to assist the engine's starter. In general, restarting and unfeathering can be accomplished within a few seconds.

Special unfeathering systems are available for certain aircraft for which restarting the engine is difficult, or for demonstrations. Such a system consists of an oil accumulator connected to the governor through a valve, as shown below in Figure 3-4-5.

In order to prevent the feathering spring from feathering the propeller when the plane is on the ground and the engine stopped, automatically removable *high-pitch stops* were incorporated into the design. These consist of spring-loaded latches fastened to the stationary hub that engage high-pitch stop plates which are bolted to the movable blade clamps. As long as the propeller is in rotation at speeds of more than 600 r.p.m., centrifugal force acts to disengage the latches from the high-pitch stop-plates so that the propeller pitch may be increased to the feathering position. At lower r.p.m., or when the engine is stopped, the latch springs engage the latches with the high-pitch stops, preventing the pitch from increasing further due to the action of the feathering spring.

One safety feature inherent in this method of feathering is that the propeller will feather if the governor oil pressure drops to zero for any reason. Because the governor obtains its supply of oil from the engine lubricating system, it follows that if the engine runs out of oil, or if oil pressure fails due to breakage of a part of the engine, the propeller will feather automatically. This action may save the engine from further damage in case the pilot is not aware of trouble.

## Reversible Propellers

*Reversible propellers,* used primarily for turboprop installations, are similar to the feathering propellers. The important exception is that the pitch is extended into the reverse range, and a hydraulic *low-pitch stop* is introduced. In order to obtain greater pitch travel, the piston and cylinder are made longer.

The hydraulic low-pitch stop, which prevents the propeller from entering the reverse range unless desired, consists of a *beta valve* that shuts off the governor oil to the propeller when the pitch reaches the low position. This valve is actuated by a linkage to the piston. In order to move the pitch into the reverse range, the pilot effectively changes the linkage to the low-pitch stop, which allows the pitch to travel into reverse.

The *beta-feedback linkage* for some propellers is external to the propeller hub (see Figure 3-4-6). Linear motion is transmitted from the rotating hub to the fixed engine by means of a *collar-carbon block assembly.* For other models, the beta-feedback linkage consists of a rod mounted

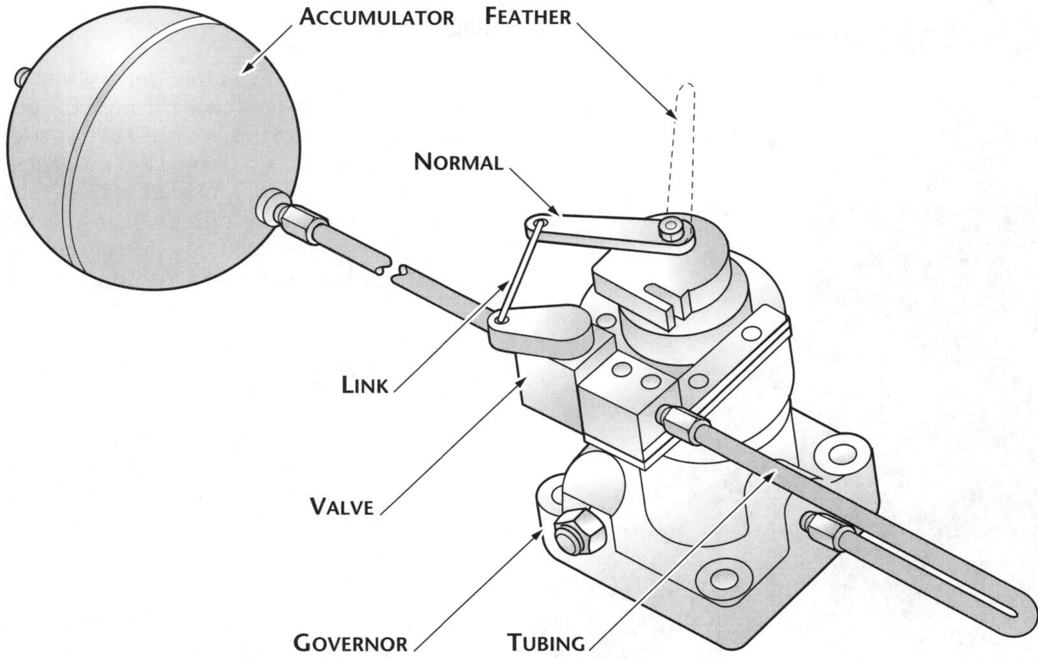


Figure 3-4-5. Unfeathering system

concentrically with the propeller shaft, actuating the beta valve at the rear.

## Compact-model Propellers

Compact-model propellers combine low weight, simplicity of design and rugged construction. In order to achieve these ends, the hub is made to be as compact as possible, utilizing aluminum alloy forgings for most of the parts. The hub shell is made in two halves, bolted together along the plane of rotation (see Figure 3-4-7, top of next page). This hub shell carries the pitch-change mechanism and blade roots internally. The hydraulic cylinder, which provides power for changing the pitch, is mounted at the front of the hub. The propel-

Figure 3-4-6. A reversible propeller, with visible beta-feedback linkage

**Figure 3-4-7. A cutaway of a compact-model propeller**

ler can only be installed on engines having flanged mounting provisions.

The constant-speed propellers utilize oil pressure from a governor to move the blades into high pitch (reduced r.p.m.). The *centrifugal twisting moment* (CTM) of the blades moves them into low pitch (high r.p.m.) in the absence of governor oil pressure.

The feathering propellers utilize oil pressure from the governor to move the blades into low pitch (high r.p.m.). The CTM of the blades also moves the blades into low pitch. Opposing these two forces is a force produced by compressed air trapped between the cylinder head and the piston, which tends to move the blades into high pitch in the absence of governor oil pressure. Thus, feathering is accomplished by the compressed air, in the absence of governor oil pressure. Feathering is accomplished by moving the governor control back to its extreme position.

The propeller is prevented from feathering, when it is stationary, by *centrifugal-responsive pins,* which engage a shoulder on the piston rod. These pins move out by centrifugal force against springs, when the propeller turns at more than 700 r.p.m.

The time necessary to feather depends upon the size of the oil passages back through the engine and governor, and the air pressure carried in the cylinder head. The larger the passages, the faster the oil from the propeller cylinder can be forced back into the engine. Also, the higher the air charge, the faster the feather action. In general, feathering can be accomplished within a few seconds.

Unfeathering can be accomplished by any of the following methods:

- Start the engine, so that the governor can pump oil back into the propeller to reduce pitch. In most light twin-engine aircraft, this procedure is considered adequate, since engine starting generally presents no problem.
- Provide an accumulator connected to the governor, with a valve to trap an air-oil charge when the propeller is feathered, but released to the propeller when the r.p.m. control is returned to normal position.
- Provide a crossover system, which allows oil from the operating engine to unfeather the propeller on the dead engine. This consists of an oil line connecting the two governors with a manual- or electric-actuated valve in between.

One safety feature inherent in this method of feathering is that the propeller will feather if the governor oil pressure drops to zero for any reason. The propeller does not require external power or oil pressure to feather; in fact, it feathers in the absence of pressure. This is very important, because external pressure or power sources often fail when most needed: during an emergency that involves engine failure. If the engine fails in such a way that the oil pressure drops, the propeller feathers automatically.

## Hartzell Propeller Governors

The governor is designed so that it may be adapted for either single-action or double-action operation. As a single-action governor, it directs oil pressure to the rear of the cylinder to decrease pitch and allows it to drain from the front of the cylinder when centrifugal force increases pitch. Propellers having counterweights use single-action governors. The counterweights and centrifugal force act together to increase pitch. For those propellers that do not use counterweights to increase pitch, oil from the governor is used to increase pitch by overcoming the centrifugal force of the blades. In this case, the plug "B" is removed and installed in passage "C" of the governor. This permits governor oil pressure to be directed to the rear of the cylinder, decreasing pitch. Oil pressure is directed to the forward side of the cylinder to increase pitch (see Figure 3-4-8).

## McCauley Propellers

McCauley Propeller Systems manufactures constant-speed propellers of both the non-feathering and feathering type. Two basic design series have been produced. These are known as the *threaded series* and the *threadless series*. The threaded series uses a retaining nut, which screws into the propeller hub and holds the blades in place. The threadless design utilizes a split retaining ring to hold the blades in the hub. (see Figure 3-4-9 on next page)

Constant-speed propellers utilize spring pressure acting on the blades to decrease blade angle. Governed oil pressure is used to increase blade angle. Blade angle is reduced by a combined force of centrifugal twisting force and spring pressure. All of the pitch-changing mechanism is inside the propeller hub.

The engine-lubricating system provides the hydraulic pressure for propeller operation. Engine oil pressure is boosted by the governor gear pump and supplied to the propeller hub through the engine shaft flange. Oil flow to the propeller is controlled by the governor-control valve.

McCauley feathering propellers utilize hydraulic pressure to oppose the force of springs and counterweights. Hydraulic pressure moves the blades toward low pitch, and springs assisted by counterweights move the blades toward high pitch. In flight, a complete loss of governor oil pressure (as caused by engine failure) will cause the springs and counterweights to automatically feather the propeller.

Some models of McCauley propellers have been modified to provide an ongoing dye-penetrant type of inspection. The hub breather holes have been sealed and the hub partially filled with oil that has been dyed red. This coloring makes the location of cracks easy to spot, and indicates that the propeller should be removed from service.

## Hamilton Standard Hydromatic® Propellers

The Hamilton Standard Hydromatic® propeller, in its many variations, represents the majority of propellers used on large radial engine-powered aircraft. They are also common today in some radial-engined agricultural aircraft, but you are not likely to encounter many in your career.

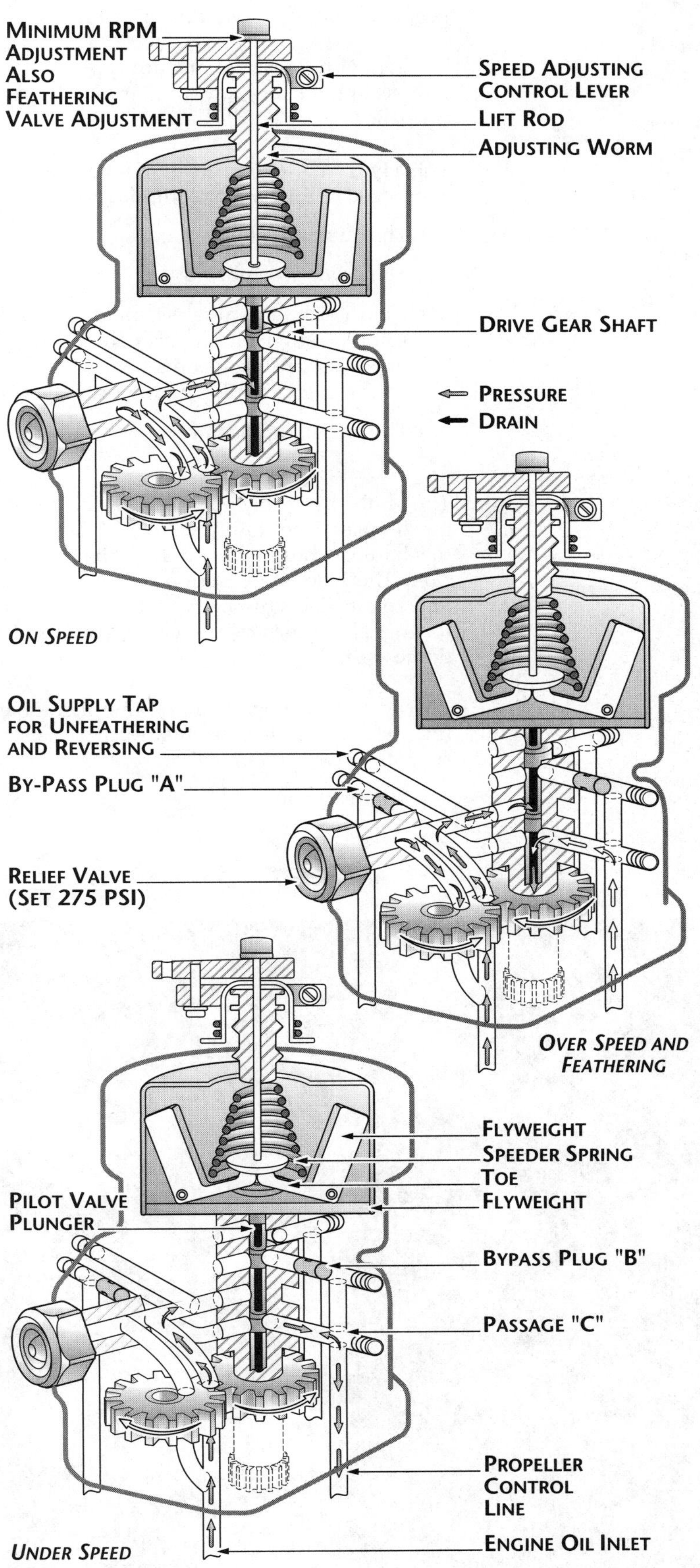


Figure 3-4-8. Woodward governor operation

They are, however, an important milestone in propeller evolution and deserve some study.

The following descriptions are typical of most models of the Hamilton Standard Hydromatic® propeller.

The Hydromatic® propeller (Figure 3-4-10) is composed of four major components:

- The hub assembly
- The dome assembly
- The distributor valve assembly (for feathering on single-acting propellers) or engine-shaft-extension assembly (for non-feathering or double-acting propellers)
- The anti-icing assembly

The hub assembly is the basic propeller mechanism. It contains both the blades and the mechanical means for holding them in position. The blades are supported by the spider and retained by the barrel. Each blade is free to turn about its axis under the control of the dome assembly.

The dome assembly contains the pitch-changing mechanism for the blades. It consists of several major components:

- Rotating cam
- Fixed cam
- Piston
- Dome shell

When the dome assembly is installed in the propeller hub, the *fixed cam* remains stationary with respect to the hub. The *rotating cam*, which can turn inside the fixed cam, meshes with gear segments on the blades.

The *piston* operates inside the *dome shell* and is the mechanism that converts engine and governor oil pressure into forces that act through the cams to turn propeller blades.

The *distributor valve*, or engine-shaft-extension assembly, provides oil passage for governor or auxiliary oil to the inboard side of the piston and for engine oil to the outboard side. During unfeathering operation, the distributor shifts under auxiliary pressure and reverses these passages so that oil from the auxiliary pump flows to the out-board side of the piston. Oil on the inboard side flows back to the engine. The engine-shaft-extension assembly is used with props that do not have feathering capabilities.

Many structural features of most Hydromatic® propellers and other constant-speed propellers are similar. The blade and hub assemblies are almost identical, and the governors are also similar, both in construction and principle of operation. The major difference is in the pitch-changing mechanism. In the Hydromatic® propeller, no counterweights are used, and the moving parts of the mechanism are completely enclosed. Oil pressure and the CTM of the blades are used together to turn the blades

Figure 3-4-9. McCauley propeller

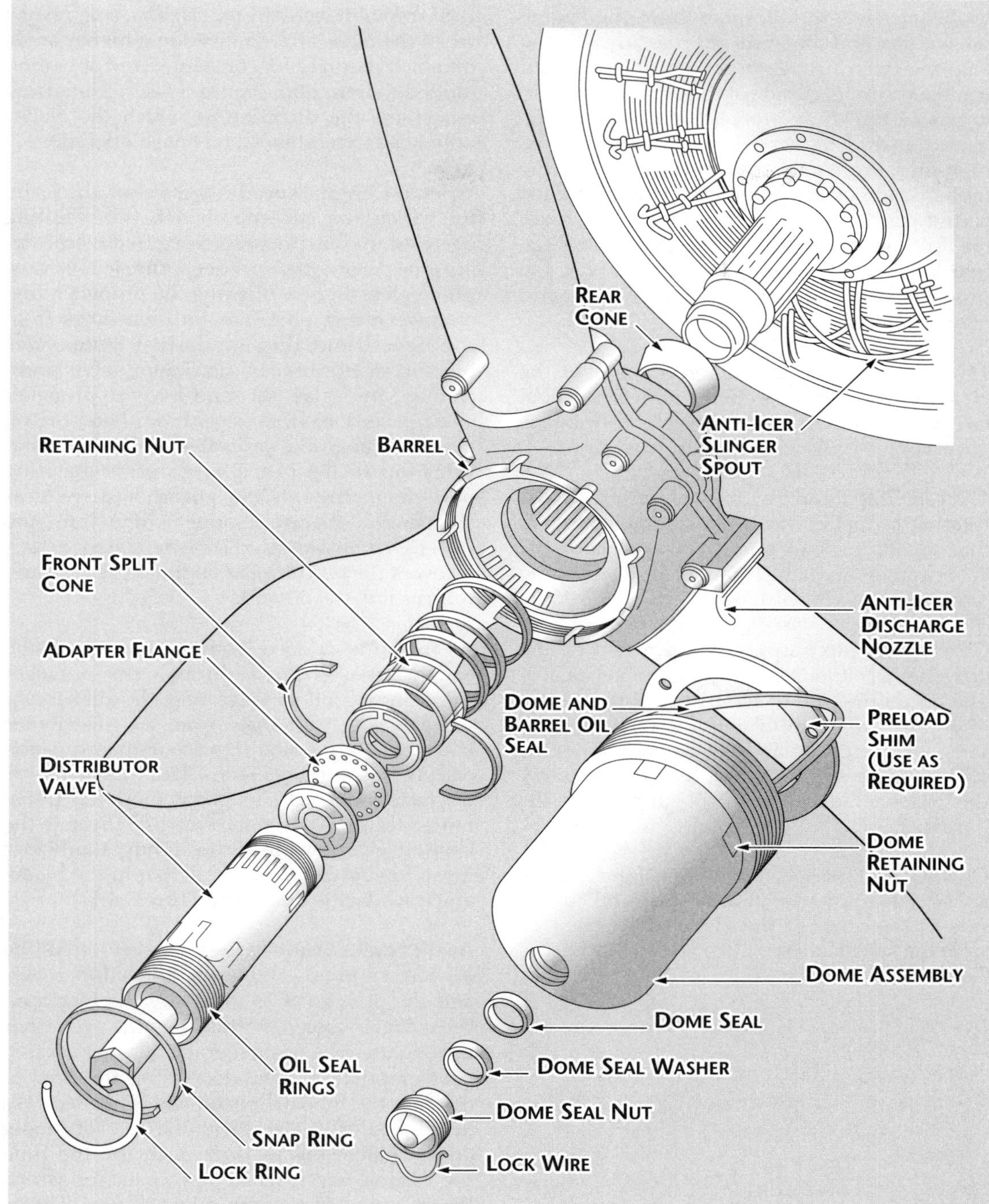


**Figure 3-4-10. Typical Hydromatic® propeller installation**

to a lower angle. The main advantages of the Hydromatic® propeller are the large blade-angle range and the feathering and reversing features.

## Principles of Operation

The pitch-changing mechanism of Hydromatic® propellers is a mechanical-hydraulic system in which hydraulic forces acting on a piston are transformed into mechanical twisting forces acting on the blades. Linear movement of the piston is converted to rotary motion by a cylindrical cam. A bevel gear on the base of the cam mates with bevel-gear segments attached to the butt ends of the blades, thereby turning the blades. This blade pitch-changing action can be understood by studying the schematic in Figure 3-4-10.

The centrifugal force acting on a rotating blade includes a component force that tends to move the blade toward low pitch. As shown in Figure 3-4-11, a second force, engine oil pressure, is supplied to the outboard side of the propeller piston to assist in moving the blade toward low pitch.

Propeller governor oil, taken from the engine oil supply and boosted in pressure by the engine-driven propeller governor, is directed against the inboard side of the propeller piston. It acts as the counterforce that can move the blades toward higher pitch. By metering this high-pressure oil to, or draining it from, the inboard side of the propeller piston by means of the constant-speed control unit, the force toward high pitch can balance and control the two forces toward low pitch. In this way, the propeller blade angle is regulated to maintain a selected r.p.m.

The basic propeller-control forces acting on the Hamilton Standard propeller are centrifugal twisting force and high-pressure oil from the governor.

The centrifugal force acting on each blade of a rotating propeller includes a component force that results in a twisting moment about the blade center line which tends, at all times, to move the blade toward low pitch.

Governor pump output oil is directed by the governor to either side of the propeller piston. The oil on the side of the piston opposite this high-pressure oil returns to the intake side of the governor pump and is used again. Engine oil at engine-supply pressure does not enter the propeller directly but is supplied to the governor.

During constant-speed operations, the double-acting governor mechanism sends oil to one side or the other of the piston, as needed, to keep the speed at a specified setting.

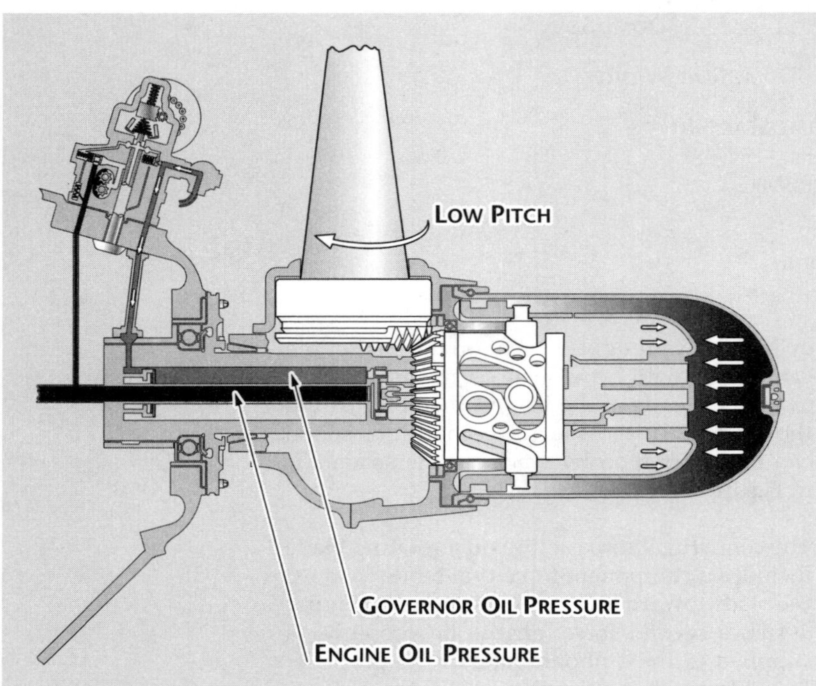


**Figure 3-4-11. Diagram of Hydromatic® propeller operational forces**

**Underspeed condition.** *Underspeed* results when the blades have moved to a higher angle than that required for constant-speed operation (dotted-line section, Figure 3-3-12). The arrow represents the direction in which the blades will move to reestablish on-speed operation.

When the engine speed drops below the r.p.m. for which the governor is set, the resulting decrease in centrifugal force exerted by the flyweights permits the speeder spring to lower the pilot valve, thereby opening the propeller governor-metering port. The oil then flows from the inboard end, through the distributor valve inboard inlet, between distributor valve lands, through the valve port and into the propeller shaft governor oil passage. From here, the oil moves through the propeller shaft oil-transfer rings, up to the propeller governor-metering port, then through the governor drive gear shaft and pilot-valve arrangement to drain into the engine nose case. The *engine scavenge pump* recovers the oil from the engine nose case and returns it to the oil tank.

As the oil is drained from the inboard piston end, engine oil flows through the propeller shaft engine oil passage and the distributor valve ports. It emerges from the distributor valve outboard outlet into the outboard piston end. With the aid of blade CTM, this oil moves the piston inboard. The piston motion is transmitted through the cam rollers and through the beveled gears to the blades. Thus, the blades move to a lower angle, as shown in the blade-angle schematic diagram (Figure 3-4-12).

As the blades assume a lower angle (dotted-line section, Figure 3-4-12), engine speed increases and the pilot valve is raised by the increased centrifugal force exerted by the governor flyweights. The propeller governor-metering port gradually closes, decreasing the flow of oil from the inboard piston end. This decrease in oil flow also decreases the rate of blade-angle change toward low pitch. By the time the engine has reached the r.p.m. for which the governor is set, the pilot valve will have assumed a neutral position (closed), which prevents any appreciable oil flow to or from the propeller. The valve is held in this position because the flyweight centrifugal force equals the speeder-spring force. The control forces are now equal, and the propeller and governor are operating on-speed.

**Overspeed condition.** If the propeller is operating above the r.p.m. for which the control is set, it is called an *overspeed* condition. The blades will be in a lower angle (solid section, Figure 3-4-13) than that required for constant speed operation (dotted lines, Figure 3-3-13). The arrow represents the direction in which the blades will move to bring the propeller to the on-speed condition.

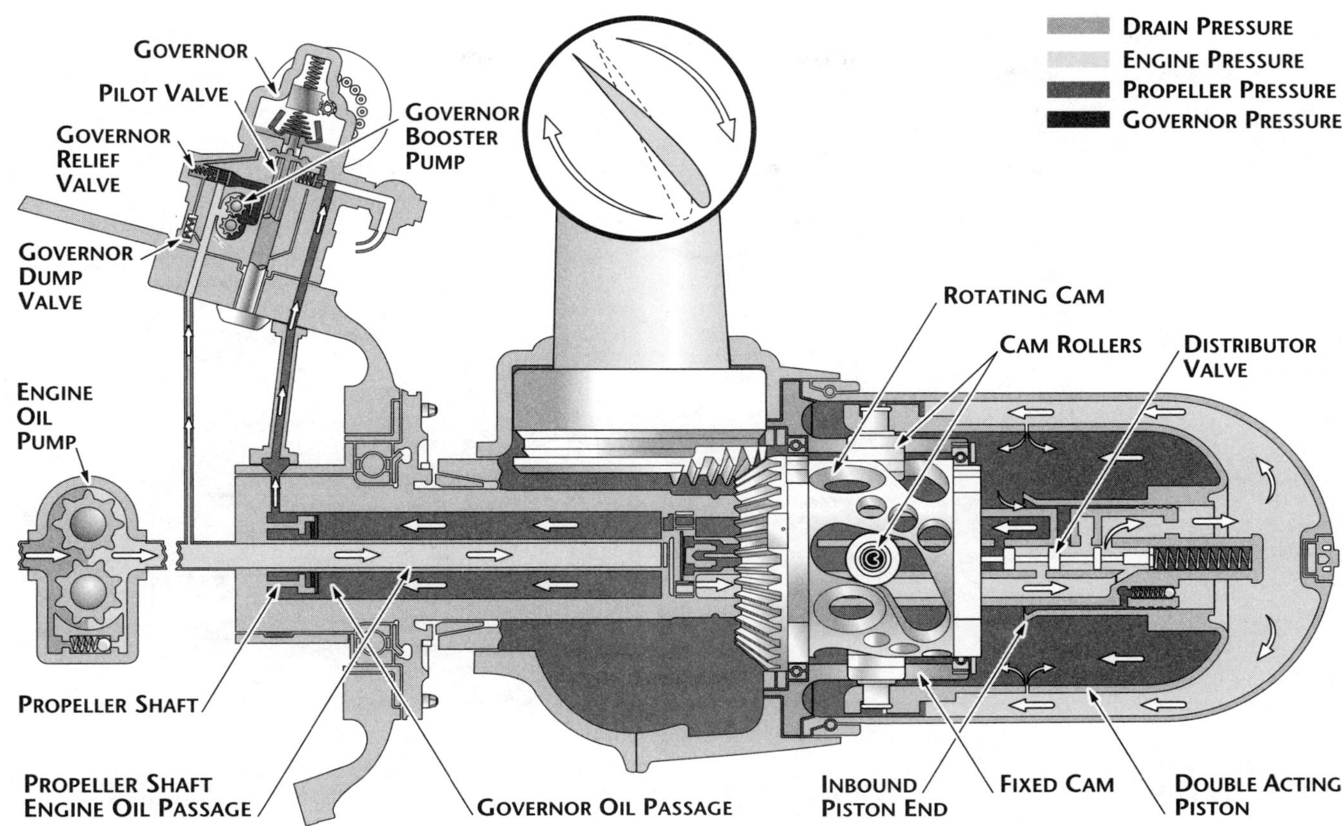


**Figure 3-4-12. Propeller operation (underspeed condition)**

When the engine speed increases above the r.p.m. for which the governor is set, note that the flyweights move outward against the force of the speeder spring, raising the pilot valve. This opens the propeller governor-metering port, allowing governor oil flow from the governor booster pump, through the propeller governor-metering port, and into the engine oil-transfer rings. From the rings, the oil passes through the propeller shaft governor oil passage, through a distributor valve port, between distributor lands and then to the inboard piston end by way of the distributor valve inboard outlet.

As a result of this flow, the piston and the attached rollers move outboard, and the rotating cam is turned by the cam track. As the piston moves outboard, oil is displaced from the outboard piston end. This oil enters the distributor valve outboard inlet, flows through the distributor valve port, past the outboard end of the valve land, through the port and into the propeller shaft engine oil passage. From that point, it is dissipated into the engine-lubricating system. The same balance forces exist across the distributor valve during overspeed as during underspeed, except that oil at governor pressure replaces oil at drain pressure on the inboard end of the valve land and between lands.

Outboard motion of the piston moves the propeller blades toward a higher angle which, in turn, decreases the engine r.p.m. A decrease in engine r.p.m. decreases the rotating speed of the governor flyweights. As a result, the flyweights are moved inward by the force of the speeder spring, the pilot valve is lowered and the propeller governor-metering port is closed. Once this port has been closed, oil flow to or from the propeller practically ceases, and the propeller and governor operate on-speed.

**Feathering operation.** A typical Hydromatic® propeller feathering installation is shown in Figure 3-4-14. When the feathering push-button switch is depressed, the low-current circuit is established from the battery through the push-button holding coil and from the battery through the solenoid relay. As long as the circuit remains closed, the holding coil keeps the push button in the depressed position. Closing the solenoid establishes the high-current circuit from the battery to the feathering motor pump unit. The feathering pump picks up engine oil from the oil supply tank, boosts its pressure, if necessary, to the relief valve setting of the pump and supplies it to the governor high-pressure transfer valve connection.

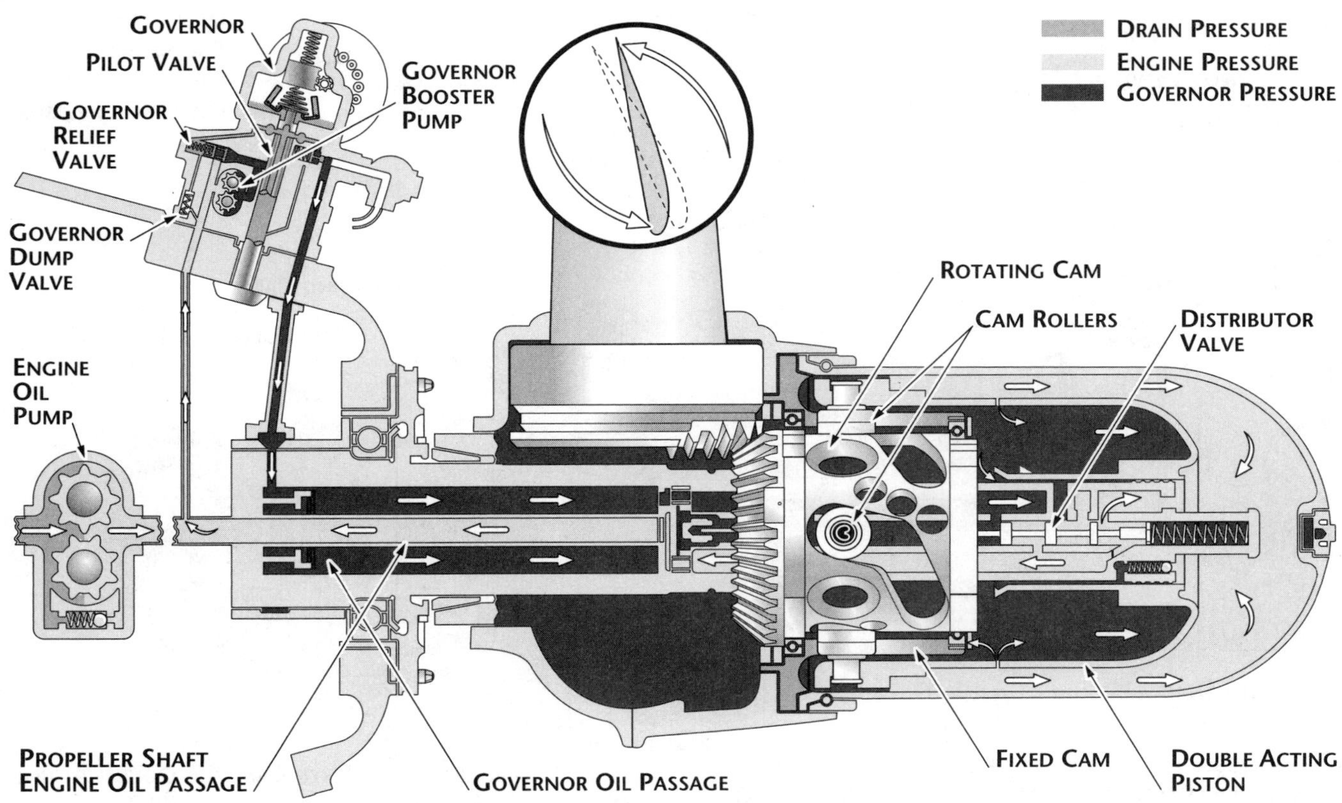


**Figure 3-4-13. Propeller operation (overspeed condition)**

Auxiliary oil entering the high-pressure transfer-valve connection shifts the governor transfer valve, which hydraulically disconnects the governor from the propeller and at the same time opens the propeller governor oil line to auxiliary oil. The oil flows through the engine transfer rings, through the propeller shaft governor oil passage, through the distributor valve port, between lands and finally to the inboard piston end by way of the valve inboard outlet.

The distributor valve does not shift during the feathering operation. It merely provides an oil passageway to the inboard piston end for auxiliary oil and the outboard piston end for engine oil. The same conditions described for underspeed operation exist in the distributor valve, except that oil at auxiliary pressure replaces drain oil at the inboard end of the land and between lands. The distributor-valve spring is backed up by engine oil pressure, which means that at all times the pressure differential

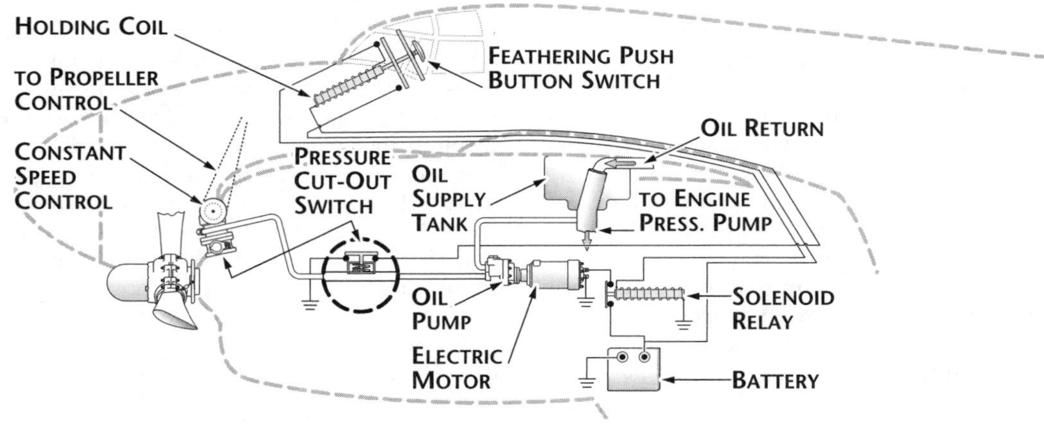


**Figure 3-4-14. Typical feathering installation**

required to move the piston will be identical with that applied to the distributor valve.

The propeller piston moves outboard under the auxiliary oil pressure at a speed proportional to the rate at which oil is supplied. This piston motion is transmitted through the piston rollers operating in the oppositely inclined cam tracks of the fixed cam and the rotating cam, and it is converted by the bevel gears into the blade-twisting moment. Only during feathering or un-feathering is the *low mechanical-advantage portion* of the cam tracks used. (The low mechanical advantage portion lies between the break and the outboard end of the track profile.)

Oil at engine pressure, displaced from the outboard piston end, flows through the distributor valve outboard inlet, past the outboard end of the valve land, through the valve port, into the propeller shaft engine oil passage and is finally delivered into the engine lubricating system. Thus, the blades move toward the full high-pitch (or feathered) angle.

Having reached the full-feathered position, further movement of the mechanism is prevented by contact between the high-angle *stop ring* in the base of the fixed cam and the *stop lugs* set in the teeth of the rotating cam. The pressure in the inboard piston end now increases rapidly, and upon reaching a set pressure, the electric cutout switch automatically opens. This cutout pressure is less than that required to shift the distributor valve.

Opening the switch de-energizes the holding coil and releases the feathering push-button control switch. Release of this switch breaks the solenoid relay circuit, which shuts off the feathering-pump motor. The pressures in both the inboard and outboard ends of the piston drop to zero, and since all the forces are balanced, the propeller blades remain in the feathered position. Meanwhile, the governor high-pressure transfer valve has shifted to its normal position as soon as the pressure in the propeller-governor line drops below that required to hold the valve open.

**Unfeathering operation.** To unfeather a Hydromatic® propeller, depress and hold the feathering switch push-button control. As in the case of feathering a propeller, the low-current control circuits from the battery through the holding coil and from the battery through the solenoid are completed when the solenoid closes. The high-current circuit from the battery starts the motor-pump unit, and oil is supplied at a high pressure to the governor transfer valve.

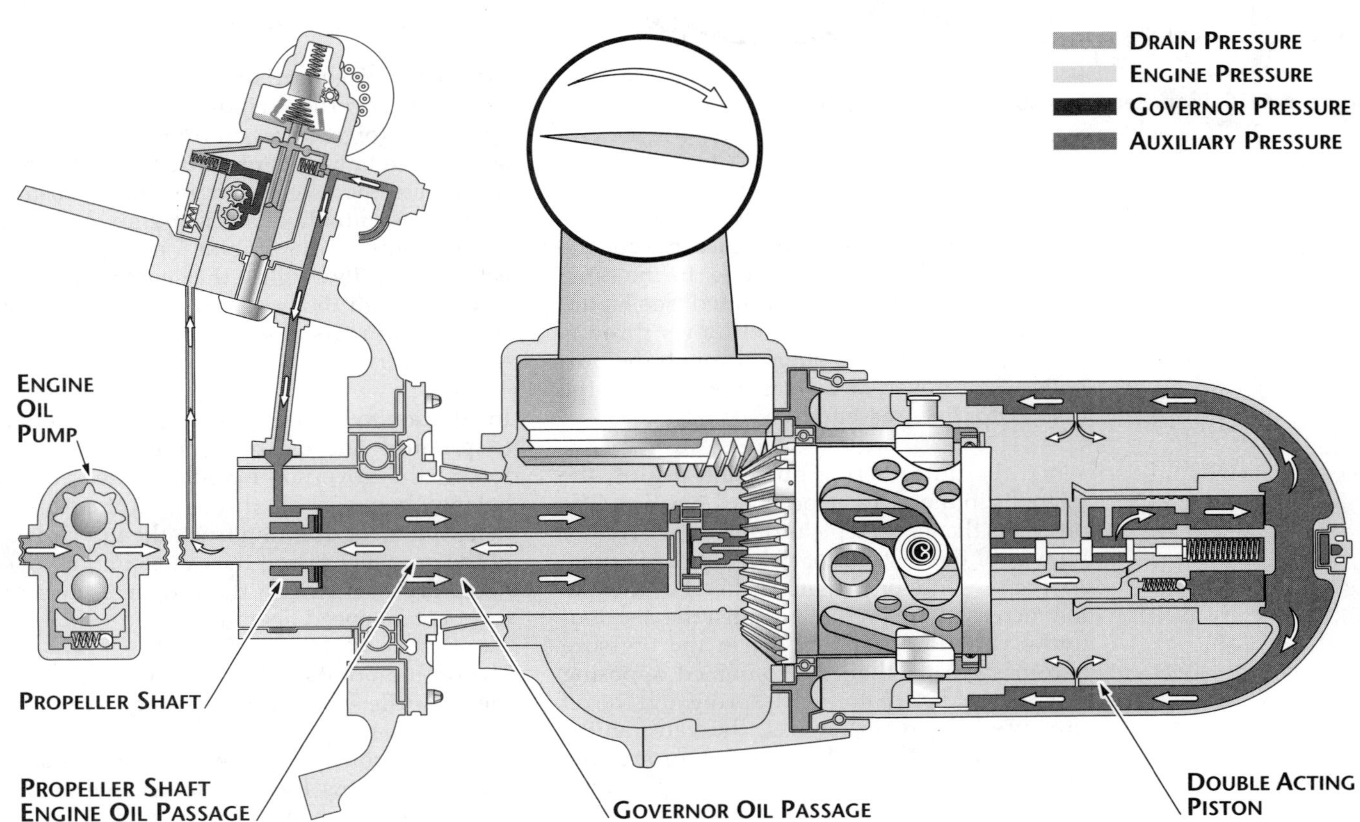


Figure 3-4-15. Propeller operation (unfeathering condition)

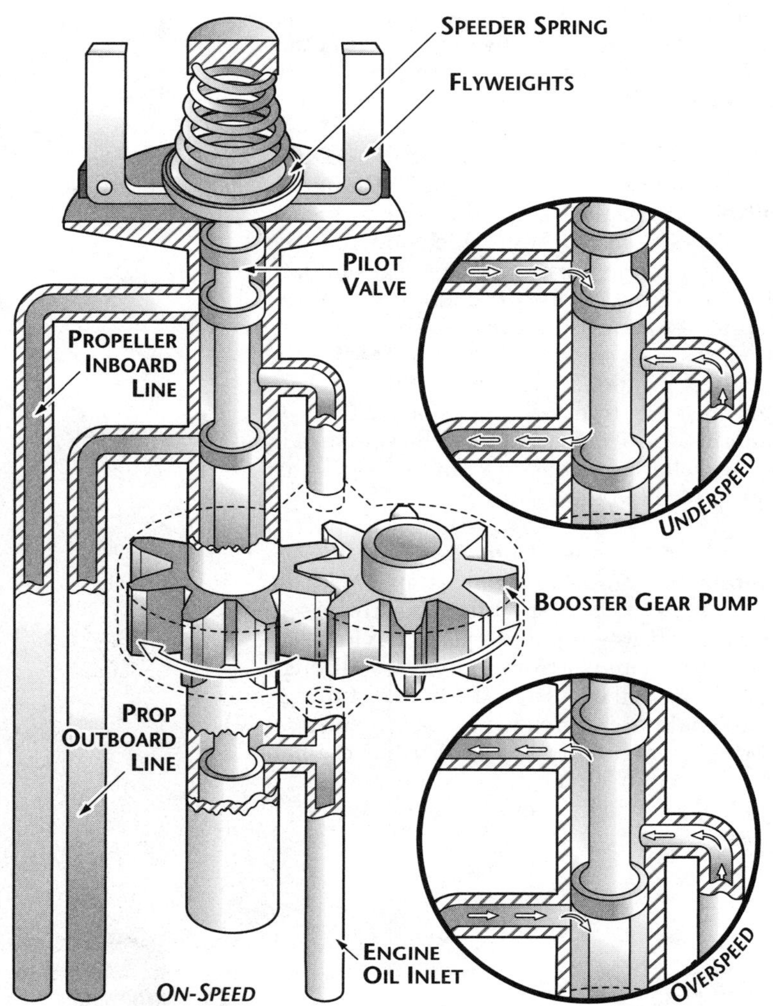


**Figure 3-4-16. Propeller-governor operating diagram**

Auxiliary oil entering through the high-pressure transfer valve connection shifts the governor transfer valve and disconnects the governor from the propeller line. In the same operation, auxiliary oil is admitted (see Figure 3-4-15, previous page). The oil flows through the engine oil transfer rings, through the propeller-shaft governor oil passage and into the distributor valve assembly.

When the unfeathering operation begins, the piston is in the extreme outboard position. The oil enters the inboard piston end of the cylinder by way of the distributor valve inboard outlet. As the pressure on the inboard end of the piston increases, the pressure against the distributor valve land builds up. When the pressure becomes greater than the combined opposing force of the distributor valve spring and the oil pressure behind this spring, the valve shifts. Once the valve shifts, the passages through the distributor valve assembly to the propeller are reversed.

A passage is opened between lands and through a port to the outboard piston end by way of the distributor valve outlet. As the piston moves inboard under the auxiliary pump oil pressure, oil is displaced from the inboard piston end through the inlet ports between the valve lands, into the propeller shaft engine oil lands and into the propeller shaft engine oil passage, where it is discharged into the engine lubricating system. At the same time. the pressure at the cutout switch increases and the switch opens. However, the circuit to the feathering pump and motor unit remains complete so long as the feathering switch is held in.

With the inboard end of the propeller piston connected to drain and auxiliary pressure flowing to the outboard end of the piston, the piston moves inboard. This unfeathers the blades, as shown in Figure 3-4-15. As the blades are unfeathered, they begin to windmill and assist the unfeathering operation by the added force toward low pitch brought about by the CTM.

When the engine speed has increased to approximately 1,000 r.p.m., the operator shuts off the feathering pump motor. The pressure in the distributor valve and at the governor transfer valve decreases, allowing the distributor valve to shift under the action of the governor high-pressure transfer valve spring. This action re-connects the governor with the propeller and establishes the same oil passages through the distributor valve that are used during constant-speed and feathering operations.

**Governor mechanism.** The engine-driven propeller governor, Figure 3-4-16, (constant-speed control) receives oil from the lubricating system and boosts its pressure to that required to operate the pitch-changing mechanism. It consists essentially of a gear pump to increase the pressure of the engine oil, a pilot valve actuated by flyweights that control the flow of oil through the governor and a relief-valve system that regulates the operating pressures in the governor.

In addition to boosting the engine oil pressure to produce one of the fundamental control forces, the governor maintains the required balance between all three control forces by metering to, or draining from, the inboard side of the propeller piston the exact quantity of oil necessary to maintain the proper blade angle for constant-speed operation.

The position of the pilot valve, with respect to the propeller-governor metering port, regulates the quantity of oil that flows through this port to or from the propeller. A spring above the rack returns the rack to an intermediate position approximating cruising r.p.m. in case of governor control failure.

Figure 3-4-17. Propeller r.p.m. adjusting screw

**Setting the propeller governor.** The propeller governor incorporates an adjustable stop, which limits the maximum speed at which the engine can run. As soon as the takeoff r.p.m. is reached, the propeller moves off the low-pitch stop. The larger propeller blade angle increases the load on the engine, thus maintaining the prescribed maximum engine speed.

At the time of propeller, propeller governor or engine installation, the following steps are normally taken to ensure that the powerplant will obtain takeoff r.p.m.

1. During ground run-up, move the throttle to takeoff position and note the resultant r.p.m. and manifold pressure.
2. If the r.p.m. obtained is higher or lower than the takeoff r.p.m. prescribed in the manufacturer's instructions, reset the adjustable stop on the governor until the prescribed r.p.m. is obtained (see Figure 3-4-17).

## Section 5

# Turboprop Propellers

Turboprop engines are used on aircraft ranging in size from large four-engine transports to medium-sized executive and relatively small single- and twin-engine aircraft (see Figure 3-5-1).

Unlike the turbojet engine, which produces thrust directly, the turboprop engine produces thrust indirectly, since the compressor and turbine assembly furnish torque to a propeller which, in turn, produces the major portion of the propulsive force that drives the aircraft. The turboprop fuel control and the propeller governor are connected and operate in coordination with each other. The power lever directs a signal from the cockpit to the fuel control for a specific amount of power from the engine. The fuel control and the propeller governor together establish the correct combination of r.p.m., fuel flow and propeller blade angle to create sufficient propeller thrust to provide the desired power.

Figure 3-5-1. A medium size commuter airplane with turbopropeller engines

The propeller-control system is divided into two types of control: one for flight and one for ground operation. For flight, the propeller blade angle and fuel flow for any given power lever setting are governed automatically, according to a predetermined schedule. Below the *flight idle* power lever position, the coordinated r.p.m. blade angle schedule becomes incapable of handling the engine efficiently. Here, the ground handling range, referred to as the *beta range*, is encountered. In the beta range of the throttle quadrant, the propeller blade angle is not governed by the propeller governor, but is controlled by the power lever position. When the power lever is moved below the start position, the propeller pitch is reversed to provide reverse thrust for rapid deceleration of the aircraft after landing.

HELICAL SPLINE COUPLING
RING GEAR COUPLING
PLANET GEAR
NTS SPLINE RING
RING GEAR
REAR PLATE
FRONT PLATE
REDUCTION GEAR CASE
PLUNGER
NTS ACTUATOR ROD


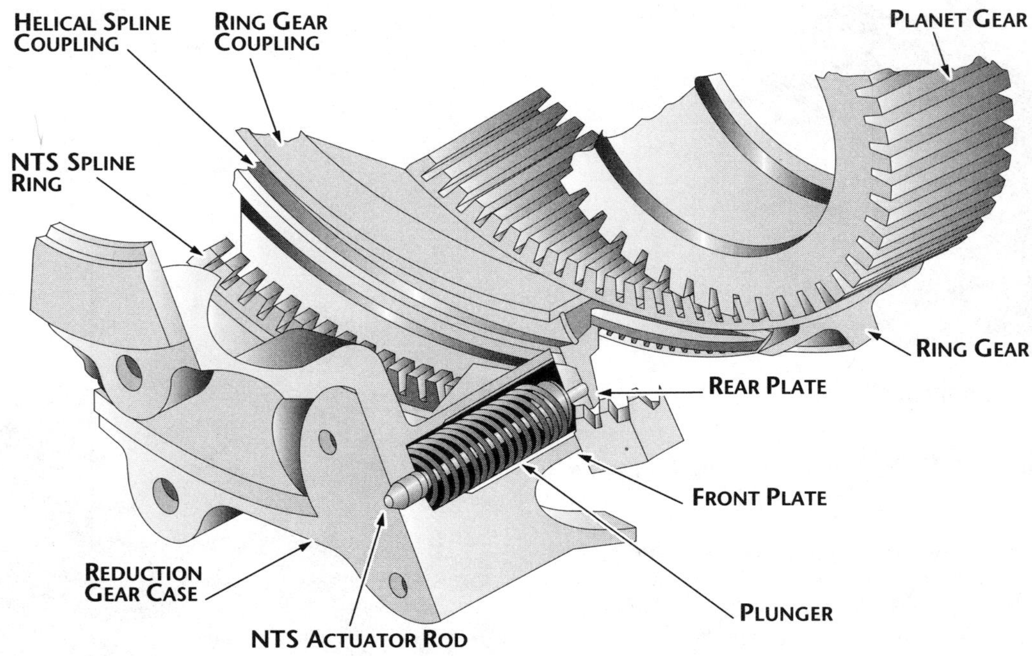

**Figure 3-5-2. Negative torque signal components**

A characteristic of the turboprop is that changes in power are not related to engine speed, but to turbine inlet temperature. During flight, the propeller maintains a constant engine speed. This speed is known as the 100-percent-rated speed of the engine, and it is the design speed at which most power and best overall efficiency can be obtained. Power changes are effected by changing the fuel flow. An increase in fuel flow causes an increase in turbine inlet temperature and a corresponding increase in energy available at the turbine. The turbine absorbs more energy and transmits it to the propeller in the form of torque. The propeller, in order to absorb the increased torque, increases blade angle, thus maintaining constant engine r.p.m.

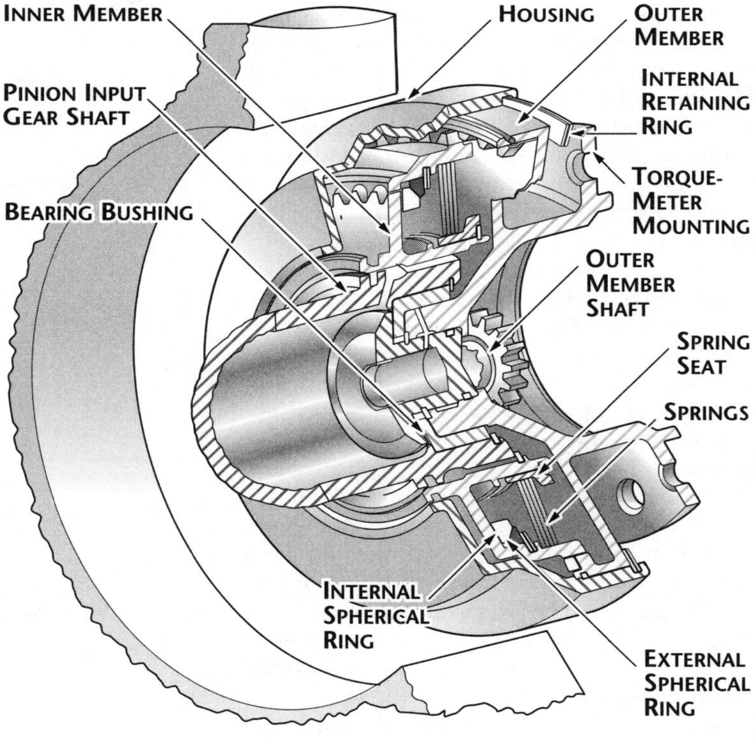

INNER MEMBER
HOUSING
OUTER MEMBER
PINION INPUT GEAR SHAFT
INTERNAL RETAINING RING
TORQUE-METER MOUNTING
BEARING BUSHING
OUTER MEMBER SHAFT
SPRING SEAT
SPRINGS
INTERNAL SPHERICAL RING
EXTERNAL SPHERICAL RING


**Figure 3-5-3. Safety coupling**

The *negative torque signal* (NTS) control system (Figure 3-5-2) provides a signal that increases propeller blade angle to limit negative shaft torque. When a predetermined negative torque is applied to the reduction gearbox, the stationary ring gear moves forward against the spring force due to a torque reaction generated by helical splines. In moving forward, the ring gear pushes two operating rods through the reduction gear nose. One or both of the rods may be used to signal the propeller and initiate an increase in propeller blade angle. This action (toward high blade angle) continues until the negative torque is relieved, resulting in the propeller returning to normal operation.

The NTS system functions when the following engine operating conditions are encountered: temporary fuel interruptions, air gust loads on the propeller, normal descents with lean fuel scheduling, high compressor air-bleed conditions at low power settings, and normal shutdowns.

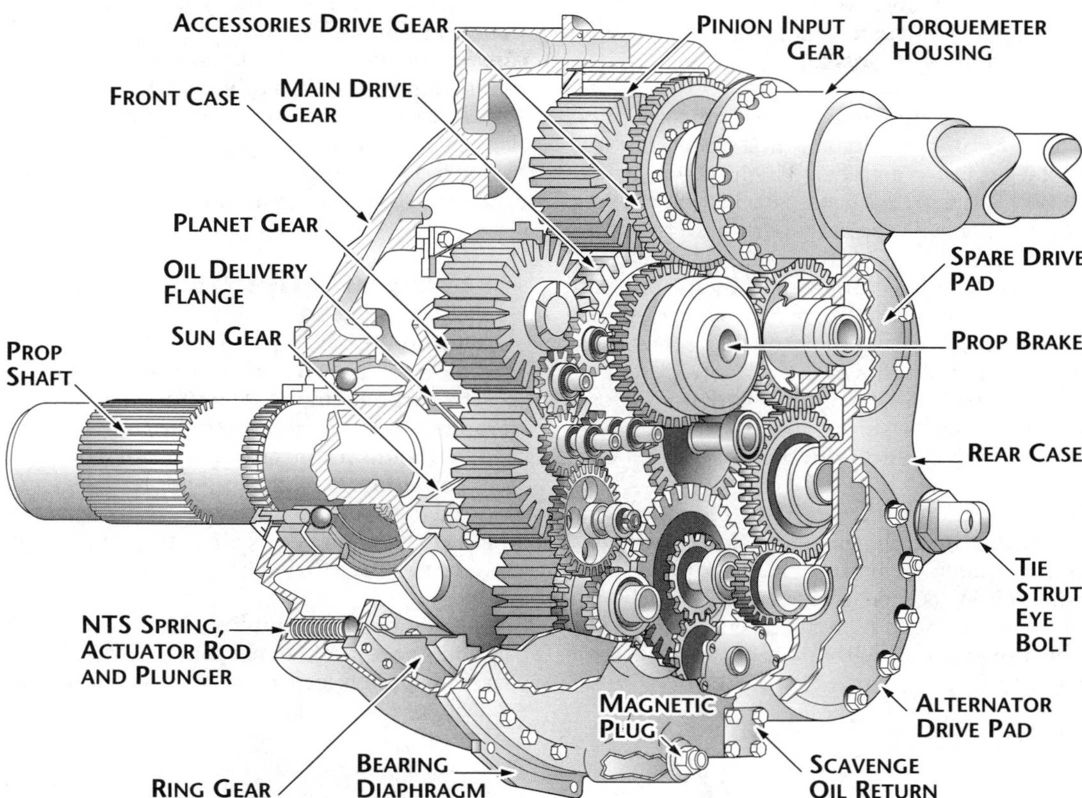


**Figure 3-5-4. Reduction gear and torquemeter assemblies**

The *thrust sensitive signal* (TSS) is a safety feature that actuates the propeller feather lever. If a power loss occurs during takeoff, propeller drag is limited to that of a feathered propeller, reducing the hazards of yawing in multi-engine aircraft. This device automatically increases blade angle and causes the propeller to feather.

The TSS system consists of an externally mounted switch assembly on the right side of the reduction gearbox. A plunger extends into the switch from the inside of the gearbox. A spring loads the plunger against the thrust signal lever mounted inside the gearbox and contacts the outer ring of the prop shaft thrust bearing. When propeller positive thrust exceeds a predetermined value, the prop shaft and ball bearing move forward, compressing two springs located between the thrust- and roller-bearing assemblies. The thrust signal lever follows the outer ring, and the TSS plunger moves into the front gearbox. The TSS system is then armed for takeoff and automatic operation. At any subsequent time when propeller thrust decreases below the predetermined value, spring force moves the prop shaft rearward. When this occurs, the TSS plunger moves outward, energizing the auto feather system. This signals the propeller to increase blade angle.

A *safety coupling* (Figure 3-5-3) disengages the reduction gear from the power unit if the power unit is operating above a preset negative torque value considerably greater than that required to actuate the NTS. The coupling consists essentially of an inner member splined to the pinion shaft, an outer member bolted to the extension shaft, an intermediate member connected to the inner member through helical teeth and to the outer member through straight teeth.

The reaction of the helical teeth moves the intermediate member forward and into mesh when positive torque is applied, rearward and out of mesh when negative torque is applied. Thus, when a predetermined negative torque is exceeded, the coupling members disengage automatically. Reengagement is also automatic during feathering or power unit shutdown. The safety coupling will operate only when negative torque is excessive.

**Reduction gear assembly.** A reduction gear assembly is shown in Figure 3-5-4. It incorporates a single propeller drive shaft, an NTS system, a TSS system, a safety coupling, a propeller brake, an independent dry-sump oil system and the necessary gearing arrangement.

The *propeller brake* is designed to prevent the propeller from windmilling when it is feathered in flight and to decrease the time for the propeller to come to a complete stop after engine shutdown.

The propeller brake is a friction-cone type, consisting of a stationary inner member and a rotating outer member which, when locked, acts upon the primary stage-reduction gearing. During normal engine operation, reduction gear oil pressure holds the brake in the released position. This is accomplished by oil pressure that holds the outer member away from the inner member. When the propeller is feathered or at engine shutdown, as reduction gear oil pressure drops off, the effective hydraulic force decreases, and a spring forcemoves the outer member into contact with the inner member.

The power unit drives the reduction-gear assembly through an extension shaft and torque meter assembly. The reduction-gear assembly is secured to the power unit by the torque-meter housing, which serves as the bottom support, and by the pair of tie struts serving as the top support.

The tie struts assist in carrying the large overhanging moments and forces produced by the propeller and reduction gear. The front ends of the struts have eccentric pins, which are splined for locking. These pins adjust the length of the strut to compensate for the manufacturing tolerances on the drive shaft housing and interconnecting parts.

## Turbo-propeller Assembly

The turbo-propeller provides an efficient and flexible means of using the power produced by the turbine engine. The propeller assembly (Figure 3-5-5), together with the control assembly, maintains a constant r.p.m. of the engine at any condition in flight idle (alpha range). For ground handling and reversing (beta range), the propeller can be operated to provide either zero or negative thrust.

The major subassemblies of the propeller assembly are the barrel, dome, low-pitch stop assembly, pitch-lock regulator assembly, blade

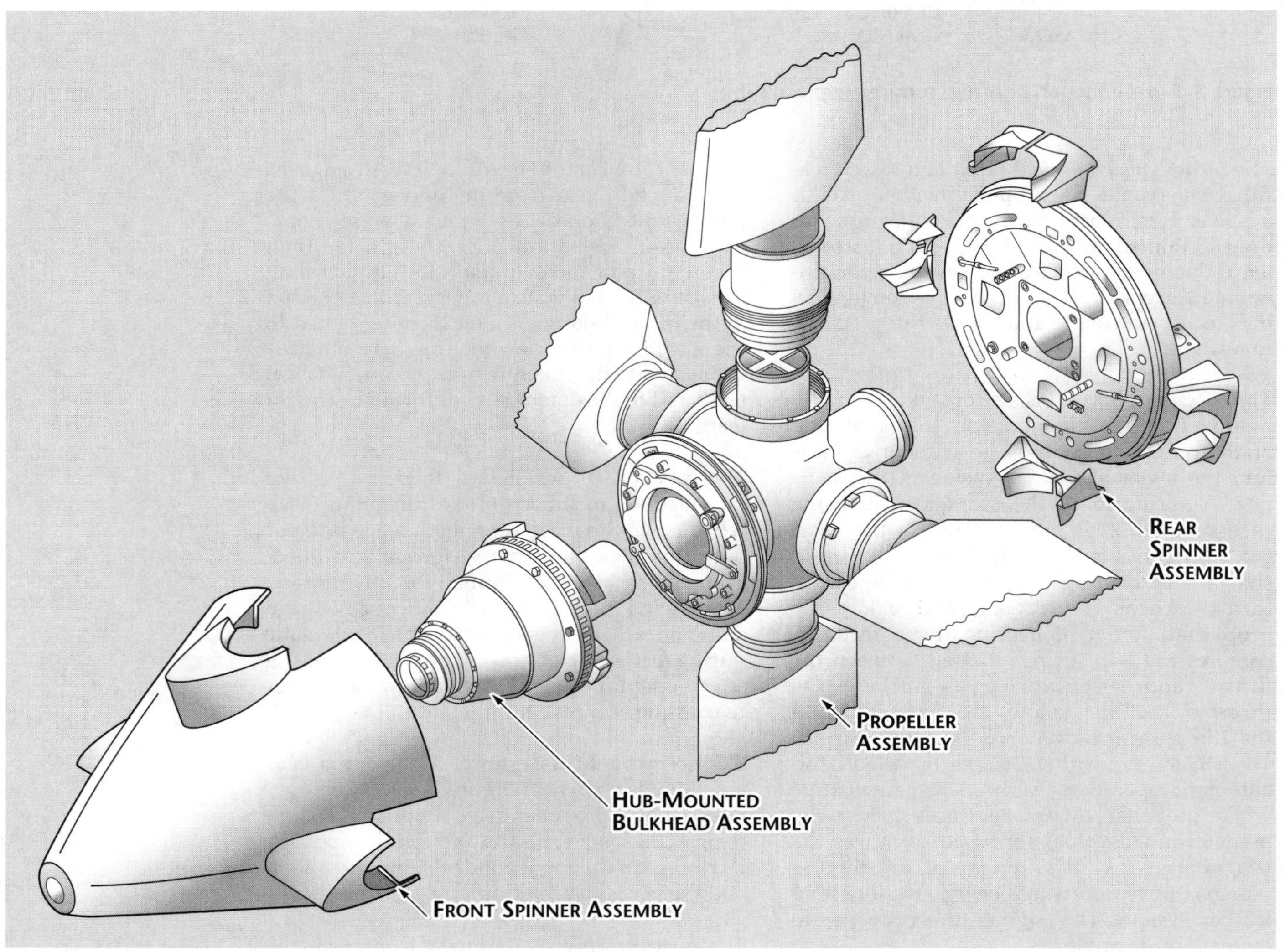


**Figure 3-5-5. Propeller assembly and associated parts for a large aircraft**

assembly and de-icing contact ring holder assembly.

The control assembly is a non-rotating assembly mounted on the aft extension of the propeller assembly barrel. It contains the oil reservoir, pumps, valves and control devices, which supply the pitch-changing mechanism with hydraulic power of proper magnitude and direction to vary pitch for the selected operating conditions. It also contains the brush housing for the electric power for the de-icer rings.

The spinner assembly is a cone-shaped configuration that mounts on the propeller and encloses the dome and barrel to reduce drag. It also provides for ram air to enter and cool the oil used in the propeller control.

The afterbody assembly is a non-rotating component mounted on the engine gearbox to enclose the control assembly. Together with the spinner, it provides a streamlined flow over the engine nacelle.

**Synchrophasing system.** The *synchrophasing system* is designed to maintain a preset angular relationship between the designated master propeller and the slave propellers. The three main units of this system are the pulse generator, the electronic synchrophaser and the speed bias servo assembly (see Figure 3-5-6).

The manual phase control provides for preselection of desired phase-angle relationship between the master and slave propellers and for vernier adjustment of the speed of the engine selected as master. This master trim provides for a master engine speed adjustment of approximately ±1 percent r.p.m.

**Propeller operation.** It is controlled by a mechanical linkage from the cockpit-mounted power lever and the emergency engine-shutdown handle (if the aircraft is provided with one) to the coordinator which, in turn, is linked to the propeller control-input lever.

The non-governing, or *taxi range,* from the reverse position to the *flight idle position* (0°-34° indexes on the coordinator, including the ground idle position) is referred to as the *beta range.* The governing or flight range from the flight-idle position to the takeoff position (34°-90° indexes on the coordinator) is referred to as the *alpha range.* The remaining portion of the coordinator segment (90° indexes to the feather position) concerns feathering only.

The beta range control for ground handling is entirely hydromechanical and is obtained by introducing a cam-and-lever system, which operates the pilot valve. One camshaft (alpha shaft) moves in response to power lever motion and establishes the desired blade-angle schedule (beta range). The other camshaft (beta shaft) is operated from the blade-feedback gearing. Its position provides a signal of actual blade angle position in the beta range. Also, the pilot valve is moved by interaction of these cams and levers to meter oil to either high or low pitch so that the actual blade angle agrees with the scheduled angle.

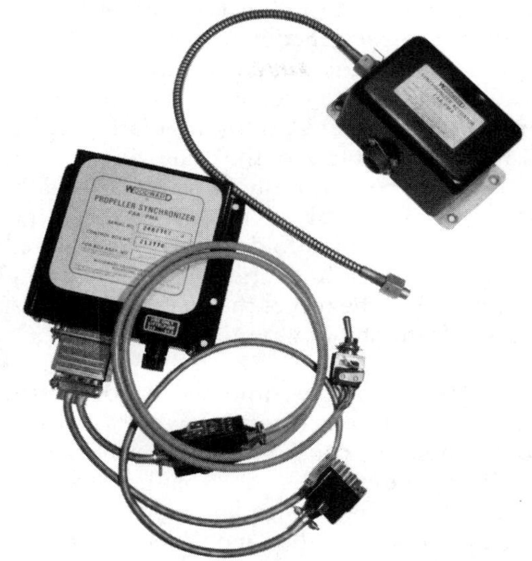

**Figure 3-5-6. The main units of a synchrophaser**

In the beta range (below flight idle), the propeller governing action is blocked out, since an overspeed would result in blade angle motion in the wrong direction of the propeller if it were in the reverse range.

When the power lever is moved to call for a blade angle below flight idle, the speed set cam (on the alpha shaft) puts additional force on the speeder spring. This holds the pilot valve in an underspeed condition against the beta lever system until the scheduled blade angle is reached.

Constant-speed governing (alpha-range control) is accomplished by a flyweight-actuated governor. The flyweights and pilot valve are driven through gearing by propeller rotation.

In the alpha range, the governor is set at its normal 100-percent r.p.m. setting by the speed-set cam (on the alpha shaft), and the pilot valve is free to move in response to off-speed conditions.

Feathering is initiated by the feather button, engine emergency shutdown handle or the *auto-feather system.* Feathering is accomplished hydraulically by a feathering valve that bypasses other control functions and routes pitch change oil directly to the propeller.

The feathering operation is separate from all normal control functions. Pressure from the pump manifold is routed through the control feather valve before going to the pilot valve and the main and standby regulating valves. Similarly, the output of the pilot valve to either low or high pitch is routed through the feather valve. When the valve is positioned for feathering, the pump manifold is connected directly to the high-pitch line. This isolates the propeller lines from the rest of the control system and closes off the standby-pump bypass.

Normal feathering is initiated by depressing the feather button. This action sends a current to the holding coil of the feathering switch, auxiliary pump and the feather solenoid, which positions the feather valve and feathers the propeller. When the propeller has been fully feathered, oil pressure buildup will operate a pressure-cutout switch that will cause the auxiliary pump and feather solenoid to become de-energized through a relay system.

Feathering may also be accomplished by pulling the engine emergency shutdown handle or switch to the shutdown position. This action mechanically positions the feather valve and electrically energizes the feathering button, sending the propeller to full feather.

The auto-feather system automatically energizes the holding coil (pulling in the feather button) when engine power loss results in a propeller thrust drop to a preset value. This system is switch-armed for use during takeoff and can function only when the power lever is near or in the takeoff position.

The NTS device mechanically moves the NTS plunger, which actuates a linkage in the propeller control when a predetermined negative torque value is sensed (when the propeller drives the engine). This plunger, working through control linkage, shifts the feather valve plunger, sending the blades toward feather.

As the blade angle increases, negative torque decreases until the NTS signal is removed, closing the feather valve. If the predetermined negative torque value is again exceeded, the NTS plunger again causes the feather valve plunger to shift.

The normal effect of the NTS is a cycling of r.p.m. slightly below the r.p.m. at which the negative torque was sensed.

Unfeathering is initiated by pulling the feather button to the unfeather position. This action supplies voltage to the auxiliary motor to drive the auxiliary pump. Because the propeller governor is in an underspeed position with the propeller feathered, the blades will move in a decreased pitch direction under auxiliary-pump pressure.

The *pitch lock* operates in the event of a loss of propeller oil pressure or an overspeed. The ratchets of the assembly become engaged when the oil pressure, which keeps them apart, is dissipated through a flyweight-actuated valve that operates at an r.p.m. slightly higher than the 100-percent r.p.m. The ratchets become disengaged when high pressure and r.p.m. settings are restored.

At the flight idle-power lever position, the control beta follow-up low-pitch stop on the beta-set cam (on the alpha shaft) is set about 2° below the flight low-pitch stop setting, acting as a secondary low-pitch stop. At the takeoff-power lever position, this secondary low-pitch stop sets a higher blade-angle stop than the mechanical flight low-pitch stop. This provides for control of overspeed after rapid power lever advance, as well as a secondary low-pitch stop.

## Propeller Synchronization

Today, virtually all multi-engined aircraft are equipped with some type of propeller-synchronization system. Synchronization systems provide a means of controlling and synchronizing engine r.p.m. Synchronization reduces vibration and eliminates the unpleasant beat produced by unsynchronized propeller operation. There are several types of synchronizer systems in use.

**Master motor synchronizer.** An early type still in use on some operating aircraft consists of a synchronizer master unit, four alternators, a tachometer, engine r.p.m. control levers, switches and wiring. These components automatically control the speed of each engine and synchronize all engines at any desired r.p.m.

A synchronizer master unit incorporates a master motor that mechanically drives four contactor units, each electrically connected to an alternator. The alternator is a small, three-phase, alternating-current (AC) generator driven by an accessory drive of the engine. The frequency of the voltage produced by the generator is directly proportional to the engine accessory speed. In automatic operation, the desired engine r.p.m. may be set by manually adjusting the r.p.m.-control lever until a master tachometer indicator on the instrument panel indicates the desired r.p.m. Any difference in r.p.m. between an engine and the master motor will cause the corresponding contactor unit to operate the pitch-changing mechanism of the propeller until the engine is on-speed (at correctly desired r.p.m.).

Figure 3-5-7. A Hartzell constant-speed, feathering propeller on a Beechcraft King Air

**One-engine master system.** Synchronizer systems are also installed in light twin-engine aircraft. Typically, such systems consist of a special propeller governor on the left-hand engine, a slave governor on the right-hand engine, a synchronizer control unit and an actuator in the right-hand engine nacelle.

The propeller governors are equipped with magnetic pickups that count the propeller revolutions and send a signal to the synchronizer unit. The synchronizer, which is usually a transistorized unit, compares the signal from the two propeller governor pickups. If the two signals are different, the propellers are out of synchronization, and the synchronizer control generates a direct-current (DC) pulse which is sent to the slave propeller unit.

The control signal is sent to an actuator, which consists of two rotary solenoids mounted to operate on a common shaft. A signal to increase the r.p.m. of the slave propeller is sent to one of the solenoids, which rotates the shaft clockwise. A signal to decrease r.p.m. is sent to the other solenoid, which moves the shaft in the opposite direction.

Each pulse signal rotates the shaft a fixed amount. This distance is called a *step.* Attached to the shaft is a flexible cable, which is connected on its other end to a trimming unit. The vernier action of the trimming unit regulates the governor arm.

## Synchrophasing

Synchrophasing systems are used with synchronization systems to further reduce noise level and vibration. While the synchronization system keeps the r.p.m. of all the propellers the same, synchrophasing systems keep the angular difference of all the propeller blades and the plane of rotation the same.

A pulse generator is keyed to the same blade on each propeller; for example, the No. 1 blade. The signal from the pulse generator is sent to a governor on the slave engine or master engine. By reading the signal from each propeller, the governor can control the synchrophasing of all of the engines. The synchrophasing system will keep the No. 1 blade of the propeller at the same angle at any given instant. This angle can be set by the pilot to compensate for varied power settings and flight conditions. These in-flight adjustments are made on the synchrophasing control panel in the cockpit.

# Hartzell Turboprop Propellers

One of the most popular propellers for smaller turboprop installations, such as Pratt & Whitney PT6-powered airplanes, is the Hartzell turboprop propeller (Figure 3-5-7). It is used on many varieties of commercial airplanes. The installation discussed here will be the one used on Beechcraft King Air 90-series airplanes. As usual, the discussion is for edu-

Figure 3-5-8. A King Air propeller in the feathered (A) and reversed (B) position

cation and familiarity. Use only the manufacturer's information for any actual repairs and maintenance.

### Turbine Reversing Propellers on Beechcraft King Air 90

A three-bladed aluminum propeller is installed on each engine. These propellers are hydraulically controlled, constant-speed, full-feathering and reversible (see Figure 3-5-8). Each propeller is controlled by engine oil acting through an engine-driven propeller governor. Feathering is accomplished by the feathering springs assisted by centrifugal force applied to the blade shank counterweights. Governor-boosted engine oil pressure moves the propeller blades to the high-r.p.m. (low-pitch) hydraulic stop and into reverse pitch. Low-pitch propeller position is determined by a mechanically actuated hydraulic stop.

**Blade angle.** In the propellers used on this aircraft, the chord (30 inches out from the propeller's center) has been selected as the position at which blade angle is measured. This position is referred to as the *30-inch station*. The sketches in Figure 3-5-9 illustrate various blade angles.

## Primary Governor

The primary governor (Figure 3-5-10) can maintain any selected propeller speed from approximately 1,800 r.p.m. to 2,200 r.p.m.

Power changes, as well as airspeed changes, cause the propeller to momentarily experience overspeed or underspeed conditions, but the governor reacts to maintain the onspeed condition.

There are times, however, when the primary governor is incapable of maintaining selected r.p.m. To help explain this situation, imagine an airplane approaching to land with its governor set at 1,900 r.p.m. As power and airspeed are both reduced, underspeed conditions exist which cause the governor to decrease blade angle to restore the onspeed condition. If blade angle could decrease all the way to 0° or reverse, the propeller would create so much drag on the airplane that the aircraft control would be dramatically reduced. As a matter of fact, the airplane would probably crash.

The propeller, acting as a large disc, would blank the airflow around the tail surfaces, and a rapid nose-down pitch change would result. To prevent these unwanted aerobatics, some device must be provided to stop the governor from selecting blade angles that are too low for safety. As the blade angle is then decreased

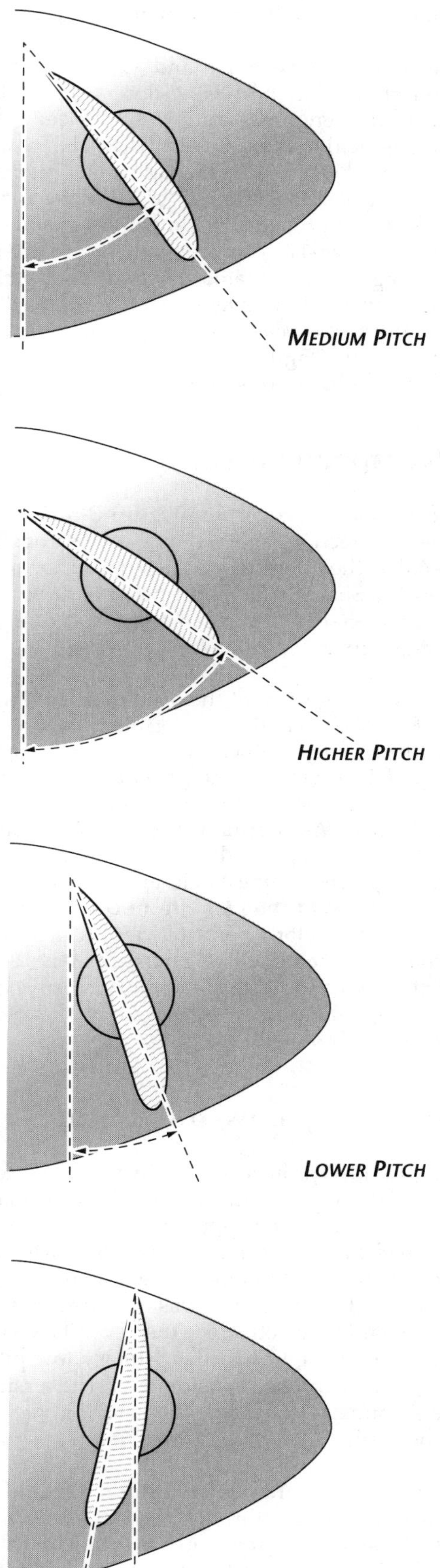


**Figure 3-5-9. A visual representation of the various propeller pitches at the 30-inch station**

**Figure 3-5-10. A primary governor**

by the governor, eventually the *low-pitch stop* is reached, and now the blade angle becomes fixed and cannot continue to a lower pitch. The governor is therefore incapable of restoring the onspeed condition, and propeller r.p.m. falls below the selected governor r.p.m. setting.

## Primary Low-pitch Stop

Whenever the propeller r.p.m. is below the selected governor r.p.m., the propeller blade angle is at the low-pitch stop.

For example, if the propeller control is set at 2,000 r.p.m. but the propeller is turning at less than 2,000 r.p.m., the blade angle is at the low-pitch stop.

On many types of airplanes, the low-pitch stop is simply at the *low-pitch limit of travel*, determined by the propeller's construction. But with a reversing propeller, the extreme travel in the low-pitch direction is past 0°, into reverse or negative blade angles. Consequently, the low-pitch stop on this propeller must be designed in such a way that it can be removed or repositioned when reversing is desired.

The low-pitch stop is created by mechanical linkage sensing the blade angle. The linkage causes a valve to close to stop the flow of oil coming into the propeller dome. Since this oil causes low-pitch and reversing, once it is blocked off a low-pitch stop has been created.

The position of the primary low-pitch stop is controlled from the cockpit by the power lever. Whenever the power lever is at idle or above, this stop is set at 15° (±1°) blade angle. Bringing the power lever aft of idle progressively repositions the stop to blade angles of less than 15° (see Figure 3-5-11).

Keep in mind that just because the primary low-pitch stop has been moved back to smaller angles than 15°, this only affects the actual blade angle when it is on the low-pitch stop. It follows, then, that as long as the propeller r.p.m. is still on selected governor setting, bringing the power lever aft of idle will not cause the propeller to reverse.

> **CAUTION:** *Though the propeller will not be made to reverse by bringing the power lever aft of idle, this action does tend to damage the linkage which provides the low-pitch stop.*

Only when the propeller r.p.m. is below the selected governor r.p.m. does reversing actually occur when the power lever is brought aft. This is because, in this condition, the blade angle is on the low-pitch stop, which is being repositioned into the reverse range.

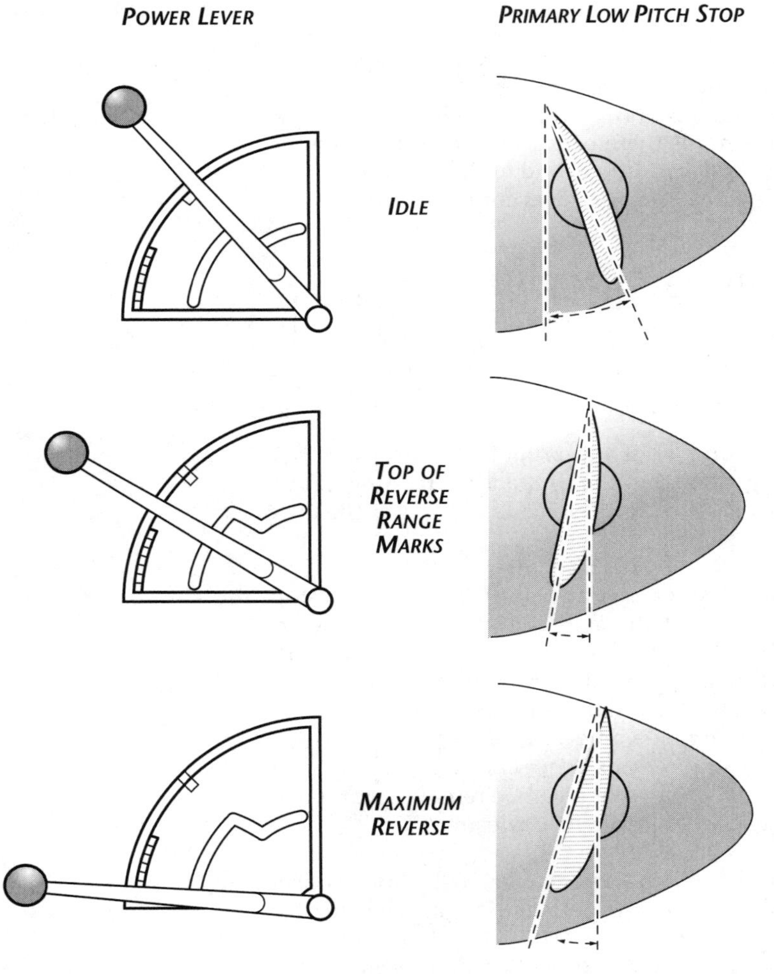

The region between 15° and -5° blade angle is referred to as the *beta-for-taxi range.* In this range, the engine's compressor speed ($N_1$) remains at the value it had when the power lever was at idle (50-70 percent) based on condition-lever position. From -5° to -11° blade angle, the $N_1$ speed progressively increases to a maximum value at -11° of approximately 85 percent. This region, designated by red and white stripes on the power-lever gate, is referred to as the *beta-plus-power range* and ends at maximum reverse. Figure 3-5-12 is an illustration of a complete governor system.



**Figure 3-5-11. The correlation between power-lever position and primary low-pitch stop position**

## Overspeed Governor

The overspeed governor (Figure 3-5-13) provides protection against excessive propeller speed in the event of primary governor malfunction. Since the PT6's propeller is driven by a *free turbine,* overspeed can rapidly occur if the primary governor fails.

The operating point of the overspeed governor is set at 4 % greater than the primary governor's maximum speed. Since the maximum propeller speed selected on the primary governor is 2,200 r.p.m., the overspeed governor is set at 2,288 r.p.m. As a runaway propeller's speed reaches 2,288 r.p.m., the overspeed governor will begin increasing blade angle to a higher pitch to prevent the r.p.m. from continuing its rise. From a pilot's point of view, a propeller tachometer stabilized at approximately 2,288 r.p.m. would indicate failure of the primary governor and proper operation of the overspeed governor.

## Secondary Low-pitch Stop

To provide protection against inadvertent reversing due to malfunction of the primary low-pitch stop, the propeller is equipped with a *secondary low-pitch stop.* Like the primary, this secondary stop blocks the flow of oil into the propeller dome. Whereas the primary stop is mechanically activated, the secondary stop is electric. And, unlike the primary low-pitch stop, which can be repositioned for reversing, the secondary low-pitch stop is set only at 12° blade angle.

Because 15° and 12° are so similar, it would be nearly impossible for the pilot to know that his primary low-pitch stop has failed and that the blade angle is at the secondary low-pitch stop if no indicating system were incorporated into the design. Thus, whenever electric power triggers the secondary low-pitch stop, a red annunciator panel light illuminates.

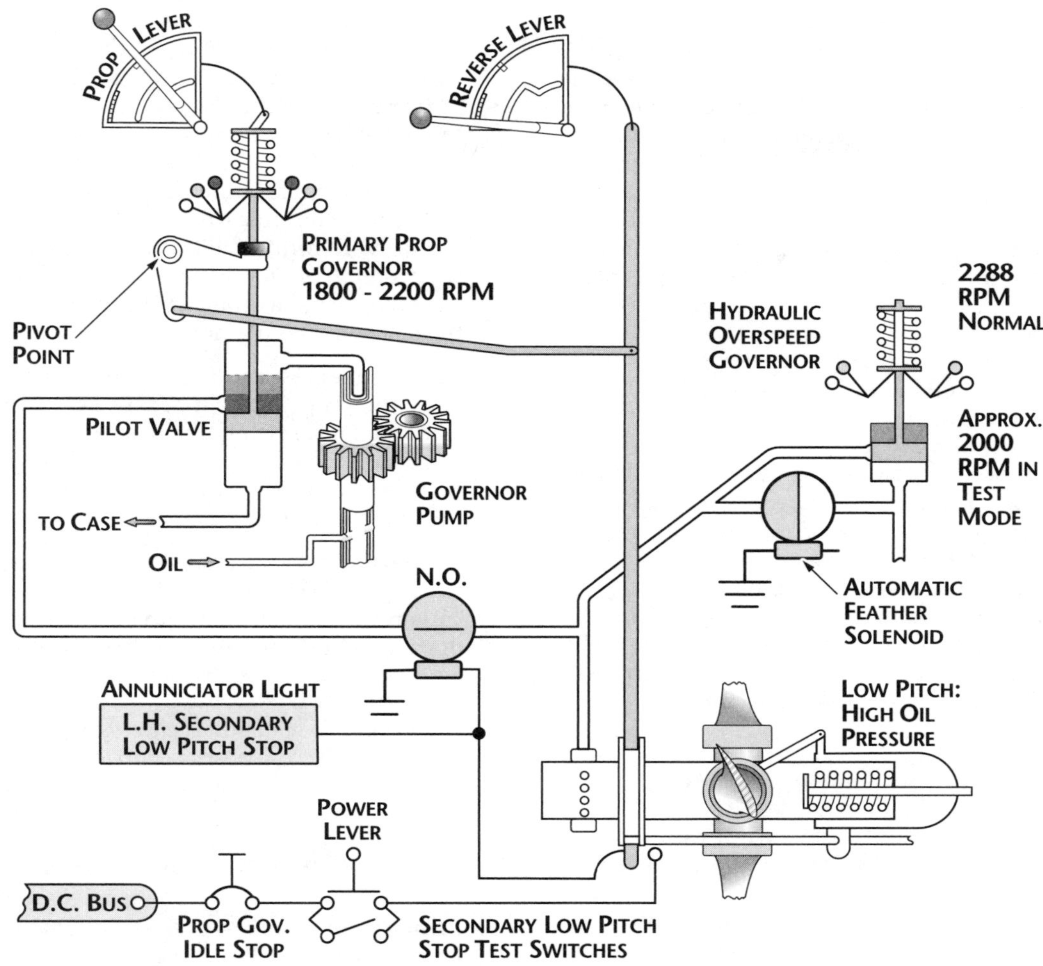


**Figure 3-5-12. Schematic of a complete governor system**

When the secondary low-pitch stop is triggered, electric power is sent to the secondary low-pitch stop solenoid, which then closes a normally open valve to prevent any oil from entering the propeller dome. However, the oil already in the dome will slowly leak out around the transfer gland, which allows the propeller to slowly move toward higher pitch. Moving away from 12°, the secondary low-pitch stop solenoid valve will open, allowing more oil to enter and sending the blade angle back to 12°. The secondary stop then triggers again and, once again, the leak moves the blade angle to more than 12°, the valve reopens, and the cycle repeats indefinitely.

When the power lever is lifted to get *aft* of idle, a switch in the pedestal opens to remove electric power from the secondary low-pitch stop system. This prevents the 12° stop from operating and allows the blade angle to continue past 12° into reverse. In summary, lifting the power lever removes the secondary low-pitch stop.

If the secondary low-pitch stop fails by not working (e.g., the solenoid valve sticks open,

**Figure 3-5-13. A PT6 overspeed propeller governor**

the electrical circuit is open or the battery and generator switches are accidentally turned off), the pilot would normally not be aware of this failure in flight. Why? It is because the secondary stop does not have to operate when the primary stop is operative. In this situation, however, there is no longer back-up protection for a primary-stop failure, so the operator should recognize the necessity for frequent pre-flight checks of his propeller system. That means ground checks before and after maintenance.

## Before Take-off Checks

In the Normal Procedures section of the Airplane Flight Manuals, the *Before Take-off Checklist* contains the propeller checks or tests by which a pilot or maintenance technician can determine the status of the propeller system. These tests may be omitted for quick turn-around at the pilot's discretion, and many pilots run them only on the day's first flight. The checks involve the following steps (see Figure 3-5-14 for reference locations):

### Testing Overspeed Governors

1. **Propeller controls — high r.p.m.** By assuring that the propeller controls are full forward (normally, they will already be there on the ground), the operator knows that the primary governor is set at 2,200 r.p.m. Since his idle r.p.m. on the ground is below this value, he knows that his blade angle is at the low-pitch stop.
2. **Power levers — below 1,900 r.p.m.** This step will usually be accomplished automatically, since low-idle propeller speed is near 1,000 r.p.m.
3. **Overspeed governor test switches — hold-to propeller governor test.** The operator reaches his left hand across to the test switches located on the pilot's right subpanel and holds them both up to the PROP GOV TEST position. This resets the overspeed governors from 2,288 r.p.m. down to approximately 2,000 r.p.m.
4. **Power levers — increase to stabilized r.p.m.** (Observe ITT and torque limits) As power is added, the propeller r.p.m. increases until the governor speed is reached, then the r.p.m. stabilizes (remains constant) even though torque and $N_1$ are still increasing. If the stabilization occurs at approximately 2,000 r.p.m., then apparently the overspeed governor is operating correctly. But if the stabilization does not occur until reaching 2,200 r.p.m., then the primary governor is working, but the overspeed governor is not testing as it should. If the test switches were released when the propeller r.p.m. has stabilized at approximately 2,000 r.p.m. on the overspeed governors, the propeller r.p.m. should increase to 2,200 r.p.m. If the switches were then held again, the r.p.m. should drop back to the test setting, but this drop occurs very abruptly. To minimize this unnecessary wear, step No. 2 (above) guarantees that the r.p.m. is below the test setting before the test switches are applied.
5. **Power levers — reduce to 1,900 r.p.m.** Once the propeller r.p.m. drops below the stabilized test setting, the blade angle has again become fixed at the low-pitch stop. In other words, the propellers are off of the overspeed governors.
6. **Propeller test switches — release.** The overspeed governors should automatically return now to 2,288 r.p.m. The r.p.m. does not rise, however, since power has been reduced to 1,900 r.p.m.

**CAUTION:** *To minimize blade erosion, this check should be accomplished on a clean run-up area, free of sand and gravel.*

## Testing Primary Governors

1. **Primary governors — exercise at 1,900 r.p.m.** As the propeller levers are moved from full forward aft to (but not into) the feather detent, the r.p.m. should decrease from 1,900 r.p.m. to 1,800 r.p.m. or below. Then, as the levers are repositioned full forward, the r.p.m. should return to 1,900. This shows that the primary governor is operating.

Since the power levers are still set to give 1,900 r.p.m., this is a convenient time to perform either or both of these checks, as described in the manuals.

### Testing Secondary Low-pitch Stops on Reversing Propellers

1. **Condition levers — high idle.** 70% $N_1$ is the reference power setting used during these checks. High idle will cause the propeller to turn with more authority (and be less affected by wind conditions) than low idle. The $N_1$ speed will remain constant, so that all propeller r.p.m. changes will be caused by blade-angle changes alone and not by $N_1$ changes.
2. **Power levers — idle.** (Read propeller r.p.m.) Normally (without Autofeather), the power levers will still be set forward

enough to give 1,900 r.p.m. during step No. 1. But now they are reduced to IDLE to let the engine's speed decrease to 70-percent $N_1$. The r.p.m. will now be below 2,200, which is the selected governor setting, and the blade angle should be at 15°, on the primary low-pitch stop (see Figure 3-5-14). By reading the r.p.m., the operator can determine the correctness of the *primary low-pitch stop setting* (PLPS). Suppose the PLPS were accidentally set to 200 instead of 15°. The higher pitch would cause a lower r.p.m. Conversely, a blade angle less than 15° would give a higher r.p.m. At 15° blade angle and 70-percent $N_1$, the propeller speed should be approximately 1,500 r.p.m. Altitude, temperature and wind conditions will vary the r.p.m. somewhat, but an r.p.m. consistently greater than 1,600 or less than 1,400 usually is an excellent indication that the PLPS is incorrectly adjusted. In this step, then, the position of the PLPS is being checked.

3. **Propeller test switches — hold-to secondary-idle stop (low-pitch stop) test.** This action completes an electrical circuit, which bypasses the power lever switches — the ones that open when the power levers are lifted. Thus, the *secondary low pitch-stop* (SLPS) will be capable of operating even after the levers are lifted.

4. **Power levers — align aft edge with top of reverse range (beta range) marks.**

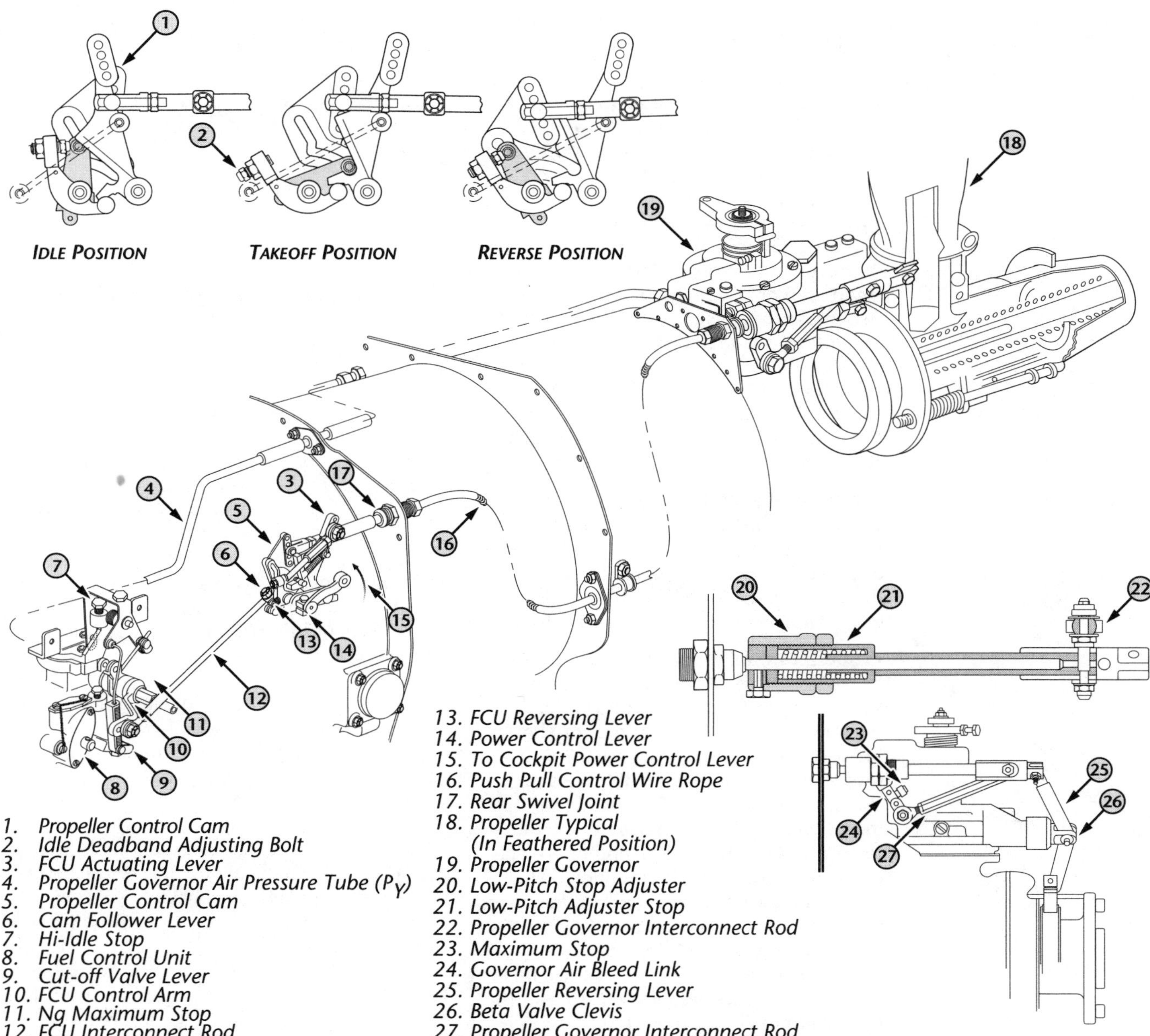


**Figure 3-5-14. To properly rig a propeller control system, all these components have to be adjusted in order**

The PLPS is now being reset prom 15° to -5° (see Figure 3-5-14). However, as the blade angle reaches 12°, the SLPS should operate, and steps No. 5 and No. 6 will check that it does. Be careful not to force the power levers too far aft into the striped range, as the PLPS linkage may be damaged.

5. **Secondary low-pitch stop lights — check on.** This is one check that the SLPS did begin operating in step No. 4, but the lights themselves give no guarantee that the SLPS works correctly. The SLPS solenoid valve could be completely inoperative, yet if the sensor operated, the lights would still illuminate. Also, the SLPS sensor could be set anywhere between 15° and -5°, and the lights could operate, even though an SLPS at 0° would be useless for true in-flight protection. Step No. 6 is much more important than step No. 5 and must not be overlooked.

6. **Propeller r.p.m. — check stabilized at 210 ±40 above r.p.m. in step No. 2.** As the blade angle decreases from 15° to 12° without a change in $N_1$, the propeller r.p.m. should increase 210 r.p.m. (with a ±40 r.p.m. tolerance). If the SLPS is set too flat, the r.p.m. rise will be excessive; too high a pitch (closer to 15°) will cause too small an increase of r.p.m. An excessive rise (more than 250 r.p.m.) indicates that the SLPS is at too flat a blade angle and, if needed in flight, it could cause excessive and dangerous drag. The SLPS sensor should be adjusted to give the proper r.p.m. rise to correct this unsatisfactory condition.

7. **Propeller test switches — release.** This step opens the bypass around the power lever switches, which are themselves open, since the power levers have been lifted. Thus, the SLPS is removed, and the blade angle should move from 12° to -5°.

8. **Propeller r.p.m. — check.** (Must increase above step No. 6) As blade angle moves from the SLPS (12°) to the PLPS (set at -5° by the power lever position), the r.p.m. should increase due to less drag on the turning propeller. If the propeller r.p.m. does not change, perhaps the SLPS is not releasing when the power levers are lifted; thus, reversing is not available. Another possible cause of this unchanged condition is that the power lever linkage is out of adjustment badly enough that the top of the striped area only corresponds to 12° instead of -5°, so there is no reason for the blades to go flatter when the switches are released. To confirm this latter possibility, move the power levers forward and aft to determine where peak r.p.m. (flattest blade angle) occurs. If it is not near the top of the striped region, the linkage is not set correctly.

9. **Power levers — idle.** Once the check described in step No. 8 is completed; the blade angle returns to 15°. On those aircraft that have a locking device on the SLPS to prevent blinking, it is now necessary to momentarily depress the propeller test switch to extinguish the light.

After this step in the checklist, instrument vacuum and pneumatic pressure are checked while the engines are still at high idle. Then, after the condition levers have been retarded to low idle, propeller feathering (manual, as compared to autofeather) is checked. In this free turbine engine, the propeller may be allowed to completely feather with the compressor operating at low idle, with no engine damage sustained. However, operation on the ground, in feather, for extended periods may overheat the fuselage and possibly damage nose-mounted avionics, because hot exhaust gases are not being blown aft by the propeller's air blast.

## Section 6

# Propeller Ice-control Systems

Effects of propeller icing. Ice formation on a propeller blade, in effect, produces a distorted blade airfoil section, which causes a loss in propeller efficiency. Generally, ice collects asymmetrically on a propeller blade and produces propeller unbalance and destructive vibration.

Several types of ice may form on propeller blades. These may include *glaze* or clear, *rime* and *glime* (a mixture of glaze and rime). Each type is described below:

**Glaze ice.** Glaze ice will appear as a hard, glossy, heavy covering. This type of ice forms when, after initial impact, the remaining portion of the raindrop flows out evenly over the surface, freezing as a smooth sheet of solid ice.

**Rime ice.** Rime ice appears with irregular shape and a rough surface. It is brittle and frostlike, and lighter than glaze ice. Rime ice forms when the raindrops are small. The liquid portion remaining after the initial impact freezes rapidly before the drop has time to spread. The small, frozen droplets trap air between them, giving the ice a white, frosty appearance.

**Glime ice.** Glime ice is a mixture of glaze and rime ice. Glime ice is a hard, rough mixture that can form rapidly. Formation occurs when raindrops vary in size or when liquid drops are mixed with snow or ice particles.

Because of the rapid deterioration of performance that may be caused by propeller icing, it is very important that the pilot be able to prevent formation of ice or to remove any that has already formed. This may be accomplished using propeller anti- or de-icing systems. *Anti-icing systems* are designed to prevent the formation or adhesion of propeller ice, while *de-icing systems* are designed to remove ice that has already formed on the propeller.

This may be accomplished using fluid or electrical systems.

**Fluid anti-icing systems.** A typical fluid system (Figure 3-6-1) includes a tank to hold a supply of anti-icing fluid. This fluid is forced to each propeller by a pump. The control system permits variation in the pumping rate so that the quantity of fluid delivered to a propeller can be varied, depending on the severity of icing. Fluid is transferred from a stationary nozzle on the engine nose case into a circular U-shaped channel (*slinger ring*) mounted on the rear of the propeller assembly. The fluid, under pressure of centrifugal force, is transferred through nozzles to each blade shank.

Because airflow around a blade shank tends to disperse anti-icing fluids to areas on which ice does not collect in large quantities, *feed shoes* or *boots* are installed on the blade leading edge. These boots are a narrow strip of rubber, extending from the blade shank to a blade station that is approximately 75 percent of the propeller radius. The feed shoes are molded with several parallel open channels in which fluid will flow from the blade shank toward the blade tip by centrifugal force. The fluid flows laterally from the channels, over the leading edge of the blade.

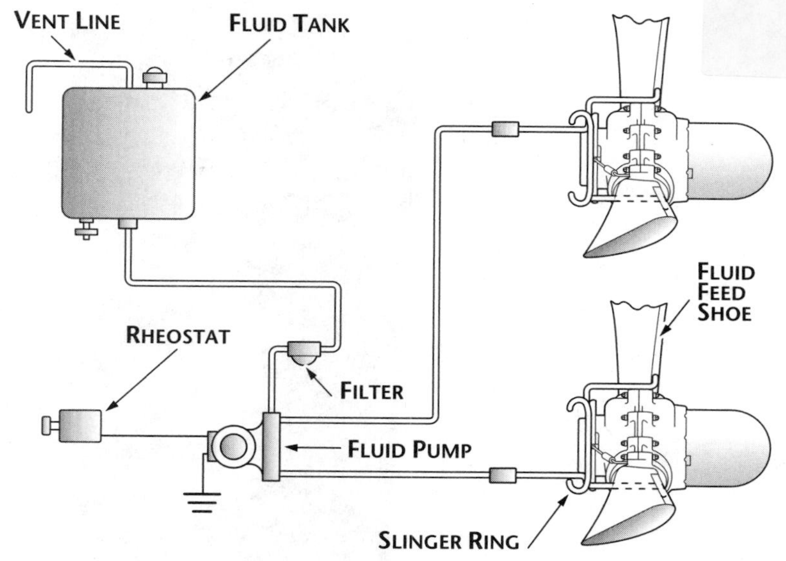


Figure 3-6-1. A typical propeller fluid anti-icing system

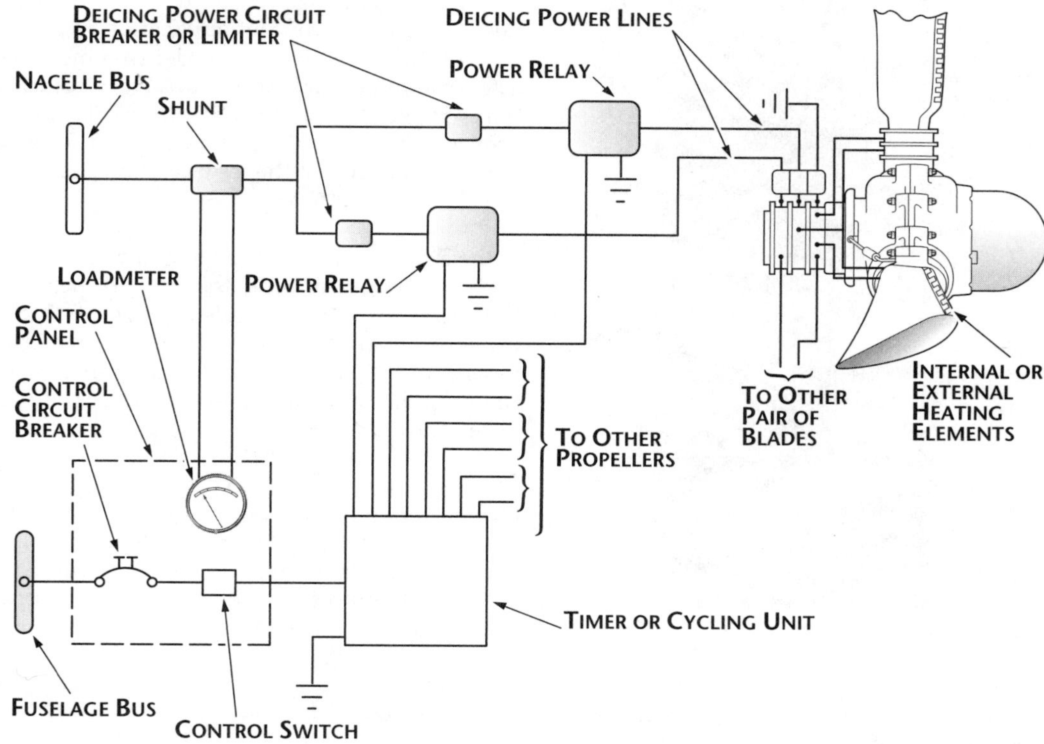


Figure 3-6-2. Typical electrical de-icing system

**Figure 3-6-3. A foil de-icer**

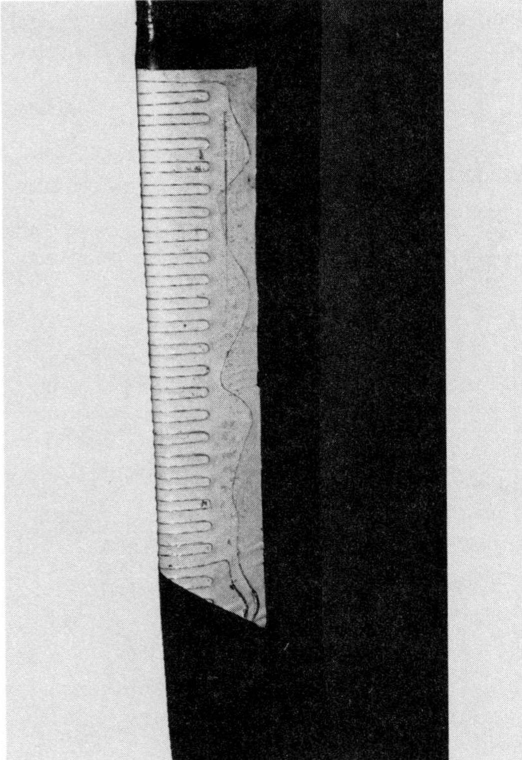

**Figure 3-6-4. A wire de-icer**

Isopropyl alcohol is used in some anti-icing systems because of its ready availability and low cost. Phosphate compounds are comparable to isopropyl alcohol in anti-icing performance and have the advantage of reduced flammability.

**Electrical de-icing systems.** Electrical de-ice systems are the current system of choice. The electrical propeller icing-control system (Figure 3-6-2 next page) is made of three basic parts:

- An electrical source
- A resistance heating element
- The system controls

The electrical source sends electrical energy through foil (see Figure 3-6-3) or wires (see Figure 3-6-4) to a slip ring-and-brush assembly located at the propeller hub. This electrical energy pushes the brushes and slip rings to the resistance heating elements, located on the prop blades and the spinner. The heating elements may be coated, either internally or externally, on the blades and spinner. Generally, the blade elements are located on the inside of rubber boots that are fastened to the blades by adhesive.

Icing control is accomplished by converting electrical energy to heat energy in the heating element. Balanced ice removal from all blades must be obtained as nearly as possible if excessive vibration is to be avoided. To obtain balanced ice removal, variation of heating current in the blade elements is controlled so that similar heating effects are obtained in opposite blades.

Electrical de-icing systems are usually designed for intermittent application of power to the heating elements, in order to remove ice after formation but before excessive accumulation. Proper control of heating intervals aids in preventing runback, since heat is applied just long enough to melt the ice face in contact with the blade.

If the heat applied to the icing surface for an excessive length of time, the ice will melt and the moisture will be pulled outward to the propeller blade tips by centrifugal force. The moisture will freeze on protected portions of the propeller and can cause severe out-of-balance conditions.

Cycling timers are used to energize the heating element circuits for periods of 15-30 seconds, with a complete cycle time of 2 minutes. A cycling timer is an electric motor-driven contactor that controls power contactors in separate sections of the circuit.

The controls for the electrical propeller de-icing system include:

- On-off switches
- Ammeters or load meters

- Protective devices, such as circuit breakers or current limiters

The *ammeters* or *load meters* allow monitoring of the individual circuits and reflect the operation of the timer.

To prevent damage caused by overheating of the element, the system should only be used in flight. Operation of the system without the engine operating can cause damage to the slip rings caused by brush arcing. When shop-testing electrical system de-icing systems, be sure to follow the manufacturer's recommended procedures. Too much heat for too long a time period will damage the boots.

## Section 7

# *Propeller Inspection and Maintenance*

The propeller inspection requirements and maintenance procedures discussed in this section are representative of those in widespread use on most of the propellers described in this chapter. No attempt has been made to include detailed maintenance procedures for a particular propeller, and all pressures, figures and sizes are solely for the purpose of illustration and do not have specific application. For maintenance information on a specific propeller, always refer to applicable manufacturer's instructions.

## Propeller Defects

The list below includes definitions for the most common defects observed during propeller inspection. These definitions are only a starting point, and the student or inexperienced AMT should always enlist the help of someone with experience in identifying and classifying propeller damage.

- **Battering.** A damaged or worn area caused by hard usage or heavy blows
- **Binding.** An area of damage caused by parts sticking together
- **Blistering.** Raised areas of plate, indicative of a lack of bond between plating and base metal or overheating above the melting temperature of the plating material
- **Bowing.** A bent blade, usually caused by foreign objects.
- **Brinelling.** Having one or more indentations on bearing races, usually caused by high-static loads or application of force during installation or removal. Indentations are rounded or spherical due to the impression left by the contacting balls or rollers of the bearing.
- **Burring.** A ragged or turned-out edge, usually resulting from machine processing
- **Chafing.** A condition caused by rubbing action between two parts under light pressure, which results in wear
- **Checking.** A condition in which there are numerous small cracks in the metal, usually caused by machine processing
- **Chipping.** The breaking away of pieces of material, usually caused by excessive stress concentration or careless handling
- **Circumferential scratching.** Damage from scratching on the external boundary or surface of an object
- **Corrosion.** Loss of metal by a chemical or electrochemical action, the products of which are generally easily removed by mechanical means (iron rust is an example of corrosion)
- **Cracking.** A partial separation of material, usually caused by vibration, overloading, internal stresses, defective assembly or fatigue — depth may be anywhere from a few thousandths of an inch to the full thickness of the piece
- **Denting.** A small, rounded depression in a surface, usually caused by the part being struck with a rounded object
- **Deposits.** Material that has accumulated in recessed areas
- **Discoloration.** Discoloration is due to the effects of heat in the presence of oxygen. Surface colors may range from light straw to gray, depending on the environmental temperature and the thickness of the oxide film that is formed on the surface.
- **Disintegration.** A breakdown or decomposition of a part into small objects
- **Distortion.** A twisting or bending of the piece from its original shape due to external forces
- **Eccentric.** Damage not having the same geometric center
- **Erosion.** Loss of metal from the surface by mechanical action of foreign objects, such as grit or fine sand. The eroded area will be rough and may be lined in the direction in which the foreign material moved relative to the surface.
- **Flaking.** The breaking loose of small

Figure 3-7-1. An up-close view of a nicked propeller blade

pieces of metal or coated surfaces, which is usually caused by defective plating or excessive loading.

- **Fracture.** A break through the entire thickness of the material
- **Fraying.** An area worn by rubbing actions
- **Fretting.** A rapid oxidation of metal surfaces caused by cyclic back-and-forth movement of closely fitted parts in the presence of oxygen, characterized by rusty surfaces ranging from black (well-bonded) to a fairly bright brick-red color (often loosely bonded or even free particles)
- **Galling.** A severe condition of chafing or fretting in which a transfer of metal from one part to another occurs, usually caused by a slight movement of mated parts having limited relative motion and under high loads
- **Glazing.** Damage which appears smooth and glossy
- **Gouging.** A furrowing condition in which a displacement of metal has occurred (a torn effect), usually caused by a piece of metal or foreign material between close moving parts
- **Grooving.** A recess or channel with rounded and smooth edges, usually caused by faulty alignment of parts
- **Indentation.** Shallow depressions in the functional surfaces where metal has been displaced but not removed, they are usually rounded, having the shape of the foreign particle, and the original surface finish can still be seen in the depressions
- **Mismatch.** Damage caused when two parts are improperly matched
- **Nick.** A sharp-sided gouge or depression with a V-shaped bottom, generally the result of careless handling of tools and parts (see Figure 3-7-1)
- **Out-of-round.** A damaged hole without the same concentric center
- **Pickup.** A buildup or rolling of metal from one area to another, usually caused by insufficient lubrication, clearances or foreign matter
- **Pitting.** Small, irregularly shaped cavities (usually with dark bottom surfaces) from which material has been removed by corrosion, which can be caused by oxidation (rust), acid attack (etching or perspiration) or electrolytic attack (non-distilled water)
- **Scoring.** A series of deep scratches caused by foreign particles between moving parts, or careless assembly or disassembly techniques (see Figure 3-7-2)
- **Scratches.** Shallow, thin lines or marks, varying in degree of depth and width, caused by the presence of fine foreign particles during operation or contact with other parts during handling (see Figure 3-7-2)

Figure 3-7-2. An example of scoring/scratching

- **Sheared.** Damage caused when a part is stressed in two different directions
- **Spalling.** The breakdown of a functional/active surface due to fatigue, usually in the form of irregularly shaped pits, the edges of which have indications of cracking and further progression
- **Stripped.** Damaged or unusable threads of a bolt or screw
- **Surface breakdown.** Breakdown of surface material caused by corrosion

## Propeller Inspection

Propellers must be inspected regularly. The exact time interval for particular propeller inspections is usually specified by the propeller manufacturer. The regular daily inspection of propellers varies little from one type to another.

Typically, it is a visual inspection of propeller blades, hubs, controls and accessories for security, safety and general condition (see Figure 3-7-3).

The following paragraphs explain general inspection procedures:

**Inspecting blades for damage.** Visual inspection of the blades does not mean a careless or casual observation. The inspection should be meticulous enough to detect any flaw or defect that may exist.

**Checking hub for looseness.** Check to make sure bolts are properly installed and tightened. If nuts can be turned, remove cotter keys and re-torque to required setting.

**Checking blades for looseness.** Check to see that blades are properly installed, in accordance with the applicable maintenance manual.

**Checking external surfaces for metal propellers.** Metal propellers and blades are susceptible to fatigue failure, resulting from concentration of stresses at the bottom of sharp nicks, cuts and scratches. It is especially necessary, therefore, to frequently and carefully inspect them for such injuries. Propeller manufacturers have published service bulletins and instructions that prescribe the manner in which these inspections are to be accomplished.

**Checking propeller governor.** Check propeller governor for leakage, security of attachment and general condition. Inspect governor oil for metal particles.

**Local etching.** Local etching or another approved non-destructive testing process may be used to detect cracks on propellers. Make certain that this is accomplished in accordance with the most recent directions provided by the propeller manufacturer.

**Figure 3-7-3. An AMT must carefully inspect all components of a propeller at regular intervals, based on manufacturer's specifications.**

Inspections performed at greater intervals of time (e.g.: 25, 50 or 100 hours) usually include a visual check of:

- Blades, spinners and other external surfaces, for excessive oil or grease deposits
- Weld and braze sections of blades and hubs, for evidence of failure
- Blade, spinner and hubs, for nicks, scratches or other flaws. Use a magnifying glass if necessary.
- Spinner or dome-shell attaching screws for tightness
- Lubricating oil levels, when applicable

If a propeller is involved in an accident and a possibility exists that internal damage may have occurred, the propeller should be disassembled and inspected. Whenever a propeller is removed from a shaft, the hub cone seats, cones and other contact parts should be examined to detect undue wear, galling or corrosion.

During major overhaul, the propeller is disassembled and all parts are inspected and checked for size, tolerances and wear. A magnetic inspection or other type of nondestructive test is usually made at this time to determine whether any fatigue cracks have developed on the steel components and assemblies.

## Repair of Aluminum Alloy Blades

Nicks, gouges and scratches in the leading or trailing edge and on the blade surface (both face and camber sections) must be removed prior to flight. Operating a propeller with this type of damage may produce a condition in which fatigue cracks will start and blade failure will occur. A small nick may be as serious as a larger one. It is important that all nicks be removed prior to each aircraft operational period. Nicks in the outer 18 inches of the propeller diameter must be treated as critical. This is the area of highest vibratory blade stress. Figure 3-7-4, photo A, shows a propeller leading edge that is badly in need of dressing. An extreme case of a nick repaired by filing is shown in Figure 3-7-4, photo B.

Propeller blades with normal damage can be repaired by a qualified technician in the field. Blades with larger nicks, gouges, etc., that may affect the structure, balance or operation of the propeller should be referred to a qualified propeller repair station for repair or replacement. There is normally sufficient material available to allow numerous repairs prior to replacement. Propeller manufacturers provide excellent reference material detailing the scope and detail of field repairs that are permitted. Make certain that all work is performed in accordance with the appropriate service manuals. AC43.13-1B provides generic repair information if no other source is available.

Local repairs may be made using small files (riffle or jeweler's files), small air grinding or polishing equipment and a variety of abrasive paper or cloth, as shown in Figure 3-7-5 (top of next page). All repairs must be accomplished parallel to the blade axis.

For damaged areas in the leading or trailing edge, begin with a round file, removing damaged material to the bottom of the damaged area. Remove material from this point out on both sides, providing a smooth, faired depression, maintaining the original airfoil concept. Using emery cloth, smooth the area to remove all traces of the initial filing and rework. Crocus cloth may then be used to polish the area. When all rework is complete, inspect the area with a 10x magnifying glass and dye penetrant, assuring no indications of damage or cracks remain.

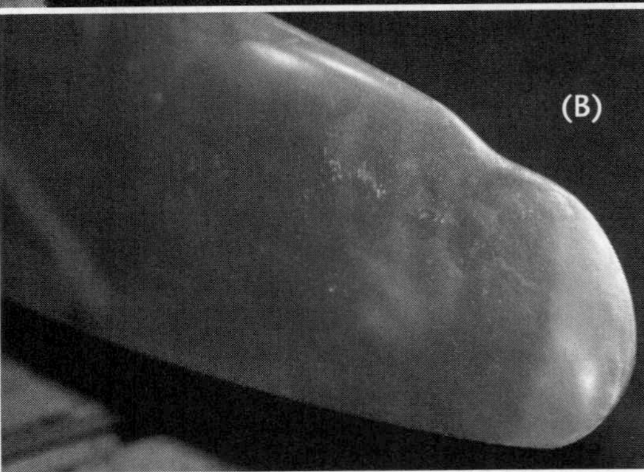


**Figure 3-7-4. A prop leading edge in this condition (A) should be dressed before its next flight. An extreme example of a file-repaired nick (B) near the tip of a propeller.**

Any repair methods, such as rolling or cold-working — which will result in moving the metal and possibly concealing damage — are not acceptable.

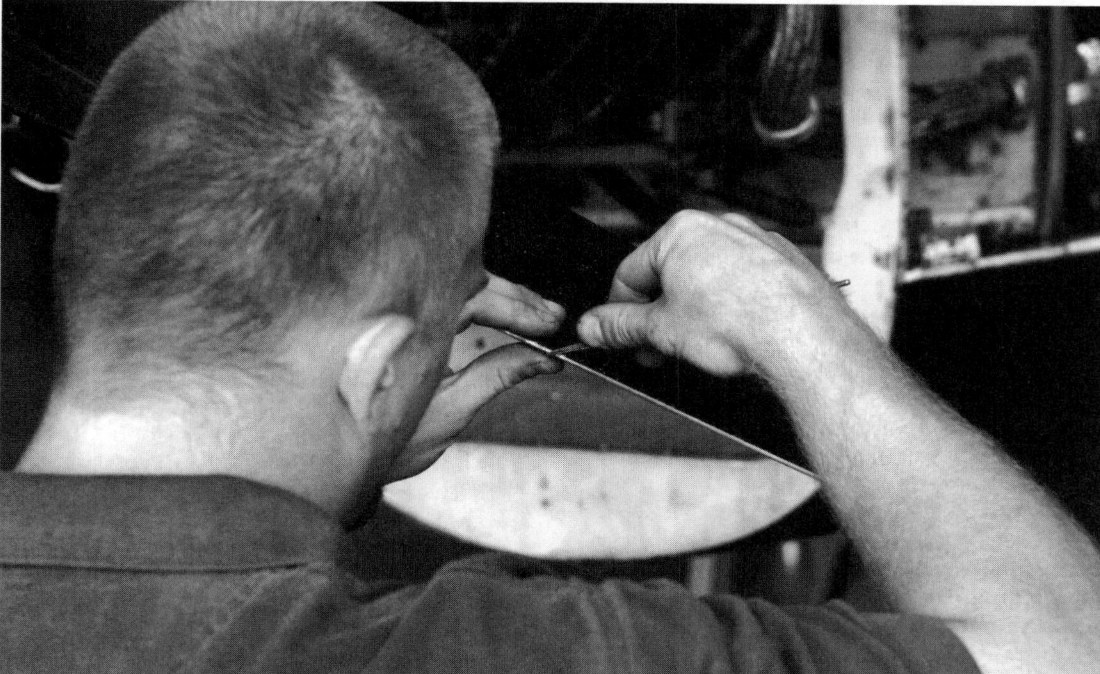

Figure 3-7-5. A technician files the damaged area of a propeller blade

## Blade Tracking

Blade tracking is the process of determining the positions of the tips of the propeller blades relative to each other. Tracking shows only the relative position of the blades, not their actual path. The blades should all track one another as closely as possible. The difference in track at like points must not exceed the tolerance specified by the propeller manufacturer.

The design and manufacture of propellers is such that the tips of the blades will give a good indication of tracking. The following method for checking tracking is normally used:

1. Install a heavy wire or small rod on the leading edge of the aircraft wing or other suitable area of the aircraft until it lightly touches the propeller blade face near the tip (See Figure 3-7-6).
2. Rotate the propeller until the next blade is in the same position as the first blade, and measure the distance between the rod and blade. Continue this process until all blades have been checked.

## Checking and Adjusting Propeller Blade Angles

When an improper blade-angle setting is found during installation or is indicated by engine performance, the following maintenance guidelines are usually followed:

1. From the applicable manufacturer's instructions, obtain the blade-angle setting and the station at which the blade angle is checked.

**CAUTION:** *Do not use metal scribes or other sharp-pointed instruments to mark the location of blade stations or to make reference lines on propeller blades, since such surface scratches can eventually result in blade failure. Tape and fineline markers work best.*

2. Use a *universal propeller protractor* to check the blade angles while the propeller is on the engine.

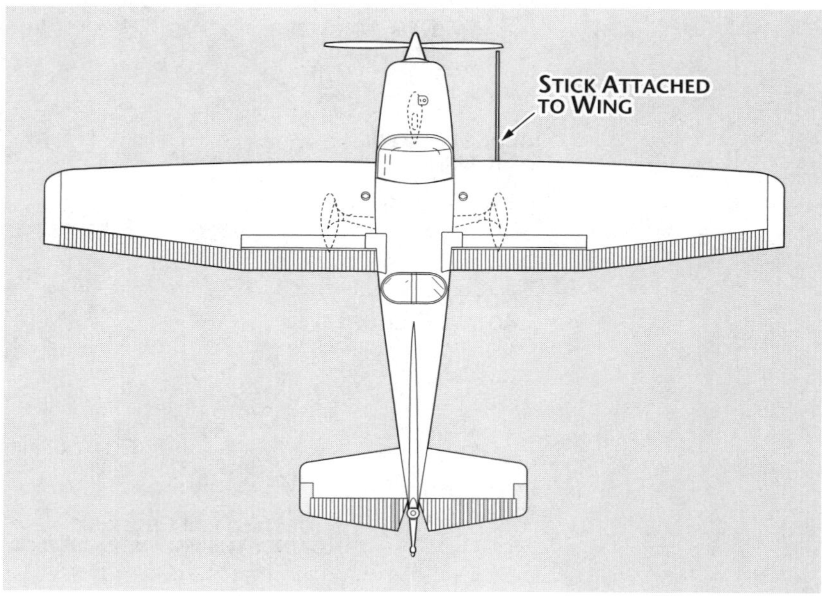


Figure 3-7-6. Checking blade track

**Universal propeller protractor.** The universal propeller protractor can be used to check propeller blade angles when the propeller is on a balancing stand or installed on the aircraft engine. Figure 3-7-7 shows the parts and adjustments of a universal propeller protractor.

The following instructions for using the protractor apply to a propeller installed on the engine:

1. Turn the propeller until the first blade to be checked is horizontal, with the leading edge up.
2. Place the *corner spirit level* (Figure 3-7-7) at right angles to the face of the protractor. Align degree and vernier scales by turning the disk adjuster before the disk is locked to the ring. The locking device is a pin that is held in the engaged position by a spring. The pin can be released by pulling it outward and turning it 90°.
3. Release the *ring-to-frame lock* (a right-hand screw with thumb nut) and turn the ring until both ring and disk zeros are at the top of the protractor.
4. Check the blade angle by determining how much the flat side of the block slants from the plane of rotation. First, locate a point to represent the plane of rotation by placing the protractor vertically against the end of the hub nut or any convenient surface known to lie in the plane of propeller rotation. Keep the protractor vertical by the corner spirit level, and turn the ring adjuster until the *center spirit level* is horizontal. This sets the zero of the vernier scale at a point representing the plane of propeller rotation. Then lock the ring to the frame.
5. Holding the protractor by the handle with the curved edge up, release the *disk-to-ring lock*. Place the *forward vertical edge* (the edge opposite the one first used) against the blade at the station specified in the manufacturer's instructions. Keep the protractor vertical by the corner spirit level, and turn the disk adjuster until the center spirit level is horizontal. The number of degrees and tenths of a degree between the two zeros indicates the blade angle.

**NOTE:** *In determining the blade angle, remember that 10 points on the vernier scale are equal to nine points on the degree scale. The graduations on the vernier scale represent tenths of a degree, but those of the degree scale represent whole degrees. The number of tenths of a degree in the blade angle is given by the number of vernier scale spaces between the zero of the vernier scale and the vernier-scale graduation line nearest to perfect alignment with a degree-scale graduation line. This reading should always be made on the vernier scale. The vernier scale increases in the same direction that the protractor scale increases. This is opposite the direction of rotation of the moving element of the protractor.*

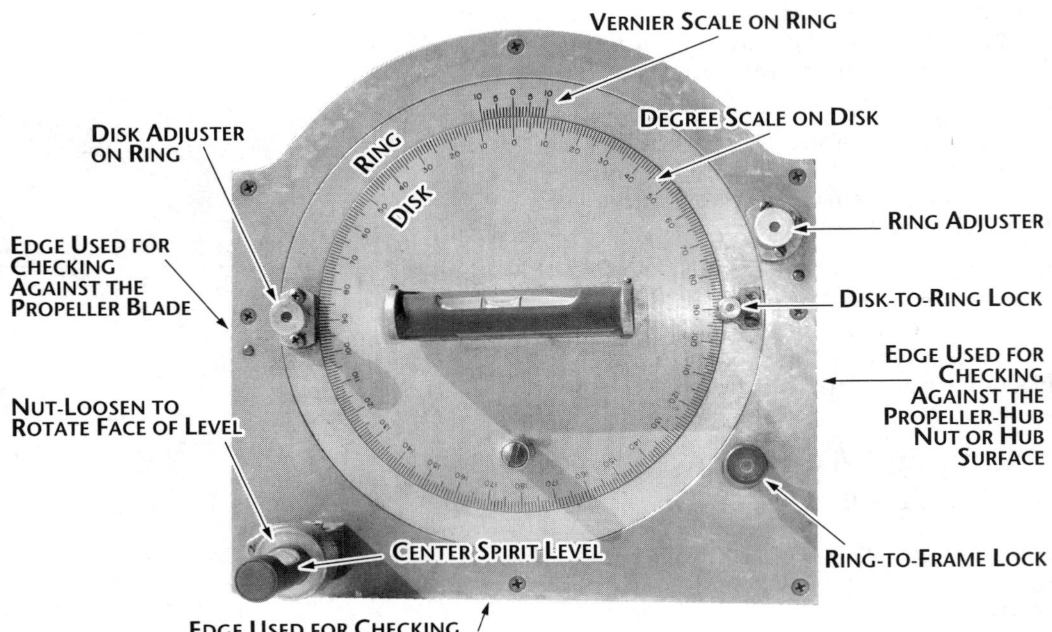


Figure 3-7-7. Universal propeller protractor

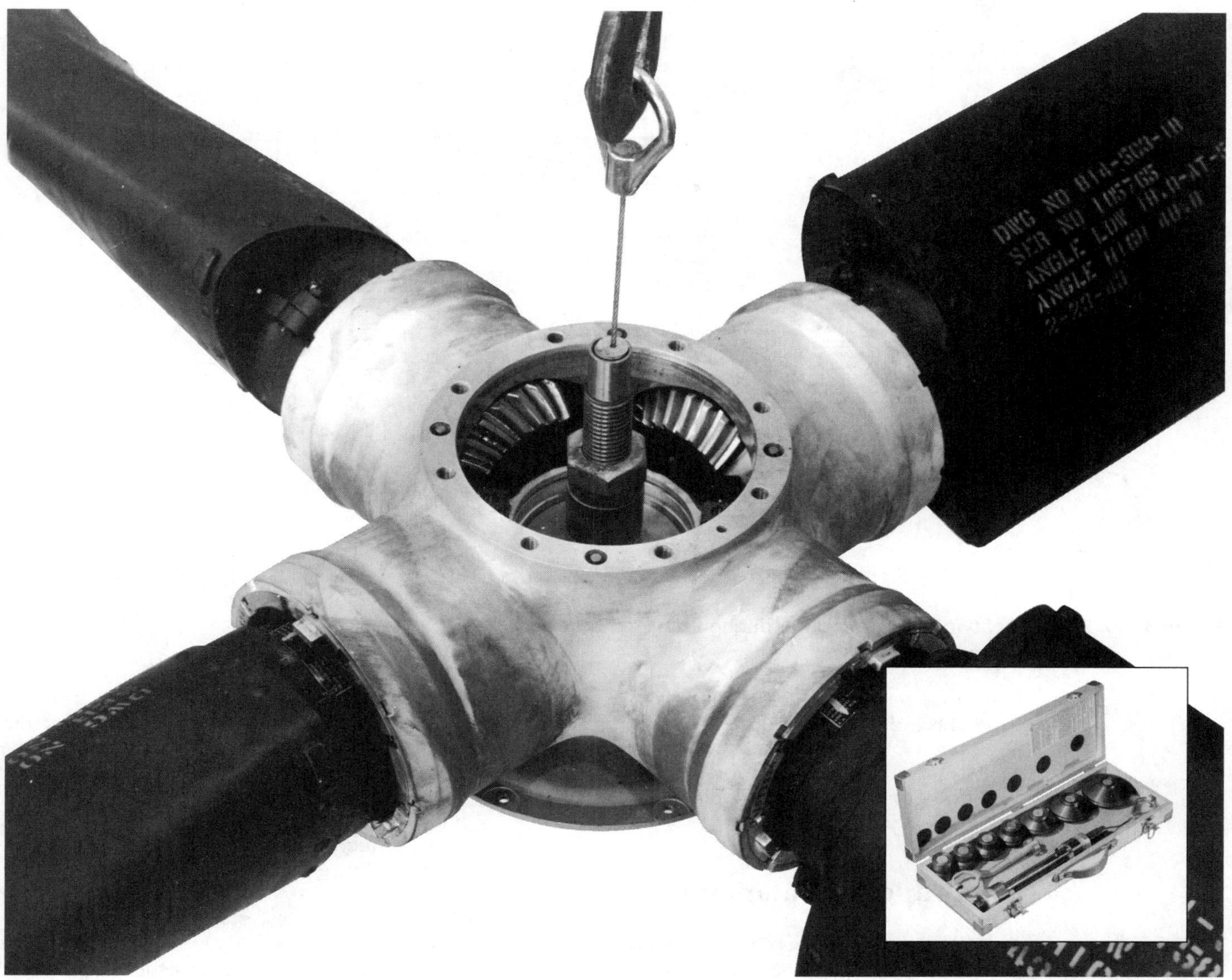

Figure 3-7-8. A suspension balance set

6. After making any necessary adjustment of the blade, lock it in position and repeat the same operations for the remaining blades of the propeller.

## Propeller Vibration

All aircraft are subject to vibration and its effects. Propeller-driven aircraft, as a rule, experience more vibration than a jet aircraft. If the propeller is driven by a reciprocating engine, the likelihood of vibration is increased. A reciprocating engine produces power as a result of controlled explosions in the cylinders. The explosions produce powerful pulses of energy that will cause the engine to vibrate. *Crankshaft dynamic dampeners* and other design features do their best to minimize this vibration but cannot eliminate it. Engine vibration is easily transmitted to the propeller because of their direct mounting.

When powerplant vibration is encountered, it is sometimes difficult to determine whether it is the result of engine vibration or propeller vibration. In most cases, the cause of the vibration can be determined by observing the propeller hub, dome or spinner while the engine is running within a 1,200- to 1,500-r.p.m. range, and determining whether or not the propeller hub rotates on an absolutely horizontal plane. If the propeller hub appears to swing in a slight orbit, the vibration will normally be caused by the propeller. If the propeller hub does not appear to rotate in an orbit, the difficulty is probably caused by engine vibration.

Propeller vibration is normally caused by one of three things:

- An unbalanced propeller blade
- Improper blade tracking
- Variation in blade angle settings

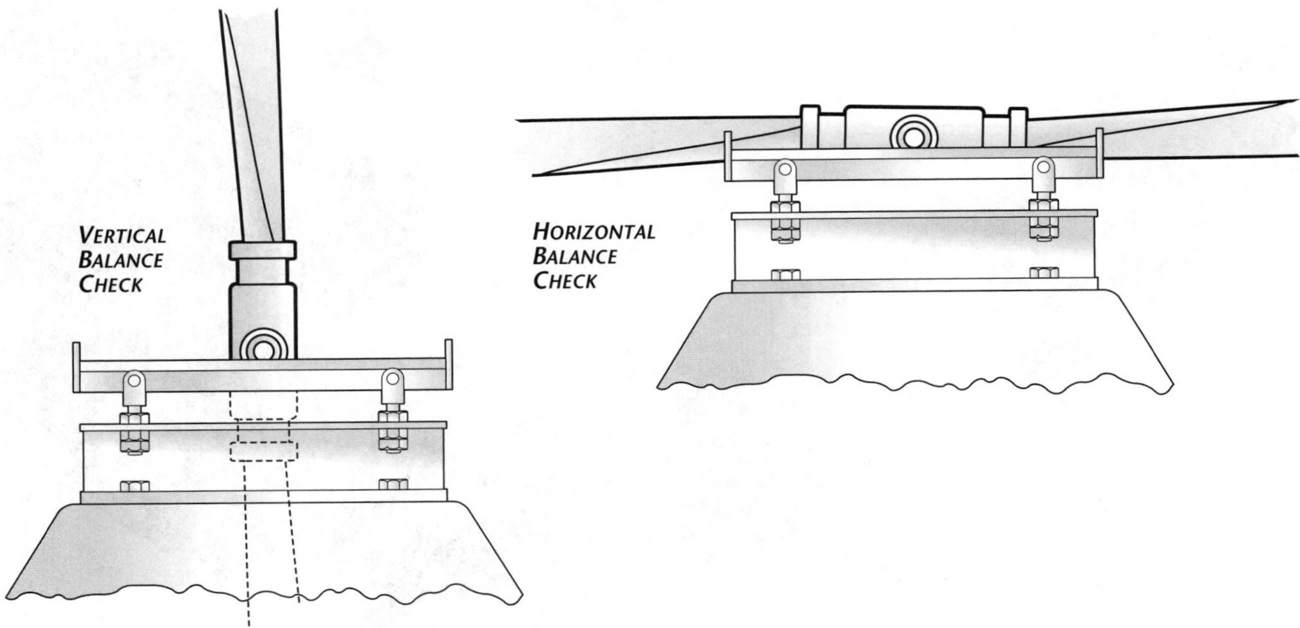


**Figure 3-7-9. Positions of two-bladed propeller during balance check**

To determine which of these is the cause, first check the blade tracking. Next, check the low-pitch blade angle setting. If both the blade tracking and blade angle setting are correct, the propeller should be removed and checked for static balance.

## Propeller Balancing

Propellers may become unbalanced over time and during use. Normal operation exerts a tremendous force on the propeller and causes it to bend and twist, while the speed of rotation imparts significant centrifugal forces. These forces will act to strain the bearings and loosen connections. Operating in dirt or gravel areas — and even the dust in the air — will erode the blades of the propeller. Routine dressing of nicks on the blades and repainting the propeller can add to an unbalanced condition. Periodically checking propeller balance and vibration is an essential part of quality maintenance. The propeller must be perfectly balanced both statically and dynamically.

Propeller unbalance, which is a source of vibration in an aircraft, may be either static or dynamic. *Static unbalance* occurs when the center of gravity of the propeller does not coincide with the axis of rotation. *Dynamic unbalance* exists when the center of gravity of similar propeller elements does not follow in the same plane of rotation.

**Static balancing.** Static balancing can be done by the *suspension method* or by the *knife-edge method*. In the suspension method, the propeller or part is hung by a cord, and any unbalance is determined by noting the eccentricity between a disk firmly attached to the cord and a cylinder attached to the assembly or part being tested. (See Figure 3-7-8 on previous page) The suspension method is used less frequently than the simpler and more accurate knife-edge method.

The knife-edge test stand (Figure 3-7-9) has two hardened steel edges mounted to allow the free rotation of an assembled propeller between them. The knife-edge test stand must be located in a room or area that is free from any air motion, preferably removed from any source of heavy vibration.

The standard method of checking propeller assembly balance involves the following sequence of operations:

1. Insert a bushing in the engine shaft hole of the propeller.
2. Insert a mandrel or arbor through the bushing.
3. Place the propeller assembly so that the ends of the arbor are supported upon the balance stand knife-edges. The propeller must be free to rotate.

If the propeller is properly balanced statically, it will remain at any position in which it is placed. Check two-bladed propeller assemblies for balance, first with the blades in a vertical position and then with the blades in a horizontal position (Figure 3-7-10). Repeat the vertical position check with the blade positions reversed; that is, with the blade which was checked in the downward position placed in the upward position.

Check a three-bladed propeller assembly with each blade placed in a downward vertical position as shown in Figure 3-7-10.

During a propeller static balance check, all blades must be at the same blade angle. Before conducting the balance check, inspect to see that each blade has been set at the same blade angle.

Unless otherwise specified by the manufacturer, an acceptable balance check requires that the propeller assembly have no tendency to rotate in any of the positions previously described. If the propeller balances perfectly in all described positions, it should also balance perfectly in all intermediate positions. When necessary, check for balance in intermediate positions to verify the check in the originally described positions.

When a propeller assembly is checked for static balance and there is a definite tendency of the assembly to rotate, the following corrections to remove the unbalance are allowed:

- The addition of permanent fixed weights at acceptable locations, when the total weight of the propeller assembly or parts is under the allowable limit
- The removal of weight at acceptable locations, when the total weight of the propeller assembly or parts is equal to the allowable limit

The location for removal or addition of weight for propeller unbalance correction has been predetermined by the propeller manufacturer. The method and point of application of unbalance corrections must be checked to see that they are in accordance with applicable drawings.

**Dynamic balancing.** Dynamic unbalance results when the center of gravity of such propeller elements as blades or counterweights do not follow in the same plane of rotation as the rest of the structure. Because the length of the propeller assembly along the engine crankshaft is short in comparison to its diameter, and since the blades are secured to the hub so they lie in the same plane perpendicular to the running axis, the dynamic unbalance resulting from improper mass distribution is usually minimal, provided the tracking requirements are met.

Modern balancing equipment has made it easy to accurately check propeller dynamic balance on most aircraft. The accuracy of this equipment allows balance corrections to be made at levels that were not previously possible. Most

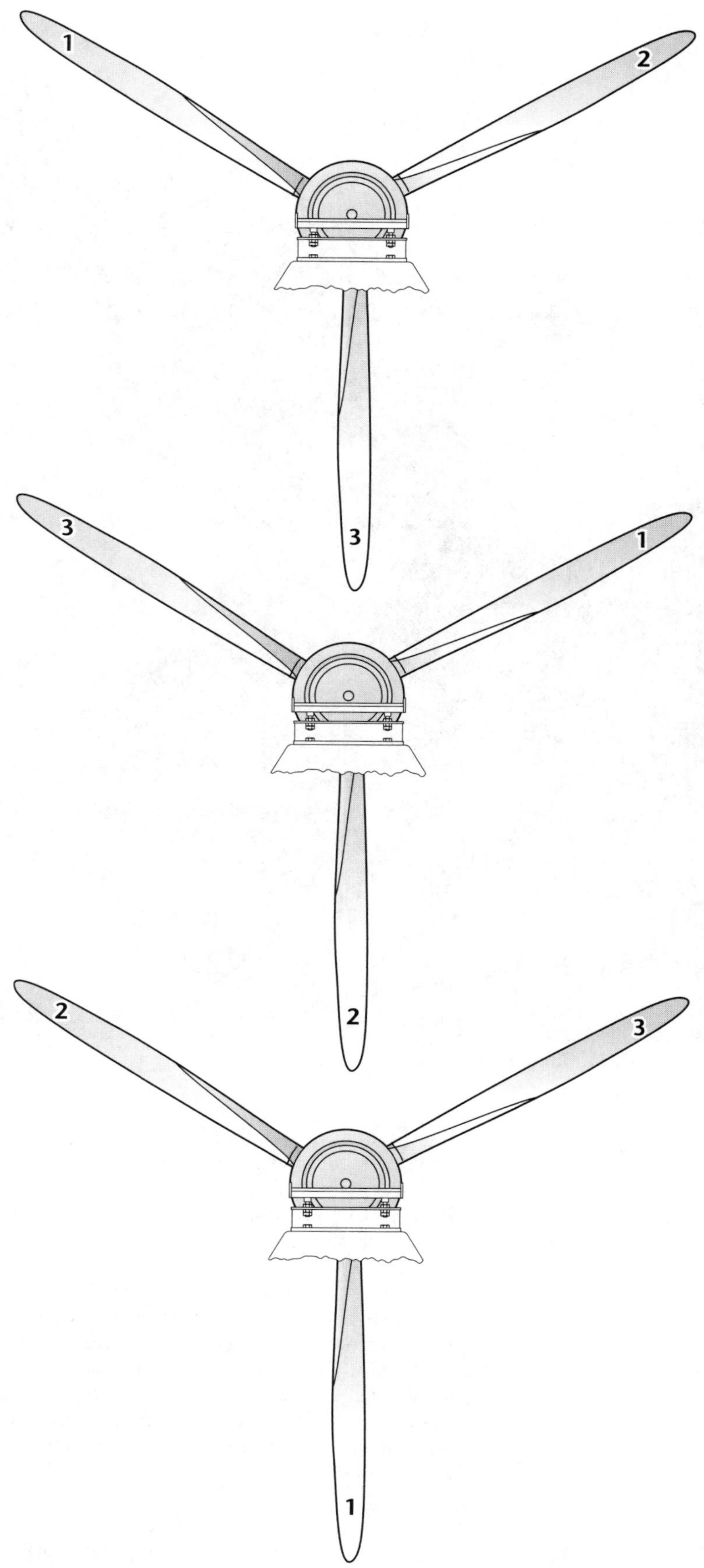


**Figure 3-7-10 Positions of three-bladed propeller during balance check**

**Figure 3-7-11. Blade cuffs**

fixed-wing aircraft could benefit from dynamic propeller balancing, yet few have ever had it done. A balanced propeller system will result in a more comfortable ride, less crew fatigue and reduced wear on many components.

The type of vibration described here occurs when shafts go round and things move about them. In order to describe a specific vibration, two parameters are necessary, *frequency* and *amplitude*. Frequency equals cycles per unit of time (usually seconds), while amplitude is measured in terms of acceleration, velocity or displacement

*Velocity, inches-per-second* (ips), is the general aviation standard in low-frequency applications, including propeller balance. The following scale may be used to demonstrate the severity of vibration in an aircraft cabin:

Excellent/ very smooth = 0.0 to 0.09 ips

Good/relatively smooth = 0.10 to 0.20 ips

Fair/somewhat bumpy = 0.21 to 0.40 ips

Poor/somewhat uncomfortable = 0.41 to 0.60 ips

Very poor/uncomfortable = 0.61 to 0.80 ips

Some studies have shown that the average airplane that has never been dynamically balanced has a vibration level of about 0.45 ips. The average level seen post-propeller balance was 0.039 ips. This generally means the operator will feel a significant difference after the procedure is accomplished.

The dynamic balancing process consists of mounting a *vibration transducer*, or *velocimeter*, to the engine and connecting the transducer by a cable to the Vibration Analyzer/Dynamic Balancer unit. The vibration transducer converts the motion produced by an out-of-balance condition to a weight-and-location solution. Trial weights are then added to the propeller as an experimental condition. The balance is checked again and again until the balance levels are acceptable. When the final weight-and-location settings have been determined, a permanent installation is made.

**Aerodynamic unbalance.** Another type of propeller unbalance, *aerodynamic unbalance*, results when the thrust (or pull) of the blades is unequal. This type of unbalance can be largely eliminated by checking blade contour and blade angle setting.

## Servicing Propellers

Propeller servicing includes cleaning, lubricating and replenishing operating oil supplies.

**Cleaning propeller blades.** Aluminum propeller blades and hubs should be cleaned regularly with soap and water. They should be wiped with a clean, soft cloth. Harsh cleaners or abrasives should never be used, as they may scratch or mar the surface.

If a high polish of the blades is desired, a commercial grade of metal polish may be used. After the desired finish is achieved, all traces of the metal polish must be removed and the blade should be covered with a film of engine oil.

To clean wooden propellers, warm water and a mild soap can be used, together with brushes or cloth.

If a propeller has been subjected to salt water, it should be flushed with fresh water until all traces of salt have been removed. This should be accomplished as soon as possible after the salt water has splashed on the propeller, regardless of whether the propeller parts are aluminum-alloy, steel or wood. After flushing, all parts should be dried thoroughly, and metal parts should be coated with clean engine oil or a suitable equivalent.

**Propeller lubrication.** Proper propeller lubrication procedures, with oil and grease specifications, are usually published in the manufacturer's instructions. Experience indicates that water sometimes gets into the propeller blade bearing assembly on some models of propellers. For this reason, the propeller manufacturer's greasing schedule must be followed to ensure proper lubrication of moving parts. Grease replacement through attached pressure fittings, or zerks, must be done in accordance with the manufacturer's instructions.

> **CAUTION:** *A grease gun can develop enough pressure to damage the propeller's internal grease seals if mishandled.*

On propellers that have self-contained hydraulic units, the reservoir oil level must be checked at specified intervals. Usually, this type of propeller must have one of the blades (generally No. 1) positioned so that the oil is visible in a sight glass on the side of the reservoir. Extreme care must be used when servicing the reservoir to avoid overfilling and servicing with the wrong specification oil.

*Hydromatic propellers* operated with engine oil do not require lubrication. Electric propellers will require oils and greases for hub lubricants and pitch-change drive mechanisms.

**Blade cuffs.** A blade cuff is a metal, wood or plastic structure designed for attachment to the shank end of the blade, with an outer surface that will transform the round shank into an airfoil section (Figure 3-7-11). The cuff is designed primarily to increase the flow of cooling air to the engine nacelle.

The cuffs are attached to the blades by mechanical clamping devices or by using bonding materials. Rubber-based adhesives and epoxy adhesives usually are used as bonding agents. Organic adhesives may cause corrosion, which results from moisture entrapment between the inner cuff surface and the outer shank surface.

# Index

## A



## B

## C

## D

## E



## F

# G



# H

## I

## J



## K



## L

## M



## N

## O



## P

## Q



## R

## S

## T

## U

## V



## W



## X



## Z